THE FINITE and DISCRETE MATH PROBLEM SOLVER®

REGISTERED TRADEMARK

Staff of Research and Education Association,

Dr. M. Fogiel, Director

Research and Education Association
505 Eighth Avenue
New York, N.Y. 10018

THE FINITE AND DISCRETE MATH PROBLEM SOLVER®

Printed in the United States of America

Library of Congress Catalog Card Number 84-61815

International Standard Book Number 0-87891-559-1

Revised Printing, 1986

PROBLEM SOLVER is a registered trademark of
Research and Education Association, New York, N.Y. 10018

WHAT THIS BOOK IS FOR

Students have generally found finite and discrete math a difficult subject to understand and learn. Despite the publication of hundreds of textbooks in this field, each one intended to provide an improvement over previous textbooks, students continue to remain perplexed as a result of the numerous conditions that must often be remembered and correlated in solving a problem. Various possible interpretations of terms used in finite and discrete math have also contributed to many of the difficulties experienced by students.

In a study of the problem, REA found the following basic reasons underlying students' difficulties with finite and discrete math taught in schools:

(a) No systematic rules of analysis have been developed which students may follow in a step-by-step manner to solve the usual problems encountered. This results from the fact that the numerous different conditions and principles which may be involved in a problem, lead to many possible different methods of solution. To prescribe a set of rules to be followed for each of the possible variations, would involve an enormous number of rules and steps to be searched through by students, and this task would perhaps be more burdensome than solving the problem directly with some accompanying trial and error to find the correct solution route.

(b) Textbooks currently available will usually explain a given principle in a few pages written by a professional who has an insight in the subject matter that is not shared by students. The explanations are often written in an abstract manner which leaves the students confused as to the application of the principle. The explanations given are not sufficiently detailed and extensive to make the student aware of the wide range of applications and different aspects of the principle being studied. The numerous possible variations of principles and their applications are usually not discussed, and it is left for the students to discover these for themselves while doing the exercises. Accordingly, the average student is expected to rediscover that which has been long known and practiced, but not published or explained extensively.

(c) The examples usually following the explanation of a topic are too few in number and too simple to enable the student to obtain a thorough grasp of the principles involved. The explanations do not provide sufficient basis to enable a student to solve problems that may be subsequently assigned for homework or given on examinations.

The examples are presented in abbreviated form which leaves out much material between steps, and requires that students derive the omitted material themselves. As a result, students find the examples difficult to understand--contrary to the purpose of the examples.

Examples are, furthermore, often worded in a confusing manner. They do not state the problem and then present the solution. Instead, they pass through a general discussion, never revealing what is to be solved for.

Examples, also, do not always include diagrams/graphs, wherever appropriate, and students do not obtain the training to draw diagrams or graphs to simplify and organize their thinking.

(d) Students can learn the subject only by doing the exercises themselves and reviewing them in class, to obtain experience in applying the principles with their different ramifications.

In doing the exercises by themselves, students find that they are required to devote considerably more time to finite and discrete math than to other subjects of comparable credits, because they are uncertain with regard to the selection and application of the theorems and principles involved. It is also often necessary for students to discover those "tricks" not revealed in their texts (or review books), that make it possible to solve problems easily. Students must usually resort to methods of trial-and-error to discover these "tricks", and as a result they find that they may sometimes spend hours in solving a single problem.

(e) When reviewing the exercises in classrooms, instructors usually request students to take turns in writing solutions on the boards and explaining them to the class. Students often find it difficult to explain in a manner that holds the interest of the class, and enables the remaining students to follow the material written on the boards. The remaining students seated in the class are, furthermore, too occupied with copying the material from the boards, to listen to the oral explanations and concentrate on the methods of solution.

This book is intended to aid students in finite and discrete math in overcoming the difficulties described, by supplying detailed illustrations of the solution methods which are usually not apparent to students. The solution methods are illustrated by problems selected from those that are most often assigned for class work and given on examinations. The problems are arranged in order of complexity to enable students to learn and understand a particular topic by reviewing the problems in sequence. The problems are illustrated with detailed step-by-step explanations, to save students the large amount of time that is often needed to fill in the gaps that are usually found between steps of illustrations in textbooks or review/outline books.

The staff of REA considers finite and discrete math a subject that is best learned by allowing students to view the methods of analysis and solution techniques themselves. This approach to learning the subject matter is similar to that practiced in various scientific laboratories, particularly in the medical fields.

In using this book, students may review and study the illustrated problems at their own pace; they are not limited to the time allowed for explaining problems on the board in class.

When students want to look up a particular type of problem and solution, they can readily locate it in the book by referring to the index which has been extensively prepared. It is also possible to locate a particular type of problem by glancing at just the material within the boxed portions. To facilitate rapid scanning of the problems, each problem has a heavy border around it. Furthermore, each problem is identified with a number immediately above the problem at the right-hand margin.

To obtain maximum benefit from the book, students should familiarize themselves with the section, "How To Use This Book", located in the front pages.

To meet the objectives of this book, staff members of REA have selected problems usually encountered in assignments and examinations, and have solved each problem meticulously to illustrate the steps which are difficult for students to comprehend. Special gratitude is expressed to them for their efforts in this area, as well as to the numerous contributors who devoted brief periods of time to this work.

Gratitude is also expressed to the many persons involved in the difficult task of typing the manuscript with its endless changes, and to the REA art staff who prepared the numerous detailed illustrations together with the layout and physical features of the book.

The difficult task of coordinating the efforts of all persons was carried out by Carl Fuchs. His conscientious work deserves much appreciation. He also trained and supervised art and production personnel in the preparation of the book for printing.

Finally, special thanks are due to Helen Kaufmann for her unique talents in rendering those difficult border-line decisions and in making constructive suggestions related to the design and organization of the book.

Max Fogiel, Ph.D.
Program Director

HOW TO USE THIS BOOK

This book can be an invaluable aid to students in finite and discrete math as a supplement to their textbooks. The book is divided into 11 chapters, each dealing with a separate topic. The subject matter is developed beginning with logic, sets, relations, and functions, vectors and matrices, graph theory, and extending through counting and binomial theorem, probability and statistics. Also included are problems in boolean algebra, linear programming and theory of games. An extensive number of applications have been included, since these appear to be more troublesome to students.

TO LEARN AND UNDERSTAND A TOPIC THOROUGHLY

1. Refer to your class text and read the section pertaining to the topic. You should become acquainted with the principles discussed there. These principles, however, may not be clear to you at that time.

2. Then locate the topic you are looking for by referring to the "Table of Contents" in the front of this book, "The Finite and Discrete Math Problem Solver".

3. Turn to the page where the topic begins and review the problems under each topic, in the order given. For each topic, the problems are arranged in order of complexity, from the simplest to the more difficult. Some problems may appear similar to others, but each problem has been selected to illustrate a different point or solution method.

To learn and understand a topic thoroughly and retain its contents, it will generally be necessary for students to review the problems several times. Repeated review is essential in order to gain experience in recognizing the principles that should be applied, and in selecting the best solution technique.

TO FIND A PARTICULAR PROBLEM

To locate one or more problems related to a particular subject matter, refer to the index. In using the index, be certain to note that the numbers given there refer to problem numbers, not to page numbers. This arrangement of the index is intended to facilitate finding a problem more rapidly, since two or more problems may appear on a page.

If a particular type of problem cannot be found readily, it is recommended that the student refer to the "Table of Contents" in the front pages, and then turn to the chapter which is applicable to the problem being sought. By scanning or glancing at the material that is boxed, it will generally be possible to find problems related to the one being sought, without consuming considerable time. After the problems have been located, the solutions can be reviewed and studied in detail. For this purpose of locating problems rapidly, students should acquaint themselves with the organization of the book as found in the "Table of Contents".

In preparing for an exam, locate the topics to be covered on the exam in the "Table of Contents", and then review the problems under those topics several times. This should equip the student with what might be needed for the exam.

CONTENTS

ix

CHAPTER 1

LOGIC

STATEMENTS, NEGATIONS, CONJUNCTIONS AND DISJUNCTIONS

Using elementary mathematics, determine the truth values of the following conjunctions:

(1) Nine is an odd number less than four.

(2) Eight is greater than one but less than fifteen.

(3) A rhombus has four equal sides, two of which are parallel.

(4) Seven and ten are prime numbers.

Solution: (1) The conjunction can be rewritten as "Nine is an odd number and nine is less than four." It is false since the second simple statement is false.

(2) The statement "Eight is greater than one and eight is less than fifteen." is equivalent to the given conjunction. This conjunction is true.

(3) The conjunction is equivalent to "A rhombus has four equal sides and A rhombus has two parallel sides." This conjunction is true.

(4) This statement has the same truth value as the statement "Seven is a prime number and ten is a prime number." It is false since the second simple statement is false.

Using elementary mathematics, evaluate the truth values of the following disjunctions:

(1) Five can be divided by one or five.

(2) The product of sixteen and four is sixty-four or the sum of sixteen and four is thirty.

(3) The difference between seven and four is greater than eight or the sum of seven and four is eleven.

(4) The number four is an odd number or $4 \div 12 = 4 + 12$.

Solution: (1) This disjunction is true since both simple statements are true.

(2) The disjunction is true because at least one of the components (the first one) is true.

(3) True, since at least one simple statement, the second statement, is true.

(4) The truth value of this disjunction is false since both simple statements are false.

Show that the implication "if a, then b" and the disjunction "(not a) or b" are logically equivalent.

The truth table is:

(1)	(2)	(3)	(4)	(5)
a	b	not a	If a, then b	(not a) or b
F	F	T	T	T
F	T	T	T	T
T	F	F	F	F
T	T	F	T	T

2

<u>Solution</u>:

Note that the implication "if a, then b" and the disjunction "(not a) or b" are logically equivalent, since the truth values in columns (4) and (5) are identical.

Evaluate the truth values of the negations of the following statements:

(1) A square has four sides.

(2) The sum of five and ten is fifty.

(3) A house is a house.

(4) France is not in Europe.

<u>Solution</u>: The negations of the given statements are

(1) A square does not have four sides.

(2) The sum of five and ten is not fifty.

(3) A house is not a house.

(4) France is in Europe.

The truth values of these statements are

(1) False
(2) True
(3) False
(4) True

Note that the negations of the given statements could have been stated alternately in the following form: "It is not the case that p."

For instance, the negation of (4) may be stated as "It is not the case that France is not in Europe."

Also note that the double negation which is defined as the negation of a negation of a simple statement gives the simple statement itself.

For example,

p : China is in Asia.

~ p : China is not in Asia

~(~p) : It is not the case that China is not in Asia. The statement ~(~p) is the same as the statement: "China is in Asia."

Finally, one must be careful in forming the negations of some statements. For example, the negation of "Four plus six is not twenty" is not "Four plus six is ten." The correct negation is "Four plus six is twenty."

Find all relations which exist between the following compound statements:

(1) ~a   (2) a ∧ b   (3) ~a ∨ b   (4) ~a ∨ ~b

(5) a → b

(1)	(2)	(3)	(4)	(5)	(6)	(7)	(8)
a	b	~b	~a	a ∧ b	~a ∨ b	~a ∨ ~b	a → b
F	F	T	T	F	T	T	T
F	T	F	T	F	T	T	T
T	F	T	F	F	F	T	F
T	T	F	F	T	T	F	T

Solution: Construct a truth table for the given statements as shown in the figure.

It can be noted that in the truth table, columns (6) and (8) are identical. Therefore the statements a → b and a ∧ b are equivalent. Columns (5) and (6) have exactly opposite truth values.

Thus, statements a ∧ b and ~a ∨ ~b are contradictory. By comparing columns (5) and (6) of the truth table, it can be seen that statement a ∧ b implies ~a ∨ b. This is because there is no case in which a ∧ b is true but ~a ∨ b is false. By the same token, it's observed that the same relation exists between statements (4) and (6), (4) and (7).

Since there is no case in which the truth values of columns (4) and (5) are both true, statements ~a and a ∧ b are inconsistent. Finally, statements ~a ∨ b and ~a ∨ ~b are subcontraries because they are never both false (see columns (6) and (7) of the truth table).

4

Show that x ∧ y is logically equivalent to y ∧ x.

(1)	(2)	(3)	(4)
x	y	x ∧ y	y ∧ x
F	F	F	F
F	T	F	F
T	F	F	F
T	T	T	T

Solution: The problem can be solved by using a truth table
as shown in the figure.

Columns (1) and (2) contain the four possible combinations
of truth values of the simple statements x and y. Since columns (3) and (4) which give the truth values of the statements x ∧ y and y ∧ x are identical, x ∧ y and y ∧ x are logically equivalent. This logical equivalence is called the
commutative property of conjunction.

● **PROBLEM** 1-7

Show that (a ∨ b) ∨ c is logically equivalent to
a ∨ (b ∨ c).

(1)	(2)	(3)	(4)	(5)	(6)	(7)
a	b	c	a∨b	(a∨b)∨c	b∨c	a∨(b∨c)
F	F	F	F	F	F	F
F	F	T	F	T	T	T
F	T	F	T	T	T	T
F	T	T	T	T	T	T
T	F	F	T	T	F	T
T	F	T	T	T	T	T
T	T	F	T	T	T	T
T	T	T	T	T	T	T

Solution: One can solve this problem by constructing a truth table as shown in the figure.

Note: If a statement consists of n simple statements, the truth table has 2^n rows.

Since the given statements have three simple statements, the truth table has $2^3 = 8$ rows. Columns (1), (2) and (3) give all eight possible combinations of truth values of the simple statements a, b, and c.

(a ∨ b) ∨ c is logically equivalent to a ∨ (b ∨ c) since columns (5) and (7), which give the truth values of (a ∨ b) ∨ c and a ∨ (b ∨ c) are identical.

This logical equivalence is called the associative property of the disjunction of statements.

TRUTH TABLE AND PROPOSITION CALCULUS

Given the following statements

(a) (x ∨ y) ∧ ~y

(b) [(a → b) ∧ (~a ∨ b)] ↔ (b → a)

construct truth tables for each.

(1)	(2)	(3)	(4)	(5)
x	y	x ∨ y	~y	(x∨y) ∧ (~y)
F	F	F	T	F
F	T	T	F	F
T	F	T	T	T
T	T	T	F	F
		W	Z	W ∧ Z

Figure 1

Solution: (a) First construct a truth table for x ∨ y.

6

Since this statement is enclosed in parentheses. The truth values of x∨ y for every logical possibility are listed in column (3), while columns (1) and (2) are filled with all possible combinations of truth values of x and y. (See Fig. 1)

Since -y is the complement of y, the truth value in each entry of column (4) must be the opposite of the truth value in column (2).

The truth value of (x∨ y)∧ ~y are listed in column (5). These truth values can be determined as follows:

Let w = x∨ y and z = ~y be the two components of the conjunction w∧ z. In the first row the truth value of column (5) is determined as follows: w = x ∨ y = T, z = -y = F. w ∧ z = T ∧ F = F.

This is because a conjunction is false if either of its components are false (or both of its components are false). A conjunction is true if all of its components are true. According to the definition above, one can easily find the other entries of column (5). The complete truth table is shown in Fig. 1.

Note that the real truth table for (x∨ y)∧ -y includes only columns (1), (2) and (5). Columns (3) and (4) are auxiliary columns used to facilitate finding the entries for column (5).

(b) The truth table is shown in Fig. 2.

(1)	(2)	(3)	(4)	(5)	(6)	(7) $(a \to b) \wedge (\text{-}a \vee b)$	(8) $[(a \to b) \wedge (\text{-}a \vee b)] \leftrightarrow (b \to a)$
a	b	-a	a → b	-a∨b	b → a		
F	F	T	T	T	T	T	T
F	T	T	T	T	F	T	F
T	F	F	F	F	T	F	F
T	T	F	T	T	T	T	T

Figure 2

To construct a truth table for

$$[(a \to b) \wedge (\text{-}a \vee b)] \leftrightarrow (b \to a),$$

first construct columns (1) and (2) which give all the possible combinations of the truth values of a and b. Next construct a column for each of the compound statements,

7

$a \rightarrow b$, $-a \lor b$, and $b \rightarrow a$. The entries in each of these columns are filled according to the definitions of "$\rightarrow$" and "$\lor$".

After obtaining all the truth values of the compound statements find the truth values of conjunction ($a \rightarrow b$) and ($-a \lor b$), which are given in column (7).

Finally, the truth values for

$$[(a \rightarrow b) \land (-a \lor b)] \leftrightarrow (b \rightarrow a)$$

is found by considering columns (6) and (7) as the truth values of the components in the definition of the biconditional.

● **PROBLEM 1-9**

Construct a truth table for the statement

$$a \land (b \lor {\sim} c)$$

(1)	(2)	(3)	(4)	(5)	(6)
a	b	c	${\sim}c$	$b \lor {\sim}c$	$a \land (b \lor {\sim}c)$
T	T	T	F	T	T
T	T	F	T	T	T
T	F	T	F	F	F
T	F	F	T	T	T
F	T	T	F	T	F
F	T	F	T	T	F
F	F	T	F	F	F
F	F	F	T	T	F

Solution: The truth table for the given statement is shown in the figure.

The truth values of the basic component statements are given in columns (1), (2) and (3).

The following method is used to assign the truth values of columns (1), (2) and (3). Divide column (1) into two equal parts, fill one part with T's and the other part with

8

F's. For this problem, the first four entries of column (1) are filled with T's. Next, divide column (2) into four equal parts and fill the first part with T's, the second part with F's, the third part with T's, and the forth part with F's.

(One can also fill the first two entries with F's, the second two entries with T's, and so on.)

Finally, fill each row of column (3) alternately with T and F.

Columns (4), (5), and (6) give the truth values of the negation ~c, the disjunction of b and ~c, and the conjunction of a and (b ∨~ c) respectively.

● **PROBLEM** 1-10

Construct a truth table for the following statements.

(1)   (x ∨ ~y) ∧ ~x

(2)   ~[(x ∧ y) ∨ (~x ∧ ~y)]

(1)	(2)	(3)	(4)	(5)	(6)
x	y	~x	~y	(x∨~y)	(x∨~y)∧~x
T	T	F	F	T	F
T	F	F	T	T	F
F	T	T	F	F	F
F	F	T	T	T	T

Fig. 1

Solution: (1)   The truth table is shown in Fig. 1.

All possible combinations of the truth values of x and y are listed in columns (1) and (2). Listed in columns (3) and (4), are the truth values of the negations of x and y respectively.

The truth values of (x ∨ ~y) are listed in column (5). The truth values of the compound statement (x ∨ ~y) ∧ ~x are listed in column (6).

9

(2) The truth table of $\sim[(x \wedge y) \vee (\sim x \wedge \sim y)]$ is shown in Fig. 2.

x	y	$\sim x$	$\sim y$	$x \wedge y$	$\sim x \wedge \sim y$	$(x \wedge y) \vee (\sim x \wedge \sim y)$	$\sim \left[(x \wedge y) \vee (\sim x \wedge \sim y)\right]$
T	T	F	F	T	F	T	F
T	F	F	T	F	F	F	T
F	T	T	F	F	F	F	T
F	F	T	T	F	T	T	F
(1)	(2)	(3)	(4)	(5)	(6)	(7)	(8)

Fig. 2

All possible combinations of the truth values of x and y are listed in the first and the second columns of the truth table.

Columns (3) and (4), contain the truth values of $\sim x$ and $\sim y$. Next, the truth values of $x \wedge y$ and $\sim x \wedge \sim y$ are listed in columns (5) and (6) respectively.

The truth values of $(x \wedge y) \vee (\sim x \wedge \sim y)$ are listed in column (7).

Finally, the negation of column (7) gives the truth values of $\sim[(x \wedge y) \vee (\sim x \wedge \sim y)]$ which are given in column (8) of the truth table.

● PROBLEM 1-11

Construct a truth table for

(1) $(\sim x) \to y$

(2) $[x \vee (\sim y)] \to z$

p	q	$p \to q$
T	T	T
T	F	F
F	T	T
F	F	T

Fig. 1

Solution: (1) The definition of the conditional statement, $P \rightarrow q$ is given by the truth table in Fig. 1.

Let $\sim x = P$ and $y = q$, then the truth table in Fig. 1 is for $(\sim x) \rightarrow y$.

One may add one more column for x as shown in Fig. 2.

x	$\sim x$	y	$\sim x \rightarrow y$
F	T	T	T
F	T	F	F
T	F	T	T
T	F	F	T

Fig. 2

(2) The statement has three components x, y, and z. A truth table for a statement of n components contains 2^n rows. Therefore, for the second statement the truth table has $2^3 = 8$ rows. The first n columns of a truth table of an n-component statement contain all the possible combinations of the truth values of the components.

The following method is used to fill the first n columns.

Step 1.   Number each row of the truth table, from top to bottom, start from 0 to 2^{n-1}

Step 2.   Fill the first n entries of each row by a n-digit binary number corresponding to the row number.

Step 3.   Replace all 1's and all 0's by T's and F's respectively. (See Fig. 3)

row	x	y	z
0	0	0	0
1	0	0	1
2	0	1	0
3	0	1	1
4	1	0	0
5	1	0	1
6	1	1	0
7	1	1	1

$\Longrightarrow$

row	x	y	z
0	F	F	F
1	F	F	T
2	F	T	F
3	F	T	T
4	T	F	F
5	T	F	T
6	T	T	F
7	T	T	T

Fig. 3

11

The truth table for [x ∨ (∿y)] → z is shown in Fig. 4.

x	y	z	∿y	x ∨ ∿y	$[x \vee (\sim y)] \rightarrow z$
F	F	F	T	T	F
F	F	T	T	T	T
F	T	F	F	F	T
F	T	T	F	F	T
T	F	F	T	T	F
T	F	T	T	T	T
T	T	F	F	T	F
T	T	T	F	T	T

Fig. 4

● **PROBLEM** 1-12

Find the truth tables of the following propositions

(1)   (A ∧ B) ∨ (A ∧ C)

(2)   (A → B) ∧ (B → C)

A	B	C	A∧B	A∧C	(A∧B) ∨ (A∧C)
T	T	T	T	T	T
T	T	F	T	F	T
T	F	T	F	T	T
T	F	F	F	F	F
F	T	T	F	F	F
F	T	F	F	F	F
F	F	T	F	F	F
F	F	F	F	F	F

Fig. 1

Solution: By repetitive use of the logical connectives one can construct compound statements which are called propositions. The truth value of a proposition depends exclusively

12

upon the truth values of its variables, that is, the truth value of a proposition is known once the truth values of its variables are known. For instance, the truth values of the statement (A ∧ B) ∨ (A ∧ C) will be known once the truth values of A, B, C are known. The truth tables for (1) and (2) are given in Fig. 1 and Fig. 2 respectively.

A	B	C	A → B	B → C	(A →B) ⋀ (B → C)
T	T	T	T	T	T
T	T	F	T	F	F
T	F	T	F	T	F
T	F	F	F	T	F
F	T	T	T	T	T
F	T	F	T	F	F
F	F	T	T	T	T
F	F	F	T	T	T

Fig. 2

● **PROBLEM** 1-13

Construct truth tables for the following statements.

(1) (a → b) ∧ (∿b → ∿a)

(2) ∿(a ∨ b) ∨ ∿(a ∧ b)

a	b	∿a	∿b	a → b	∿b→∿a	(a →b) ⋀ (∿b →∿a)
T	T	F	F	T	T	T
T	F	F	T	F	F	F
F	T	T	F	T	T	T
F	F	T	T	T	T	T
1	2	3	4	5	6	7

Fig. 1

Solution: (1) The truth table is given in Fig. 1.

13

Note that columns (3) and (4) are just the complements of columns (1) and (2) respectively. Furthermore, note that columns (5) and (6) are identical, that is, $a \to b \equiv \sim b \to \sim a$. It is because $a \to b$ is equivalent to $\sim b \to \sim a$, that $(a \to b) \wedge (\sim b \to \sim a)$ must have the same truth values as either $a \to b$ or $\sim b \to \sim a$. Since, by the definition of "$\wedge$", $a \wedge b$ is true only if a and b are both true. Hence, column (7) is identical to columns (5) and (6).

(2) The truth table is shown in Fig. 2.

a	b	a∨b	a∧b	~(a∨b)	~(a∧b)	~(a∨b) ∨ ~(a∧b)
T	T	T	T	F	F	F
T	F	T	F	F	T	T
F	T	T	F	F	T	T
F	F	F	F	T	T	T

Fig. 2

● **PROBLEM 1-14**

Construct a truth table for $[(a \vee b) \wedge \sim b] \to a$.

Fig.1

(1)	(2)	(3)	(4)	(5)	(6)	(7)
a	b	[(a ∨ b)	∧	~b]	→	a
T	T	T	F	F	T	T
T	F	T	T	T	T	T
F	T	T	F	F	T	F
F	F	F	F	T	T	F

Solution: The following truth table is constructed in a form that is different from the truth tables in the previous problems. The truth table is shown in the figure.

    Columns (1) to (7) are filled in the following order: (1), (2), (5), (7), (3), (4), (6). Column (3) gives the truth values of a ∨ b according to the definition of "∨". Column (4) gives the truth values of (a ∨ b) ∧ ~b. Finally, column (6) gives the truth values of [(a ∨ b) ∧ ~b] → a.

14

[(a∨ b) ∧ ~b] → a is a tautology since column (6) consists of T's only.

CONDITIONAL AND BICONDITIONAL STATEMENTS

Symbolize and determine the validity of the following argument:

"The disjunction of x and y is true, and x is true. Therefore, y must be false."

x	y	~y	x $\bigvee$ y	(x $\bigvee$ y) $\bigwedge$ x	$\left[(x\bigvee y)\bigwedge x\right] \rightarrow \, \sim y$
T	T	F	T	T	F
T	F	T	T	T	T
F	T	F	T	F	T
F	F	T	F	F	T

Solution: The symbolized argument is

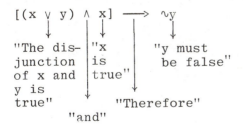

The argument's validity can be determined by using a truth table. An argument is said to be valid when it is impossible for all the premises to be true and the conclusion false. Thus, to determine whether an argument is valid, the argument must be symbolized in the following form:

$$\left[\text{Premise}_1 \wedge \text{Premise}_2 \wedge \ldots \wedge \text{Premise}_n\right] \rightarrow \text{Conclusion}$$

The argument proposition must be a tautology if the argument is to be valid.

15

The truth table for $[(x \lor y) \land x] \to \sim y$ is shown in the figure.

Since the argument proposition is not a tautology, the argument is not valid.

Given the statement "Every tested television set produced by this factory is assigned a serial number." Write the converse, inverse, and contrapositive of this statement.

Solution: By definition, the statement formed when the hypothesis and the conclusion of a conditional statement are interchanged is called the converse of the conditional. In symbols, $a \to b$ is the converse of $b \to a$.

Hence, the converse of the given statement is: "If a television set produced from this factory has a serial number, it has been tested."

The inverse of a statement is a statement in which the premise and conclusion are each replaced by their respective negations. In symbols, $\sim a \to \sim b$ is the inverse of the statement $a \to b$.

Therefore, the inverse of the given statement is: "If a television set produced from this factory has not been tested, it does not have a serial number."

The contrapositive of a statement is defined as the following: $\sim a \to \sim b$ is the contrapositive form of the conditional $b \to a$.

Thus, the contrapositive of the given statement is: "If a television set produced from this factory does not have a serial number, then it has not been tested."

Find the hypothesis, conclusion, converse, inverse, and contrapositive of the following argument:

"If it is raining, then the streets are wet."

Solution: The hypothesis of the statement is: "It is raining."

Its conclusion is: "The streets are wet."

Its converse is: "If the streets are wet, then it is raining."

Its inverse is: "If it is not raining, then the streets are not wet."

Its contrapositive is: "If the streets are not wet, then it is not raining."

(1) Find the hypothesis and conclusion of the following statements:

   (i)   If John does not go, then we shall cancel the trip.

  (ii)   If you go, we won't go.

 (iii)   I will join the army, if I can't find a job.

(2) Do the following statements have the same meaning as the statement, "If Jean is not in her room, then she is at the library."?

   (i)   If Jean is at the library, then she is not in her room.

  (ii)   If Jean is at her room, then she is not at the library.

 (iii)   If Jean is not at the library, then she is in her room.

 (iv)   Jean is in her room, or she is at the library.

Are the statements above related to each other? Explain.

(1)	(2)	(3)	(4)	(5)	(6)	(7)	(8)	(9)
p	q	$\sim p$	$\sim q$	$p \rightarrow q$	$q \rightarrow p$	$\sim p \rightarrow \sim q$	$\sim q \rightarrow \sim p$	$\sim p \vee q$
T	T	F	F	T	T	T	T	T
T	F	F	T	F	T	T	F	F
F	T	T	F	T	F	F	T	T
F	F	T	T	T	T	T	T	T

Fig. 1

Solution: (1) (i) Hypothesis: John doesn't go.
                     Conclusion: We shall cancel the trip.

           (ii) Hypothesis: You go.
                 Conclusion: We won't go.

           (iii) Hypothesis: I can't find a job.
                 Conclusion: I will join the army.

(2) Symbolize all the given statements as the following:
     let p be the statement "Jean is not in her room.", and
     q be the statement "Jean is at the library."

     Then the original statement, in symbols, is $p \rightarrow q$.

     Other statements in symbols are,

       (i) $q \rightarrow p$

       (ii) $\sim p \rightarrow \sim q$

       (iii) $\sim q \rightarrow \sim p$

       (iv) $\sim p \lor q$

     where $\sim p$ is "Jean is in her room ", and $\sim q$ is "Jean is
     not at the library."

     Next, construct a truth table as shown in the figure.

By examining the truth table in Figure 1, it is seen that
columns (5), (8) and (9) are identical. That is, the state-
ments $\sim q \rightarrow \sim p$ and $\sim p \lor q$ have the same truth values as the
statement $p \rightarrow q$. Therefore, statements (iii) and (iv) have
the same meaning as the original statement.

     However, all the statements are related. Statement
(i) is the converse of the original statement; statement
(ii) is called the inverse of the original statement and
statements (iii) and (iv) are just different forms of its
contrapositive.

● PROBLEM 1-19

Use a truth table to determine whether $\sim A$ implies
$\sim(A \lor B)$.

A	~ A	B	A ∨ B	~ (A ∨ B)	~ A → ~ (A ∨ B)
T	F	T	T	F	T
T	F	F	T	F	T
F	T	T	T	F	F
F	T	F	F	T	T

Fig. 1

Solution: The symbolized statement is ~A → ~(A ∨ B). The truth table for ~A → ~(A ∨ B) is shown in the figure.

~A is said to imply ~(A ∨ B) only if ~A → ~(A ∨ B) is a tautology. Since ~A → ~(A ∨ B) is not a tautology, ~A does not imply ~(A ∨ B).

● PROBLEM 1-20

Rewrite the following statements in the if - then form:

(1) We'll pay the postage, if you buy our products.

(2) George gets up at 10 a.m. only if he should go to school at 8 a.m.

(3) A baseball game is a sufficient reason for Philip to miss school.

(4) A grade point average of 4.0 is necessary for a student to receive this scholarship.

(5) The fourth day of July is Independence Day.

Solution: (1) If you buy our products, then we will pay the postage.

(2) If George gets up at 10 a.m., then he should go to school at 8 a.m.

(3) If there is a baseball game, then Philip misses school.

(4) If a student receives this scholarship, then he has a grade point average of 4.0.

(5) If it's the fourth day of July, then it is Independence Day.

Using elementary mathematics, determine the truth values of the converses of the following implications:

(1)   If $10 < 15$, then $10 + 5 = 15$.

(2)   If $4 + 5 = 20$, then $4 \times 5 = 20$.

(3)   If $15 - 3 = 12$, then $3 = 12$.

(4)   If the geometric figure is a square, then it is not a circle.

p	q	p → q
T	T	T
T	F	F
F	T	T
F	F	T

Fig. 1

Solution: First find the converses of the given implications.   By definition, $p \to q$ is the converse of $q \to p$, where p and q are simple statements.   Hence, the converses of implications (1) thru (4) are:

(1)   If $10 + 5 = 15$, then $10 < 15$.

(2)   If $4 \times 5 = 20$, then $4 + 5 = 20$.

(3)   If $3 = 12$, then $15 - 3 = 12$.

(4)   If a geometric figure is not a circle, then it is a square.

The truth values of the above statements are:

(1)   True
(2)   False
(3)   True
(4)   False

The results can be checked by using truth tables.   The truth table in the figure is constructed according to the definition of " → ".   In the truth table p and q are the hypothesis and conclusion of the implication respectively.

The truth table states that an implication is false if and only if its hypothesis is true , but its conclusion is false.

For example, the truth table for implication (3) is,

p	q	p → q
T	F	F

, and the truth table of its converse is,

p	q	q → p
T	F	T

Note that implications (2) and (4) are true, but their converses are false, whereas implication (3) is false but its converse is true.

● **PROBLEM 1-22**

Verify the following argument,

"If he is a college student, then he is intelligent.

He is a college student.

Therefore, he is intelligent."

p	q	p → q	(p → q) ∧ P	[(P → q) ∧ P] → q
T	T	T	T	T
T	F	F	F	T
F	T	T	F	T
F	F	T	F	T

Fig. 1

Solution: Verify this argument by using a truth table.

First, write the argument in the following form:

$$\left[(Premise\ 1)\ \wedge\ (Premise\ 2)\right] \to Conclusion$$

For the given argument, the two premises are: "If he is a college student, then he is intelligent." and "He is a college student." The conclusion is: "He is intelligent."

Symbolize the statement "He is a college student." by p and "He is intelligent." by q. The argument in symbols is:

$$p \rightarrow q$$

$$p$$

$$\overline{}$$

$$\therefore \quad q$$

The truth table for $[(p \rightarrow q) \wedge p] \rightarrow q$ is given in the figure.

The truth table for implication $[(p \rightarrow q) \wedge p] \rightarrow q$ represents a tautology, i.e. there is no case in which both premises of the argument are true, but the conclusion is false. Therefore, the argument is valid.

Is the following a valid argument?

"All dogs are animals.

A horse is an animal.

Therefore, a horse is a dog."

p	q	p → q	$(p \rightarrow q) \bigwedge q$	$\left[(p \rightarrow q) \bigwedge q\right] \rightarrow p$
T	T	T	T	T
T	F	F	F	T
F	T	T	T	F
F	F	T	F	T

Fig. 1

Solution: The first premise can be restated as "If it is a dog then it is an animal." Further, represent the statements "It is a dog," and "it is an animal." by p and q respectively. Then the argument, in symbols, is

$$p \rightarrow q$$

$$q$$

$$\overline{}$$

$$\therefore \quad p$$

The argument is tested by constructing a truth table for $[(p \rightarrow q) \wedge q] \rightarrow p$. If the truth table represents a tautology then the argument is valid. Otherwise, it is invalid. This is because a valid argument is defined as an argument for which one can find no case in which all the premises of the argument are true but the conclusion of the argument is false. It is also known that a → b is false if and only if a is true while b is false. By writing an argument in the form of

$$\left[\text{Premise}_1 \land \text{Premise}_2 \land \ldots \land \text{Premise}_n \right] \to \text{Conclusion}$$

the proposition can be false if and only if all premises are true but the conclusion is false. Therefore, if the proposition above is a tautology, then there is no case in which all the premises of the argument are true but the conclusion is false.

The truth table is shown in the figure.

This truth table does not represent a tautology. Hence, the argument is invalid.

● **PROBLEM 1-24**

Is the following argument valid?

"If it is raining, then the streets are wet.

Either it is snowing, or the streets are not wet.

It is not snowing.

Therefore, it is not raining."

Solution: First symbolize the given statements. Let

a be "It is raining."

b be "Streets are wet."

and c be "It is snowing."

Then the argument is symbolized as the following:

Premise 1: a → b
Premise 2: c ∨ (~b)
Premise 3: ~c
_____
Conclusion: ∴ ~a

There are several methods one may use to verify an argument; the method of natural deduction is used here.

Use the rule of syllogism

$$p \lor q$$

$$\frac{\sim p}{\therefore \quad q}$$

and replace the second and the third premises by ~b. The original argument then becomes:

$$a \to b$$

$$\frac{\sim b}{\therefore \quad \sim a}$$

This is a valid argument since it has the form of modus tollens.

$$p \to q$$

$$\frac{\sim q}{\therefore \quad \sim p}$$

Therefore, the original argument is valid.

Determine whether the following argument is valid:

"I can graduate only if I have a grade point average 3.5. Either I am smart or I do not have a G.P.A. of 3.5. I did not graduate. Therefore, I am not smart."

Solution: Use the method of natural deduction to determine the validity of the given argument. First, symbolize the given argument. Let

m be "I can graduate."

n be "I have a G.P.A. of 3.5."

and ℓ be "I am smart."

The argument can be put in symbols as follows:

Premise₁: $m \to n$

Premise₂: $\ell \lor (\sim n)$

Premise₃: $\sim m$

Conclusion: $\sim \ell$

Logically, premise$_2$ is equivalent to $(\sim n) \vee \ell$ since the disjunction operation "$\vee$" is commutative. $(\sim n) \vee \ell$ is logically equivalent to the implication $n \rightarrow \ell$. Now, the original argument is replaced by

$$m \rightarrow n$$
$$n \rightarrow \ell$$
$$\underline{\sim m}$$
$$\sim \ell$$

By the rule of hypothetical syllogism,

$$p \rightarrow q$$
$$\underline{q \rightarrow r}$$
$$p \rightarrow r$$;

the first and second premises are replaced by $m \rightarrow \ell$. So, the original argument becomes,

$$m \rightarrow \ell$$
$$\underline{\sim m}$$
$$\sim \ell$$

This is an invalid argument because it is an example of the fallacy of denying the antecedent.

● PROBLEM 1-26

Indicate the necessary and sufficient conditions of the following implications:

(1) I go to work late only if I get up late.

(2) A's are hard to get only if you don't study hard.

(3) Jobs are hard to find if the unemployment rate is high.

(4) If you make more money, then you can spend more.

Solution:

	Sufficient condition	Necessary condition
(1)	I go to work late.	I get up late.
(2)	A's are hard to get.	You don't study hard.
(3)	The unemployment rate is high.	Jobs are hard to find.
(4)	You make more money.	You can spend more.

Construct truth tables for the following:

(1)   ~(x ∧ y) ↔ (~x ∨ ~y).

(2)   ~(x ∨ y) ↔ (~x ∧ ~y).

p	q	p ↔ q
T	T	T
T	F	F
F	T	F
F	F	T

Fig. 1

Solution: A statement of the form "p if and only if q", de-noted by p ↔ q, is called a biconditional statement. The truth table for p ↔ q is in Fig. 1.

By examining the truth table, it is seen that p ↔ q is true only if p and q have the same truth values.

Therefore, according to the definition of " ↔ ", the truth tables for ~(x ∧ y) ↔ (~x ∨ ~y) and ~(x ∨ y) ↔ (~x ∧ ~y) are constructed as shown in Fig. 2 and Fig. 3 respectively.

x	y	~x	~y	x ∧ y	~ (x ∧ y)	~x ∨ ~y	~ (x ∧ y) ↔ ~x ∨ ~y
T	T	F	F	T	F	F	T
T	F	F	T	F	T	T	T
F	T	T	F	F	T	T	T
F	F	T	T	F	T	T	T

Fig. 2

x	y	~x	~y	x ∨ y	~ (x ∨ y)	~x ∧ ~y	~ (x ∨ y) ↔ ~x ∧ ~y
T	T	F	F	T	F	F	T
T	F	F	T	T	F	F	T
F	T	T	F	T	F	F	T
F	F	T	T	F	T	T	T

Fig. 3

Note that for the truth tables in Fig. 2 and Fig. 3, the last columns of these two tables contain T's only. A statement is called a tautology if it has all T's in its truth column. Hence, the given biconditional statements are tautologies.

• PROBLEM 1-28

Construct a truth table to determine the validity of the following argument proposition,

$$[(p \vee q) \wedge \sim q] \rightarrow p.$$

(1)	(2)	(3)	(4)	(5)	(6)
p	q	$p \vee q$	$\sim q$	$(p \vee q) \wedge \sim q$	$[(p \vee q) \wedge \sim q] \rightarrow p$
T	T	T	F	F	T
T	F	T	T	T	T
F	T	T	F	F	T
F	F	F	T	F	T

Fig. 1

Solution: An argument is valid only if it is impossible to have a false conclusion while all the premises are true. In other words, the argument proposition must be a tautology if the argument is to be true.

For the argument $[(p \vee q) \wedge \sim q] \rightarrow p$, the premises are $p \vee q$ and $\sim q$; the conclusion is p, and the truth table is shown in the figure.

The argument is valid since $[(p \vee q) \wedge \sim q] \rightarrow p$ is a tautology. Note that the validity of the given argument could be determined without the construction of column (6). Examining column (5), the two premises are both true only in the second row. In this row, however, the conclusion p (in the first column), is also true. Therefore, the argument is valid.

• PROBLEM 1-29

(1) Show that $a \rightarrow b$ and $\sim b \rightarrow \sim a$ are logically equivalent.

(2) Show that the biconditional statement $(a \rightarrow b) \leftrightarrow (\sim b \rightarrow \sim a)$ is a tautology.

Solution: (1) The truth table for $a \rightarrow b$ and $\sim b \rightarrow \sim a$ is shown in Fig. 1.

a	b	~a	~b	a → b	~b → ~a
T	T	F	F	T	T
T	F	F	T	F	F
F	T	T	F	T	T
F	F	T	T	T	T
(1)	(2)	(3)	(4)	(5)	(6)

Fig. 1

Since the columns (5) and (6) are identical, a → b and ~b → ~a are logically equivalent. That is, a → b and ~b → ~a have the same truth values for any given a and b. In symbols this relation is represented by (a → b) ↔ (~b → ~a) or (a → b) ≡ ~b → ~a.

(2)   From part (1), we know that a → b and ~b → ~a have the same truth values for any given a and b. Furthermore, it is known that a ↔ b is true only if a and b have the same truth values. Therefore, (a → b) ↔ (~b → ~a) must be a tautology. (One can, of course, arrive at the same conclusion by using a truth table.)

MATHEMATICAL INDUCTION

● **PROBLEM 1-30**

Use the principle of mathematical induction to prove that for every positive integer n,

$$1 + 2 + 3 + 4 + \ldots + n = \frac{n(n+1)}{2} .$$

Solution: The principle of mathematical induction can be stated as the following:

"Let P be a proposition defined on the positive integers N, i.e., P(n) is either true or false for each n in N. Suppose P has the following two properties:

(1) P(1)   is true.

(2) P(n+1) is true whenever P(n) is true.

Then P is true for every positive integer." Another way of stating the same principle is the following:

"Let P be a proposition defined on the positive integers N such that

(1)   P(1)   is true.

(2)   P(n)   is true whenever P(k) is true, for all
        $1 \le k < n$.

Then P is true for every positive integer."

The proof of the given proposition,

$$P = 1 + 2 + 3 + \ldots + n = \frac{n(n+1)}{2}$$

is the following.

First, prove that P(1) is true.

$$P(1) = 1 = \frac{n(n+1)}{2} = \frac{1 \cdot (1+1)}{2} = 1$$

Next, prove that the proposition P(n) implies the proposition P(n+1). (Note that the proof which one gives must be valid for every $n \varepsilon N$.) Therefore, assume that P(n) is true, that is,

$$1 + 2 + 3 + \ldots + n = \frac{n(n+1)}{2} \ .$$

Then, add n+1 to both sides of the equation above, one obtains

$$1 + 2 + 3 + \ldots + n + (n+1) = \frac{n(n+1)}{2} + (n+1)$$

$$= \frac{(n+1)(n+2)}{2}$$

This equation is the statement of P(n+1).

Hence, from the above it is seen that the truth of P(n) implies that P(n+1) is true.   The proof is completed.

● **PROBLEM 1-31**

Prove by mathematical induction
$$1^2 + 2^2 + 3^2 + \ldots + n^2 = \frac{1}{6} n(n+1)(2n+1).$$

Solution: Mathematical induction is a method of proof. The steps are:
(1)   The verification of the proposed formula or theorem for the smallest value of n. It is desirable, but not necessary, to verify it for several values of n.
(2)   The proof that if the proposed formula or theorem is true for n = k, some positive integer, it is true also for n = k+1. That is, if the proposition is true for any particular value of n, it must be true for

the next larger value of n.

(3) A conclusion that the proposed formula holds true for all values of
n.

Proof: Step 1. Verify:

For   n = 1: $1^2 = \frac{1}{6}(1)(1+1)[2(1)+1] = \frac{1}{6}(1)(2)(3) = \frac{1}{6}(6) = 1$

$$1 = 1 ✓$$

For   n = 2: $1^2 + 2^2 = \frac{1}{6}(2)(2+1)[2(2)+1] = \frac{1}{6}(2)(3)(5) = \frac{1}{6}(6)(5)$

$$1 + 4 = (1)(5)$$

$$5 = 5 ✓$$

For   n = 3: $1^2 + 2^2 + 3^2 = \frac{1}{6}(3)(3+1)[2(3)+1]$

$$1 + 4 + 9 = \frac{1}{6}(3)(4)(7) = \frac{1}{6}(12)(7) = 14$$

$$14 = 14 ✓$$

Step 2. Let k represent any particular value of n. For
n = k, the formula becomes

$$1^2 + 2^2 + 3^2 + \ldots + k^2 = \frac{1}{6}k(k+1)(2k+1). \tag{A}$$

For n = k+1, the formula is

$$1^2 + 2^2 + 3^2 + \ldots + k^2 + (k+1)^2 = \frac{1}{6}(k+1)[(k+1) + 1][2(k+1) + 1]$$

$$= \frac{1}{6}(k+1)(k+2)(2k+3). \tag{B}$$

We must show that if the formula is true for n = k, then it must be true
for n = k+1. In other words, we must show that (B) follows from (A).
The left side of (A) can be converted into the left side of (B) by merely
adding $(k+1)^2$. All that remains to be demonstrated is that when $(k+1)^2$
is added to the right side of (A), the result is the right side of (B).

$$1^2 + 2^2 + \ldots + k^2 + (k+1)^2 = \frac{1}{6}k(k+1)(2k+1) + (k+1)^2$$

Factor out (k+1):

$$1^2 + 2^2 + 3^2 + \ldots + k^2 + (k+1)^2 = (k+1)\left[\frac{1}{6}k(2k+1) + (k+1)\right]$$

$$= (k+1)\left[\frac{k(2k+1)}{6} + \frac{(k+1)6}{6}\right]$$

$$= (k+1)\frac{2k^2 + k + 6k + 6}{6}$$

$$= \frac{(k+1)(2k^2 + 7k + 6)}{6}$$

$$= \frac{1}{6}(k+1)(k+2)(2k+3),$$

since $2k^2 + 7k + 6 = (k+2)(2k+3).$

Thus, we have shown that if we add $(k+1)^2$ to both sides of the equation
for n = k, then we obtain the equation **or** formula for n = k+1. We have
thus established that if (A) is true, then (B) must be true; that is, if
the formula is true for n = k, then it must be true for n = k+1. In
other words, we have proved that if the proposition is true for a certain
positive integer k, then it is also true for the next greater integer
k+1.

Step 3. The proposition is true for n = 1,2,3 (Step 1). Since
it is true for n = 3, it is true for n = 4 (Step 2, where k = 3 and
k+1 = 4). Since it is true for n = 4, it is true for n = 5, and so on,
for all positive integers n.

Prove by mathematical induction that, for all positive integral values of n,

$$\frac{1}{1\cdot3} + \frac{1}{3\cdot5} + \frac{1}{5\cdot7} + \cdots + \frac{1}{(2n-1)(2n+1)} = \frac{n}{2n+1} \; .$$

Solution: Step 1. The formula is true for n = 1, since

$$\frac{1}{(2-1)(2+1)} = \frac{1}{2+1} = \frac{1}{3} \; .$$

Step 2. Assume that the formula is true for n = k. Then

$$\frac{1}{1\cdot3} + \frac{1}{3\cdot5} + \frac{1}{5\cdot7} + \cdots + \frac{1}{(2k-1)(2k+1)} = \frac{k}{2k+1} \; .$$

Add the (k+1)th term, which is $\frac{1}{(2k+1)(2k+3)}$, to both sides of the above

equation. Then

$$\frac{1}{1\cdot3} + \frac{1}{3\cdot5} + \frac{1}{5\cdot7} + \cdots + \frac{1}{(2k-1)(2k+1)} + \frac{1}{(2k+1)(2k+3)} = \frac{k}{2k+1} + \frac{1}{(2k+1)(2k+3)} \; .$$

The right hand side of this equation $= \frac{k(2k+3)+1}{(2k+1)(2k+3)} = \frac{k+1}{2k+3}$, which is the

value of $\frac{n}{2n+1}$ when n is replaced by (k+1).

Hence if the formula is true for n = k, it is true for n = k + 1.

But the formula holds for n = 1; hence it holds for n = 1 + 1 = 2.
Then, since it holds for n = 2, it holds for n = 2 + 1 = 3, and so
on. Thus the formula is true for all positive integral values of n.

Using mathematical induction, prove that
$$x^{2n} - y^{2n} \text{ is divisible by } x + y.$$

Solution:

(1) The theorem is true for n = 1, since $x^2 - y^2 = (x-y)(x+y)$ is
divisible by x + y.

(2) Let us assume the theorem true for n = k, a positive integer;
that is, let us assume

(A) $x^{2k} - y^{2k}$ is divisible by x + y.

We wish to show that, when (A) is true,

(B) $x^{2k+2} - y^{2k+2}$ is divisible by x + y.

Now $x^{2k+2} - y^{2k+2} = \left(x^{2k+2} - x^2 y^{2k}\right) + \left(x^2 y^{2k} - y^{2k+2}\right)$

$$= x^2\left(x^{2k} - y^{2k}\right) + y^{2k}\left(x^2 - y^2\right).$$

In the first term $\left(x^{2k} - y^{2k}\right)$ is divisible by $(x+y)$ by assumption, and in the second term $\left(x^2 - y^2\right)$ is divisible by $(x+y)$ by Step (1); hence, if the theorem is true for $n = k$, a positive integer, it is true for the next one $n = k + 1$.

(3) Since the theorem is true for $n = k = 1$, it is true for $n = k + 1 = 2$; being true for $n = k = 2$, it is true for $n = k + 1 = 3$; and so on, for every positive integral value of n.

● **PROBLEM** 1-34

Prove by mathematical induction that

$$\frac{5}{1 \cdot 2 \cdot 3} + \frac{6}{2 \cdot 3 \cdot 4} + \frac{7}{3 \cdot 4 \cdot 5} + \cdots + \frac{n+4}{n(n+1)(n+2)} = \frac{n(3n+7)}{2(n+1)(n+2)} \quad .$$

Solution:

(1) The formula is true for $n = 1$, since $\dfrac{5}{1 \cdot 2 \cdot 3} = \dfrac{1(3+7)}{2 \cdot 2 \cdot 3} = \dfrac{5}{6}$.

(2) Assume the formula to be true for $n = k$, a positive integer; that is, assume

(A) $\quad \dfrac{5}{1 \cdot 2 \cdot 3} + \dfrac{6}{2 \cdot 3 \cdot 4} + \cdots + \dfrac{k+4}{k(k+1)(k+2)} = \dfrac{k(3k+7)}{2(k+1)(k+2)}$.

Under this assumption we wish to show that

(B) $\quad \dfrac{5}{1 \cdot 2 \cdot 3} + \dfrac{6}{2 \cdot 3 \cdot 4} + \cdots + \dfrac{k+4}{k(k+1)(k+2)} + \dfrac{k+5}{(k+1)(k+2)(k+3)}$

$$= \frac{(k+1)(3k+10)}{2(k+2)(k+3)} \quad .$$

When $\dfrac{k+5}{(k+1)(k+2)(k+3)}$ is added to both members of (A), we have

on the right

$$\frac{k(3k+7)}{2(k+1)(k+2)} + \frac{k+5}{(k+1)(k+2)(k+3)} = \frac{1}{(k+1)(k+2)}\left[\frac{k(3k+7)}{2} + \frac{k+5}{k+3}\right]$$

$$= \frac{1}{(k+1)(k+2)} \frac{k(3k+7)(k+3)+2(k+5)}{2(k+3)} = \frac{1}{(k+1)(k+2)} \frac{3k^3 + 16k^2 + 23k + 10}{2(k+3)}$$

$$= \frac{1}{(k+1)(k+2)} \frac{(k+1)^2(3k+10)}{2(k+3)} = \frac{(k+1)(3k+10)}{2(k+2)(k+3)} \quad ;$$

32

hence, if the formula is true for $n = k$ it is true for $n = k + 1$.

(3) Since the formula is true for $n = k = 1$ (Step 1), it is true for $n = k + 1 = 2$; being true for $n = k = 2$, it is true for $n = k + 1 = 3$; and so on, for all positive integral values of n.

● PROBLEM 1-35

Prove that the sum of the cubes of the first n natural numbers is equal to $\left\{ \dfrac{n(n+1)}{2} \right\}^2$.

Solution: We note by trial that the statement is true when n =1, or 2, or 3 (when n = 1,

$$1^3 = 1 \quad \text{and} \quad \left[\frac{1(1+1)}{2}\right]^2 = \left[\frac{1(2)}{2}\right]^2 = \left(\frac{2}{2}\right)^2 = 1^2 = 1,$$

when $n = 2$,
$$1^3 + 2^3 = 1 + 8 = 9 \quad \text{and} \quad \left[\frac{2(2+1)}{2}\right]^2 = \left[\frac{2(3)}{2}\right]^2$$
$$= 3^2 = 9$$

etc.) Assume that it is true when n terms are taken; that is, suppose

$$1^3 + 2^3 + 3^3 + \ldots \text{ to n terms} = \left\{\frac{n(n+1)}{2}\right\}^2 .$$

Add the $(n+1)^{th}$ term, that is, $(n+1)^3$ to each side; then

$$1^3 + 2^3 + 3^3 + \ldots \text{ to n+1 terms} = \left\{\frac{n(n+1)}{2}\right\}^2 + (n + 1)^3$$

$$= \frac{n^2(n+1)^2}{2^2} + (n+1)(n+1)^2$$

$$= (n+1)^2\left(\frac{n^2}{4}\right) + (n+1)^2(n+1)$$

$$= (n+1)^2\left(\frac{n^2}{4} + n + 1\right)$$

$$= (n+1)^2\left(\frac{n^2}{4} + \frac{4n}{4} + \frac{4}{4}\right)$$

$$= (n+1)^2\left(\frac{n^2 + 4n + 4}{4}\right)$$

$$= \frac{(n+1)^2(n^2 + 4n + 4)}{4}$$

$$= \frac{(n+1)^2(n+2)^2}{2^2}$$

$$= \left\{\frac{(n+1)(n+2)}{2}\right\}^2 ;$$

$$= \left\{ \frac{(n+1)\left[(n+1) + 1\right]}{2} \right\}^2$$

which is of the same form as the result we assumed to be true for n terms, $n + 1$ taking the place of n; in other words, if the result is true when we take a certain number of terms, whatever that number may be, it is true when we increase that number by one; but we see that it is true when 3 terms are taken; therefore it is true when 4 terms are taken; it is therefore true when 5 terms are taken; and so on. Thus the result is true universally.

Using mathematical induction, prove the binomial formula

$$(a+x)^n = a^n + na^{n-1}x + \frac{n(n-1)}{2!}a^{n-2}x^2 + \ldots + \frac{n(n-1)\ldots(n-r+2)}{(r-1)!}a^{n-r+1}x^{r-1}$$

$$+ \ldots + x^n$$

for positive integral values of n.

<u>Solution:</u> Step 1. The formula is true for $n = 1$.

Step 2. Assume the formula is true for $n = k$. Then

$$(a+x)^k = a^k + ka^{k-1}x + \frac{k(k-1)}{2!}a^{k-2}x^2 + \ldots + \frac{k(k-1)\ldots(k-r+2)}{(r-1)!}a^{k-r+1}x^{r-1}$$

$$+ \ldots + x^k .$$

Multiply both sides by $a+x$. The multiplication on the right may be written

$$a^{k+1} + ka^k x + \frac{k(k-1)}{2!}a^{k-1}x^2 + \ldots + \frac{k(k-1)\ldots(k-r+2)}{(r-1)!}a^{k-r+2}x^{r-1}$$

$$+ \ldots + ax^k$$

$$+ a^k x + ka^{k-1}x^2 + \ldots + \frac{k(k-1)\ldots(k-r+3)}{(r-2)!}a^{k-r+2}x^{r-1} + \ldots + x^{k+1} .$$

Since

$$\frac{k(k-1)\ldots(k-r+2)}{(r-1)!}a^{k-r+2}x^{r-1} + \frac{k(k-1)\ldots(k-r+3)}{(r-2)!}a^{k-r+2}x^{r-1}$$

$$= \frac{k(k-1)\ldots(k-r+3)}{(r-2)!}a^{k-r+2}x^{r-1}\left\{ \frac{k-r+2}{r-1} + 1 \right\} = \frac{(k+1)k(k-1)\ldots(k-r+3)}{(r-1)!}a^{k-r+2}x^{r-1},$$

the produc may be written

$$(a+x)^{k+1} = a^{k+1} + (k+1)a^k x + \ldots + \frac{(k+1)k(k-1)\ldots(k-r+3)}{(r-1)!}a^{k-r+2}x^{r-1}$$

$$+ \ldots + x^{k+1}$$

which is the binomial formula with n replaced by $k+1$.

Hence if the formula is true for $n = k$, it is true for $n = k + 1$. But the formula holds for $n = 1$; hence it holds for $n = 1 + 1 = 2$, and so on. Thus the formula is true for all positive integral values of n.

CHAPTER 2

SET THEORY

SETS AND SUBSETS

> List all the subsets of C = {1,2}.

Solution: {1}, {2}, {1,2}, $\emptyset$, where $\emptyset$ is the empty set.
Each set listed in the solution contains at least one
element of the set C. The set {2,1} is identical to {1,2}
and therefore is not listed. $\emptyset$ is included in the solution because $\emptyset$ is a subset of every set.

> Find four proper subsets of P = {n: n ε I, -5 < n ≤ 5}.

Solution: P = {-4, -3, -2, -1, 0, 1, 2, 3, 4, 5}. All
these elements are integers that are either less than or
equal to 5 or greater than -5. A set A is a proper subset
of P if every element of A is an element of B and in addition
there is an element of B which is not in A.
   (a) B = {-4, -2, 0, 2, 4} is a subset because each
element of B is an integer greater than -5 but less than or
equal to 5. B is a proper subset because 3 is an element
of P but not an element of B. We can write 3 ε P but
3 ∉ B.
   (b) C = {3} is a subset of P, since 3 ε P. However,
5 ε P but 5 ∉ C. Hence, C ⊂ P.
   (c) D = {-4, -3, -2, -1, 1, 2, 3, 4, 5} is a proper
subset of P, since each element of D is an element of P,
but 0 ε P and 0 ∉ D.
   (d) φ ⊂ P, since φ has no elements. Note that φ is

the empty set. φ is a proper subset of every set except
itself.

> If U = the set of whole numbers and E = the set of even whole
> numbers: find Ē.

<u>Solution</u>: Ē is called the complement of E. Ē is the set of all
elements in the universal set, U, that are not elements of E. Therefore,

$$\bar{E} = \{1,3,5,\ldots\},$$

the set of odd whole numbers.

> If U = { 1, 2, 3, 4, 5} and A = { 2, 4}, find A'.

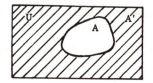

<u>Solution</u>: The complement of a set A in U is the set of all
elements of U that do not belong to A. The symbol A' (or,
sometimes, Ā, ⌐A, or Â) denotes the complement of A in U.
The figure gives a representation of A', the complement of
A in U. In this problem, since,

    U = { 1, 2, 3, 4, 5}

and

    A = { 2, 4},

    A' = { 1, 3, 5}.

> U = {1,2,3,4,5,6,7,8,9,10}, P = {2,4,6,8,10}, Q = {1,2,3,4,5}.
> Find (a) P̄ and (b) Q̄ .

<u>Solution</u>: P̄ and Q̄ are the complements of P and Q respectively.

That is, P̄ is the set of all elements in the universal set, U, that
are not elements of P, and Q̄ is the set of elements in U that are
not in Q. Therefore,

    (a) P̄ = {1,3,5,7,9}; (b) Q̄ = {6,7,8,9,10}

Given the universal set U = {2,4,6,...,12}

   (a) Find the set S = {x ∈ U|x²−5x+6 = 0}

   (b) Find the set S if U is changed to be
       U = {0,1,2,3,...,10}.

Solution: (a) Solving the equation $x^2 - 5x + 6 = 0$,

$$(x-2)(x-3) = 0$$

$$x = 2$$

$$x = 3$$

Thus, the solution set is {2,3}. Since the universal set is

$$U = \{2,4,6,\ldots,12\}, \quad S = \{2\}.$$

Note that 3 ∉ S, although 3 is one of the two solutions of the equation, since S cannot have any element that is not in U.

(b) The solution set of the given equation is {2,3}. Since both elements in the solution set are in U, S = {2,3}.

Sketch the graph of the subset S of the universal set U, U = {All real numbers}, where

   S = {(x,y) | y ≤ x and 1 < x < 8}.

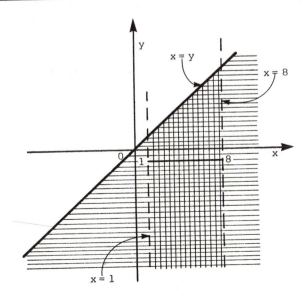

Solution: The elements of S are the points, (x,y), where
x ≥ y and 1 < x < 8.

They are all the points on or below the line x = y and
between the lines x = 1 and x = 8.

Thus, the graph of the set S is the area bound by the
lines x = 1, x = 8, and x = y as indicated in the figure by
the double-hatched area.

● PROBLEM 2-8

Given the set S = {1,2,3,4,5,6} find a partition of S.

Solution: A partition of a set S is a subdivision of the set
into subsets that are disjoint and exhaustive, i.e., every
element of S must belong to one and only one of the subsets.
Each subset in the partition is also called a cell. Therefore, $(S_1, S_2, \ldots, S_n)$ is a partition of S if

(a) $S_i \cap S_j = \phi$ (where ϕ is the empty set) for all $i \neq j$

(the cells are disjoint), and

(b) $S_1 \cup S_2 \cup S_3 \cup \ldots \cup S_n = S$ (the cells are exhaustive).

Hence, one of the partitions of S is

$$\{ \{1,2,3\}, \{4\}, \{5,6\} \}.$$

The partition { {1},{2},{3},{4},{5},{6} } is a partition
into unit sets.

● PROBLEM 2-9

Prove that the power set P(A) of any set A of n elements contains exactly 2^n elements.

Solution: If A is an empty set then it has no elements,
n = 0. A = $\phi \in$ P(A) and the power set of A = ϕ consists of
only one element, namely, the empty set ϕ = A such that, 2^n
= 2^0 = 1. Therefore, the theorem is true for A = ϕ. For a
nonempty set A = $\{a_1, a_2, a_3, a_4, \ldots, a_n\}$, P(A) is the set of

all subsets of A. There are C(n,1) = n singleton subsets belonging to P(A), $\{a_1\}, \{a_2\}, \ldots \{a_n\}$. There are exactly

C(n,2) subsets of A containing exactly two elements of A,

$\{a_1, a_2\}, \{a_1, a_3\}, \ldots \{a_1, a_n\}$, $\{a_2, a_3\}, \{a_2, a_4\} \ldots \{a_2, a_n\}$,

$\{a_3, a_4\}, \ldots \{a_3, a_n\} \ldots, \{a_{n-1}, a_n\}$.

Continuing, there are C(n,3) subsets of A containing exactly three elements, C(n,4) subsets of A containing exactly four, etc. Finally, there is exactly one, C(n,n) = 1, subset of A containing all n elements of A (A itself), and one empty set ϕ (C(n,0) = 1), $\phi \in P(A)$. Therefore, the total number of subsets of A of n elements is given by

$$C(n,0) + C(n,1) + C(n,2) + \ldots + C(n,n).$$

Using the binomial expansion for $(1+1)^n$, we have

$$(1+1)^n = C(n,0) + C(n,1) + C(n,2) + \ldots + C(n)$$

$$= 2^n$$

Hence, the number of elements in P(A) is 2^n.

● **PROBLEM 2-10**

Find the power set of the "tripleton" set S = {a,b,c}, and the number of elements of P(S).

Solution: The subsets of S are:

the empty set ϕ,   {a},{b},{c},{a,b}, {a,c}, {b,c}, and the

set S itself.

By definition, the power set of the set S is the set of all subsets of S.

Therefore, the power set is

P(S) = {ϕ,{a},{b},{c},{a,b},{a,c},{b,c},{a,b,c} }

The number of elements of set P(S) is found according to the following:

if A is a set with n(A) = m, then n(P(A)) = 2^m.

For this problem, n(S) = 3, m = 3, and n(P(S)) = $2^m = 2^3 = 8$. Hence, P(S) has 8 elements.

● **PROBLEM 2-11**

Give the definitions of a finite and an infinite set, and two examples of finite sets and infinte sets.

Solution: By definition, a set S is infinite provided that it has a proper subset P such that there exists a one-to-one correspondence between S and P. A set is finite if it is not infinite.

The empty set and the singleton set are two examples of the finite set. This is because the empty set does not have any proper subset, and the singleton set (a set which consists of one element alone) has only one proper subset, namely the empty set which cannot be put into one-to-one correspondence with the singleton set itself.

The set of all real numbers and the set of all positive odd integers are two examples of the infinite set.

Determine whether the following sets are inductive sets.

(1)   A = {1,2,3,4,...}

(2)   B = {2} ∪ {x|x³ - 17x² + 19x - 12 = 0}

Solution: (1) An inductive set S must have the following properties:

   (1) 1 ∈ S

   (2) If a ∈ S then (a+1) ∈ S

Since 1 ∈ A and for every element a ∈ A,(a+1) ∈ A (because A is the set of all positive integers), A is an inductive set.

(2)   Set {x|x³ - 17x² + 19x - 12 = 0} can be expressed in tabular form as {1,3,4}. This is because

$$x^3 - 17x^2 + 19x - 12 = 0$$

$$(x-1)(x-3)(x-4) = 0 \quad \text{and, solving for x, one has:}$$

$$x_1 = 1, \quad x_2 = 3, \quad \text{and} \quad x_3 = 4.$$

Therefore,   B = {2} ∪ {x|x³ - 17x² + 19x - 12 = 0}

= {2} ∪ {1,3,4} = {1,2,3,4}. We have 1 ∈ B. However, 4 ∈ B and 4 + 1 ∉ B.

Therefore, B is not an inductive set.

Prove that P ⊂ Q, where

Q = {(x,y) | |x+y| < 21} and

P = {(x,y) | |x-1| < 10   and |y-1| < 9}.

Solution: By definition, A is a subset of a set B if every element in A is also an element of B. This relationship is written $A \subseteq B$. If A is a non-empty subset of B and $A \neq B$, then it is written $A \subset B$, and A is called a proper subset of B. Hence, to prove $P \subset Q$, it must be shown that every element of P is an element of Q and that Q contains at least one element which does not belong to P. This can be done by proving that every ordered pair (x,y) which satisfies the conditions $|x-1| < 10$ and $|y-1| < 9$ also satisfies the condition $|x+y| < 21$, and the existence of at least one ordered pair (x_0,y_0) which satisfies the condition $|x+y| < 21$, but does not satisfy

$$|x-1| < 10 \quad \text{and} \quad |y-1| < 9.$$

The absolute value $|x+y|$ can be written as the following:

$$|x+y| = \left| \left[(x-1)+1 \right] + \left[(y-1)+1 \right] \right|$$
$$\leq |(x-1)+1| + |(y-1)+1|$$
$$\leq |x-1| + 1 + |y-1| + 1$$
$$\leq |x-1| + |y-1| + 2$$

Substitute $|x-1| < 10$ and $|y-1| < 9$, which are the conditions that every element of P must satisfy, into the above inequality.

$$|x+y| \leq |x-1| + |y-1| + 2 < 10 + 9 + 2 = 21$$
$$|x+y| < 10 + 9 + 2 = 21$$
$$|x+y| < 21$$

Thus, all elements of P also satisfy the condition $|x+y| < 21$. In other words, every element of P is also an element of Q.

Furthermore, by inspection, $(12,-9)$ belongs to Q, but it does not belong to P. Hence, there exists at least one ordered pair $(x_0,y_0) = (12,-9)$ which satisfies the condition $|x+y| < 21$, but does not satisfy the conditions $|x-1| < 10$ and $|y-1| < 9$. Since $|12-9| = 3 < 21$, but $|12-1| = 11 \not< 10$, and $|-9-1| = 10 \not< 9$, $P \subset Q$.

● PROBLEM 2-14

If $A = \{(x,y) \mid |x-4| \leq 1 \text{ and } |y-3| \leq 1\}$ and

$B = \{(x,y) \mid x^2+y^2-8x-6y+21 \leq 0\}$. Show that $A \subset B$.

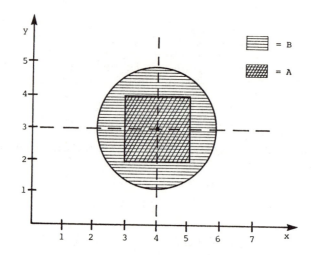

Solution: $x^2 + y^2 - 8x - 6y + 21 \le 0$ can be rewritten as $(x-4)^2 + (y-3)^2 \le 4$, and

$$(x - 4)^2 + (y - 3)^2$$

$$= |x - 4|^2 + |y - 3|^2 \le 4$$

Substitute $|x-4| \le 1$ and $|y-3| \le 1$ which expresses the set A in terms of the above inequality,

$$|x - 4|^2 + |y - 3|^2 \le 1 + 1 = 2 < 4.$$

Therefore, $A \subseteq B$ since all ordered pairs (x,y) which satisfy the definition of set A also satisfy the definition of set B.

It can be observed that there is at least one ordered pair that is in B but not in A. For instance, $(2,3)$ is in B. $(2-4)^2 + (3-3)^2 = 4$; but it is not in A. $|2-4| = 2 \not\le 1$. Thus, $A \subset B$.

The graphs of sets A and B are shown in the figure.

● **PROBLEM 2-15**

For this problem we shall use the following definition:

If Y is a subset of X then $P(X:Y) = \{A \in P(X) | Y \subseteq A\}$

Let   Y = {1,2},

X = {1,2,3,4,5},

Z = {a,1}.

(A) List all the elements of the set $P(X:Y)$.

(B) List the elements of the set $P(Y:Z)$.

(C) Show that for every set A, $P(A:\phi) = P(A)$.

Solution: First we shall list the elements of P(Y) and P(X).
Recall that the power set of a set is the set containing all
subsets of that set (including the null set). Recall also
that if a set has n elements the power set of that set will
have 2^n elements.

We see that P(Y) has $2^2 = 4$ elements:

$$P(Y) = \left[\phi, \{1\}, \{2\}, \{1,2\} \right] .$$

Similarly, P(X) has $2^5 = 32$ elements which are listed
below.

$$P(X) = \left[\phi, \{1\}, \{2\}, \{3\}, \{4\}, \{5\}, \{1,2\}, \{1,3\}, \{1,4\}, \right.$$

$$\{1,5\}, \{2,3\}, \{2,4\}, \{2,5\}, \{3,4\}, \{3,5\}, \{4,5\},$$

$$\{1,2,3\}, \{1,2,4\}, \{1,2,5\}, \{1,3,4\}, \{1,3,5\},$$

$$\{1,4,5\}, \{2,3,4\}, \{2,3,5\}, \{2,4,5\}, \{3,4,5\},$$

$$\{1,2,3,4\}, \{2,3,4,5\}, \{1,3,4,5\}, \{1,2,3,5\},$$

$$\left. \{1,2,4,5\}, \{1,2,3,4,5\} \right] .$$

(A) Now, by definition, P(X:Y) consists of those elements
of P(X) for which Y is contained as a subset of that
element. These are:

$$P(X:Y) = \left[\{1,2\}, \{1,2,3\}, \{1,2,4\}, \{1,2,5\}, \{1,2,3,4\}, \right.$$

$$\left. \{1,2,3,5\}, \{1,2,4,5\}, \{1,2,3,4,5\} \right] .$$

(B) Similarly, we obtain P(Y:Z) = ϕ. This follows from the
fact that no element of P(Y) contains the element a as
a member. Hence, for no element of P(Y) will Z be
contained in it (since a $\in$ Z).

(C) By definition given a set A we have:

$$P(A:\phi) = \{B \in P(A) \mid \phi \subseteq B\}.$$ But, notice that ϕ is

contained in every set as a subset. This implies that
for every B $\in$ P(A), $\phi \in$ B. In other words,

$$P(A:\phi) = P(A).$$

● PROBLEM 2-16

Prove that the empty set ϕ is a subset of every set.

P	Q	P→Q
T	T	T
T	F	F
F	T	T
F	F	T

Fig. 1

Solution: First recall the definition of subset. Given two sets X and Y, we say X is a subset of Y, or X ⊆ Y, if and only if all elements of X are also elements of Y. For the above proof we must also remember the truth table for the logical connective "implies". This is shown in Figure 1.

Observe that the only time "implies", (⇒), has a value of "False" is when the antecedent has a value of "True" and the consequence has a value of "False. Now lets return to our definition of subset. If we express the definition of X ⊆ Y using symbolic logic we get:

$$X \subseteq Y \Leftrightarrow \left[(x \in X) \Rightarrow (x \in Y) \right]$$

We can derive a false statement on the right-hand side of the equivalence only by having x ∈ X true and x ∈ Y false, The statement that the null set is a subset of every set is translated into

$$\phi \subseteq Y \Leftrightarrow \left[(x \in \phi) \Rightarrow (x \in Y) \right]$$

for every set Y. Notice that the antecedent of the right-hand statement is always false, (i.e., x ∈ φ), since the null set contains no elements. Hence, the implication is always true regardless of the truth or falsity of the consequence.

Therefore, φ ⊆ Y is true for any set Y thereby completing the proof.

SET OPERATIONS

● **PROBLEM 2-17**

Give the definition of

(1)   The complement of a subset S of the universal set U.

(2)   The intersection of two subsets A and B of the universal set U.

(3)   Disjoint sets.

(4)   The inclusive union of two subsets A and B of the

universal set U.

(5) The exclusive union of two subsets A and B of the universal set U.

(6) The difference of two subsets A and B of the universal set U.

Solution: (1) The complement of a subset S of the universal set U is the set of elements which belong to U but do not belong to S. It is denoted as S^c, or $\sim S$, or S'.

(2) The intersection of two subsets A and B of the universal set U, denoted by A ∩ B, is the set of elements which belong to both A and B.

$$A \cap B = \{X \mid X \in A \text{ and } X \in B\}$$

(3) If A ∩ B = φ, that is, if A and B do not have any elements in common, then A and B are said to be disjoint, or non-intersecting.

(4) The inclusive union of two subsets A and B of the universal set U, denoted by A ∪ B, is the set of elements which belong to A or B, or both:

$$A \cup B = \{X \mid X \in A, \text{ or } X \in B, \text{ or both.}\}$$

(5) The exclusive union of two subsets A and B of the universal set U is the set of all elements of U that are elements of A or B, but not both. This set is denoted by A ∪ B,

$$A \underset{\smile}{\cup} B = \{X \mid X \in A, \text{ or } X \in B, \text{ but not both}\}$$

(6) The difference of A and B is also called the relative complement of a set B with respect to a set A, denoted by A \ B or A - B, the relative complement of B with respect to A is the set of elements which belong to A but do not belong to B:

$$A \setminus B = \{X \mid X \in A, X \notin B\}.$$

● **PROBLEM 2-18**

State the laws of set operations.

Solution: If S is an algebra of sets, and if A,B,C,...,∅, U, are elements of S, then the following hold for U, ∩, and '.

IDENTITY LAWS

1a. A ∪ ∅ = A   1b. A ∩ ∅ = ∅

2a. A ∪ U = U   2b. A ∩ U = A

IDEMPOTENT LAWS

3a. A ∪ A = A   3b. A ∩ A = A

COMPLEMENT LAWS

4a. A ∪ A' = U   4b. A ∩ A' = φ
5a. (A')' = A   5b. ∅' = U; U' = ∅

COMMUTATIVE LAWS

6a. A ∪ B = B ∪ A   6b. A ∩ B = B ∩ A

ASSOCIATIVE LAWS

7a. (A ∪ B) ∪ C = A ∪ (B ∪ C)
7b. (A ∩ B) ∩ C = A ∩ (B ∩ C)

DISTRIBUTIVE LAWS

8a. A ∪ (B ∩ C) = (A ∪ B) ∩ (A ∪ C)
8b. A ∩ (B ∪ C) = (A ∩ B) ∪ (A ∩ C)

DE MORGAN'S LAWS

9a. (A ∪ B)' = A' ∩ B'   9b. (A ∩ B)' = A' ∪ B'

● **PROBLEM 2-19**

Show that the complement of the complement of a set is the
set itself.

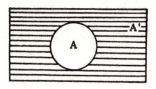

Solution: The complement of set A is given by A'. Therefore, the complement of the complement of a set is given by
(A')'. This set, (A')', must be shown to be the set A;
that is, that (A')' = A. In the figure the complement of
the set A, or A', is the set of all points not in set A;
that is, all points in the rectangle that are not in the
circle. This is the shaded area in the figure. Therefore,
this shaded area is A'. The complement of this set, or
(A')', is the set of all points of the rectangle that are
not in the shaded area; that is, all points in the circle,
which is the set A. Therefore, the set (A')' is the same
as set A; that is,

$$(A')' = A.$$

If a = { 1, 2, 3, 4, 5} and b = { 2, 3, 4, 5, 6}, find a ∪ b.

Solution: The symbol ∪ is used to denote the union of sets. Thus a ∪ b (which is read the union of a and b) is the set of all elements that are in either a or b or both. In this problem, if,

then   a = { 1, 2, 3, 4, 5} and b = { 2, 3, 4, 5, 6},

a ∪ b = { 1, 2, 3, 4, 5, 6}.

If a = { 1, 2, 3, 4, 5) and b = { 2, 3, 4, 5, 6}, find a ∩ b.

Solution: The intersection of two sets a and b is the set of all elements that belong to both a and b; that is, all elements common to a and b. In this problem, if

a = { 1, 2, 3, 4, 5} and b = { 2, 3, 4, 5, 6},

then   a ∩ b = { 2, 3, 4, 5}.

If A = {2,3,5,7} and B = {1,-2,3,4,-5,$\sqrt{6}$}, find
(a) A ∪ B and (b) A ∩ B.

Solution: (a) A ∪ B is the set of all elements in A or in B or in both A and B, with no element included twice in the union set.

A ∪ B = {1,2,-2,3,4,5,-5,$\sqrt{6}$,7}

(b) A ∩ B is the set of all elements in both A and B.

A ∩ B = {3}

Sometimes two sets have no elements in common. Let S = {3,4,7} and T = {2,-4,6}. What is the intersection of S and T? In this case S ∩ T has no elements. Hence S ∩ T = ∅, the empty set. In that case, the sets are said to be disjoint.

The set of all elements entering a discussion is called the universal set, U, When the universal set is not given, we assume it to be the set of real numbers. The set of all elements in the universal set that are not elements of A is called the complement of A, written Ā .

Given U = {1,2,3,4,5,6,7}, A = {1,2}, B = {3,4}, and C = {7}. Find U', A ∪ C, A ∩ B, (A')', (B ∪ C)', and (A ∪ B) ∩ C.

<u>Solution</u>:   U' = empty set φ.

$$A' = \{a \mid a \in U \text{ but } a \notin A\}$$

$$= \{3,4,5,6,7\}.$$

$$A \cup C = \{x \mid x \in A \text{ or } x \in C\}$$

$$= \{1,2,7\}.$$

$$A \cap B = \{x \mid x \in A \text{ and } X \in B\}$$

$$= \phi$$

$$(A')' = A = \{1,2\}.$$

$$(B \cup C) = \{x \mid x \in B \text{ or } X \in C\}$$

$$= \{3,4,7\}, \text{ hence}$$

$$(B \cup C)' = \{1,2,5,6\}.$$

$$A \cup B = \{1,2,3,4\}, \quad \text{thus}$$

$$(A \cup B) \cap C = \{x \mid x \in (A \cup B) \text{ and } x \in C\}$$

$$= \phi.$$

Given U = {a,b,c,d,e,f,g}, A = {a,b}, B = {c,d,e,f,g}, C = {a,d,e,g}, and D = {a}.

Find:

(1) A^C

(2) D^C

(3) $A^C \backslash B^C$

(4) $\left[A \cup B \cap C \cup U \backslash D \right]^C$

(5) $(A \cap B)^C$

<u>Solution</u>:   (1) By definition, $A^C = \{a \mid a \in U, a \notin A\}$ i.e. A^C is the complement of A. Thus, $A^C = \{c,d,e,f,g\}$.

(2) $D^C = \{b,c,d,e,f,g\}$.

(3) By definition, $X\backslash Y = \{x \mid x \in X$ and $x \notin Y\}$. $X\backslash Y$, (or $X - Y$), is the set difference of X and Y or the relative complement of a set Y with respect to a set X.

Since $A^C = \{c,d,e,f,g\}$, and

$B^C = \{a,b\}$, we have:

$A^C\backslash B^C = \{c,d,e,f,g\}$.

$X\backslash Y$ can also be writen as $X\backslash Y = \{x \mid x \in X$ and $x \in Y^C\}$ which implies that $X\backslash Y = X \cap Y^C$. Observe that $(B^C)^C = B = \{c,d,e,f,g\}$ so that $A^C \cap (B^C)^C = A^C \cap B = \{c,d,e,f,g\}$

$$= A^C\backslash B^C.$$

Furthermore, note that since $A^C \cap B^C = \phi$ we can directly write $A^C\backslash B^C = A^C$. This is justified since $A^C\backslash B^C = A^C \cap (B^C)^C = A^C \cap B$ and $A^C = B$, thus $A^C \cap B = A^C = B$. That is $A^C \cap B = A^C\backslash B^C = A^C$.

(4) $A \cup B = U$,

$U \cap C = C$,

$C \cup U = U$,

$U\backslash D = U \cap D^C = D^C$,

$(D^C)^C = D$.

Therefore,

$$\left[A \cup B \cap C \cup U \backslash D\right]^C = D = \{a\}.$$

(5) Since $A \cap B = \phi$,

$$\phi^C = U = \{a,b,c,d,e,f,g\}.$$

Find $A - B$ and $A - (A \cap B)$ for

$A = \{1,2,3,4\}$ and

$B = \{2,4,6,8,10\}$.

Solution: The relative complement of subsets A and B of the universal set U is defined as the set,

$$A - B = \{x \mid x \in A$ and $x \in B\}.$$

Note that it is not assumed that $B \subseteq A$. Hence,

$$A - B = \{1,3\}.$$

To find $A - (A \cap B)$, first find $A \cap B$. The set $A \cap B$ is a set of elements that are common to both A and B, so $A \cap B = \{2,4\}$.

Therefore, $A - (A \cap B) = \{1,3\}$.

● **PROBLEM** 2-26

Let the universal set

$$U = \{0,1,2,3,4,5,6,7,8\}.$$

Let $X = \{1,4,7\}$,

        $B = \{3,4,8\}.$

(A)   Find $X \cup B$

(B)   Find $X \cap B$

(C)   Find $X - B$

(D)   Find $B - X$

Solution: (A)   Recall that, by definition, $X \cup B$ is the subset of U given that

$$X \cup B = \{a \mid a \in X \text{ or } a \in B\}.$$

In this example:

$$X \cup B = \{1,3,4,7,8\}.$$

(B) $X \cap B$ is defined by

$$X \cap B = \{a \mid a \in X \text{ and } a \in B\}.$$

In this example:

$$A \cap B = \{4\}$$

(C)   Recall that the difference of two sets $X - B$ is the relative complement of the second set with respect to the first, i.e.:

$$X - B = X \cap B'$$

$$= \{a \mid a \in X \text{ and } a \notin B\}.$$

In our example we have

$$X - B = \{1,7\}.$$

(D)   Finally, B - X

$$= \{a \mid a \in B \text{ and } a \notin X\}$$

$$= \{3,8\}.$$

Given: U = {a,b,c,1,2,3},

          A = {a,b,3} and B = {a,1,2,3},

find:

(1)   A' ∪ B'

(2)   A ∩ B

(3)   (A ∩ B') ∪ (A' ∩ B)

Solution: (1)   By definition, the complement of a subset S of the universal set U is the set of elements of U that are not elements of S.   This set is denoted by S',

$$S' = \{x \in U \mid x \notin S\}.$$

Thus,   A' = {c,1,2}, B' = {b,c}.

The inclusive union of two sets S_1 and S_2 (S_1 and S_2 are the subsets of the universal set U) is defined as

$$S_1 \cup S_2 = \{x \in U \mid x \in S_1 \text{ or } x \in S_2 \text{ or both}\}.$$

So,   A' ∪ B' = {b,c,1,2}.

(2)   The intersection of two subsets S_1 and S_2 of the universal set U is defined as the set of elements of U that are elements of both S_1 and S_2.

$$S_1 \cap S_2 = \{x \in U \mid x \in S_1 \text{ and } x \in S_2\}.$$

Hence,   A ∩ B = {a,3}.

(3)   Since   A ∩ B' = {b} and A' ∩ B = {1,2},

$$(A \cap B') \quad (A' \cap B) = \{b\} \cup \{1,2\}$$

$$= \{b,1,2\}.$$

Given sets $S_1 = \{(x,y) \mid 4 \le x^2 + y^2 \le 16\}$

$S_2 = \{(x,y) \mid -4 \le x \le 2\}$

and the universal set U,

$U = \{\text{All real numbers}\}$

sketch the graph of S_1, S_2, and $S_1 \cap S_2$.

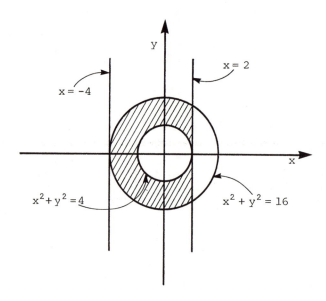

<u>Solution</u>: Sketching the graphs of $x^2+y^2 = 4$, $x^2+y^2 = 16$, $x = -4$, and $x = 2$ (see Figure 1).

The graph of S_1 is the area between the two circles $x^2+y^2 = 4$ and $x^2+y^2 = 16$. The graph of S_2 is the area between the lines $x = -4$ and $x = 2$.

By definition, the intersection of two subsets A and B of the universal set U is the set of all elements of U that are elements of both A and B.

That is, $A \cap B = \{x \in U \mid x \in A \text{ and } x \in B\}$.

Hence, the graph of $S_1 \cap S_2$ is the area bound by $x^2+y^2 = 4$, $x^2+y^2 = 16$, $x = -4$ and $x = 2$; this region is indicated in Figure 1 by the single-hatched area.

(1) Find set   S = A ∪ (B ∩ C)   where

      U = {2,4,6,8,10,x,y,z},

      A = {2,4,x,y},   B = {2,4,6,8,10},

and   C = {6,8,z}.

(2) Draw the Venn Diagram of the set A ∪ (B ∩ C).

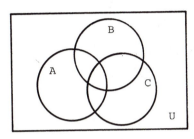

Fig. (a)

Solution: (1) First, find B ∩ C since this operation is in parentheses, B ∩ C = {6,8}. Since A = {2,4,x,y},

    A ∪ (B ∩ C) = {2,4,6,8,x,y} = S

(2) Draw three intersecting circles labeled A, B and C as shown in Fig.(a). The rectangle is used to represent the universal set U. Note that the universal set is partitioned into eight subsets (regions) labeled $R_1, R_2, R_3, \ldots R_8$ (see Fig.(b)).

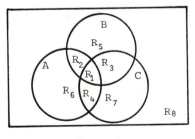

Fig. (b)

Furthermore,

      R_1 = A ∩ B ∩ C, R_2 ∪ R_1 = A ∩ B

      R_1 ∪ R_3 = B ∩ C, R_1 ∪ R_4 = A ∩ C

      R_1 ∪ R_2 ∪ R_4 ∪ R_6 = A, R_1 ∪ R_2 R_3 ∪ R_5 = B

and R_1 ∪ R_3 ∪ R_4 ∪ R_7 = C.

One may also have

$$R_6 = A - B - C, \quad R_5 = B - A - C, \quad \ldots \ldots \text{etc.}$$

To obtain the Venn diagram of $A \cup (B \cap C)$, shade A (regions R_1, R_2, R_4 and R_6) with strokes in one direction and shade $B \cap C$ (regions R_1 and R_3) with strokes in another direction; the total shaded area is $A \cup (B \cap C)$. See Fig.(c). The complete Venn diagram is shown in Fig.(d) where set $S = A \cup (B \cap C)$ is indicated by the shaded area.

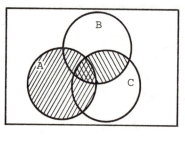

Fig. (c)

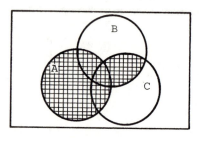

Fig. (d)

● PROBLEM 2-30

Given $A = \{1,2,4\}$, $B = \{3,4\}$, $C = \{4,5,6\}$ and

         $D = \{4,10\}$,

find   (1) $A \cup B \cup C \cup D$

       (2) $A \cap B \cap C \cap D$.

Solution: (1)   In general, the union of n (where n is a positive integer, $n \geq 2$) sets is defined as the set of elements which belong to any one (or more) of the n sets.

Therefore, $A \cup B \cup C \cup D = \{1,2,3,4,5,6,10\}$.

(2)   The intersection of n sets, ($n \geq 2$), is the set of elements which are common to all n sets. Hence,

       $A \cap B \cap C \cap D = \{4\}$.

● PROBLEM 2-31

Find $\underset{i=I}{U} A_i$ if $I = N$ – the set of natural numbers,

and

       $A_i = \{a \in R \mid 7 < a < i^2 + 10\}$,

where R is the set of real numbers.

Solution: $\underset{i=I}{U} A_i = \{a \in R \mid 7 < a < 1^2 + 10\} U \{a \in R \mid 7 < a < 2^2 + 10\}$

$$U \{a \in R \mid 7 < a < 3^2 + 10\} U \ldots$$

$$U \{a \in R \mid 7 < a < n^2 + 10\} U \ldots$$

$n \in N$, $0 < n < \infty$.

Therefore, $\underset{i=I}{U} A_i$ is the union of all sets $A_1, A_2, A_3,$
$\ldots \circ A_i \ldots$ where A_i is as given. The result is an infinite set of real numbers,

$$S = \underset{i=I}{U} A_i = \{x \in R \mid 7 < x < \infty\}.$$

● **PROBLEM** 2-32

Using the definitions of union and intersection show that, when viewed as binary operations on sets, union and intersection demonstrate, for all sets X, Y and Z:

(A) Commutativity

(B) Associativity

(C) Distributivity of intersection over union.

Solution: (A) Symbolically, the commutative law for union is:

$$X \cup Y = Y \cup X.$$

Using the definition of union and the commutativity of logical disjunction we get

$$a \in (X \cup Y) <=>$$

$$(a \in X) \vee (a \in Y) <=>$$

$$(a \in Y) \vee (a \in X) <=>$$

$$a \in (Y \cup X).$$

By once more using the definition of union, $X \cup Y = Y \cup X$.

Similarly, $(X \cap Y) = (Y \cap X)$

since $(a \in X) \wedge (a \in Y) <=>$

$$(a \in Y) \wedge (a \in X).$$

(B) Symbolically, the associative law for intersection is:

$$(X \cap Y) \ Z = X \cap (Y \cap Z).$$

55

Using the logical definition of intersection and the associative law of logical conjunction we get:

$$a \in \left[(X \cap Y) \cap Z\right] \iff$$

$$a \in (X \cap Y) \wedge a \in Z \iff$$

$$\left[(a \in X) \wedge (a \in Y)\right] \wedge (a \in Z) \iff$$

$$a \in X \wedge \left[(a \in Y) \wedge (a \in Z)\right] \iff$$

$$a \in X \cap (Y \cap Z).$$

The associative proof for union is similar.

(C) The distributive law of intersection over union may be written symbolically as:

$$X \cap (Y \cup Z) = (X \cap Y) \cup (X \cap Z)$$

To prove this we apply the logical definitions of union and intersection along with the rules for the logical connective conjunction and disjunction. Thus,

$$a \in \left[X \cap (Y \cup Z)\right] \iff$$

$$(a \in X) \wedge a \in (Y \cup Z) \iff$$

$$(a \in X) \wedge \left[(a \in Y) \cup (a \in Z)\right] \iff$$

$$\left[(a \in X) \wedge (a \in Y)\right] \cup \left[(a \in X) \wedge (a \in Z)\right] \iff$$

$$\left[a \in (X \cap Y)\right] \cup \left[a \in (X \cap Z)\right] \iff$$

$$a \in \left[(X \cap Y) \cup (X \cap Z)\right].$$

● **PROBLEM 2-33**

Illustrate one of De Morgan's Theorems with the use of Venn Diagrams.

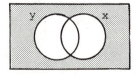

Fig. 1.

Solution: Recall that De Morgan's Theorems state that:

(1) The complement of the union of two sets is equal

56

to the intersection of the complements of the two sets.

(2) The complement of the intersection of two sets is equal to the union of the complements of the two sets.

$$(A \cup B)' = A' \cap B'$$

$$(A \cap B)' = A' \cup B'$$

De Morgan's first theorem for sets can be illustrated as in Figs. 1 and 2.

In Figure 1 we represent the set corresponding to the complement of the union of two sets, X and Y.

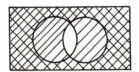

Fig. 2.

In Figure 2 we have the complement of X represented by the following hatched area: \\\\ . Similarly the complement of Y is represented by ////. Finally, the intersection of the complements of the two sets is depicted by the cross-hatched area:XXXX.

Observe that this cross-hatched region corresponds with the entire shocked region of Figure 1.

● **PROBLEM 2-34**

Prove that $(A \cap B) \cap C = A \cap (B \cap C)$.

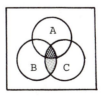

 - A∩B

- (A∩B)∩C

Fig. 1.

<u>Solution</u>: <u>Method 1</u>: Using the Venn diagram.

The Venn diagrams for (A ∩ B) ∩ C and A ∩ (B ∩ C) are shown in Figs. 1 and 2. Observe that the double shaded area in Figs. 1 and 2 are identical. Hence, (A ∩ B) ∩ C = A ∩ (B ∩ C).

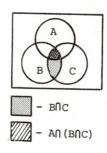

▨ – B∩C

▨ – A∩(B∩C)

Fig. 2.

<u>Method 2</u>: Let x be an element of set (A ∩ B) ∩ C and y be an element of set A ∩ (B ∩ C). x ∈ [(A ∩ B) ∩ C], i.e., x ∈ A ∩ B and x ∈ C.

If x ∈ A ∩ B, then x ∈ A and x ∈ B. Therefore, x ∈ A and x ∈ B and x ∈ C, which implies that x ∈ A and x ∈ (B ∩ C), i.e. x ∈ A ∩ (B ∩ C). Hence,

x ∈ [(A ∩ B) ∩ C] implies x ∈ [A ∩ (B ∩ C)],

which means (A ∩ B) ∩ C is a subset of

A ∩ (B ∩ C), $\left[(A \cap B) \cap C\right] \le \left[A \cap (B \cap C)\right]$.

By a similar procedure one proves that A ∩ (B ∩ C) ≤ (A ∩ B) ∩ C as follows:

y ∈ A ∩ (B ∩ C), i.e. y ∈ A and y ∈ (B ∩ C).

If y ∈ (B ∩ C) then y ∈ B and y ∈ C. Thus y ∈ A and y ∈ B and y ∈ C which implies

y ∈ A ∩ B and y ∈ C, i.e., y ∈ (A ∩ B) ∩ C.

Therefore, A ∩ (B ∩ C) ⊆ (A ∩ B) ∩ C.

By definition, if X ⊆ Y and Y ⊆ X, then X = Y.

Therefore, (A ∩ B) ∩ C = A ∩ (B ∩ C).

58

(1) Prove that $(A \cup B \cup C)' = A' \cap B' \cap C'$ where A, B, and C are subsets of the universal set U.

(2) Prove the dual of the equality in (1).

Solution: (1) $(A \cup B \cup C)' = [A \cup (B \cup C)]'$

Associative Law:

$P \cup (Q \cup S) = (P \cup Q) \cup S.$

$= A' \cap (B \cap C)'$

De Morgan's Law (treating B C as a single set.)

$(P \cup Q)' = P' \cap Q'.$

$= A' \cap (B' \cap C')$ De Morgan's Law

$= A' \cap B' \cap C'$ Associative Law:

$P \cap (Q \cap S) = (P \cap Q) \cap S.$

Therefore, $(A \cup B \cup C)' = A' \cap B' \cap C'$

(2) The dual of $(A \cup B \cup C)' = A' \cap B' \cap C'$ is $(A \cap B \cap C)' = A' \cup B' \cup C'$.

The dual of a set equality is obtained by replacing $\cup$'s by $\cap$'s and $\cap$'s by $\cup$'s, ϕ's by U's and U's by ϕ's, while all other elements in the equality are unchanged.

The proof is the following:

$(A \cap B \cap C)'$

$= [A \cap (B \cap C)]'$ Associative Law:

$P \cap (Q \cap S) = (P \cap Q) \cap S.$

$= A' \cup (B \cap C)'$ De Morgan's Law:

$(P \cap Q)' = P' \cup Q'.$

$= A' \cup (B' \cup C')$ De Morgan's Law

$= A' \cup B' \cup C'$ Associative Law

It can be shown that the proof of any set equality will consitute the proof of the dual equality, and hence, a proof of either quality guarantees the proof of both equalities. This is called the principle of duality.

If X, Y, and Z are any three subsets of the universal set U, prove that

$$(X \cap Y) - (X \cap Z) = X \cap (Y - Z).$$

Solution: If S is an algebra of sets and if A, B, C, . . ., $\emptyset$, U are elements of S, then the following hold for $\cup, \cap$, and '.

IDENTITY LAWS

1a. $A \cup \emptyset = A$ 1b. $A \cap \emptyset = \emptyset$
2a. $A \cup U = U$ 2b. $A \cap U = A$

IDEMPOTENT LAWS

3a. $A \cup A = A$ 3b. $A \cap A = A$

COMPLEMENT LAWS

4a. $A \cup A' = U$ 4b. $A \cap A' = \emptyset$
5a. $(A')' = A$ 5b. $\emptyset' = U; \ U' = \emptyset$

COMMUTATIVE LAWS

6a. $A \cup B = B \cup A$ 6b. $A \cap B = B \cap A$

ASSOCIATIVE LAWS

7a. $(A \cup B) \cup C = A \cup (B \cup C)$
7b. $(A \cap B) \cap C = A \cap (B \cap C)$

DISTRIBUTIVE LAWS

8a. $A \cup (B \cap C) = (A \cup B) \cap (A \cup C)$
8b. $A \cap (B \cup C) = (A \cap B) \cup (A \cap C)$

DE MORGAN'S LAWS

9a. $(A \cup B)' = A' \cap B'$ 9b. $(A \cap B)' = A' \cup B'$

By using the laws above, one obtains,

	LAWS OR THEOREMS USED:
$(X \cap Y) - (X \cap Z)$	
$= (X \cap Y) \cap (X \cap Z)'$	$A - B = A \cap B'$
$= (X \cap Y) \cap (X' \cup Z')$	9b.
$= (X \cap Y \cap X') \cup (X \cap Y \cap Z')$	8b.
$= (X \cap X' \cap Y) \cup (X \cap Y \cap Z')$	6b.
$= (\emptyset \cap Y) \cup (X \cap Y \cap Z')$	4b.

$= \emptyset \cup (X \cap Y \cap Z')$ 1b.

$= (X \cap (Y \cap Z'))$ 1a. and 7b.

$= X \cap (Y - Z)$ Theorem: $A - B = A \cap B'$

Hence, $(X \cap Y) - (X \cap Z) = X \cap (Y - Z)$

● **PROBLEM** 2-37

Simplify the following expressions:

(a) $(P \cup Q)' \cup (P' \cap Q)$

(b) $Q \cup [(P' \cup Q) \cap P]'$

Solution: If S is an algebra of sets and if A, B, C, ..., $\emptyset$, U are elements of S, then the following hold for $\cup, \cap$, and '.

IDENTITY LAWS

1a. $A \cup \emptyset = A$ 1b. $A \cap \emptyset = \emptyset$
2a. $A \cup U = U$ 2b. $A \cap U = A$

IDEMPOTENT LAWS

3a. $A \cup A = A$ 3b. $A \cap A = A$

COMPLEMENT LAWS

4a. $A \cup A' = U$ 4b. $A \cap A' = \emptyset$
5a. $(A')' = A$ 5b. $\emptyset' = U; \ U' = \emptyset$

COMMUTATIVE LAWS

6a. $A \cup B = B \cup A$ 6b. $A \cap B = B \cap A$

ASSOCIATIVE LAWS

7a. $(A \cup B) \cup C = A \cup (B \cup C)$
7b. $(A \cap B) \cap C = A \cap (B \cap C)$

DISTRIBUTIVE LAWS

8a. $A \cup (B \cap C) = (A \cup B) \cap (A \cup C)$
8b. $A \cap (B \cup C) = (A \cap B) \cup (A \cap C)$

DE MORGAN'S LAWS

9a. $(A \cup B)' = A' \cap B'$ 9b. $(A \cap B)' = A' \cup B'$

By applying the laws above, one obtains:

61

(1)   (P∪Q)'∪(P'∩Q)   <u>LAW USED</u>:

    = (P'∩Q')∪(P'∩Q)   Law 9a.

    = P'∩(Q'∪Q)   Law 8b.

    = P'∩U   Law 4a.

    = P'   Law 2b.

(2)   Q∪[(P'∪Q)∩P]'

    = Q∪[(P'∪Q)'∪P']   Law 9b.

    = Q∪⟨[(P')'∩Q']∪P'⟩   Law 9a.

    = Q∪[(P∩Q')∪P']   Law 5a.

    = Q∪[P'∪(P∩Q')]   Law 6a.

    = Q∪[(P'∪P)∩(P'∪Q')]   Law 8a.

    = Q∪[U∩(P'∪Q')]   Law 4a.

    = Q∪(P'∪Q')   Law 6b. and 2b.

    = Q∪(Q'∪P')   Law 6a.

    = (Q∪Q')∪P'   Law 7a.

    = U∪P'   Law 4a.

    = U   Laws 6b. and 2a.

VENN DIAGRAM

● **PROBLEM** 2-38

Draw the Venn diagram of sets

    U, A∩B, (A∪B)∩C and A'∩(B'∩C),

where A, B, C are subsets of the universal set U.

<u>Solution</u>:

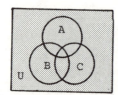

The shaded area
represents U

Fig. 1.

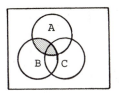

The shaded region
is A∩B

Fig. 2.

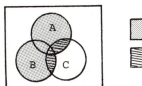

- A∪B

- (A∪B)∩C

Fig. 3.

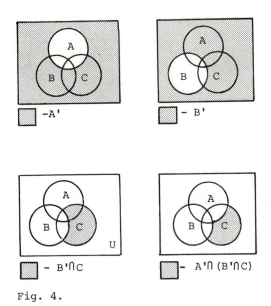

□ -A'

□ - B'

□ - B'∩C

□- A'∩ (B'∩C)

Fig. 4.

● **PROBLEM** 2-39

Prove that if A⊆B, then B'⊆A'. Draw the Venn diagram
of A⊆B, A' and B'.

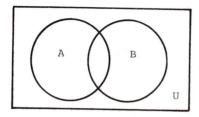 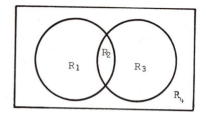

Fig. 1

Solution: If $A \subseteq B$, then every element a, $a \in A$, is also an element of B, whereas it is not necessary for every element of B to be in A.   Let x be any element of B', then $x \in B'$ and $x \notin B$.   Therefore, $x \notin A$, which implies that $x \in A'$.   Since any element x in B' is also in A', $B' \subseteq A'$.   The proof is now complete.

To draw the Venn diagram of this problem, first draw a rectangle for the universal set U, then draw two overlapping circles for A and B which partition the universal set into four subsets labeled R_1, R_2, R_3 and R_4, as shown in Fig. 1. $A \subseteq B$ means that A is contained in B.   Therefore, set $A - B$ (region R_1) is empty since $A - B$ is defined as the set of elements that are in A but not in B (see Fig. 2).

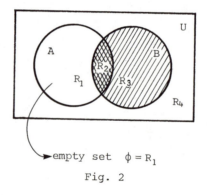

empty set $\phi = R_1$

Fig. 2

The Venn diagrams for A' and B' are shown in Fig. 3. Note that region R_1 is empty thus it's not shaded even though it belongs to B' (see Fig. 3(c)).   It is observed that the area shaded in Fig. 3(c) is also shaded in Fig. 3(b), and the only non-empty area (region R_3) that is not shaded in Fig. 3(c) is shaded in Fig. 3(b).

Therefore, $B' \subseteq A'$ is also verified by the Venn diagram.

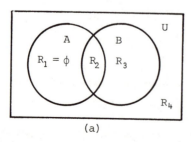 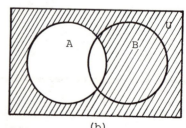

(a)   (b)

Venn diagram for A'

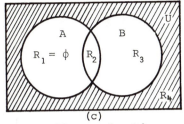

(c)

Venn diagram for B'

Fig. 3

● PROBLEM 2-40

Given the universal set U and its subsets A, B and C
where B ⊆ A and B ∩ C = φ. Draw the Venn diagram of

(1) A ∪ B ∩ C

(2) (A ∩ C)' ∩ B\C

(3) U \ A \ C

(4) A \ C \ B

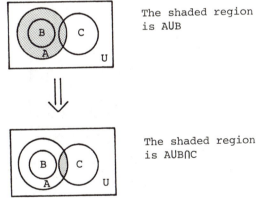

The shaded region
is A∪B

The shaded region
is A∪B∩C

Fig. 1.

Solution: (1) B ⊆ A, that is B is a subset of A, and
B ∩ C = φ, that is B and C are disjoint sets. In the Venn
diagram in Fig. 1 for A ∪ B ∩ C, the circle which represents B
is inside the circle that represents A and the circles representing B and C don't overlap.

(2) The Venn diagram is shown in Fig. 2.

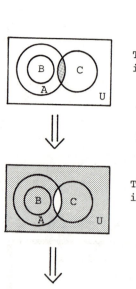

The shaded region is A∩C

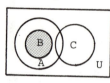

The shaded region is (A∩C)'

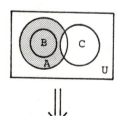

The shaded region is (A∩C)'∩B and it also represents (A∩C)'∩B∖C

Fig. 2.

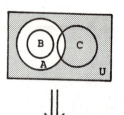

Venn diagram for U∖A

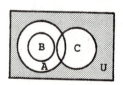

Venn diagram for U∖A∖C

Fig. 3.

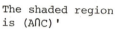

A∖C

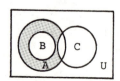

A∖C∖B

Fig. 4.

66

Prove $A \cup B' = (A' \cap B)'$ by using a Venn diagram.

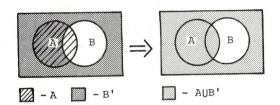

Fig. 1.

Solution: The Venn diagrams for $A \cup B'$ and $(A' \cap B)'$ are shown in Fig. 1 and Fig. 2, respectively. Observe that the shaded area in Fig. 1 is identical to that of Fig. 2, hence, $A \cup B' = (A' \cap B)'$.

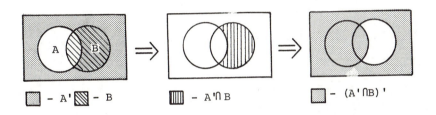

Fig. 2.

Simplify the set

$$(X \cup Y) \cup (X \cap Y)$$

and then use the Venn diagram to verify the result.

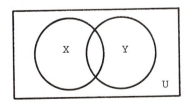

Fig. 1

Solution: If S is an algebra of sets and if A, B, C, . . .,

ϕ, U are elements of S, then the following hold for U, $\cap$, and '.

IDENTITY LAWS

1a. $A \cup \phi = A$ 1b. $A \cap \phi = \phi$
2a. $A \cup U = U$ 2b. $A \cap U = A$

IDEMPOTENT LAWS

3a. $A \cup A = A$ 3b. $A \cap A = A$

COMPLEMENT LAWS

4a. $A \cup A' = U$ 4b. $A \cap A' = \phi$
5a. $(A')' = A$ 5b. $\phi' = U;\ U' = \phi$

COMMUTATIVE LAWS

6a. $A \cup B = B \cup A$ 6b. $A \cap B = B \cap A$

ASSOCIATIVE LAWS

7a. $(A \cup B) \cup C = A \cup (B \cup C)$
7b. $(A \cap B) \cap C = A \cap (B \cap C)$

DISTRIBUTIVE LAWS

8a. $A \cup (B \cap C) = (A \cup B) \cap (A \cup C)$
8b. $A \cap (B \cup C) = (A \cap B) \cup (A \cap C)$

DE MORGAN'S LAWS

9a. $(A \cup B)' = A' \cap B'$ 9b. $(A \cap B)' = A' \cup B'$

$(X \cup Y) \cup (X \cap Y)$

$= [(X \cup Y) \cup X] \cap [(X \cup Y) \cup Y]$ Distributive Law 8a., treating $X \cup Y$ as a single term.

$= [(X \cup Y) \cup X] \cap [X \cup (Y \cup Y)]$ Associative Law 7a.

$= [(X \cup Y) \cup X] \cap (X \cup Y)$ Idempotent Law 3a.

$= [X \cup (X \cup Y)] \cap (X \cup Y)$ Commutative Law 6a.

$= [(X \cup X) \cup Y] \cap (X \cup Y)$ Law 7a.

$= (X \cup Y) \cap (X \cup Y)$ Law 3a.

$= X \cup Y$ Law 3b.

To verify the result by using Venn diagrams, first draw a rectangle to represent the universal set U, then draw two intersecting circles to represent sets X and Y. (see Fig. 1). The Venn diagrams for $X \cap Y$, $X \cup Y$, and

68

(X ∪ Y) ∪ (X∩ Y) are shown in Fig. (2),(3), and (4), respectively. It is observed that the shaded areas in
Fig.(3) and (4) are precisely the same. Hence, we have
verified the result: (X ∪ Y) ∪ (X∩Y) = X ∪ Y.

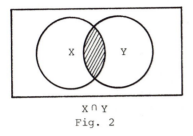

X ∩ Y

Fig. 2

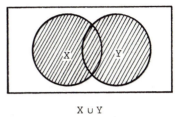

X ∪ Y

Fig. 3

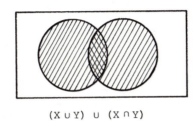

(X ∪ Y) ∪ (X ∩ Y)

Fig. 4

● **PROBLEM 2-43**

Given the Venn diagram in Fig. 1 with subsets A, B and
C of the universal set U as shown.   Shade the set
(A' ∩ B) \ C ∩ U (A ∩ B) ∪ (B ∩ C).

Solution: First, find A' and B, then A'∩ B as shown in Fig. 2.

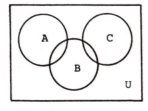

Fig. 1.

Set (A' ∩ B) \ C is shown in Fig. 3.

69

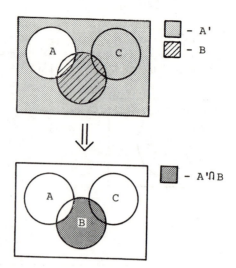

$-$ A'

$-$ B

$-$ A'$\cap$B

Fig. 2.

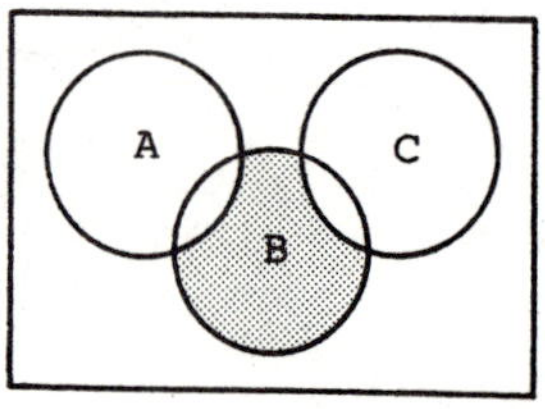

Fig. 3.

Since $X \cap U = X$, the intersection of any subset X of the universal set U with U is equal to X.

$$A' \cap B \setminus C \cap U = A' \cap B \setminus C.$$

The sets $A \cap B$ and $B \cap C$ are shown in Fig. 4 and 5, respectively.

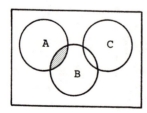

Fig. 4.

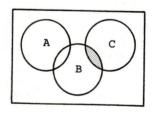

Fig. 5.

70

The Venn diagram for the union of the sets A' ∩ B \ C,
A ∩ B, and B ∩ C is shown in Fig. 6.

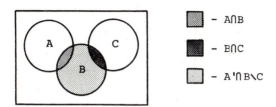

Fig. 6.

Therefore, we get the Venn diagram of the set
(A' ∩ B) \ C ∩ U(A ∩ B) ∪ (B ∩ C) as shown in Fig. 7.

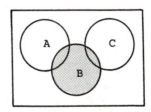

Fig. 7.

Observe that the given set is equivalent to set B, i.e.

A' ∩ B \ C ∩ U(A ∩ B) ∪ (B ∩ C) = B.

Thus, we see that the Venn diagram may be used to simplify a set expression.

● PROBLEM 2-44

Given the Venn diagrams in Figs. 1, 2 and 3 with subsets
A, B, C, D, E, F, G, H of the universal set U. Shade
the sets given in (1), (2), and (3) on the Venn diagrams
in Figs. 1, 2, and 3, respectively.

(1) A \ B \ C ∪ (D ∪ E)

(2) D ∪ B \ C ∪ (E ∩ F)

(3) (B \ A \ C) ∪ (C \ B \ E \ D) ∪ (G ∩ H) ∪ (H ∩ F)

        ∪ (G ∩ E\F)∪(E ∩ F\G)

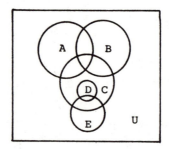

Fig. 1.

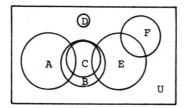

Fig. 2.

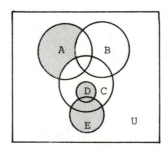

Fig. 3.

Solution:

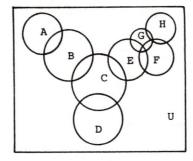

Fig. 4.

72

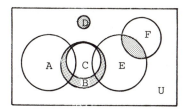

Fig. 5.

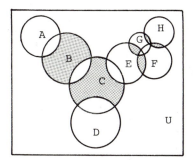

Fig. 6.

● PROBLEM 2-45

Write the expression which represents the set given by its Venn diagram in Fig. 1.

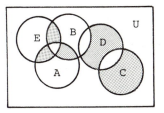

The set is represented
by the shaded region.

Fig. 1.

Solution: One of the straight forward ways of writing this expression is to find the expression for the small shaded regions first. If a certain region can't be represented by a simple expression with one set operation on two sets, one can try to break that region into smaller subregions for which a simple expression can be found.

Finally, use the union to join all the expressions for each shaded region and subregions.

73

Hence, the expression for the set given by the Venn diagram in Fig. 1 is found as shown in Fig. 2.

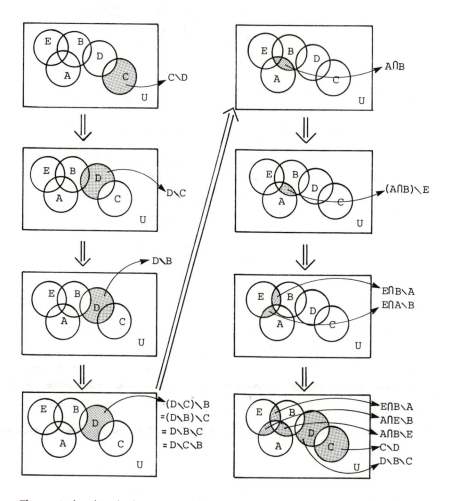

The set is $(C\backslash D)\cup(D\backslash B\backslash C)\cup(A\cap B\backslash E)\cup(A\cap E\backslash B)\cup(E\cap B\backslash A)$

Fig. 2.

● **PROBLEM 2-46**

Write the expressions for the sets represented by the Venn diagrams in Figs. 1, 2 and 3.

Solution: The set represented by the Venn diagram in Fig. 1 is

$$(A\backslash B\backslash D)\cup(B\cap D\backslash A\backslash C).$$

74

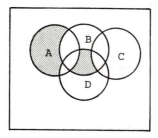

Fig. 1.

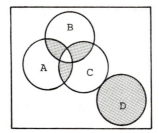

Fig. 2.

The Venn diagram in Fig. 2 represents the set

$$(A \cap B \setminus C) \cup (B \cap C \setminus A) \cup (A \cap C \setminus B) \cup D$$

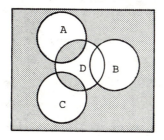

Fig. 3.

and, finally, the Venn diagram in Fig. 3 represents the set:

$$U \setminus A \setminus B \setminus C \setminus D \cup (A \cap D) \cup (C \cap D).$$

Note that the set represented by a Venn diagram is unique. However the expression which represents the set may not be unique. For example, the expression which represents the set represented by the Venn diagram in Fig. 3 may be written differently as

$$A' \setminus B \setminus C \setminus D \cup (A \cap D) \cup (D \cap C).$$

75

CARTESIAN PRODUCT

Find $A \times B$ and $B \times A$, for

$$A = \{1,2,3\} , B = \{a,b\}.$$

Solution: The cartesian-product set $A \times B$ is the set of all ordered pairs whose first components are chosen from A and whose second components are chosen from B.

$$A \times B = \{(x,y) \mid x \in A \text{ and } y \in B\}$$

The scheme for formation of all possible ordered pairs is as follows:

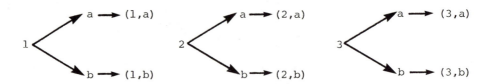

Hence,

$$A \times B = \{(1,a),(1,b),(2,a),(2,b),(3,a),(3,b)\}$$

To find $B \times A$, the first components of the ordered pairs are chosen from B and the second components from A.

Thus, $B \times A = \{(a,1),(a,2),(a,3),(b,1),(b,2),(b,3)\}$

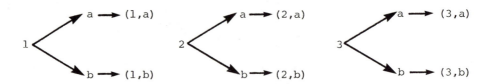

Here, $A \times B \neq B \times A$. This is generally true since the elements of $A \times B$ are usually different from those of $B \times A$.

Write the cartesian product $U \times U$ for $U = \{a,b\}$.

Solution: By definition, $U \times U = \{(x,y) \mid x \in U \text{ and } y \in U\}$.

Hence,

$$U \times U = \{(a,b),(b,a),(a,a),(b,b)\}.$$

The scheme for the formation of all the possible ordered pairs yields

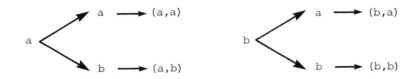

If a = { 1, 2, 3} and b = { 5, 6}, find a x b and b x a.

Solution: The Cartesian product of two sets a and b, denoted by a x b, is the set of all ordered pairs (x, y) such that x ε a and y ε b. In this problem, if a = { 1, 2, 3} and b = { 5, 6}, then the Cartesian product a x b is:

a x b = { (1, 5), (1, 6), (2,5), (2, 6), (3, 5),

(3, 6)}.

The Cartesian product b x a is the set of all ordered pairs (x, y) such that x ε b and y ε a. Hence, the Cartesian product b x a is:

b x a = { (5, 1), (5, 2), (5, 3), (6, 1), (6, 2),

(6, 3)}.

Let M = {1, 2} and N = {p, q}. Find (a) M × N, (b) N × M, and (c) M × M.

Solution: (a) M×N is the set of all ordered pairs in which the first component is a member of M and the second component is a member of N. Thus,

M × N = {(1,p), (1, q), (2, p), (2, q)}.

Note that the number of elements in M is 2,

the number of elements in N is 2,

and the number of elements in M × N = 2 × 2 = 4.

(b) N ×M is the set of all ordered pairs in which the first component is a member of N and the second component is a member of M. Thus,

$$N \times M = \{(p, 1), (q, 1), (p, 2), (q, 2)\}.$$

Once again note that the number of elements in $N \times M$ is $2 \times 2 = 4$.

(c) $M \times M$ is the set of all ordered pairs in which both components are members of M. Thus,

$$M \times M = \{(1, 1), (1, 2), (2, 1), 2, 2)\}.$$

Here too, the number of elements in $M \times M$ is $2 \times 2 = 4$.

● **PROBLEM** 2-51

Given $A = \{a,b,c\}$ and $B = \{1,2,3,4\}$; find $A \times B$, $B \times A$, $n(A \times B)$ and $n(B \times A)$.

Solution: By definition, $A \times B = \{(x,y) \mid x \in A \text{ and } y \in B\}$. Thus,

$$A \times B = \{(a,1),(a,2),(a,3),(a,4),(b,1),(b,2),(b,3),$$

$$(b,4),(c,1),(c,2),(c,3),(c,4)\}$$

Similarly,

$$B \times A = \{(1,a),(1,b),(1,c),(2,a),(2,b),(2,c),(3,a),$$

$$(3,b),(3,c),(4,a),(4,b),(4,c)\}$$

In general, if A and B are finite sets, A with $n(A)$ elements and B with $n(B)$ elements, then $A \times B$ contains $n(A \times B)$ elements and $n(A \times B) = n(A) \cdot n(B)$. Here, $n(A) = 3$, $n(B) = 4$. Therefore,

$$n(A \times B) = n(A) \cdot n(B) = 3 \cdot 4 = 12$$

$$n(B \times A) = n(B) \cdot n(A) = 4 \cdot 3 = 12$$

Note that even though normally $A \times B \neq B \times A$, $n(A) \cdot n(B) = n(B) \cdot n(A)$, so $n(A \times B) = n(B \times A)$.

This is understandable since there exists a one-to-one correspondence relationship between $A \times B$ and $B \times A$.

● **PROBLEM** 2-52

Use an example to show that $A \cup (B \times C) \neq (A \cup B) \times (A \cup C)$.

Solution: Let A = {a}, B = {b}, C = {c}, then

$$A \cup (B \times C) = \{a\} \cup \{(b,c)\}$$

$$= \{a, (b,c)\},$$

and (A ∪ B) × (A ∪ C) = {a,b} × {a,c}

$$= \{(a,a),(a,c),(b,a),(b,c)\}$$

Hence, $A \cup (B \times C) \neq (A \cup B) \times (A \cup C)$.

● PROBLEM 2-53

Verify that P×(Q ∪ R) = (P×Q) ∪ (P×R).

Solution: In order to prove the given equation, it must be shown that every element of set P×(Q ∪ R) is also an element in (P×Q) ∪ (P×R) and vice versa.

(a) P×(Q ∪ R) = {(x,y) | x ∈ P, and y ∈ (Q ∪ R)}

$$= \{(x,y) \mid x \in P, \text{ and } (y \in Q \text{ or } y \in R)\}$$

(b) (P×Q) ∪ (P×R)

$$= \{(x,y) \mid (x,y) \in [(P \; Q) \cup (P \times R)]\}$$

$$= \{(x,y) \mid (x,y) \in P \times Q \text{ or } (x,y) \in P \times R\}$$

$$= \{(x,y) \mid (x \in P \text{ and } y \in Q) \text{ or } (x \in P \text{ and } y \in R)\}$$

$$= \{(x,y) \mid x \in P \text{ and } (y \in Q \text{ or } y \in R)\}$$

Since P×(Q ∪ R) = {(x,y) | x ∈ P and (y ∈ Q or y ∈ R)},

and (P×Q) ∪ (P×R) is also defined as the set {(x,y) | x ∈ P, and (y ∈ Q or y ∈ R)}.

Therefore,

$$P \times (Q \cup R) = (P \times Q) \cup (P \times R).$$

● PROBLEM 2-54

If S is any subset of the universal set U, and ϕ is the empty set, find $S \times \phi$ and $\phi \times S$.

Solution: By definition, if A, B are any sets, A × B = {(x,y) | x ∈ A and y ∈ B}. Therefore, $S \times \phi$ = {(x,y) | x ∈ S, y ∈ φ}. Since y ∈ φ, and φ contains no elements, y doesn't exist.

In other words, there is no such ordered pair (x,y), where x ε S and y ε φ since y doesn't exist. Hence, S × φ = φ. By the same token, φ × S = φ.

Let: A = {1,3,5}.

Let: B = {2,x}.

Find: (A)   A × B

          (B)   B × A

          (C)   A × A

          (D)   B × B.

Solution: First recall that if X and Y are sets then the cartesian product of X and Y is:

$$X × Y = \{(a,b) \mid a ε X \text{ and } b ε Y\}.$$

In this example, we list the products by using the above definition applied to the specific sets A and B:

(A)   A × B = {(1,2),(1,x),(3,2),(3,x),(5,2),(5,x)}.

(B)   B × A = {(2,1),(2,3),(2,5),(x,1),(x,3),(x,5)}.

(C)   A × A = {(1,1),(1,3),(1,5),(3,1),(3,3),(3,5),

          (5,1),(5,3),(5,5)}.

(D)   B × B = {(2,2),(2,x),(x,2),(x,x)}.

Given the intervals

    A = {a | 0≤ a ≤4},   B = {b | 1≤ b ≤2}, and

    C = {c | -1 < c < 2};

find A × B and B × C.   Also graph A × C and (-∞,0] × [0,∞).

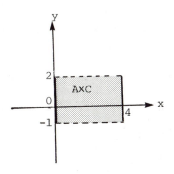

The set AxC is represented by the shaded
region. The solid line segments indicate
that the points on those two line segments
belong to AxC whereas the dotted line segments indicate that the points on those
segments are not elements of AxC.

Fig. 1.

<u>Solution</u>: Note that sets A, B and C are not sets of discrete
numbers but sets of continuous intervals. Cartesian products
can also be formed by using intervals.

$A \times B = \{(a,b) \mid a \in A$ and $b \in B\}$

$\quad\quad = \{(a,b) \mid 0 \le a \le 4,\ 1 \le b \le 2\}$

$B \times C = \{(b,c) \mid b \in B$ and $c \in C\}$

$\quad\quad = \{(b,c) \mid 1 < b \le 2,\ -1 < c < 2\}$

$A \times C = \{(a,c) \mid a \in A$ and $c \in C\}$

$\quad\quad = \{(a,c) \mid 0 \le a \le 4,\ -1 < c < 2\}$

The graph of $A \times C$ and the graph of the cartesian product of
intervals $(-\infty,0]$ and $[0,\infty)$ are shown in Fig. 1 and Fig. 2,
respectively.

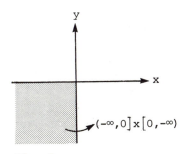

Fig. 2.

Given   A = {4,9} and B = {1,2},   find all the subsets
of A × B.

Solution: A × B = {(4,1),(4,2),(9,1),(9,2)}

The subsets of A × B are:

$\quad$ {(4,1)},{(4,2)},{(9,1)},{(9,2)},{(4,1),(4,2)},

$\quad$ {(4,1),(9,1)},{(4,1),(9,2)},{(4,2),(9,1)},{(4,2),(9,2)},

$\quad$ {(9,1),(9,2)},{(4,1),(4,2),(9,1)},{(4,1),(4,2),(9,2)},

$\quad$ {(4,1),(9,1),(9,2)},{(4,2),(9,1),(9,2)}, set A × B itself

$\quad$ and empty set φ = { }.

There are total 16 subsets.

Given A = {1,2}, B = {3,4}, and C ={5,6}. Find A x B x C.

Solution: The concept of cartesian product can be extended
to more than two sets. For example, the cartesian-product
set A × B × C can be defined as the set of all ordered triples
(a,b,c) where a ϵ A, b ϵ B, and c ϵ C.

$\quad$ That is,   A × B × C = {(a,b,c) | a ϵ A,b ϵ B, and c ϵ C}

Therefore,

$\quad$ A × B × C = {(1,3,5),(1,3,6),(1,4,5),(1,4,6),(2,3,5),

$\quad\quad\quad$ (2,3,6),(2,4,5),(2,4,6)}.

$\quad$ Schematically,   A × B × C   could be obtained as follows:

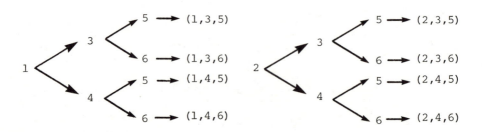

82

Given A = {2,4,6} and B = {y | 1 ≤ y ≤ 4 and y ∈ R}, sketch
the graph of A × B.

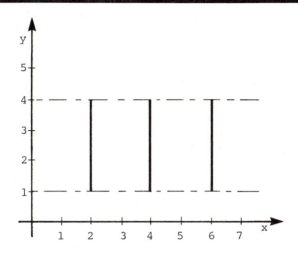

Solution: By definition, A × B = {(x,y) | x ∈ A, y ∈ B}. In the
set A × B, the first coordinate of each ordered pair which
represents a point on a segment is 2, 4, or 6; the second coordinate is any real number y, such that 1 ≤ y ≤ 4. The graph
of A × B consists of three vertical segments whose end points
are (2,0) and (2,4), (4,1) and (4,4), (6,1) and (6,4).

Find   A × B ∪ C \ B   where A = {1,2}, B = {3,4},

and   C = {1,2,3,4}.

Solution: A × B = {(a,b) | a ∈ A and b ∈ B}

                        = {(1,3),(1,4),(2,3),(2,4)}

                A × B ∪ C = {x | x ∈ A × B or x ∈ C}

                        = {(1,3),(1,4),(2,3),(2,4),(1,2,3,4)}

        A × B ∪ C \ B = {x | x ∈ A × B ∪ C and x ∉ B}

                        = {(1,3),(1,4),(2,3),(2,4),(1,2)}.

Prove that the cartesian product is non-associative.

Solution: One can prove that the cartesian product is not associative by using an example. For instance, let

$$A = \{1,2\}, \quad B = \{3,4\}, \quad C = \{5,6\}.$$

$$A \times B = \{(1,3),(1,4),(2,3),(2,4)\}$$

$$(A \times B) \times C = \{((1,3),5),((1,3),6),((1,4),5),((1,4),6),$$

$$((2,3),5),((2,3),6),((2,4),5),((2,4),6)\}$$

whereas,

$$A \times (B \times C) = \{(1,(3,5)),(1,(3,6)),(1,(4,5)),(1,(4,6)),$$

$$(2,(3,5)),(2,(3,6)),(2,(4,5)),(2,(4,6))\}.$$

It's obvious that

$$(A \times B) \times C \neq A \times (B \times C).$$

Given three sets X, Y and Z, prove the following facts regarding the cartesian product.

(A) $X \times (Y \cap Z) = (X \times Y) \cap (X \times Z)$

(B) $X \times (Y \cup Z) = (X \times Y) \cup (X \times Z)$

Solution: First, recall how the cartesian product is defined using symbolic logic. Given two sets A and B then:

$$A \times B = \{(a,b) \mid a \in A \wedge b \in B\}.$$

Our method of proof in both parts (A) and (B) will be to translate our original expression into symbolic logic and then use the laws and definitions of logic to derive an equivalent expression which is the statement of what we are trying to prove.

(A) From our definition of cartesian product we see that making the claim that

$$(a,x) \in X \times (y \cap Z)$$

translates into the following logically equivalent form:

$$(a \in X) \wedge (X \in Y \cap Z)$$

We may now apply the definition of intersection to obtain:

$$(a \in X) \wedge (x \in Y \wedge x \in Z)$$

We now use the idempotency of logical conjunction (i.e. the fact that $P \wedge P <=> P$), as well as the associativity of logical conjunction, yielding the following equivalent statement:

$$(a \in X) \wedge (a \in X) \wedge (x \in Y) \wedge (x \in Z)$$

From this we may use the commutative and associative laws of logical conjunction, yielding:

$$\left[(a \in X) \wedge (x \in Y)\right] \wedge \left[(a \in X) \wedge (x \in Z)\right]$$

Next we re-apply the definition of cartesian product (this time in reverse) and the two expressions in square brackets give:

$$\left[(a,x) \in X \times Y\right] \wedge \left[(a,x) \in Y \times Z\right]$$

Finally, for the last step of the proof we use the definition of intersection and get:

$$(a,x) \in (X \times Y) \cap (X \times Z)$$

Since, we have shown that an ordered pair is an element of $X \times (Y \cap Z)$ if and only if it is an element of $(X \times Y) \cap (X \times Z)$ we have shown, by the definition of equal sets

$$X \times (Y \cap Z) = (X \times Y) \cap (X \times Z)$$

(B)   We prove this by assuming that if an ordered pair is in $X \times (Y \cup Z)$ then this is equivalent to saying that the ordered pair is in $(X \times Y) \cup (X \times Z)$.   By assumption:

$$(a,x) \in X \times (Y \cup Z).$$

From the definition of cartesian product this gives:

$$(a \in X) \wedge \left[X \in (Y \cup Z)\right]$$

The definition of union gives:

$$(a \in X) \wedge \left[(x \in Y) \vee (x \in Z)\right]$$

To this expression we apply the law of logical distributivity of conjunction over disjunction (i.e.

$$p \wedge (q \vee r) <=> (p \wedge q) \vee (p \wedge r))$$

This yields from above

$$\left[(a \in X) \wedge (x \in Y)\right] \vee \left[(a \in X) \wedge (x \in Z)\right]$$

To this expression we apply the definition of cartesian product once again, giving us:

$$\left[(a,x) \in X \times Y\right] \lor \left[(a,x) \in (X \times Z)\right]$$

Finally, from the definition of union, we get

$$(a,x) \in (X \times Y) \cup (Y \times Z)$$

Hence, we have:

$$X \times (Y \cup Z) = (X \times Y) \cup (Y \times Z)$$

● **PROBLEM** 2-63

State the definitions of cardinality and cardinal number. Let x be an arbitrary cardinal number, find $1 \cdot x$ and $0 \cdot x$.

<u>Solution</u>: Two sets A and B are said to have the same cardinality if there exists a one-to-one correspondence $f : A \to B$.

The cardinal numbers are considered as symbols assigned to sets in such a way that two sets are assigned the same symbol if and only if they have the same cardinality.

Given any cardinal numbers x and y, the cardinal product xy is defined to be the cardinal number of the cartesian product $X \times Y$, where card X = x and card Y = y. In addition, cardinal multiplication is commutative, associative and distributive.

Let A = {a} with cardinal number 1, and let card X = x. Then,

$$A \times X = \{(a,x) \mid a \in A \text{ and } x \in X\}.$$

Set $A \times X$ has the same cardinality as set X because there is a one-to-one correspondence $f : X \to A \times X$, where $f(x) = (a,x)$ for all $x \in X$ and $(a,x) \in A \times X$. Therefore, $1 \cdot x = x$.

Similarly, if ϕ is the empty set, card $\phi = 0$, since

$$\phi \times A = \phi, \quad 0 \cdot x = 0.$$

● **PROBLEM** 2-64

Find the cardinal sum $4 + 5$ of the two finite cardinal numbers 4 and 5.

Solution: The cardinal number of a finite set is the number of elements in that set.   The cardinal sum of x and y denoted by $x + y$, is the cardinal number card($A \cup B$), where A and B are disjoint sets such that, card A = x and card B = y.   Let $A = \{a_1, a_2, a_3, a_4\}$ with card A = 4, $B = \{b_1, b_2, b_3, b_4, b_5\}$ with card B = 5 and $A \cap B = \phi$.

Then, $A \cup B = \{a_1, a_2, a_3, a_4, b_1, b_2, b_3, b_4, b_5\}$.

Hence, card($A \cup B$) = 9 which is the cardinal sum $4 + 5$ of the two finite cardinal numbers 4 and 5.

● **PROBLEM** 2-65

Calculate the cardinal sum $N_o + N_o$ where N_o = card N and N is the set of natural numbers.

Solution: Let N_e and N_o be the sets of even and odd non-negative integers, respectively.   Note N_o = card N_e = card N_o. Then,

$$N = N_e \cup N_o \quad \text{and,}$$

$$N_o + N_o = \text{card } N_e + \text{card } N_o$$

$$= \text{card}(N_e \cup N_o)$$

$$= \text{card } N$$

$$= N_o$$

The result above is one of the distinctive properties of the cardinal numbers of the infinite sets.   For finite sets, the cardinal numbers $a + b = a$ if and only if $b = 0$.

APPLICATIONS

● **PROBLEM** 2-66

Find the greatest common divisor (GCD) of 32 and 48 by using set theory.

Solution: Let sets A and B be the sets of positive divisors of 32 and 48 respectively.

$$A = \{1, 2, 4, 8, 16, 32\}$$

$$B = \{1, 2, 3, 4, 6, 8, 12, 16, 24, 48\}$$

The GCD of 32 and 48 must belong to the set A ∩ B which is the set of positive common divisors of 32 and 48.

$$A \cap B = \{1,2,4,8,16\}$$

By inspection, we see that GCD of 32 and 48 is 16.

● PROBLEM 2-67

In a survey carried out in a school snack shop, the following results were obtained. Of 100 boys questioned, 78 liked sweets, 74 ice-cream, 53 cake, 57 liked both sweets and ice-cream. 46 liked both sweets and cake while only 31 boys liked all three. If all the boys interviewed liked at least one item, draw a Venn diagram to illustrate the results. How many boys liked both ice-cream and cake?

Solution: A Venn diagram is a pictorial representation of the relationship between sets. A set is a collection of objects. The number of objects in a particular set is the cardinality of a set.

To draw a Venn diagram we start with the following picture:

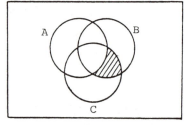

Each circle represents set A, B or C respectively. Let

A = set of boys who like ice-cream

B = set of boys who like cake

C = set of boys who like sweets.

The sections of overlap between circles represents the members of one set who are also members of another set. For example, the shaded region in the picture indicates the set of boys who are in sets B and C but not A. This is the set of boys who like both cake and sweets but not ice-cream. The inner section common to all three circles indicates the set of boys who belong to all three sets simultaneously.

We wish to find the number of boys who liked both ice-cream and cake. Let us label the sections of the diagram with the cardinality of these sections. The cardinality of the region common to all three sets is the number of boys who liked all three items or 31.

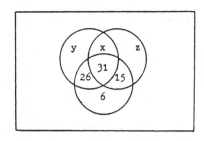

The number of boys who liked ice-cream and sweets
was 57. Of these 57, 31 like all three leaving 26 boys in
set A and set C but not set B. Similarly there are 15 boys
in B and C, but not in A. 78 - 26 - 31 - 15 = 6 boys in C
but not in A or B.

Let x = number of boys who are in A and B but not C

     y = number of boys are in A but not B or C

     z = number of boys who are in B but not A or C.

We know that the sum of all the labeled areas is 100 or

26 + 31 + 15 + 6 + x + y + z = 100

           78 + x + y + z = 100.

Also, there are 74 boys total in set A or

              x + y + 31 + 26 = 74   and 53 total in

set B or   x + z + 46 = 53.

Combining: x + y + z = 100 - 78 = 22

                x + y = 74 - 57 = 17

                x + z = 53 - 46 = 7 .

Substracting the second equation from the first gives
z = 5 implying x = 2 and y = 15. Our answer is the number
of boys in sets A and B = x + 31 = 33.

● PROBLEM 2-68

Of 37 men and 33 women, 36 are teetotalers. Nine of the
women are non-smokers and 18 of the men smoke but do not
drink. 13 of the men and seven of the women drink but do
not smoke. How many, at most,   both drink and
smoke.

Solution: A = set of all smokers

             B = set of all drinkers

             C = set of all women

             D = set of all men.

   We construct two Venn diagrams and label it in the following way:

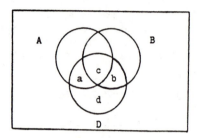 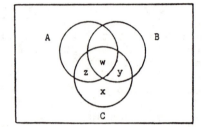

   Each section on the graph indicates a subset of the group of men and women. For example, the section labeled "z" is the subset including all women who smoke but do not drink. The section labeled "b" is the subset including all men who drink but do not smoke.

   In addition to labels, these letters also will indicate the cardinality, the number of objects, in the subset. We are told there are 37 men; thus, $a + b + c + d = 37$. There are 33 women; thus, $x + y + z + w = 33$. There are 9 women non-smokers which includes $x + y$. $d = 18$ the number of non-drinking, smoking men.

Similarly, $x + z + a + d = 36$, the teetotalers

          $b = 13$, the drinking, non-smoking men

          $y = 7$, the drinking, non-smoking women.

   Collecting all these equations, we wish to find the maximum value of $c + w$, the number of drinkers and smokers.

$x + z + a + d = 36$, $a + b + c + d = 37$

$b = 13$, $d = 18$, $x + y + z + w = 33$

     $y = 7$, $x + y = 9$.

Substituting we see that:

$x + y = x + 7 = 9$ or $x = 2$ from this we have

$2 + z + a + 18 = 36$ $a + 13 + c + 18 = 37$

          $a + z = 16$ $a + c = 6$

  $2 + 7 + z + w = 33$

          $z + w = 24$.

We now solve for c + w:

$a = 6 - c$ and thus $z + 6 - c = 16$ or $z - c = 10$

$c = z - 10$ and $w = 24 - z$

thus, $c + w = z - 10 + 24 - z = 14$.

The maximum number of drinkers and smokers is 14.

● PROBLEM 2-69

In the fall semester at a certain university, 60 percent of the freshman students registered to take General Physics I, and 70 percent of the freshman students registered to take General Chemistry I. If every freshman had to register for at least one of the courses, how many freshman students registered to take both courses?

Solution: Let A and B be the sets of students who registered to take General Physics I and General Chemistry I, respectively. Let n(A) denote the number of elements in the set A. Then

$n(A) = 60\%$

$n(B) = 70\%$,   and

$n(A \cup B) = 100\%$

The set of students who registered to take both courses can be represented by $A \cap B$.   Since $n(A \cup B) = n(A) + n(B) - n(A \ B)$,

Since $n(A \cap B) = n(A) + n(B) - n(A \cap B)$

$= 130\% - 100\%$

$= 30\%$

Hence, 30% of the freshman students registered to take both courses.

● PROBLEM 2-70

In a certain high school 600 students purchased tickets to a dance, 300 purchased tickets to a basketball game, and 173 students purchased tickets to both events. How many students purchased tickets to either of the two events?

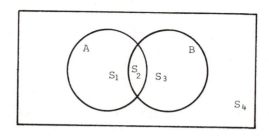

Solution: The problem requires us to find the number of
elements in the set P ∪ Q where P is the set of students who
purchased tickets to a dance and Q is the set of students who
purchased tickets to the basketball game.   Since some students purchased tickets to both events

$$P \cap Q \neq \phi.$$

If $P \cap Q = \phi$ then, $n(P \cup Q) = n(P) + n(Q)$.   However, for
$P \cap Q \neq \phi$, $n(P \cup Q)$ is determined by finding the sum of the
number of elements in P and the number of elements in Q and
subtracting the number of elements in the intersection of A
and B from the result.   This subtraction is necessary because
the elements in $P \cap Q$ have been counted twice during the process of finding $n(P \cup Q)$.

$$n(P) = 600 \quad , \quad n(Q) = 300 \quad , \quad n(P \cap Q) = 173$$

$$n(P \cup Q) = n(P) + n(Q) - n(P \cap Q)$$

$$= 600 + 300 - 173$$

$$= 727$$

Therefore, a total of 727 students purchased tickets for
either of the two events.

A Venn diagram is often useful for solving this kind of
problem.   For example, the formula for $n(P \cup Q)$ may be derived
by using a Venn diagram as follows: (see Figure for the   Venn
diagram)

The Venn diagram in the figure is partitioned into four
sets S_1, S_2, S_3 and S_4.   By examining the Venn diagram, one
obtains, $A = S_1 \cup S_2$, $B = S_2 \cup S_3$, $A \cup B = S_1 \cup S_2 \cup S_3$ and
$A \cap B = S_2$.   Hence, $n(A) = n(S_1) + n(S_2)$, $n(B) = n(S_2) + n(S_3)$,
$n(A \cup B) = n(S_1) + n(S_2) + n(S_3)$ and $n(A \cap B) = n(S_2)$.   Since

$$n(A) + n(B) = \left[n(S_1) + n(S_2) \right] + \left[n(S_2) + n(S_3) \right]$$

$$= \left[n(S_1) + n(S_2) + n(S_3) \right] + n(S_2)$$

$$= n(A \cup B) + n(S_2)$$

$$= n(A \cup B) + n(A \cap B),$$

$$n(A \cup B) = n(A) + n(B) - n(A \cap B)$$

The same procedure may be used to find $n(A \cup B \cup C)$.

In a freshman class of 200 students of a certain college, records indicate that 80 students registered to take Biology I, 90 registered to take Calculus I, 55 registered to take General Physics I, 32 registered to take both Biology I and Calculus I, 23 registered to take both Calculus I and General Physics I, 16 registered to take both Biology I and General Physics I, and 8 registered to take all three courses. Is the record from the registrar's office accurate? (Assume that each of the 200 students registered for at least one course.)

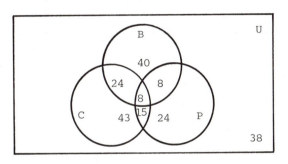

Solution: Let B = set of students who registered to take Biology I

C = set of students who registered to take Calculus I

P = set of students who registered to take General Physics I

then

$$n(B) = 80, \quad n(C) = 90$$

$$n(P) = 55, \quad n(B \cap C) = 32$$

$$n(C \cap P) = 23, \quad n(B \cap P) = 16$$

and $n(B \cap C \cap P) = 8$

If the record is to be accurate, $n(B \cup C \cup P)$ must equal 200.

By definition, $n(B \cup C \cup P)$

$$= n(B) + n(C) + n(P) - n(B \cap C)$$

$$- n(B \cap P) - n(P \cap C) + n(B \cap C \cap P)$$

thus $n(B \cup C \cup P) = 80 + 90 + 55 - 32 - 16$

$$- 23 + 8$$

$$= 162$$

Since $n(B \cup C \cup P) \neq 200$, the record is not accurate. The record indicates that there are 200 - 162 = 38 students that didn't register for any of the three courses, which is inconsistent with the initial assumption.

In solving problems of this type, the analysis should begin with the set $(B \cap C \cap P)$ and then extend outwardly in all directions.

In addition, the Venn diagram of this problem is shown in Figure 1.

● PROBLEM 2-72

A survey of 100 people was conducted to determine the popularity of three local radio stations; V.P.H.K, B.A.P.C and W.P.Q.W. The results were as follows:

42 people liked VPHK

48 people liked BAPC

41 people liked WPQW

15 people liked both VPHK and BAPC

17 people liked both VPHK and WPQW

18 people liked both BAPC and WPQW

10 people liked all the three radio stations

Find the number of people who liked none of the three stations.

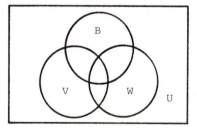 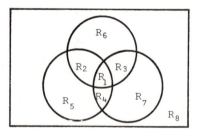

Fig. 1

Solution: The universal set is the set of 100 people. The number of people who like none of the three stations can be found by using set operations and/or Venn diagrams. Let the sets of people who liked VPHK, BAPC and WPQW be V, B and W respectively. Then

$$n(U) = 100, \quad n(V) = 42, \quad n(B) = 48$$

94

$n(W) = 41$, $n(V \cap B) = 15$,

$n(V \cap W) = 17$, $n(B \cap W) = 18$, and

$n(V \cap B \cap W) = 10$.

The Venn diagram from the given information is shown in Fig. 1. The set of people who liked none of the three stations is represented by the set $U - (V \cup B \cup W)$ or $V' \cap B' \cap W'$, which is also labeled as R_8. To find $n(R_8)$, one needs to find $n(R_1), n(R_2). . . n(R_7)$ first. In solving problems of this type, the analysis should begin with the set $(V \cap B \cap W)$ and then extend outwardly in all directions. This enables one to examine first those elements common to all three sets, then those elements common to two of the sets, and finally those elements that appear in each set that are not contained in any of the others. To proceed, first find $n(R_1)$,

$$n(R_1) = n(V \cap B \cap W)$$

$$= 10$$

Since $R_4 \cup R_1 = V \cap W$ and $n(R_4 \cup R_1) = n(V \cap W) = 17$,

$$n(R_4) = 17 - 10$$

$$= 7$$

Similarly, $n(R_2) = n(V \cap B) - n(R_1)$

$$= 15 - 10 = 5$$

and $n(R_3) = n(B \cap W) - n(R_1)$

$$= 18 - 10 = 8$$

Further, $R_5 \cup R_4 \cup R_2 \cup R_1 = V$,

$n(R_5) = n(V) - n(R_4) - n(R_2) - n(R_1)$

$n(R_5) = 42 - 7 - 5 - 10 = 20$.

By the same token, $n(R_6)$

$$= n(B) - n(R_1) - n(R_2) - n(R_3)$$

$$= 48 - 10 - 5 - 8$$

$$= 25$$

and $n(R_7) = n(W) - n(R_1) - n(R_3) - n(R_4)$

$$= 41 - 10 - 8 - 7$$

$$= 16$$

Finally, $n(R_8) = n(U) - n(R_1) - n(R_2) - n(R_3) - n(R_4)$

$$- n(R_5) - n(R_6) - n(R_7)$$

$$= 100 - 10 - 5 - 8 - 7 - 20 - 25 - 16$$

$$= 9 = n(V' \cap B' \cap W').$$

Therefore, the number of people who liked none of the three radio stations is 9. The results are shown in Fig. 2.

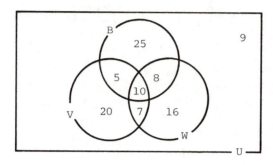

Fig. 2

CHAPTER 3

RELATIONS

RELATIONS AND GRAPHS

Find the relation defined by $y^2 = 25 - x^2$, where x belongs to $D = \{0,3,4,5\}$.

<u>Solution:</u> x takes on the values 0,3,4, and 5. Replacing x by these values in the equation $y^2 = 25 - x^2$ we obtain the corresponding values of y:

x	$y^2 = 25 - x^2$	y
0	$y^2 = 25 - 0$ $y^2 = 25$ $y = \sqrt{25}$ $y = \pm 5$	± 5
3	$y^2 = 25 - 3^2$ $y^2 = 25 - 9$ $y^2 = 16$ $y = \sqrt{16}$ $y = \pm 4$	± 4
4	$y^2 = 25 - 4^2$ $y^2 = 25 - 16$ $y^2 = 9$ $y = \sqrt{9}$ $y = \pm 3$	± 3
5	$y^2 = 25 - 5^2$ $y^2 = 25 - 25$ $y^2 = 0$ $y = 0$	0

Hence the relation defined by $y^2 = 25 - x^2$ where x belongs to $D = \{0,3,4,5\}$ is

$$\{(0,5),(0,-5),(3,4),(3,-4),(4,3),(4,-3),(5,0)\} .$$

Find the relation M over set $S = \{1,2,3\}$ if
$$M = \{ (x,r(x)) : r(x) = 2x - 1 \}$$

Solution: x takes on the values 1,2, and 3. Replacing x by these values in the equation $r(x) = 2x - 1$ we obtain the corresponding values of $r(x)$:

x	$r(x) = 2x - 1$	$r(x)$
1	$r(x) = 2(1)-1$ $= 2 - 1$ $= 1$	1
2	$r(x) = 2(2)-1$ $= 4 - 1$ $= 3$	3
3	$r(x) = 2(3)-1$ $= 6 - 1$ $= 5$	5

Thus the rule of correspondence $r(x) = 2x - 1$ determines the set of ordered pairs
$$\{(1,1),(2,3),(3,5)\}$$

But the relation must be a subset of $S \times S$. Since $(3,5)$ is not a subset of $S \times S$ (5 is not a member of S) we eliminate this pair. Hence,
$$M = \{(1,1),(2,3)\}$$

Find the relation Q over $S \times T$ if $S = \{1,2,3\}$, $T = \{4,5\}$, and the rule of correspondence is

$$r(x) = x + 2.$$

Solution: We first find the image of each element in S by substituting each element for x in the rule of correspondence $r(x) = x + 2$.

$$r(1) = 1 + 2 = 3 \qquad r(2) = 2 + 2 = 4$$

$$r(3) = 3 + 2 = 5.$$

Thus, the rule of correspondence determines the following set of ordered pairs:

$$\{ (1,3), (2,4), (3,5) \}.$$

However, the relation Q must be a subset of S × T, which equals { (1,4), (1,5), (2,4), (2,5), (3,4), (3,5) }. Therefore the point (1,3) won't appear in Q because (1,3) doesn't appear in S × T. Therefore, the relation over S × T determined by $r(x) = x + 2$ is

$$Q = \{ (2,4), (3,5) \}.$$

We can use set-builder notation to describe the relation discussed in the above example. In the example, a set of ordered pairs was determined by a rule of correspondence. The first component, x, was chosen from the domain, S. The second component, $r(x)$, was the corresponding image from the range, T. Thus, we can describe the relation Q in the following manner:

$$Q = \{ \left(x, r(x) \right): r(x) = x + 2 \}.$$

This notation refers to all ordered pairs [x, r(x)], such that $r(x) = x + 2$.

● PROBLEM 3-4

Find the set of ordered pairs { (x,y) } if $y = x^2 - 2x - 3$ and D = { x | x is an integer and $1 \le x \le 4$ }.

Solution: We first note that D = { 1,2,3,4 }. Substituting these values of x in the equation

$$y = x^2 - 2x - 3,$$

we find the corresponding y values. Thus,

for x = 1, $y = 1^2 - 2(1) - 3 = 1 - 2 - 3 = -4$

for x = 2, $y = 2^2 - 2(2) - 3 = 4 - 4 - 3 = -3,$

for x = 3, $y = 3^2 - 2(3) - 3 = 9 - 6 - 3 = 0,$ and

for x = 4, $y = 4^2 - 2(4) - 3 = 16 - 8 - 3 = 5$.

Hence { (x,y) } = { (1,-4), (2,-3), (3,0), (4,5) }.

List all the relations from set A to set B where

   A = {α,β}   and   B = {x}.

Solution: A relation R from set A to set B is defined as a subset of A × B.

$$A × B = \{(α,x), (β,x)\}$$

Therefore, the relations from A to B are:

$$R_1 = \{(α,x)\}$$

$$R_2 = \{(β,x)\}$$

$$R_3 = \text{empty set } φ$$

$$R_4 = A × B$$

Given relations

   $R_1 = \{(1,2),(3,4),(5,6),(7,8),(9,10)\}$

and

   $R_2 = \{(a,b),(1,2),(c,d),(3,4)\}$

Find relations $R_1 \cap R_2$ and $R_1 \cup R_2$.

Solution: By definition, a relation R from set A to set B is a subset of A × B.   Hence, relations are also sets and operations like union and intersection can be performed on them. Therefore,

$$R_1 \cap R_2 = \{(1,2),(3,4)\} \quad \text{and}$$

$$R_1 \cup R_2 = \{(1,2),(3,4),(5,6),(7,8),(9,10),$$

$$(a,b),(c,d)\}.$$

Find the coordinate diagram of the following relations:

  (1) R_1 = {(x,a),(y,a),(z,b)}

  (2) R_2 = {(1,2),(2,3),(3,4),(6,3),(5,2),(4,1),(4,3)}

Solution: We will represent the first element of each ordered pair in R_1 as points on a horizontal axis, and the second element in each ordered pair in R_1 by points on a vertical axis. Thus we have created a coordinate system, and each element of R_1 is represented by a point in this coordinate system as shown in Fig. 1.

(1)

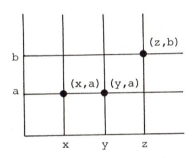

Fig. 1.

(2) The coordinate diagram for (2) is shown in Fig. 2.

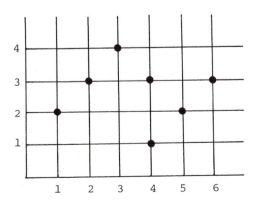

Fig. 2.

Write the matrix of the following relations:

 (1) R_1 = {(a,1),(b,1),(c,4),(d,15),(a,4)}

 (2) R_2 = {(2,4),(2,5),(2,8),(4,1),(4,3)}

Solution: A matrix of a relation R from set A to set B is a rectangular array whose rows are labeled by the elements of A and whose columns are labeled by the elements of B. A "1" or a "0" is placed in each entry of the array according to whether a ε A is or is not related to b ε B.

  The matrices for R_1 and R_2 are shown in Figs. 1 and 2, respectively.

	1	4	15
a	1	1	0
b	1	0	0
c	0	1	0
d	0	0	1

Fig. 1

	1	3	4	5	8
2	0	0	1	1	1
4	1	1	0	0	0

Fig. 2

Draw the arrow diagram of the relation R over the set

 A = {1,2,3,4,5,6,7,8,9,10,11,12,13,14}

where R = {(x,R(x)) : R(x) = 2x + 6}.

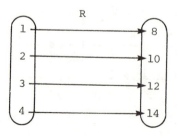

R

Fig. 1.

Solution: x takes on the values 1, 2, 3 and 4. Replacing x by these values in the equation R(x) = 2x + 6, we obtain:

$$R(1) = 2 \cdot 1 + 6 = 8$$

$$R(2) = 2 \cdot 2 + 6 = 10$$

$$R(3) = 2 \cdot 3 + 6 = 12$$

$$R(4) = 2 \cdot 4 + 6 = 14$$

Thus,   R = {(1,8),(2,10),(3,12),(4,14)}.

The arrow diagram of this relation is shown in Fig. 1.

● PROBLEM 3-10

Draw the graphs of the following relations:

(1) R_1: $x^2 + y^2 = 4$

(2) R_2: $\dfrac{x^2}{4} - \dfrac{y^2}{4} = 1$

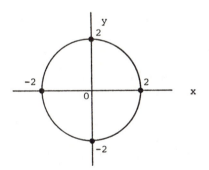

Fig. 1.

Solution: (1)   The equation which describes R_1 is the equation of a circle.   The graph is shown in Fig. 1.   The graph for R_2 is shown in Fig. 2.

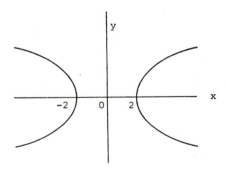

Fig. 2.

Define a binary operation "*" on the set R^+ of all positive real numbers by the formula

$$a*b = \frac{ab}{a+b}$$

Determine whether the operation "*" is associative.

Solution: Note first that the operation "*" is at least well defined on R^+, since if a and b are positive then so are ab and a+b and hence so is ab/(a+b); i.e. ab/(a+b) belongs to R^+.

To test the associative law, one must compute (a*b)*c and a*(b*c) and compare them.

$$(a*b)*c = \left(\frac{ab}{a+b}\right)*c \qquad \text{(definition of "*")}$$

$$= \frac{\left(\frac{ab}{a+b}\right)c}{\left(\frac{ab}{a+b}\right)+c} \qquad \text{(definition of "*")}$$

$$= \frac{\left(\frac{ab}{a+b}\right)c(a+b)}{\left[\left(\frac{ab}{a+b}\right)+c\right](a+b)} \qquad \text{(multiplying both numerator and denominator by a+b to remove inner fractions)}$$

$$= \frac{(ab)c}{ab + c(a+b)} \qquad \text{(simplifying)}$$

$$= \frac{abc}{ab+ac+bc} \qquad \text{(expanding and rearranging terms)}$$

$$a*(b*c) = a * \left(\frac{bc}{b+c}\right) \qquad \text{(definition of "*")}$$

$$= \frac{a\left(\frac{bc}{b+c}\right)}{a + \left(\frac{bc}{b+c}\right)} \qquad \text{(definition of "*")}$$

$$= \frac{a\left(\frac{bc}{b+c}\right)(b+c)}{\left(a + \left(\frac{bc}{b+c}\right)\right)(b+c)} \qquad \text{(multiplying numerator and denominator by b+c)}$$

$$= \frac{a(bc)}{a(b+c) + bc} \qquad \text{(simplifying)}$$

$$= \frac{abc}{ab+ac+bc}$$

Since (a*b)*c and a*(b*c) are equal, "*" is associative.

INVERSE RELATIONS AND COMPOSITION OF RELATIONS

Given A = {9,10,11} and R is a relation in A × A, defined as R = {(a,b) | a = b}

  (1) find R, and the complementary relation to R;

  (2) draw the graphs of A × A, R, and R'.

Solution: A × A = {(9,9),(10,10),(11,11),(9,10),(9,11),

                              (10,9)(10,11)(11,9)(11,10)}

From the definition of R one obtains R = {(a,b) | a = b}

                                    = {(9,9),(10,10),

                                       (11,11)}

when R is restricted to A × A.

    By definition, if R is a relation in a cartesian-product set S × S, then

      R' = {(x,y) | (x,y) ε S × S and (x,y) ∉ R}

is the complementary relation to R. In other words, R' is the set containing those ordered pairs (x,y) of A × A not in R.

    Therefore,

      R' = {(9,10),(9,11),(10,9)(10,11),(11,9),(11,10)}

(2) The graphs of A × A, R, and R' are shown in the figure.

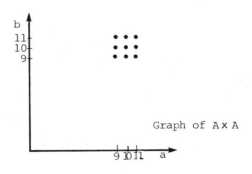

Graph of A × A

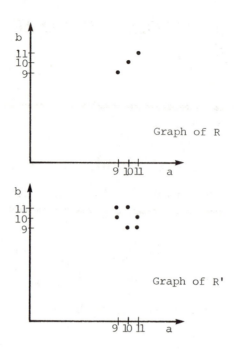

Graph of R

Graph of R'

Given the relation R = {(9,8),(10,9),(11,10)}

in the set S x S, where S = {8,9,10,11}

   (1)   find the inverse of R and the complementary re-
       lation to R;

   (2)   find the domains and the ranges of R and R^{-1};

   (3)   sketch R, R^{-1}, and R'.

Solution: (1) Let R be a relation from set A to set B. The
inverse of R, denoted by R^{-1}, is the relation from B to A
consisting of those ordered pairs which when reversed belong
to R.

$$R^{-1} = \{(y,x) \mid (x,y) \; \varepsilon \; R\}.$$

If R is a relation in a cartesian-product set A × A, then
the complementary relation to R, denoted by R', is the set
containing those ordered pairs (x,y) of A × A not in R:

$$R' = \{(x,y) \mid (x,y) \notin R \text{ and } (x,y) \; \varepsilon \; A \times A\}.$$

Note that R^{-1}, the inverse of R, is defined for a relation
from any set A to any other set B. However, the complementary relation to R is defined for relations in a cartesian-
product only!

The given R is R = {(9,8),(10,9),(11,10)}

$$= \{(x,y) \mid x = y + 1\}$$

If a relation R is represented by a certain defining condition, then the corresponding condition defining R^{-1} is obtained by replacing x for y and y for x. Hence,

$$R^{-1} = \{(x,y) \mid y = x + 1\}$$

$$= \{(8,9),(9,10),(10,11)\}$$

$$R' = \{(8,8),(8,9),(8,10),(8,11),(9,9),(9,10),$$

$$(9,11),(10,10),(10,11),(10,8),(11,8)$$

$$(11,9),(11,11)\}$$

$$= \{(x,y) \mid (x,y) \; \varepsilon \; S \times S \text{ and } (x,y) \notin R\}$$

(2) The domain of a relation R is the set of all first elements of the ordered pairs which belong to R, and the range of R is the set of second elements. The domain of the given R is {9,10,11}, and the range of R is {8,9,10}. For R^{-1}, the domain is {9,10,11} and the range is {8,9,10}. It is seen that, as a result of the process of interchanging the components of the ordered pairs, the domain of R is the range of R^{-1}, and the range of R is the domain of R^{-1}.

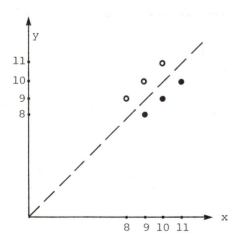

Each • represents an element of R.

Each O represents an element of R^{-1}.

Fig. 1. The graphs of R and R^{-1}.

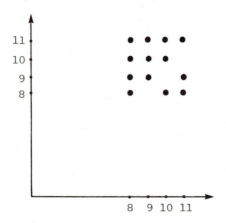

$$11$$
$$10$$
$$9$$
$$8$$

$$8 \quad 9 \quad 10 \quad 11$$

Fig. 2. Graph of R'

(3)

The process of obtaining R^{-1} yields mirror images of the original points and creates correspondingly a set of ordered pairs which are the inverse relation. This is clearly observed from Fig. 1.

● PROBLEM 3-14

Draw the arrow diagram of the composition of relations $R_1 \circ R_2 \circ R_3$, where R_1, R_2 and R_3 are given as follows:

$R_1 = \{(1,a),(2,b),(3,a),(4,b),(5,c)\}$

$R_2 = \{(a,x),(b,y),(b,z)\}$

$R_3 = \{(x,10),(y,10),(z,20),(z,30)\}$

Solution: The arrow diagrams for R_1, R_2, and R_3 are shown in Figures 1, 2, and 3, respectively.

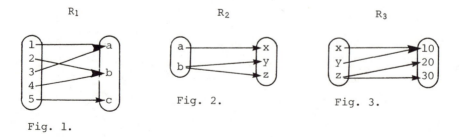

R_1

R_2

R_3

Fig. 1.

Fig. 2.

Fig. 3.

The arrow diagram for $R_1 \circ R_2 \circ R_3$ is shown in Fig. 4.

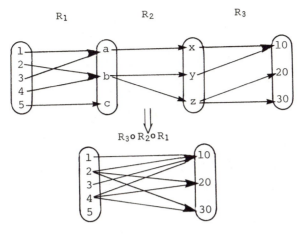

R_1 R_2 R_3

$R_3 \circ R_2 \circ R_1$

Fig. 4.

Given two relations R_1 and R_2 such that R_1 is a relation from a set A to a set B and R_2 is a relation from B to another set C show that the inverse of the composition,

$$(R_1 \circ R_2)^{-1}$$

is equal to the composition of the inverses taken in the reverse order,

$$R_2^{-1} \circ R_1^{-1}.$$

Solution: Symbolically what we are being asked to prove is that

$$(R_1 \circ R_2)^{-1} = R_2^{-1} \circ R_1^{-1}$$

We use the definition of inverse relation to show that claiming

$$(c,a) \; \varepsilon \; (R_1 \circ R_2)^{-1}$$

is equivalent to making the claim that

$$(a,c) \; \varepsilon \; R_1 \circ R_2.$$

From the definition of the composition of relations this yields that there is a $b \varepsilon A \cap C$ such that $(a,b) \varepsilon R_1$ and $(b,c) \varepsilon R_2$. From this we can use the definition of inverse relation applied to both R_1 and R_2 to see that

$$(b,a) \; \varepsilon \; R_1^{-1} \text{ and } (c,b) \; \varepsilon \; R_2^{-1}$$

109

Applying the definition of composition of relations to R_2^{-1} and R_1^{-1}, we have

$$(c,a) \; \varepsilon \; R_2^{-1} o \, R_1^{-1}$$

as an equivalent statement.   This sequence of logical equivalences demonstrated that

$$(c,a) \; \varepsilon \; (R_1 o \, R_2)^{-1}$$

if and only if

$$(c,a) \; \varepsilon \; R_2^{-1} o \, R_1^{-1}$$

Thus, we have

$$R_2^{-1} o \, R_1^{-1} \; = \; (R_1 o \, R_2)^{-1}$$

and the proof is complete.

PROPERTIES OF RELATIONS

Given   A = {1,2,3,4,5,6,7} and relations:

   R_1 = {(1,1),(2,2),(1,2),(2,3),(3,2),(3,3),

         (4,4),(5,5),(7,6),(6.7),(6,6),(2,1),(7,7)}

   R_2 = {(1,2),(2,3),(1,3),(1,4),(2,4)}

   R_3 = {(1,4),(2,5),(3,7),(6,6)}

   R_4 = {(2,4),(4,2),(4,6),(2,6),(6,4),(6,2),(2,2),

         (4,4),(6,6)}

Determine which of the given relations from set A to itself are:

   (1)   Reflexive
   (2)   Symmetric
   (3)   Transitive
   (4)   Anti-symmetric

Solution: First, we recall the following definitions defined for relations from a set S to itself:

Definition 1:   A relation R is reflexive if and only if $s\bar{R}s$ for every $s \in S$.

Definition 2:   A relation R is symmetric if $(xRy) => (yRx)$ where $x,y \in S$.

Definition 3:   R is transitive if xRy and yRz implies xRz.

Definition 4:   R is anti-symmetric if s_1Rs_2 and s_2Rs_1 implies $s_1 = s_2$.

Therefore, only R_1 is reflexive since for every $s_i \in S$, there is an ordered pair containing s_i with itself in R_1. R_1 and R_4 are symmetric. Observe that every time an ordered pair occurs in R_1 (or R_4) the ordered pair consisting of the elements in the reverse order also appears.

R_2, R_3 and R_4 are transitive. R_2 and R_3 are anti-symmetric.

● **PROBLEM 3-17**

Given a relation R from set $A = \{x,y,z\}$ to itself, determine whether $R = \phi$ is

(1)   symmetric,
(2)   reflexive,
(3)   transitive.

Solution: (1) By definition, R is symmetric if $(a,b) \in A$ and $(a,b) \in R => (b,a) \in R$.

Since R is an empty set, it has no elements. Thus, this relation satisfies the condition of a symmetric relation. Therefore, $R = \phi$ is symmetric.

(2) $R = \phi$ is obviously not reflexive since it contains no element; it can't satisfy the condition for a reflexive relation (for all $a \in A$, $(a,a) \in R$).

(3) $R = \phi$ is transitive since it satisfies the condition for a transitive relation, i.e. R is transitive if aRb and bRc implies aRc where (a,b) and $(b,c) \in R$.

111

Given a relation R on a set S and a relation R_0,

$R_0 = \{(a,b) \mid (a,b) \in S \times S \text{ and } a = b\}$

Prove   (1)   R is reflexive if and only if $R_0 \subset R$.

                (2)   R is transitive if and only if $(RoR) \subset R$.

                (3)   R is symmetric if and only if $R = R^{-1}$.

                (4)   R is anti-symmetric if and only if (or iff) $R \cap R^{-1} \subset R_0$.

Solution: (1) R is a subset of $S \times S$, if R is reflexive, then $(a,a) \in R$ for all $(a,a) \in S \times S$. However, by the definition of R_0, $(a,a) \in R_0$ for all $(a,a) \in S \times S$. Hence, $R_0 \subset R$. Further, if $R_0 \subset R$, then since $(a,a) \in R_0$, (a,a) must also belong to R. Therefore, R is reflexive. The proof is now complete.

(2) Let A, B and C be sets, and let R_1 be a relation from A to B; R_2 be a relation from B to C. Then $R_1 o R_2$ is called the composition of R_1 and R_2, such that

$R_1 o R_2 = \{(a,c) \mid$ there exists $b \in B$ for which

$(a,b) \in R_1$ and $(b,c) \in R_2\}$

Let R be transitive, that is, if $(a,b) \in R$ and $(b,c) \in R$, then $(a,c) \in R$. Let $(x,z) \in RoR$, $(x,y) \in R$ and $(y,z) \in R$ for some $y \in S$. Since R is transitive, $(x,z) \in R$. Hence, $Ro R \subset R$. Conversely assume $RoR \subset R$. If $(x,y) \in R$ and $(y,z) \in R$, then $(x,z) \in RoR$. Since $Ro R \subset R$, $(x,z) \in R$. Thus, R is transitive.

(3) R is symmetric if $(a,b) \in R$ implies $(b,a) \in R$. Let $R = \{(x,y) \mid (x,y) \in S \times S\}$. Then $R^{-1} = \{(y,x) \mid (y,x) \in S \times S\}$. Assume $R^{-1} = R$, one has $(x,y) \in R$ and $(y,x) \in R$. Thus R is symmetric. Assume R is symmetric in order to prove the converse. If R is symmetric, then $(x,y) \in R$ implies $(y,x) \in R$. $(y,x) \in R^{-1}$ hence $R^{-1} \subset R$. Similarly, if R is symmetric so is R^{-1}. $(x,y) \in R^{-1}$ implies $(y,x) \in R^{-1}$. $(y,x) \in R$ thus $R \subset R^{-1}$. So, $R = R^{-1}$ since $R \subset R^{-1}$ and $R^{-1} \subset R$.

(4) Let R be anti-symmetric. Assume $(x,y) \in R \cap R^{-1}$, then (x,y) belongs to both R and R^{-1}. If $(x,y) \in R^{-1}$, then $(y,x) \in R$, and $x = y$ since R is anti-symmetric. Hence, $(x,y) \in R_0$, which implies $R \cap R^{-1} \subset R_0$.

To complete the proof, let $R \cap R^{-1} \subset R_0$, $(x,y) \in R$, $(y,x) \in R$ and $(x,y) \in R^{-1}$. Then $(x,y) \in R \cap R^{-1}$.

Since $R \cap R^{-1} \subset R_0$, $(x,y) \in R_0$, i.e. $x = y$. Thus, $(x,y) = (y,x)$, which implies that R is anti-symmetric.

Give examples of relations R on A = {1,2,3,4} which are:

   (1)   symmetric, reflexive and transitive;

   (2)   symmetric and reflexive but not transitive;

   (3)   reflexive and transitive but not symmetric;

   (4)   reflexive, but not symmetric nor transitive.

Solution: First, R must be a subset of

$$A \times A = \{(1,1),(1,2),(1,3),(1,4),(2,1),(2,2),(2,3),(2,4),$$
$$(3,1),(3,2),(3,3),(3,4),(4,1),(4,2),(4,3),(4,4)\}$$

(1) The universal set $A \times A$ is an example of a relation that is symmetric, reflexive and transitive.

(2) R = {(1,1),(2,2),(3,3),(4,4),(1,2),(2,4),(2,1),(4,2)} is an example of a relation that is both symmetric and reflexive. But R is not transitive since 1R2 and 2R4 but 1$\not{R}$4.

(3) R = {(1,1),(2,2),(3,3),(4,4),(1,2),(2,3),(1,3)} is both reflexive and transitive. But R is not symmetric since, for instance, 1R2 but 2$\not{R}$1.

(4) R = {(1,1),(2,2),(3,3),(4,4),(1,2),(2,1)} is reflexive since we have aRa for all a ε A. However, it is not symmetric since 4R2 but 2$\not{R}$4, and it's not transitive since $(4,2) \varepsilon$ R, $(2,1) \varepsilon$ R but $(4,1) \not\in$ R.

EQUIVALENCE RELATIONS

(1) Find all the partitions of A = {1,2,3}.

Solution: (1) By definition, a partition of a set X is a collection of disjoint subsets of X whose union equals X.

A has $2^3 = 8$ subsets. They are:

{1},{2},{3},{1,2},{1,3},{2,3},{1,2,3}

and the empty set ϕ.

The partitions of A are:

[{1},{2},{3}],

[{1},{2,3}], [{2},{1,3}], [{3},{1,2}],

and   [{1,2,3}]

● **PROBLEM** 3-21

(a) Prove that equality for sets is an equivalence relation.

(b) Prove that inclusion of sets is reflexive, anti-
symmetric and transitive.

Solution: Let R be a relation on a set A.

(1) R is reflexive if aRa for every a in A.

(2) R is symmetric if aRb implies bRa.

(3) R is anti-symmetric if aRb and bRa implies a = b.

(4) R is transitive if aRb and bRc implies aRc.

Observe that these properties are only defined for relations
on a set.

In order to show that equality of sets is an equivalence
relation we must show it is (i) Reflexive, (ii) Symmetric and
(iii) Transitive.

(a) (i) By the axiom of extension, (which states that A = B
means $x \in A$ if and only if $x \in B$) $x \in S$ iff $x \in S$, so, S = S for
set S. Hence, equality for sets is reflexive.

(ii) If S = P, then P = S for all sets S and P since "$x \in S$
iff $x \in P$" is equivalent to "$x \in P$ iff $x \in S$". Thus, equality
for sets is symmetric since S = P implies P = S for all sets
S and P.

(iii) If S = P and P = Q, then "$x \in S$ iff $x \in P$" and "$x \in P$ iff
$x \in Q$"; this implies "$x \in S$ iff $x \in Q$." Therefore, S = P and
P = Q imply S = Q for all sets S, P and Q.

Thus, equality for sets is transitive. Therefore,
equality for sets is an equivalence relation.

(b) (1)   Inclusion of sets is reflexive since $x \in S$ implies $x \in S$.

(2)   Since "If $x \in S$, then $x \in P$" (i.e. $S \subseteq P$) and "If $x \in P$, then $x \in S$", (i.e. $P \subseteq S$) implies "$x \in S$ if and only if $x \in P$" (i.e. $S = P$).   Inclusion of sets is therefore anti-symmetric.

(3)   Inclusion of sets is transitive because "If $x \in S$, then $x \in P$" (i.e. $S \subseteq P$) and "If $x \in P$, then $x \in Q$" (i.e. $P \subset Q$) implies "If $x \in S$, then $x \in Q$" (i.e. $S \subseteq Q$).

● **PROBLEM 3-22**

Show that the relation a ≡ b (modulo n) is an equivalence relation defined on the set of integers.

Solution: First we recall how congruence modulo n is defined. Two integers a and b are said to be congruent modulo n if and only if

$$a - b = \ell \cdot n$$

for some integer ℓ.   Next, we must show that this relation is reflexive, symmetric, transitive, and hence satisfies the requirements for an equivalence relation.

(A)   Reflexive: Observe that for any integer $c : c - c = 0$ and $0 = 0 \cdot n$ (i.e. we let $\ell = 0$).   Thus $c \equiv c(\mod n)$.

(B)   Symmetric: Let $a \equiv b(\text{modulo } n)$.   In this case, from the definition of congruence modulo n, it follows that $a - b = \ell \cdot n$ for some integer ℓ.   This implies

$$b - a = -\ell \cdot n$$

(since $a - b = -1(b-a)$).   Since ℓ is an integer we know that $-\ell$ must also be an integer.   Thus, by definition,

$$b \equiv a(\mod n)$$

and we have demonstrated symmetry.

(C)   Transitive: Let

$$a \equiv b(\text{modulo } n) \quad \text{and}$$

$$b \equiv c(\text{modulo } n).$$

From the definition of congruence modulo n these statements imply that:

(1) $a - b = \ell_1 \cdot n$ and

(2) $b - c = \ell_2 \cdot n$

115

for some integers ℓ_1 and ℓ_2. Adding equations one and two gives:

$$(a-b) + (b-c) = (\ell_1 \cdot n) + (\ell_2 \cdot n)$$

or $$a - c = (\ell_1 + \ell_2)n.$$

Since ℓ_1 and ℓ_2 are both integers, their sum must also be an integer. Hence, by the definition of equivalence modulo n, we have that

$$a \equiv c(\bmod\ n).$$

We have shown that the relation congruence modulo n is reflexive, symmetric and transitive. Therefore it meets the requirements for an equivalence relation.

● **PROBLEM 3-23**

If we let R be an equivalence relation and are given a set A with $a \in A$, prove that each a/R is a non-empty subset of A.

Solution: First recall the definition of a set modulo on an equivalence relation.

$$A/R = \{a/R \mid a \in A \text{ where } a/R = \{b \in A \mid bRa\} \}$$

From the reflexivity of R we see that for every $a \in A$, aRa implies that $a \in a/R$ (from the above definition) and hence a is a non-empty subset of A/R

● **PROBLEM 3-24**

Let R be the relation on $A = N \times N$, defined by:

$(x,y)\ R\ (p,q)$ if and only if $x + q = y + p$, where

N = {nonnegative integers}. Prove that R is an equivalence relation.

Solution: By definition, a relation R on a set S is called an equivalent relation if R is reflexive, symmetric and transitive. Furthermore, R is reflexive if aRa for every a in S, R is symmetric if aRb implies bRa, and R is transitive if aRb and bRc implies aRc.

The given relation R is

(a) reflexive since (x,y) R (x,y). In other words, since $x+y = y+x$, (x,y) R (x,y) for all (x,y) in A.

(b) symmetric: If (x,y) R (p,q) then $x+q = y+p$ which can be rewritten as $p+y = q+x$, this implies that (p,q) R (x,y). Hence, (x,y) R (p,q) implies (p,q) R (x,y).

(c) transitive: Let (x,y) R (p,q) and (p,q) R (a,b), then $x+q = y+p$ and $p+b = q+a$. By addition: $(x+q) + (p+b) = (y+p) + (q+a)$; rearranging the terms: $(x+b) + (p+q) = (y+a) + (p+q)$; and subtracting $p+q$ from both sides: $x+b = y+a$. Hence, (x,y) R (a,b).

Therefore, R is an equivalence relation.

CHAPTER 4

FUNCTIONS

FUNCTIONS AND GRAPHS

Determine whether or not each of the following diagrams defines a function from

$$X = \{1,3,5,7\} \text{ into } Y = \{2,4,10,26,50\}.$$

If the answer is "No" explain why.

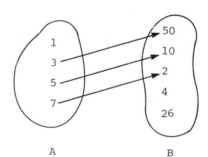

Fig. 1.   A   B

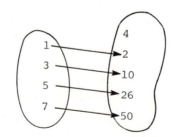

Fig. 2.

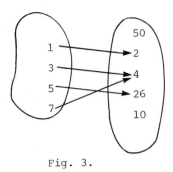

Fig. 3.

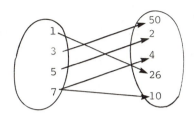

Fig. 4.

Solution: Definition: A function (mapping) f:A → B is a subset of A × B where each member of A appears exactly once as the first component of an ordered pair.

Therefore, the diagram in (1) is not a function since we have

$$f(3) = 50$$

$$f(5) = 10$$

$$f(7) = 2$$

but $1 \in X$, $1 \notin f$.

The diagram in Fig. 2 defines a function $f : X \rightarrow Y$, and $y = f(x) = x^2 + 1$. The diagram in (3) defines a function, too.

The diagram in Fig. 4 does not define a function since $7 \in X$ appears more than once; observe $f(7) = 4$ and $f(7) = 10$.

● PROBLEM 4-2

Given the mapping depicted in the figure, does this mapping represent a function? Explain.

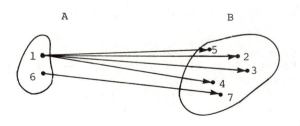

A   B

119

Solution: No. This mapping does not represent a function
because a function is a set of ordered pairs no two of which
have the same first coordinate, i.e. a function f associates
with each element x of the domain exactly one element y of
the range. Hence, a function cannot have more than one image
for each element of the domain.

In the given mapping from A to B, the element 1 in A has
four images in B.

● **PROBLEM 4-3**

Define a function f from A to B and give two examples.

Solution: A function from A to B is a binary relation such
that with each element of A there is associated in some way
exactly one element of another set B. For example, x = y
and x = 5y + 12 are functions.

● **PROBLEM 4-4**

Given that $\phi : A \to B$ is a function with a set C containing the image of ϕ, prove that $\phi : A \to C$ is also a
function.

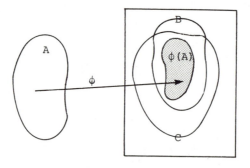

Fig. 1.

Solution: We first illustrate this problem with the use of
a diagram.

In the diagram of Figure 1 we see that ϕ is a function
which maps the set A on the left to its image, $\phi(A)$, which is
the shaded region on the right. Observe that the set B,
which is the range of $\phi : A \to B$, contains the image. The
theorem makes the claim that if another set, C, also contains

120

$\phi(A)$ then $\phi : A \rightarrow C$ is a function. Before giving the proof, we restate the set-theoretic definition of a function: A function is a triple (Φ, A, B) where A and B are sets and Φ is a relation, such that the domain of Φ is A and given that $(a, b_1) \in \Phi$ and $(a, b_2) \in \Phi$ we know that $b_1 = b_2$.

It is clear that ϕ of our problem is a relation from A to B. By the definition of Image, given that $(a,b) \in \phi$, we must have $a \in A$ and $b \in (Image(\phi))$. Also since we are given that $IM(\phi) \subseteq C$, it follows that $a \in A$ and $b \in C$. Hence, by the definition of Cartesian product, we have $(a,b) \in A \times C$.

Starting from the assumption that an ordered pair (a,b) is in ϕ, we have shown that (a,b) must also be in $A \times B$. Thus ϕ is a relation from A to C. Since $\phi : A \rightarrow B$ is a function we see that the domain of f is A. Hence, the second requirement for a function $\phi : A \rightarrow C$ is met.

● **PROBLEM** 4-5

If $X = \{1,2,3,4,5,6,7,8\}$ and $Y = \{2,4,6,8\}$, use $f(x)$ notation to indicate the image of each element of X in the following mapping.

Set X   1   2   3   4   5   6   7   8

Set Y   2   4   6   8

Solution: The mapping of an element x in the set X to an element y in the set Y may be written in $f(x)$ notation as $f(x) = y$ when f is a function mapping x to y. That is, $f: x \rightarrow y$. Therefore,

$f(1) = 2, \ f(2) = 2, \ f(3) = 4, \ f(4) = 4, \ f(5) = 6, \ f(6) = 6,$

$f(7) = 8, \ f(8) = 8$

● **PROBLEM** 4-6

Given the following functions state the domain, the range, and the image of each.

(A) $f_1 : R \rightarrow R$ given by

     $f_1(x) = x^2$

(B) $f_2 : R^+ + \{0\} \rightarrow R^+ + \{0\}$

     $f_2(x) = +\sqrt{x}$

(C) $f_3 : R \rightarrow R$

$$f_3(x) = [x] \equiv \text{the greatest integer less than or}$$
$$\text{equal to x.}$$

(D) $f_4 : R \to R$

$$f_4(x) = \pi.$$

Solution: Recall that a function $f : A \to B$ may be defined equivalently as the triple (f,A,B). Here, f is a relation from A to B. The domain of f is A. Also given $(a,b) \in f$ and $(a,c) \in f$ then $(b = c)$. The range of f is B. Finally, the image under f is

$$\{f(a) \mid a \in A\}$$

(A) In this case both the domain and the range are the set of real numbers. The image under f_1 is:

$$Im(f_1) = \{f_1(x) \mid x \in R\}$$
$$= \{x^2 \mid x \in R\} = \{y \mid y \geq 0\}.$$

(B) In this case the domain and range are both the set of non-negative real numbers. The image under f_2 is:

$$Im(f_2) = \{f(x) \mid x \in R^+ + \{0\}\}$$
$$= \{+\sqrt{x} \mid x \in R^+ + \{0\}\}$$
$$= \{y \in R^+ + \{0\}\}.$$

(C) In this example the domain and range of f_3 are the set of reals. The image is:

$$Im(f_3) = \{f_3(x) \mid x \in R\}$$
$$= \{[x] \mid x \in R\}$$
$$= \{\text{integers}\}.$$

(D) The domain and range are again the set of reals. The image is:

$$Im(f_4) = \{f_4(x) \mid x \in R\}$$
$$= \{\pi\}.$$

The singleton set containing the number $\pi = 3.14159...$

Draw the graphs of:

(1) $f(x) = 4x$

(2) $f(x) = 10$

(3) $f(x) = x^2$

(4) $f(x) = |x|$

Solution:

(1)

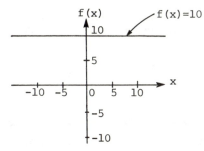

(2)

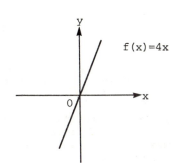

(3)

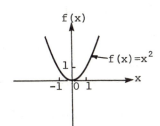

(4)

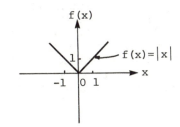

Let the domain of $M = \{(x,y): y = x\}$ be the set of real numbers. Is M a function?

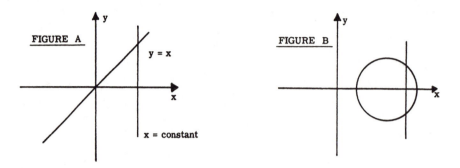

FIGURE A

$y = x$

$x = $ constant

FIGURE B

Solution: The range is also the set of real numbers since $y = \{y \mid y = x\}$. The graph of $y = x$ is the graph of a line ($y = mx + b$ where $m = 1$ and $b = 0$). See fig. A. If for every value of x in the domain, there corresponds only one y value then y is said to be a function of x. Since each element in the domain of M has exactly one element for its image, M is a function. Also notice that a vertical line (x = constant) crosses the graph $y = x$ only once. Whenever this is true the graph defines a function. Consult figure B.

The vertical line (x = constant) crosses the graph of the circle twice; i.e., for each x,y is not unique, therefore the graph does not define a function.

Which of the following sets are functions of x?

$A = \{(5,1),(4,2),(4,3),(6,4)\}$,

$B = \{(x,y) \mid y = |x|\}$,

$C = \{(x,y) \mid x = |y|\}$?

Solution: A function is a relation having the property that each member of its domain is paired with exactly one member of its range. Thus, set A is not a function, for it contains the pairs (4,2) and (4,3) - that is, one member of its domain, 4, is paired with more than one member of its range, 2 and 3. If each x value has only one corresponding y value, any vertical line only intersects the graph of a function at one point. Thus, from figure 1 we note that set B is a function. Notice that a function may contain two pairs with the same second member; for example, our function B contains the pairs (1,1) and (- 1,1).

Fig. 1 $y=|x|$

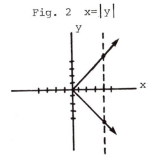

Fig. 2 $x=|y|$

From figure 2 we note that a vertical line intersects the graph of C in two places, thus there are x
values of C which have more than one corresponding y
value, and C is not a function.

● **PROBLEM** 4-10

(a) Sketch the following binary relations in $A \times A$,
     $A = \{$All real numbers$\}$.

   (1) $R_1 : x^2 + y^2 = 4$

   (2) $R_2 : x = 4y^2$

   (3) $R_3 : \dfrac{x^2}{4} - y^2 = 1$

(b) Are R_1, R_2, and R_3 functions?

Solution: (a) (1)

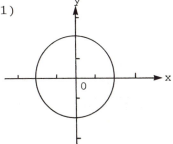

(2)

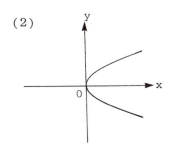

(3)

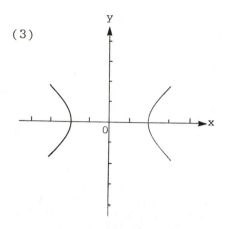

(b) A binary relation R, from set A to set B is a subset of A × B, and a function from A to B is a binary relation such that with each element of A there is associated in some way exactly one element of another set B. Thus to determine whether the given relations are also functions is equivalent to determining whether the set of ordered pairs of a given relation has any two ordered pairs with the same first coordinate. This is done by drawing vertical lines on the graph of a relation. If no vertical lines have more than one intersection with the graph i.e. there are no two ordered pairs in the relation having the same first coordinate, then, the given relation is a function. The procedure described above is called the vertical-line test.

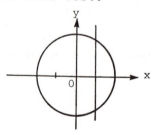

By the vertical-line test, it's seen that none of the given relations are functions from X to Y, where X and Y are the set of x-coordinates and y-coordinates, respectively. Note that relation (2) is a function from Y to X since no horizontal lines can have more than one intersection with the graph of relations (2) whereas relations (1) and (3) are neither functions from X to Y nor functions from Y to X.

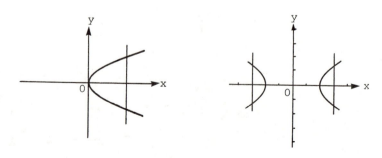

Which of the following graphs represent:

  (a)   relations between x and y?

  (b)   a function f : x → y?

  (c)   a function f⁻¹ : y → x?

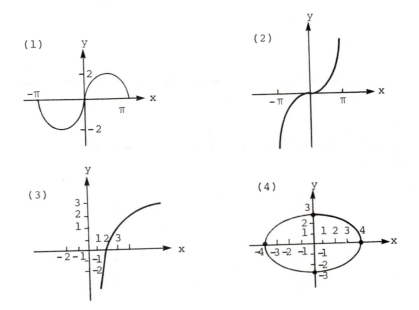

(1)   (2)   (3)   (4)

Solution: (a)   If A and B are sets, then a binary relation, R, from A to B is a subset of A × B.   If (x,y) ∈ R, it is said that x is R-related to y and denoted by xRy.   Here, A and B are the set of x-coordinates and y-coordinates, respectively of all the points in the coordinate plane.   Since each of the graphs represents a certain set of points in the coordinate plane, each graph represents a subset of A × B.   Therefore, all the given graphs represent relations between x and y.

(b)   By definition, if each element of a set A there is assigned to a unique element of a set B, then the collection of such assignments is called a function from A into B.   Hence, graphs (1), (2) and (3) are functions f : x → y.

(c)   Graph (2) and graph (3) represent functions f⁻¹: y → x. Note that graph (1) does not represent a function f⁻¹ : y → x since for certain values of y the associated element x is not unique.   Similarly, graph (4) is not a function at all since for certain x (or y) the associated element y (or x) is not unique.   This is illustrated by drawing a vertical and/or horizontal line on the graphs as shown in the figure.

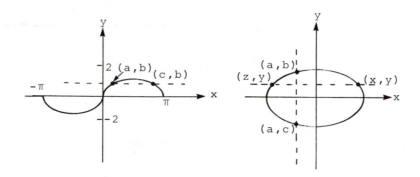

Given the following graphs, indicate: (a) which graph represents a binary relation? (b) which are graphs of functions?

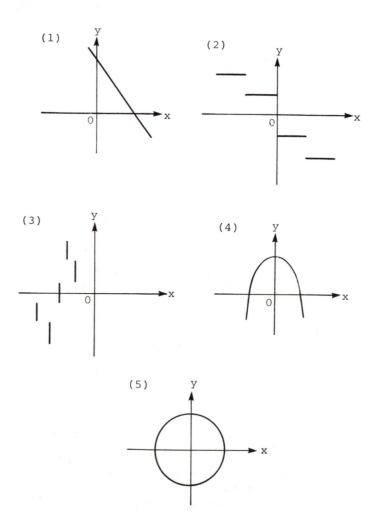

Solution: (a) By definition, a binary relation in A is a set of ordered pairs whose first and second coordinates are both members of set A. Here, A is the set of real numbers. Therefore, all the given graphs are graphs of binary relations.

(b) A function f from set A into set B is defined as a set of ordered pairs {(a,b) | a ∈ A and b ∈ B} no two of which have the same first coordinate. Hence, by the vertical-line test (see Figure), it's found that graph (1), (2) and (4) are graphs of functions from X to Y where X and Y are sets of x-coordinates and y-coordinates respectively.

Graphs (1) and (3) are graphs of functions from Y to X by the horizontal-line test (see Fig.).

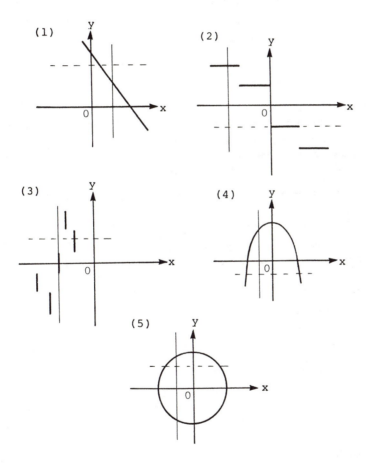

● **PROBLEM** 4-13

Let f be a mapping with the rule of correspondence

$$f(x) = 3x^2 - 2x + 1.$$

Find f(1), f(-3), f(-b).

Solution: In order to find $f(1)$, $f(-3)$, $f(-b)$, we replace x by 1, (-3), and $(-b)$ respectively in our equation for $f(x)$, $f(x) = 3x^2 - 2x + 1$. Thus

$$f(1) = 3(1)^2 - 2(1) + 1$$
$$= 3 - 2 + 1$$
$$= 1 + 1$$
$$= 2$$
$$f(-3) = 3(-3)^2 - 2(-3) + 1$$
$$= 3(9) + 6 + 1$$
$$= 27 + 7$$
$$= 34$$
$$f(-b) = 3(-b)^2 - 2(-b) + 1$$
$$= 3(b^2) + 2b + 1$$
$$= 3b^2 + 2b + 1$$

● **PROBLEM** 4-14

Let f be the function whose domain is the set of all real numbers, whose range is the set of all numbers greater than or equal to 2, and whose rule of correspondence is given by the equation $f(x) = x^2 + 2$. Find $3f(0) + f(-1)f(2)$.

Solution: The rule of correspondence in this example is expressed by the equation $f(x) = x^2 + 2$. To find the number in the range that is associated with any particular number in the domain, we merely replace the letter x wherever it appears in the equation $f(x) = x^2 + 2$ by the given number. Thus

$$f(0) = 0^2 + 2 = 2 , \qquad f(-1) = (-1)^2 + 2 = 1 + 2 = 3 ,$$

$$f(2) = 2^2 + 2 = 4 + 2 = 6 , \quad \text{and}$$

$$3f(0) + f(-1)f(2) = 3(2) + (3)(6) = 6 + 18 = 24.$$

● **PROBLEM** 4-15

If $f(x) = 3x + 4$ and $D = \{x| -1 \le x \le 3\}$, find the range of $f(x)$.

Solution: We first prove that the value of $3x + 4$ increases when x increases. If $X > x$, then we may multiply both sides of the inequality by a positive number to obtain an equivalent inequality. Thus, $3X > 3x$. We may also add a number to both sides of the inequality to obtain an equivalent in-

130

equality. Thus

$$3X + 4 > 3x + 4.$$

Hence, if x belongs to D, the function value $f(x) = 3x + 4$ is least when $x = -1$ and greatest when $x = 3$. Consequently, since $f(-1) = -3 + 4 = 1$ and $f(3) = 9 + 4 = 13$, the range is all y from 1 to 13; that is,

$$R = \{y \mid 1 \leq y \leq 13\}.$$

● **PROBLEM** 4-16

Find the domain D and the range R of the function $\left(x, \dfrac{x}{|x|}\right)$.

Solution: Note that the y-value of any coordinate pair (x,y) is $\dfrac{x}{|x|}$. We can replace x in the formula $\dfrac{x}{|x|}$ with any number except 0, since the denominator, $|x|$, can not equal 0, (i.e. $|x| \neq 0$) which is equivalent to $x \neq 0$. This is because division by 0 is undefined. Therefore, the domain D is the set of all real numbers except 0. If x is negative, i.e. $x < 0$, then $|x| = -x$ by definition. Hence, if x is negative, then $\dfrac{x}{|x|} = \dfrac{x}{-x} = -1$. If x is positive, i.e. $x > 0$, then $|x| = x$ by definition. Hence, if x is posi-tive, then $\dfrac{x}{|x|} = \dfrac{x}{x} = 1$. (The case where $x = 0$ has already been found to be undefined). Thus, there are only two numbers -1 and 1 in the range R of the function; that is, $R = \{-1, 1\}$.

● **PROBLEM** 4-17

Find the zeros of the function

$$\frac{2x + 7}{5} + \frac{3x - 5}{4} + \frac{33}{10}.$$

Solution: Let the function $f(x)$ be equal to $\dfrac{2x + 7}{5} + \dfrac{3x - 5}{4} + \dfrac{33}{10}$. A number, a, is a zero of a function $f(x)$ if $f(a) = 0$. A zero of $f(x)$ is a root of the equation $f(x) = 0$. Thus, the zeros of the function are the roots of the equation

$$\frac{2x + 7}{5} + \frac{3x - 5}{4} + \frac{33}{10} = 0.$$

The least common denominator, LCD, of the denominators of 5, 4, and 10 is 20. This is a fractional equation which can be solved by multiplying both members of the equation by the LCD.

$$20\left(\frac{2x+7}{5} + \frac{3x-5}{4} + \frac{33}{10}\right) = (20)(0)$$

$$4(2x+7) + 5(3x-5) + (2 \cdot 33) = 0.$$

Distributing,

$$8x + 28 + 15x - 25 + 66 = 0.$$

$$23x + 69 = 0$$

$$23x = -69$$

$$x = -3$$

Hence x = -3 is the zero of the given function.

● **PROBLEM** 4-18

Find the image of each element in

$$A = \{1,2,3,4,5,6,7,8,9\}$$

under the following mapping:

$$f(x) = \begin{cases} 2x, & \text{if } x < 5 \\ 8, & \text{if } x \geq 5 \end{cases}$$

Solution: The image of each element in $A = \{1,2,3,4,5,6,7,8,9\}$ under the mapping $f(x)$, is $f(1), f(2), f(3), f(4), f(5), f(6), f(7), f(8), f(9)$. $f(x)$ has two corresponding values, depending on the value of x. If $x < 5$, $f(x) = 2x$, thus for

$$x = 1, \quad f(1) = 2(1) = 2$$
$$x = 2, \quad f(2) = 2(2) = 4$$
$$x = 3, \quad f(3) = 2(3) = 6$$
$$x = 4, \quad f(4) = 2(4) = 8$$

and if

$$x \geq 5, \quad f(x) = 8, \qquad \text{thus for}$$
$$x = 5, \quad f(5) = 8$$
$$x = 6, \quad f(6) = 8$$
$$x = 7, \quad f(7) = 8$$
$$x = 8, \quad f(8) = 8$$
$$x = 9, \quad f(9) = 8$$

● **PROBLEM** 4-19

Given: $f : x \rightarrow y$, $f(x) = x+4$, and $Y = \{y \mid 4 \leq y < 10\}$; find $f^{-1}(Y)$.

Solution: Since $Y = \{y \mid 4 < y < 10\}$ and $y = x + 4$, then

$$4 \leq (y = x+4) < 10, \text{ that is } 4 \leq x + 4 < 10.$$

Adding -4 to all terms of $4 \leq x + 4 < 10$ one has $0 \leq x < 6$.

Therefore, $f^{-1}(Y) = \{x \mid 0 \leq x < 6\}$.

Graphically, f is first drawn and Y is located on the y-axis. Next project Y onto f, and the corresponding subset of f is projected onto the x-axis, which is $f^{-1}(Y)$ (see Fig. 1).

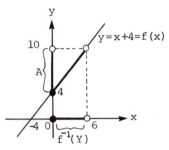

Fig. 1.

● **PROBLEM 4-20**

Let $f : R \to R$ and $g : R \to R$ be two functions given by $f(x) = 2x + 5$ and $g(x) = 4x^2$ respectively for all x in R, where R is the set of real numbers. Find expressions for the compositions $(f \circ g)(x)$ and $(g \circ f)(x)$.

Solution: Consider functions $f : A \to B$ and $g : B \to C$, that is, where the codomain of f is the domain of g. Then the function $g \circ f$ is defined as $g \circ f : A \to C$ where $(g \circ f)(x) = g(f(x))$ for all x in A, and it is called the composition of f and g.

In another notation,

$g \circ f = \{(x,z) \in A \times C \mid$ for all $y \in B$ such that

$(x,y) \in f$ and $(y,z) \in g\}$.

Therefore,

$$(f \circ g)(x) = f(g(x))$$

$$= f(4x^2)$$

$$= 2[4x^2] + 5$$

$$= 8x^2 + 5 \quad \text{and}$$

$$(g \circ f)(x) = g(f(x))$$

133

$$= g(2x + 5)$$

$$= 4[2x + 5]^2$$

$$= 4(4x^2 + 20x + 25)$$

$$= 16x^2 + 80x + 100$$

It's seen from the results that f∘g ≠ g∘f. This is true in general, i.e. function composition is not always a commutative operation.

● **PROBLEM** 4-21

Prove that composition of functions is an associative operation.

Solution: Let f : A → B, g : B → C and h : C → D be functions. Then to prove that functional composition is associative is the same as proving that (h∘g)∘f = h∘(g∘f). Note that both (h∘g)∘f and h∘(g∘f) are functions from A to D. Thus, to prove that (h∘g)∘f = h∘(g∘f) one needs only to show that ((h∘g)∘f)(x) = (h∘(g∘f))(x) for all x in A, (because if f : X → Y and g : X → Y are functions then f = g if and only if f(x) = g(x) for all x in X.)

By the definition of functional composition,

((h∘g)∘f)(x) = (h∘g)(f(x)) = h(g(f(x))) and

(h∘(g∘f))(x) = h(g∘f(x)) = h(g(f(x)))

for all x in A. Therefore,

(h∘g)∘f = h∘(g∘f).

● **PROBLEM** 4-22

Let   Φ : R → R given by

Φ = {(x,x²) | x ∈ R}.

Let   Γ : R → R given by

Γ = {(x,2x²+9) | x ∈ R}.

Compute

Φ∘Γ   and   Γ∘Φ.

Solution: Using the above notation for the functions Φ and Γ, we have that

$$\Phi(x) = x^2 \quad \text{and} \quad \Gamma(x) = 2x^2 + 9.$$

Now

$$(\Phi \circ \Gamma)(x) = \Phi(\Gamma(x))$$

$$= \Phi(2x^2+9)$$

$$= (2x^2+9)^2$$

$$= 4x^4 + 36x^2 + 81$$

Also, we have that:

$$(\Gamma \circ \Phi)(x) = \Gamma(\Phi(x))$$

$$= \Gamma(x^2)$$

$$= 2(x^2)^2 + 9 = 2x^4 + 9.$$

Note that $\quad \Phi \circ \Gamma \neq \Gamma \circ \Phi.$

● **PROBLEM** 4-23

The functions $f : A \to B$, $g : B \to C$ and $h : C \to A$ are defined by the diagram in Fig. 1. Draw the diagram which defines the composition function $h \circ g \circ f : A \to A$.

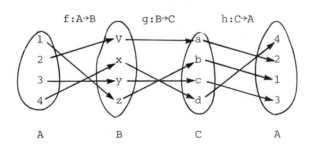

Fig. 1.

Solution: Definition: Let $f_1 : X \to Y$ and $f_2 : Y \to Z$. Then the composition function $f_2 \circ f_1$ is a function from X to Z defined by $f_2(f_1(x))$.

Therefore, we have

$$h \circ g \circ f = h(g(f(a))).$$

For example,

$$f(1) = z, \quad g(z) = b, \quad h(b) = 1,$$

thus,

$$h(g(f(1)))$$

$$= h(g(z))$$

$$= h(b) = 1.$$

$$h(g(f(2)))$$

$$= h(g(V))$$

$$= h(a)$$

$$= 2.$$

$$h(g(f(3)))$$

$$= h(g(y))$$

$$= h(c)$$

$$= 3.$$

$$h(g(f(4)))$$

$$= h(g(x))$$

$$= h(d)$$

$$= 4.$$

The diagram that defines $h \circ g \circ f$ is shown in Fig. 2.

h∘g∘f:A→A

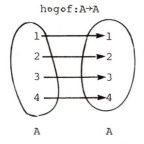

Fig. 2.   A   A

One can, of course, obtain the same diagram simply by inspection. As an example, Fig. 3 shows how we can find $h(g(f(1)))$ directly through the diagram in Fig. 1.

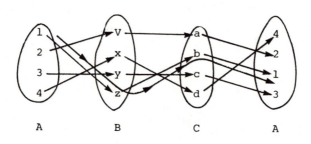

Fig. 3.   A   B   C   A

136

Given $f(x) = x + 10$, $g(x) = x^2$ and $h(x) = \sqrt[3]{x+1}$. Find

(1) $f \circ g(x)$

(2) $g \circ h(x)$

(3) $h \circ f \circ g(x)$

(4) $h \circ g \circ f(x)$

Solution: (1) $f \circ g(x) = f(g(x))$

$= f(x^2)$

$= (x^2) + 10$

$= x^2 + 10$

(2) $g \circ h(x) = g(h(x))$

$= g(\sqrt[3]{x} + 1)$

$= (\sqrt[3]{x} + 1)^2$

$= x^{\frac{2}{3}} + 2\sqrt[3]{x} + 1$

(3) $h \circ f \circ g(x)$

$= h(f(g(x)))$

$= h(f(x^2))$

$= h(x^2 + 10)$

$= \sqrt[3]{(x^2+10)} + 1$

$= (x^2 + 10)^{\frac{1}{3}} + 1$

(4) $h \circ g \circ f(x)$

$= h(g(f(x)))$

$= h(g(x+10))$

$= h((x+10)^2)$

$= h(x^2 + 20x + 100)$

$= \sqrt{x^2+20x+100} + 1$

Let E be a set, and let f, g be mappings of E → E.
Prove: if f and g are each one-to-one, then the compo-
site mapping f∘g is one-to-one. (f∘g is defined by the
formula (f∘g)(x) = f(g(x))).

Solution: One must show that if (f∘g)(x) = (f∘g)(y), then
x = y. One does so as follows:

$$(f∘g)(x) = (f∘g)(y) \qquad \text{(given)}$$

$$f(g(x)) = f(g(y)) \qquad \text{(definition of f∘g)}$$

$$g(x) = g(y) \qquad \text{(because f is one-to-one)}$$

$$x = y \qquad \text{(because g is one-to-one)}$$

Draw the graph of $f(x) = x^3$ and find its inverse,
$f^{-1}(x)$, if it exists. Draw the graph of $f^{-1}(x)$.

Solution: The graph of $f(x) = x^3$ is shown in Fig. 1.

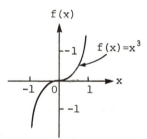

Fig. 1.

The inverse of $f(x) = x^3$ is $f^{-1}(x) = g(x) = \sqrt[3]{x}$; its
graph is shown in Fig. 2.

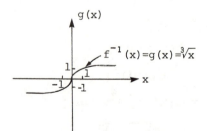

Fig. 2.

Show that the inverse of the function $y = x^2 + 4x - 5$ is not a function.

Solution: Given the function f such that no two of its ordered pairs have the same second element, the inverse function f^{-1} is the set of ordered pairs obtained from f by interchanging in each ordered pair the first and second elements. Thus, the inverse of the function

$y = x^2 + 4x - 5$ is $x = y^2 + 4y - 5$.

The given function has more than one first component corresponding to a given second component. For example, if $y = 0$, then $x = -5$ or 1. If the elements $(-5,0)$ and $(1,0)$ are reversed, we have $(0,-5)$ and $(0,1)$ as elements of the inverse. Since the first component 0 has more than one second component, the inverse is not a function (a function can have only one y value corresponding to each x value).

Given a function φ which maps a set A to itself show that if the identity map from A to A is contained in φ then φ is the identity map.

Solution: We are given that I_A is contained in our map φ, to show equality it is necessary to show inclusion the other way (i.e. $\phi \subset I_A$). First recall two definitions:

A function f from x_0 to y_0 is a subset of $X \times Y$ such that for all $x \in X_0$ there is a $y \in Y_0$ such that $(x,y) \in f$. Also if $(x,y) \in f$ and $(x,y_2) \in f$ then $y_1 = y_2$.

The identity map $I_X : X \to X$ is by definition the set: $\{(x,x)\}$ for every $x \in X$.

Assume that $(a_1,b_2) \in \phi$, then by definition of φ we have $a_1 \in A$. Also by the definition of I_A, $(a_1,a_1) \in I_A$. By our assumption that $I_A \subset \phi$, we have $(a_1,a_1) \in \phi$. Also by the definition of I_A, if $(a_1,a_2) \in \phi$, then $a_1 = a_2$.

Thus $(a_1,a_2) = (a_1,a_1) \in I_A$. Since, if we are given any $(a_1,a_2) \in \phi$, it follows that (a_1,a_2) must be an element of I_A, we have shown that $\phi \subset I_A$.

Thus, $\phi = I_A$.

SURJECTIVE, INJECTIVE AND BIJECTIVE FUNCTIONS

● **PROBLEM** 4-29

Give the definition of an injective function, a surjective function and a bijective function. Give an example
for each of the above functions.

Solution: A function f : A → B is said to be one-to-one or
injective if distinct elements in the domain A have distinct
images, i.e. if f(x) = f(y) implies x = y. For an example:
y = f(x) = x² defined over the domain $\{x \in R \mid x \geq 0\}$ is an in-
jection or an injective function.

A function f : A → B is said to be surjective or an onto
function if each element of B is the image of some element
of A. i.e. f(A) = B. For instance, y = x sinx, is a sur-
jection or a surjective function.

A function f : A → B is said to be bijective or a bijection if f is both injective and surjective. f is also
called a one-to-one correspondence between A and B. An example of such function would be y = x. The definitions
given above are also illustrated by the graphs shown in the
figure.

f:A→B is injective

f:A→B is surjective

f:A→B is bijective

Let $\phi(a,b) = a + b^2$ where a and b are any real numbers.

   (A)   Give the domain of ϕ.

   (B)   Give the co-domain of ϕ.

   (C)   Give the range of ϕ.

   (D)   Give the inverse image under ϕ of 0.

   (E)   Is ϕ injective?

   (F)   Is ϕ surjective?

Solution: Note that we can use the following notation:

$$\phi : R \times R \rightarrow R \quad \text{given by}$$

$$\phi(a,b) = a + b^2 \text{ for all a and b which are real numbers.}$$

(A)   From the above notation we see that the domain of ϕ is the set $R \times R$. (The plane of ordered pairs of real numbers.)

(B)   The co-domain of ϕ is the set R of real numbers.

(C)   The range of ϕ is also the set R of real numbers.

(D)   The inverse image of 0 is

$$\phi^{-1}(0) = \{(a,b) \in R \times R \mid \phi(a,b) = 0\}$$

$$= \{(a,b) \in R \times R \mid a + b^2 = 0\}$$

$$= \{(a,b) \mid a = -b^2\}$$

   This is a parabola with vertex at the origin and opening downward.

(E)   Note that ϕ is not injective since $\phi(0,0) = 0$ and $\phi(1,-1) = 0$ but $(0,0) \neq (1,-1)$.

(F) ϕ is surjective since for any $C \in R$ there exists an a and b in R such that

$$C = \phi(a,b) = a + b^2$$

Which of the following functions are injective, surjective or bijective?

(A) $f_1 : R \rightarrow [-1,1]$ given by:

$f_1(x) = \sin(x)$

(B) $f_2 : R \rightarrow R$ given by:

$f_2(x) = \sin(x)$

(C) $f_3 : R \rightarrow R$ given by:

$f_3(x) = x^3$

(D) $f_4 : R \rightarrow R$ given by;

$f_4(x) = x^2 - x^3.$

Solution: (A) This function is surjective since for every $y \in [-1,1]$ there is an $x \in R$ such that $\sin x = y$. The function is not injective since there are x_1 and x_2 with $x_1 \neq x_2$ and $\sin x_1 = \sin x_2$ (i.e. $x_1 = 2\pi$ and $x_2 = 4\pi$ in which case $\sin x_1 = 0 = \sin x_2$).

(B) The function $f_2 : R \rightarrow R$ given by $f_2(x) = \sin x$ is neither injective nor surjective. Note that the same example as was used above will show that f_2 is not injective. It is not surjective since given any $|y| > 1$, there is no $x \in R$ such that $f_2(x) = y$.

(C) This function, $f_3 : R \rightarrow R$, is injective since given an $x_1 \neq x_2$, $x_1^3 \neq x_2^3$. It is surjective since for every $y \in R$ there is an x such that $y = x^3$ (namely $y^{\frac{1}{3}}$). Hence, f_3 is also bijective.

(D) This map is surjective since for every $y \in R$ there is an $x \in R$ such that $y = x^2 - x^3$. It is not injective since for $x_1 = 0$ and $x_2 = 1$, $f_4(x_1) = f_4(x_2) = 0$ but $x_1 \neq x_2$.

We see that the only bijection is (C).

● **PROBLEM** 4-32

Show that the mapping $\Phi : R \rightarrow R$ by $\Phi = \{(x,y) \mid y = 5x-1\}$ is a surjective mapping.

Solution: By definition a map $\phi : A \rightarrow B$ is surjective if, given any $b \in B$, there exists an $a \in A$ such that $b = \phi(a)$. In our example, suppose we are given an arbitrary real number q; since the reals are the field, the expression

$\frac{q+1}{5}$ is also a real number.

(All we did was add one and multiply by the inverse of 5.) Now, observe that the ordered pair

$$\left(\frac{q+1}{5}, q \right) \in \phi$$

for every real number q. Hence, Φ is a surjective function.

● PROBLEM 4-33

Let ϕ be a mapping from set A to set B. Show:

(A) If there exists another map γ from B to A such that $\gamma \circ \phi = I_A$ (where I_A is the identity map from A to A) then $\phi : A \rightarrow B$ is an injective map.

(B) If there is a mapping θ from B to A such that $\phi \circ \theta = I_B$ then $\phi : A \rightarrow B$ is a surjective map.

Solution: (A) First we assume that there exists a function $\gamma : B \rightarrow A$ with $\gamma \circ \phi = I_A$. Now suppose we have $a_1, a_2 \in A$ such that $\phi(a_1) = \phi(a_2)$, in this case $a_1 = (\gamma \circ \phi)(a_1)$ which by definition of composition becomes $a_1 = \gamma(\phi(a_1))$. From our assumption that $a_1 = a_2$, we have $a_1 = \gamma(\phi(a_2))$, and, applying the definition of composition of functions once again, this gives $a_1 = (\gamma \circ \phi)(a_2) = a_2$ by our other assumption $(\gamma \circ \phi = I_A)$. Thus, we have shown that $\phi : A \rightarrow B$ is an injective map, since by definition, an injective, or one-to-one, map is a map $f : A \rightarrow B$ such that $f(x_1) = f(x_2)$ implies that $x_1 = x_2$ for any $x_1, x_2 \in A$.

(B) First we assume that there exists a function $\theta : B \rightarrow A$ with $\phi \circ \theta = I_B$. This implies that given any $b \in B$ there is an $a \in A$ with

$$a = \theta(b) \quad \text{so that}$$

$$\phi(a) = \phi(\theta(b))$$

$$= \phi \circ \theta(b)$$

$$= I_B(b) = b.$$

By definition a surjective (or an onto) map is a map $f : A \rightarrow B$ such that every element $b \in B$ is an image of some element of A. Thus we have shown that $\phi : A - B$ satisfies the criterion of a surjective map.

Define the function $\phi : R \rightarrow R$ by $\phi = \{(x,y) \mid y = 2x - 3\}$. Show that this is an injective map.

Solution: Let $(x_1,y) \in \phi$ and $(x_2,y) \in \phi$ for some $x_1, x_2 \in R$. From the definition of ϕ we have:

$$y = \phi(x_1) = 2x_1 - 3,$$

$$y = \phi(x_2) = 2x_2 - 3.$$

Thus, we have

$$2x_1 - 3 = 2x_2 - 3 \quad \text{and}$$

$$x_1 = x_2.$$

Therefore, we have shown that ϕ is injective.

Let function f be a subset of $A \times A$. Prove that, for every function f_1 and f_2 which are subsets of $A \times A$, if $f \circ f_1 = f \circ f_2$ then $f_1 = f_2$ if and only if f is injective.

Solution: First, we prove the implication from right to left, that is to show that if f is not injective then, given $f \circ f_1 = f \circ f_2$, $f_1 \neq f_2$.

Thus, assume f is not injective, which implies that there are at least two elements a_1, $a_2 \in A$, $a_1 \neq a_2$ but $f(a_1) = f(a_2)$. Next let's define $f_1 : A \rightarrow A$, $f_1(a) = a_1$ for all $a \in A$, and $f_2 : A \rightarrow A$, $f_2(a) = a_2$ for all $a \in A$.

If $f \circ f_1 = f \circ f_2$ then

$$f \circ f_1(a) = f(f_1(a)) = f(a_1),$$

$$f \circ f_2(a) = f(f_2(a)) = f(a_2). \quad \text{By our assumption}$$

$f(a_1) = f(a_2)$; we have

$$f \circ f_1 = f \circ f_2 \quad \text{but}$$

$$a_1 = f_1(a) \neq a_2 = f_2(a), \quad \text{i.e.} \quad f_1 \neq f_2.$$

We now prove that if $f \circ f_1 = f \circ f_2$ and $f_1 \neq f_2$ then f is not injective.

Let $f \circ f_1 = f \circ f_2$ and f_1, f_2 be defined as before. Furthermore, let $f_1(a_0) \neq f_2(a_0)$ for some $a_0 \in A$. Then:

$$f \circ f_1 = f(f(a_0)) = f(a_1),$$

$$f \circ f_2 = f(f_2(a_0)) = f(a_2).$$

We have $\quad f(a_1) = f(a_2), \quad$ however

$$a_1 = f_1(a_0) \neq a_2 = f_2(a_0).$$

Therefore, f is not injective.

Hence, the proof is completed.

● **PROBLEM** 4-36

Given the bijective mapping $\phi : A \rightarrow B$, prove that $\phi^{-1} : B \rightarrow A$ is also a bijection.

Solution: We must first show that ϕ^{-1} from B to A is a function. Note that because $\phi : A \rightarrow B$ is surjective we know that the domain of the inverse relation ϕ^{-1} is $\text{Im}(\phi) = B$. And, so the first requirement for a function is met.

Next, let (b, a_1) and (b, a_2) be elements of ϕ^{-1}. Then we have that (a_1, b) and (a_2, b) are both elements of ϕ. Hence by change of notation, $\phi(a_1) = b = \phi(a_2)$.

Since $\phi : A \rightarrow B$ is injective, we must have that $a_1 = a_2$. Hence the second requirement for $\phi^{-1} : B \rightarrow A$ to be a function is met.

Now we show that $\phi^{-1} : B \rightarrow A$ is injective. Let $b_1, b_2 \in B$ with $\phi^{-1}(b_1) = \phi^{-1}(b_2) = a$. Thus, we have that $\phi(a) = b_1$ and $\phi(a) = b_2$. Since ϕ is a function we must have that $b_1 = b_2$.

Thus we have shown that ϕ^{-1} is an injective function.

The final requirement for a bijection is that $\phi^{-1} : B \rightarrow A$ is surjective. We know that the image of $\phi^{-1} = \text{Dom}(\phi) = A$, and hence ϕ^{-1} is surjective. The proof is thus complete.

Given the characteristic function

$$\lambda_X : A \to \{0,1\} = 2$$

by
$$\lambda_X(a) = \begin{cases} 0 \text{ if } u \notin A \\ 1 \text{ if } u \in A \end{cases}$$

where $X \subseteq A$. Show that there is a bijection

$$\beta : \rho(A) \to \{\lambda_X\}$$

given by: $\beta(A) = \lambda_A$.

Solution: It is clearly true that $\beta : A \to \lambda_A$ is a function. Next we must show that β is injective. We begin by assuming that

$$\beta(X) = \beta(Y).$$

Note if $\lambda_X = \lambda_Y$ then

$$\lambda_X(a) = \lambda_Y(a) \quad \text{for all } a \in A.$$

Note $b \in X$ if and only if $\lambda_x(b) = 1$ (by the definition of the characteristic function). Equivalently:

$$\lambda_y(b) = 1 \quad \text{if and only if } b \in Y.$$

Thus $X = Y$ and we have shown that β is injective.

Next we prove that β is surjective. First let $\lambda \in \{\lambda_A\}$, define X by:

$$x = \{a \mid a \in A \quad \text{and } \phi(a) = 1\},$$

and thus,

$$X \in \rho(A) \quad \text{and } \rho(A) = \lambda.$$

The proof is complete since β is both injective and surjective and hence bijective.

Let γ be a function from a set A to itself. Show that if there is a positive integer r such that if

$$\gamma^r : A \to A, \text{ and}$$

$$\gamma^r(a) = a$$

for every $a \in A$ (i.e. $\gamma^r = I_A$) then γ is a bijective map.

Solution: First we must define what is meant by the power of a function. The power of a function is defined using reason: given a map $\delta : X \to X$ then $\delta' = \delta$ and $\delta^{n+1} = \delta \circ \delta^n$.

For the proof, first assume $r = 1$. Then, trivially, $\gamma' = \gamma = I_A$ which is a bijective map.

Next we assume $r > 1$. Assume that γ is not injective. Then $\gamma(a_1) = \gamma(a_2)$ for some a_1 and a_2 in A with $a_1 \neq a_2$. By our definition of γ^r we see that $\gamma^{r-1}\gamma(a_1) = \gamma^{r-1}\gamma(a_2)$ for all $a_1 \neq a_2$ in A. Thus γ^r is not an injective map (and hence not I_A in particular).

Next assume that γ is not surjective. Then, by definition there is an $a_2 \in A$ such that $\gamma(a_1) = a_2$ for all $a_1 \in A$. Thus,

$$\gamma(\gamma^{r-1}(a_3)) \neq a_2$$

for all $a_3 \in A$. We conclude that γ^r is not surjective (and hence not the surjective map $I_A : A \to A$). The proof is complete.

(A) Use the analogy of the citizens of the U.S. who have Social Security numbers and the set of Social Security numbers to illustrate bijective maps.

(B) Give an analogy for the set of maps which are surjective but not injective.

Solution: A mapping $\delta : A \to B$ is injective if and only if, given a_1 and a_2 in A with $a_1 \neq a_2$, $\delta(a_1) \neq \delta(a_2)$. A mapping $\phi : A \to B$ is surjective if given any $b \in B$ there is an $a \in A$ such that $\phi(a) = b$. Finally, a map which is both injective and surjective is bijective.

(A) Let C be the set of citizens of the United States who have Social Security numbers. Let S be the set of Social Security numbers. Let $\phi: C \to S$ be given by the map which produces the Social Security number of any given citizen. In order to show that ϕ is bijective we must show that it is both injective and surjective. Injective: Assume that we have two distinct citizens of the United States, that is c_1 and c_2 with $c_1 \neq c_2$. Every citizen of the U.S. is given a unique Social Security number, hence, the Social Security number of c_1, $\phi(c_1)$, is not the same as the Social Security number, $\phi(c_2)$, of citizen c_2. Thus $c_1 \neq c_2$ implies $\phi(c_1) \neq \phi(c_2)$. Hence ϕ is an injective map. Surjective: It is true that every citizen of the United States in our set C has a Social Security number. Thus given any Social Security number s (i.e. $s \in S$) there is some citizen ($c \in C$) such that $(c) = s$. Since ϕ is both injective and surjective, ϕ is a bijective map.

(B) For the analogy let C be the set of citizens of the U.S. and let A be the set of ages of people in the U.S. Let $\alpha : C \to A$ by $\alpha(c)$ be the age of the citizen denoted by c.

α is not injective since there are distinct citizens c_1, $c_2 \in C$ ($c_1 \neq c_2$) such that $\alpha(c_1) = \alpha(c_2)$ (i.e. they have the same age). α is surjective. This is true because given any age of a U.S. citizen, a, there is some citizen c, such that the age of c is a (i.e., for every $a \in A$ there is a $c \in C$ such that $\alpha(c) = a$).

● **PROBLEM 4-40**

Given that R is the set of real numbers find:

(A) A function $\phi : R \to R$ such that ϕ is injective but not surjective.

(B) A function $\gamma : R \to R$ such that γ is surjective but not injective.

Solution: Recall the following definitions:

(1) A mapping $\phi : A \to B$ is injective if, given a_1 and a_2 on A such that $a_1 \neq a_2$, $\phi(a_1) \neq \phi(a_2)$.

(2) A mapping $\phi : A \to B$ is surjective if for every $b \in B$ there is an $a \in A$ such that $\phi(a) = b$.

(A) The mapping

$$\phi : R \to R \quad \text{given by}$$

$$\phi(x) = e^x$$

is an injective mapping which is not surjective. ϕ is injective since, if we are given x_1 and x_2 in R with $x_1 \neq x_2$,

$$e^{x_1} \neq e^{x_2}$$

Observe that ϕ is not surjective since for any $y \in R$ such that $y \leq 0$ there is no $x \in R$ with $y = e^x$.

(B)   Consider the mapping

$$\gamma : R \rightarrow R \quad \text{given by}$$

$$\gamma(x) = x^2 - x^3.$$

Observe that this function is surjective since given $y \in R$ there is an x such that $x^2 - x^3 = y$. Note that γ is not injective since, for instance, $0 \neq 1$ but $\gamma(0) = 0^2 - 0^3 = 0$ and $\gamma(1) = 1^2 - 1^3 = 0$. So, $\gamma(0) = \gamma(1)$.

● **PROBLEM 4-41**

Prove that if ϕ is a function from a set A to a set B then ϕ has an inverse if and only if ϕ is bijective.

Solution: By definition a function is bijective if it is both injective and surjective. Assume ϕ is not injective. Then there exist a_1 and a_2 in A with $a_1 \neq a_2$ but $\phi(a_1) = \phi(a_2)$. Let $\phi(a_1) = \phi(a_2) = b \in B$. If ϕ^{-1} exists then $\phi^{-1}(b) = a_1$ and $\phi^{-1}(b) = a_2$. In this case ϕ^{-1} does not satisfy the requirements of a function, namely ϕ^{-1} maps one element of its domain into two distinct elements of its image. Thus, we have a contradiction and ϕ must be injective. Next assume ϕ is not surjective. In this case there must exist a $b \in B$ such that there is no $a \in A$ with $\phi(a) = b$. However, in this case $\phi^{-1}(b)$ does not exist and so ϕ^{-1} is not a function. We have reached another contradiction and hence must assume that ϕ is surjective; thus ϕ must be bijective.

Next, we prove the converse. Assume ϕ is bijective. In this case given any $a \in A$, there is a unique $b \in B$ such that $\phi(a) = b$ and hence $\phi^{-1}(b) = a$. Also since every $b \in B$ is in the image of ϕ, ϕ^{-1} is defined on every element of B.

● **PROBLEM 4-42**

Given the parallelogram ABCD shown in Fig. 1

justify the statement: The set of points of the segment AB is equipotent with the set of points of CD.

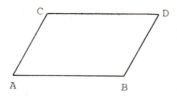

Fig. 1.

Solution: By definition two sets, X and Y, are equipotent if and only if there exists a map $\gamma : X \rightarrow Y$ such that γ is a bijection. We may construct this bijection by referring to the diagram of Figure 2.

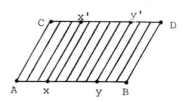

Fig. 2.

We have drawn line segments from side AB to side CD parallel to the other two sides (AC and BD). Our mapping is defined as follows: Let the points on AB be the elements of set S_1; the points on CD be the elements of set S_2. Then the bijective mapping $f : S_1 \rightarrow S_2$ is defined as the map where $f(S_1) = S_2$ is the endpoint of a line segment which is parallel to either AC or BD and connects S_1 and S_2.

From the diagram one sees that our mapping is bijective. It is injective since given any two distinct points X and Y of AB, their images will be distinct points. It is surjective since for any line segment (parallel to A (or BD)) with endpoint X' or CD there is another endpoint on AB, namely X.

Hence, every $x' \in S_2$ is an image of some element $x \in S_1$.

● **PROBLEM 4-43**

Show that the set of maps from set Z to set Y to set X is equipotent to the set of maps from $Y \times Z$ to X.

Solution: Symbolically what we want to prove is that $\left(X^Y\right)^Z \approx X^{Y \times Z}$. In order to do this we need to show that there is a map Δ from $(X^Y)^Z$ to $X^{Y \times Z}$ such that Δ is a bijection. Define

$$\Delta^{-1} : X^{Y \times Z} \rightarrow (X^Y)^Z \quad \text{by}$$

$$\Delta^{-1}(\gamma) = \{(x_1, a_2) \mid (x_1, a_2) \in Z \times X^Y \text{ and}$$

$$a_2(y) = \gamma(y, z)\},$$

where γ is a map from $Y \times Z$ to X. Δ^{-1} is easily shown to be both injective and surjective, hence it is a bijection. From this and the definition of equipotency it follows that

$$X^{Y \times Z} \approx (X^Y)^Z$$

which is what we needed to show.

● PROBLEM 4-44

Given three sets: X, Y and Z, such that $Y \cap Z = \phi$. Prove that the set of functions from $Y \cup Z$ to X is equipotent to the set of functions from Y to X crossed with the set of functions from Z to X.

Solution: Symbolically what we are being asked to show is
$$X^{Y \cup Z} \approx X^Y \times X^Z$$

Assume $\gamma \in X^{Y \cup Z}$, that is $\gamma : Y \cup Z \to X$. It follows that

$$\gamma = \gamma' \cup \gamma'' \quad \text{where}$$

$$\gamma' = \gamma \cap (Y \times X) \quad \text{and}$$

$$\gamma'' = \gamma \cap (Z \times X).$$

Note that $\gamma' \cap \gamma''$ is the empty set since $Y \cap Z = \phi$.

From our definitions we have that

$$\gamma' : Y \to X \quad \text{and that}$$

$$\gamma'' : Z \to X.$$

Define the map $\Delta : X^{Y \cup Z} \to X^Y \times X^Z$

by
$$\Delta(\gamma) = \Delta(\gamma' \cup \gamma'')$$

$$= (\gamma', \gamma'').$$

Note that Δ is a well defined function. Since it is clearly injective and surjective, it follows that

$$X^{Y \cup Z} \quad \text{is equipotent with}$$

$$X^Y \cup X^Z.$$

The proof is thus complete.

Show diagrammatically that the set of points of a circle contained inside a convex polygon are equipotent with the set of points of a convex polygon. Explain your reasoning fully.

Solution: Recall that by definition two sets are equipotent if and only if there exists a bijective map from one set to the other. In this example we illustrate our bijective map through the use of the diagram of Figure 1.

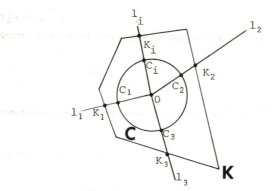

Fig. 1.

In Figure 1 we have a convex polygon K. (Recall that a set Q is convex if and only if given q_1 and q_2 in Q, every point of the line segment $\overline{(q_1 q_2)}$ is contained in Q.) In this case, the set K' of the polygon K together with all points inside K is the convex set we are referring to. The circle C lies entirely inside of K and the center of C is the point 0. The map which we would like to show is a bijection is

$$\phi : C \to K \text{ given by}$$

$$\phi(c_i) = k_i ,$$

where k_i is the point of intersection of the ray l_i, originating at 0 and passing through c_i, with K. By inspection, we see that given any point c_i on the circle C it would not be possible for $\phi(c_i)$ to be more than single valued (i.e. every ray originating from 0 and passing through c_i will intersect K at only one point and at one point only). Thus the map $\phi : C \to K$ is well defined. That ϕ is injective is seen by the fact that given two distinct points $c_i \neq c_j$ with c_i, $c_j \in C$, the images under ϕ will be distinct points of K. That ϕ is surjective is seen by choosing an arbitrary point of K, say k_i, and drawing a ray from 0 extending through k_i.

It is clear that this ray must intersect the circle C in a point c_i. Thus for each k_i of K there will be a point c_i of C such that $\phi(c_i) = k_i$. Hence, ϕ is a surjective map. Thus we have shown that ϕ is bijective and, by definition, the set of points of C must be equipotent with the set of points of K. This finishes the proof.

CHAPTER 5

VECTORS AND MATRICES

VECTORS

Which of the following vectors are equal to $\overrightarrow{MN}$ if
M = (2, 1) and N = (3, − 4)?
(a) $\overrightarrow{AB}$, where A = (1, − 1) and B = (2, 3)
(b) $\overrightarrow{CD}$, where C = (− 4, 5) and D = (− 3, 10)
(c) $\overrightarrow{EF}$, where E = (3, − 2) and F = (4, − 7).

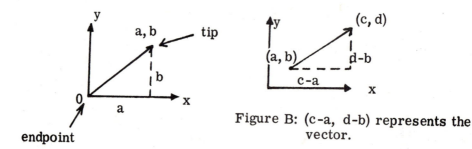

Figure A: (a-0, b-0) represents
the vector.

Figure B: (c-a, d-b) represents the
vector.

Solution: With each ordered pair in the plane there can
be associated a vector from the origin to that point.

The vector is determined by subtracting the coordinates of the endpoint from the corresponding coordinates of the tip. As for $\overrightarrow{MN}$, the tip is the point
corresponding to the second letter of the alphabetical
notation, N, while the endpoint is the point corresponding to the first, M. In this problem the vectors
are of a general nature wherein their endpoints do not
lie at the origin.

First find the ordered pair which represents $\overrightarrow{MN}$.

$$\overrightarrow{MN} = (3 - 2, - 4 - 1) = (1, - 5)$$

Then, find the ordered pair representing each vector.

(a) $\overrightarrow{AB} = (2 - 1, 3 - (- 1)) = (1, 4)$

(b) $\overrightarrow{CD} = ((- 3) - (- 4), 10 - 5) = (1, 5)$

(c) $\overrightarrow{EF} = (4 - 3, - 7 - (- 2)) = (1, - 5)$

Only $\overrightarrow{EF}$ and $\overrightarrow{MN}$ are equal.

● **PROBLEM 5-2**

What is the angle between a diagonal of a cube and one of its edges?

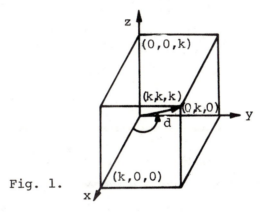

Fig. 1.

Solution: We assume the cube is centered at the origin of a three dimensional coordinate system and let k be the length of an edge. Now, find the length of the diagonal. It can be seen from Fig. 1 that the vector d (diagonal) has coordinates (k,k,k). But (k,k,k) can be written as a linear combination of the standard basis vectors for R^3 :

$$d = (k,k,k) = k(1,0,0) + k(0,1,0) + k(0,0,1).$$
$$= ku_1 + ku_2 + ku_3 .$$

Choose one of the edges for example, ku_1 . The angle between the diagonal and this edge can be found by the angle formulation of the dot product. That is, if v_1, v_2 are two vectors in R^n , then

$$\cos \theta = \frac{v_1 \cdot v_2}{\|v_1\| \|v_2\|} \tag{1}$$

where $\|v_1\|$ and $\|v_2\|$ denote the lengths of v_1 and v_2, respectively. Thus,

155

$$\cos \theta = \frac{ku_1 \cdot d}{\|ku_1\| \, \|d\|} . \tag{2}$$

In n-dimensional Euclidean space, the length of a vector $(x_1, x_2, \ldots, x_n)$ is given by

$$\sqrt{x_1^2 + x_2^2 + \ldots + x_n^2} = \left(\sum_{i=1}^{n} x_i^2 \right)^{\frac{1}{2}} .$$

Hence, $\|ku_1\| = k\|u_1\| = k(1 + 0 + 0)^{\frac{1}{2}} = k$; $\|d\| = \sqrt{k^2 + k^2 + k^2} = \sqrt{3k^2} = \sqrt{3} \, k.$

The dot product of two vectors $v, w \in R^n$ is defined as

$$v \cdot w = (v_1, v_2, \ldots, v_n) \cdot (w_1, w_2, \ldots, w_n)$$

$$= v_1 w_1 + v_2 w_2 + \ldots + v_n w_n = \sum_{i=1}^{n} v_i w_i .$$

Hence, $ku_1 \cdot d = (k, 0, 0) \cdot (k, k, k) = k^2 .$

We substitute the above results into (2) yielding

$$\cos \theta = \frac{k^2}{\sqrt{3} \, k^2} = \frac{1}{\sqrt{3}} .$$

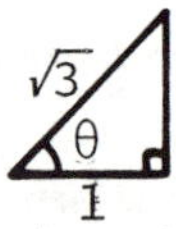

Fig. 2

The angle whose cosine is $1/\sqrt{3}$ is given by $\cos^{-1}\left(\frac{1}{\sqrt{3}}\right)$ and

$$\theta = \cos^{-1}\left(\frac{1}{\sqrt{3}}\right) = 54° \ 41' .$$

● **PROBLEM 5-3**

Compute $u \cdot v$ where i) $u = (2, -3, 6)$, $v = (8, 2, -3)$; ii) $u = (1, -8, 0, 5)$, $v = (3, 6, 4)$; iii) $u = (3, -5, 2, 1)$, $v = (4, 1, -2, 5)$.

Solution: In the vector space R^n, the dot product is defined as follows: for $X = (x_1, x_2, \ldots, x_n)$, $Y = (y_1, y_2, \ldots, y_n)$:

$$X \cdot Y = (x_1 y_1 + x_2 y_2 + \ldots + x_n y_n). \tag{1}$$

Thus, we compute the dot product of two vectors from R^n, multiply corresponding components and add. The result of taking the dot product is a scalar.

i) $u \cdot v = (2, -3, 6) \cdot (8, 2, -3) = 2(8) - 3(2) + 6(-3) = -8 .$

ii) $u \cdot v = (1, -8, 0, 5) \cdot (3, 6, 4)$. But, here the dot product is not defined since $u \in R^4$ while $v \in R^3$.

iii) $u \cdot v = (3, -5, 2, 1) \cdot (4, 1, -2, 5) = 3(4) - 5(1) + 2(-2) + 1(5) = 8.$

● **PROBLEM 5-4**

Find the distance between the vectors u and v where i) $u = (1, 7)$, $v = (6, -5)$; ii) $u = (3, -5, 4)$, $v = (6, 2, -1)$; iii) $u = (5, 3, -2, -4, 1)$. $v = (2, -1, 0, -7, 2)$.

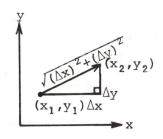

Solution: The Pythagorean theorem provides the foundation for the concept
of distance in Euclidean space.
In R^2, $d(x,y) = \sqrt{(\Delta x)^2 + (\Delta y)^2}$ where $\Delta x = x_2 - x_1$ and $\Delta y = y_2 - y_1$.
Generalizing to R^n, let $x = (x_1, x_2, \ldots, x_n)$, $y = (y_1, y_2, \ldots, y_n)$ be two
vectors in R^n. Then $d(x,y) = (\ (\Delta x_1)^2 + (\Delta x_2)^2 + \ldots + (\Delta x_n)^2\ )^{\frac{1}{2}}$ where
$\Delta x_1 = y_1 - x_1$, $\Delta x_2 = y_2 - x_2, \ldots, \Delta x_n = y_n - x_n$.
i) $d(u,v) = \sqrt{(1-6)^2 + (7+5)^2} = \sqrt{169} = 13$.

ii) $d(u,v) = \sqrt{9 + 49 + 24} = \sqrt{83}$.

iii) $d(u,v) = \sqrt{9 + 16 + 4 + 9 + 1} = \sqrt{39}$.

Note that $d(u,v) = d(v,u)$, i.e., the distance between two points is a
symmetric function of the points.

● **PROBLEM 5-5**

Show that the dot product can be derived from the theorem of Pythagoras and
the law of cosines.

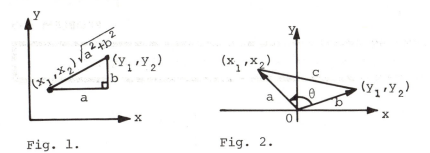

Fig. 1.   Fig. 2.

Solution: The Pythagorean theorem is used to derive the notion of distance
between two points in the plane while the law of cosines enables angle measurements to be made.
Definition: $d(\alpha, \beta) = \sqrt{\alpha^2 + \beta^2}$, the distance between the origin and the
point (α, β).
Notice that $a = |x_1 - y_1|$ and $b = |x_2 - y_2|$.

From Fig. 1, the distance between points (x_1, x_2) and (y_1, y_2) is
$$((x_1 - y_1)^2 + (x_2 - y_2)^2)^{\frac{1}{2}} = d((x_1 - y_1), (x_2 - y_2)).$$

We can express any distance in R^2 in terms of the function d.
In Fig. 2, the distance of (x_1, x_2) from the origin is given by $a =$

$d(x_1,x_2)$ while the distance of (y_1,y_2) from 0 is given by $b = d(y_1,y_2)$. The distance from (x_1,x_2) to (y_1,y_2) is given by c. Then, the law of cosines states that

$$\cos \theta = \frac{a^2 + b^2 - c^2}{2ab} \qquad (1)$$

Since $a^2 + b^2 - c^2 = x_1^2 + x_2^2 + y_1^2 + y_2^2 - (x_1-y_1)^2 - (x_2-y_2)^2$

$$= 2(x_1y_1 + x_2y_2) \ , \ (1) \text{ may be rewritten as}$$

$$\cos \theta = \frac{x_1y_1 + x_2y_2}{d(x_1,x_2)d(y_1,y_2)} \qquad (2)$$

The dot product of two vectors $u = (x_1,x_2)$ and $v = (y_1,y_2)$ is defined as; $u \cdot v = x_1y_1 + x_2y_2$. Thus,

$$\cos \theta = \frac{u \cdot v}{d(x_1,x_2)d(y_1,y_2)}$$

Now notice that $d(x_1,x_2) = \sqrt{x_1^2 + x_2^2} = \sqrt{(x_1,x_2)\cdot(x_1,x_2)} = \sqrt{u \cdot u} = (u \cdot u)^{\frac{1}{2}}$.
Similarly, $d(y_1,y_2) = (v \cdot v)^{\frac{1}{2}}$. Hence,

$$\cos \theta = \frac{u \cdot v}{(u \cdot u)^{\frac{1}{2}}(v \cdot v)^{\frac{1}{2}}} \qquad (3)$$

From (2), the dot product we can express in terms of the angle between two vectors and the distance between them:

$$u \cdot v = x_1y_1 + x_2y_2 = (\cos \theta)(d(x_1,x_2)d(y_1,y_2)) \ .$$

From (3), the length or norm of a vector can be expressed through the dot product.

● **PROBLEM 5-6**

Find a vector orthogonal to $A = (2,1,-1)$ and $B = (1,2,1)$.

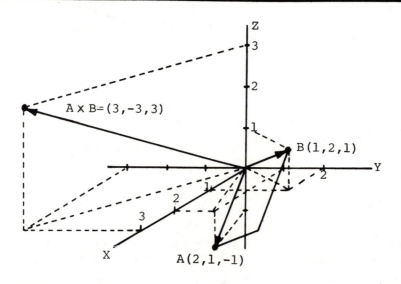

Solution: From the definition of orthogonality of vectors, a vector $V=(v_1,v_2,v_3)$ is said to be orthogonal to A and B if $A \cdot V = 0$ and $B \cdot V = 0$. In other words,

$$(v_1,v_2,v_3) \cdot (2,1,-1) = 2v_1 + v_2 - v_3 = 0 \qquad (1)$$

and

$$(v_1,v_2,v_3) \cdot (1,2,1) = v_1 + 2v_2 + v_3 = 0 \qquad (2)$$

Adding (1) and (2)

$$3v_1 + 3v_2 = 0 .$$

Therefore $v_1 = -v_2$. Let $v_2 = -3$. From $v_1 = -v_2$ and $v_3 = -(v_1+2v_2)$, the result is $v_1 = 3$ and $v_3 = -(3-6) = 3$. Therefore, the vector $(3,-3,3)$ is orthogonal to $A = (2,1,-1)$ and $B = (1,2,1)$.

● **PROBLEM 5-7**

1.)   Find $A \times B$ where $A = (1,2,-2)$ and $B = (3,0,1)$.

2.)   Verify directly that $A \cdot (A \times B) = 0$ and $B \cdot (A \times B) = 0$ where $A = (1,2,-2)$ and $B = (3,0,1)$.

3.)   Show that $A \cdot (A \times B) = 0$ and $B \cdot (A \times B) = 0$ where A,B are any vectors in R^3.

Solution: If $A = (a_1,a_2,a_3) \neq 0$ and $B = (b_1,b_2,b_3) \neq 0$ are vectors,

we define the vector $A \times B$, by:

$$A \times B = (a_2 b_3 - a_3 b_2, a_3 b_1 - a_1 b_3, a_1 b_2 - a_2 b_1),$$

$A \times B$ is called the cross product or vector product of A and B. If $A = (a_1,a_2,a_3)$ and $B = (b_1,b_2,b_3)$ are vectors in R^3, then the dot product is defined by: $A \cdot B = a_1 b_1 + a_2 b_2 + a_3 b_3$.

1.) $A = (1,2,-2)$ and $B = (3,0,1)$. By the definition of the cross product above, we compute $A \times B = (1,2,-2) \times (3,0,1) = (2(1)-(-2)(0),(-2)(3)-(1)(1), (1)(0) - 2(3)) = (2,-7,-6)$.

2.)   In 1) we found $A \times B = (2,-7,-6)$. Therefore, the dot product $A \cdot (A \times B) = (1,2,-2) \cdot (2,-7,-6) = 2 - 14 + 12 = 0$, and the dot product $B \cdot (A \times B) = (3,0,1) \cdot (2,-7,-6) = 6 - 0 - 6 = 0$.

3.)   By the definition of cross product, $A \times B = (a_2 b_3 - a_3 b_2, a_3 b_1 - a_1 b_3, a_1 b_2 - a_2 b_1$ if $A = (a_1,a_2,a_3)$ and $B = (b_1,b_2,b_3)$. By the definition of dot product,

$$A \cdot (A \times B) = (a_1,a_2,a_3) \cdot (a_2 b_3 - a_3 b_2, a_3 b_1 - a_1 b_3, a_1 b_2 - a_2 b_1) =$$

$$= a_1(a_2 b_3 - a_3 b_2) + a_2(a_3 b_1 - a_1 b_3) + a_3(a_1 b_2 - a_2 b_1) = 0$$

$$B \cdot (A \times B) = (b_1,b_2,b_3) \cdot (a_2 b_3 - a_3 b_2, a_3 b_1 - a_1 b_3, a_1 b_2 - a_2 b_1)$$

$$= b_1(a_2 b_3 - a_3 b_2) + b_2(a_3 b_1 - a_1 b_3) + b_3(a_1 b_2 - a_2 b_1) = 0 .$$

Find the area of the triangle determined by the points $P_1(2,2,0)$, $P_2(-1,0,1)$ and $P_3(0,4,3)$ by using the cross-product.

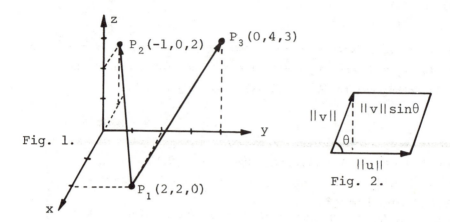

Fig. 1.

Fig. 2.

Solution: Fig. 1.   The triangle whose area is to be found is given in the figure above.

We relate the area of a triangle to the notion of a cross-product by using the following argument:

Lagrange's Identity states that

$$\|u \times v\|^2 = \|u\|^2 \|v\|^2 - (u \cdot v)^2 \tag{1}$$

where $\|w\|$ denotes the length of a vector in R^3. Thus, if $w = (w_1, w_2, w_3)$, $\|w\| = [\sum_{i=1}^{3} w_i^2]^{\frac{1}{2}}$. $(u \cdot v)$ is the dot product, $(u \cdot v) = \|u\| \|v\| \cos \theta$ where θ is the angle between u and v.   Equation (1) may be rewritten as

$$\|u \times v\|^2 = \|u\|^2 \|v\|^2 - \|u\|^2 \|v\|^2 \cos^2 \theta$$

$$= \|u^2\| \|v^2\| (1 - \cos^2 \theta)$$

$$= \|u^2\| \|v^2\| (\sin^2 \theta)$$

and, hence,

$$\|u \times v\| = \|u\| \|v\| \sin \theta \tag{2}$$

From Fig.2, observe that $\|v\| \sin \theta$ is the altitude of the parallelogram determined by u and v. The base of the parallelogram is u. Since

Area of parallelogram = (base)(altitude),

it can be seen from (2) that the norm of $u \times v$ equals the area of the parallelogram determined by u and v.

We utilize this result in solving the given problem. From Fig. 1, the area of the triangle is $\frac{1}{2}$ the area of the parallelogram formed by the vectors $\overrightarrow{P_1P_2}$ and $\overrightarrow{P_1P_3}$. $\overrightarrow{P_1P_2}$ is given by $P_2(-1,0,2) - P_1(2,2,0) = (-3,-2,2)$, and $\overrightarrow{P_1P_3} = P_3(0,4,3) - P_1(2,2,0) = (-2,2,3)$. In R^3, the cross-product is defined as follows:

$$X \times Y = \begin{vmatrix} e_1 & e_2 & e_3 \\ x_1 & x_2 & x_3 \\ y_1 & y_2 & y_3 \end{vmatrix} \qquad \text{where}$$

$X = (x_1, x_2, x_3)$, $Y = (y_1, y_2, y_3)$ and $\{e_1, e_2, e_3\}$ is the standard basis for R^3. Now, $\overrightarrow{P_1 P_2} \times \overrightarrow{P_1 P_3} = (-3, -2, 2) \times (-2, 2, 3) =$

$$\begin{vmatrix} e_1 & e_2 & e_3 \\ -3 & -2 & 2 \\ -2 & 2 & 3 \end{vmatrix} = e_1 \begin{vmatrix} -2 & 2 \\ 2 & 3 \end{vmatrix} - e_2 \begin{vmatrix} -3 & 2 \\ -2 & 3 \end{vmatrix} +$$

$$e_3 \begin{vmatrix} -3 & -2 \\ -2 & 2 \end{vmatrix}$$

$= -10e_1 + 5e_2 - 10e_3$. This is $(-10, 5, -10)$ when written as a point in R^3. Since $\|(-10, 5, -10)\| = \sqrt{100+25+100} = 15$

$$\text{Area (Triangle)} = \tfrac{1}{2} \|\overrightarrow{P_1 P_2} \times \overrightarrow{P_1 P_3}\|$$

$$= \tfrac{1}{2}(15) = 7.5 .$$

● **PROBLEM** 5-9

(1) Define an eigenvalue.
(2) Show that if u and v are eigenvectors of a linear operator f which belong to λ and if a is a real number, then **(a)** u + v and **(b)** au are also eigenvectors of f which belong to λ .

Solution: (1) A real number, λ, is an eigenvalue of a linear mapping f if and only if there exists a non-zero vector u in V such that $f(u) = \lambda u$.

Thus, an eigenvalue is a number which acts as a scalar multiple of some non-zero vector to give its f-image. If λ is an eigenvalue and $f(u) = \lambda u$, then u is called an eigenvector of f belonging to λ . The set of eigenvectors of f which belong to a given eigenvalue, λ, constitutes a subspace of R^n which is called the eigenspace of λ .

Let P be the graph point of an eigenvector u which belongs to an eigenvalue λ of an operator f on R^2 . If Q is the graph point of λu, then $\overrightarrow{OP}$ and $\overrightarrow{OQ}$ are collinear. The eigenvalue, λ, signifies an extension, contraction or reversal in direction of its eigenspace according to whether its value is greater than 1, between 0 and 1 or less than 0. (See Figure 1).

 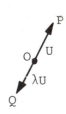

Fig. 1.

(2a) $f(u+v) = f(u) + f(v) = \lambda u + \lambda v = \lambda(u+v)$. Thus, u+v are also eigenvectors of f which belong to λ .

(b) $f(au) = af(u) = a(\lambda u) = \lambda(au)$. Thus, each scalar multiple au is an eigenvector.

● **PROBLEM** 5-10

Find the real eigenvalues of A and their associated eigenvectors when

$$A = \begin{bmatrix} 1 & 1 \\ -2 & 4 \end{bmatrix} .$$

Solution: The problem requires that we find all real numbers λ and all non-zero vectors $X = \begin{pmatrix} x_1 \\ x_2 \end{pmatrix}$ such that $AX = \lambda X$:

$$\begin{bmatrix} 1 & 1 \\ -2 & 4 \end{bmatrix} \begin{bmatrix} x_1 \\ x_2 \end{bmatrix} = \lambda \begin{bmatrix} x_1 \\ x_2 \end{bmatrix}$$

The above matrix equation is equivalent to the homogeneous system,

$$x_1 + x_2 = \lambda x_1$$

$$-2x_1 + 4x_2 = \lambda x_2$$

or,

$$(\lambda-1)x_1 - x_2 = 0 \tag{1}$$

$$+2x_1 + (\lambda-4)x_2 = 0 .$$

Recall that a homogeneous system has a non-zero solution if and only if the determinant of the matrix of coefficients is zero. Thus,

$$\begin{vmatrix} \lambda-1 & -1 \\ 2 & \lambda-4 \end{vmatrix} = 0$$

or,

$$(\lambda-1)(\lambda-4) + 2 = 0 .$$

Therefore,

$$\lambda^2 - 5\lambda + 6 = 0$$

or,

$$(\lambda-3)(\lambda-2) = 0 .$$

Hence, $\lambda_1 = 2$ and $\lambda_2 = 3$ are the eigenvalues of A. We find an eigenvector of A associated with $\lambda_1 = 2$, by forming the linear system:

$$AX = 2X$$

or,

$$\begin{bmatrix} 1 & 1 \\ -2 & 4 \end{bmatrix} \begin{bmatrix} x_1 \\ x_2 \end{bmatrix} = 2 \begin{bmatrix} x_1 \\ x_2 \end{bmatrix}$$

This gives

$$x_1 + x_2 = 2x_1 \qquad \qquad (2-1)x_1 - x_2 = 0$$
$$\text{or}$$
$$-2x_1 + 4x_2 = 2x_2 \qquad \qquad 2x_1 + (2-4)x_2 = 0$$

or,

$$x_1 - x_2 = 0$$
$$2x_1 - 2x_2 = 0 \qquad \text{or, simply, } x_1 - x_2 = 0 \ .$$

Observe that one obtains this last linear system by substituting $\lambda=2$ in (1). It can be seen that any vector in R^2 of the form

$$x = k \begin{bmatrix} 1 \\ 1 \end{bmatrix}, \quad k \quad \text{a scalar, is an eigenvector of } A$$

associated with $\lambda_1 = 2$. Thus,

$$x_1 = \begin{bmatrix} 1 \\ 1 \end{bmatrix} \quad \text{is an eigenvector of } A \text{ associated with}$$

$\lambda_1 = 2$. Similarly, for $\lambda_2 = 3$, we obtain from (1):

$$(3-1)x_1 - x_2 = 0 \qquad \qquad 2x_1 = x_2 = 0$$
$$\qquad \qquad \qquad \text{or,}$$
$$2x_1 + (3-4)x_2 = 0 \qquad \qquad 2x_1 - x_2 = 0 \ .$$

Thus, $x_2 = \begin{bmatrix} 1 \\ 2 \end{bmatrix}$ is an eigenvector of A associated with $\lambda_2 = 3$.

● **PROBLEM** 5-11

Given

$$A = \begin{bmatrix} 1 & 0 & -1 & 0 \\ 0 & 1 & 0 & -1 \\ 1 & 0 & -1 & 0 \\ 0 & 1 & 0 & -1 \end{bmatrix} \quad \text{and} \quad B = \begin{bmatrix} 4 & -1 & -1 & 0 \\ -1 & 4 & 0 & -1 \\ 1 & 0 & 2 & -1 \\ 0 & 1 & -1 & 2 \end{bmatrix},$$

Find a characteristic vector common to A and B.

Solution: If $AB = BA$, then A and B have a characteristic vector in common.

We prove this assertion as follows: Let λ_1 be a characteristic root of A and let $x_1, \ldots, x_k$ be a basis for the null space of $\lambda_1 I_n - A$. Then, any non-zero linear combination of $x_1, \ldots, x_k$ is a characteristic vector corresponding to λ_1. Conversely, we express any characteristic vector of A corresponding to λ_1 as a linear combination of $(x_1, x_2, \ldots, x_k)$. Suppose that $Bx_i \neq 0$, $i=1, \ldots k$. then

$$A(Bx_i) = BAx_i$$
$$= B(\lambda_1 x_i)$$
$$= \lambda_1 (Bx_i) \ ;$$

thus,

$$Bx_i \in Sp(x_1, \ldots, x_k) \ , \quad i = 1, \ldots, k \ .$$

Let X be the matrix whose jth column is x_j, $j = 1, \ldots, k$. Therefore, the system of equations,

$$Xy_i = Bx_i \ , \quad i = 1, \ldots, k \ , \tag{1}$$

is the matrix whose ith column is y_i, $i = 1, \ldots, k$. The relation (1) can be written compactly as

$$XY = BX \ . \tag{2}$$

Let μ_1 be any characteristic root of Y and $z = (\alpha_1, \ldots, \alpha_k)$, a characteristic vector of Y corresponding to μ_1 . Then, by (2),

$$B(xz) = XYz$$

and since

$$Yz = \mu_1 z \ ,$$

$$B(xz) = \mu_1 (Xz) \ .$$

Now,

$$Xz = \sum_{t=1}^{k} \alpha_t x_t \neq 0$$

because $z = (\alpha_1, \ldots, \alpha_k) \neq 0$ and $x_1, \ldots, x_k$ are linearly independent. Thus, Xz is a characteristic vector of B . But,

$$Xz = \sum_{t=1}^{k} \alpha_t x_t \in (x_1, \ldots, x_k)$$

and, therefore, Xz is a characteristic vector of A . Now, find the product of AB and BA.

$$AB = \begin{bmatrix} 3 & -1 & -3 & 1 \\ -1 & 3 & 1 & -3 \\ 3 & -1 & -3 & 1 \\ -1 & 3 & 1 & -3 \end{bmatrix} = BA \ .$$

Thus, $AB = BA$. Therefore, A and B have a common characteristic vector. The matrix A is clearly singular and, therefore, $\det(I\lambda - A) = 0$ implies $\lambda = 0$ (since $\det A = 0$). Since the rank of $\lambda_1 I_4 - A$ is 2, the dimension of its null space is $4-2 = 2$. The vector $x_1 = (1,0,1,0)$ and $x_2 = (0,1,0,1)$ form a basis for this null space

$$Bx_1 = (3,-1,1,3); \ Bx_2 = (-1,3,-1,3)$$

Thus,

$$Bx_1 = 3x_1 - x_2 \quad \text{and} \quad Bx_2 = -x_1 + 3x_2 \ .$$

Let

$$X = \begin{bmatrix} 1 & 0 \\ 0 & 1 \\ 1 & 0 \\ 0 & 1 \end{bmatrix} \quad \text{and} \quad Y = \begin{bmatrix} 3 & -1 \\ -1 & 3 \end{bmatrix} \ .$$

Then, as in (2),

$$XY = BX \ .$$

To find the characteristic root of Y, solve the equation, $\det(\lambda I - Y) = 0$. Then,

$$\det \begin{vmatrix} \lambda-3 & 1 \\ 1 & \lambda-3 \end{vmatrix} = 0$$

or,

$$(\lambda-3)^2 - 1 = 0$$

$$(\lambda-4)(\lambda-2) = 0 \ .$$

Therefore, $\mu_1 = 4$ is the characteristic root of Y, and $z = (1,-1)$ is the corresponding characteristic vector. Then,

$$Xz = \begin{bmatrix} 1 & 0 \\ 0 & 1 \\ 1 & 0 \\ 0 & 1 \end{bmatrix} \begin{bmatrix} 1 \\ -1 \end{bmatrix} = \begin{bmatrix} 1 \\ -1 \\ 1 \\ -1 \end{bmatrix}$$

Thus, $Xz = (1,-1,1,-1)$ is a characteristic vector common to A and B.

● **PROBLEM** 5-12

Show for the following matrix, A, that any column eigenvector corresponding to a particular eigenvalue is orthogonal to all row eigenvectors corresponding to other eigenvalues and vice versa.

$$A = \begin{bmatrix} -1 & 2 & 2 \\ -8 & 7 & 4 \\ -13 & 5 & 8 \end{bmatrix}$$

Solution:By definition if λ is an eigenvalue of a square matrix A and X is a corresponding eigenvector (column eigenvector), they satisfy the relation

$$AX = \lambda X .$$

The eigenvalues can be obtained as the roots of the characteristic equation $\det(A - \lambda I) = 0$.

The eigenvectors corresponding to the various eigenvalues can be found by solving the equations

$$(A - \lambda I)X = 0$$

after substituting the appropriate values for λ . A similar course of action can be carried out to find the row eigenvectors of the matrix.

A row eigenvector is a non-zero vector Y which satisfies the relation:

$$YA = \lambda Y .$$

Thus, if $Y = (y_1, y_2, y_3)$, the row eigenvector satisfies

$$[y_1, y_2, y_3] \begin{bmatrix} -1 & 2 & 2 \\ -8 & 7 & 4 \\ -13 & 5 & 8 \end{bmatrix} = \lambda [y_1, y_2, y_3] .$$

Hence,

$$(-y_1 - 8y_2 - 13y_3, \ 2y_1 + 7y_2 + 5y_3, \ 2y_1 + 4y_2 + 8y_3) = (\lambda y_1, \ \lambda y_2, \ \lambda y_3)$$

or

$$-y_1 - 8y_2 - 13y_3 = \lambda y_1$$

$$+2y_1 + 7y_2 + 5y_3 = \lambda y_2$$

$$2y_1 + 4y_2 + 8y_3 = \lambda y_3 .$$

Rearranging,

$$(-1-\lambda)y_1 - 8y_2 - 13y_3 = 0$$

$$2y_1 + (7-\lambda)y_2 + 5y_3 = 0 \qquad (1)$$

$$2y_1 + 4y_2 + (8-\lambda)y_3 = 0 .$$

It is known that a set of homogeneous linear equations has a solution in which the unknowns are not all zero if and only if the determinant of the

165

coefficients is zero. One has

$$\begin{vmatrix} -1-\lambda & -8 & -13 \\ 2 & 7-\lambda & 5 \\ 2 & 4 & 8-\lambda \end{vmatrix} = 0$$

or,

$$(-1-\lambda) \begin{vmatrix} 7-\lambda & 5 \\ 4 & 8-\lambda \end{vmatrix} - (-8) \begin{vmatrix} 2 & 5 \\ 2 & 8-\lambda \end{vmatrix} -13 \begin{vmatrix} 2 & 7-\lambda \\ 2 & 4 \end{vmatrix}$$

$$=(-1-\lambda)[(7-\lambda)(8-\lambda)- 20] + 8[2(8-\lambda) - 10] - 13[8-2(7-\lambda)]$$

or,

$$\lambda^3 - 14\lambda^2 + 63\lambda - 90 = 0$$

$$\lambda^2 (\lambda-3) - 11\lambda(\lambda-3) + 30(\lambda-3) = 0$$

$$(\lambda-3)(\lambda-6)(\lambda-5) = 0 .$$

Therefore, the eigenvalues are $\lambda = 3$, $\lambda = 5$ and $\lambda = 6$.
When $\lambda = 3$, equations (1) become

$$-4y_1 - 8y_2 - 13y_3 = 0$$
$$-2y_1 + 4y_2 + 5y_3 = 0$$
$$2y_1 + 4y_2 + 5y_3 = 0 .$$

These have solutions

$$y_1 = -2, \ y_2 = 1, \ y_3 = 0 .$$

Thus, $Y_1 = (-2,1,0)$ is the row eigenvector associated with eigenvalue $\lambda = 3$. We let $\lambda=5$ in equations (1) yielding

$$Y_2 = (7,-2,-2)$$

and, for $\lambda = 6$, $y_3 = (-3,1,1)$. The column eigenvectors are obtained by solving the equation $(A - \lambda I)X = 0$ for X.

For $\lambda = 3$,

$$(A - 3I)X = 0$$

$$\begin{bmatrix} -4 & 2 & 2 \\ -8 & 4 & 4 \\ -13 & 5 & 5 \end{bmatrix} \begin{bmatrix} x_1 \\ x_2 \\ x_3 \end{bmatrix} = \begin{bmatrix} 0 \\ 0 \\ 0 \end{bmatrix}$$

Thus, $X_1 = \begin{bmatrix} 0 \\ 1 \\ -1 \end{bmatrix}$ is an eigenvector corresponding to $\lambda = 3$.

Similarly, $X_2 = \begin{bmatrix} 1 \\ 2 \\ 1 \end{bmatrix}$ with eigenvalue $\lambda = 5$, $X_3 = \begin{bmatrix} 2 \\ 4 \\ 3 \end{bmatrix}$ with eigen-

value $\lambda = 6$. Now,

$$y_1 x_2 = [-2,1,0] \begin{bmatrix} 1 \\ 2 \\ 1 \end{bmatrix} = 0$$

$$y_1 x_3 = [-2,1,0] \begin{bmatrix} 2 \\ 4 \\ 3 \end{bmatrix} = 0$$

and, likewise, $Y_r X_s = 0$ where $r \neq s$. Thus, the column eigenvector

166

corresponding to a particular eigenvalue is orthogonal to all row eigen-vectors corresponding to other eigenvalues.

MATRIX ARITHMETIC

● **PROBLEM** 5-13

If $A = \begin{bmatrix} 2 & -2 & 4 \\ -1 & 1 & 1 \end{bmatrix}$ and $B = \begin{bmatrix} 0 & 1 & -3 \\ 1 & 3 & 1 \end{bmatrix}$, find $2A + B$.

Solution: For an $m \times n$ matrix, $A = (a_{ij})$, we know $cA = (ca_{ij})$. Hence,

$$2A = 2 \begin{bmatrix} 2 & -2 & 4 \\ -1 & 1 & 1 \end{bmatrix} = \begin{bmatrix} 2 \cdot 2 & 2 \cdot (-2) & 2 \cdot 4 \\ 2 \cdot (-1) & 2 \cdot 1 & 2 \cdot 1 \end{bmatrix} = \begin{bmatrix} 4 & -4 & 8 \\ -2 & 2 & 2 \end{bmatrix}.$$

For two $m \times n$ matrices, $A = (\alpha_{ij})$ and $B = (\beta_{ij})$, the ith row of the matrix $A + B$ is computed by $e_i \cdot (A+B) = (\alpha_{i1} + \beta_{i1}, \ldots, \alpha_{in} + \beta_{in})$.

Thus,

$$2A + B = \begin{bmatrix} 4 & -4 & 8 \\ -2 & 2 & 2 \end{bmatrix} + \begin{bmatrix} 0 & 1 & -3 \\ 1 & 3 & 1 \end{bmatrix} = \begin{bmatrix} 4+0 & -4+1 & 8-3 \\ -2+1 & 2+3 & 2+1 \end{bmatrix}$$

$$2A + B = \begin{bmatrix} 4 & -3 & 5 \\ -1 & 5 & 3 \end{bmatrix}.$$

● **PROBLEM** 5-14

Find $A - B$ if $A = \begin{bmatrix} 3 & -2 & 5 \\ -1 & 2 & 3 \end{bmatrix}$ and $B = \begin{bmatrix} 2 & 3 & 2 \\ -3 & 4 & 6 \end{bmatrix}$.

Solution: Let A and B be two matrices with the same size, i.e., the same number of rows and of columns, say $m \times n$.

$$A = \begin{bmatrix} a_{11} & a_{12} & \cdots & a_{1n} \\ \vdots & \vdots & & \vdots \\ a_{m1} & a_{m2} & \cdots & a_{mn} \end{bmatrix}$$

and

$$B = \begin{bmatrix} b_{11} & b_{12} & \cdots & b_{1n} \\ \vdots & \vdots & & \vdots \\ b_{m1} & b_{m2} & \cdots & b_{mn} \end{bmatrix}.$$

The sum of A and B, written $A + B$, is the matrix obtained by adding corresponding components.

$$A+B = \begin{bmatrix} a_{11}+b_{11} & a_{12}+b_{12} & \cdots & \cdots & a_{1n}+b_{1n} \\ a_{21}+b_{21} & a_{22}+b_{22} & \cdots & \cdots & a_{2n}+b_{2n} \\ \cdots & \cdots & \cdots & \cdots & \cdots \\ a_{m1}+b_{m1} & a_{m2}+b_{m2} & & & a_{mn}+b_{mn} \end{bmatrix}$$

Also, define $A - B \equiv A + (-B)$. Here $-B$ represents the matrix $-1 \cdot B$.

In the given problem,

$$A = \begin{bmatrix} 3 & -2 & 5 \\ -1 & 2 & 3 \end{bmatrix} ; \quad B = \begin{bmatrix} 2 & 3 & 2 \\ -3 & 4 & 6 \end{bmatrix} .$$

Then,

$$-B = -1 \cdot B = \begin{bmatrix} -1 \cdot 2 & -1 \cdot 3 & -1 \cdot 2 \\ -1 \cdot (-3) & -1 \cdot 4 & -1 \cdot 6 \end{bmatrix} = \begin{bmatrix} -2 & -3 & -2 \\ 3 & -4 & -6 \end{bmatrix} .$$

Thus,

$$A - B = A + (-B) = \begin{bmatrix} 3 & -2 & 5 \\ -1 & 2 & 3 \end{bmatrix} = \begin{bmatrix} -2 & -3 & -2 \\ 3 & -4 & -6 \end{bmatrix} .$$

● **PROBLEM 5-15**

Show that

a) $A + B = B + A$ where

$$A = \begin{bmatrix} 3 & 1 & 1 \\ 2 & -1 & 1 \end{bmatrix} ; \quad B = \begin{bmatrix} 4 & 2 & -1 \\ 0 & 0 & 2 \end{bmatrix} .$$

b) $(A+B) + C = A + (B+C)$ where

$$A = \begin{bmatrix} -2 & 6 \\ 2 & 1 \end{bmatrix} , \quad B = \begin{bmatrix} 2 & 1 \\ -0 & 3 \end{bmatrix} \quad \text{and} \quad C = \begin{bmatrix} -1 & 0 \\ 7 & 2 \end{bmatrix} .$$

c) If A and the zero matrix (0_{ij}) have the same size, then $A + 0 = A$ where

$$A = \begin{bmatrix} 2 & 1 \\ 1 & 2 \end{bmatrix} .$$

d) $A + (-A) = 0$ where

$$A = \begin{bmatrix} 2 & 1 \\ 1 & 2 \end{bmatrix} .$$

e) $(ab)A = a(bA)$ where $a = -5$, $b = 3$ and

$$A = \begin{bmatrix} 6 & -1 & 0 \\ 1 & 2 & 1 \end{bmatrix} .$$

f) Find B if $2A - 3B + C = 0$ where

$$A = \begin{bmatrix} -1 & 3 \\ 0 & 0 \end{bmatrix} \quad \text{and} \quad C = \begin{bmatrix} -2 & -1 \\ -1 & 1 \end{bmatrix} .$$

Solution: a) By the definition of matrix addition, we compute:

$$A + B = \begin{bmatrix} 3 & 1 & 1 \\ 2 & -1 & 1 \end{bmatrix} + \begin{bmatrix} 4 & 2 & -1 \\ 0 & 0 & 2 \end{bmatrix}$$

$$= \begin{bmatrix} 3+4 & 1+2 & 1+(-1) \\ 2+0 & -1+0 & 1+2 \end{bmatrix}$$

$$= \begin{bmatrix} 7 & 3 & 0 \\ 2 & -1 & 3 \end{bmatrix}$$

and

$$B + A = \begin{bmatrix} 4 & 2 & -1 \\ 0 & 0 & 2 \end{bmatrix} + \begin{bmatrix} 3 & 1 & 1 \\ 2 & -1 & 1 \end{bmatrix}$$

$$= \begin{bmatrix} 4+3 & 2+1 & -1+1 \\ 0+2 & 0+(-1) & 2+1 \end{bmatrix} + \begin{bmatrix} 7 & 3 & 0 \\ 2 & -1 & 3 \end{bmatrix}$$

Thus, $A + B = B + A$.

b)

$$A + B = \begin{bmatrix} -2 & 6 \\ 2 & 1 \end{bmatrix} + \begin{bmatrix} 2 & 1 \\ 0 & 3 \end{bmatrix}$$

$$= \begin{bmatrix} -2+2 & 6+1 \\ 2+0 & 1+3 \end{bmatrix} = \begin{bmatrix} 0 & 7 \\ 2 & 4 \end{bmatrix}$$

and

$$(A+B) + C = \begin{bmatrix} 0 & 7 \\ 2 & 4 \end{bmatrix} + \begin{bmatrix} -1 & 0 \\ 7 & 2 \end{bmatrix} = \begin{bmatrix} 0+(-1) & 7+0 \\ 2+7 & 4+2 \end{bmatrix} = \begin{bmatrix} -1 & 7 \\ 9 & 6 \end{bmatrix}.$$

$$B + C = \begin{bmatrix} 2 & 1 \\ 0 & 3 \end{bmatrix} + \begin{bmatrix} -1 & 0 \\ 7 & 2 \end{bmatrix} = \begin{bmatrix} 2+(-1) & 1+0 \\ 0+7 & 3+2 \end{bmatrix} = \begin{bmatrix} 1 & 1 \\ 7 & 5 \end{bmatrix}$$

and

$$A + (B+C) = \begin{bmatrix} -2 & 6 \\ 2 & 1 \end{bmatrix} + \begin{bmatrix} 1 & 1 \\ 7 & 5 \end{bmatrix} = \begin{bmatrix} -2+1 & 6+1 \\ 2+7 & 1+5 \end{bmatrix} = \begin{bmatrix} -1 & 7 \\ 9 & 6 \end{bmatrix}.$$

Thus, $(A+B) + C = A + (B+C)$.

c)   An m×n matrix all of whose elements are zeros is called a zero matrix and is usually denoted by $0_{m\ n}$.

$$A = \begin{bmatrix} 2 & 1 \\ 1 & 2 \end{bmatrix} \qquad 0 = \begin{bmatrix} 0 & 0 \\ 0 & 0 \end{bmatrix}.$$

Thus,

$$A + 0 = \begin{bmatrix} 2 & 1 \\ 1 & 2 \end{bmatrix} + \begin{bmatrix} 0 & 0 \\ 0 & 0 \end{bmatrix} = \begin{bmatrix} 2+0 & 1+0 \\ 1+0 & 2+0 \end{bmatrix} = \begin{bmatrix} 2 & 1 \\ 1 & 2 \end{bmatrix}.$$

Hence, $A + 0 = A$.

d) $-A = -1 \cdot \begin{bmatrix} 2 & 1 \\ 1 & 2 \end{bmatrix} = \begin{bmatrix} -1 \cdot 2 & -1 \cdot 1 \\ -1 \cdot 1 & -1 \cdot 2 \end{bmatrix} = \begin{bmatrix} -2 & -1 \\ -1 & -2 \end{bmatrix}.$

Thus,

$$A + (-A) = \begin{bmatrix} 2 & 1 \\ 1 & 2 \end{bmatrix} + \begin{bmatrix} -2 & -1 \\ -1 & -2 \end{bmatrix} = \begin{bmatrix} 2+(-2) & 1+(-1) \\ 1+(-1) & 2+(-2) \end{bmatrix} = \begin{bmatrix} 0 & 0 \\ 0 & 0 \end{bmatrix}$$

Therefore, $A + (-A) = 0$.

e)   If $A = \begin{bmatrix} a_1 & b_1 \\ c_1 & d_1 \end{bmatrix}$ and a is any scalar from a field, aA is

defined by

$$aA = \begin{bmatrix} aa_1 & ab_1 \\ ac_1 & ad_1 \end{bmatrix}.$$

169

So, $bA = 3 \begin{bmatrix} 6 & -1 & 0 \\ 1 & 2 & 1 \end{bmatrix} = \begin{bmatrix} 3\cdot 6 & 3\cdot(-1) & 3\cdot 0 \\ 3\cdot 1 & 3\cdot 2 & 3\cdot 1 \end{bmatrix}$

$$= \begin{bmatrix} 18 & -3 & 0 \\ 3 & 6 & 3 \end{bmatrix}$$

and

$a(bA) = -5 \begin{bmatrix} 18 & -3 & 0 \\ 3 & 6 & 3 \end{bmatrix} = \begin{bmatrix} -90 & 15 & 0 \\ -15 & -30 & -15 \end{bmatrix}$

$(ab)A = ((-5)(3)) \begin{bmatrix} 6 & -1 & 0 \\ 1 & 2 & 1 \end{bmatrix} = -15 \begin{bmatrix} 6 & -1 & 0 \\ 1 & 2 & 1 \end{bmatrix} = \begin{bmatrix} -90 & 15 & 0 \\ -15 & -30 & -15 \end{bmatrix}$

Thus, $(ab)A = a(bA)$.

f) $2A - 3B + C = 2A + C - 3B = 0$ since matrix addition is commutative.

Now, add $3B$ to both sides of the equation,

$$2A + C - 3B = 0 ,$$

to obtain $2A + C - 3B + 3B = 0 + 3B$.   (1)

Using the laws that were exemplified in parts a) through d),
(1) becomes $2A + C = 3B$. Now,

$$\tfrac{1}{3}(2A + C) = \tfrac{1}{3}(3B)$$

which implies $B = \tfrac{1}{3}(2A + C)$.

$2A + C = \begin{bmatrix} 2(-1) & 2(3) \\ 2(0) & 2(0) \end{bmatrix} + \begin{bmatrix} -2 & -1 \\ -1 & 1 \end{bmatrix} = \begin{bmatrix} -4 & 5 \\ -1 & 1 \end{bmatrix}$.

Thus,

$$\tfrac{1}{3}(2A + C) = \tfrac{1}{3}\begin{bmatrix} -4 & 5 \\ -1 & 1 \end{bmatrix} = \begin{bmatrix} -4/3 & 5/3 \\ -1/3 & 1/3 \end{bmatrix} .$$

● **PROBLEM 5-16**

If $A = \begin{bmatrix} 1 & 2 & 4 \\ 2 & 6 & 0 \end{bmatrix}$; $B = \begin{bmatrix} 4 & 1 & 4 & 3 \\ 0 & -1 & 3 & 1 \\ 2 & 7 & 5 & 2 \end{bmatrix}$,

find AB .

Solution: Since   A   is a 2×3 matrix and   B   is a 3×4 matrix, the
product   AB   is a 2×4 matrix.

$$AB = \begin{bmatrix} 1 & 2 & 4 \\ 2 & 6 & 0 \end{bmatrix} \begin{bmatrix} 4 & 1 & 4 & 3 \\ 0 & -1 & 3 & 1 \\ 2 & 7 & 5 & 2 \end{bmatrix}$$

$$= \begin{bmatrix} 1\cdot 4+2\cdot 0+4\cdot 2 & 1\cdot 1+2\cdot(-1)+4\cdot 7 & 1\cdot 4+2\cdot 3+4\cdot 5 & 1\cdot 3+2\cdot 1+4\cdot 2 \\ 2\cdot 2+6\cdot 0+0\cdot 2 & 2\cdot 1+6\cdot(-1)+0\cdot 7 & 2\cdot 4+6\cdot 3+0\cdot 5 & 2\cdot 3+6\cdot 1+0\cdot 2 \end{bmatrix}$$

$$= \begin{bmatrix} 4+0+8 & 1-2+28 & 4+6+20 & 3+2+8 \\ 8+0+0 & 2-6+0 & 8+18+0 & 6+6+0 \end{bmatrix}$$

$$AB = \begin{bmatrix} 12 & 27 & 30 & 13 \\ 8 & -4 & 26 & 12 \end{bmatrix}$$

a) Suppose $A = \begin{bmatrix} 1 & 3 \\ 2 & -1 \end{bmatrix}$ and $B = \begin{bmatrix} 2 & 0 & -4 \\ 3 & -2 & 6 \end{bmatrix}$. Find i) AB and

ii) BA .

b) Suppose $A = [2,1]$ and $B = \begin{bmatrix} 1 & -2 & 0 \\ 4 & 5 & -3 \end{bmatrix}$. Find i) AB, and

ii) BA.

<u>Solution</u>: Suppose $A = (a_{ij})$ and $B = (b_{jk})$ are matrices such that the number of columns of A equals the number of rows of B; also suppose A is an $m \times n$ matrix and B is an $n \times s$ matrix.

$$A = \begin{bmatrix} a_{11} & \cdots & a_{1n} \\ \vdots & & \vdots \\ a_{m1} & \cdots & a_{mn} \end{bmatrix}$$

$$B = \begin{bmatrix} b_{11} & \cdots & b_{1s} \\ \vdots & & \vdots \\ b_{n1} & \cdots & b_{ns} \end{bmatrix} .$$

Then the product AB is the $m \times s$ matrix whose ik-element is
$$\sum_{j=1}^{n} a_{ij} b_{jk} = a_{i1} b_{ik} + a_{i2} b_{2k} + \ldots + a_{in} b_{nk} .$$ If $A_1, \ldots, A_m$ are the row vectors of the matrix A, and if $B^1, \ldots, B^s$ are the column vectors of the matrix B, then the ik-element of the product AB is equal to $A_1, \ldots, B^k$. Thus,

$$\begin{matrix} A_1 \cdot B^1 & \cdots & A_1 \circ B^s \\ \vdots & & \vdots \\ A_m B^1 & \cdots & A_m \cdot B^s \end{matrix}$$

a) i) $A = \begin{bmatrix} 1 & 3 \\ 2 & -1 \end{bmatrix}$ and $B = \begin{bmatrix} 2 & 0 & -4 \\ 3 & -2 & 6 \end{bmatrix}$. Here, A is 2×2 and

B is 2×3 . The product AB is a 2×3 matrix. To obtain the components in the first row of AB, one multiplies the first row (1,3) of A by the columns $\begin{bmatrix} 2 \\ 3 \end{bmatrix}$, $\begin{bmatrix} 0 \\ -2 \end{bmatrix}$ and $\begin{bmatrix} -4 \\ 6 \end{bmatrix}$ of B, respectively.

$$\begin{bmatrix} 1 & 3 \\ 2 & -1 \end{bmatrix} \begin{bmatrix} 2 & 0 & -4 \\ 3 & -2 & 6 \end{bmatrix} = [1 \cdot 2 + 3 \cdot 3 \quad 1 \cdot 0 + 3 \cdot (-2) \quad 1 \cdot (-4) + 3 \cdot 6]$$

$$= [2 + 0 \quad 0 - 6 \quad -4 + 18] = [11 \quad -6 \quad 14]$$

To obtain the components in the second row of AB, multiply the second row (2,01) of A by the columns of B, respectively.

$$\begin{bmatrix} 1 & 3 \\ 2 & -1 \end{bmatrix} \begin{bmatrix} 2 & 0 & -4 \\ 3 & -2 & 6 \end{bmatrix} = [(2 \cdot 2 + (-1) \cdot 3) \quad (2 \cdot 0 + (-1) \cdot (-2)) \quad (2 \cdot (-4) + (-1) \cdot 6)]$$

$$= [4 - 3 \quad 0 + 2 \quad -8 - 6] = [1 \quad 2 \quad -14].$$

Thus,

$$AB = \begin{bmatrix} 11 & -6 & 14 \\ 1 & 2 & -14 \end{bmatrix}.$$

ii)   Here B is 2×3 and A is 2×2 . Since the number of columns of B is not equal to the number of rows of A, the product BA is not defined.

b) i)   Since A is a 1×2 and B is a 2×3, the product AB is a 1×3 matrix.

$$AB = \begin{bmatrix} 2,1 \end{bmatrix} \begin{bmatrix} 1 & -2 & 0 \\ 4 & 5 & -3 \end{bmatrix} = \begin{bmatrix} (2 \cdot 1 + 1 \cdot 4, & 2 \cdot (-2) + 1 \cdot 5, & 2 \cdot 0 + 1 \cdot (-3) \end{bmatrix}$$

$$= \begin{bmatrix} 6 & 1 & -3 \end{bmatrix}$$

ii)   In this case, B is 2×3 and A is 1×2 . Since the number of columns of B is not equal to the number of rows of A, the product BA is not defined.

● **PROBLEM** 5-18

Let A = $\begin{bmatrix} 2 & -1 \\ 1 & 0 \\ -3 & 4 \end{bmatrix}$ and B = $\begin{bmatrix} 1 & -2 & -5 \\ 3 & 4 & 0 \end{bmatrix}$.

Find (i) AB, (ii) BA .

Solution: Since A is 3×2 and B is 2×3, the product AB is a 3×3 matrix. To obtain the first row of AB, multiply the first row of A by each column of B, respectively. The first row of

$$\begin{bmatrix} 2 & -1 \\ 1 & 0 \\ -3 & 4 \end{bmatrix} \begin{bmatrix} 1 & -2 & -5 \\ 3 & 4 & 0 \end{bmatrix} \quad \text{is} \quad [2 \cdot 1 + (-1) \cdot 3 \quad 2 \cdot (-2) + (-1)(4) \quad 2 \cdot (-5) + -1 \cdot 0]$$

$$= [2 - 3 \quad -4 - 4 \quad -10 + 0] = [-1 \quad -8 \quad -10].$$

To obtain the second row of AB, multiply the second row of A by each column of B, respectively.

$$\begin{bmatrix} 2 & -1 \\ 1 & 0 \\ -3 & 4 \end{bmatrix} \begin{bmatrix} 1 & -2 & -5 \\ 3 & 4 & 0 \end{bmatrix} = [1 \cdot 1 + 0 \cdot 3 \quad 1 \cdot (-2) + 0 \cdot 4 \quad 1 \cdot (-5) + 0 \cdot 0]$$

$$= [1+0 \quad -2+0 \quad -5+0] = [1 \quad -2 \quad -5]$$

To obtain the third row of AB, multiply the third row of A by each column of B, respectively. Then, AB =

$$\begin{bmatrix} 2 & -1 \\ 1 & 0 \\ -3 & 4 \end{bmatrix} \begin{bmatrix} 1 & -2 & -5 \\ 3 & 4 & 0 \end{bmatrix} = \begin{bmatrix} -1 & -8 & -10 \\ 1 & -2 & -5 \\ (-3)\cdot1+4\cdot3 & (-3)\cdot(-2)+4\cdot4 & (-3)(-5)+4\cdot0 \end{bmatrix}$$

$$= \begin{bmatrix} -1 & -8 & -10 \\ 1 & -2 & -5 \\ -3+12 & 6+16 & 15+0 \end{bmatrix} = \begin{bmatrix} -1 & -8 & -10 \\ 1 & -2 & -5 \\ 9 & 22 & 15 \end{bmatrix} .$$

Thus,

$$AB = \begin{bmatrix} -1 & -8 & -10 \\ 1 & -2 & -5 \\ 9 & 22 & 15 \end{bmatrix} .$$

ii) Since B is 2x3 and A is 3X2, the product BA is a 2x2 matrix. To obtain the first row of BA, multiply the first row of B by each column of A, respectively.

$$\begin{bmatrix} 1 & -2 & -5 \\ 3 & 4 & 0 \end{bmatrix} \begin{bmatrix} 2 & -1 \\ 1 & 0 \\ -3 & 4 \end{bmatrix}$$

$$= \quad [1\cdot2+(-2)\cdot1+(-5)\cdot(-3) \quad 1\cdot(-1)+(-2)\cdot0+(-5)\cdot4]$$

$$= \quad [2-2+15 \quad -1+0-20] = [15 \quad -21] .$$

To obtain the second row of BA, multiply the second row of B by each column of A, respectively. Then, BA =

$$\begin{bmatrix} 1 & -2 & -5 \\ 3 & 4 & 0 \end{bmatrix} \begin{bmatrix} 2 & -1 \\ 1 & 0 \\ -3 & 4 \end{bmatrix}$$

$$= \begin{bmatrix} 15 & -21 \\ 3\cdot2+4\cdot1+0\cdot(-3) & 3\cdot(-1)+4\cdot0+0\cdot4 \end{bmatrix}$$

$$= \begin{bmatrix} 15 & -21 \\ -6+4+0 & -3+0+0 \end{bmatrix} = \begin{bmatrix} 15 & -21 \\ 10 & -3 \end{bmatrix} .$$

Thus,

$$BA = \begin{bmatrix} 15 & -21 \\ 10 & -3 \end{bmatrix} .$$

Remark: In this case, observe that both AB and BA are defined but they are not equal. In fact, they do not even have the same shape.

● PROBLEM 5-19

If $A = \begin{bmatrix} 2 & -1 & 0 \\ 1 & 0 & -3 \end{bmatrix}$ and $B = \begin{bmatrix} 1 & -4 & 0 & 1 \\ 2 & -1 & 3 & -1 \\ 4 & 0 & -2 & 0 \end{bmatrix}$

(1) Determine the shape of AB .

(2) If c_{ij} denotes the element in the ith row and jth column of the product matrix AB, find c_{23} c_{14} and c_{21} .

Solution: (1) A matrix with m rows and n columns is called an m by n matrix or an m×n matrix. The pair of numbers (m,n) is called its size or shape. Since A is 2×3 and B is 3×4 , the product AB is a 2×4 matrix. Thus, the shape of AB is (2×4).

(2) Now, c_{23} is the element in the 2nd row and 3rd column of the product matrix AB . To obtain c_{23} , multiply the second row of A by the third column of B . Hence,

$$c_{23} = [1 \cdot 0 \quad -3] \begin{bmatrix} 0 \\ 3 \\ -2 \end{bmatrix}$$

$$= 1 \cdot 0 + 0 \cdot 3 + (-3) \cdot (-2) = 0 + 0 + 6 = 6 ,$$

and

$$c_{14} = [2 \quad -1 \quad 0] \begin{bmatrix} 1 \\ -1 \\ 0 \end{bmatrix}$$

$$= 2 \cdot 1 + (-1) \cdot (-1) + 0 \cdot 0 = 2 + 1 + 0 = 3 .$$

$$c_{21} = [1 \quad 0 \quad -3] \begin{bmatrix} 1 \\ 2 \\ 4 \end{bmatrix} = 1 \cdot 1 + 0 \cdot 2 + (-3) \cdot 4$$

$$= 1 + 0 - 12 = -11 .$$

● **PROBLEM** 5-20

Prove (AB)C = A(BC) where $A = \begin{bmatrix} 5 & 2 & 3 \\ 2 & -3 & 4 \end{bmatrix}$, $B = \begin{bmatrix} 2 & -1 & 1 & 0 \\ 0 & 2 & 2 & 2 \\ 3 & 0 & -1 & 3 \end{bmatrix}$

and $C = \begin{bmatrix} 1 & 0 & 2 \\ 2 & -3 & 0 \\ 0 & 0 & 3 \\ 2 & 1 & 0 \end{bmatrix}$

Solution: First, find (AB)C and then A(BC).

$$AB = \begin{bmatrix} 5 & 2 & 3 \\ 2 & -3 & 4 \end{bmatrix} \begin{bmatrix} 2 & -1 & 1 & 0 \\ 0 & 2 & 2 & 2 \\ 3 & 0 & -1 & 3 \end{bmatrix}$$

$$= \begin{bmatrix} 10+0+9 & -5+4+0 & 5+4-3 & 0+4+9 \\ 4+0+12 & -2-6+0 & 2-6-4 & 0-6+12 \end{bmatrix}$$

$$= \begin{bmatrix} 19 & -1 & 6 & 13 \\ 16 & -8 & -8 & 6 \end{bmatrix}$$

Now,

$$(AB)C = \begin{bmatrix} 19 & -1 & 6 & 13 \\ 16 & -8 & -8 & 6 \end{bmatrix} \begin{bmatrix} 1 & 0 & 2 \\ 2 & -3 & 0 \\ 0 & 0 & 3 \\ 2 & 1 & 0 \end{bmatrix}$$

174

$$= \begin{bmatrix} 19-2+0+26 & 0+3+0+13 & 38+0+18+0 \\ 16-16+0+12 & 0+24+0+6 & 32+0-24+0 \end{bmatrix} = \begin{bmatrix} 43 & 16 & 56 \\ 12 & 30 & 8 \end{bmatrix}.$$

Thus,
$$(AB)C = \begin{bmatrix} 43 & 16 & 56 \\ 12 & 30 & 8 \end{bmatrix}.$$

$$BC = \begin{bmatrix} 2 & -1 & 1 & 0 \\ 0 & 2 & 2 & 2 \\ 3 & 0 & -1 & 3 \end{bmatrix} \begin{bmatrix} 1 & 0 & 2 \\ 2 & -3 & 0 \\ 0 & 0 & 3 \\ 2 & 1 & 0 \end{bmatrix}$$

$$= \begin{bmatrix} 2-2+0+0 & 0+3+0+0 & 4+0+3+0 \\ 0+4+0+4 & 0-6+0+2 & 0+0+6+0 \\ 3+0+0+6 & 0+0+0+3 & 6+0-3+0 \end{bmatrix}$$

$$BC = \begin{bmatrix} 0 & 3 & 7 \\ 8 & -4 & 6 \\ 9 & 3 & 3 \end{bmatrix}$$

and
$$A(BC) = \begin{bmatrix} 5 & 2 & 3 \\ 2 & -3 & 4 \end{bmatrix} \begin{bmatrix} 0 & 3 & 7 \\ 8 & -4 & 6 \\ 9 & 3 & 3 \end{bmatrix}$$

$$= \begin{bmatrix} 0+16+27 & 15-8+9 & 35+12+9 \\ 0-24+36 & 6+12+12 & 14-18+12 \end{bmatrix} = \begin{bmatrix} 43 & 16 & 56 \\ 12 & 30 & 8 \end{bmatrix}$$

Thus, $(AB)C = A(BC)$.

● **PROBLEM** 5-21

Find $A(B+C)$ and $AB + AC$ if $A = \begin{bmatrix} 2 & 2 & 3 \\ 3 & -1 & 2 \end{bmatrix}$, $B = \begin{bmatrix} 1 & 0 \\ 2 & 2 \\ 3 & -1 \end{bmatrix}$

and $C = \begin{bmatrix} -1 & 2 \\ 1 & 0 \\ 2 & -2 \end{bmatrix}$.

Solution: We perform the indicated matrix operations:

$$B + C = \begin{bmatrix} 1 & 0 \\ 2 & 2 \\ 3 & -1 \end{bmatrix} + \begin{bmatrix} -1 & 2 \\ 1 & 0 \\ 2 & -2 \end{bmatrix}$$

$$= \begin{bmatrix} 1+(-1) & 0+2 \\ 2+1 & 2+0 \\ 3+2 & -1+(-2) \end{bmatrix} = \begin{bmatrix} 0 & 2 \\ 3 & 2 \\ 5 & -3 \end{bmatrix}$$

then,
$$A(B+C) = \begin{bmatrix} 2 & 2 & 3 \\ 3 & -1 & 2 \end{bmatrix} \begin{bmatrix} 0 & 2 \\ 3 & 2 \\ 5 & -3 \end{bmatrix}$$

$$= \begin{bmatrix} 2 \cdot 0+2 \cdot 3+3 \cdot 5 & 2 \cdot 2+2 \cdot 2+3 \cdot (-3) \\ 3 \cdot 0+(-1) \cdot 3+2 \cdot 5 & 3 \cdot 2+(-1) \cdot 2+2 \cdot (-3) \end{bmatrix}.$$

$$A(B+C) = \begin{bmatrix} 0+6+15 & 4+4-9 \\ 0-3+10 & 6-2-6 \end{bmatrix} = \begin{bmatrix} 21 & -1 \\ 7 & -2 \end{bmatrix}.$$

$$AB = \begin{bmatrix} 2 & 2 & 3 \\ 3 & -1 & 2 \end{bmatrix} \begin{bmatrix} 1 & 0 \\ 2 & 2 \\ 3 & -1 \end{bmatrix}$$

$$= \begin{bmatrix} 2 \cdot 1+2 \cdot 2+3 \cdot 3 & 2 \cdot 0+2 \cdot 2+3 \cdot (-1) \\ 3 \cdot 1+(-1) \cdot 2+2 \cdot 3 & 3 \cdot 0+(-1) \cdot 2+2 \cdot (-1) \end{bmatrix}$$

$$= \begin{bmatrix} 2+4+9 & 0+4+(-3) \\ 3-2+6 & 0-2-2 \end{bmatrix} = \begin{bmatrix} 15 & 1 \\ 7 & -4 \end{bmatrix}.$$

$$AC = \begin{bmatrix} 2 & 2 & 3 \\ 3 & -1 & 2 \end{bmatrix} \begin{bmatrix} -1 & 2 \\ 1 & 0 \\ 2 & -2 \end{bmatrix}$$

$$= \begin{bmatrix} 2 \cdot (-1)+2 \cdot 1+3 \cdot 2 & 2 \cdot 2+2 \cdot 0+3 \cdot (-2) \\ 3 \cdot (-1)+(-1) \cdot 1+2 \cdot 2 & 3 \cdot 2+(-1) \cdot 0+2 \cdot (-2) \end{bmatrix}$$

$$AC = \begin{bmatrix} -2+2+6 & 4+0-6 \\ -3-1+4 & 6+0-4 \end{bmatrix} = \begin{bmatrix} 6 & -2 \\ 0 & 2 \end{bmatrix}.$$

Then,

$$AB + AC = \begin{bmatrix} 15 & 1 \\ 7 & -4 \end{bmatrix} + \begin{bmatrix} 6 & -2 \\ 0 & 2 \end{bmatrix} = \begin{bmatrix} 15+6 & 1+(-2) \\ 7+0 & -4+2 \end{bmatrix}$$

$$= \begin{bmatrix} 21 & -1 \\ 7 & -2 \end{bmatrix}$$

Remark: $A(B+C) = \begin{bmatrix} 21 & -1 \\ 7 & -2 \end{bmatrix}$ and $AB + AC = \begin{bmatrix} 21 & -1 \\ 7 & -2 \end{bmatrix}$.

Thus, $A(B+C) = AB + AC$. This is called the left distributive law.

● **PROBLEM** 5-22

Find (i) A^2 (ii) A^3 (iii) A^4 when $A = \begin{bmatrix} 1 & 2 \\ -1 & 1 \end{bmatrix}$.

Solution: We compute the indicated powers of the given matrix by using the rules for matrix multiplication:

$$A^2 = AA = \begin{bmatrix} 1 & 2 \\ -1 & 1 \end{bmatrix} \begin{bmatrix} 1 & 2 \\ -1 & 1 \end{bmatrix} = \begin{bmatrix} 1-2 & 2+2 \\ -1-1 & -2+1 \end{bmatrix} = \begin{bmatrix} -1 & 4 \\ -2 & -1 \end{bmatrix}.$$

$$A^3 = AAA = A^2A = \begin{bmatrix} -1 & 4 \\ -2 & -1 \end{bmatrix} \begin{bmatrix} 1 & 2 \\ -1 & 1 \end{bmatrix} = \begin{bmatrix} -1-4 & -2+4 \\ -2+1 & -4-1 \end{bmatrix} = \begin{bmatrix} -5 & 2 \\ -1 & -5 \end{bmatrix}.$$

The usual laws for exponents are $A^mA^n = A^{m+n}$ and $(A^m)^n = A^{mn}$. Thus, $A^4 = A^3A$ or, $A^4 = (A^2)^2 = A^2A^2$.

$$A^4 = A^3A = \begin{bmatrix} -5 & 2 \\ -1 & -5 \end{bmatrix} \begin{bmatrix} 1 & 2 \\ -1 & 1 \end{bmatrix} = \begin{bmatrix} -5-2 & -10+2 \\ -1+5 & -2-5 \end{bmatrix} = \begin{bmatrix} -7 & -8 \\ 4 & -7 \end{bmatrix}.$$

Observe that

$$A^4 = A^2 A^2 = \begin{bmatrix} -1 & 4 \\ -2 & -1 \end{bmatrix} \begin{bmatrix} -1 & 4 \\ -2 & -1 \end{bmatrix} = \begin{bmatrix} 1-8 & -4-4 \\ 2+2 & -8+1 \end{bmatrix}$$

$$= \begin{bmatrix} -7 & -8 \\ 4 & -7 \end{bmatrix}$$

• **PROBLEM** 5-23

a)   Show that:

   (i)   A0 = 0   (ii)   0A = 0   (iii)   AI = A   (iv)   IA = A

where 0 and I denote the zero and identity matrices respectively, and

$$A = \begin{bmatrix} 2 & 1 & 3 \\ 4 & -1 & -1 \end{bmatrix}$$

b)   Give examples of the following rules: (i) if A has a row of
zeros, the same row of AB consists of zeros. (ii) if B has a
column of zeros, the same column of AB consists of zeros.

Solution: a) By definition, the mxn matrix whose entries are all zero
is called the zero martix and is denoted by 0. By definition, the nxn
matrix with 1's on the diagonal and 0's elsewhere, denoted by I, is
called the unit or identity matrix; e.g., in R^3,

$$I = \begin{bmatrix} 1 & 0 & 0 \\ 0 & 1 & 0 \\ 0 & 0 & 1 \end{bmatrix}$$

Performing the indicated
matrix operations:
  (i)

$$A = \begin{bmatrix} 2 & 1 & 3 \\ 4 & -1 & -7 \end{bmatrix}$$

$$0 = \begin{bmatrix} 0 & 0 & 0 & 0 \\ 0 & 0 & 0 & 0 \\ 0 & 0 & 0 & 0 \end{bmatrix}$$

$$A0 = \begin{bmatrix} 2 & 1 & 3 \\ 4 & -1 & -7 \end{bmatrix} \begin{bmatrix} 0 & 0 & 0 & 0 \\ 0 & 0 & 0 & 0 \\ 0 & 0 & 0 & 0 \end{bmatrix}$$

$$= \begin{bmatrix} 0+0+0 & 0+0+0 & 0+0+0 & 0+0+0 \\ 0+0+0 & 0+0+0 & 0+0+0 & 0+0+0 \end{bmatrix} = \begin{bmatrix} 0 & 0 & 0 & 0 \\ 0 & 0 & 0 & 0 \\ 0 & 0 & 0 & 0 \end{bmatrix} .$$

(ii) Likewise,

$$0A = \begin{bmatrix} 0 & 0 \\ 0 & 0 \end{bmatrix} \begin{bmatrix} 2 & 1 & 3 \\ 4 & -1 & -7 \end{bmatrix}$$

$$= \begin{bmatrix} 0+0 & 0+0 & 0+0 \\ 0+0 & 0+0 & 0+0 \end{bmatrix} = \begin{bmatrix} 0 & 0 & 0 \\ 0 & 0 & 0 \end{bmatrix}$$

(iii)

$$I = \begin{bmatrix} 1 & 0 & 0 \\ 0 & 1 & 0 \\ 0 & 0 & 1 \end{bmatrix}$$

177

$$AI = \begin{bmatrix} 2 & 1 & 3 \\ 4 & -1 & -7 \end{bmatrix} \begin{bmatrix} 1 & 0 & 0 \\ 0 & 1 & 0 \\ 0 & 0 & 1 \end{bmatrix}$$

$$= \begin{bmatrix} 2+0+0 & 0+1+0 & 0+0+3 \\ 4+0+0 & 0-1+0 & 0+0-7 \end{bmatrix}$$

$$= \begin{bmatrix} 2 & 1 & 3 \\ 4 & -1 & -7 \end{bmatrix}.$$

Thus, $AI = A$.

(iv) $$I = \begin{bmatrix} 1 & 0 \\ 0 & 1 \end{bmatrix}$$

$$IA = \begin{bmatrix} 1 & 0 \\ 0 & 1 \end{bmatrix} \begin{bmatrix} 2 & 1 & 3 \\ 4 & -1 & -7 \end{bmatrix}$$

$$= \begin{bmatrix} 2+0 & 1+0 & 3+0 \\ 0+4 & 0+(-1) & 0+(-7) \end{bmatrix}$$

$$= \begin{bmatrix} 2 & 1 & 3 \\ 4 & -1 & -7 \end{bmatrix}.$$

Thus, $IA = A$.

Observe that the zero and identity matrices must have the appropriate size in order for the products to be defined. For example,

$$\begin{bmatrix} 0 & 0 & 0 \\ 0 & 0 & 0 \end{bmatrix} \begin{bmatrix} 2 & 1 & 3 \\ 4 & -1 & -7 \end{bmatrix}$$

is not defined.

b) Let $A = \begin{bmatrix} -2 & -3 & 1 \\ 0 & 0 & 0 \\ 1 & -1 & 0 \end{bmatrix}$ and $B = \begin{bmatrix} 6 & -2 & 1 \\ 3 & 1 & 2 \\ -1 & 1 & 1 \end{bmatrix}$.

Then,

$$AB = \begin{bmatrix} -2 & -3 & 1 \\ 0 & 0 & 0 \\ 1 & -1 & 0 \end{bmatrix} \begin{bmatrix} 6 & -2 & 1 \\ 3 & 1 & 2 \\ -1 & 1 & 1 \end{bmatrix}$$

$$AB = \begin{bmatrix} -12-9-1 & 4-3+1 & -2-6+1 \\ 0+0+0 & 0+0+0 & 0+0+0 \\ 6-3+0 & -2-1+0 & 1-2+0 \end{bmatrix}$$

$$= \begin{bmatrix} -22 & 2 & -7 \\ 0 & 0 & 0 \\ 3 & -3 & -1 \end{bmatrix}.$$

(ii) Let $A = \begin{bmatrix} 4 & 1 & 0 \\ -1 & 2 & 3 \\ 1 & 0 & 1 \end{bmatrix}$ and $B = \begin{bmatrix} 1 & 0 & 1 \\ -2 & 0 & 1 \\ 1 & 0 & 1 \end{bmatrix}$.

Then,

$$AB = \begin{bmatrix} 4 & 1 & 0 \\ -1 & 2 & 3 \\ 1 & 0 & 1 \end{bmatrix} \begin{bmatrix} 1 & 0 & 1 \\ -2 & 0 & 1 \\ 1 & 0 & 1 \end{bmatrix}$$

178

$$= \begin{bmatrix} 4-2+0 & 0+0+0 & 4+1+0 \\ -1-4+3 & 0+0+0 & -1+2+3 \\ 1+0+1 & 0+0+0 & 1+0+1 \end{bmatrix} = \begin{bmatrix} 2 & 0 & 5 \\ -2 & 0 & 4 \\ 2 & 0 & 2 \end{bmatrix} .$$

● **PROBLEM 5-24**

Give examples to show that in matrix arithmetic one can have the following:
(a) $AB \neq BA$.
(b) $A \neq 0$, $B \neq 0$ and yet, $AB = 0$.
(c) $A \neq 0$ and $A^2 = 0$.
(d) $A \neq 0$, $A^2 \neq 0$ and $A^3 = 0$.
(e) $A^2 = A$ with $A \neq 0$ and $A \neq I$.
(f) $A^2 = I$ with $A \neq I$ and $A \neq -I$.

Solution: (a) $A = \begin{bmatrix} 2 & 1 \\ -1 & 0 \end{bmatrix}$ $B = \begin{bmatrix} 1 & 0 \\ 3 & 1 \end{bmatrix}$

$$AB = \begin{bmatrix} 2+3 & 0+1 \\ -1+0 & 0 \end{bmatrix} = \begin{bmatrix} 5 & 1 \\ -1 & 0 \end{bmatrix}$$

$$BA = \begin{bmatrix} 1 & 0 \\ 3 & 1 \end{bmatrix} \begin{bmatrix} 2 & 1 \\ -1 & 0 \end{bmatrix} = \begin{bmatrix} 2+0 & 1+0 \\ 6-1 & 3+0 \end{bmatrix} = \begin{bmatrix} 2 & 1 \\ 5 & 3 \end{bmatrix}$$

Thus, $AB \neq BA$.
(b)   Let $A = \begin{bmatrix} 0 & 1 \\ 0 & 0 \end{bmatrix}$; $B = \begin{bmatrix} 0 & 4 \\ 0 & 0 \end{bmatrix}$.

$$AB = \begin{bmatrix} 0 & 1 \\ 0 & 0 \end{bmatrix} \begin{bmatrix} 0 & 4 \\ 0 & 0 \end{bmatrix} = \begin{bmatrix} 0+0 & 0+0 \\ 0+0 & 0+0 \end{bmatrix}$$
$$= \begin{bmatrix} 0 & 0 \\ 0 & 0 \end{bmatrix}$$

Thus, $AB = 0$, $A \neq 0$ and $B \neq 0$.
(c)   Let $A = \begin{bmatrix} 0 & 1 \\ 0 & 0 \end{bmatrix}$

$$A^2 - AA = \begin{bmatrix} 0 & 1 \\ 0 & 0 \end{bmatrix} \begin{bmatrix} 0 & 1 \\ 0 & 0 \end{bmatrix} = \begin{bmatrix} 0+0 & 0+0 \\ 0+0 & 0+0 \end{bmatrix} = \begin{bmatrix} 0 & 0 \\ 0 & 0 \end{bmatrix} .$$

Let $A = \begin{bmatrix} 1 & -1 \\ 1 & -1 \end{bmatrix}$,

then

$$A^2 = AA = \begin{bmatrix} 1 & -1 \\ 1 & -1 \end{bmatrix} \begin{bmatrix} 1 & -1 \\ 1 & -1 \end{bmatrix} = \begin{bmatrix} 1-1 & -1+1 \\ 1-1 & -1+1 \end{bmatrix} = \begin{bmatrix} 0 & 0 \\ 0 & 0 \end{bmatrix} .$$

Thus, $A \neq 0$ and $A^2 = 0$.

(d)   Let $A = \begin{bmatrix} 0 & 1 & 1 \\ 0 & 0 & 1 \\ 0 & 0 & 0 \end{bmatrix}$

$$A^2 = AA = \begin{bmatrix} 0 & 1 & 1 \\ 0 & 0 & 1 \\ 0 & 0 & 0 \end{bmatrix} \begin{bmatrix} 0 & 1 & 1 \\ 0 & 0 & 1 \\ 0 & 0 & 0 \end{bmatrix}$$

$$= \begin{bmatrix} 0+0+0 & 0+0+0 & 0+1+0 \\ 0+0+0 & 0+0+0 & 0+0+0 \\ 0+0+0 & 0+0+0 & 0+0+0 \end{bmatrix} = \begin{bmatrix} 0 & 0 & 1 \\ 0 & 0 & 0 \\ 0 & 0 & 0 \end{bmatrix} .$$

179

$$A^3 = A^2 A = \begin{bmatrix} 0 & 0 & 1 \\ 0 & 0 & 0 \\ 0 & 0 & 0 \end{bmatrix} \begin{bmatrix} 0 & 1 & 1 \\ 0 & 0 & 1 \\ 0 & 0 & 0 \end{bmatrix}$$

$$= \begin{bmatrix} 0 & 0 & 0 \\ 0 & 0 & 0 \\ 0 & 0 & 0 \end{bmatrix}.$$

(e)   Let $A = \begin{bmatrix} \frac{1}{2} & \frac{1}{2} \\ \frac{1}{2} & \frac{1}{2} \end{bmatrix}$

$$A^2 = AA = \begin{bmatrix} \frac{1}{2} & \frac{1}{2} \\ \frac{1}{2} & \frac{1}{2} \end{bmatrix} \begin{bmatrix} \frac{1}{2} & \frac{1}{2} \\ \frac{1}{2} & \frac{1}{2} \end{bmatrix} = \begin{bmatrix} \frac{1}{4}+\frac{1}{4} & \frac{1}{4}+\frac{1}{4} \\ \frac{1}{4}+\frac{1}{4} & \frac{1}{4}+\frac{1}{4} \end{bmatrix} = \begin{bmatrix} \frac{1}{2} & \frac{1}{2} \\ \frac{1}{2} & \frac{1}{2} \end{bmatrix}.$$

Thus, $A^2 = A$ with $A \neq 0$ and $A \neq I$.

(f)   Let $A = \begin{bmatrix} -1 & 0 \\ 0 & 1 \end{bmatrix}$

$$A^2 = AA = \begin{bmatrix} -1 & 0 \\ 0 & 1 \end{bmatrix} \begin{bmatrix} -1 & 0 \\ 0 & 1 \end{bmatrix}$$

$$A^2 = \begin{bmatrix} 1+0 & 0+0 \\ 0+0 & 0+1 \end{bmatrix} = \begin{bmatrix} 1 & 0 \\ 0 & 1 \end{bmatrix}.$$

Hence, $A^2 = I$ with $A \neq I$ and $A \neq -I$.

● **PROBLEM 5-25**

Define the transpose of a matrix. Find the transpose of the following matrices:

$$A = \begin{bmatrix} 4 & -2 & 3 \\ 0 & 5 & -2 \end{bmatrix} \qquad B = \begin{bmatrix} 6 & 2 & -4 \\ 3 & -1 & 2 \\ 0 & 4 & 3 \end{bmatrix}$$

$$C = \begin{bmatrix} 5 & 4 \\ 3 & 2 \\ 2 & -3 \end{bmatrix} \qquad D = \begin{bmatrix} 3 & -5 & 1 \end{bmatrix} \qquad E = \begin{bmatrix} 2 \\ -1 \\ 3 \end{bmatrix}$$

Solution: Definition: If $A = [a_{ij}]$ is an $m \times n$ matrix, then the $n \times m$ matrix $A^t = [a^t_{ij}]$ where

$$a^t_{ij} = a_{ji} \quad [1 \le i \le m . \quad 1 \le j \le n]$$

is called the transpose of A. Thus, the transpose of A is obtained by interchanging the rows and columns of A.

$$A = \begin{bmatrix} 4 & -2 & 3 \\ 0 & 5 & -2 \end{bmatrix}.$$

Then,

$$A^t = \begin{bmatrix} 4 & 0 \\ -2 & 5 \\ 3 & -2 \end{bmatrix}$$

$$B = \begin{bmatrix} 6 & 2 & -4 \\ 3 & -1 & 2 \\ 0 & 4 & 3 \end{bmatrix} .$$

$$B^t = \begin{bmatrix} 6 & 3 & 0 \\ 2 & -1 & 4 \\ -4 & 2 & 3 \end{bmatrix} .$$

$$C = \begin{bmatrix} 5 & 4 \\ -3 & 2 \\ 2 & -3 \end{bmatrix} ; \quad \text{thus,} \quad C^t = \begin{bmatrix} 5 & -3 & 2 \\ 4 & 2 & -3 \end{bmatrix} .$$

$$D = [3 \quad -5 \quad 1] .$$

Then,

$$D^t = \begin{bmatrix} 3 \\ -5 \\ 1 \end{bmatrix} .$$

$$E = \begin{bmatrix} 2 \\ -1 \\ 3 \end{bmatrix} ; \quad \text{hence,} \quad E^t = [2 \quad -1 \quad 3] .$$

● **PROBLEM** 5-26

Let $A = \begin{bmatrix} 1 & 2 & 0 \\ 3 & -1 & 4 \end{bmatrix}$. Find (i) AA^t , (ii) A^tA .

Solution: The transpose of A, denoted by A^t, is the matrix obtained from A by interchanging the rows and columns of A. For example, if

$$A = \begin{bmatrix} a_{11} & a_{12} & \cdots & a_{1n} \\ a_{21} & a_{22} & \cdots & a_{2n} \\ \vdots & & \ddots & \vdots \\ a_{m1} & a_{m2} & & a_{mn} \end{bmatrix} ,$$

then

$$A^t = \begin{bmatrix} a_{11} & a_{21} & \cdots & a_{m1} \\ a_{12} & a_{22} & \cdots & a_{m2} \\ \vdots & & \ddots & \vdots \\ a_{1n} & a_{2n} & \cdots & a_{mn} \end{bmatrix}$$

Observe that if A is an $m \times n$ matrix, then A^t is an $n \times m$ matrix. Hence, the products AA^t and A^tA are always defined.

$$A = \begin{bmatrix} 1 & 2 & 0 \\ 3 & -1 & 4 \end{bmatrix}$$

then,

$$A^t = \begin{bmatrix} 1 & 3 \\ 2 & -1 \\ 0 & 4 \end{bmatrix} .$$

181

(i) $\quad AA^t = \begin{bmatrix} 1 & 2 & 0 \\ 3 & -1 & 4 \end{bmatrix} \begin{bmatrix} 1 & 3 \\ 2 & -1 \\ 0 & 4 \end{bmatrix}$

$\quad\quad = \begin{bmatrix} 1\cdot 1+2\cdot 2+0\cdot 0 & 1\cdot 3+2\cdot(-1)+0\cdot 4 \\ 3\cdot 1+(-1)\cdot 2+(4)\cdot 0 & 3\cdot 3+(-1)\cdot(-1)+4\cdot 4 \end{bmatrix}$

$\quad\quad = \begin{bmatrix} 1+4+0 & 3-2+0 \\ 3-2+0 & 9+1+16 \end{bmatrix} = \begin{bmatrix} 5 & 1 \\ 1 & 26 \end{bmatrix}$.

(ii) $\quad A^t A = \begin{bmatrix} 1 & 3 \\ 2 & -1 \\ 0 & 4 \end{bmatrix} \begin{bmatrix} 1 & 2 & 0 \\ 3 & -1 & 4 \end{bmatrix}$

$\quad\quad = \begin{bmatrix} 1\cdot 1+3\cdot 3 & 1\cdot 2+3\cdot(-1) & 1\cdot 0+3\cdot 4 \\ 2\cdot 1+(-1)\cdot 3 & 2\cdot 2+(-1)\cdot(-1) & 2\cdot 0+(-1)\cdot 4 \\ 0\cdot 1+4\cdot 3 & 0\cdot 2+4\cdot(-1) & 0\cdot 0+4\cdot 4 \end{bmatrix}$

$\quad\quad = \begin{bmatrix} 1+9 & 2-3 & 0+12 \\ 2-3 & 4+1 & 0-4 \\ 0+12 & 0-4 & 0+16 \end{bmatrix}$

$\quad\quad = \begin{bmatrix} 10 & -1 & 12 \\ -1 & 5 & -4 \\ 12 & -4 & 16 \end{bmatrix}$

● **PROBLEM** 5-27

Compute AB using block multiplication where

$$A = \left[\begin{array}{cc:c} 1 & 2 & 1 \\ 3 & 4 & 0 \\ \hdashline 0 & 0 & 2 \end{array} \right] \quad \text{and} \quad B = \left[\begin{array}{ccc:c} 1 & 2 & 3 & 1 \\ 4 & 5 & 6 & 1 \\ \hdashline 0 & 0 & 0 & 1 \end{array} \right] .$$

Solution: Using a system of horizontal and vertical lines, we partition a matrix A into smaller "submatrices" of A. The matrix A is then called a block matrix. A given matrix may be divided into blocks in different ways. For example,

$$\begin{bmatrix} 1 & -2 & 0 & 1 \\ 2 & 3 & 5 & 7 \\ 3 & 1 & 4 & 5 \end{bmatrix} = \left[\begin{array}{cc:cc} 1 & -2 & 0 & 1 \\ 2 & 3 & 5 & 7 \\ \hdashline 3 & 1 & 4 & 5 \end{array} \right]$$

$$= \left[\begin{array}{ccc:c} 1 & -2 & 0 & 1 \\ \hdashline 2 & 3 & 5 & 7 \\ \hdashline 3 & 1 & 4 & 5 \end{array} \right] .$$

$$A = \begin{bmatrix} 1 & 2 & | & 1 \\ 3 & 4 & | & 0 \\ - & - & - & | & - \\ 0 & 0 & | & 2 \end{bmatrix} .$$

Let $E \begin{bmatrix} 1 & 2 \\ 3 & 4 \end{bmatrix}$, $F = \begin{bmatrix} 1 \\ 0 \end{bmatrix}$ and $G[2]$ then,

$$A = \begin{bmatrix} E & | & F \\ 0 & | & G \end{bmatrix}$$

$$B = \begin{bmatrix} 1 & 2 & 3 & | & 1 \\ 4 & 5 & 6 & | & 1 \\ - & - & - & - & | & - \\ 0 & 0 & 0 & | & 1 \end{bmatrix} .$$

Let $R = \begin{bmatrix} 1 & 2 & 3 \\ 4 & 5 & 6 \end{bmatrix}$ $S = \begin{bmatrix} 1 \\ 1 \end{bmatrix}$ $T = [1]$.

Then,

$$B = \begin{bmatrix} R & | & S \\ 0 & | & T \end{bmatrix} .$$

After partitioning the matrices into block matrices, multiplication of the
matrices is the usual matrix multiplication with each entire block con--
sidered as a unit entry of the matrix. If two matrices can be multiplied,
then they can be multiplied as block matrices if they are each partitioned
into blocks similarly; that is, into an equal number of blocks so that
corresponding blocks have the same size. Suppose

$$A = \begin{bmatrix} A_1 & | & A_2 \\ - & - & | & - - \\ A_3 & | & A_4 \end{bmatrix} \quad \text{and} \quad B = \begin{bmatrix} B_1 & | & B_2 \\ - & - & | & - - \\ B_3 & | & B_4 \end{bmatrix}$$

where A_1 and B_1, A_2 and B_2, A_3 and B_3, A_4 and B_4 are the same

sizes, respectively. Then AB is given by

$$A_1 B_1 + A_2 B_3 \qquad A_1 B_2 + A_2 B_4$$
$$A_3 B_1 + A_3 B_3 \qquad A_3 B_2 + A_4 B_4$$

In the problem,

$$AB = \begin{bmatrix} E & F \\ 0 & G \end{bmatrix} \begin{bmatrix} R & S \\ 0 & T \end{bmatrix} = \begin{bmatrix} ER + F \cdot 0 & ES + FT \\ OR + G \cdot 0 & 0S + GT \end{bmatrix}$$

$$= \begin{bmatrix} ER & ES + FT \\ 0 & GT \end{bmatrix} =$$

$$= \begin{bmatrix} \begin{bmatrix} 1 & 2 \\ 3 & 4 \end{bmatrix} \begin{bmatrix} 1 & 2 & 3 \\ 4 & 5 & 6 \end{bmatrix} & \begin{bmatrix} 1 & 2 \\ 3 & 4 \end{bmatrix} \begin{bmatrix} 1 \\ 1 \end{bmatrix} + \begin{bmatrix} 1 \\ 0 \end{bmatrix} [1] \\ 0 & [2] \, [1] \end{bmatrix}$$

$$= \begin{bmatrix} \begin{bmatrix} 1+8 & 2+10 & 3+12 \\ 3+16 & 6+20 & 9+24 \end{bmatrix} & \begin{bmatrix} 1+2 \\ 3+4 \end{bmatrix} + \begin{bmatrix} 1 \\ 0 \end{bmatrix} \\ [0 \quad 0 \quad 0] & [2] \end{bmatrix}$$

$$= \begin{bmatrix} 9 & 12 & 15 & 4 \\ 19 & 26 & 33 & 7 \\ 0 & 0 & 0 & 2 \end{bmatrix}$$

● **PROBLEM** 5-28

Define elementary row operations and give an example.

Solution: The three elementary row operations on a matrix A are:
1. Interchange the i-th and the j-th row of A.
2. Add the i-th row of A to the j-th row of A, $i \neq j$.
3. Multiply the i-th row of A by a non-zero scalar k. Let

$$A = \begin{bmatrix} 1 & 6 & 3 & 4 \\ 1 & 2 & 1 & 1 \\ -1 & 2 & 1 & 2 \end{bmatrix}.$$

The following row operations are performed on the matrix A:
(1) Interchange the first and the second rows

$$\begin{bmatrix} 1 & 2 & 1 & 1 \\ 1 & 6 & 3 & 4 \\ -1 & 2 & 1 & 2 \end{bmatrix}.$$

(2) Add the first row to the third row and -1 times the first row to the second row. Adding the first row to the third row,

$$\begin{bmatrix} 1 & 2 & 1 & 1 \\ 1 & 6 & 3 & 4 \\ -1+1 & 2+2 & 1+1 & 2+1 \end{bmatrix} = \begin{bmatrix} 1 & 2 & 1 & 1 \\ 1 & 6 & 3 & 4 \\ 0 & 4 & 2 & 3 \end{bmatrix};$$

adding -1 times the first row to the second,

$$\begin{bmatrix} 1 & 2 & 1 & 1 \\ (-1 \cdot 1)+1 & (-1 \cdot 2)+6 & (-1 \cdot 1)+3 & (-1 \cdot 1)+4 \\ 0 & 4 & 2 & 3 \end{bmatrix}$$

$$= \begin{bmatrix} 1 & 2 & 1 & 1 \\ 0 & 4 & 2 & 3 \\ 0 & 4 & 2 & 3 \end{bmatrix}.$$

Add -1 times the second row to the third row.

$$\begin{bmatrix} 1 & 2 & 1 & 1 \\ 0 & 4 & 2 & 3 \\ 0 & 0 & 0 & 0 \end{bmatrix}$$

Divide the second row by 4.

$$\begin{bmatrix} 1 & 2 & 1 & 1 \\ 0 & 1 & \frac{1}{2} & 3/4 \\ 0 & 0 & 0 & 0 \end{bmatrix}$$

184

Add -2 times the second row to the first row.

$$\begin{bmatrix} (-2\cdot0)+1 & (-2\cdot1)+2 & (-2\cdot\frac{1}{2})+1 & (-2\cdot3/4)+1 \\ 0 & 1 & \frac{1}{2} & 3/4 \\ 0 & 0 & 0 & 0 \end{bmatrix}$$

$$= \begin{bmatrix} 1 & 0 & 0 & -\frac{1}{2} \\ 0 & 1 & \frac{1}{2} & 3/4 \\ 0 & 0 & 0 & 0 \end{bmatrix} .$$

Note that this matrix is in row-reduced echelon form. The elementary row operations can be applied to reduce a matrix to echelon form, and this technique is used in solving systems of linear equations.

● **PROBLEM 5-29**

Reduce the following matrices to echelon form and then to row reduced echelon form.

(a)
$$A = \begin{bmatrix} 0 & 1 & 3 & -2 \\ 2 & 1 & -4 & 3 \\ 2 & 3 & 2 & -1 \end{bmatrix}$$

(b)
$$A = \begin{bmatrix} 6 & 3 & -4 \\ -4 & 1 & -6 \\ 1 & 2 & -5 \end{bmatrix}$$

Solution: In echelon form, the first non-zero entry of any row is a 1, and any row of zeros lies below the rows with non-zero entries. Furthermore, the first non-zero entry of any row is in a column to the left of the first non-zero entry in the next row. In addition, in reduced echelon form, the entire column containing the first non-zero entry of any row is all zeros except for that entry.

(a)   We put A into row reduced echelon form by using a series of elementary row operations. First we interchange the first and the second rows.

$$\begin{bmatrix} 2 & 1 & -4 & 3 \\ 0 & 1 & 3 & -2 \\ 2 & 3 & 2 & -1 \end{bmatrix}$$

Add -1 times the first row to the third row.

$$\begin{bmatrix} 2 & 1 & -4 & 3 \\ 0 & 1 & 3 & -2 \\ 0 & 2 & 6 & -4 \end{bmatrix}$$

Add -2 times the second row to the third row.

$$\begin{bmatrix} 2 & 1 & -4 & 3 \\ 0 & 1 & 3 & -2 \\ 0 & 0 & 0 & 0 \end{bmatrix}$$

Finally, to obtain the echelon form, multiply the first column by $\frac{1}{2}$.
Hence,

$$\begin{bmatrix} 1 & \frac{1}{2} & 2 & 3/2 \\ 0 & 1 & 3 & -2 \\ 0 & 0 & 0 & 0 \end{bmatrix} .$$

Now add $-\frac{1}{2}$ times the second row to the first row, to obtain the row reduced echelon form

$$\begin{bmatrix} 1 & 0 & -7/2 & 5/2 \\ 0 & 1 & 3 & -2 \\ 0 & 0 & 0 & 0 \end{bmatrix} .$$

(b) First interchange the first and third rows.

$$\begin{bmatrix} 1 & 2 & -5 \\ -4 & 1 & -6 \\ 6 & 3 & -4 \end{bmatrix} .$$

Add 4 times the first row to the second row and -6 times the first row to the third row.

$$\begin{bmatrix} 1 & 2 & -5 \\ 0 & 9 & -26 \\ 0 & -9 & 26 \end{bmatrix}$$

Now add the second row to the third row.

$$\begin{bmatrix} 1 & 2 & -5 \\ 0 & 9 & -26 \\ 0 & 0 & 0 \end{bmatrix}$$

Divide the second row by 9 to obtain the echelon form.

$$\begin{bmatrix} 1 & 2 & -5 \\ 0 & 1 & -26/9 \\ 0 & 0 & 0 \end{bmatrix}$$

Add -2 times the second row to the first row to obtain the row-reduced echelon form.

$$\begin{bmatrix} 1 & 0 & 7/9 \\ 0 & 1 & -26/9 \\ 0 & 0 & 0 \end{bmatrix} .$$

● **PROBLEM 5-30**

Find f(A) where $A = \begin{pmatrix} 1 & -2 \\ 4 & 5 \end{pmatrix}$

and $f(t) = t^2 - 3t + 7$.

Solution: The general polynomial of a scalar variable t is denoted by

$$f(t) = a_n t^n + a_{n-1} t^{n-1} + \ldots + a_1 t + a_0 .$$

The general polynomial of a square matrix A of order n is

$$f(A) = a_n A^n + a_{n-1} A^{n-1} + \ldots + a_1 A + a_0 I ,$$

where I is an nxn identity matrix.

Given that

$$f(t) = t^2 - 3t + 7$$

then,

$$f(A) = A^2 - 3A + 7I .$$

$$A^2 = \begin{pmatrix} 1 & -2 \\ 4 & 5 \end{pmatrix} \begin{pmatrix} 1 & -2 \\ 4 & 5 \end{pmatrix} = \begin{pmatrix} 1-8 & -2-10 \\ 4+20 & -8+25 \end{pmatrix} = \begin{pmatrix} -7 & -12 \\ 24 & 17 \end{pmatrix}$$

Hence, we can use this to compute

$$f(A) = \begin{pmatrix} -7 & -12 \\ 24 & 17 \end{pmatrix} - 3 \begin{pmatrix} 1 & -2 \\ 4 & 5 \end{pmatrix} + 7 \begin{pmatrix} 1 & 0 \\ 0 & 1 \end{pmatrix}$$

$$= \begin{pmatrix} -3 & -6 \\ 12 & 9 \end{pmatrix} .$$

● **PROBLEM 5-31**

Given

$$A = \begin{pmatrix} 1 & e^t \\ t^2 & t \end{pmatrix} .$$

find $\int_0^1 A(t)\, dt.$

Solution: When the elements of the matrix A, $[a_{ij}]$ are functions of a scalar variable t, the integral of A with respect to t taken between the limits a and b is defined as that matrix which has for its (i, j)th element

$$\int_a^b a_{ij}(t)\,dt. \quad \text{Therefore,}$$

$$\int_a^b A(t)\,dt = [c_{ij}]$$

where $[c_{ij}] = \int_a^b a_{ij}(t)\,dt.$

In defining matrix integration, it has been assumed that the elements $[a_{ij}]$ of the matrix A are continuous real-valued functions of a real variable t.

Now consider the given matrix

$$A = \begin{bmatrix} 1 & e^t \\ t^2 & t \end{bmatrix}.$$

Then,

$$\int_0^1 A(t)\,dt = c_{ij}, \quad c_{ij} = \int_0^1 a_{ij}(t)\,dt.$$

Therefore,

$$c_{11} = \int_0^1 a_{11}(t)\,dt = \int_0^1 1\,dt.$$

$$c_{12} = \int_0^1 a_{12}(t)\,dt = \int_0^1 e^t\,dt$$

$$c_{21} = \int_0^1 a_{21}(t)\,dt = \int_0^1 t^2\,dt$$

$$c_{22} = \int_0^1 a_{22}(t)\,dt = \int_0^1 t\,dt.$$

Thus, the integral over (0, 1) of A(t) is

$$\int_0^1 A(t)\,dt = \begin{pmatrix} \int_0^1 1\,dt & \int_0^1 e^t\,dt \\[2em] \int_0^1 t^2\,dt & \int_0^1 t\,dt \end{pmatrix}$$

Upon integration of each element of the above matrix we get

$$\int_0^1 A(t)\,dt = \begin{pmatrix} t\Big|_0^1 & e^t\Big|_0^1 \\[2em] \dfrac{t^3}{3}\Big|_0^1 & \dfrac{t^2}{2}\Big|_0^1 \end{pmatrix}$$

$$\int_0^1 A(t)\,dt = \begin{pmatrix} 1 & e-1 \\[1em] 1/3 & 1/2 \end{pmatrix}$$

● **PROBLEM 5-32**

Find $e^A = f(A)$, where

$$A = \begin{pmatrix} 3 & -3 & 3 \\ -1 & 5 & -2 \\ -1 & 3 & 0 \end{pmatrix}$$

Solution: We know that e^x can be defined by the following power series:

$$e^x = 1 + \frac{x}{1!} + \frac{x^2}{2!} + \frac{x^3}{3!} + \cdots + \frac{x^n}{n!} + \cdots .$$

Now the exponential function of a square matrix A is defined by the same power series as the exponential function of a scalar. Thus

$$e^A = I + \frac{A}{1!} + \frac{A^2}{2!} + \frac{A^3}{3!} + \cdots + \frac{A^n}{n!} + \cdots \tag{1}$$

To compute e^A, the expansion (1), is inconvenient. It is possible to reduce (1) to the polynomial form. Thus, if A is a square matrix of order n, then

$$e^A = \alpha_{n-1} A^{n-1} + \alpha_{n-2} A^{n-2} + \cdots + \alpha_1 A + \alpha_0 I \tag{2}$$

where α_{n-1}, α_{n-2}, $\cdots$ α_0 are constants.

To find α_0, α_1, $\cdots$, α_{n-1} we first compute the eigenvalues of A, λ_1, λ_2, $\cdots$ λ_n. For each eigenvalue we have

$$e^{\lambda_i} = r(\lambda_i)$$

where $r(\lambda) = \alpha_{n-1} \lambda^{n-1} + \cdots + \alpha_2 \lambda^2 + \alpha_1 \lambda + \alpha_0$. $\tag{3}$

The characteristic equation of the given matrix A is det. $[\lambda I - A] = 0$. Therefore,

$$\det \begin{bmatrix} \lambda - 3 & 3 & -3 \\ 1 & \lambda - 5 & 2 \\ 1 & -3 & \lambda \end{bmatrix} = 0$$

or $\lambda - 3 \begin{vmatrix} \lambda-5 & 2 \\ -3 & \lambda \end{vmatrix} - 3 \begin{vmatrix} 1 & 2 \\ 1 & \lambda \end{vmatrix} - 3 \begin{vmatrix} 1 & \lambda-5 \\ 1 & -3 \end{vmatrix} = 0$

$\lambda - 3 \ [\lambda(\lambda-5) + 6] -3 \ [\lambda-2] -3 \ [-3 - (\lambda-5)] = 0$

$(\lambda-2) \ [(\lambda-3)^2 - 3 + 3] = 0$

or $(\lambda-2)(\lambda-3)^2 = 0$.

Then, the eigenvalues are $\lambda = 2$, $\lambda = 3$, and $\lambda = 3$.

Substitute these values into (3) to obtain

$$r(2) = \alpha_2 4 + \alpha_1 2 + \alpha_0.$$

Since $e^{\lambda_i} = r(\lambda_i)$, $e^2 = 4\alpha_2 + 2\alpha_1 + \alpha_0$. $\tag{i}$

Similarly,

$r(3) = 9\alpha_2 + 3\alpha_1 + \alpha_0$ $\tag{ii}$

190

or $e^3 = 9\alpha_2 + 3\alpha_1 + \alpha_0$.

Note that there are only two equations and three unknowns.
The third equation is given by

$$e^3 = \frac{d}{d\lambda}\left(r(\lambda)\right)\big|_{\lambda=3}$$

$$e^3 = \frac{d}{d\lambda}\left(\alpha_2\lambda^2 + \lambda\,\alpha_1 + \alpha_0\right)\big|_{\lambda=3}$$

$$= 2\alpha_2\lambda + \alpha_1\big|_{\lambda=3}$$

$$= 6\alpha_2 + \alpha_1 \quad . \tag{iii}$$

one has

$$e^2 = 4\alpha_2 + 2\alpha_1 + \alpha_0 \tag{i}$$

$$e^3 = 9\alpha_2 + 3\alpha_1 + \alpha_0 \tag{ii}$$

$$e^3 = 6\alpha_2 + \alpha_1 \tag{iii}$$

or $\quad E = A\alpha \qquad$ where $\qquad E = \begin{pmatrix} e^2 \\ e^3 \\ e^3 \end{pmatrix} \qquad$ and

$$A = \begin{bmatrix} 4 & 2 & 1 \\ 9 & 3 & 1 \\ 6 & 1 & 0 \end{bmatrix} \qquad \text{and} \qquad \alpha = \begin{pmatrix} \alpha_2 \\ \alpha_1 \\ \alpha_0 \end{pmatrix}$$

Use Cramer's rule to solve this system yields

$$\alpha_2 = \frac{\begin{vmatrix} e^2 & 2 & 1 \\ e^3 & 3 & 1 \\ e^3 & 1 & 0 \end{vmatrix}}{\begin{vmatrix} 4 & 2 & 1 \\ 9 & 3 & 1 \\ 6 & 1 & 0 \end{vmatrix}} \quad . \tag{4}$$

Now

$$\begin{vmatrix} 4 & 2 & 1 \\ 9 & 3 & 1 \\ 6 & 1 & 0 \end{vmatrix} = -1 \qquad \text{and}$$

$$\begin{vmatrix} e^2 & 2 & 1 \\ e^3 & 3 & 1 \\ e^3 & 1 & 0 \end{vmatrix} = \begin{vmatrix} e^3 & 3 \\ e^3 & 1 \end{vmatrix} - \begin{vmatrix} e^2 & 2 \\ e^3 & 1 \end{vmatrix} = -e^2$$

So (4) implies that $\alpha_2 = \dfrac{-e^2}{-1} = e^2$.

Similarly,

$$\alpha_1 = \frac{\begin{vmatrix} 4 & e^2 & 1 \\ 9 & e^3 & 1 \\ 6 & e^3 & 0 \end{vmatrix}}{\begin{vmatrix} 4 & 2 & 1 \\ 9 & 3 & 1 \\ 6 & 1 & 0 \end{vmatrix}} = \frac{-e^3 + 6e^2}{-1}$$

and

$$\alpha_0 = \frac{\begin{vmatrix} 4 & 2 & e^2 \\ 9 & 3 & e^3 \\ 6 & 1 & e^3 \end{vmatrix}}{\begin{vmatrix} 4 & 2 & 1 \\ 9 & 3 & 1 \\ 6 & 1 & 0 \end{vmatrix}} = \frac{-9e^2 + 2e^3}{-1}.$$

Thus solving equations (i), (ii), and (iii), yields

$$\alpha_2 = e^2 , \quad \alpha_1 = e^3 - 6e^2 , \quad \alpha_0 = 9e^2 - 2e^3 .$$

In the given problem, since the matrix is of order 3 x 3, n = 3. From (2) one obtains

$$e^A = \alpha_2 A^2 + \alpha_1 A + \alpha_0 I .$$

$$A^2 = \begin{bmatrix} 3 & -3 & 3 \\ -1 & 5 & -2 \\ -1 & 3 & 0 \end{bmatrix} \begin{bmatrix} 3 & -3 & 3 \\ -1 & 5 & -2 \\ -1 & 3 & 0 \end{bmatrix}$$

$$= \begin{bmatrix} 9+3-3 & -9-15+9 & 9+6+0 \\ -3-5+2 & 3+25-6 & -3-10+0 \\ -3-3+0 & 3+15+0 & -3-6+0 \end{bmatrix}$$

$$= \begin{bmatrix} 9 & -15 & 15 \\ -6 & 22 & -13 \\ -6 & 18 & -9 \end{bmatrix} .$$

Thus,

$$e^A = \alpha_2 \begin{bmatrix} 9 & -15 & 15 \\ -6 & 22 & -13 \\ -6 & 18 & -9 \end{bmatrix} + \alpha_1 \begin{bmatrix} 3 & -3 & 3 \\ -1 & 5 & -2 \\ -1 & 3 & 0 \end{bmatrix}$$

$$+ \alpha_o \begin{bmatrix} 1 & 0 & 0 \\ 0 & 1 & 0 \\ 0 & 0 & 1 \end{bmatrix}$$

$$e^A = \begin{bmatrix} 9\alpha_2 + 3\alpha_1 + \alpha_o & -15\alpha_2 - 3\alpha_1 & 15\alpha_2 + 3\alpha_o \\ 6\alpha_2 - \alpha_1 & 22\alpha_2 + 5\alpha_1 + \alpha_o & -13\alpha_2 - 2\alpha_1 \\ -6\alpha_2 - \alpha_1 & 18\alpha_2 + 3\alpha_1 & -9\alpha_2 + \alpha_o \end{bmatrix}$$

Now substitute the values of α_2 , α_1 and α_o into the above matrix. This yields

$$e^A = \begin{bmatrix} 9e^2+3e^3-18e^2+9e^2-2e^3 & -15e^2-3e^3+18e^2 & 15e^2+3e^3-18e^2 \\ -6e^2-e^3+6e^2 & 22e^2+5e^3-30e^2+9e^2-2e^3 & -13e^2-2e^3+12e^2 \\ -6e^2-e^3+6e^2 & 18e^2+3e^3-18e^2 & -9e^2+9e^2-2e^3 \end{bmatrix}$$

or,

$$e^A = \begin{bmatrix} e^3 & -3e^3 + 3e^2 & 3e^3 - 3e^2 \\ -e^3 & 3e^3 + e^2 & -2e^3 - e^2 \\ -e^3 & 3e^3 & -2e^3 \end{bmatrix}$$

THE INVERSE AND RANK OF A MATRIX

Find the inverse of the matrix A where

$$A = \begin{bmatrix} 1 & 1 & 1 & 1 \\ 0 & 1 & 1 & 1 \\ 0 & 0 & 1 & 1 \\ 0 & 0 & 0 & 1 \end{bmatrix}$$

Show that the inverse of a diagonal matrix is obtained by inverting the diagonal entries.

Solution: We compute the inverse by row reducing the following augmented matrix:

$$[A : I] = \left[\begin{array}{cccc:cccc} 1 & 1 & 1 & 1 & 1 & 0 & 0 & 0 \\ 0 & 1 & 1 & 1 & 0 & 1 & 0 & 0 \\ 0 & 0 & 1 & 1 & 0 & 0 & 1 & 0 \\ 0 & 0 & 0 & 1 & 0 & 0 & 0 & 1 \end{array}\right]$$

Subtract the second row from the first row:

$$\left[\begin{array}{cccc:cccc} 1 & 0 & 0 & 0 & 1 & -1 & 0 & 0 \\ 0 & 1 & 1 & 1 & 0 & 1 & 0 & 0 \\ 0 & 0 & 1 & 1 & 0 & 0 & 1 & 0 \\ 0 & 0 & 0 & 1 & 0 & 0 & 0 & 1 \end{array}\right]$$

Subtract the third row from the second row, and the fourth row from the third row:

$$\left[\begin{array}{cccc:cccc} 1 & 0 & 0 & 0 & 1 & -1 & 0 & 0 \\ 0 & 1 & 0 & 0 & 0 & 1 & -1 & 0 \\ 0 & 0 & 1 & 0 & 0 & 0 & 1 & -1 \\ 0 & 0 & 0 & 1 & 0 & 0 & 0 & 1 \end{array}\right]$$

Hence

$$A^{-1} = \begin{bmatrix} 1 & -1 & 0 & 0 \\ 0 & 1 & -1 & 0 \\ 0 & 0 & 1 & -1 \\ 0 & 0 & 0 & 1 \end{bmatrix}$$

A diagonal matrix is a square matrix whose non-diagonal entries are all zero. Let A be a diagonal matrix whose diagonal entries are all non-zero, and let

$$A = \begin{bmatrix} a_{11} & 0 & \cdots & & 0 \\ 0 & a_{22} & \cdots & & 0 \\ \cdot & & 0 & a_{kk} & \cdot & \cdot \\ \cdot & & & & & \cdot \\ 0 & \cdots & \cdots & \cdots & a_{nn} \end{bmatrix}$$

with $a_{ii} \neq 0$, $i = 1, \ldots, n$.

Now apply the same procedure as above for finding the inverse of a matrix.
Then

$$[A : I] = \begin{bmatrix} a_{11} & & 0 & \vert & 1 & 0 & \cdots & 0 \\ 0 & a_{22} & 0 & \vert & 0 & 1 & \cdots & 0 \\ \cdot & & & \vert & & \cdot & & \cdot \\ \cdot & & & \vert & & \cdot & & \cdot \\ \cdot & & & \vert & & \cdot & & \cdot \\ 0 & \cdots & \cdot a_{nn} & \vert & 0 & \cdots & & \cdot 1 \end{bmatrix}$$

Multiply the first row by $\dfrac{1}{a_{11}}$, the second row by $\dfrac{1}{a_{22}}$ $\ldots$ and
the nth row by $\dfrac{1}{a_{nn}}$, to obtain

$$[I : B] = \begin{bmatrix} 1 & 0 & \cdots & 0 & \vert & 1/a_{11} & 0 & \cdots & 0 \\ 0 & 1 & \cdots & 0 & \vert & 0 & 1/a_{22} & \cdots & 0 \\ \cdot & & & & \vert & & \cdot & & \cdot \\ \cdot & & & & \vert & & \cdot & & \cdot \\ 0 & \cdots & & 1 & \vert & 0 & \cdots & & 1/a_{nn} \end{bmatrix}$$

Hence

$$A^{-1} = \begin{bmatrix} 1/a_{11} & 0 & \cdots & 0 \\ 0 & 1/a_{22} & \cdots & 0 \\ \cdot & & & \cdot \\ \cdot & & & \cdot \\ 0 & \cdots & & 1/a_{nn} \end{bmatrix}$$

Thus the inverse of a diagonal matrix is obtained by inverting the diagonal entries.

Observe that if one of the diagonal entries is zero, the matrix is not invertible. For example,

$$\begin{bmatrix} 1 & 0 & 0 \\ 0 & 0 & 0 \\ 0 & 0 & 3 \end{bmatrix}$$

is not invertible.

Find the inverses of the following matrices.

(1)   (2)

$$A = \begin{bmatrix} 3 & 1 \\ -1 & 6 \end{bmatrix}$$

$$A = \begin{bmatrix} 1 & -7 & -14 \\ 2 & 1 & -1 \\ 1 & 3 & 4 \end{bmatrix}$$

(3)

$$A = \begin{bmatrix} 3 & 1 & 0 \\ 1 & -1 & 2 \\ 1 & 1 & 1 \end{bmatrix}.$$

Solution: The method of solution is the same in all three cases, namely, forming the block matrix [A : I]   where I is the n × n identity matrix, and using elementary row operations to reduce it to [I : A⁻¹].

(1) $A = \begin{bmatrix} 3 & 1 \\ -1 & 6 \end{bmatrix}.$

Now $[A : I] = \begin{bmatrix} 3 & 1 & \vdots & 1 & 0 \\ -1 & 6 & \vdots & 0 & 1 \end{bmatrix}.$

To row reduce this block matrix, first multiply the first row by 6:

$$\begin{bmatrix} 18 & 6 & \vdots & 6 & 0 \\ -1 & 6 & \vdots & 0 & 1 \end{bmatrix}$$

Subtract the second row from the first row:

$$\begin{bmatrix} 19 & 0 & \vdots & 6 & -1 \\ -1 & 6 & \vdots & 0 & 1 \end{bmatrix}$$

Multiply the second row by 19:

$$\begin{bmatrix} 19 & 0 & \vdots & 6 & -1 \\ -19 & 114 & \vdots & 0 & 19 \end{bmatrix}$$

Add the first row to the second row:

$$\begin{bmatrix} 19 & 0 & \vdots & 6 & -1 \\ 0 & 114 & \vdots & 6 & 18 \end{bmatrix}$$

Divide the first and second rows by 19:

$$\begin{bmatrix} 1 & 0 & \vdots & 6/19 & -1/19 \\ 0 & 6 & \vdots & 6/19 & 18/19 \end{bmatrix}$$

Divide the second row by 6:

196

$$\begin{bmatrix} 1 & 0 & \vdots & 6/19 & -1/19 \\ 0 & 1 & \vdots & 1/19 & 3/19 \end{bmatrix}$$

Therefore

$$A^{-1} = \begin{bmatrix} 6/19 & -1/19 \\ 1/19 & 3/19 \end{bmatrix}$$

(2)

$$A = \begin{bmatrix} 1 & -7 & -14 \\ 2 & 1 & -1 \\ 1 & 3 & 4 \end{bmatrix}$$

$$[A : I] = \begin{bmatrix} 1 & -7 & -14 & \vdots & 1 & 0 & 0 \\ 2 & 1 & -1 & \vdots & 0 & 1 & 0 \\ 1 & 3 & 4 & \vdots & 0 & 0 & 1 \end{bmatrix}$$

Subtract the first row from the third row:

$$\begin{bmatrix} 1 & -7 & -14 & \vdots & 1 & 0 & 0 \\ 2 & 1 & -1 & \vdots & 0 & 1 & 0 \\ 0 & 10 & 18 & \vdots & -1 & 0 & 1 \end{bmatrix}$$

Divide the third row by 2:

$$\begin{bmatrix} 1 & -7 & -14 & \vdots & 1 & 0 & 0 \\ 2 & 1 & -1 & \vdots & 0 & 1 & 0 \\ 0 & 5 & 9 & \vdots & -1/2 & 0 & 1/2 \end{bmatrix}$$

Add -2 times the first row to the second row:

$$\begin{bmatrix} 1 & -7 & 9 & \vdots & 1 & 0 & 0 \\ 0 & 15 & 27 & \vdots & -2 & 1 & 0 \\ 0 & 5 & 9 & \vdots & -1/2 & 0 & 1/2 \end{bmatrix}$$

Divide the second row by 3:

$$\begin{bmatrix} 1 & -7 & -14 & \vdots & 1 & 0 & 0 \\ 0 & 5 & 9 & \vdots & -2/3 & 1/3 & 0 \\ 0 & 5 & 9 & \vdots & -1/2 & 0 & 1/2 \end{bmatrix}$$

Subtract the second row from the third row:

$$\begin{bmatrix} 1 & -7 & -14 & \vdots & 1 & 0 & 0 \\ 0 & 5 & 9 & \vdots & -2/3 & 1/3 & 0 \\ 0 & 0 & 0 & \vdots & 1/6 & -1/3 & 1/2 \end{bmatrix}$$

At this point A is row equivalent to

197

$$F = \begin{bmatrix} 1 & -7 & -14 \\ 0 & 5 & 9 \\ 0 & 0 & 0 \end{bmatrix}$$

The matrix A is singular and therefore A does not have an inverse.

(3)

$$A = \begin{bmatrix} 3 & 1 & 0 \\ 1 & -1 & 2 \\ 1 & 1 & 1 \end{bmatrix}$$

$$[A : I] = \begin{bmatrix} 3 & 1 & 0 & \vdots & 1 & 0 & 0 \\ 1 & -1 & 2 & \vdots & 0 & 1 & 0 \\ 1 & 1 & 1 & \vdots & 0 & 0 & 1 \end{bmatrix}$$

Interchange the first and third rows:

$$\begin{bmatrix} 1 & 1 & 1 & \vdots & 0 & 0 & 1 \\ 1 & -1 & 2 & \vdots & 0 & 1 & 0 \\ 3 & 1 & 0 & \vdots & 1 & 0 & 0 \end{bmatrix}$$

Subtract the first row from the second row and add -3 times the first row to the third row:

$$\begin{bmatrix} 1 & 1 & 1 & \vdots & 0 & 0 & 1 \\ 0 & -2 & 1 & \vdots & 0 & 1 & -1 \\ 0 & -2 & -3 & \vdots & 1 & 0 & -3 \end{bmatrix}$$

Divide the second row by -2:

$$\begin{bmatrix} 1 & 1 & 1 & \vdots & 0 & 0 & 1 \\ 0 & 1 & -1/2 & \vdots & 0 & -1/2 & 1/2 \\ 0 & -2 & -3 & \vdots & 1 & 0 & -3 \end{bmatrix}$$

Subtract the second row from the first row:

$$\begin{bmatrix} 1 & 0 & 3/2 & \vdots & 0 & 1/2 & 1/2 \\ 0 & 1 & -1/2 & \vdots & 0 & -1/2 & 1/2 \\ 0 & -2 & -3 & \vdots & 1 & 0 & -3 \end{bmatrix}$$

Add 2 times the second row to the third row:

$$\begin{bmatrix} 1 & 0 & 3/2 & \vdots & 0 & 1/2 & 1/2 \\ 0 & 1 & -1/2 & \vdots & 0 & -1/2 & 1/2 \\ 0 & 0 & -4 & \vdots & 1 & -1 & -2 \end{bmatrix}$$

Divide the third row by -4:

$$\begin{bmatrix} 1 & 0 & 3/2 & \vdots & 0 & 1/2 & 1/2 \\ 0 & 1 & -1/2 & \vdots & 0 & -1/2 & 1/2 \\ 0 & 0 & 1 & \vdots & -1/4 & 1/4 & +2/4 \end{bmatrix}$$

Add −3/2 times the third row to the first row and add 1/2 times the third row to the second row:

$$\begin{bmatrix} 1 & 0 & 0 & \vdots & 3/8 & +1/8 & -2/8 \\ 0 & 1 & 0 & \vdots & -1/8 & -3/8 & 6/8 \\ 0 & 0 & 1 & \vdots & -1/4 & 1/4 & +2/4 \end{bmatrix}$$

Thus

$$A^{-1} = \begin{bmatrix} 3/8 & 1/8 & -2/8 \\ -1/8 & -3/8 & 6/8 \\ -1/4 & 1/4 & 2/4 \end{bmatrix}$$

● **PROBLEM** 5-35

Find the inverse of A where
$$A = \begin{bmatrix} 2 & 3 \\ 3 & 5 \end{bmatrix}.$$

Solution: We know that

$$AA^{-1} = I$$

where I is the identity matrix. Let

$$A^{-1} = \begin{bmatrix} a & b \\ c & d \end{bmatrix}.$$

Since $A A^{-1} = I$, we have

$$\begin{bmatrix} 2 & 3 \\ 3 & 5 \end{bmatrix} \begin{bmatrix} a & b \\ c & d \end{bmatrix} = \begin{bmatrix} 1 & 0 \\ 0 & 1 \end{bmatrix}$$

Performing the matrix multiplication, we obtain:

$$\begin{bmatrix} 2a + 3c & 2b + 3d \\ 3a + 5c & 3b + 5d \end{bmatrix} = \begin{bmatrix} 1 & 0 \\ 0 & 1 \end{bmatrix}.$$

Now this matrix equality is equivalent to the following system of equations to be satisfied by a,b,c,d:

$$2a + 3c = 1 \qquad\qquad 2b + 3d = 0$$

$$3a + 5c = 0 \qquad\qquad 3b + 5d = 1 \ .$$

The pair of equations on the left yields a = 5 and c = −3, while the pair on the right yields b = −3 and d = 2. Hence

$$A^{-1} = \begin{bmatrix} 5 & -3 \\ -3 & 2 \end{bmatrix}.$$

The method used in this example reduces a matrix inversion problem to one of solving a system of linear equations, and it may be applied to a square matrix of any order.

● **PROBLEM** 5-36

Use the classical adjoint to find A^{-1} where

$$A = \begin{bmatrix} 1 & 0 & -1 \\ 0 & 2 & 2 \\ 1 & 1 & -1 \end{bmatrix}$$

<u>Solution:</u>   Recall some definitions: If $A = (a_{ij})$, then a co-factor of an entry a_{ij} is denoted A_{ij} and is given by $(-1)^{i+j}$ times the determinant of the $(n-1) \times (n-1)$ minor matrix obtained from A by deleting its ith row and jth column.

By the matrix of cofactors, we mean the matrix

$$C = \begin{bmatrix} A_{11} & \cdots & \cdots & A_{1n} \\ \cdot & & & \cdot \\ \cdot & & & \cdot \\ \cdot & & & \cdot \\ A_{n1} & \cdots & \cdots & A_{nn} \end{bmatrix} .$$

Then the adjoint of A is C^T, i.e.,

$$\text{adj } A = \begin{bmatrix} A_{11} & A_{21} & \cdots & A_{n1} \\ A_{12} & A_{22} & \cdots & \\ \cdot & \cdot & & \cdot \\ \cdot & \cdot & & \cdot \\ \cdot & \cdot & & \cdot \\ A_{1n} & A_{2n} & \cdots & A_{nn} \end{bmatrix}$$

Recall that A^{-1} exists if and only if $\det A = |A| \neq 0$. The rule for obtaining A^{-1} is then

$$A^{-1} = \frac{1}{|A|} [\text{adj } A]$$

where $|A| = $ determinant of the $n \times n$ square matrix. Let us first compute the determinant of matrix A

$$A = \begin{bmatrix} 1 & 0 & -1 \\ 0 & 2 & 2 \\ 1 & 1 & -1 \end{bmatrix}$$

$$|A| = \begin{vmatrix} 1 & 0 & -1 \\ 0 & 2 & 2 \\ 1 & 1 & -1 \end{vmatrix}$$

$$= 1 \begin{vmatrix} 2 & 2 \\ 1 & -1 \end{vmatrix} - 0 \begin{vmatrix} 0 & 2 \\ 1 & -1 \end{vmatrix} + (-1) \begin{vmatrix} 0 & 2 \\ 1 & 1 \end{vmatrix}$$

$$= 1(-2-2) - 0(0-2) - 1(0-2)$$

$$= -4-0+2 = -2 .$$

We find that $|A| \neq 0$. Therefore A^{-1} exists. The classical adjoint of A is found by replacing each element of A by its cofactor and taking the transpose of the resulting matrix.

Let us now compute the cofactors of the entries of A.

$$A = \begin{bmatrix} 1 & 0 & -1 \\ 0 & 2 & 2 \\ 1 & 1 & -1 \end{bmatrix}$$

To find A_{11}, we delete the first row and first column of A to obtain the matrix

$$\begin{bmatrix} 2 & 2 \\ 1 & -1 \end{bmatrix} .$$

The cofactor A_{11} is then $(-1)^{1+1}$ times the determinant of the above matrix, i.e.,

$$A_{11} = (1)^2 \cdot \begin{vmatrix} 2 & 2 \\ 1 & -1 \end{vmatrix} = \begin{vmatrix} 2 & 2 \\ 1 & -1 \end{vmatrix}$$

$$= (-2-2) = -4 .$$

We find the cofactors of the remaining elements of A by the same method. The cofactors of the nine elements of A are

$$A_{11} = + \begin{vmatrix} 2 & 2 \\ 1 & -1 \end{vmatrix}, \quad A_{12} = - \begin{vmatrix} 0 & 2 \\ 1 & -1 \end{vmatrix}, \quad A_{13} = + \begin{vmatrix} 0 & 2 \\ 1 & 1 \end{vmatrix}$$

$$\qquad = (-2-2) \qquad\qquad = -(0-2) \qquad\qquad = (0-2)$$

$$\qquad = -4 \qquad\qquad\quad = 2 \qquad\qquad\qquad = -2$$

$$A_{21} = - \begin{vmatrix} 0 & -1 \\ 1 & -1 \end{vmatrix}, \quad A_{22} = + \begin{vmatrix} 1 & -1 \\ 1 & -1 \end{vmatrix}, \quad A_{23} = - \begin{vmatrix} 1 & 0 \\ 1 & 1 \end{vmatrix}$$

$$\qquad = -(0+1) \qquad\qquad = (-1+1) \qquad\qquad = -(1-0)$$

$$\qquad = -1 \qquad\qquad\quad = 0 \qquad\qquad\qquad = -1$$

$$A_{31} = + \begin{vmatrix} 0 & -1 \\ 2 & 2 \end{vmatrix}, \quad A_{32} = - \begin{vmatrix} 1 & -1 \\ 0 & 2 \end{vmatrix}, \quad A_{33} = + \begin{vmatrix} 1 & 0 \\ 0 & 2 \end{vmatrix}$$

$$\qquad = (0+2) \qquad\qquad = -(2-0) \qquad\qquad = (2-0)$$

$$\qquad = 2 \qquad\qquad\quad = -2 \qquad\qquad\qquad = 2$$

The matrix of cofactors C is given by

$$C = \begin{bmatrix} -4 & 2 & -2 \\ -1 & 0 & -1 \\ 2 & -2 & 2 \end{bmatrix} .$$

We form the transpose of the matrix of cofactors to obtain the classical adjoint of A:

$$Adj\ A = \begin{bmatrix} -4 & -1 & 2 \\ 2 & 0 & -2 \\ -2 & -1 & 2 \end{bmatrix}$$

Now,

$$A^{-1} = \frac{1}{|A|} [adj\ A]$$

$$= -\frac{1}{2} \begin{bmatrix} -4 & -1 & 2 \\ 2 & 0 & -2 \\ -2 & -1 & 2 \end{bmatrix} ,$$

So

$$A^{-1} = \begin{bmatrix} 2 & 1/2 & -1 \\ -1 & 0 & 1 \\ 1 & 1/2 & -1 \end{bmatrix} .$$

It is easy to check the computation by verifying that

$$AA^{-1} = I$$

$$\begin{bmatrix} 1 & 0 & -1 \\ 0 & 2 & 2 \\ 1 & 1 & -1 \end{bmatrix} \begin{bmatrix} 2 & 1/2 & -1 \\ -1 & 0 & 1 \\ 1 & 1/2 & -1 \end{bmatrix}$$

$$= \begin{bmatrix} 2+0-1 & 1/2+0-1/2 & -1+0+1 \\ 0-2+2 & 0+0+1 & 0+2-2 \\ 2-1-1 & 1/2+0-1/2 & -1+1+1 \end{bmatrix}$$

$$= \begin{bmatrix} 1 & 0 & 0 \\ 0 & 1 & 0 \\ 0 & 0 & 1 \end{bmatrix} = I .$$

● **PROBLEM 5-37**

Find the rank of the matrix A where $A = \begin{bmatrix} 1 & 3 & 2 \\ 2 & 6 & 1 \end{bmatrix} .$

Solution: If A is a matrix, then the rank of A, written r(A), is the maximum number of linearly independent columns or, equivalently,

202

rows. Since A is 2×3 , the rank must be two or less. First, check the rows for linear independence. Set

$$c_1(1,3,2) + c_3(2,6,1) = 0$$

to obtain the system of equations:

$$L_1 : \quad c_1 + 2c_2 = 0$$

$$L_2 : \quad 3c_1 + 6c_2 = 0$$

$$L_3 : \quad 2c_1 + c_2 = 0 \ .$$

By solving this sytem of equations, we find that it has only a trivial solution:

$$-3L_1 : \quad -3c_1 - 6c_2 = 0$$

$$+L_2 : \quad \underline{3c_1 + 6c_2 = 0}$$

$$0c_1 + 0c_2 = 0$$

$$-2L_1 : \quad -2c_1 - 4c_2 = 0$$

$$+L_3 : \quad \underline{2c_1 + c_2 = 0}$$

$$0c_1 - 3c_2 = 0$$

$$c_2 = 0$$

$$\text{From } L_1 : \quad c_1 + 2c_2 = 0$$

$$c_1 = 0$$

Solution: $c_1 = c_2 = 0$.

Thus, the two rows are independent and $r(A) = 2$.

We also could have found $r(A)$ by checking the maximum number of linearly independent columns. The two column vectors

$$\begin{bmatrix} 1 \\ 2 \end{bmatrix} \quad \text{and} \quad \begin{bmatrix} 2 \\ 1 \end{bmatrix}$$

are linearly independent since

$$c_1 \begin{bmatrix} 1 \\ 2 \end{bmatrix} + c_2 \begin{bmatrix} 2 \\ 1 \end{bmatrix} = 0$$

implies $c_1 = c_2 = 0$. Obtaining the system of equations,

$$L_1 : \quad c_1 + 2c_2 = 0$$

$$+L_2 : \quad 2c_1 + c_2 = 0 \ ,$$

and solving this system of equations, the result is the trivial solution $c_1 = c_2 = 0$.

$$-2L_1 : \quad -2c_1 - 4c_2 = 0$$

$$L_2 : \quad \underline{2c_1 + c_2 = 0}$$

$$-3c_2 = 0$$

$$c_2 = 0$$

From L_1 : $c_1 + 2c_2 = 0$

$c_1 = 0$

Solution: $c_1 = c_2 = 0$.   Furthermore, since the columns are vectors in R^2 , the maximum number of linearly independent columns can only equal two (dim $R^2 = 2$).   Thus, again, $r(A) = 2$.

● **PROBLEM** 5-38

Find the rank of the matrix   A   where:

(i)
$$A = \begin{bmatrix} 1 & 3 & 1 & -2 & -3 \\ 1 & 4 & 3 & -1 & -4 \\ 2 & 3 & -4 & -7 & -3 \\ 3 & 8 & 1 & -7 & -8 \end{bmatrix}$$

(ii)
$$A = \begin{bmatrix} 1 & 2 & -3 \\ 2 & 1 & 0 \\ -2 & -1 & 3 \\ -1 & 4 & -2 \end{bmatrix}$$

(iii)
$$A = \begin{bmatrix} 1 & 3 \\ 0 & -2 \\ 5 & -1 \\ -2 & 3 \end{bmatrix}$$

Solution: (i)   First, reduce the matrix   A   to echelon form using the elementary row operations.
(a)   Add   -1   times the first row to the second row.
(b)   Add   -2   times the first row to the third row.
(c)   Add   -3   times the first row to the third row.

$$A = \begin{bmatrix} 1 & 3 & 1 & -2 & -3 \\ 0 & 1 & 2 & 1 & -1 \\ 0 & -3 & -6 & -3 & 3 \\ 0 & -1 & -2 & -1 & 1 \end{bmatrix}$$

Add   +3 times the second row to the third row.
Add the second row to the fourth row.   Then,

$$A = \begin{bmatrix} 1 & 3 & 1 & -2 & -3 \\ 0 & 1 & 2 & 1 & -1 \\ 0 & 0 & 0 & 0 & 0 \\ 0 & 0 & 0 & 0 & 0 \end{bmatrix} .$$

Since the echelon matrix has two nonzero rows, rank (A) = 2.

(ii)   Since row rank equals column rank it is easier to form the transpose of   A   and then row reduce to echelon form.

204

$$\begin{bmatrix} 1 & 2 & -2 & -1 \\ 2 & 1 & -1 & 4 \\ -3 & 0 & 3 & -2 \end{bmatrix}$$

Add -2 times the first row to the second row, and add 3 times the first row to the third row.

$$A = \begin{bmatrix} 1 & 2 & -2 & -1 \\ 0 & -3 & 3 & 6 \\ 0 & 6 & -3 & -5 \end{bmatrix}$$

Add 2 times the second row to the third row.

$$A = \begin{bmatrix} 1 & 2 & -2 & -1 \\ 0 & -3 & 3 & 6 \\ 0 & 0 & 3 & 7 \end{bmatrix}$$

Since the echelon matrix has three nonzero rows, rank (A) = 3.

(iii)   The two columns are linearly independent since one is not a multiple of the other.   Hence, rank [A] = 2 .

● **PROBLEM** 5-39

Let A be the matrix

$$\begin{bmatrix} 0 & 1 & 3 & -2 & -1 & 2 \\ 0 & 2 & 6 & -4 & -2 & 4 \\ 0 & 1 & 3 & -2 & 1 & 4 \\ 0 & 2 & 6 & 1 & -1 & 0 \end{bmatrix}$$

Find the determinant rank of A .

Solution: If A is an m✕n   matrix, the determinant rank of A is defined as follows: The order of the largest non-zero determinant which is obtainable by the possible deletion of rows and columns from the matrix.
    The standard method of computing the determinant rank is the one shown below. First, use elementary row operations to reduce the matrix to echelon form (the leading coefficient of each equation equals one). Then, from the echelon matrix, select the largest upper triangular matrix which has one's along the main diagonal. The determinant of this matrix is the product of the diagonal elements, and, hence, the determinant rank is the order of this determinant. Applying the three elementary row operations on A, we obtain the equivalent matrix:

$$\begin{bmatrix} 0 & 1 & 3 & -2 & -1 & 2 \\ 0 & 0 & 0 & 1 & 1/5 & -4/5 \\ 0 & 0 & 0 & 0 & 1 & 1 \\ 0 & 0 & 0 & 0 & 0 & 0 \end{bmatrix} \qquad (1)$$

Examining (1), it can be seen that the second, fourth and fifth columns form the largest possible upper triangular matrix. Thus,

$$\begin{bmatrix} 1 & -2 & -1 \\ 0 & 1 & 1/5 \\ 0 & 0 & 1 \end{bmatrix}$$

205

has determinant equal to one, and the determinant rank of A is three. Since (1) contains three non-zero rows, the row-rank of A is also three, i.e., determinant rank = row rank. The last statement is always true.

Given:
$$A = \begin{bmatrix} 1 & 1 \\ 0 & 1 \end{bmatrix} \quad \text{and} \quad P = \begin{bmatrix} 1 & 1 \\ 1 & -1 \end{bmatrix}.$$

(a) Find P^{-1}.

(b) Find $P^{-1}AP$.

(c) Verify that, if B is similar to A, then A is similar to B.

(d) Show that $B^k = P^{-1}A^kP$ if $B = P^{-1}AP$ where k is any positive integer.

Solution: (a) It is known that $P^{-1} = \dfrac{1}{\det P} \, \mathrm{adj}\, P$.

$$\det P = \begin{vmatrix} 1 & 1 \\ 1 & -1 \end{vmatrix} = -1-1 = -2 .$$

$$\mathrm{adj}\, P = \begin{bmatrix} -1 & -1 \\ -1 & 1 \end{bmatrix} .$$

Therefore,
$$P^{-1} = \begin{bmatrix} \frac{1}{2} & \frac{1}{2} \\ \frac{1}{2} & -\frac{1}{2} \end{bmatrix} .$$

(b)
$$P^{-1}AP = \begin{bmatrix} \frac{1}{2} & \frac{1}{2} \\ \frac{1}{2} & -\frac{1}{2} \end{bmatrix} \begin{bmatrix} 1 & 1 \\ 0 & 1 \end{bmatrix} \begin{bmatrix} 1 & 1 \\ 1 & -1 \end{bmatrix}$$

$$= \begin{bmatrix} \frac{1}{2} & \frac{1}{2} \\ \frac{1}{2} & -\frac{1}{2} \end{bmatrix} \begin{bmatrix} 2 & 0 \\ 1 & -1 \end{bmatrix}$$

$$= \begin{bmatrix} 3/2 & -\frac{1}{2} \\ \frac{1}{2} & \frac{1}{2} \end{bmatrix} .$$

We say that the matrix B is similar to the matrix A if there is an invertible matrix P such that
$$B = P^{-1}AP .$$
Therefore, let
$$B = \begin{bmatrix} 3/2 & -\frac{1}{2} \\ \frac{1}{2} & \frac{1}{2} \end{bmatrix}$$
and then B is similar to A.

(c) If P is invertible and $B = P^{-1}AP$, then
$$PBP^{-1} = P(P^{-1}AP)P^{-1}$$

$$= (PP^{-1})A(PP^{-1})$$

$$= A \quad \text{since} \quad PP^{-1} = I .$$

Let $Q = P^{-1}$ so that $Q^{-1} = P$. Then $Q^{-1}BQ = A$.

$$Q = P^{-1} = \begin{bmatrix} \frac{1}{2} & \frac{1}{2} \\ \frac{1}{2} & -\frac{1}{2} \end{bmatrix} .$$

Thus,

$$Q^{-1} = \frac{1}{\det Q} \operatorname{adj} Q$$

$$Q^{-1} = \frac{1}{-\frac{1}{2}} \begin{bmatrix} -\frac{1}{2} & -\frac{1}{2} \\ -\frac{1}{2} & \frac{1}{2} \end{bmatrix} = \begin{bmatrix} 1 & 1 \\ 1 & -1 \end{bmatrix}$$

$$Q^{-1}BQ = \begin{bmatrix} 1 & 1 \\ 1 & -1 \end{bmatrix} \begin{bmatrix} 3/2 & -\frac{1}{2} \\ \frac{1}{2} & \frac{1}{2} \end{bmatrix} \begin{bmatrix} \frac{1}{2} & \frac{1}{2} \\ \frac{1}{2} & -\frac{1}{2} \end{bmatrix}$$

$$= \begin{bmatrix} 1 & 1 \\ 1 & -1 \end{bmatrix} \begin{bmatrix} \frac{1}{2} & 1 \\ \frac{1}{2} & 0 \end{bmatrix}$$

$$= \begin{bmatrix} 1 & 1 \\ 0 & 1 \end{bmatrix}$$

$$= A .$$

Thus, if B is similar to A, then A is similar to B.

(d) Let $K = 2$; check that $P^{-1}A^2 P = B^2$. Then

$$B^2 = \begin{bmatrix} 3/2 & -\frac{1}{2} \\ \frac{1}{2} & \frac{1}{2} \end{bmatrix} \begin{bmatrix} 3/2 & -\frac{1}{2} \\ \frac{1}{2} & \frac{1}{2} \end{bmatrix}$$

$$= \begin{bmatrix} 2 & -1 \\ 1 & 0 \end{bmatrix} .$$

$$A^2 = \begin{bmatrix} 1 & 1 \\ 0 & 1 \end{bmatrix} \begin{bmatrix} 1 & 1 \\ 0 & 1 \end{bmatrix} = \begin{bmatrix} 1 & 2 \\ 0 & 1 \end{bmatrix} .$$

Then,

$$P^{-1}A^2 P = \begin{bmatrix} \frac{1}{2} & \frac{1}{2} \\ \frac{1}{2} & -\frac{1}{2} \end{bmatrix} \begin{bmatrix} 1 & 2 \\ 0 & 1 \end{bmatrix} \begin{bmatrix} 1 & 1 \\ 1 & -1 \end{bmatrix}$$

$$= \begin{bmatrix} \frac{1}{2} & \frac{1}{2} \\ \frac{1}{2} & -\frac{1}{2} \end{bmatrix} \begin{bmatrix} 3 & -1 \\ 1 & -1 \end{bmatrix}$$

$$= \begin{bmatrix} 2 & -1 \\ 1 & 0 \end{bmatrix}$$

$$= B^2 .$$

Suppose $K = 3$:

$$A^3 = A^2 A = \begin{bmatrix} 1 & 2 \\ 0 & 1 \end{bmatrix} \begin{bmatrix} 1 & 1 \\ 0 & 1 \end{bmatrix} = \begin{bmatrix} 1 & 3 \\ 0 & 1 \end{bmatrix} \ ,$$

$$B^3 = B^2 B = \begin{bmatrix} 2 & -1 \\ 1 & 0 \end{bmatrix} \begin{bmatrix} 3/2 & -\frac{1}{2} \\ \frac{1}{2} & \frac{1}{2} \end{bmatrix} = \begin{bmatrix} 5/2 & -3/2 \\ 3/2 & -\frac{1}{2} \end{bmatrix} \ .$$

Then,

$$P^{-1} A^3 P = \begin{bmatrix} \frac{1}{2} & \frac{1}{2} \\ \frac{1}{2} & -\frac{1}{2} \end{bmatrix} \begin{bmatrix} 1 & 3 \\ 0 & 1 \end{bmatrix} \begin{bmatrix} 1 & 1 \\ 1 & -1 \end{bmatrix}$$

$$= \begin{bmatrix} \frac{1}{2} & \frac{1}{2} \\ \frac{1}{2} & -\frac{1}{2} \end{bmatrix} \begin{bmatrix} 4 & -2 \\ 1 & -1 \end{bmatrix}$$

$$= \begin{bmatrix} 5/2 & -3/2 \\ 3/2 & -\frac{1}{2} \end{bmatrix}$$

$$= B^3 \ .$$

In general for any positive integer k, $B^k = P^{-1} A^k P$ if $B = P^{-1} A P$. To prove this rigorously for any matrices A and B and an invertible matrix P, use an inductive argument. Given $P^{-1} A P = B$, show $P^{-1} A^k P = B^k$. Take $n = 1$; $P^{-1} A P = B$. When $n = 2$; $P^{-1} A P = B$ gives $B^2 = P^{-1} A P P^{-1} A P$ so $B^2 = P^{-1} A^2 P$ since $P P^{-1} = I$.

Assume $P^{-1} A^k P = B^k$ is true for $k = n$; show that it is true for $k = n+1$. $P^{-1} A^n P = B^n$ so, since $B^{n+1} = B^n \cdot B = P^{-1} A^n P B = P^{-1} A^n P P^{-1} A P = P^{-1} A^{n+1} P$, $B^{n+1} = P^{-1} A^{n+1} P$.

From this it follows that if B is similar to A, then B^k is similar to A^k. Observe that the powers of A are easy to find. Direct calculation gives:

$$A^3 = \begin{bmatrix} 1 & 3 \\ 0 & 1 \end{bmatrix} \ ; \qquad A^4 = \begin{bmatrix} 1 & 4 \\ 0 & 1 \end{bmatrix} \ .$$

In general, we obtain the formula

$$A^k = \begin{bmatrix} 1 & k \\ 0 & 1 \end{bmatrix} \ .$$

Again, to be rigorous, one would need to use an inductive argument. To find B^k, use the formula $B = P^{-1} A^k B$. Thus,

$$B^k = P^{-1} \begin{bmatrix} 1 & k \\ 0 & 1 \end{bmatrix} \begin{bmatrix} 1 & 1 \\ 1 & -1 \end{bmatrix}$$

$$= P^{-1} \begin{bmatrix} 1+k & 1-k \\ 1 & -1 \end{bmatrix}$$

$$= \begin{bmatrix} \frac{1}{2} & \frac{1}{2} \\ \frac{1}{2} & -\frac{1}{2} \end{bmatrix} \begin{bmatrix} 1+k & 1-k \\ 1 & -1 \end{bmatrix}$$

$$= \begin{bmatrix} 1+k/2 & -k/2 \\ k/2 & 1-k/2 \end{bmatrix} \quad .$$

DETERMINANTS

● **PROBLEM** 5-41

Define permutations. Find the permutations of order 3.

Solution: A permutation of order n, n = 1,2,..., is an ordered set
$<i_1, i_2,...,i_n>$ of integers in which each of the integers 1,2,...,n
occurs exactly once.
   We denote permutation σ by:

$$\sigma = \begin{pmatrix} 1 & 2 & . & . & . & n \\ i_1 & i_2 & . & . & . & i_n \end{pmatrix}$$

or $\sigma = i_1 i_2 \ldots i_n$. The permutations of order 3 are <123> , <132>
<213> , <231> , <312> , <321> . Note that the number of such permuta-
tions is n! .
   This follows from the reasoning below:
   Let {1,2,...,n} be the first n integers. There are n values to
which 1 can be sent, (n-1) values for 2 and, proceeding, 1 value for
n(n-1)...1 = n!

● **PROBLEM** 5-42

Define an inversion of a permutation.
Define the sign of a permutation.
Find the inversion of <3,1,4,2>.
Find the sign of <3,1,4,2>.

Solution: An inversion of a permutation $<i_1, i_2,...,i_n>$ is a pair
(p,q) of integers from among 1,2,...,n such that p $>$ q and p
occurs before q in the list $i_1, i_2,...,i_n$. $I<i_1, i_2,...,i_n>$ will
denote the number of inversions of $<i_1, i_2,...,i_n>$.

   The inversions of <3, 1, 4, 2> are: (3,1), (3,2) and (4,2).
Hence, I<3, 1, 4, 2> = 3.   The number,

$$\text{sgn} <i_1, i_2,...,i_n> = (-1)^{I<i_1, i_2,...,i_n>}$$

is called the sign of the permutation $<i_1, i_2,...,i_n>$. Hence,
sgn<3, 1, 4, 2> = $(-1)^3$ = -1 .

Determine the parity of σ = 542163.

<u>Solution</u>: Consider an arbitrary permutation, σ , in S_n : $\sigma = j_1 j_2 \cdots j_n$.
We say σ is even or odd according as to whether there is an even or odd
number of pairs (i K) for which

$$i > K \text{ but } i \text{ precedes } K \text{ in } \sigma \text{ .}$$

Define the sign or parity of σ , written sgn σ , by:

(a)   sgn $\sigma = (-1)^n = +1$ if n is even ,

(b)   sgn $\sigma = (-1)^n = -1$ if n is odd.

Method 1: σ = 542163.
It is necessary to obtain the number of pairs (i,j) for which i > j
and i precedes j in σ . 5, 4, and 2 precede and are greater than
1. Hence, the number of pairs is 3, {(5,1), (4,2), (2,1). 5 and 4
precede and are greater than 2, hence the number of pairs is 2,
{(5,2), (4,2)} . 5, 4 and 6 precede, and are greater than 3. Hence,
the number of pairs is 3, {(5,3), (4,3), (6,3)}. 5 precedes, and is
greater than 4. Hence, the number of pairs is 1 , {(5,4)} . Since
3 + 2 + 3 + 1 = 9 is odd, σ is an odd permutation. Therefore,
sgn σ = -1 .

Method 2:
   Transpose 1 to the first position as follows:

        5 4 2 1 6 3 to 1 5 4 2 6 3

   Transpose 2 to the second position:

        1 5 4 2 6 3 to 1 2 5 4 6 3

   Transpose 3 to the third position:

        1 2 5 4 6 3 to 1 2 3 5 4 6

   Transpose 4 to the fourth position:

        1 2 3 5 4 6 to 1 2 3 4 5 6

Note that 5 and 6 are in the correct positions. Add the numbers
"jumped": 3 + 2 + 3 + 1 = 9. Since 9 is odd, σ is an odd permutation.
Hence, sgn σ = -1.

Method 3:
   An interchange of two numbers in a permutation is equivalent to multiplying the permutation by a transposition. Therefore, transform σ to
the identity permutation using transpositions such as:

Thus, the number of transpositions is 5. Since 5 is odd, σ is an

210

odd permutation. Hence, sgn σ = -1.

Let σ = 24513, and τ = 41352, be permutations in S_5. Find: (i) the composition permutations $\tau\sigma$ and $\sigma\tau$; (ii) σ^{-1}.

Solution: σ = 24513 τ = 41352. We can write,

$$\sigma = \begin{pmatrix} 1 & 2 & 3 & 4 & 5 \\ 2 & 4 & 5 & 1 & 3 \end{pmatrix} \qquad \tau = \begin{pmatrix} 1 & 2 & 3 & 4 & 5 \\ 4 & 1 & 3 & 5 & 2 \end{pmatrix}$$

which means,

$\sigma(1)$ = 2 , $\sigma(2)$ = 4 , $\sigma(3)$ = 5, $\sigma(4)$=1
and
$\sigma(5)$ = 3. Also,

$\tau(1)$ = 4 , $\tau(2)$ = 1 , $\tau(3)$ = 3 , $\tau(4)$ = 5 and $\tau(5)$ = 2 .

Now,
$\tau\sigma = \tau\sigma(i) = \tau(\sigma(i))$,
so,
$\tau\sigma(1) = \tau(\sigma(1)) = \tau(2) = 1$
$\tau\sigma(2) = \tau(\sigma(2)) = \tau(4) = 5$
$\tau\sigma(3) = \tau(\sigma(3)) = \tau(5) = 2$
$\tau\sigma(4) = \tau(\sigma(4)) = \tau(1) = 4$
$\tau\sigma(5) = \tau(\sigma(5)) = \tau(3) = 3$.
Thus, $\tau\sigma$ = 15243.

$\sigma\tau = \sigma\tau(i) = \sigma(\tau(i))$:
$\sigma(\tau(1)) = \sigma(4) = 1$
$\sigma(\tau(2)) = \sigma(1) = 2$
$\sigma(\tau(3)) = \sigma(3) = 5$
$\sigma(\tau(4)) = \sigma(5) = 3$
$\sigma(\tau(5)) = \sigma(2) = 4$.

Hence, $\sigma\tau$ = 12534.
(ii)

$$\sigma = \begin{pmatrix} 1 & 2 & 3 & 4 & 5 \\ 2 & 4 & 5 & 1 & 3 \end{pmatrix} .$$

We know that,
$$\sigma\sigma^{-1} = e$$
where e is the identity permutation,

$$e = \begin{pmatrix} 1 & 2 & 3 & \ldots & n \\ 1 & 2 & 3 & \ldots & n \end{pmatrix} .$$

To obtain e from σ , permute 2 4 5 1 3 to 1 2 3 4 5. Thus,

$$\sigma^{-1} = \begin{pmatrix} 2 & 4 & 5 & 1 & 3 \\ 1 & 2 & 3 & 4 & 5 \end{pmatrix} .$$

Then,
$\sigma^{-1}(1)$ = 4 , $\sigma^{-1}(2)$ = 1 , $\sigma^{-1}(3)$ = 5 , $\sigma^{-1}(4)$ = 2

and
$$\sigma^{-1}(5) = 3 .$$

Thus,
$$\sigma^{-1} = \begin{pmatrix} 1 & 2 & 3 & 4 & 5 \\ 4 & 1 & 5 & 2 & 3 \end{pmatrix}$$

So, $\sigma^{-1} = 41523$.

Check the result by showing: $\sigma\sigma^{-1} = e$

$$\sigma\sigma^{-1} = \begin{pmatrix} 1 & 2 & 3 & 4 & 5 \\ 2 & 4 & 5 & 1 & 3 \end{pmatrix} \begin{pmatrix} 1 & 2 & 3 & 4 & 5 \\ 4 & 1 & 5 & 2 & 3 \end{pmatrix} .$$

$$\sigma\sigma^{-1}(1) = \sigma(\sigma^{-1}(1)) = \sigma(4) = 1$$

$$\sigma\sigma^{-1}(2) = 2 , \ \sigma\sigma^{-1}(3) = 3 , \ \sigma\sigma^{-1}(4) = 4 , \ \sigma\sigma^{-1}(5) = 5 .$$

Thus,
$$\sigma\sigma^{-1} = \begin{pmatrix} 1 & 2 & 3 & 4 & 5 \\ 1 & 2 & 3 & 4 & 5 \end{pmatrix} = e .$$

● PROBLEM 5-45

Find the product of two permutations σ and φ, where,

(i) $\quad \sigma = \begin{pmatrix} 1 & 2 & 3 & 4 & 5 \\ 2 & 4 & 1 & 5 & 3 \end{pmatrix} \qquad \varphi = \begin{pmatrix} 1 & 2 & 3 & 4 & 5 \\ 4 & 1 & 2 & 5 & 3 \end{pmatrix}$

(ii) $\quad \sigma = \begin{pmatrix} 1 & 2 & 3 & 4 & 5 \\ 4 & 1 & 2 & 5 & 3 \end{pmatrix} \qquad \varphi = \begin{pmatrix} 4 & 1 & 2 & 5 & 3 \\ 1 & 2 & 3 & 4 & 5 \end{pmatrix}$

(iii) $\quad \sigma = \begin{pmatrix} 1 & 2 & 3 & 4 & 5 \\ 5 & 2 & 4 & 3 & 1 \end{pmatrix} \qquad \varphi = \begin{pmatrix} 4 & 1 & 2 & 5 & 3 \\ 1 & 2 & 3 & 4 & 5 \end{pmatrix}$

(iv) $\quad \sigma = \begin{pmatrix} 1 & 2 & 3 & 4 & 5 \\ 4 & 1 & 2 & 5 & 3 \end{pmatrix} \qquad \varphi = \begin{pmatrix} 1 & 2 & 3 & 4 & 5 \\ 2 & 4 & 1 & 5 & 3 \end{pmatrix}$

<u>Solution</u>: The product of two permutations σ and φ is defined in terms of function composition:

$$(\sigma\varphi)(i) = \sigma[\varphi(i)] , \ i = 1,\ldots,n .$$

The set of all permutations on n objects, together with this operation of multiplication is called the symmetric group of degree n, and is denoted by S_n .

(i) $\quad (\sigma\varphi)(i) = \sigma[\varphi(i)] , \quad i = 1,\ldots,n;$
then
$$(\sigma\varphi)(1) = \sigma[\varphi(1)] = \sigma(4) = 5$$
$$(\sigma\varphi)(2) = \sigma[\varphi(2)] = \sigma(1) = 2$$
$$(\sigma\varphi)(3) = \sigma[\varphi(3)] = \sigma(2) = 4 .$$

Similarly, we find $\sigma\varphi(4) = 3$ and $\sigma\varphi(5) = 1$. Thus,

(i)
$$\sigma\phi = \begin{pmatrix} 1 & 2 & 3 & 4 & 5 \\ 2 & 4 & 1 & 5 & 3 \end{pmatrix} \begin{pmatrix} 1 & 2 & 3 & 4 & 5 \\ 4 & 1 & 2 & 5 & 3 \end{pmatrix} = \begin{pmatrix} 1 & 2 & 3 & 4 & 5 \\ 5 & 2 & 4 & 3 & 1 \end{pmatrix}$$

(ii)
$$\sigma\phi = \begin{pmatrix} 1 & 2 & 3 & 4 & 5 \\ 4 & 1 & 2 & 5 & 3 \end{pmatrix} \begin{pmatrix} 4 & 1 & 2 & 5 & 3 \\ 1 & 2 & 3 & 4 & 5 \end{pmatrix} = \begin{pmatrix} 1 & 2 & 3 & 4 & 5 \\ 1 & 2 & 3 & 4 & 5 \end{pmatrix}$$

(iii)
$$\sigma\phi = \begin{pmatrix} 1 & 2 & 3 & 4 & 5 \\ 5 & 2 & 4 & 3 & 1 \end{pmatrix} \begin{pmatrix} 4 & 1 & 2 & 5 & 3 \\ 1 & 2 & 3 & 4 & 5 \end{pmatrix} = \begin{pmatrix} 1 & 2 & 3 & 4 & 5 \\ 2 & 4 & 1 & 5 & 3 \end{pmatrix}$$

(iv)
$$\sigma\phi = \begin{pmatrix} 1 & 2 & 3 & 4 & 5 \\ 4 & 1 & 2 & 5 & 3 \end{pmatrix} \begin{pmatrix} 1 & 2 & 3 & 4 & 5 \\ 2 & 4 & 1 & 5 & 3 \end{pmatrix} = \begin{pmatrix} 1 & 2 & 3 & 4 & 5 \\ 1 & 5 & 4 & 3 & 2 \end{pmatrix}$$

Observe that the examples (i) and (iv) show that multiplication of permutations is not always commutative, (i.e., $\sigma\phi \neq \phi\sigma$), in general.

● **PROBLEM 5-46**

Define det A and find the determinant of the following matrices:

(a) $[a_{11}]$ (b) $\begin{bmatrix} a_{11} & a_{12} \\ a_{21} & a_{22} \end{bmatrix}$ (c) $\begin{bmatrix} 0 & 0 & 0 \\ 0 & 0 & 0 \\ 0 & 0 & 0 \end{bmatrix}$

(d) $\begin{bmatrix} a_{11} & a_{12} & a_{13} \\ a_{21} & a_{22} & a_{23} \\ a_{31} & a_{32} & a_{33} \end{bmatrix}$

Solution: Determinants are formally defined as:

$$Det(A) = |A| = \sum_{\sigma} sgn(\sigma) \prod_{j=1}^{n} a_{\sigma(j)j}$$

$$= \sum_{\sigma} sgn(\sigma)\, a_{\sigma(1)1}\, a_{\sigma(2)2} \cdots a_{\sigma(n)n}$$

where summation extends over the $n!$, different permutations, σ, of the n symbols $1,2,\ldots,n$ and $sgn(\sigma) = +1$, σ even
-1, σ odd

$|A|$ is also known as an $n{\times}n$ determinant or a determinant of order n.

(a)   det A = a_{11}

(b)   det A = $\begin{vmatrix} a_{11} & a_{12} \\ a_{21} & a_{22} \end{vmatrix}$ = $a_{11}a_{22} - a_{21}a_{12}$

(c)   det A = $|0|$ = 0

213

(d)
$$\det A = \begin{vmatrix} a_{11} & a_{12} & a_{13} \\ a_{21} & a_{22} & a_{23} \\ a_{31} & a_{32} & a_{33} \end{vmatrix}$$

The permutations of S_3 and their signs are:

Permutation	Sign	Permutation	Sign
1 2 3	+	2 1 3	-
1 3 2	-	3 1 2	+
2 3 1	+	3 2 1	-

Then, $\det A = a_{11}a_{22}a_{33} - a_{11}a_{32}a_{23} + a_{21}a_{32}a_{13} - a_{21}a_{12}a_{33} + a_{31}a_{12}a_{23}$
$$- a_{31}a_{22}a_{13} .$$

● **PROBLEM 5-47**

a) Find the determinant of an arbitrary 3×3 matrix.
b) Find $\det A$ where:

$$A = \begin{bmatrix} -5 & 0 & 2 \\ 6 & 1 & 2 \\ 2 & 3 & 1 \end{bmatrix}$$

Solution: Let

$$A = \begin{bmatrix} b_{11} & b_{12} & b_{13} \\ b_{21} & b_{22} & b_{23} \\ b_{31} & b_{32} & b_{33} \end{bmatrix}.$$

$$\det A = \det \begin{bmatrix} b_{11} & b_{12} & b_{13} \\ b_{21} & b_{22} & b_{23} \\ b_{31} & b_{32} & b_{33} \end{bmatrix}.$$

Expand the above determinant by minors, using the first column.

$$\det A = +b_{11} \begin{vmatrix} b_{22} & b_{23} \\ b_{32} & b_{33} \end{vmatrix} - b_{21} \begin{vmatrix} b_{12} & b_{13} \\ b_{32} & b_{33} \end{vmatrix}$$
$$+ b_{31} \begin{vmatrix} b_{12} & b_{13} \\ b_{22} & b_{23} \end{vmatrix}$$

$\det A = b_{11}(b_{22}b_{33} - b_{32}b_{23}) - b_{21}(b_{12}b_{33} - b_{32}b_{13}) + b_{31}(b_{12}b_{23} - b_{22}b_{13}).$

Now expand the determinant by minors, using the second row:

214

$$\det A = -b_{21} \begin{vmatrix} b_{12} & b_{13} \\ b_{32} & b_{33} \end{vmatrix} + b_{22} \begin{vmatrix} b_{11} & b_{13} \\ b_{31} & b_{33} \end{vmatrix}$$

$$- b_{23} \begin{vmatrix} b_{11} & b_{12} \\ b_{31} & b_{32} \end{vmatrix} .$$

$$\det A = -b_{21}(b_{12}b_{33} - b_{32}b_{13}) + b_{22}(b_{11}b_{33} - b_{31}b_{13}) - b_{23}(b_{11}b_{32} - b_{31}b_{12})$$

$$= b_{22}b_{11}b_{33} - b_{22}b_{31}b_{13} - b_{23}b_{11}b_{32} + b_{23}b_{31}b_{12} - b_{21}(b_{12}b_{33} - b_{32}b_{13})$$

$$= b_{11}(b_{22}b_{33} - b_{32}b_{23}) - b_{21}(b_{12}b_{33} - b_{32}b_{13}) + b_{31}(b_{12}b_{23} - b_{22}b_{13})$$

Clearly, this is the same as the first answer. Note, also, that det A can be rearranged algebraically until it can be written as:

$$\det A = b_{11}b_{22}b_{33} + b_{12}b_{23}b_{31} + b_{13}b_{32}b_{21}$$

$$- [b_{13}b_{22}b_{31} + b_{23}b_{32}b_{11} + b_{33}b_{21}b_{12}] .$$

It is easy to remember this result (det of a 3x3 matrix) using the following mnemonic device: Figure 1

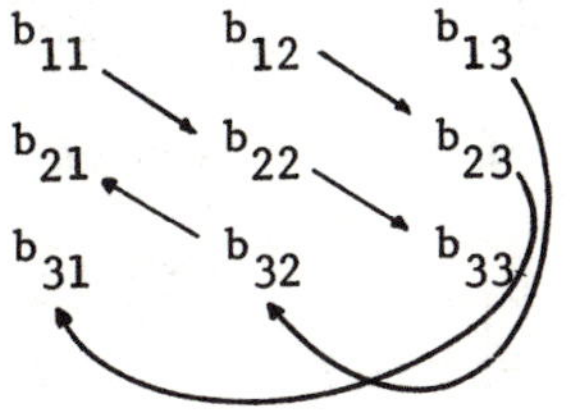

$$\left(b_{11}b_{22}b_{33} + b_{12}b_{23}b_{31} + b_{13}b_{32}b_{21} \right)$$

Figure 2:

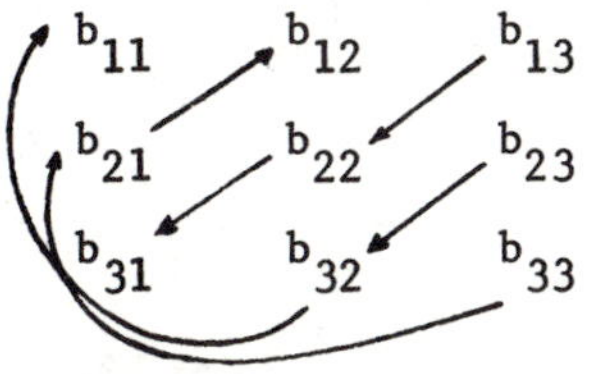

$$\left(b_{13}b_{22}b_{31} + b_{23}b_{32}b_{11} + b_{33}b_{21}b_{12} \right)$$

$$\det A = b_{11}b_{22}b_{33} + b_{12}b_{23}b_{31} + b_{13}b_{32}b_{21}$$

$$- [b_{13}b_{22}b_{31} + b_{23}b_{32}b_{11} + b_{33}b_{21}b_{12}] .$$

This makes taking 3x3 determinants simpler.

b) **Expand the determinant by minors, using the first column.**

$$\det A = -5 \begin{vmatrix} 1 & 3 \\ 3 & 1 \end{vmatrix} - 6 \begin{vmatrix} 0 & 2 \\ 3 & 1 \end{vmatrix} + 2 \begin{vmatrix} 0 & 2 \\ 1 & 2 \end{vmatrix}$$

$$= -5(1-6) - 6(0-6) + 2(0-2) = +25 + 36 - 4 = 57.$$

Find the determinant of the following matrix:

$$A = \begin{bmatrix} 2 & 0 & 3 & 0 \\ 2 & 1 & 1 & 2 \\ 3 & -1 & 1 & -2 \\ 2 & 1 & -2 & 1 \end{bmatrix}$$

Solution: Use the method of expansion by minors.

$$A = \begin{bmatrix} 2 & 0 & 3 & 0 \\ 2 & 1 & 1 & 2 \\ 3 & -1 & 1 & -2 \\ 2 & 1 & -2 & 1 \end{bmatrix}$$

Expanding along the first row:

$$\det A = 2 \begin{vmatrix} 1 & 1 & 2 \\ -1 & 1 & -2 \\ 1 & -2 & 1 \end{vmatrix} + 3 \begin{vmatrix} 2 & 1 & 2 \\ 3 & -1 & -2 \\ 2 & 1 & 1 \end{vmatrix}$$

Note that the minors, whose multiplying factors were zero, have been eliminated. This illustrates the general principle that, when evaluating determinants, expansion along the row (or column) containing the most zeros is the optimal procedure.

Add the second row to the first row for each of the 3 by 3 determinants:

$$\det A = 2 \begin{vmatrix} 0 & 2 & 0 \\ -1 & 1 & -2 \\ 1 & -2 & 1 \end{vmatrix} + 3 \begin{vmatrix} 5 & 0 & 0 \\ 3 & -1 & -2 \\ 2 & 1 & 1 \end{vmatrix}$$

Now expand the above determinants by minors using the first row.

$$\det A = 2(-2) \begin{vmatrix} -1 & -2 \\ 1 & 1 \end{vmatrix} + 3(5) \begin{vmatrix} -1 & -2 \\ 1 & 1 \end{vmatrix}$$

$$= (-4)(-1+2) + 15(-1+2)$$

$$= -4+15 = 11 .$$

a)   Compute the determinant of the matrix A, where:

$$A = \begin{bmatrix} 2 & 1 & -3 \\ 4 & 1 & 1 \\ 2 & 0 & -2 \end{bmatrix}$$

b)   Prove that the determinant of a lower-triangular matrix A is the product of the diagonal entries of A.

Solution: a) To find the determinant of A, apply row operations until an upper-triangular matrix is obtained. We then use the fact that the determinant of an upper triangular matrix is equal to the product of its diagonal entries.

$$A = \begin{bmatrix} 2 & 1 & -3 \\ 4 & 1 & 1 \\ 2 & 0 & -2 \end{bmatrix}$$

Now, add -2 times the first row to the second row and subtract the first row from the third row.

$$B = \begin{bmatrix} 2 & 1 & -3 \\ 0 & -1 & 7 \\ 0 & -1 & 1 \end{bmatrix}$$

B has the same determinant as A. Subtract the second row from the third row

$$C = \begin{bmatrix} 2 & 1 & -3 \\ 0 & -1 & 7 \\ 0 & 0 & -6 \end{bmatrix}$$

C has the same determinant as B. C is an upper triangular matrix and therefore the determinant of C is the product of the diagonal entries of C. Hence,

$$\det A = \det B = \det C = (2) \cdot (-1) \cdot (-6) = 12 \ .$$

b) Proof:

$$\text{Let } A = \begin{bmatrix} a_{11} & 0 & 0 \ldots 0 \\ a_{21} & a_{22} & 0 \ldots 0 \\ & & \cdots \\ a_{n1} & a_{n2} & a_{n3} \cdot a_{nn} \end{bmatrix} .$$

Suppose that $<i_1, i_2, \ldots, i_n>$ is a permutation for which $a_{1i_1} a_{2i_2} \cdots$ $a_{ni_n} \neq 0$. Since $a_{1i_1} \neq 0$, $i_1 = 1$.

Notice that a_{11} is the only non-zero element in the first row, i.e. $a_{1j} = 0$, $j \neq 1$.

Since $a_{2i_2} \neq 0$ and $i_2 \neq i_1 = 1$, $i_2 = 2$. Similarly, $i_3 = 3$, $i_4 = 4, \ldots, i_n = n$. Thus, the only non-zero term in

$$\det A = \Sigma \text{ sgn } <i_1, \ldots, i_n> a_{1i_1} a_{2i_2} \cdots a_{ni_n} ,$$

is $a_{11} a_{22} \cdots a_{nn}$. So, $\det A = a_{11} a_{22} \cdots a_{nn}$ equals the product of the diagonal entries.

Evaluate det A where:

$$A = \begin{bmatrix} 0 & 1 & 5 \\ 3 & -6 & 9 \\ 2 & 6 & 1 \end{bmatrix}$$

Solution: Interchange the first and second rows of matrix A, obtaining matrix

$$B = \begin{bmatrix} 3 & -6 & 9 \\ 0 & 1 & 5 \\ 2 & 6 & 1 \end{bmatrix} \quad ;$$

and by the properties of the function,

$$\det A = -\det B .$$

$$= - \det \begin{bmatrix} 3 & -6 & 9 \\ 0 & 1 & 5 \\ 2 & 6 & 1 \end{bmatrix}$$

or

$$\det A = -3 \det \begin{bmatrix} 1 & -2 & 3 \\ 0 & 1 & 5 \\ 2 & 6 & 1 \end{bmatrix} .$$

A common factor of 3 from the first row of the matrix B was taken out. Add -2 times the first row to the third row. The value of the determinant of A will remain the same.
Thus,

$$\det A = -3 \det \begin{bmatrix} 1 & -2 & 3 \\ 0 & 1 & 5 \\ 0 & 10 & -5 \end{bmatrix} .$$

Add -10 times the second row to the third row.
Thus,

$$\det A = -3 \det \begin{bmatrix} 1 & -2 & 3 \\ 0 & 1 & 5 \\ 0 & 0 & -55 \end{bmatrix} .$$

As we know, the determinant of a triangular matrix is equal to the product of the diagonal elements.
Thus,

$$\det A = (-3) \cdot (1) \cdot (1) \cdot (-55) = 165 .$$

Compute the determinant of

$$A = \begin{bmatrix} 1 & 0 & 0 & 3 \\ 2 & 7 & 0 & 6 \\ 0 & 6 & 3 & 0 \\ 7 & 3 & 1 & -5 \end{bmatrix}.$$

Solution: First list the basic properties of the determinant. Let A be a square matrix.

(1)   If we interchange two rows (columns) of A, the determinant changes by a factor of (-1).

(2)   By adding to one row (column) of A a multiple of another row (column), the determinant is not changed.

(3)   If A is triangular, i.e., A has zeros above or below the diagonal, the determinant of A is the product of the diagonal elements of A.

(4)   If a row (column) of A is multiplied by a scalar K, but all of the other rows (columns) are left unchanged, then the determinant of the resulting matrix is K times the determinant of A.

Given:

$$A = \begin{bmatrix} 1 & 0 & 0 & 3 \\ 2 & 7 & 0 & 6 \\ 0 & 6 & 3 & 0 \\ 7 & 3 & 1 & -5 \end{bmatrix}.$$

Now add -3 times the first column to the fourth column. But the value of the determinant A will not be changed.

Thus,

$$A = \begin{bmatrix} 1 & 0 & 0 & 0 \\ 2 & 7 & 0 & 0 \\ 0 & 6 & 3 & 0 \\ 7 & 3 & 1 & -26 \end{bmatrix}$$

Now the above matrix A is a lower triangular matrix. Using property (3), yields:

$$\det A = (1) \cdot (3) \cdot (-26) = -546 .$$

Find the cofactors of the matrix A where:

$$A = \begin{bmatrix} 1 & 2 & 3 \\ 3 & 2 & 1 \\ 2 & 0 & 2 \end{bmatrix}.$$

Solution: If A is an $n \times n$ matrix, then M_{ij} will denote the $(n-1) \times (n-1)$ matrix obtained from A by deleting its ith row and jth column. The determinant $|M_{ij}|$ is called the minor of the element a_{ij} of A, and we define the cofactor of a_{ij}; denoted by A_{ij}, as:

$$A_{ij} = (-1)^{i+j} |M_{ij}| .$$

$$A = \begin{bmatrix} 1 & 2 & 3 \\ 3 & 2 & 1 \\ 2 & 0 & 2 \end{bmatrix} .$$

Then

$$A_{11} = (-1)^2 \begin{vmatrix} 2 & 1 \\ 0 & 2 \end{vmatrix} = 4 - 0 = 4$$

$$A_{12} = (-1)^{1+2} \begin{vmatrix} 3 & 1 \\ 2 & 2 \end{vmatrix} = -(6-2) = -4$$

$$A_{13} = (-1)^{1+3} \begin{vmatrix} 3 & 2 \\ 2 & 0 \end{vmatrix} = +(0-4) = -4$$

$$A_{21} = (-1)^{2+1} \begin{vmatrix} 2 & 3 \\ 0 & 2 \end{vmatrix} = -(4-0) = -4$$

$$A_{22} = (-1)^{2+2} \begin{vmatrix} 1 & 3 \\ 2 & 2 \end{vmatrix} = +(2-6) = -4$$

$$A_{23} = (-1)^{2+3} \begin{vmatrix} 1 & 2 \\ 2 & 0 \end{vmatrix} = -(0-4) = 4$$

$$A_{31} = (-1)^{3+1} \begin{vmatrix} 2 & 3 \\ 2 & 1 \end{vmatrix} = +(2-6) = -4$$

$$A_{32} = (-1)^{3+2} \begin{vmatrix} 1 & 3 \\ 3 & 1 \end{vmatrix} = -1(1-9) = 8$$

$$A_{33} = (-1)^{3+3} \begin{vmatrix} 1 & 2 \\ 3 & 2 \end{vmatrix} = +(2-6) = -4$$

Thus, the cofactors of A are:

$$A_{11} = 4 \qquad A_{12} = -4 \qquad A_{13} = -4$$

$$A_{21} = -4 \qquad A_{22} = -4 \qquad A_{23} = 4$$

$$A_{31} = -4 \qquad A_{32} = 8 \qquad A_{33} = -4 .$$

● **PROBLEM** 5-53

a) Given:
$$A = \begin{bmatrix} 5 & 2 & -1 \\ 3 & 1 & 2 \\ 2 & 7 & 4 \end{bmatrix} , \quad \text{find det A .}$$

b) If
$$A = \begin{bmatrix} 2 & 3 & 1 \\ -2 & 4 & 5 \\ 2 & 0 & 7 \end{bmatrix} , \quad \text{find adj A .}$$

<u>Solution</u>:   The definition of the determinant in terms of permutations is a mathematical one. For computational purposes, more efficient evaluative procedures are available.

Define the cofactor aij of an $n \times n$ determinant $|A|$ as $(-1)^{i+j} |C|$ where $|C|$ is the $(n-1) \times (n-1)$ sub-determinant obtained by deleting the ith row and jth column of $|A|$.

Suppose that $A = (a_{ij})$ is an $n \times n$ matrix, and let A_{ij} denote the cofactor of a_{ij} , $i = 1, 2, \ldots, n$; $j = 1, 2, \ldots, n$. Then,

$$\det A = \sum_{k=1}^{n} a_{kj} A_{kj} , \quad j = 1, 2, \ldots, n ,$$

and

$$\det A = \sum_{k=1}^{n} a_{ik} A_{ik} , \quad i = 1, 2, \ldots, n .$$

Note that we sum across k to obtain det A. Since $j = 1, 2, \ldots, n$, there are 2n ways of computing det A, one for each row (or column) of elements.

$$A = \begin{bmatrix} 5 & 2 & -1 \\ 3 & 1 & 2 \\ 2 & 7 & 4 \end{bmatrix}$$

The cofactors along the first column are:

$$A_{11} = (-1)^{1+1} \begin{vmatrix} 1 & 2 \\ 7 & 4 \end{vmatrix} = + (4-14) = -10$$

$$A_{21} = (-1)^{2+1} \begin{vmatrix} 2 & -1 \\ 7 & 4 \end{vmatrix} = -(8+7) = -15$$

$$A_{31} = (-1)^{3+1} \begin{vmatrix} 2 & -1 \\ 1 & 2 \end{vmatrix} = +(4+1) = 5 .$$

Hence,

$$\det A = \sum_{k=1}^{n} a_{k1} A_{k1} = a_{11} A_{11} + a_{21} A_{21} + a_{31} A_{31}$$

$$= 5(-10) + 3(-15) + 2(5)$$
$$= -50 - 45 + 10$$
$$= -85 .$$

Also, expanding along the 2nd row,

$$A_{21} = -15, \; A_{22} = 22, \; A_{23} = -31 .$$

Therefore,

$$\det A = a_{21} A_{21} + a_{22} A_{22} + a_{23} A_{23}$$

$$= 3(-15) + 1(22) + 2(-31)$$
$$= -45 + 22 - 62$$
$$= -85 .$$

b) Let A be a $n \times n$ matrix. Then the adjoint of A, adj A, is defined as the matrix of transposed cofactors. Let $C_{11}, C_{12}, \ldots, C_{1n}$,

$C_{21}, \ldots, C_{2n}, \ldots, C_{n1}, \ldots, C_{nn}$ denote the cofactors of $a_{11}, a_{12}, \ldots, a_{nn}$, respectively. Then

$$\text{adj. A} = \begin{bmatrix} C_{11} & C_{21} & \cdots & C_{n1} \\ C_{12} & & & \\ \cdot & & & \\ \cdot & & & \\ \cdot & & & \\ C_{1n} & \cdots & & C_{nn} \end{bmatrix}$$

The adjoint is useful in finding the inverse of a non-singular matrix. The cofactors of A are:

$$A_{11} = + \begin{vmatrix} 4 & 5 \\ 0 & 7 \end{vmatrix} = (28.0) = 28$$

$$A_{12} = - \begin{vmatrix} -2 & 5 \\ 2 & 7 \end{vmatrix} = -(-14-10) = 24$$

$$A_{13} = + \begin{vmatrix} -2 & 4 \\ 2 & 0 \end{vmatrix} = (0-8) = -8$$

$$A_{21} = - \begin{vmatrix} 3 & 1 \\ 0 & 7 \end{vmatrix} = -(21-0) = -21$$

$$A_{22} = + \begin{vmatrix} 2 & 1 \\ 2 & 7 \end{vmatrix} = (14-2) = 12$$

$$A_{23} = - \begin{vmatrix} 2 & 3 \\ 2 & 0 \end{vmatrix} = -(0-6) = 6$$

$$A_{31} = + \begin{vmatrix} 3 & 1 \\ 4 & 5 \end{vmatrix} = (15-4) = 11$$

$$A_{32} = - \begin{vmatrix} 2 & 1 \\ -2 & 5 \end{vmatrix} = -(10+2) = -12$$

$$A_{33} = + \begin{vmatrix} 2 & 3 \\ -2 & 4 \end{vmatrix} = (8+6) = 14 \quad .$$

The matrix of cofactors is:

$$\begin{bmatrix} 28 & 24 & -8 \\ -21 & 12 & 6 \\ 11 & -12 & 14 \end{bmatrix}$$

and the adj of A is:

$$\text{Adj } [A] = \begin{bmatrix} 28 & -21 & 11 \\ 24 & 12 & -12 \\ -8 & 6 & 14 \end{bmatrix} .$$

● PROBLEM 5-54

Find the adjoint of the matrix A, where:

$$A = \begin{bmatrix} 3 & 2 & -1 \\ 1 & 6 & 3 \\ 2 & -4 & 0 \end{bmatrix} .$$

Solution: The transpose of the matrix of cofactors of the elements a_{ij} of A, which is denoted by adj. A, is called the adjoint of A.

The cofactor of a_{ij} of A, denoted by A_{ij} is

$$A_{ij} = (-1)^{i+j} |M_{ij}| ,$$

where M_{ij} is the (n-1) square submatrix of A, obtained by deleting its ith row and jth column. The matrix of cofactors is:

$$\begin{bmatrix} A_{11} & A_{12} \cdots A_{1n} \\ A_{21} & A_{22} \cdots A_{2n} \\ A_{n1} & A_{n2} \cdots A_{nn} \end{bmatrix} .$$

Thus,

$$adj. A = \begin{bmatrix} A_{11} & A_{21} \cdots A_{n1} \\ A_{12} & A_{22} \cdots A_{n2} \\ \vdots \\ A_{1n} & A_{2n} \cdots A_{nn} \end{bmatrix} .$$

Now,

$$A = \begin{bmatrix} 3 & 2 & -1 \\ 1 & 6 & 3 \\ 2 & -4 & 0 \end{bmatrix} .$$

The cofactors of A are:

$$A_{11} = (-1)^{1+1} \begin{vmatrix} 6 & 3 \\ -4 & 0 \end{vmatrix} = + (0+12) = 12$$

$$A_{12} = (-1)^{1+2} \begin{vmatrix} 1 & 3 \\ 2 & 0 \end{vmatrix} = -(0-6) = 6$$

$$A_{13} = (-1)^{1+3} \begin{vmatrix} 1 & 6 \\ 2 & -4 \end{vmatrix} = +(-4-12) = -16$$

$$A_{21} = (-1)^{2+1} \begin{vmatrix} 2 & -1 \\ -4 & 0 \end{vmatrix} = -(0-4) = 4$$

$$A_{22} = (-1)^{2+2} \begin{vmatrix} 3 & -1 \\ 2 & 0 \end{vmatrix} = +(0+2) = +2$$

$$A_{23} = (-1)^{2+3} \begin{vmatrix} 3 & 2 \\ 2 & -4 \end{vmatrix} = -(-12-4) = 16$$

$$A_{31} = (-1)^{3+1} \begin{vmatrix} 2 & -1 \\ 6 & 3 \end{vmatrix} = +(6+6) = 12$$

$$A_{32} = (-1)^{3+2} \begin{vmatrix} 3 & -1 \\ 1 & 3 \end{vmatrix} = -(9+1) = -10$$

$$A_{33} = (-1)^{3+3} \begin{vmatrix} 3 & 2 \\ 1 & 6 \end{vmatrix} = +(18-2) = 16 \ .$$

The matrix of cofactors is:

$$\begin{bmatrix} 12 & 6 & -16 \\ 4 & 2 & 16 \\ 12 & -10 & 16 \end{bmatrix} .$$

The adjoint of A is:

$$\text{adj.}(A) = \begin{bmatrix} 12 & 4 & 12 \\ 6 & 2 & -10 \\ -16 & 16 & 16 \end{bmatrix}$$

● **PROBLEM** 5-55

Find the adjoint A of the following matrices:

(a) $\quad A = \begin{bmatrix} a_{11} & a_{12} \\ a_{21} & a_{22} \end{bmatrix}$ (b) $\quad A = \begin{bmatrix} 1 & 0 & 5 \\ 2 & 1 & 0 \\ 0 & 4 & 0 \end{bmatrix}$

(c) $\quad A = \begin{bmatrix} \lambda_1 & \cdots & 0 \\ & \ddots & \\ 0 & & \lambda_n \end{bmatrix}$, $\quad \lambda_i \neq 0$, $\quad i = 1,2,\ldots,n$

Solution:a)The transpose of the matrix of cofactors of the elements, a_{ij} of A, is called the adjoint of A.

$$A = \begin{bmatrix} a_{11} & a_{12} \\ a_{21} & a_{22} \end{bmatrix} .$$

The cofactors of the four elements are:

$$A_{11} = a_{22} , \ A_{12} = -a_{21}$$
$$A_{21} = -a_{12}, \ A_{22} = a_{11} .$$

The matrix of the cofactors is:

$$\begin{bmatrix} a_{22} & -a_{21} \\ -a_{12} & a_{11} \end{bmatrix}$$

Thus,

$$\text{adj } A = \begin{bmatrix} a_{22} & -a_{12} \\ -a_{21} & a_{11} \end{bmatrix} .$$

224

(b)

$$A = \begin{bmatrix} 1 & 0 & 5 \\ 2 & 1 & 0 \\ 0 & 4 & 0 \end{bmatrix}$$

The cofactors of the nine elements are:

$$A_{11} = + \begin{vmatrix} 1 & 0 \\ 4 & 0 \end{vmatrix} = 1(0-0) = 0$$

$$A_{12} = - \begin{vmatrix} 2 & 0 \\ 0 & 0 \end{vmatrix} = 0$$

$$A_{13} = + \begin{vmatrix} 2 & 1 \\ 0 & 4 \end{vmatrix} = (8-0) = +8$$

$$A_{21} = - \begin{vmatrix} 0 & 5 \\ 4 & 0 \end{vmatrix} = -(0-20) = +20$$

$$A_{22} = + \begin{vmatrix} 1 & 5 \\ 0 & 0 \end{vmatrix} = 0$$

$$A_{23} = - \begin{vmatrix} 1 & 0 \\ 0 & 4 \end{vmatrix} = -(4-0) = -4$$

$$A_{31} = + \begin{vmatrix} 0 & 5 \\ 1 & 0 \end{vmatrix} = (0-5) = -5$$

$$A_{32} = - \begin{vmatrix} 1 & 5 \\ 2 & 0 \end{vmatrix} = -(0-10) = +10$$

$$A_{33} = + \begin{vmatrix} 1 & 0 \\ 2 & 1 \end{vmatrix} = (1-0) = 1$$

The matrix of the cofactors is:

$$\begin{bmatrix} 0 & 0 & 8 \\ 20 & 0 & -4 \\ -5 & +10 & 1 \end{bmatrix}$$

Thus,

$$\text{adj } A = \begin{bmatrix} 0 & 20 & -5 \\ 0 & 0 & +10 \\ 8 & -4 & 1 \end{bmatrix}$$

(c)

$$A = \begin{bmatrix} \lambda_1 & & 0 \\ & \ddots & \\ 0 & & \lambda_n \end{bmatrix}$$

Here, the matrix A is a diagonal matrix. We know,

$$A^{-1} = \frac{1}{\det A} (\text{adj } A)$$

or

$$\text{adj } A = |A| A^{-1}.$$

We also know that the inverse of a diagonal matrix is simply made up of the inverses of its non-diagonal elements. Now,

$$A = \begin{bmatrix} \lambda_1 & & 0 \\ & \ddots & \\ 0 & & \lambda_n \end{bmatrix}.$$

Then,

$$A^{-1} = \begin{bmatrix} \lambda_1^{-1} & 0 & \cdots & 0 \\ 0 & \lambda_2^{-1} & & \vdots \\ \vdots & & \ddots & \\ 0 & & & \lambda_n^{-1} \end{bmatrix},$$

providing $\lambda_i \neq 0$ $(i = 1, \ldots, n)$. Hence,

$$\text{adj } A = |A| \begin{bmatrix} \lambda_1^{-1} & 0 & \cdots & 0 \\ 0 & \lambda_2^{-1} & \cdots & 0 \\ \vdots & & \ddots & \\ 0 & 0 & & \lambda_n^{-1} \end{bmatrix}$$

Since $|A| = \lambda_1 \lambda_2 \cdots \lambda_n = \prod_{i=1}^{n} \lambda_i$,

$$\text{adj } A = \prod_{i=1}^{n} \lambda_i \begin{bmatrix} \dfrac{1}{\lambda_1} & 0 & \cdots & 0 \\ 0 & \dfrac{1}{\lambda_2} & & 0 \\ \vdots & & \ddots & \\ 0 & \cdots & & \dfrac{1}{\lambda_n} \end{bmatrix}$$

● **PROBLEM** 5-56

Given:
$$A = \begin{bmatrix} 3 & 1 & 2 \\ 0 & 1 & 1 \\ -1 & 1 & 0 \end{bmatrix}.$$

Show that $(\text{adj } A) \cdot A = (\det A)I$ where I is the identity matrix.

<u>Solution</u>: It is known that the classical adjoint, or adj A, is the transpose of the matrix of cofactors of the elements a_{ij} of A. The cofactors of the nine elements of the given matrix A^{-1} are:

$$A_{11} = + \begin{vmatrix} 1 & 1 \\ 1 & 0 \end{vmatrix} = (0-1) = -1$$

$$A_{12} = - \begin{vmatrix} 0 & 1 \\ -1 & 0 \end{vmatrix} = -(0+1) = -1$$

$$A_{13} = + \begin{vmatrix} 0 & 1 \\ -1 & 1 \end{vmatrix} = (0+1) = 1$$

$$A_{21} = - \begin{vmatrix} 1 & 2 \\ 1 & 0 \end{vmatrix} = -(0-2) = +2$$

$$A_{22} = + \begin{vmatrix} 3 & 2 \\ -1 & 0 \end{vmatrix} = (0+2) = 2$$

$$A_{23} = - \begin{vmatrix} 3 & 1 \\ -1 & 1 \end{vmatrix} = -(3+1) = -4$$

$$A_{31} = + \begin{vmatrix} 1 & 2 \\ 1 & 1 \end{vmatrix} = (1-2) = -1$$

$$A_{32} = - \begin{vmatrix} 3 & 2 \\ 0 & 1 \end{vmatrix} = -(3-0) = -3$$

$$A_{33} = + \begin{vmatrix} 3 & 1 \\ 0 & 1 \end{vmatrix} = (3-0) = 3 \ .$$

Then the matrix of the cofactors is:

$$\begin{bmatrix} -1 & -1 & 1 \\ 2 & 2 & -4 \\ -1 & -3 & 3 \end{bmatrix} \ .$$

Hence, from the above definition of adjoint

$$\text{adj } A = \begin{bmatrix} -1 & 2 & -1 \\ -1 & 2 & -3 \\ 1 & -4 & 3 \end{bmatrix} \ .$$

$$(\text{adj } A) \cdot A = \begin{bmatrix} -1 & 2 & -1 \\ -1 & 2 & -3 \\ 1 & -4 & 3 \end{bmatrix} \begin{bmatrix} 3 & 1 & 2 \\ 0 & 1 & 1 \\ -1 & 1 & 0 \end{bmatrix}$$

$$= \begin{bmatrix} -3+0+1 & -1+2-1 & -2+2+0 \\ -3+0+3 & -1+2-3 & -2+2+0 \\ 3+0-3 & 1-4+3 & 2-4+0 \end{bmatrix}$$

$$= \begin{bmatrix} -2 & 0 & 0 \\ 0 & -2 & 0 \\ 0 & 0 & -2 \end{bmatrix} \ .$$

$$\text{adj } A \cdot = -2 \begin{bmatrix} 1 & 0 & 0 \\ 0 & 1 & 0 \\ 0 & 0 & 1 \end{bmatrix} = -2I$$

$$\det A = \begin{vmatrix} 3 & 1 & 2 \\ 0 & 1 & 1 \\ -1 & 1 & 0 \end{vmatrix} \ .$$

$$\det A = 3 \begin{vmatrix} 1 & 1 \\ 1 & 0 \end{vmatrix} - 0 \begin{vmatrix} 1 & 2 \\ 1 & 0 \end{vmatrix} - 1 \begin{vmatrix} 1 & 2 \\ 1 & 1 \end{vmatrix}$$

$$= 3(0-1) -1 (1-2)$$

$$= -3 + 1 = -2 \ .$$

Hence,

$$\text{adj } A \cdot A = -2I = (\det A)I \ .$$

Compute the determinants of each of the following matrices and find which of the matrices are invertible.

(a) $\begin{bmatrix} 3 & 1 & 2 \\ 1 & 0 & 6 \\ -1 & 1 & 1 \end{bmatrix}$ (b) $\begin{bmatrix} -1 & 1 & 3 \\ 2 & 1 & 1 \\ 4 & 2 & 2 \end{bmatrix}$ (c) $\begin{bmatrix} 2 & 1 & 1 \\ 0 & 0 & 0 \\ 4 & 3 & 1 \end{bmatrix}$

Solution: We can evaluate determinants by using the basic properties of the determinant function.

Properties of Determinants:

(1) If each element in a row (or column) is zero, the value of the determinant is zero.

(2) If two rows (or columns) of a determinant are identical, the value of the determinant is zero.

(3) The determinant of a matrix A and its transpose A^t are equal: $|A| = |A^t|$.

(4) The matrix A has an inverse if and only if det $A \neq 0$.

$$\det A = \begin{vmatrix} 3 & 1 & 2 \\ 1 & 0 & 6 \\ -1 & 1 & 1 \end{vmatrix}$$

$$= 3\begin{vmatrix} 0 & 6 \\ 1 & 1 \end{vmatrix} - 1\begin{vmatrix} 1 & 6 \\ -1 & 1 \end{vmatrix} + 2\begin{vmatrix} 1 & 0 \\ -1 & 1 \end{vmatrix}$$

$$= 3(0-6) - 1(1+6) + 2(1-0)$$

$$= -18-7+2 = -23 .$$

Since det $A = -23 \neq 0$, this matrix is invertible.

(b)

$$A = \begin{bmatrix} -1 & 1 & 3 \\ 2 & 1 & 1 \\ 4 & 2 & 2 \end{bmatrix}$$

Here det $A = 0$, since the third row is a multiple of the second row. Since det $A = 0$, the matrix is not invertible.

(c)

$$A = \begin{bmatrix} 2 & 1 & 1 \\ 0 & 0 & 0 \\ 4 & 3 & 1 \end{bmatrix}$$

$$\det A = \begin{vmatrix} 2 & 1 & 1 \\ 0 & 0 & 0 \\ 4 & 3 & 1 \end{vmatrix} = 0.$$

Here, each element in the second row is zero, therefore, the value of the determinant is zero. Since det $A = 0$, the matrix is not invertible.

Find the value of $\begin{vmatrix} 67 & 19 & 21 \\ 39 & 13 & 14 \\ 81 & 24 & 26 \end{vmatrix}$

Solution: Our aim in this problem is to break down the given determinant into one that is easier to evaluate. We can therefore rewrite our determinant as:

$$\begin{vmatrix} 67 & 19 & 21 \\ 39 & 13 & 14 \\ 81 & 24 & 26 \end{vmatrix} = \begin{vmatrix} 10+57 & 19 & 21 \\ 0+39 & 13 & 14 \\ 9+72 & 24 & 26 \end{vmatrix}.$$

Now we can make use of one of the well-known properties of determinants; that is, if each element of a column of a determinant is expressed as the sum of two terms, the determinant can be expressed as the sum of two determinants. Thus,

$$\begin{vmatrix} 10+57 & 19 & 21 \\ 0+39 & 13 & 14 \\ 9+72 & 24 & 26 \end{vmatrix} = \begin{vmatrix} 10 & 19 & 21 \\ 0 & 13 & 14 \\ 9 & 24 & 26 \end{vmatrix} + \begin{vmatrix} 57 & 19 & 21 \\ 39 & 13 & 14 \\ 72 & 24 & 26 \end{vmatrix}.$$

The determinant can again be simplified further. Let us examine the second determinant in the above sum. Remember that multiplying each element in a column of a determinant by a number and adding that product to the corresponding elements in another column does not change the value of the determinant. Therefore, we can perform this on the determinant using -3 as the number, and adding the product of -3 and the elements of column two to the corresponding elements of column one. Thus, we obtain:

$$\begin{vmatrix} 57 & 19 & 21 \\ 39 & 13 & 14 \\ 72 & 24 & 26 \end{vmatrix} = \begin{vmatrix} 57+(-3)(19) & 19 & 21 \\ 39+(-3)(13) & 13 & 14 \\ 72+(-3)(24) & 24 & 26 \end{vmatrix}$$

$$= \begin{vmatrix} 0 & 19 & 21 \\ 0 & 13 & 14 \\ 0 & 24 & 26 \end{vmatrix}.$$

Now, since each element in a column of a determinant is zero, the value of the determinant is zero. Thus, the value of the second determinant in the above sum is zero, and we have:

$$\begin{vmatrix} 67 & 19 & 21 \\ 39 & 13 & 14 \\ 81 & 24 & 26 \end{vmatrix} = \begin{vmatrix} 10 & 19 & 21 \\ 0 & 13 & 14 \\ 9 & 24 & 26 \end{vmatrix} .$$

But, this can be rewritten as:

$$\begin{vmatrix} 10 & 19 & 19+2 \\ 0 & 13 & 13+1 \\ 9 & 24 & 24+2 \end{vmatrix} = \begin{vmatrix} 10 & 19 & 19 \\ 0 & 13 & 13 \\ 9 & 24 & 24 \end{vmatrix} + \begin{vmatrix} 10 & 19 & 2 \\ 0 & 13 & 1 \\ 9 & 24 & 2 \end{vmatrix} .$$

If two columns of a determinant have the same elements, then its value is zero. Thus the first determinant in the above sum is zero, and we are left with:

$$\begin{vmatrix} 10 & 19 & 2 \\ 0 & 13 & 1 \\ 9 & 24 & 2 \end{vmatrix} .$$

We now use minors to determine the value of the determinant. Let us choose column one, and call its elements a_1, a_2, a_3. Then their corresponding minors are A_1, A_2, A_3. We form the products a_1A_1, a_2A_2, a_3A_3. Since a_1 is in the first row and the first column, and $1 + 1 = 2$, which is even, the sign of a_1A_1 is positive. Similarly, the sign of a_2A_2 is negative, and that of a_3A_3 is positive. Thus, we have:

$a_1A_1 - a_2A_2 + a_3A_3$, and substituting we obtain:

$10A_1 - 0A_2 + 9A_3$. The second term vanishes. We find the minors A_1 and A_3 by eliminating from the determinant the row and column that a_1 and a_3 are found in. Thus,

$$A_1 = \begin{vmatrix} 13 & 1 \\ 24 & 2 \end{vmatrix}, \qquad A_3 = \begin{vmatrix} 19 & 2 \\ 13 & 1 \end{vmatrix}$$

and

$$\begin{vmatrix} 67 & 19 & 21 \\ 39 & 13 & 14 \\ 81 & 24 & 26 \end{vmatrix} = \begin{vmatrix} 10 & 19 & 2 \\ 0 & 13 & 1 \\ 9 & 24 & 2 \end{vmatrix}$$

$$= 10 \begin{vmatrix} 13 & 1 \\ 24 & 2 \end{vmatrix} + 9 \begin{vmatrix} 19 & 2 \\ 13 & 1 \end{vmatrix} .$$

Now, these two determinants are easily evaluated.
The first,

$$\begin{vmatrix} 13 & 1 \\ 24 & 2 \end{vmatrix} = (13)(2) - (1)(24) = 26 - 24 = 2,\text{ and the second}$$

$$\begin{vmatrix} 19 & 2 \\ 13 & 1 \end{vmatrix} = (19)(1) - (2)(13) = 19 - 26 = -7.$$

Thus,

$$\begin{vmatrix} 67 & 19 & 21 \\ 39 & 13 & 14 \\ 81 & 24 & 26 \end{vmatrix} = 10(2) + 9(-7) = 20 - 63 = -43.$$

Another way to approach this problem is to use the expansion scheme for determinants of third order. Using this method we rewrite the given determinant as follows:

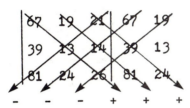

We multiply the elements falling on the same diagonal, thus obtaining six terms. The three terms on the lines sloping downward from left to right have a positive value, and the three on the lines sloping downward from right to left have a negative value. Upon expanding we obtain:

$$(67)(13)(26) + (19)(14)(81) + (21)(39)(24) - (21)(13)(81)$$

$$- (67)(14)(24) - (19)(39)(26).$$

Performing the indicated operations should give us the same value obtained using our previous method, that is - 43.

The advantage of the first method is that it does not involve a long multiplication process.

● **PROBLEM** 5-59

Without expanding, prove that

$$D = \begin{vmatrix} x & y & 2x \\ z & w & 2z \\ u & v & 2u \end{vmatrix} = 0$$

<u>Solution:</u> If the elements of a column (row) of a determinant are multiplied by any number m, the determinant is multiplied by m. Therefore,

$$D = \begin{vmatrix} x & y & 2x \\ z & w & 2z \\ u & v & 2u \end{vmatrix} = 2 \begin{vmatrix} x & y & x \\ z & w & z \\ u & v & u \end{vmatrix}$$

since the elements of the third column of D are multiples of 2 .
Since two rows (two columns) of this determinant are identical, the value of

$$D = \begin{vmatrix} x & y & 2x \\ z & w & 2z \\ u & v & 2u \end{vmatrix} = 2 \begin{vmatrix} x & y & x \\ z & w & z \\ u & v & u \end{vmatrix} = 0$$

(the first and third columns are identical)

● **PROBLEM** 5-60

If
$$D_1 = \begin{vmatrix} a & b & c \\ d & e & f \\ g & h & k \end{vmatrix}, \quad D_2 = \begin{vmatrix} a & g & x \\ b & h & y \\ c & k & z \end{vmatrix}$$

and d = tx, e = ty, f = tz, prove without expanding that $D_1 = -tD_2$.

<u>Solution:</u> In D_1 if we replace d by tx, e by ty and f by tz, we have

$$D_1 = \begin{vmatrix} a & b & c \\ tx & ty & tz \\ g & h & k \end{vmatrix}$$

If the elements of a row (a column) of a determinant are multiplied by any number t, the determinant is multiplied by t. Therefore

$$D_1 = \begin{vmatrix} a & b & c \\ tx & ty & tz \\ g & h & k \end{vmatrix} = \begin{vmatrix} a & b & c \\ t(x) & t(y) & t(z) \\ g & h & k \end{vmatrix}$$

$$= t \begin{vmatrix} a & b & c \\ x & y & z \\ g & h & k \end{vmatrix}$$

$$= t \begin{vmatrix} a & x & g \\ b & y & h \\ c & z & k \end{vmatrix} \quad \text{since the rows of this deter-minant are the columns of the one just above.}$$

Now, since interchanging two columns (rows) of a determinant changes the sign of the determinant, interchanging columns 2 and 3 gives us:

$$D_1 = -t \begin{vmatrix} a & g & x \\ b & h & y \\ c & k & z \end{vmatrix} \qquad \text{But} \qquad \begin{vmatrix} a & g & x \\ b & h & y \\ c & k & z \end{vmatrix} = D_2;$$

hence, $D_1 = -tD_2$.

● **PROBLEM** 5-61

Show that $\begin{vmatrix} b + c & a - b & a \\ c + a & b - c & b \\ a + b & c - a & c \end{vmatrix} = 3abc - a^3 - b^3 - c^3.$

Solution: The following is a known property of deter-
minants: $\begin{vmatrix} a_1 + \overline{a}_1 & b_1 & c_1 \\ a_2 + \overline{a}_2 & b_2 & c_2 \\ a_3 + \overline{a}_3 & b_3 & c_3 \end{vmatrix} =$

$\begin{vmatrix} a_1 & b_1 & c_1 \\ a_2 & b_2 & c_2 \\ a_3 & b_3 & c_3 \end{vmatrix} + \begin{vmatrix} \overline{a}_1 & b_1 & c_1 \\ \overline{a}_2 & b_2 & b_2 \\ \overline{a}_3 & b_3 & c_3 \end{vmatrix}.$ Notice that in our given

determinant there are two columns with their elements expressed as the sum of two terms. We will first apply the
above property to the first column. We thus obtain:

$$\begin{vmatrix} b + c & a - b & a \\ c + a & b - c & b \\ a + b & c - a & c \end{vmatrix} = \begin{vmatrix} b & a - b & a \\ c & b - c & b \\ a & c - a & c \end{vmatrix} + \begin{vmatrix} c & a - b & a \\ a & b - c & b \\ b & c - a & c \end{vmatrix}.$$

Now, applying this property to the second column
of both determinants on the right side of the equal
sign we obtain:

$$\begin{vmatrix} b & a & a \\ c & b & b \\ a & c & c \end{vmatrix} + \begin{vmatrix} b & - b & a \\ c & - c & b \\ a & - a & c \end{vmatrix} + \begin{vmatrix} c & a & a \\ a & b & b \\ b & c & c \end{vmatrix} + \begin{vmatrix} c & - b & a \\ a & - c & b \\ b & - a & c \end{vmatrix}.$$

But, if each element in a column of a determinant is
multiplied by a number p, in this case p = - 1, then
the value of the determinant is multiplied by p. That
is,

233

$$\begin{vmatrix} a_1 & pb_1 & c_1 \\ a_2 & pb_2 & c_2 \\ a_3 & pb_3 & c_3 \end{vmatrix} = p \begin{vmatrix} a_1 & b_1 & c_1 \\ a_2 & b_2 & c_2 \\ a_3 & b_3 & c_3 \end{vmatrix}.$$

Thus, our above determinants become,

$$= \begin{vmatrix} b & a & a \\ c & b & b \\ a & c & c \end{vmatrix} - \begin{vmatrix} b & b & a \\ c & c & b \\ a & a & c \end{vmatrix} + \begin{vmatrix} c & a & a \\ a & b & b \\ b & c & c \end{vmatrix} - \begin{vmatrix} c & b & a \\ a & c & b \\ b & a & c \end{vmatrix}.$$

Recall that when two columns of a determinant are identical, the value of the determinant is zero. Thus, the first three determinants vanish and we are left with

$$- \begin{vmatrix} c & b & a \\ a & c & b \\ b & a & c \end{vmatrix}.$$

To evaluate this third order determinant we employ the following method: rewrite the first two columns of the determinant next to the third column, obtaining:

$$- \begin{vmatrix} c & b & a \\ a & c & b \\ b & a & c \end{vmatrix} \begin{matrix} c & b \\ a & c \\ b & a \end{matrix}.$$

Draw three diagonal lines sloping downward from left to right, each of which encompasses three elements of the determinant. Do this also from right to left.

The diagram now looks like: -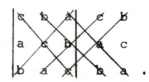

We now form the products of the elements in each of the six diagonals, preceding each of the terms in the left to right diagonals by a positive sign, and each of the terms in the right to left diagonals by a negative sign. The sum of the six products is the required expansion of the determinant. Thus, we obtain:

$$- (c \cdot c \cdot c + b \cdot b \cdot b + a \cdot a \cdot a - acb - cba - bac) =$$

$$- c^3 - b^3 - a^3 + 3abc = 3abc - a^3 - b^3 - c^3.$$

i Define the absolute determinant.

ii If
    (a) $A = \begin{bmatrix} a_{11} \\ a_{21} \end{bmatrix}$,   (b) $A = \begin{bmatrix} 1 & 3 \\ 2 & 0 \\ 1 & 4 \end{bmatrix}$,

find the absolute determinant of A .

Solution: Definition: The absolute determinant, A, of a m×n matrix A, n ≤ m, is given by the equation:

$$\text{absolute} \quad \det A = (\det A^t A)^{\frac{1}{2}} .$$

The absolute determinant formula is not applied to the case n > m , because, then, it would always give a value of zero.

(a) $A = \begin{bmatrix} a_{11} \\ a_{21} \end{bmatrix}$ then $A^t = \begin{bmatrix} a_{11} & a_{21} \end{bmatrix}$ and,

$$A^t A = \begin{bmatrix} a_{11} & a_{21} \end{bmatrix} \begin{bmatrix} a_{11} \\ a_{21} \end{bmatrix} = a_{11}^2 + a_{21}^2 .$$

Hence,
$$\text{absolute } \det A = (\det A^t A)^{\frac{1}{2}}$$
$$= (a_{11}^2 + a_{21}^2)^{\frac{1}{2}} .$$

(b) $A = \begin{bmatrix} 1 & 3 \\ 2 & 0 \\ 1 & 4 \end{bmatrix}$.

Then,
$$A^t = \begin{bmatrix} 1 & 2 & 1 \\ 3 & 0 & 4 \end{bmatrix}$$

$$A^t A = \begin{bmatrix} 1 & 2 & 1 \\ 3 & 0 & 4 \end{bmatrix} \begin{bmatrix} 1 & 3 \\ 2 & 0 \\ 1 & 4 \end{bmatrix}$$

$$= \begin{bmatrix} 1+4+1 & 3+0+4 \\ 3+0+4 & 9+0+16 \end{bmatrix}$$

$$= \begin{bmatrix} 6 & 7 \\ 7 & 25 \end{bmatrix} .$$

Hence,
$$\text{absolute } \det A = (\det A^t A)^{\frac{1}{2}}$$
$$= \left(\det \begin{vmatrix} 6 & 7 \\ 7 & 25 \end{vmatrix} \right)^{\frac{1}{2}}$$
$$= [150 - 49]^{\frac{1}{2}}$$
$$= (101)^{\frac{1}{2}} = \sqrt{101} .$$

Define submatrix and subdeterminant.

Solution: The totality of $m \times n$ matrices (i.e., m rows and n columns) with entries in R will be designated by $M_{m,n}(R)$.

Let r and n be positive integers. Define $\Gamma_{r,n}$ to be the totality of sequences α, $\alpha = (\alpha_1, \ldots, \alpha_r)$, in which each α_1 is an integer satisfying $1 \le \alpha_1 \le n$. Next, let $G_{r,n}$ be the subset of $\Gamma_{r,n}$ consisting of precisely those sequences α for which $1 \le \alpha_1 \le \alpha_2 \le \ldots \le \alpha_r \le n$. Finally, if $r \le n$, then $Q_{r,n}$ will denote the subset of $G_{r,n}$ consisting of those α for which $1 \le \alpha_1 < \ldots < \alpha_r \le n$. Thus, $\Gamma_{r,n}$ is the set of all sequences of length r chosen from $1, \ldots, n$; $G_{r,n}$ is the set of non-decreasing sequences in $\Gamma_{r,n}$ and $Q_{r,n}$ is the set of strictly increasing sequences in $G_{r,n}$.

Let R be a field, and $A \in M_{m,n}(R)$. Let $\alpha \in \Gamma_{r,m}$ and $\beta \in \Gamma_{s,n}$. Then, $A[\alpha|\beta]$ is the matrix in $M_{r,s}(R)$ whose (ij) entry is $a_{\alpha_i \beta_j}$; $i = 1, \ldots, r$, $j = 1, \ldots, s$. **Definition**: If α and β are in $Q_{r,m}$ and $Q_{s,n}$ respectively, then $A[\alpha|\beta]$ is called a submatrix of A. The determinant of a square submatrix is called a subdeterminant. For $\alpha \in Q_{r,m}$ and $\beta \in Q_{s,n}$, let α' and β' denote the sequences in $Q_{m-r,m}$ and $Q_{n-s,n}$ whose integers are complementary to α, and β, respectively.

Then, define: $A(\alpha|\beta) = A[\alpha'|\beta']$, $A[\alpha|\beta) = A[\alpha|\beta']$, $A(\alpha|\beta) = A[\alpha'|\beta]$. If $m = n$, $r = s$ and $\alpha \in Q_{r,n}$ then $A[\alpha|\alpha]$ is called a principal submatrix of A. The determinant of a principal submatrix is called a principal subdeterminant. For example, let:

$$A = \begin{bmatrix} a_{11} & a_{12} & a_{13} \\ a_{21} & a_{22} & a_{23} \\ a_{31} & a_{32} & a_{33} \end{bmatrix}$$

and $\alpha = (2,2,3) \in \Gamma_{3,3}$, $\beta = (1,1) \in \Gamma_{2,3}$. Then

$$A[\alpha|\beta] = A[2,2,3|1,1] = \begin{bmatrix} a_{21} & a_{21} \\ a_{21} & a_{21} \\ a_{31} & a_{31} \end{bmatrix}$$

Again, if $\alpha = (1,2) \in Q_{2,3}$, $\beta = (2,3) \in Q_{2,3}$, then

$$A[\alpha|\beta] = \begin{bmatrix} a_{12} & a_{13} \\ a_{22} & a_{23} \end{bmatrix},$$

$$A[\alpha|\beta] = A[\alpha'|\beta] = A[3|2,3] = [a_{32} \quad a_{33}] ,$$

and

$$A(\alpha|\beta) = A[\alpha'|\beta'] = A[3|1] = [a_{31}] .$$

Define the Wronskian.

Solution: The Wronskian of two functions at $x = x_0$ is defined as the following determinant:

$$W(f,g,x_0) = \det \begin{bmatrix} f(x_0) & g(x_0) \\ f'(x_0) & g'(x_0) \end{bmatrix}$$

If f and g are dependent functions, then there are numbers c_1 and c_2 not both zero such that

$$c_1 f + c_2 g = 0 .$$

Differentiation gives:

$$c_1 f' + c_2 g' = 0 .$$

c_1 and c_2 are not both 0 so we know the system of homogeneous equations

$$c_1 f + c_2 g = 0$$
$$c_1 f' + c_2 g' = 0$$

has a non-zero solution.

But a system of homogeneous equations has a non-zero solution if and only if the determinant of the coefficient matrix is 0. Therefore,

$$W(f,g,x_0) = \det \begin{bmatrix} f(x_0) & g(x_0) \\ f'(x_0) & g'(x_0) \end{bmatrix} = 0$$

so we obtain the following result:

If f and g are dependent, then for each x_0, $W(f,g,x_0) = 0$. In general, the converse of this is not true. The Wronskian for more than two functions is defined analogously. For example,

$$w(f,g,h,x_0) = \det \begin{bmatrix} f(x_0) & g(x_0) & h(x_0) \\ f'(x_0) & g'(x_0) & h'(x_0) \\ f''(x_0) & g''(x_0) & h''(x_0) \end{bmatrix} .$$

MATRICES AND SYSTEMS OF EQUATIONS, CRAMMER'S RULE

● **PROBLEM** 5-65

Show that the following homogeneous system of equations has a unique solution:

$$x_1 + 2x_2 + x_3 = 0$$
$$x_2 - 3x_3 = 0$$
$$-x_1 + x_2 - x_3 = 0 .$$

Solution: A homogeneous system of linear equations has either   a)   the unique trivial solution $x_1 = x_2 = \ldots = x_n = 0$ or   b)   an infinite number of non-trivial solutions, plus the trivial solution.

First, write the above equations in matrix notation:

$$\begin{bmatrix} 1 & 2 & 1 \\ 0 & 1 & -3 \\ -1 & 1 & -1 \end{bmatrix} \begin{bmatrix} x_1 \\ x_2 \\ x_3 \end{bmatrix} = \begin{bmatrix} 0 \\ 0 \\ 0 \end{bmatrix}$$

or, $AX = 0$.

$$\det A = 1 \begin{vmatrix} 1 & -3 \\ 1 & -1 \end{vmatrix} - 2 \begin{vmatrix} 0 & -3 \\ -1 & -1 \end{vmatrix} + 1 \begin{vmatrix} 0 & 1 \\ -1 & 1 \end{vmatrix}$$

$$= \left[(-1+3) - 2(0-3) + (0+1) \right]$$
$$= 2 + 6 + 1$$
$$= 9 .$$

Hence, $\det A \neq 0$, and according to Cramer's Rule, the above system has a unique solution.

Therefore, $X = 0$, i.e., $x_1 = x_2 = x_3 = 0$ since the homogeneous system $AX = 0$ has a non-zero solution if and only if $\det A = 0$.

● **PROBLEM** 5-66

Solve the following system of equations by forming the matrix of coefficients and reducing it to echelon form.

$$3x + 2y - z = 0$$
$$x - y + 2z = 0 \qquad\qquad (1)$$
$$x + y - 6z = 0$$

Solution: The most general linear system of   m   equations in   n   unknowns (variables) $x_1, x_2, \ldots, x_n$ is of the form

$$a_{11}x_1 + a_{12}x_2 + \ldots + a_{1n}x_n = c_1$$
$$a_{21}x_1 + a_{22}x_2 + \ldots + a_{2n}x_n = c_2 \qquad\qquad (2)$$
$$\cdot \quad \cdot \quad \cdot \quad \cdot \quad \cdot \quad \cdot \quad \cdot \quad \cdot \quad \cdot \quad \cdot \quad \cdot \quad \cdot \quad \cdot$$
$$a_{m1}x_1 + a_{m2}x_2 + \ldots + a_{mn}x_n = c_m ,$$

where $a_{11}, a_{12}, \ldots, a_{1n}$, $a_{21}, a_{22}, \ldots, a_{2n}$, $a_{m1}, a_{m2}, \ldots, a_{mn}$ are called the coefficients of the system (2), and $c_1, c_2, \ldots, c_m$ are called the constants of the system (2).

If $c_1 = c_2 = \ldots = c_m = 0$, a system of the form (2) is called a homogeneous system. Let $A = (a_{ij})$ be a matrix of size $m \times n$ where the entries a_{ij} are the same as the coefficients of the system given by (2). Then we say that A is the matrix of coefficients of (2). In order to solve the homogeneous system

$$a_{11}x_1 + a_{12}x_2 + \ldots + a_{1n}x_n = 0$$
$$a_{21}x_1 + a_{22}x_2 + \ldots + a_{2n}x_n = 0 \qquad (3)$$
$$\cdot$$
$$\cdot$$
$$\cdot$$
$$a_{m1}x_1 + a_{m2}x_2 + \ldots + a_{mn}x_n = 0 \quad ,$$

first form the matrix of coefficients $A = (a_{ij})$. Next, one performs a series of elementary row operations on A which transform it into a row-echelon matrix $R = (r_{ij})$. Since the solutions to (3) are the same as the solutions to

$$r_{11}x_1 + r_{12}x_2 + \ldots + r_{1n}x_n = 0$$
$$r_{21}x_1 + r_{22}x_2 + \ldots + r_{2n}x_n = 0$$
$$\cdot$$
$$\cdot \qquad (4)$$
$$\cdot$$
$$r_{m1}x_1 + r_{m2}x_2 + \ldots + r_{mn}x_n = 0$$

where the coefficients r_{ij} are identically the entries of R (i.e., R is the coefficient matrix of the homogeneous system (4)) we solve the original system (3) by solving the reduced system (4). Now solve the given homogeneous system (1).

The matrix of coefficients of the system (1) is

$$\begin{bmatrix} 3 & 2 & -1 \\ 1 & -1 & 2 \\ 1 & 1 & -6 \end{bmatrix} .$$

Reduce the above matrix to echelon form. Add -3 times the second row to the first row, and add -1 times the second row to the third row:

$$\begin{bmatrix} 0 & 5 & -7 \\ 1 & -1 & 2 \\ 0 & 2 & -8 \end{bmatrix}$$

Divide the third row by 2.

$$\begin{bmatrix} 0 & 5 & -7 \\ 1 & -1 & 2 \\ 0 & 1 & -4 \end{bmatrix}$$

Add -5 times the third row to the first row, and add the third row to

239

the second row.

$$\begin{bmatrix} 0 & 0 & 13 \\ 1 & 0 & -2 \\ 0 & 1 & -4 \end{bmatrix}$$

Divide row one by 13; then add 2 times the resulting row one to row two. Next, add 4 times the resulting row one to row three:

$$\begin{bmatrix} 0 & 0 & 1 \\ 1 & 0 & 0 \\ 0 & 1 & 0 \end{bmatrix}$$

Interchange rows one and two, then rows two and three:

$$\begin{bmatrix} 1 & 0 & 0 \\ 0 & 1 & 0 \\ 0 & 0 & 1 \end{bmatrix}$$

This matrix is reduced and gives the system

$$x = 0$$
$$y = 0$$
$$z = 0 \quad .$$

Thus, the unique solution to the original system is $x = y = z = 0$, the solution that is called the trivial solution.

● **PROBLEM** 5-67

By forming the augmented matrix and row reducing, determine the solutions of the following system

$$\begin{array}{rcl} 2x \ - \ y \ + \ 3z &=& 4 \\ 3x \qquad\ + \ 2z &=& 5 \\ -2x \ + \ y \ + \ 4z &=& 6 \ . \end{array} \qquad (1)$$

Solution: The system of equations

$$a_{11}x_1 + a_{12}x_2 + \cdots + a_{1n}x_n = c_1$$

$$a_{21}x_1 + a_{22}x_2 + \cdots + a_{2n}x_n = c_2$$

$$\vdots \qquad\qquad\qquad\qquad (2)$$

$$a_{m1}x_1 + a_{m2}x_2 + \cdots + a_{mn}x_n = c_m$$

is called a non-homogeneous linear system if the constants $c_1, c_2, \ldots, c_m$ are not all zero.

We form the $m \times (n+1)$ matrix A' defined by

$$A' = \begin{bmatrix} a_{11} & a_{12} & \cdots & a_{1n} & \vdots & c_1 \\ a_{21} & a_{22} & \cdots & a_{2n} & \vdots & c_2 \\ \vdots & & & & \vdots & \\ a_{m1} & a_{m2} & \cdots & a_{mn} & \vdots & c_m \end{bmatrix} \quad .$$

This matrix is called the augmented matrix of the system (2). The first n columns of A′ consist of the coefficient matrix of (2), and the last column of A′ consists of the corresponding constants.

To solve the non-homogeneous linear system, form the augmented matrix A′. Apply row operations to A′ to reduce it to echelon form. Now, the augmented matrix of the system (1) is

$$\begin{bmatrix} 2 & -1 & 3 & | & 4 \\ 3 & 0 & 2 & | & 5 \\ -2 & 1 & 4 & | & 6 \end{bmatrix} .$$

To begin the row reduction, add the first row to the third row

$$\begin{bmatrix} 2 & -1 & 3 & | & 4 \\ 3 & 0 & 2 & | & 5 \\ 0 & 0 & 7 & | & 10 \end{bmatrix} .$$

This is the augmented matrix of

$$\begin{array}{rcrcrcl} 2x & - & y & + & 3z & = & 4 \\ 3x & & & + & 2z & = & 5 \\ & & & & 7z & = & 10 \end{array} .$$

The system has been sufficiently simplified now so that the solution can be found.

From the last equation we have $z = 10/7$. Substituting this value into the second equation and solving for x gives $x = 5/7$. Substituting $x = 5/7$ and $z = 10/7$ into the first equation and solving for y yields $y = 12/7$. The solution to system (1) is, therefore,

$$x = 5/7 \ , \ y = 12/7 \ , \ z = 10/7 \ .$$

Note: We could have further reduced the matrix to row-reduced echelon form and solved the system directly from the reduced matrix. That is, by adding $-2/3$ times the second row to the first row, we have

$$\begin{bmatrix} 0 & -1 & 5/3 & | & 2/3 \\ 3 & 0 & 2 & | & 5 \\ 0 & 0 & 7 & | & 10 \end{bmatrix} .$$

Multiplying the second row by $1/3$ and the first row by -1 and interchanging the two, then multiplying the third row by $1/7$ results in

$$\begin{bmatrix} 1 & 0 & 2/3 & 5/3 \\ 0 & 1 & -5/3 & -2/3 \\ 0 & 0 & 1 & 10/7 \end{bmatrix} .$$

Then, adding $5/3$ times the third row to the second and $-2/3$ times the third row to the first gives

$$\begin{bmatrix} 1 & 0 & 0 & 5/7 \\ 0 & 1 & 0 & 12/7 \\ 0 & 0 & 1 & 10/7 \end{bmatrix} .$$

The solution to a non-homogeneous system found in the above manner is called the particular solution. The non-homogeneous system will be satisfied by any sum of the particular solution and a solution to the corresponding homogeneous system. In this case, the only solution to the homogeneous system is the trivial solution. Therefore, the only solution to the non-homogeneous problem is the particular solution.

Solve the following system by Gauss-Jordan elimination

$$x_1 + 3x_2 - 2x_3 + 2x_5 = 0$$
$$2x_1 + 6x_2 - 5x_3 - 2x_4 + 4x_5 - 3x_6 = -1$$
$$5x_3 + 10x_4 + 15x_6 = 5$$
$$2x_1 + 6x_2 + 8x_4 + 4x_5 + 18x_6 = 6 .$$

Solution: The augmented matrix for the system is

$$\begin{bmatrix} 1 & 3 & -2 & 0 & 2 & 0 & | & 0 \\ 2 & 6 & -5 & -2 & 4 & -3 & | & -1 \\ 0 & 0 & 5 & 10 & 0 & 15 & | & 5 \\ 2 & 6 & 0 & 8 & 4 & 18 & | & 6 \end{bmatrix} .$$

Reduce this matrix to row-reduced echelon form. Add -2 times the first row to the second and fourth rows

$$\begin{bmatrix} 1 & 3 & -2 & 0 & 2 & 0 & | & 0 \\ 0 & 0 & -1 & -2 & 0 & -3 & | & -1 \\ 0 & 0 & 5 & 10 & 0 & 15 & | & 5 \\ 0 & 0 & 4 & 8 & 0 & 18 & | & 6 \end{bmatrix} .$$

Add 5 times the second row to the third row and 4 times the second row to the fourth row.

$$\begin{bmatrix} 1 & 3 & -2 & 0 & 2 & 0 & | & 0 \\ 0 & 0 & -1 & -2 & 0 & -3 & | & -1 \\ 0 & 0 & 0 & 0 & 0 & 0 & | & 0 \\ 0 & 0 & 0 & 0 & 0 & 6 & | & 2 \end{bmatrix} .$$

Multiply the second row by -1 and the fourth row by 1/6. Then, interchange the third and fourth rows.

$$\begin{bmatrix} 1 & 3 & -2 & 0 & 2 & 0 & | & 0 \\ 0 & 0 & 1 & 2 & 0 & 3 & | & 1 \\ 0 & 0 & 0 & 0 & 0 & 1 & | & 1/3 \\ 0 & 0 & 0 & 0 & 0 & 0 & | & 0 \end{bmatrix} .$$

Add -3 times the third row to the second row. Then add 2 times the second row to the first row.

$$\begin{bmatrix} 1 & 3 & 0 & 4 & 2 & 0 & | & 0 \\ 0 & 0 & 1 & 2 & 0 & 0 & | & 0 \\ 0 & 0 & 0 & 0 & 0 & 1 & | & 1/3 \\ 0 & 0 & 0 & 0 & 0 & 0 & | & 0 \end{bmatrix} .$$

Now the corresponding system of equations is

$$x_1 + 3x_2 + 4x_4 + 2x_5 = 0$$
$$x_3 + 2x_4 = 0$$
$$x_6 = 1/3 .$$

Then, solving for the leading variables results in

$$x_1 = -3x_2 - 4x_4 - 2x_5$$
$$x_3 = -2x_4$$
$$x_6 = 1/3 .$$

If we assign x_2, x_4 and x_5 the arbitrary values r, s, and t, respectively, the solution set is given by the formulas,

$$x_1 = -3r - 4s - 2t, \quad x_2 = r, \quad x_3 = -2s, \quad x_4 = s$$
$$x_5 = t, \quad x_6 = 1/3 .$$

● **PROBLEM** 5-69

For the following system, find the augmented matrix; then, by reducing, determine whether the system has a solution.

$$\begin{array}{rcl} 3x - y + z &=& 1 \\ 7x + y - z &=& 6 \\ 2x + y - z &=& 2 \end{array} \qquad (1)$$

Solution: The augmented matrix is the matrix of coefficients with an additional column which corresponds to the right hand side of the equalities in the system of equations.

Reduction of the augmented matrix corresponds to reducing the system to a simpler equivalent form. If reduction of the matrix shows a contradiction the system has no solution.

The augmented matrix for the system is

$$\begin{bmatrix} 3 & -1 & 1 & | & 1 \\ 7 & 1 & -1 & | & 6 \\ 2 & 1 & -1 & | & 2 \end{bmatrix} .$$

This can be reduced by performing the following row operations. Divide the first row by 3

$$\begin{bmatrix} 1 & -1/3 & 1/3 & | & 1/3 \\ 7 & 1 & -1 & | & 6 \\ 2 & 1 & -1 & | & 2 \end{bmatrix} .$$

Now add -7 times the first row to the second row and -2 times the first row to the third row

$$\begin{bmatrix} 1 & -1/3 & 1/3 & | & 1/3 \\ 0 & 10/3 & -10/3 & | & 11/3 \\ 0 & 5/3 & -5/3 & | & 4/3 \end{bmatrix} .$$

Divide the second row by 10/3, and add -5/3 times the second row to the third row

$$\begin{bmatrix} 1 & -1/3 & 1/3 & | & 1/3 \\ 0 & 1 & -1 & | & 11/10 \\ 0 & 0 & 0 & | & -1/2 \end{bmatrix} .$$

This is the augmented matrix of the system

$$x - 1/3 \, y + 1/3 \, z = 1/3$$

$$y \quad - \quad z \quad = \quad 11/10 \qquad\qquad (2)$$
$$0 \quad = \quad -1/2 \quad .$$

The last equation cannot hold for any choice of x,y, and z. Thus, system (2) has no solution. Therefore, the system (1) has no solution. Observe that a homogeneous system always has at least one solution (namely, the one in which all the variables are equal to zero) while a non-homogeneous system may have no solution.

● **PROBLEM** 5-70

Show that the following system has more than one solution.

$$
\begin{aligned}
3x &- y + 7z = 0 \\
2x &- y + 4z = \tfrac{1}{2} \\
x &- y + z = 1 \\
6x &- 4y + 10z = 3 \quad .
\end{aligned}
\qquad (1)
$$

Solution: The augmented matrix for the system (1) is

$$
\left[
\begin{array}{ccc|c}
3 & -1 & 7 & 0 \\
2 & -1 & 4 & \tfrac{1}{2} \\
1 & -1 & 1 & 1 \\
6 & -4 & 10 & 3
\end{array}
\right] .
$$

Reduce it to row reduced echelon form. Add -3 times the third row to the first row, add -2 times the third row to the second row and add -6 times the third row to the fourth row

$$
\left[
\begin{array}{ccc|c}
0 & 2 & 4 & -3 \\
0 & 1 & 2 & -3/2 \\
1 & -1 & 1 & 1 \\
0 & 2 & 4 & -3
\end{array}
\right] .
$$

Add -1 times the first row to the fourth row; add the second row to the third row

$$
\left[
\begin{array}{ccc|c}
0 & 2 & 4 & -3 \\
0 & 1 & 2 & -3/2 \\
1 & 0 & 3 & -1/2 \\
0 & 0 & 0 & 0
\end{array}
\right] .
$$

Add -2 times the second row to the first row

$$
\left[
\begin{array}{ccc|c}
0 & 0 & 0 & 0 \\
0 & 1 & 2 & -3/2 \\
1 & 0 & 3 & -1/2 \\
0 & 0 & 0 & 0
\end{array}
\right] .
$$

Interchanging row one and row three gives

$$
\left[
\begin{array}{ccc|c}
1 & 0 & 3 & -1/2 \\
0 & 1 & 2 & -3/2 \\
0 & 0 & 0 & 0 \\
0 & 0 & 0 & 0
\end{array}
\right] .
$$

This is the augmented matrix for the system

$$
\begin{aligned}
x \quad + \quad 3z &= -\tfrac{1}{2} \\
y + 2z &= -3/2 \\
0 &= 0 \\
0 &= 0 \quad.
\end{aligned}
$$

We can write the above system as

$$
\begin{aligned}
x &= -3z - \tfrac{1}{2} \\
y &= -2z - 3/2 \quad.
\end{aligned}
$$

Since z can be assigned any arbitrary value, there are infinitely many solutions, one for each value of z. Thus, the system (1) has more than one solution.

In general, let R be a row reduced echelon form of the augmented matrix of the given system. Let r be the number of non-zero rows of this R and n be the number of unknowns of the system. If n > r, then the system has more than one particular solution.

When n = r, the system may or may not have a solution, and there cannot be more than one particular solution. That is, if there is no row whose only non-zero entry is in the last column, then the particular solution exists and is unique.

● **PROBLEM** 5-71

Use determinants to show that the following system is inconsistent.

$$
\begin{aligned}
x + y &= 3(x - 2y) + 5 && (1) \\
14y - 4x &= 11 && (2)
\end{aligned}
$$

Solution: The method of solving a system of equations by determinants is based upon Cramer's rule. Cramer's rule is stated:

In a system of n linear equations in n variables, if the determinant of the coefficients is not zero, the system has a unique solution. The value of each variable is a fraction whose denominator is the determinant of the coefficients and whose numerator is the same determinant, with the coefficients of that variable replaced by the corresponding constants.

Thus, if we have two equations arranged in standard form ax + by = c and dx + ey = f, then

$$
x = \frac{\begin{vmatrix} c & b \\ f & e \end{vmatrix}}{\begin{vmatrix} a & b \\ d & e \end{vmatrix}} = \frac{ce - fb}{ae - db}
$$

$$
y = \frac{\begin{vmatrix} a & c \\ d & f \end{vmatrix}}{\begin{vmatrix} a & b \\ d & e \end{vmatrix}} = \frac{af - dc}{ae - db}
$$

If the numerator is not zero and the denominator is zero, the system is inconsistent. One can apply Cramer's rule to solve this particular problem as follows: The determinant can be obtained more readily if the terms are arranged in the standard form, ax+by=c. Equation (1) becomes 2x-7y=-5 and equation (2) becomes -4+14y=11. Hence,

$$x = \frac{\begin{vmatrix} -5 & -7 \\ 11 & 14 \end{vmatrix}}{\begin{vmatrix} 2 & -7 \\ -4 & 14 \end{vmatrix}} = \frac{-5(14) - 11(-7)}{2(14) - (-4)(-7)} = \frac{-70 + 77}{28 - 28} = \frac{7}{0}$$

$$y = \frac{\begin{vmatrix} 2 & -5 \\ -4 & 11 \end{vmatrix}}{\begin{vmatrix} 2 & -7 \\ -4 & 14 \end{vmatrix}} = \frac{2(11) - (-4)(-5)}{0} = \frac{22 - 20}{0} = \frac{2}{0}$$

Since both x and y are of the form $\frac{a}{0}$, $a \neq 0$, the solution set is the empty set, and the system is inconsistant.

If both numerator and denominator are zero, the values of x and y are indeterminate; any (x,y) pair that satisfies one equation will satisfy the other also. Since the two equations have the same solution set, they are dependent equations.

Suppose that the augmented matrix for a system of linear equations has been reduced by row operations to the given reduced row echelon forms. Solve the systems.

(a) $\begin{bmatrix} 1 & 0 & 0 & | & 5 \\ 0 & 1 & 0 & | & -2 \\ 0 & 0 & 1 & | & 4 \end{bmatrix}$ (b) $\begin{bmatrix} 1 & 0 & 0 & 4 & | & -1 \\ 0 & 1 & 0 & 2 & | & 6 \\ 0 & 0 & 1 & 3 & | & 2 \end{bmatrix}$

(c) $\begin{bmatrix} 1 & 6 & 0 & 0 & 4 & | & -2 \\ 0 & 0 & 1 & 0 & 3 & | & 1 \\ 0 & 0 & 0 & 1 & 5 & | & 2 \\ 0 & 0 & 0 & 0 & 0 & | & 0 \end{bmatrix}$ (d) $\begin{bmatrix} 1 & 0 & 0 & | & 0 \\ 0 & 1 & 0 & | & 0 \\ 0 & 0 & 0 & | & 1 \end{bmatrix}$

Solution: (a) If the number of non-zero rows, r, of the reduced matrix is equal to the number of unknowns, n, and there is no inconsistent equation $0 = k$, then the system has a unique particular solution.

(a) The corresponding system of equations is

$$\begin{aligned} x_1 & & & = 5 \\ & x_2 & & = -2 \\ & & x_3 & - 4 \end{aligned}.$$

Therefore, the solution to the system is

$$x_1 = 5, \quad x_2 = -2 \quad \text{and} \quad x_3 = 4.$$

(b) Since $n > r$, we have more than one solution. The corresponding system of equations is

$$\begin{aligned} x_1 & & & + 4x_4 & = -1 \\ & x_2 & & + 2x_4 & = 6 \\ & & x_3 & + 3x_4 & = 2 \end{aligned}.$$

The above system can be written as

$$x_1 = -1 - 4x_4$$
$$x_2 = 6 - 2x_4 \tag{1}$$
$$x_3 = 2 - 3x_4$$

Assign x_4 any value, and then compute x_1, x_2 and x_3 from (1). Thus, we have many solutions, one for each value of x_4.

(c) $n > r$; therefore, the system has many solutions. The corresponding system of equations is

$$x_1 + 6x_2 \qquad\qquad + 4x_5 = -2$$
$$x_3 \quad + 3x_5 = 1$$
$$x_4 + 5x_5 = 2 .$$

The above equations can be written as

$$x_1 = -2 - 4x_5 - 6x_2$$
$$x_3 = 1 - 3x_5$$
$$x_4 = 2 - 5x_5 .$$

Since x_5 can be assigned an arbitrary value, t, and x_2 can be assigned an arbitrary value, s, there are infinitely many solutions. The solution set is given by the formula

$$x_1 = -2 - 4t - 6s, \; x_2 = s, \; x_3 = 1 - 3t ,$$
$$x_4 = 2 - 5t, \; x_5 = t.$$

(d) This system has no solution since the row-reduced echelon form has a row in which the first non-zero entry is in the last column.

● **PROBLEM** 5-73

Solve the following system

$$x_1 - 2x_2 - 3x_3 = 3$$
$$2x_1 - x_2 - 4x_3 = 7 \tag{1}$$
$$3x_1 - 3x_2 - 5x_3 = 8 .$$

Solution: The matrix of coefficients for the system (1) is

$$A = \begin{bmatrix} 1 & -2 & -3 \\ 2 & -1 & -4 \\ 3 & -3 & -5 \end{bmatrix} .$$

The system (1) may be written in matrix form as

$$\begin{bmatrix} 1 & -2 & -3 \\ 2 & -1 & -4 \\ 3 & -3 & -5 \end{bmatrix} \begin{bmatrix} x_1 \\ x_2 \\ x_3 \end{bmatrix} = \begin{bmatrix} 3 \\ 7 \\ 8 \end{bmatrix} . \tag{2}$$

Let

$$X = \begin{bmatrix} x_1 \\ x_2 \\ x_3 \end{bmatrix}, \qquad b = \begin{bmatrix} 3 \\ 7 \\ 8 \end{bmatrix}.$$

Then equation (2) is written

$$\vec{AX} = \vec{b} . \tag{3}$$

A solution vector $\vec{x}$ can be found by multiplying both sides of equation (3) by A^{-1}. Then we have

$$A^{-1}AX = A^{-1}b ,$$

but $A^{-1}A = I$. Hence,

$$IX = A^{-1}b$$

or

$$X = A^{-1}b . \tag{4}$$

Thus, the solutions of a system of linear equations can be obtained by finding the inverse matrix of the coefficient matrix of the system and then solving equation (4). To find A^{-1}, first form the matrix $[A : I]$, and reduce this matrix, by applying row operations, to the form $[I : B]$. Then, $B = A^{-1}$. Now,

$$[A : I] = \left[\begin{array}{ccc|ccc} 1 & -2 & -3 & 1 & 0 & 0 \\ 2 & -1 & -4 & 0 & 1 & 0 \\ 3 & -3 & -5 & 0 & 0 & 1 \end{array} \right] .$$

Add -2 times the first row to the second row and -3 times the first row to the third row

$$\left[\begin{array}{ccc|ccc} 1 & -2 & -3 & 1 & 0 & 0 \\ 0 & 3 & 2 & -2 & 1 & 0 \\ 0 & 3 & 4 & -3 & 0 & 1 \end{array} \right] .$$

Now add -1 times the second row to the third row

$$\left[\begin{array}{ccc|ccc} 1 & -2 & -3 & 1 & 0 & 0 \\ 0 & 3 & 2 & -2 & 1 & 0 \\ 0 & 0 & 2 & -1 & -1 & 0 \end{array} \right] .$$

Divide the third row by 2

$$\left[\begin{array}{ccc|ccc} 1 & -2 & -3 & 1 & 0 & 0 \\ 0 & 3 & 2 & -2 & 1 & 0 \\ 0 & 0 & 1 & -\frac{1}{2} & -\frac{1}{2} & \frac{1}{2} \end{array} \right] .$$

Add -2 times the third row to the second row and 3 times the third row to the first row

$$\left[\begin{array}{ccc|ccc} 1 & -2 & 0 & -\frac{1}{2} & -3/2 & 3/2 \\ 0 & 3 & 0 & -1 & 2 & -1 \\ 0 & 0 & 1 & -\frac{1}{2} & -\frac{1}{2} & \frac{1}{2} \end{array} \right] .$$

Divide the second row by 3; then add 2 times the resulting second row to the first row.

$$\left[\begin{array}{ccc|ccc} 1 & 0 & 0 & -7/6 & -1/6 & 5/6 \\ 0 & 1 & 0 & -1/3 & 2/3 & -1/3 \\ 0 & 0 & 1 & -\frac{1}{2} & -\frac{1}{2} & \frac{1}{2} \end{array} \right] .$$

Thus,

$$A^{-1} = \begin{bmatrix} -7/6 & -1/6 & 5/6 \\ -1/3 & 2/3 & -1/3 \\ -\frac{1}{2} & -\frac{1}{2} & \frac{1}{2} \end{bmatrix} .$$

Then equation (4) becomes

$$\begin{bmatrix} x_1 \\ x_2 \\ x_3 \end{bmatrix} = \begin{bmatrix} -7/6 & -1/6 & 5/6 \\ -1/3 & 2/3 & -1/3 \\ -\frac{1}{2} & -\frac{1}{2} & \frac{1}{2} \end{bmatrix} \begin{bmatrix} 3 \\ 7 \\ 8 \end{bmatrix} .$$

Multiplying, we have

$$\begin{bmatrix} x_1 \\ x_2 \\ x_3 \end{bmatrix} = \begin{bmatrix} -21/6 - 7/6 + 40/6 \\ -1 + 14/3 - 8/3 \\ -3/2 - 7/2 + 4 \end{bmatrix} = \begin{bmatrix} 2 \\ 1 \\ -1 \end{bmatrix} .$$

Thus,

$$x_1 = 2, \; x_2 = 1, \; x_3 = -1 .$$

It is interesting to note that the calculation of the inverse matrix is closely related to the solution of simultaneous equations. Indeed, the two processes are essentially the same. To show this, we can use the method of successive elimination. Eliminating x_1 from the second and third equations,

$$x_1 - 2x_2 - 3x_3 = 3$$
$$3x_2 + 2x_3 = 1$$
$$3x_2 + 4x_3 = -1 .$$

Then, eliminating x_2 from the third equation,

$$x_1 - 2x_2 - 3x_3 = 3$$
$$3x_2 + 2x_3 = 1$$
$$2x_3 = -2 .$$

We obtain

$$x_1 - 2x_2 - 3x_3 = 3$$
$$3x_2 + 2x_3 = 1$$
$$x_3 = -1 ,$$

$$x_1 - 2x_2 = 0$$
$$3x_2 = 3$$
$$x_3 = -1$$

and, finally,

$$x_1 = 2$$
$$x_2 = 1$$
$$x_3 = -1 .$$

It can be seen that the solution to the system $A\vec{x} = \vec{b}$, calculated directly, is the same solution we obtained by calculating A^{-1} and finding $A^{-1}\vec{b}$. Observe that the inverse matrix can also be found by using the formula

$$A^{-1} = \frac{1}{\det A} \text{ adj } A .$$

Show, by example, how the method of Gauss elimination is related to row echelon matrices.

<u>Solution:</u> Gauss elimination is used to solve systems of linear equations. Consider the following set of equations:

$$2x_1 - 3x_2 + 2x_3 + 5x_4 = 3$$
$$x_1 - x_2 + x_3 + 2x_4 = 1 \tag{1}$$
$$3x_1 + 2x_2 + 2x_3 + x_4 = 0$$
$$x_1 + x_2 - 3x_3 - x_4 = 0 .$$

According to the method of Gauss elimination, use one of the equations to eliminate one of the unknowns from the remaining three equations. In (1) above, choose the first equation for use and x_1 for elimination.

By suitably multiplying this equation, x_1 can be successively eliminated from the other equations by subtraction. For example, multiplying the first equation by ·5 and subtracting from the second eliminates x_1 from the second. In this manner we obtain

$$2x_1 - 3x_2 + 2x_3 + 5x_4 = 3$$
$$.5x_2 - .5x_4 = - .5 \tag{2}$$
$$6.5x_2 - x_3 - 6.5x_4 = - 4.5$$
$$2.5x_2 - 4x_3 - 3.5x_4 = - 1.5 .$$

Now use another equation from (2) and eliminate another unknown from the remaining two equations. Here, choose the second equation and x_2 for elimination. The result is

$$2x_1 - 3x_2 + 2x_3 + 5x_4 = 3$$
$$.5x_2 + 0x_3 - .5x_4 = -.5 \tag{3}$$
$$x_3 + 0x_4 = -2$$
$$4x_3 + x_4 = -1 .$$

Now use the third equation to eliminate x_3 from the last equation. Thus,

$$2x_1 - 3x_2 + 2x_3 + 5x_4 = 3$$
$$.5x_2 + 0x_3 - .5x_4 = - .5 \tag{4}$$
$$x_3 + 0x_4 = - 2$$
$$- x_4 = - 7 .$$

System (4) is in triangular form and is amenable to back-substitution. From the last equation, $x_4 = 7$; substituting this into the equation above, $x_3 = -2$. Proceeding to the next equation, $x_2 = 6$ and finally $x_1 = -5$. It is more convenient to work with the coefficient matrix of system (1). Form the augmented matrix

$$\begin{bmatrix} 2 & -3 & 2 & 5 & | & 3 \\ 1 & -1 & 1 & 2 & | & 1 \\ 3 & 2 & 2 & 1 & | & 0 \\ 1 & 1 & -3 & -1 & | & 0 \end{bmatrix} \qquad (5)$$

Using elementary row operations, we obtain the equivalent row echelon matrix

$$\begin{bmatrix} 1 & -1.5 & 1 & 2.5 & | & 1.5 \\ 0 & 1 & 0 & -1 & | & -1 \\ 0 & 0 & 1 & 0 & | & -2 \\ 0 & 0 & 0 & 1 & | & 7 \end{bmatrix} . \qquad (6)$$

Comparing (4) and (6), note that they differ only in that (6) has the leading coefficients of the unknowns equal to one. We conclude that the method of Gauss elimination is equivalent to using elementary row operations to reduce a matrix to row echelon form. In fact, the steps of a Gauss elimination process, in which we eliminate coefficients from the equations, correspond to the row operations used to reduce the matrix of coefficients or the augmented matrix.

● **PROBLEM** 5-75

Solve the following linear equations by using Cramer's Rule:

$$-2x_1 + 3x_2 - x_3 = 1$$
$$x_1 + 2x_2 - x_3 = 4$$
$$-2x_1 - x_2 + x_3 = -3 .$$

Solution: Consider a system of n linear equations in n unknowns:

$$a_{11}x_1 + a_{12}x_2 + \ldots + a_{1n}x_n = b_1$$

$$a_{21}x_1 + a_{22}x_2 + \ldots + a_{2n}x_n = b_2$$

$$\cdot \quad \cdot \quad \cdot \quad \cdot \quad \cdot \quad \cdot \quad \cdot \quad \cdot \quad \cdot \quad \cdot \quad \cdot \quad \cdot$$

$$a_{n1}x_1 + a_{n2}x_2 + \ldots + a_{nn}x_n = b_n .$$

Write the above equations in matrix notation.

$$\begin{bmatrix} a_{11} & a_{12} & \cdots & a_{1n} \\ a_{21} & a_{22} & \cdots & a_{2n} \\ \vdots & & & \vdots \\ a_{n1} & a_{n2} & \cdots & a_{nn} \end{bmatrix} \begin{bmatrix} x_1 \\ x_2 \\ \vdots \\ x_n \end{bmatrix} = \begin{bmatrix} b_1 \\ b_2 \\ \vdots \\ b_n \end{bmatrix}$$

or, AX = B .

Let A be an $n \times n$ matrix over the field F such that det A $\neq$ 0. If $b_1, b_2, \ldots, b_n$ are any scalars in F, the unique solution of the system of equations AX = B is given by:

$$x_i = \frac{\det A_i}{\det A} \qquad i = 1, 2, \ldots, n ,$$

where A_i is the $n \times n$ matrix obtained from A by replacing the ith column of A by the column vector

$$\begin{bmatrix} b_1 \\ b_2 \\ \cdot \\ \cdot \\ \cdot \\ b_n \end{bmatrix}$$

The above theorem is known as "Cramer's Rule" for solving systems of linear equations. Cramer's Rule applies only to systems of n linear equations in n unknowns with non-zero determinants.

We now apply Cramer's rule in solving a specific system of linear equations. Consider the given system:

$$-2x_1 + 3x_2 - x_3 = 1$$
$$x_1 + 2x_2 - x_3 = 4$$
$$-2x_1 - x_2 + x_3 = -3$$

or,

$$\underbrace{\begin{bmatrix} -2 & 3 & -1 \\ 1 & 2 & --1 \\ -2 & -1 & 1 \end{bmatrix}}_{A} \quad \underbrace{\begin{bmatrix} x_1 \\ x_2 \\ x_3 \end{bmatrix}}_{X} = \underbrace{\begin{bmatrix} 1 \\ 4 \\ 3 \end{bmatrix}}_{B}$$

$$\text{Det } A = \begin{vmatrix} -2 & 3 & -1 \\ 1 & 2 & -1 \\ -2 & -1 & 1 \end{vmatrix}$$

$$\text{Det } A = -2 \begin{vmatrix} 2 & -1 \\ -1 & 1 \end{vmatrix} - 3 \begin{vmatrix} 1 & -1 \\ -2 & 1 \end{vmatrix} - 1 \begin{vmatrix} 1 & 2 \\ -2 & -1 \end{vmatrix}$$

$$= -2(2-1) - 3(1-2) - (-1+4)$$

$$= -2 + 3 - 3 = -2 .$$

Since $\det A \neq 0$, the system has a unique solution. Now,

$$x_1 = \frac{\det A_1}{\det A} , \quad x_2 = \frac{\det A_2}{\det A} , \quad x_3 = \frac{\det A_3}{\det A} .$$

$\text{Det } A_1$ is the determinant of the matrix obtained by replacing the 1st column of A by the column of B. Thus,

$$\det A_1 = \begin{vmatrix} 1 & 3 & -1 \\ 4 & 2 & -1 \\ 3 & -1 & 1 \end{vmatrix}$$

$$= -4 .$$

Then,

$$x_1 = \frac{-4}{-2} = 2 .$$

$$x_2 = \frac{\begin{vmatrix} -2 & 1 & -1 \\ 1 & 4 & -1 \\ -2 & -3 & 1 \end{vmatrix}}{|A|} = \frac{-6}{-2} = 3 .$$

$$x_3 = \frac{\det A_3}{\det A} = \frac{\begin{vmatrix} -2 & 3 & 1 \\ 1 & 2 & 4 \\ -2 & -1 & -3 \end{vmatrix}}{-2} = \frac{-8}{-2} = 4 .$$

Thus,

$$x_1 = 2 , \; x_2 = 3 , \; x_3 = 4 ,$$

is the unique solution to the given system.

SPECIAL KINDS OF MATRICES

● **PROBLEM** 5-76

Define (1) An upper triangular matrix.
       (2) A lower triangular matrix.
       (3) A properly triangular matrix.
Give examples.

Solution: A triangular matrix is an $n \times n$ matrix whose non-zero elements lie on the diagonal and all are either above or below the diagonal.

Definition: (1) An upper triangular matrix is an $n \times n$ matrix all of whose non-zero entries lie on its diagonal and above. Example:

$$\begin{bmatrix} 1 & 2 & 3 \\ 0 & 0 & 0 \\ 0 & 0 & 6 \end{bmatrix}$$

(2) A lower triangular matrix is an $n \times n$ matrix all of whose non-zero entries lie on its diagonal and below. Example:

$$\begin{bmatrix} 1 & 0 & 0 \\ 2 & 0 & 0 \\ 3 & 0 & 6 \end{bmatrix} .$$

(3) A properly triangular matrix is an $n \times n$ matrix in which all of the diagonal entries are zero. Example:

$$\begin{bmatrix} 0 & 1 & 2 \\ 0 & 0 & 3 \\ 0 & 0 & 0 \end{bmatrix}$$

It should be noted that the zero matrix and all diagonal matrices are triangular matrices.

If A and B are both diagonal matrices having n rows and n columns, they commute. Demonstrate this in the specific case where

$$A = \begin{bmatrix} 2 & 0 & 0 \\ 0 & -1 & 0 \\ 0 & 0 & 3 \end{bmatrix} \quad B = \begin{bmatrix} -2 & 0 & 0 \\ 0 & 4 & 0 \\ 0 & 0 & -6 \end{bmatrix} .$$

ie. show that AB=BA

Solution: A diagonal matrix is a square matrix whose non-diagonal entries are all zero.

$$AB = \begin{bmatrix} 2 & 0 & 0 \\ 0 & -1 & 0 \\ 0 & 0 & 3 \end{bmatrix} \begin{bmatrix} -2 & 0 & 0 \\ 0 & 4 & 0 \\ 0 & 0 & -6 \end{bmatrix}$$

$$= \begin{bmatrix} 2 \cdot (-2) & 0 & 0 \\ 0 & (-1) \cdot 4 & 0 \\ 0 & 0 & 3 \cdot (-6) \end{bmatrix} = \begin{bmatrix} -4 & 0 & 0 \\ 0 & -4 & 0 \\ 0 & 0 & -18 \end{bmatrix}$$

$$BA = \begin{bmatrix} -2 & 0 & 0 \\ 0 & 4 & 0 \\ 0 & 0 & -6 \end{bmatrix} \begin{bmatrix} 2 & 0 & 0 \\ 0 & -1 & 0 \\ 0 & 0 & 3 \end{bmatrix}$$

$$= \begin{bmatrix} (-2) \cdot 2 & 0 & 0 \\ 0 & 4 \cdot (-1) & 0 \\ 0 & 0 & (-6) \cdot 3 \end{bmatrix} = \begin{bmatrix} -4 & 0 & 0 \\ 0 & -4 & 0 \\ 0 & 0 & -18 \end{bmatrix}$$

Thus, AB = BA .

A matrix P is called idempotent if $P^2 = P$. Show that the matrices

$$\begin{bmatrix} 25 & -20 \\ 30 & -24 \end{bmatrix}, \quad \begin{bmatrix} -26 & -18 & -27 \\ 21 & 15 & 21 \\ 12 & 8 & 13 \end{bmatrix} \quad \text{and} \quad \begin{bmatrix} 1 & 0 & 0 \\ 0 & 1 & 0 \\ 0 & 0 & 0 \end{bmatrix}$$

are idempotent.

Solution: $P^2 = PP = \begin{bmatrix} 25 & -20 \\ 30 & -24 \end{bmatrix} \begin{bmatrix} 25 & -20 \\ 30 & -24 \end{bmatrix}$

$$= \begin{bmatrix} 25 \cdot 25 + (-20) \cdot 30 & 25 \cdot (-20) + (-20) \cdot (-24) \\ 30 \cdot 25 + (-24) \cdot 30 & 30 \cdot (-20) + (-24) \cdot (-24) \end{bmatrix}$$

$$= \begin{bmatrix} 625 - 600 & -500 + 480 \\ 750 - 720 & -600 + 576 \end{bmatrix}$$

$$= \begin{bmatrix} 25 & -20 \\ 30 & -24 \end{bmatrix} = P \ .$$

Thus, $P^2 = P$.

Thus, the matrix $\begin{pmatrix} 25 & -20 \\ 30 & -24 \end{pmatrix}$

is idempotent.

$$P^2 = PP = \begin{bmatrix} -26 & -18 & -27 \\ 21 & 15 & 21 \\ 12 & 8 & 13 \end{bmatrix} \begin{bmatrix} -26 & -18 & -27 \\ 21 & 15 & 21 \\ 12 & 8 & 13 \end{bmatrix}$$

$$=$$

$\begin{array}{l} (-26)\cdot(-26)+(-18)\cdot(21)+-27\cdot12, \quad (-26)\cdot(-18)+(-18)\cdot15+(-27)\cdot8 \ , \\ \qquad\qquad\qquad\qquad\qquad\qquad\qquad (-26)\cdot(-27)+(-18)\cdot21+(-27)\cdot13 \end{array}$

$21\cdot(-26)+15\cdot21\cdot12, \quad 21\cdot(-18)+15\cdot15+21\cdot8, \quad 21\cdot(-27)+15\cdot21+21\cdot13$

$21\cdot(-26)+8\cdot21+13\cdot12, \quad 12\cdot(-18)+8\cdot15+13\cdot8, \quad 12\cdot(-27)+8\cdot21+13\cdot13$

$$= \begin{bmatrix} 676-378-325 & 468-270-216 & 702-378-351 \\ -546+315+252 & -378+225+168 & -567+315+273 \\ -312+168+156 & -216+120+104 & -324+168+169 \end{bmatrix}$$

$$= \begin{bmatrix} -26 & -18 & -27 \\ 21 & 15 & 21 \\ 12 & 8 & 13 \end{bmatrix} = P \ .$$

Thus, the matrix $\begin{bmatrix} -26 & -18 & -27 \\ 21 & 15 & 21 \\ 12 & 8 & 13 \end{bmatrix}$ is idempotent.

$$P^2 = PP = \begin{bmatrix} 1 & 0 & 0 \\ 0 & 1 & 0 \\ 0 & 0 & 0 \end{bmatrix} \begin{bmatrix} 1 & 0 & 0 \\ 0 & 1 & 0 \\ 0 & 0 & 0 \end{bmatrix}$$

$$= \begin{bmatrix} 1\cdot1+0\cdot0+0\cdot0 & 1\cdot0+0\cdot1+0\cdot0 & 1\cdot0+0\cdot0+0\cdot0 \\ 0\cdot1+1\cdot0+0\cdot0 & 0\cdot0+1\cdot1+0\cdot0 & 0\cdot0+1\cdot0+0\cdot0 \\ 0\cdot1+0\cdot0+0\cdot0 & 0\cdot0+0\cdot1+0\cdot0 & 0\cdot0+0\cdot0+0\cdot0 \end{bmatrix}$$

$$= \begin{bmatrix} 1 & 0 & 0 \\ 0 & 1 & 0 \\ 0 & 0 & 0 \end{bmatrix} = P \ .$$

Thus, the matrix

$$\begin{bmatrix} 1 & 0 & 0 \\ 0 & 1 & 0 \\ 0 & 0 & 0 \end{bmatrix}$$

is idempotent.

Define row-reduced echelon form and give examples.

Solution: A row-reduced echelon form is a matrix such that
(a) Each row that consists entirely of zeros is below each row which contains a non-zero entry.
(b) The first non-zero entry in each row is a 1.
(c) The first non-zero entry in each row is to the right of the first non-zero entry of the preceding row.
(d) Each column that contains the first non-zero entry of some row has zeros everywhere else.
    If a matrix satisfies only properties (a)-(c), it is said to be in row echelon form or just echelon form.
    Note that, together (b) and (d) are equivalent to saying that the column in which the leading entry of a non-zero row "i" occurs is $\vec{e}_i$,

i.e., $\begin{bmatrix} 0 \\ \vdots \\ 0 \\ \vdots \\ 1 \\ \vdots \\ 0 \\ \vdots \\ 0 \end{bmatrix}$, where the 1 occurs in the ith place.

Examples: The following matrices are in reduced row-echelon form.

(1) $\begin{bmatrix} 1 & 0 & 0 & 4 \\ 0 & 1 & 0 & 7 \\ 0 & 0 & 1 & -1 \end{bmatrix}$ (2) $\begin{bmatrix} 1 & 0 & 0 \\ 0 & 1 & 0 \\ 0 & 0 & 1 \end{bmatrix}$

(3) $\begin{bmatrix} 0 & 1 & -2 & 0 & 1 \\ 0 & 0 & 0 & 1 & 3 \\ 0 & 0 & 0 & 0 & 0 \\ 0 & 0 & 0 & 0 & 0 \end{bmatrix}$ (4) $\begin{bmatrix} 0 & 0 \\ 0 & 0 \end{bmatrix}$

The following matrices are not in row-echelon form.

(1) $\begin{bmatrix} 1 & 1 & 3 & 7 \\ 0 & 1 & 6 & 2 \\ 0 & 0 & 1 & 5 \end{bmatrix}$ (2) $\begin{bmatrix} 1 & 1 & 0 \\ 0 & 1 & 0 \\ 0 & 0 & 0 \end{bmatrix}$

(3) $\begin{bmatrix} 0 & 1 & 2 & 6 & 0 \\ 0 & 0 & 1 & -1 & 0 \\ 0 & 0 & 0 & 0 & 1 \end{bmatrix}$

Note that in (1) and (2), although the first non-zero member of the second row occurs in the second column, there are not zeros everywhere else in that column.

1. Which of the following matrices are in reduced-row echelon form?

(a) $\begin{bmatrix} 1 & 0 & 0 \\ 0 & 0 & 0 \\ 0 & 0 & 1 \end{bmatrix}$ (b) $\begin{bmatrix} 0 & 1 & 0 \\ 1 & 0 & 0 \\ 0 & 0 & 0 \end{bmatrix}$ (c) $\begin{bmatrix} 1 & 1 & 0 \\ 0 & 1 & 0 \\ 0 & 0 & 0 \end{bmatrix}$

(d) $\begin{bmatrix} 1 & 2 & 0 & 3 & 0 \\ 0 & 0 & 1 & 1 & 0 \\ 0 & 0 & 0 & 0 & 1 \\ 0 & 0 & 0 & 0 & 0 \end{bmatrix}$ (e) $\begin{bmatrix} 1 & 0 & 0 & 5 \\ 0 & 0 & 1 & 3 \\ 0 & 1 & 0 & 4 \end{bmatrix}$

(f) $\begin{bmatrix} 1 & 0 & 3 & 1 \\ 0 & 1 & 2 & 4 \end{bmatrix}$

2. Which of the following matrices are in row-echelon form?

(a) $\begin{bmatrix} 1 & 2 & 3 \\ 0 & 0 & 0 \\ 0 & 0 & 1 \end{bmatrix}$ (b) $\begin{bmatrix} 1 & -7 & 5 & 5 \\ 0 & 1 & 3 & 2 \end{bmatrix}$ (c) $\begin{bmatrix} 1 & 1 & 0 \\ 0 & 1 & 0 \\ 0 & 0 & 0 \end{bmatrix}$

(d) $\begin{bmatrix} 1 & 3 & 0 & 2 & 0 \\ 1 & 0 & 2 & 2 & 0 \\ 0 & 0 & 0 & 0 & 1 \\ 0 & 0 & 0 & 0 & 0 \end{bmatrix}$ (e) $\begin{bmatrix} 2 & 3 & 4 \\ 0 & 1 & 2 \\ 0 & 0 & 3 \end{bmatrix}$

(f) $\begin{bmatrix} 0 & 0 & 0 \\ 0 & 0 & 0 \\ 0 & 0 & 0 \end{bmatrix}$

Solution: A matrix is in row-reduced echelon form if it satisfies the following conditions:

(i) Any zero row lies below the non-zero rows.
(ii) The leading entry of any non-zero row is 1.
(iii) The leading entry of any non-zero row is to the right of the leading entry of each preceding row.
(iv) The column that contains the leading entry of any row has zero for all other entries.

The matrices (d) and (f) satisfy all the necessary conditions; therefore, they are in reduced row-echelon form.
The matrix (a) is not in row-reduced echelon form because it violates condition (i). The matrices (b) and (e) do not satisfy condition (iii) while the matrix (c) does not satisfy condition (iv).

2. If a matrix satisfies the conditions (i), (ii) and (iii), then it is in row echelon form. Only the matrices (b), (c) and (f) satisfy these conditions so they are in row-echelon form.

1) Define a column-reduced matrix and give an example.
2) Define column-reduced echelon form and give an example.

Solution: The matrix A is said to be column-reduced if
a) The leading entry of each non-zero column is 1.
b) Every row containing the leading entry of some non-zero column has all its other entries zero.

A non-zero column of a matrix A means a column of A whose entries are not all zero. By the leading entry of a non-zero column of A, we mean the first non-zero entry of that column.

Example: The following matrices are column-reduced:

$$A = \begin{bmatrix} 1 & 0 & 0 \\ 0 & 1 & 0 \\ 0 & 0 & 0 \\ 0 & 0 & 0 \end{bmatrix} \qquad B = \begin{bmatrix} 1 & 0 & 0 \\ 0 & 1 & 0 \\ 0 & 0 & 0 \\ 0 & 0 & 0 \end{bmatrix}.$$

2) A matrix B is said to be in column-reduced echelon form if it satisfies (a) and (b) above and also the following:

c) Each of its zero columns lies to the right of all its non-zero columns.
d) The leading non-zero entry in any column is in the row that lies above the leading entry in the next column.

Note that a matrix in column-reduced echelon form is the transpose of a matrix in row-reduced echelon form.

Example:

$$B = \begin{bmatrix} 1 & 0 & 0 & 0 & 0 \\ 0 & 1 & 0 & 0 & 0 \\ 0 & 2 & 0 & 0 & 0 \\ 0 & 0 & 1 & 0 & 0 \end{bmatrix}$$

Define the following types of symmetric matrices:
  (a) Positive - definite.
  (b) Positive - semi-definite.
  (c) Negative-definite.
  (d) Negative-semi-definite.
  (e) Indefinite.

Solution: A matrix A is symmetric if its respective row and column vectors are equal. Symbolically, $A = A^T$.
  A symmetric matrix A is:

(a) Positive-definite if and only if all the eigenvalues are positive.
(b) Positive - semi-definite if and only if all the eigenvalues are non-negative.
(c) Negative - definite if and only if all the eigenvalues are negative.
(d) Negative - semi-definite if and only if all the eigenvalues are non-positive.
(e) Indefinite if and only if it has at least one positive and at least one negative eigenvalue. For example, let

$$A = \begin{bmatrix} 1 & 2 & 0 \\ 2 & 1 & 0 \\ 0 & 0 & 3 \end{bmatrix} .$$

The matrix A has eigenvalues -1 and 3 and is, therefore, indefinite. Let

$$A = \begin{bmatrix} 4 & 0 \\ 0 & 3 \end{bmatrix} .$$

The matrix A has eigenvalues 3 and 4 and is, therefore, positive-definite.

● **PROBLEM** 5-83

Show that the matrix A is not diagonalizable where

$$A = \begin{bmatrix} -3 & 2 \\ -2 & 1 \end{bmatrix} .$$

Solution: The characteristic equation of A is $\det(\lambda I - A) = 0$; therefore,

$$\begin{vmatrix} \lambda+3 & -2 \\ +2 & \lambda-1 \end{vmatrix} = 0 ,$$

or

$$(\lambda+3)(\lambda-1) + 4 = 0$$

$$(\lambda+1)^2 = 0 .$$

Thus, $\lambda = -1$ is the only eigenvalue of A; the eigenvectors corresponding to $\lambda = -1$ are the solutions of $(-I-A)x = 0$. Thus,

$$\begin{bmatrix} 2 & -2 \\ 2 & -2 \end{bmatrix} \begin{bmatrix} x_1 \\ x_2 \end{bmatrix} = \begin{bmatrix} 0 \\ 0 \end{bmatrix} ,$$

or

$$2x_1 - 2x_2 = 0$$
$$2x_1 - 2x_2 = 0 .$$

The solutions of this system are $x_1 = t$, $x_2 = t$. Hence, the eigenspace consists of all vectors of the form

$$\begin{bmatrix} t \\ t \end{bmatrix} = t \begin{bmatrix} 1 \\ 1 \end{bmatrix} .$$

Since this space is 1-dimensional, A does not have two linearly independent eigenvectors and, therefore is not diagonalizable.

● **PROBLEM** 5-84

Demonstrate the following rules by giving examples:

(a) If A is an $n \times n$ diagonal matrix and B is an $n \times n$ matrix, each row of AB is then just the product of the diagonal entry of A times the corresponding row of B.

Solution: (a) Let

$$A = \begin{bmatrix} 2 & 0 & 0 \\ 0 & -1 & 0 \\ 0 & 0 & 3 \end{bmatrix} \quad \text{and} \quad B = \begin{bmatrix} 4 & 2 & 1 \\ -1 & 0 & 6 \\ 2 & 1 & -3 \end{bmatrix}.$$

Then

$$AB = \begin{bmatrix} 2 & 0 & 0 \\ 0 & -1 & 0 \\ 0 & 0 & 3 \end{bmatrix} \begin{bmatrix} 4 & 2 & 1 \\ -1 & 0 & 6 \\ 2 & 1 & -3 \end{bmatrix}$$

$$= \begin{bmatrix} 8+0+0 & 4+0+0 & 2+0+0 \\ 0+1+0 & 0+0+0 & 0-6+0 \\ 0+0+6 & 0+0+3 & 0+0-9 \end{bmatrix} = \begin{bmatrix} 8 & 4 & 2 \\ 1 & 0 & -6 \\ 6 & 3 & -9 \end{bmatrix}.$$

This shows that each row of AB is the product of the diagonal element of A and the corresponding row of B.
Let

$$AB = \begin{bmatrix} 4 & 0 & 0 & 0 \\ 0 & 0 & 0 & 0 \\ 0 & 0 & 3 & 0 \\ 0 & 0 & 0 & -2 \end{bmatrix} \quad \text{and} \quad B = \begin{bmatrix} 3 & 0 & -1 & 2 \\ -1 & 1 & 0 & 1 \\ 4 & -1 & -2 & 1 \\ 0 & 1 & 3 & -4 \end{bmatrix}$$

Then

$$AB = \begin{bmatrix} 12 & 0 & -4 & 8 \\ 0 & 0 & 0 & 0 \\ 12 & -3 & -6 & 3 \\ 0 & -2 & -6 & 8 \end{bmatrix}$$

Now use this rule to take the powers of a diagonal matrix. Find A^2 where A is the first matrix above.

$$\begin{bmatrix} 2 & 0 & 0 \\ 0 & -1 & 0 \\ 0 & 0 & 3 \end{bmatrix}^2 = \begin{bmatrix} 2 & 0 & 0 \\ 0 & -1 & 0 \\ 0 & 0 & 3 \end{bmatrix} \begin{bmatrix} 2 & 0 & 0 \\ 0 & -1 & 0 \\ 0 & 0 & 3 \end{bmatrix}$$

$$= \begin{bmatrix} 4 & 0 & 0 \\ 0 & 1 & 0 \\ 0 & 0 & 9 \end{bmatrix}$$

First find A^3:

$$A^3 = \begin{bmatrix} 2 & 0 & 0 \\ 0 & -1 & 0 \\ 0 & 0 & 3 \end{bmatrix}^3 = \begin{bmatrix} 2 & 0 & 0 \\ 0 & -1 & 0 \\ 0 & 0 & 3 \end{bmatrix} \begin{bmatrix} 2 & 0 & 0 \\ 0 & -1 & 0 \\ 0 & 0 & 3 \end{bmatrix}^2$$

$$= \begin{bmatrix} 2 & 0 & 0 \\ 0 & -1 & 0 \\ 0 & 0 & 3 \end{bmatrix} \begin{bmatrix} 4 & 0 & 0 \\ 0 & 1 & 0 \\ 0 & 0 & 9 \end{bmatrix} = \begin{bmatrix} 8 & 0 & 0 \\ 0 & -1 & 0 \\ 0 & 0 & 2 \end{bmatrix}$$

Observe how easy it is to take the powers of a diagonal matrix. Now continue with the further powers.

$$\begin{bmatrix} 2 & 0 & 0 \\ 0 & -1 & 0 \\ 0 & 0 & 3 \end{bmatrix}^{10} = \begin{bmatrix} 2^{10} & 0 & 0 \\ 0 & 1 & 0 \\ 0 & 0 & 3^{10} \end{bmatrix} = \begin{bmatrix} 1024 & 0 & 0 \\ 0 & 1 & 0 \\ 0 & 0 & 59049 \end{bmatrix}.$$

(b)
$$BA = \begin{bmatrix} 4 & 2 & 1 \\ -1 & 0 & 6 \\ 2 & 1 & 3 \end{bmatrix} \begin{bmatrix} 2 & 0 & 0 \\ 0 & -1 & 0 \\ 0 & 0 & 3 \end{bmatrix} = \begin{bmatrix} 8 & -2 & 3 \\ -2 & 0 & 18 \\ 4 & -1 & -9 \end{bmatrix}$$

$$BA = \begin{bmatrix} 3 & 0 & -1 & 2 \\ -1 & 1 & 0 & 1 \\ 4 & -1 & -2 & 1 \\ 0 & 1 & 3 & -4 \end{bmatrix} \begin{bmatrix} 4 & 0 & 0 & 0 \\ 0 & 0 & 0 & 0 \\ 0 & 0 & 3 & 0 \\ 0 & 0 & 0 & -2 \end{bmatrix}$$

$$= \begin{bmatrix} 12 & 0 & -3 & -4 \\ -4 & 0 & 0 & -2 \\ 16 & 0 & -6 & -2 \\ 0 & 0 & 9 & 8 \end{bmatrix}.$$

Thus, each column of BA is the product of a column of B and the corresponding diagonal element of A.

● **PROBLEM** 5-85

Define a symmetric matrix. Is every symmetric matrix similar to a diagonal matrix?

Solution: The transpose A^t of A is the matrix obtained from A by interchanging the rows and columns of A. A matrix A is symmetric if $A^t = A$. For example

$$A = \begin{bmatrix} 2 & 1 \\ 1 & 3 \end{bmatrix} \quad \text{is symmetric since} \quad A^t = \begin{bmatrix} 2 & 1 \\ 1 & 3 \end{bmatrix},$$

while

$$A = \begin{bmatrix} 2 & 1 \\ 2 & 3 \end{bmatrix} \quad \text{is not symmetric since} \quad A^t = \begin{bmatrix} 2 & 2 \\ 1 & 3 \end{bmatrix} \neq A.$$

If a matrix is symmetric, then it is similar to a diagonal matrix. The characteristic polynomial of a symmetric matrix has only real roots and for each root of multiplicity k, one can find k independent characteristic vectors.

If A is symmetric we can actually find an orthogonal matrix P such that $P^{-1}AP$ is diagonal. The orthogonal matrix is a matrix whose columns are orthonormal. Thus, a symmetric matrix is always similar to a diagonal matrix.

Find an orthogonal matrix P that diagonalizes

$$A = \begin{bmatrix} 4 & 2 & 2 \\ 2 & 4 & 2 \\ 2 & 2 & 4 \end{bmatrix} .$$

Solution: The matrix A is symmetric. Construct a matrix P whose column vectors form an orthonormal set of eigenvectors of A. This can be done as follows:

1) Find a basis for each eigenspace of A .
2) Apply the Gram-Schmidt process to each of these bases to obtain an orthonormal basis for each eigenspace.
3) Form the matrix P whose columns are the basis vectors constructed in Step 2; this matrix orthogonally diagonalizes A. The characteristic equation of A is

$$\det(\lambda I - A) = \det \begin{vmatrix} \lambda-4 & -2 & -2 \\ -2 & \lambda-4 & -2 \\ -2 & -2 & \lambda-4 \end{vmatrix} = 0 ,$$

or

$$(\lambda-4)[(\lambda-4)^2 - 4] - (-2)[-2(\lambda-4) - 4] + (-2)[4 + 2(\lambda-4)] = 0 .$$

$$(\lambda-4)(\lambda-6)(\lambda-2) - 4(\lambda-2) - 4(\lambda-2) = 0 .$$

$$(\lambda-2)[\lambda^2 -10\lambda + 24 - 4 - 4] = (\lambda-2)^2 (\lambda-8) = 0 .$$

Thus, the eigenvalues of A are $\lambda = 2$ and $\lambda = 8$. To find the eigenvectors, solve the equation $(\lambda I - A)x = 0$ for x . First, with $\lambda = 2$, $(2I - A)x = 0$, or

$$\begin{bmatrix} -2 & -2 & -2 \\ -2 & -2 & -2 \\ -2 & -2 & -2 \end{bmatrix} \begin{bmatrix} x_1 \\ x_2 \\ x_3 \end{bmatrix} = \begin{bmatrix} 0 \\ 0 \\ 0 \end{bmatrix} .$$

Solving this system gives $x_1 + x_2 + x_3 = 0$. So, $X_1 = \begin{bmatrix} -1 \\ 1 \\ 0 \end{bmatrix}$ and $X_2 = \begin{bmatrix} -1 \\ 0 \\ 1 \end{bmatrix}$ are two linearly independent vectors of this form. X_1 and

X_2 form a basis for the eigenspace corresponding to $\lambda = 2$. Before proceeding, define the Gram-Schmidt process which enables us to obtain an orthonormal basis from any given basis. Let $\{u_1, u_2, \ldots, u_n\}$ be a basis of R^n . First, select any one of the original vectors u_1, for example, then set $v_1 = \frac{1}{|u_1|} u_1$, so v_1 has unit length. Thus, $v_1 \cdot v_1 = 1$. Set

$w_2 = u_2 - (u_2 \cdot v_1)v_1$. Now $w_2 \neq 0$. The second member v_2 of the desired set of orthogonal unit vectors is obtained by dividing w_2 by its length. Thus, $v_2 = \frac{1}{|w_2|} \cdot w_2$: then $v_2 \cdot v_2 = 1$, $v_1 \cdot v_2 = 0$. In the

third step write $w_3 = u_3 - (u_3 \cdot v_1)v_3 - (u_3 \cdot v_2)v_2$ $(w_3 \neq 0)$. The third required vector v_3 is then given by $v_3 = \dfrac{1}{|w_3|} w_3$; then $v_3 \cdot v_3 = 1$,

$v_1 \cdot v_3 = v_2 \cdot v_3 = 0$. A continuation of this process finally determines the nth member of the required set in the form

$$v_n = \frac{1}{|w_n|} w_n$$

where $w_n = u_n - \sum\limits_{k=1}^{n-1} (u_n \cdot v_k) v_k$. Applying the Gram-Schmidt process to $[X_1; X_2]$ yields the orthonormal eigenvectors. First, recall $|X_1|$ is defined to be $\sqrt{X_1 \cdot X_1} = \sqrt{x_a^2 + x_b^2 + x_c^2}$ where x_a, x_b, x_c are the components of X_1 . So, in this case $|X| = \sqrt{(-1)^2 + (1)^2 + 0^2} = \sqrt{2}$.
Therefore,

$$v_1 = \frac{X_1}{|X_1|} = \frac{1}{\sqrt{2}} \begin{bmatrix} -1 \\ 1 \\ 0 \end{bmatrix} = \begin{bmatrix} -1/\sqrt{2} \\ 1/\sqrt{2} \\ 0 \end{bmatrix} ,$$

and,

$$w_2 = X_2 - (X_2 \cdot v_1) v_1 .$$

Therefore,

$$w_2 = \begin{bmatrix} -1 \\ 0 \\ 1 \end{bmatrix} - 1/\sqrt{2} \begin{bmatrix} -1/\sqrt{2} \\ 1/\sqrt{2} \\ 0 \end{bmatrix} , \qquad w_2 = \begin{bmatrix} -\tfrac{1}{2} \\ -\tfrac{1}{2} \\ 1 \end{bmatrix} ,$$

and, hence,

$$v_2 = \frac{w_2}{|w_2|} = \frac{1}{\sqrt{6}} \begin{bmatrix} -1 \\ -1 \\ 2 \end{bmatrix} = \begin{bmatrix} -1/\sqrt{6} \\ -1/\sqrt{6} \\ 2/\sqrt{6} \end{bmatrix} .$$

Now let $\lambda = 8$. Then $(8I - A)x = 0$, or

$$\begin{bmatrix} 4 & -2 & -2 \\ -2 & 4 & -2 \\ -2 & -2 & 4 \end{bmatrix} \begin{bmatrix} x_1 \\ x_2 \\ x_3 \end{bmatrix} = \begin{bmatrix} 0 \\ 0 \\ 0 \end{bmatrix} .$$

Thus, $X_3 = \begin{bmatrix} 1 \\ 1 \\ 1 \end{bmatrix}$ forms a basis for the eigenspace corresponding to $\lambda = 8$.

Applying the Gram-Schmidt process to X_3 yields

$$v_3 = \begin{bmatrix} 1/\sqrt{3} \\ 1/\sqrt{3} \\ 1/\sqrt{3} \end{bmatrix} .$$

By construction $\langle v_1, v_2 \rangle = 0$; further, $\langle v_1 \cdot v_3 \rangle = \langle v_2 \cdot v_3 \rangle = 0$ so that $\{v_1, v_2, v_3\}$ is an orthonormal set of eigenvectors. Thus,

$$P = \begin{bmatrix} -1/\sqrt{2} & -1/\sqrt{6} & 1/\sqrt{3} \\ 1/\sqrt{2} & -1/\sqrt{6} & 1/\sqrt{3} \\ 0 & 2/\sqrt{6} & 1/\sqrt{3} \end{bmatrix}$$

orthogonally diagonalizes A. Thus, P is an orthonormal set of eigen-vectors and $P^{-1}AP$ is a diagonal matrix.

CHAPTER 6

GRAPH THEORY

GRAPHS AND DIRECTED GRAPHS

Give examples of the following concepts

a)   Graph

b)   Digraph

c)   Matrix of a digraph

d)   Incidence matrix.

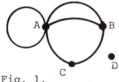

Fig. 1.

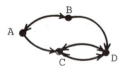

Fig. 2.

Solution: a) We first define the notion of a graph. A graph is a collection of a finite number of vertices P_1, P_2, ..., P_n together with a finite number of edges P_iP_j joining a pair of vertices P_i and P_j

In Fig. 1 there are four vertices A, B, C, D. There are 5 edges, i.e., lines joining some of the vertices. Note that a vertex can be the starting and final point of an edge as the loop from A to A demonstrates. Furthermore, two ver-

264

tices may be joined by more than one edge (as AB) and a vertex need not have any edge.

b) A digraph is a directed graph. This means that the edges joining vertices have direction.

In Fig. 2, A → B and B → A. Also, A → C. Note that we cannot have a loop in a digraph since a vertex cannot be directed toward itself (a point has no direction). Also, the directed edge AB is different from the directed edge BA.

c) Every digraph has a matrix representation. To determine the matrix representative of a digraph, use a square matrix in which the entry in row A column B is the number of directed edges from A to B. For example, the digraph in Fig. 2 has the matrix

	A	B	C	D
A	0	1	1	0
B	1	0	0	1
C	0	0	0	2
D	0	0	2	0

d) An incidence is the matrix of a digraph that has only the entries 0 or 1. Consider the following digraph.

Fig. 3.

The matrix representative is

	A	B	C
A	0	1	0
B	1	0	1
C	1	0	0

This is an incidence matrix.

265

Give an example of

    (1)   a multigraph;
    (2)   a graph;
    (3)   a loop-free multigraph;
    (4)   a connected graph.

<u>Solution</u>: (1) A multiple-graph is a graph with multiple edges and/or loops. An example is given in Fig. 1.

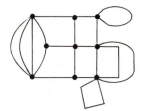

Fig. 1.

(2) A graph is a multigraph without any multiple-edges or loops. An example is shown in Fig. 2.

Fig. 2.

(3) A loop-free multigraph is shown in Fig. 3.

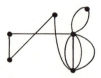

Fig. 3.

(4) A graph is said to be connected if there is a path between any two of its vertices. An example is shown in Fig. 4.

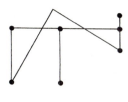

Fig. 4.

Indicate which of the following figures are multigraphs
and which are graphs.

(1)

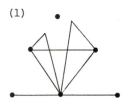

Fig. 1.

(2)

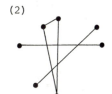

Fig. 2.

(3)

Fig. 3.

(4)

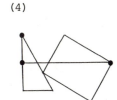

Fig. 4.

Solution: (1) is a multigraph with 4 multiple-edges.

(2) is a graph.

(3) is a multigraph.

(4) is a multigraph with one loop.

267

Determine whether the structures in the figure are graphs.

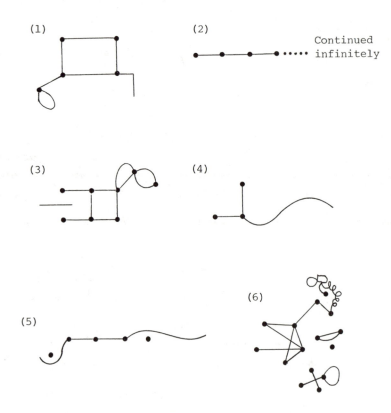

(1)

(2)

Continued
••••• infinitely

(3)

(4)

(6)

(5)

Solution: **Definition:** A graph is an ordered triple (N,A,f) where

N = 1 non-empty finite set of nodes or vertices

A = a finite set of arcs or edges

f = a function associating with each arc a an un-ordered pair x - y of nodes called the end-points of a.

(1) is not a graph since one of its arcs has only one end-point. For the same reason (4) and (5) are not graphs.

(2) is not a graph because there are infinitely many vert-icies and edges.

(3) is not a graph because there is an arc with no endpoints.

(6) is a graph.

Classify the following graphs according to the following
criteria:

a) directed or undirected

b) connected or unconnected

c) planar or nonplanar

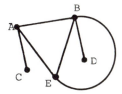

Fig. 1.

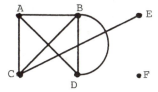

Fig. 2.

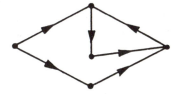

Fig. 3.

Solution: Figure 1 is an example of an undirected, connected, planar graph. Undirected graphs contain no arrows:
there is no particular direction associated with any of the
branches. This graph is connected, since each node has at
least one branch leading to it. Also, the graph is planar,
because we can draw it on a plane so that none of the
branches overlaps.

Figure 2 is an undirected, unconnected, nonplanar graph.
It is unconnected because node F is not attached to any
branch. In addition, it is a nonplanar graph because of
the overlapping of the various branches. There is no way
to redraw this graph on a plane without running into overlaps.

Finally, Figure 3 is a connected, directed, and planar
graph. Notice the arrow along the branches, giving direc-
tionality to it. We say it is a digraph, an abbreviated
form of directed graph.

● **PROBLEM** 6-6

Find the diameters of all the complete bipartite graphs.

Solution: Though there is an infinite number of different
complete bipartite graphs, this problem is still solveable.
A complete bipartite graph is a bipartite graph G(V,E) such

that every vertex in one subset of V is connected to every vertex in the other subset of V. The diameter of a connected graph is the maximum distance between any two vertices of that graph. The diameter for K_{11}, the simplest complete bipartite graph, is 1. The graph is shown in Fig. 1.

Fig. 1.

For all other complete bipartite graphs the diameters are all the same - 2. For example, K_3, as shown in Fig. 2. The graph has two subsets of vertices A and B where

A = {1,2,3}   and   B = {4,5,6,7,8}.

Fig. 2.

The maximum distance between a vertex in A and a vertex in B is 1 since every vertex in A is connected with every vertex in B and vice versa. Furthermore, the maximum distance between any two vertices in A, (or B), is 2. For instance, the distance from 4 to 8 is 2, as shown in Fig. 3.

Fig. 3.

The distance between two vertices a and b of a connected graph G is the length of the shortest path between a and b. There is no case in which the length of the shortest path between any two vertices of a complete bipartite graph is greater than two. Therefore, the diameter for K_{11} is 1, and the diameter for any other complete bipartite graph is 2.

(1) Find the degree of each vertex of the following graphs:

    (a) G(V,E), V = {a,b,c,d,e,f,g}

        E = [{a,b},{b,c},{a,d},{b,d},{c,d},{d,e},

              {e,f},{d,f},{c,f},{d,g},{c,g}]

    (b) G(V,E), V = {1,2,3,4,5,6,7,8,9,10}

        E = [{1,2},{2,3},{3,4},{1,10},{2,5},{2,6},

              {2,10},{3,9},{3,8},{4,7},{9,10},{5,9},

              {6,10},{7,10},{8,10}]

(2) Find the diameter of the following graphs:

(3) Identify all edges, nodes, and loops of the following graph G.

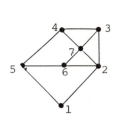

Fig. 1.

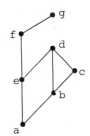

Fig. 2.

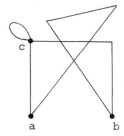

Fig. 3.

Solution: (a) Recall that for a graph G(V,E), V is the set of vertices or nodes and E is the set of unordered pairs of distinct vertices called edges. Therefore, the graph given by G(V,E) with V = {a,b,c,d,e,f,g} and E = [{a,b},{b,c}, {a,d},{b,d},{c,d},{d,e},{e,f},{d,f},{c,f},{d,g},{c,g}] is the graph shown in Fig. 5.

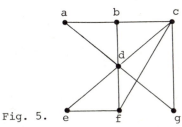

Fig. 5.

271

The graph is obtained by first placing all seven vertices, and then connecting all pairs of vertices as given in
E.   For example, {b,c}   E means that there is an edge that
joins vertices b and c.   Note that the graph given is unique
but its shape is not necessarily unique.   For instance, the
given graph can be drawn as shown in Fig. 6.

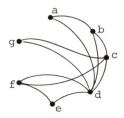

Fig. 6.

(b)   The graph is shown in Fig. 7.

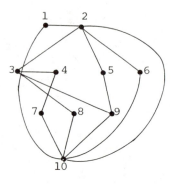

Fig. 7.

(2)   The diameter of a connected graph is defined as the maximum distance between any two vertices of the graph.   The
distance between two vertices a and b of a connected graph G
is the length of the shortest path between a and b.   A path
in G consists of an alternating sequence of vertices and
edges of the form $v_0, e_1, v_1, e_2, v_2, \ldots, e_n, v_n$ where each edge
e_i is incident on v_{i-1} and v_i, and all vertices are distinct.
Hence, the diameter of the graph in Fig. 1 is 2.   The diameter of the graph in Fig. 2 is 4.

(3)   The given graph can be redrawn as shown in Fig. 4.   It
has edges {a,c},{a,b},{b,c} and {c,c}; and nodes a, b and c.

272

Fig. 4.

A loop is an edge whose endpoints are the same vertex. Thus, edge {c,c} is a loop. Therefore, we have G(V,E) with V = {a,b,c} and E = [{a,b},{a,c},{b,c},{c,c}].

● **PROBLEM** 6-8

Find all the subgraphs of the graph in Fig. 1.

Fig. 1.

Solution: The given graph can be described by G(V,E) where V = {a,b,c} and E = [{a,b},{a,c},{b,c}]. By definition, a subgraph of a graph G(V,E) is G*(V*,E*) where V* ⊆ V, E* ⊆ E and the endpoints of E* belong to V*. Thus, the subgraphs are shown in Fig. 2. There is a total of 14 subgraphs.

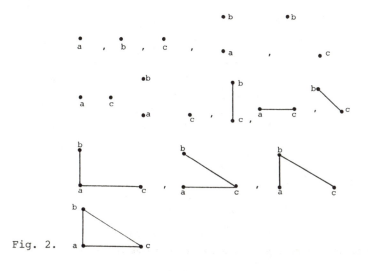

Fig. 2.

All the possible subsets of V are $\{\phi\}, \{a\}, \{b\}, \{c\}, \{a,b\},$
$\{a,c\}, \{b,c\}$ and $\{a,b,c\}$. All the possible subsets of E are
$[\{\phi\}], [\{a,b\}], [\{a,c\}], [\{b,c\}], [\{a,b\},\{a,c\}], [\{a,b\},\{b,c\}],$
$[\{a,c\},\{b,c\}],$ and $[\{a,b\},\{a,c\},\{b,c\}].$ ϕ is the empty set.
The subgraphs in Fig. 2 may be represented as follows:

$$G_1(V_1,E_1) = G_1(\{u\},[\{\phi\}])$$

$$G_2(V_2,E_2) = G_2(\{b\},[\{\phi\}])$$

$$G_3(V_3,E_3) = G_3(\{c\},[\{\phi\}])$$

$$G_4(V_4,E_4) = G_4(\{a,b\},[\{\phi\}])$$

$$G_5(V_5,E_5) = G_5(\{a,c\},[\{\phi\}])$$

$$G_6(V_6,E_6) = G_6(\{b,c\},[\{\phi\}])$$

$$G_7(V_7,E_7) = G_7(\{a,b,c\},[\{\phi\}])$$

$$G_8(V_8,E_8) = G_8(\{a,b\},[\{a,b\}])$$

$$G_9(V_9,E_9) = G_9(\{a,c\},[\{a,c\}])$$

$$G_{10}(V_{10},V_{10}) = G_{10}(\{b,c\},[\{b,c\}])$$

$$G_{11}(V_{11},V_{11}) = G_{11}(\{a,b,c\},[\{a,b\},\{a,c\}])$$

$$G_{12}(V_{12},V_{12}) = G_{12}(\{a,b,c\},[\{b,c\},\{a,c\}])$$

$$G_{13}(V_{13},V_{13}) = G_{13}(\{a,b,c\},[\{a,b\},\{b,c\}])$$

$$G_{14}(V_{14},V_{14}) = G(V,E)$$

● **PROBLEM** 6-9

Draw the following graphs.

   (a) $G_1(V_1,E_1)$

       $V_1 = \{1,2,3,4,5\}$

       $E_1 = [\{1,2\},\{1,3\},\{2,3\},\{4,5\}]$

   (b) $G_2(V_2,E_2)$

       $V_2 = \{a,b,c,d,e,f\}$

       $E_2 = [\{a,b\},\{b,c\},\{a,f\},\{b,f\},\{f,c\},\{c,d\},\{d,f\},$
              $\{e,f\},\{d,e\}]$

   (c) $G_3(V_3,E_3)$

       $V_3 = \{r,s,t,u,v,w,x,y,z\}$

       $E_3 = [\{s,t\},\{s,v\},\{r,z\},\{y,r\},\{w,r\},\{x,w\},\{x,y\}]$

Solution:

(a)

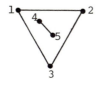

Fig. 1.

(b)

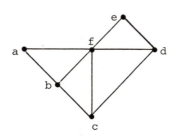

Fig. 2.

(c)

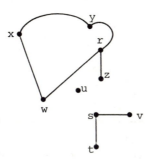

Fig. 3.

• PROBLEM 6-10

Construct a graph with at least 9 edges, for which every edge is a bridge. In addition, give a graph with no bridges.

Solution: Definition: An edge E is called a bridge of G-E has more components than G, where G is any graph and $G - [x_i, x_j]$ (or G-E where E is the edge $[x_i, x_j]$) will denote the graph obtained from G by the deletion of the edge $[x_i, x_j]$.

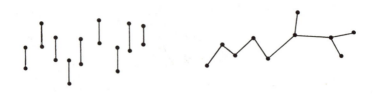

Fig. 1.

The graphs in Fig. 1 are graphs with at least nine edges for which every edge is a bridge. A graph with no bridges is shown in Fig. 2.

Fig. 2.

Draw three regular graphs with 10 vertices.

Solution: A 3-regular graph is a graph in which every vertice has exactly 3 edges incident upon it (degree-3). Using this definition, the solutions are as follows:

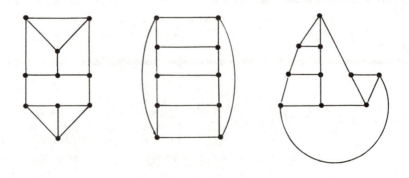

Draw the following graphs.

   (1) K_4

   (2) $K_{4,4}$

   (3) K_2

   (4) K_3

   (5) K_8

<u>Solution</u>: K_n is used to denote a complete graph of n verti-
ces. $K_{m,n}$ denotes a complete bipartite graph. A complete
graph is a graph in which each vertex is connected to every
other vertex. A graph is said to be bipartite if its vertices V can be partitioned into two subsets A and B, such that
each edge of the graph connects a vertex of A to a vertex of
B. A complete bipartite graph $K_{m,n}$ is a bipartite graph such
that each vertex of A is connected to each vertex in B; m is
the number of vertices in A and n is the number of vertices
in B.

(1)

(2)

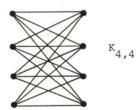

$K_{4,4}$

(3)

K_2

(4)

K_3

277

(5)

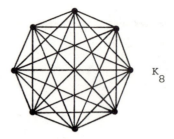

K_8

● **PROBLEM** 6-13

Draw the following:

   (1) K_1

   (2) a connected **regular** graph of degree 0

   (3) a graph that is 1-regular

   (4) $K_{2,8}$

   (5) a bipartite graph of 10 vertices

   (6) a connected regular graph of degree 3 with 8 vertices.

Solution: (1)   ● K_1

Fig. 1.

(2) A graph G is **regular** of degree n or n-regular if **every** vertex has degree n. The required graph is shown in Fig. 2. ●

Fig. 2.

(3)   ●————●

Fig. 3.

(4)

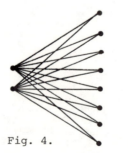

$K_{2,8}$

Fig. 4.

278

(5)   A bipartite graph is shown in Fig. 5.

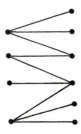

Fig. 5.

(6)

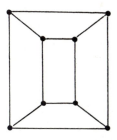

Fig. 6.

Given the following sequences of vertices, determine, by referring the graph in the figure, which ones are paths. Paths repeat no edges or vertices.

(1)   a,b,c,d,c,h,g,f,g

(2)   h,b,c,g,h,b

(3)   h,b,c,g,h

(4)   a,b,h,j,i,h,c,g,f,e

(5)   h,i,j,h

(6)   j,i,h,c,g,f,e

(7)   a,b,c,d,c,h,k,h,g,c,b,a

(8)   c,a,k,i,j

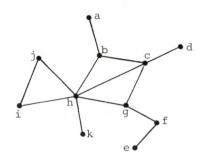

Solution: Definition: A path in a graph (or digraph) from vertex x_1 to vertex x_n is a sequence $x_1, e_1, x_2, e_2, x_3 \ldots x_{n-1}$, e_{n-1}, x_n of vertices and edges where for each i, the endpoints of edge e_i are x_i and x_{i+1}.

For digraph e_i with endpoints x_i and x_{i+1} is the edge from x_i to x_{i+1}. A path can also be represented by a sequence of vertices (or nodes) or edges (or arcs) alone.

(1)   A path
(2)   A path
(3)   A path without repeating any edges. It's also called a cycle since there is only one vertex has been repeated and it is both the beginning vertex and the ending vertex.
(4)   A path repeats no edges. This kind of path sometimes is called a trial.
(5)   A path repeats no edges. It's also a cycle.
(6)   A path repeats no vertices (thus repeats no edges).
(7)   A path whose first and last vertices are the same; it is also called a circuit.
(8)   Not a path.

● **PROBLEM** 6-15

Determine which of the following multigraphs are traversable.

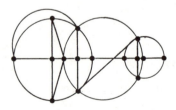

Fig. 1.

Fig. 2.

Fig. 3.

Solution: Any finite connected graph with either two odd
vertices (vertices whose degrees are odd numbers) or all even
vertices is traversable.

A traversable trail may begin at either odd vertex and
will end at the other odd vertex. The multigraphs in Figures
1 and 3 are traversable, that is, we can draw them without
any breaks in the curve and without repeating any edge. This
is shown in Figures 4 and 5.

For the graph in Fig. 1, a traversable trail is shown
below:

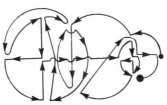

Fig. 4.

As seen from the figure above the traversable trail
starts from one of the two odd vertices and terminates at
the other odd vertex.

Fig. 5.

The graph in Fig. 2 is not traversable since it has 5
odd vertices.

Given graphs G_1, G_2, and G_3 as shown in Figs. 1, 2 and 3, respectively; for each graph find:

(1) all paths from vertex 1 to vertex 4;

(2) all bridges; and

(3) all cut points if any exist.

Fig. 1.

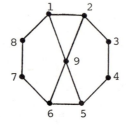

Fig. 2.

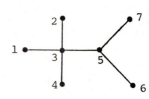

Fig. 3.

Solution: (1) A path in a multigraph consists of an alternating sequence of vertices and edges of the form $v_0, e_1, v_1,$ $e_2, v_2, \ldots e_n, v_n$; where each edge e_i is incident on v_{i-1} and v_i, and all vertices are distinct.

Therefore, there is no path between vertex 1 and vertex 4 for G .

The paths from vertex 1 to vertex 4 for G_2 are:

(i)1,2,3,4; (ii)1,8,7,6,5,4; (iii)1,9,5,4; (iv)1,9,2,3,4;

(v)1,9,6,5,4; (vi)1,8,7,6,9,5,4; (vii)1,8,7,6,9,2,3,4;

(viii)1,2,9,6,5,4; (ix)1,2,9,5,4 and (x)1,8,7,6,5,9,2,3,4.

The only path from vertices 1 to 4 for G_3, is 1,3,4.

(2) A bridge in a connected graph is an edge e such that the subgraph obtained from the graph by deleting the edge e is disconnected. The graph in Fig. 1 is not a connected graph. There is no bridge in the graph in Fig. 2 since no disconnected graph may be obtained by the removing of any edge of the graph. The graph in Fig. 3 is called a tree which is a connected graph with no cycles. A cycle is a closed walk such that all vertices are distinct except the starting and ending vertices, which are the same. (A walk in a graph consists of an alternating sequence of vertices and edges of the form $v_0, e_1, v_1, e_2, v_2 \ldots e_n, v_n$ where each edge e_i is incident on v_{i-1} and v_i.) Every edge in a tree must be a bridge.

(3)   A cut point, by definition, is a vertex x such that a disconnected subgraph is obtained if x and all edges incident on x are deleted from a connected graph G.   Again, the graph in Fig. 1 is not connected; there is no cut point in the graph in Fig. 2; and Nodes 3 and 5 of the graph in Fig. 3 are the cut points.   The subgraph obtained by deleting vertex 3 and 5 are shown in Figs. 4 and 5, respectively.

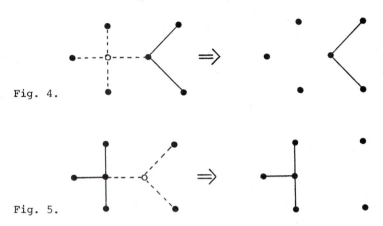

Fig. 4.

Fig. 5.

● **PROBLEM** 6-17

Find the critical path in Figure 1.

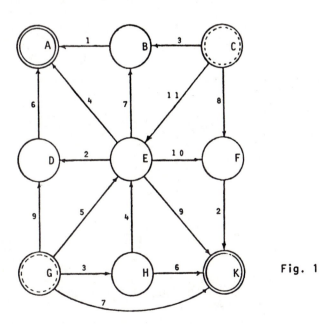

Fig. 1

Acyclic network in which C and G are start events and A and K are end events.

<u>Solution:</u> c_{ij}, s_i, t_i stand for: duration of activity

between i and j, earliest time event i can occur and latest time of event i, respectively. The algorithm used for finding the critical path is as follows:

1. Index the nodes $C = 1$, $G = 2$, etc.

2. Let s_1 and $S_C = 0$.

   Let $t^* = \max\limits_i s_i$ after using

$$s_j = \max_{i<j} \{s_i + c_{ij}, 0\} \quad \text{for} \quad j = 2,3,\ldots,p.$$

For example, $s_8 = s_A = \max (1 + 18, 6 + 13, 4 + 11, 0) = 19$.

3. Let

$$t_p = t_K = t^* = 23$$

and use

$$t_i = \min_{j>i} \{t_j - c_{ij}, t^*\} \quad \text{for } i = p - 1,\ldots,2,1.$$

Thus

$$t_2 = t_G = \min (17-9,\ 11-5,\ 7-3, 23-7, 23) = 4.$$

4. The set Z of critical nodes is $\{i: s_i = t_i\} = \{C,E,F,K\}$.

5. Critical arcs (i,j) have $i \in z$, $j \in z$, $t_j = s_i + c_{ij}$.

   CE, EF, FK are critical; thus the critical path is CEFK.

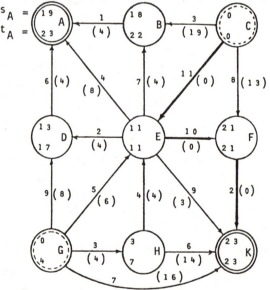

Critical path CEFK in an acyclic network. Earliest times s_i in order CG,H,E,BDF, AK; then the latest times t_j in the reverse order. Start with $s_C = s_G = 0$, $s_H = 3 + 0$, $s_E = \max (11 + 0,\ 5 + 0, 4+3)$ $=11,\ldots, t_A = t_K = t^* = 23$, $t_F = 23 - 2,\ldots, t_H = \min (11 - 4, 23 - 6) = 7,\ldots$ ..The value of the total float $t_j - s_i - c_{ij}$ is in parentheses; it is zero if and only if the arc is critical.

The critical path, earliest times and latest times for each node are demonstrated in Figure 2.

For the digraphs in Fig. 1 and Fig. 2, find:

(1)   the indegree and outdegree of vertices 1 and 4;

(2)   whether there are any sources or sinks.

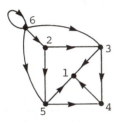

Fig. 1.

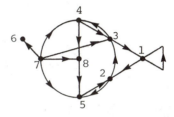

Fig. 2.

Solution: For the digraph in Fig. 1,

outdegree of vertex 1 is 0,

indegree of vertex 1 is 3.

A vertex with zero outdegree is called a sink, hence vertex 1 is a sink. A vertex with zero indegree is called a source. There are no sources in this digraph.

The outdegree and indegree of vertex 4 are 1 and 2, respectively.

For the digraph in Fig. 2,

outdegree of vertex 1 is 2,

indegree of vertex 1 is 2,

outdegree of vertex 4 is 2   and

indegree of vertex 4 is 2.

There is one source (vertex 7) and vertex 6 is a sink.

(a) Given the directed graph in Fig. 1, determine

    (1) whether the digraph is unilaterally connected;

    (2) whether it's strongly connected.

(b) Find the matrix of this digraph.

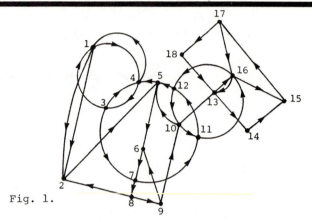

Fig. 1.

Solution: (a) The digraph is strongly connected since for any two vertices a and b of this digraph there exists a path from a to b and a path from b to a. The digraph is also unilaterally connected.

(b) The matrix is shown below.

Vertex	1	2	3	4	5	6	7	8	9	10	11	12	13	14	15	16	17	18
Vertex 1	0	2	0	1	0	0	0	0	0	0	0	0	0	0	0	0	0	0
2	0	0	0	0	1	0	0	0	0	0	0	0	0	0	0	0	0	0
3	1	0	0	1	0	0	1	0	1	0	0	0	0	0	0	0	0	0
4	1	0	1	0	0	0	0	0	0	0	0	0	0	0	0	0	0	0
5	0	0	0	1	0	1	0	1	0	0	0	0	0	0	0	0	0	0
6	0	0	0	0	0	0	0	1	0	0	0	0	0	0	0	0	0	0
7	0	0	0	0	0	0	0	0	1	0	0	1	0	0	0	0	0	0
8	0	1	0	0	0	0	0	0	1	0	0	0	0	0	0	0	0	0
M = 9	0	0	0	0	0	1	0	0	0	1	0	0	0	0	0	0	0	0
10	0	0	0	0	1	0	0	0	0	0	1	0	1	0	0	0	0	0
11	0	0	0	0	0	0	0	0	0	0	0	1	0	0	0	1	0	0
12	0	0	0	0	1	0	0	0	0	1	0	0	0	0	0	0	0	0
13	0	0	0	0	0	0	0	0	0	0	0	0	0	1	0	1	0	0
14	0	0	0	0	0	0	0	0	0	0	0	0	0	0	1	0	0	0
15	0	0	0	0	0	0	0	0	0	0	0	0	0	0	0	1	1	0
16	0	0	0	0	0	0	0	0	0	0	0	0	1	0	0	0	0	0
17	0	0	0	0	0	0	0	0	0	0	0	0	0	0	0	1	0	1
18	0	0	0	0	0	0	0	0	0	0	0	0	1	0	0	0	0	0

M is a n × n matrix with entry

$$a_{ij} = \begin{cases} \text{the number of arcs from vertex} \\ \quad \text{i to vertex j} \\ 0 \quad \text{otherwise} \end{cases}$$

MATRICES AND GRAPHS

● **PROBLEM** 6-20

Find the connection matrix C and the N-connection matrix D for each of the following graphs.

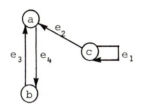
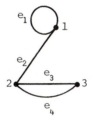

Solution: The Boolean connection matrix C of a graph G with m vertices is the m × m array of 0's and 1's such that

$$c_{ij} = \begin{cases} 1 \quad \text{if there is an edge e with endpoints} \\ \quad x_i \text{ and } x_j \text{ or in the case of a dia-} \\ \quad \text{graph, there is an edge e from vertex} \\ \quad x_i \text{ to vertex } x_j \\ \\ 0 \quad \text{if no edge joins } x_i \text{ to } x_j \end{cases}$$

The N-connection matrix D of G is the m × m array of natural numbers such that

d_{ij} = the number of different edges e with endpoints
   x_i and x_j or, in the case of a diagraph, the
   number of different edges from x_i to x_j case.

Therefore the C and D matrices of the given graphs are:

(1)

$$C = \begin{vmatrix} 0 & 1 & 0 \\ 1 & 0 & 0 \\ 1 & 0 & 1 \end{vmatrix} \qquad D = \begin{vmatrix} 0 & 1 & 0 \\ 1 & 0 & 0 \\ 1 & 0 & 1 \end{vmatrix}$$

(2)

$$C = \begin{vmatrix} 1 & 1 & 0 \\ 1 & 0 & 1 \\ 0 & 1 & 0 \end{vmatrix} \qquad D = \begin{vmatrix} 1 & 1 & 0 \\ 1 & 0 & 2 \\ 0 & 2 & 0 \end{vmatrix}$$

● **PROBLEM** 6-21

Find the N-connection matrix of the following graph.

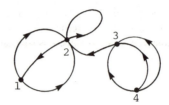

Solution: The matrix is found as follows:

$a_{11} = 0$ there are no edges from vertex 1 to itself

$a_{12} = 2$ there are two edges from vertex 1 to vertex 2.

$a_{13} = 0$ there are no edges from vertex 1 to vertex 3

$a_{14} = 0$ there are no edges from vertex 1 to vertex 4

$a_{21} = 1$ there is one edge from vertex 2 to vertex 1

$a_{22} = 1$ there is an edge from vertex 2 to itself

.
°
°

The resulting matrix is

$$M = \begin{vmatrix} 0 & 2 & 0 & 0 \\ 1 & 1 & 0 & 0 \\ 0 & 1 & 0 & 0 \\ 0 & 0 & 3 & 0 \end{vmatrix}$$

288

Find the edge matrix, incidence matrix and adjacency
matrix of the given graph.

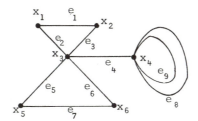

Solution: For a graph G with m vertices $x_1, x_2, \ldots, x_m$ and n
edges $e_1, e_2, \ldots, e_n$ the edge matrix of G is an nxz integer
matrix where each row of the matrix denotes an edge.   Therefore, the edge matrix for the given graph is

$$
E = \begin{array}{|cc|}
1 & 2 \\
1 & 3 \\
3 & 3 \\
3 & 4 \\
3 & 5 \\
3 & 6 \\
5 & 6 \\
4 & 4 \\
4 & 4 \\
\end{array}
$$

where the first row denotes
the edge e_1 or $\{x_1, x_2\}$,
for example

The adjacency matrix of G is an $m \times m$ matrix defined by

$$a_{ij} = \text{the number of different edges e}$$
$$\text{with endpoints } x_i \text{ and } x_j.$$

It is also called the N-connection matrix of G.   Hence,
the adjacency matrix for the given graph is

$$
A = \begin{array}{c|cccccc}
 & x_1 & x_2 & x_3 & x_4 & x_5 & x_6 \\
\hline
x_1 & 0 & 1 & 1 & 0 & 0 & 0 \\
x_2 & 1 & 0 & 1 & 0 & 0 & 0 \\
x_3 & 1 & 1 & 0 & 1 & 1 & 1 \\
x_4 & 0 & 0 & 1 & 2 & 0 & 0 \\
x_5 & 0 & 0 & 1 & 0 & 0 & 1 \\
x_6 & 0 & 0 & 1 & 0 & 1 & 0 \\
\end{array}
$$

The incidence matrix of G is an **m** × **n** matrix defined by:

$$m_{ij} = \begin{cases} 1 & \text{if the vertex } x_i \text{ is incident} \\ & \text{on the edge } e_j. \\ 0 & \text{otherwise} \end{cases}$$

Thus, the incidence matrix of the given graph is therefore:

$$I = \begin{array}{c} \\ x_1 \\ x_2 \\ x_3 \\ x_4 \\ x_5 \\ x_6 \end{array} \begin{array}{ccccccccc} e_1 & e_2 & e_3 & e_4 & e_5 & e_6 & e_7 & e_8 & e_9 \\ 1 & 1 & 0 & 0 & 0 & 0 & 0 & 0 & 0 \\ 1 & 0 & 1 & 0 & 0 & 0 & 0 & 0 & 0 \\ 0 & 1 & 1 & 1 & 1 & 1 & 0 & 0 & 0 \\ 0 & 0 & 0 & 1 & 0 & 0 & 0 & 1 & 1 \\ 0 & 0 & 0 & 0 & 1 & 0 & 1 & 0 & 0 \\ 0 & 0 & 0 & 0 & 0 & 1 & 1 & 0 & 0 \end{array}$$

● PROBLEM 6-23

Find the incidence matrix of the graph below:

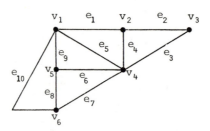

Solution: An incidence matrix contains a "1" in the i,j entry if edge (j) is incident (i.e., connected to) vertice (i). Otherwise, a "0" is placed in that i,j entry.

$$M = \begin{pmatrix} 1 & 0 & 0 & 0 & 1 & 0 & 0 & 0 & 1 & 1 \\ 1 & 1 & 0 & 1 & 0 & 0 & 0 & 0 & 0 & 0 \\ 0 & 1 & 1 & 0 & 0 & 0 & 0 & 0 & 0 & 0 \\ 0 & 0 & 1 & 1 & 1 & 1 & 1 & 0 & 0 & 0 \\ 0 & 0 & 0 & 0 & 0 & 1 & 0 & 1 & 1 & 0 \\ 0 & 0 & 0 & 0 & 0 & 0 & 1 & 1 & 0 & 1 \end{pmatrix}$$

Find the adjacency matrices for the following graphs:

(1)

(2)

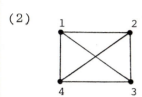

Solution: Since, by definition, an adjacency matrix contains a "1" in the i, j entry if its graph has an edge (i,j), and a "0" otherwise, the solutions are as follows:

(1)

$$M = \begin{pmatrix} 0 & 0 & 0 & 1 & 1 \\ 0 & 0 & 0 & 1 & 1 \\ 0 & 0 & 0 & 1 & 1 \\ 1 & 1 & 1 & 0 & 0 \\ 1 & 1 & 1 & 0 & 0 \end{pmatrix}$$

(2)

$$M = \begin{pmatrix} 0 & 1 & 1 & 1 \\ 1 & 0 & 1 & 1 \\ 1 & 1 & 0 & 1 \\ 1 & 1 & 1 & 0 \end{pmatrix}$$

Find the adjacency matrices for the following graphs:

(1)

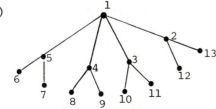

(2)

Solution: An adjacency matrix is formed by placing a "1" in the i,j entry if there is an edge between points i and j. Otherwise, a "0" is placed in that i,j entry.

(1)

$$M = \begin{pmatrix}
0 & 1 & 1 & 1 & 1 & 0 & 0 & 0 & 0 & 0 & 0 & 0 & 0 \\
1 & 0 & 0 & 0 & 0 & 0 & 0 & 0 & 0 & 0 & 0 & 1 & 1 \\
1 & 0 & 0 & 0 & 0 & 0 & 0 & 0 & 0 & 1 & 1 & 0 & 0 \\
1 & 0 & 0 & 0 & 0 & 0 & 0 & 1 & 1 & 0 & 0 & 0 & 0 \\
1 & 0 & 0 & 0 & 0 & 1 & 1 & 0 & 0 & 0 & 0 & 0 & 0 \\
0 & 0 & 0 & 0 & 1 & 0 & 0 & 0 & 0 & 0 & 0 & 0 & 0 \\
0 & 0 & 0 & 0 & 1 & 0 & 0 & 0 & 0 & 0 & 0 & 0 & 0 \\
0 & 0 & 0 & 1 & 0 & 0 & 0 & 0 & 0 & 0 & 0 & 0 & 0 \\
0 & 0 & 0 & 1 & 0 & 0 & 0 & 0 & 0 & 0 & 0 & 0 & 0 \\
0 & 0 & 1 & 0 & 0 & 0 & 0 & 0 & 0 & 0 & 0 & 0 & 0 \\
0 & 0 & 1 & 0 & 0 & 0 & 0 & 0 & 0 & 0 & 0 & 0 & 0 \\
0 & 1 & 0 & 0 & 0 & 0 & 0 & 0 & 0 & 0 & 0 & 0 & 0 \\
0 & 1 & 0 & 0 & 0 & 0 & 0 & 0 & 0 & 0 & 0 & 0 & 0
\end{pmatrix}$$

(2)

$$M = \begin{pmatrix}
0 & 1 & 1 & 1 & 1 & 1 \\
1 & 0 & 1 & 1 & 1 & 1 \\
1 & 1 & 0 & 1 & 1 & 1 \\
1 & 1 & 1 & 0 & 1 & 1 \\
1 & 1 & 1 & 1 & 0 & 1 \\
1 & 1 & 1 & 1 & 1 & 0
\end{pmatrix}$$

● **PROBLEM 6-26**

Determine the digraph that is described by the following matrix.

$$A = \begin{vmatrix}
2 & 1 & 0 & 2 \\
1 & 0 & 3 & 0 \\
1 & 0 & 0 & 0 \\
0 & 1 & 1 & 0
\end{vmatrix}$$

Since matrix A is defined by

$$a_{ij} = \text{the number of different edges} \\ e \text{ from vertex } x_i \text{ to } x_j$$

the graph which is described by the given matrix is found as the following:

$a_{11} = 2$ there are two edges from
          vertex 1 to itself

$a_{12} = 1$ there is a single edge from
          vertex x_1 to vertex x_2

$a_{13} = 0$ there are no edges from
          vertex x_1 to vertex x_3

     °
     °
     •

Repeat this procedure until all entires of matrix A have been taken care of.

The digraph is shown in the figure.

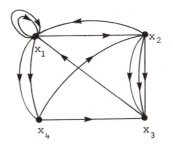

● **PROBLEM 6-27**

Given

(a)   Edge

Edge	
e_1	(1,1)
e_2	(1,2)
e_3	(3,2)
e_4	(2,3)
e_5	(4,5)

and

(b)

Edge	
e_1	{1,2}
e_2	{2,2}
e_3	{3,2}
e_4	{3,1}

Draw the graphs of (a) and (b).

Solution: The graphs are:

(a)

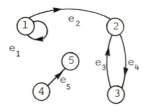

(b)

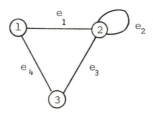

Sketch the digraphs that have the following matrix representations.

(a)
$$\begin{vmatrix} 0 & 0 & 2 & 1 \\ 1 & 0 & 0 & 1 \\ 0 & 1 & 1 & 0 \\ 0 & 0 & 0 & 0 \end{vmatrix}$$

(b)
$$\begin{vmatrix} 1 & 3 & 0 & 2 \\ 0 & 0 & 0 & 0 \\ 1 & 1 & 0 & 0 \\ 0 & 1 & 0 & 0 \end{vmatrix}$$

Solution: (a)

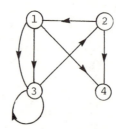

(b)

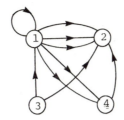

Draw the directed graph with vertices

$$V = \{v_1, v_2, \ldots, v_6\}, \text{ edges}$$

$$E = \{e_1, e_2, \ldots, e_{10}\},$$

and f defined as the following:

e	f(e)
e_1	(v_1, v_2)
e_2	(v_2, v_1)
e_3	(v_2, v_2)
e_4	(v_2, v_3)
e_5	(v_4, v_3)
e_6	(v_4, v_4)
e_7	(v_3, v_5)
e_8	(v_3, v_6)
e_9	(v_3, v_6)
e_{10}	(v_5, v_6)

Solution:

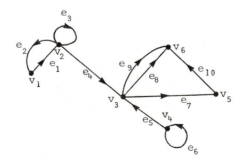

295

Draw the multigraphs whose adjacency matrices are given below:

(1)

$$G_1 = \begin{pmatrix} 1 & 1 & 1 & 1 & 2 \\ 1 & 3 & 1 & 3 & 1 \\ 1 & 1 & 0 & 1 & 1 \\ 1 & 3 & 1 & 0 & 1 \\ 2 & 1 & 1 & 1 & 0 \end{pmatrix}$$

(2)

$$G_2 = \begin{pmatrix} 0 & 2 & 2 & 3 \\ 2 & 0 & 3 & 2 \\ 2 & 3 & 0 & 0 \\ 3 & 2 & 0 & 0 \end{pmatrix}$$

Solution: (1) The adjacency matrix for a multigraph contains the number of edges incident on vertices (i,j) for every i,j entry. Hence, to reconstruct a multigraph from its adjacency matrix one must set up points corresponding to each "i" (or "j") in the matrix, and then fill in the proper edges incident on the vertices as dictated by the matrix.

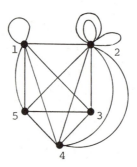

Fig. 1.

(2)

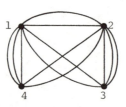

Fig. 2.

Draw the graphs whose incidence matrices are given below:

(1)

$$A = \begin{array}{c} \\ v_1 \\ v_2 \\ v_3 \\ v_4 \\ v_5 \\ v_6 \\ v_7 \end{array} \begin{array}{c} \begin{array}{cccccccccccc} e_1 & e_2 & e_3 & e_4 & e_5 & e_6 & e_7 & e_8 & e_9 & e_{10} & e_{11} & e_{12} \end{array} \\ \left[\begin{array}{cccccccccccc} 1 & 0 & 0 & 1 & 0 & 0 & 1 & 0 & 0 & 1 & 0 & 0 \\ 0 & 1 & 0 & 0 & 1 & 0 & 0 & 1 & 0 & 0 & 1 & 0 \\ 0 & 0 & 1 & 0 & 0 & 1 & 0 & 0 & 1 & 0 & 0 & 1 \\ 1 & 1 & 1 & 0 & 0 & 0 & 0 & 0 & 0 & 0 & 0 & 0 \\ 0 & 0 & 0 & 1 & 1 & 1 & 0 & 0 & 0 & 0 & 0 & 0 \\ 0 & 0 & 0 & 0 & 0 & 0 & 1 & 1 & 1 & 0 & 0 & 0 \\ 0 & 0 & 0 & 0 & 0 & 0 & 0 & 0 & 0 & 1 & 1 & 1 \end{array}\right] \end{array}$$

(2)

$$B = \left[\begin{array}{cccccccccc} 1 & 0 & 0 & 0 & 1 & 1 & 1 & 0 & 0 & 0 \\ 1 & 1 & 0 & 0 & 0 & 0 & 0 & 1 & 0 & 1 \\ 0 & 1 & 1 & 0 & 0 & 1 & 0 & 0 & 1 & 0 \\ 0 & 0 & 1 & 1 & 0 & 0 & 1 & 0 & 0 & 1 \\ 0 & 0 & 0 & 1 & 1 & 0 & 0 & 1 & 1 & 0 \end{array}\right]$$

Solution: (1)

Fig. 1.

(2)

Fig. 2.

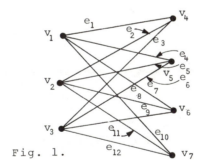

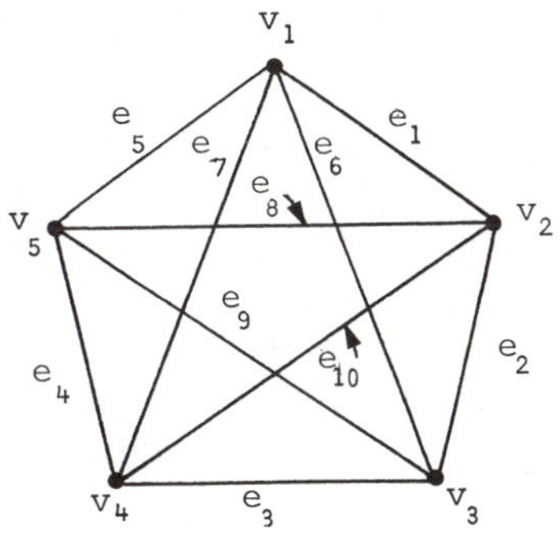

A sociologist is interested in determining which one in a group of 5 people has the most influence over the others. The group is asked to fill out the following questionnaire:

Your name	Person whose opinion you value most

The answers are tabulated as shown in Fig. 1. The people in the group are labeled as A, B, C, D, and E. It is assumed that the person whose opinion a member of the group values most is the one who influences that member most.

Use the given data and graph theory to determine which one of the group has most influence over the others.

Group Number	Person Whose Opinion He or She Values Most
A	D
B	A
C	A
D	B
E	D

Fig. 1

Solution: Represent the data given in Fig. 1 by a digraph as shown in Fig. 2. Here the arcs correspond to direct influence and the arrows indicate direction of influence. For instance, C values the opinion of A most, hence, C is influenced by A and there is a direct arc from A to C. (See Fig. 2.)

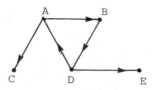

Fig. 2.

In the graph in Fig. 2, one-paths correspond to direct influence. Paths involving two, three or more arcs correspond to indirect influences: the longer the path, the more remote, and presumably the smaller the influence.

A table is constructed according to the graph, it is shown in Fig. 3.

Person	Number of one-path	Number of two-path	Number of three-path
A	$2\begin{pmatrix}A{\to}B\\A{\to}C\end{pmatrix}$	$1\,(A{\to}B{\to}D)$	$1\,(A{\to}B{\to}D{\to}E)$
B	$1\,(B{\to}D)$	$2\begin{pmatrix}B{\to}D{\to}E\\B{\to}D{\to}A\end{pmatrix}$	$1\,(B{\to}D{\to}A{\to}C)$
C	0	0	0
D	$2\begin{pmatrix}D{\to}A\\D{\to}E\end{pmatrix}$	$2\begin{pmatrix}D{\to}A{\to}B\\D{\to}A{\to}C\end{pmatrix}$	0
E	0	0	0

Fig. 3.

We conclude from the table in Fig. 3 that D has the most influence over the others. A and D have the same number of direct influences (two), the strongest influence over the others. However, D has two two-path influences which is not as strong as the one-path influence, but stronger than the three-path influence, whereas A has only one two-path influence. Therefore, D has the edge over A as far as influence goes.

● PROBLEM 6-33

The figure below shows a schematic map of the interconnections between the airports in three different countries a, b and c.

The figures beside the links denote the number of choices along the links. For example, a 4 indicates that four airlines fly services along that route.

Construct a matrix showing the number of choices of routes between the airports in country a and country c.

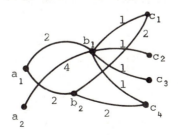

299

<u>Solution:</u> First tabulate the information given in the figure.

P:	b_1	b_2		Q:	c_1	c_2	c_3	c_4
a_1	2	2		b_1	1	1	1	1
a_2	4	0		b_2	2	0	0	2

Now, a route from a_1 to c_1 may pass through b_1 or b_2. In the first case there are 2x1 choices and in the second case 2x2 choices. This means 6 choices in all and the other entries for the required matrix may be worked out in the same way, giving

	c_1	c_2	c_3	c_4
a_1	6	2	2	6
a_2	4	4	4	4

But this is simply the matrix product of P and Q:

$$PQ = \begin{bmatrix} 2 & 2 \\ 4 & 0 \end{bmatrix} \begin{bmatrix} 1 & 1 & 1 & 1 \\ 2 & 0 & 0 & 2 \end{bmatrix} = \begin{bmatrix} 6 & 2 & 2 & 6 \\ 4 & 4 & 4 & 4 \end{bmatrix}$$

The natural interpretation of the product matrix is as follows: the two rows of P represent ways of flying from a_1 and a_2 to b_1 and b_2 respectively. The columns of Q represent ways in which c_i can be reached from b_1 and b_2. The product of P and Q therefore gives the number of choices of route between the airports in country a and country c.

● **PROBLEM 6-34**

Consider an abstract map giving the routes between cities as shown in Figure 1.
Show how matrix methods may be used to count the number of different routes between two points on a map.

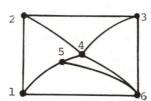

Fig. 1.

Solution: The map in Fig. 1 is a graph. The six numbered dots are called vertices and the connecting lines are called edges. Two vertices are adjacent if there is an edge directly connecting them. For example, vertex 2 and vertex 4 are adjacent whereas vertices 3 and 5 are not. A route in a graph is a sequence of edges such that the terminal point of one is the initial point of the next edge in the sequence. For example, $6 \to 3 \to 2 \to 4$ denotes the route starting at vertex 6 and proceeding to vertex 3, then continuing to vertex 2, and finally ending at vertex 4. The same edge can appear more than once in a route; for example, the route $3 \to 4 \to 2 \to 4 \to 5 \to 1$ uses the edge joining 2 and 4 twice.

A graph can be numerically described by its incidence matrix. The size of this nxn matrix is determined by the number of vertices in the graph. In the present example, the incidence matrix is a 6x6 matrix where the i,j entry a_{ij} equals the number of edges joining vertex i to vertex j. For example, $a_{16} = 1$ since vertices 1 and 6 have one edge joining them, while $a_{52} = 0$ since there is no edge joining vertex 5 to vertex 2. Thus, the incidence matrix is:

$$\begin{bmatrix} 0 & 1 & 0 & 0 & 1 & 1 \\ 1 & 0 & 1 & 1 & 0 & 0 \\ 0 & 1 & 0 & 1 & 0 & 1 \\ 0 & 1 & 1 & 0 & 1 & 1 \\ 1 & 0 & 0 & 1 & 0 & 1 \\ 1 & 0 & 1 & 1 & 1 & 0 \end{bmatrix}$$

The length of a route is the number of edges in the sequence that forms the route. If the same edge is used more than once, it is counted more than once in determining the length of the route. For example, the route $4 \to 2 \to 3 \to 4 \to 2 \to 1$ has length 5. The i,j entry in the incidence matrix can be interpreted as the number of routes of length 1 from vertex i to vertex j, since there is a route of length 1 from i to j exactly when there is an edge joining i and j.

If we form the product

$$A^2 = AA = \begin{bmatrix} 3 & 0 & 2 & 3 & 1 & 1 \\ 0 & 3 & 1 & 1 & 2 & 3 \\ 2 & 1 & 3 & 2 & 2 & 1 \\ 3 & 1 & 2 & 4 & 1 & 2 \\ 1 & 2 & 2 & 1 & 3 & 2 \\ 1 & 3 & 1 & 2 & 2 & 4 \end{bmatrix}$$

observe that the i,j entry of this matrix gives the number of routes from i to j of length 2. This follows since

$$\text{i, j entry of } A^2 = a_{i1}\,a_{1j} + a_{i2}a_{2j} + \ldots + a_{i6}a_{6j}.$$

The first product $a_{i1}a_{1j}$, counts the number of routes that run from i to 1 to j; the second product $a_{i2}a_{2j}$ counts the number of routes that run from i to 2 to j and so on. For example, the entry 3 in the 1, 4 position of A^2 counts the three routes $1 \to 2 \to 4$, $1 \to 5 \to 4$, $1 \to 6 \to 4$.

In general, if k is any positive integer, the i, j entry of A^k is the number of routes of length k from vertex i to vertex j. Using k = 3, we have

$$A^3 = \begin{bmatrix} 2 & 8 & 4 & 4 & 7 & 9 \\ 8 & 2 & 7 & 9 & 4 & 4 \\ 4 & 7 & 4 & 7 & 5 & 9 \\ 4 & 9 & 7 & 6 & 9 & 10 \\ 7 & 4 & 5 & 9 & 4 & 7 \\ 9 & 4 & 9 & 10 & 7 & 6 \end{bmatrix}$$

The entry 9 in the 4, 2 position counts the nine routes of length 3 from vertex 4 to vertex 2: $4 \to 2 \to 1 \to 2$, $4 \to 2 \to 3 \to 2$, $4 \to 2 \to 4 \to 2$, $4 \to 3 \to 4 \to 2$, $4 \to 5 \to 1 \to 2$, $4 \to 5 \to 4 \to 2$, $4 \to 6 \to 1 \to 2$, $4 \to 6 \to 3 \to 2$, $4 \to 6 \to 4 \to 2$.

● **PROBLEM** 6-35

The results of a six-person round-robin backgammon tournament are given in the graph below.

An arrow pointing from player i to player j means that i beat j in their match. For example, player 2 beat player 1 but player 6 beat player 2. Who is the best player?

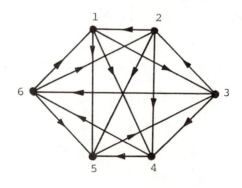

Solution: A matrix can be used to describe the results of the tournament. Let A be the 6x6 matrix where $a_{ij} = 1$ if player i beat player j in their match and $a_{ij} = 0$ if j beat i. Hence,

$$A = \begin{bmatrix} 0 & 0 & 1 & 1 & 1 & 0 \\ 1 & 0 & 0 & 1 & 1 & 0 \\ 0 & 1 & 0 & 1 & 0 & 1 \\ 0 & 0 & 0 & 0 & 1 & 1 \\ 0 & 0 & 1 & 0 & 0 & 0 \\ 1 & 1 & 0 & 0 & 1 & 0 \end{bmatrix}$$

To find how many matches each player won, add across each row. We have

Player #	# wins
1	3
2	3
3	3
4	2
5	1
6	3

and hence the ranking of the players is

Place	Player #
1st	1, 2, 3, 6
5th	4
6th	5

One method of breaking the four way tie for first place is as follows:

First form the product of A with itself.

$$A^2 = AA = \begin{bmatrix} 0 & 0 & 1 & 1 & 1 & 0 \\ 1 & 0 & 0 & 1 & 1 & 0 \\ 0 & 1 & 0 & 1 & 0 & 1 \\ 0 & 0 & 0 & 0 & 1 & 1 \\ 0 & 0 & 1 & 0 & 0 & 0 \\ 1 & 1 & 0 & 0 & 1 & 0 \end{bmatrix} \begin{bmatrix} 0 & 0 & 1 & 1 & 1 & 0 \\ 1 & 0 & 0 & 1 & 1 & 0 \\ 0 & 1 & 0 & 1 & 0 & 1 \\ 0 & 0 & 0 & 0 & 1 & 1 \\ 0 & 0 & 1 & 0 & 0 & 0 \\ 1 & 1 & 0 & 0 & 1 & 0 \end{bmatrix}$$

$$
= \begin{bmatrix} 0 & 1 & 1 & 1 & 1 & 2 \\ 0 & 0 & 2 & 1 & 2 & 1 \\ 2 & 1 & 0 & 1 & 3 & 1 \\ 1 & 1 & 1 & 0 & 1 & 0 \\ 0 & 1 & 0 & 1 & 0 & 1 \\ 1 & 0 & 2 & 2 & 2 & 0 \end{bmatrix} \tag{1}
$$

The matrix (1) gives the two step wins for i over j. To illustrate the meaning of this statement consider the following: Multiplying the first row of A by its second column yields $0 \cdot 0 + 0 \cdot 0 + 1 \cdot 1 + 1 \cdot 0 + 1 \cdot 0 + 0 \cdot 1$

Player 1 who did not play with himself, lost to player 2. But he beat player 3 who beat player 2. Thus player 1 had a two step win over player 2 and hence the (1,2) element of A^2 has 1.

Similarly player 3 had 3 two-step wins over player 5 since he beat player 2 who beat player 5; he also beat player 4 who beat player 5 and finally, beat player 6 who beat player 5.

Now let $B = A + A^2$:

$$
B = \begin{bmatrix} 0 & 1 & 2 & 2 & 2 & 2 \\ 1 & 0 & 2 & 2 & 3 & 1 \\ 2 & 2 & 0 & 2 & 3 & 2 \\ 1 & 1 & 1 & 0 & 2 & 1 \\ 0 & 1 & 1 & 1 & 0 & 1 \\ 2 & 1 & 2 & 2 & 3 & 0 \end{bmatrix}
$$

The i, j entry of B equals the number of direct wins plus the number of two-step wins that player i has over player j. Adding across rows yields

Player #	# of direct wins + # of 2-step wins
1	9
2	9
3	11
4	6
5	4
6	10

Now the ranking becomes

Place	Player #
1st	3
2nd	6
3rd	1, 2
5th	4
6th	5

Assuming that this method measures a player's strength, observe that player 3 is the strongest player.

Note that there are other ways of measuring relative strength. It is possible to use $2A + A^2$ which gives twice as much weight to a direct win as a two step win.

● **PROBLEM 6-36**

The digraph given below represents relationships between people in a group. An arrow denotes "is friendly to" while a double arrow indicates mutual friendship.

Find the number of cliques and the people belonging to each clique.

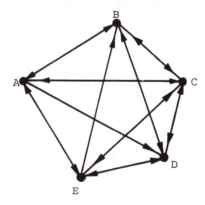

Solution: The incidence matrix of the given digraph is

	A	B	C	D	E
A	0	1	1	1	1
B	1	0	1	1	0
C	1	1	0	1	1
D	0	1	1	0	1
E	1	1	1	1	0.

Next define clique. A clique is the largest collection of three or more individuals with the property that any two of them are mutual friends.

This means that if three people are mutual friends and if each is a mutual friend of a fourth person, the three do not form a clique since they do not form the largest collection of mutual friends.

From the given incidence matrix we can obtain a symmetric matrix by considering only pairs connected by a double arrow, namely

$$A \leftrightarrow B, \ A \leftrightarrow C, \ A \leftrightarrow E, \ C \leftrightarrow E$$

$$B \leftrightarrow C \quad B \leftrightarrow D \quad C \leftrightarrow D \quad D \leftrightarrow E.$$

The resulting symmetric matrix S is

$$S = \begin{array}{c|ccccc} & A & B & C & D & E \\ \hline A & 0 & 1 & 1 & 0 & 1 \\ B & 1 & 0 & 1 & 1 & 0 \\ C & 1 & 1 & 0 & 1 & 1 \\ D & 0 & 1 & 1 & 0 & 1 \\ E & 1 & 0 & 1 & 1 & 0 \end{array}$$

To find cliques in a group examine the entries in the matrix S^3. The reason for this is that the entries along the diagonal of S^3 give the number of three stage relations between a person and himself. This means mutual friendships with at least two other people. We find S^3:

$$S^3 = \begin{array}{c|ccccc} & A & B & C & D & E \\ \hline A & 4 & 8 & 8 & 4 & 8 \\ B & 8 & 4 & 8 & 8 & 4 \\ C & 8 & 8 & 8 & 8 & 8 \\ D & 4 & 8 & 8 & 4 & 8 \\ E & 8 & 4 & 8 & 8 & 4 \end{array}$$

Now use the following facts

1. If $S_{ii}^{(3)}$ is positive, the person P_i belongs to at least one clique.

2. If $S_{ii}^{(3)} = 0$, the person belongs to no clique.

Since all the diagonal entries in S are positive every person in the group belongs to a clique. The size and number of cliques is found by using the theorem below:

306

Let $S = (s_{ij})$ be the symmetric matrix associated with the incidence matrix of a clique digraph. An individual is a member of exactly one clique with k members if and only if his diagonal entry in S^3 equals $(k - 1)(k - 2)$ i.e., $s_{ii}^{(3)} = (k - 1)(k - 2)$.

For the given digraph the number of possible members in a clique can be 3, 4, or 5. Since none of the diagonal entries is 2, 6, or 12 each person must belong to more than one clique.

Consider A whose diagonal entry is 4. Since $4 = 2 + 2$, A belongs to two cliques each with three members. Similarly B, D and E belong to two cliques each containing three people. The diagonal entry 8 in S^3 can only be obtained from 2, 6 and 12 by $2 + 2 + 2 + 2$ or $2 + 6$. Thus C belongs to four cliques of three persons each or else to two cliques one containing 3 persons and the other containing 4 persons. But since no one else belongs to a 4-person clique, C must belong to 4 cliques of 3 persons each.

Consulting the matrix S we can determine the composition of the 4 cliques. They are: {A, B, C}, {A, C, E}, {B, C, D} and {C, D, E}.

● **PROBLEM** 6-37

Suppose that six individuals have been meeting in group therapy for a long time and their leader, who is not part of the group, has drawn the digraph G below to describe the influence relations among the various individuals.

Who is the leader of the group? Who is the person who influences no one else?

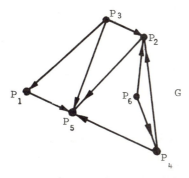

Solution: Every directed graph or digraph has a matrix representation, called the incidence matrix. The incidence matrix for the digraph G is given below:

	P_1	P_2	P_3	P_4	P_5	P_6
P_1	0	0	0	0	1	0
P_2	0	0	0	0	1	0
P_3	1	1	0	0	1	0
P_4	0	1	0	0	1	0
P_5	0	0	0	0	0	0
P_6	0	1	0	1	0	0

A 1 in the i, j position indicates that the ith person dominates (has influence over) the jth person.

Define the leader of the group to be the person who dominates the greatest number of people. Let $P(i)$ denote the number of people that the ith person dominates.

$$P(1) = 1, \quad P(2) = 1 \quad P(3) = 3 \quad P(4) = 2$$

$$P(5) = 0 \quad P(6) = 2.$$

Note that no persons dominate themselves. P_3 influences three people - more than any other individual. Thus P_3 is called the leader of the group. P_5 is the person who influences no one.

● PROBLEM 6-38

Which of the following matrices can be interpreted as perfect communications matrices? For those that are, find the two stage and three stage communication lines that are feedbacks.

	A	B	C	D
A	0	1	1	2
B	1	0	1	0
C	1	1	0	1
D	2	0	1	0

	A	B	C
A	0	1	1
B	1	0	1
C	1	1	0

Solution: In a perfect communication model we have a group of vertices and a mode of communication between them. These could be telephone lines or highways. In such a model if vertex A communicates with vertex B, then necessarily vertex B communicates with vertex A.

First, draw the digraph of the given matrix.

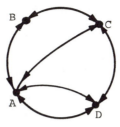

Fig. 1.

Observe that the matrix for Fig. 1 is a perfect communication matrix. Examining the matrix itself, observe that it is symmetric. This follows from the fact that if A is connected by an edge to B, then B is connected by an edge to A.

If the original matrix is squared, we obtain the number of lines of communication between two vertices passing through exactly one other vertex.

$$\text{Now, } M^2 = \begin{bmatrix} 6 & 1 & 3 & 1 \\ 1 & 2 & 1 & 3 \\ 3 & 1 & 3 & 2 \\ 1 & 3 & 2 & 5 \end{bmatrix}$$

The entries given in M^2 can be verified from Fig. 1. For example A can communicate with C by the lines A → B → C, A → D → C, A → D → C. Similarly B can communicate with itself through B → A → B; B → C → B.

The matrix M^3 gives the number of lines of communication between two vertices passing through exactly two other vertices.

$$M^3 = \begin{bmatrix} 6 & 1 & 3 & 1 \\ 1 & 2 & 1 & 3 \\ 3 & 1 & 3 & 2 \\ 1 & 3 & 2 & 5 \end{bmatrix} \begin{bmatrix} 0 & 1 & 1 & 2 \\ 1 & 0 & 1 & 0 \\ 1 & 1 & 0 & 1 \\ 2 & 0 & 1 & 0 \end{bmatrix} = \begin{bmatrix} 6 & 9 & 8 & 15 \\ 9 & 2 & 6 & 3 \\ 8 & 6 & 6 & 9 \\ 15 & 3 & 9 & 4 \end{bmatrix}$$

The feedback from a vertex to itself is given by the entries along the main diagonal. To find out how many ways A can obtain two-stage or three-stage feedback, it is necessary to only look at the value in the diagonal of $M + M^2 + M^3$.

	A	B	C	D
A	12	11	12	18
B	11	4	8	6
C	12	8	9	12
D	18	6	12	9

$$M + M^2 + M^3 =$$

Thus, A has 12 ways to get feedback. Similarly B has 4, C has 9 and D has 9.

The digraph for the second matrix is

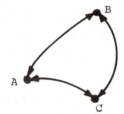

Fig. 2.

Thus, the matrix is a perfect communication matrix. Observe that it is symmetric with only ones or zeros as elements. It is an incidence matrix. The square of the matrix is

	A	B	C
A	2	1	1
B	1	2	1
C	1	1	2

$$M^2 =$$

This matrix illustrates the number of two stage communication lines between pairs of vertices in the group. The 2's on the diagonal indicate that there are two ways in which each vertex can get information back to itself through one other vertex. The matrix M^3 will give the number of three-stage communication lines between pairs of vertices. The values along the diagonal give the number of ways a vertex can get information back to itself through two other vertices. Here, A has two ways of obtaining feedback information, namely A → B → C → A and A → C → B → A.

	A	B	C
A	2	3	3
B	3	2	3
C	3	3	2

$$M^3 =$$

Again, to find out how many ways A, B, or C can obtain

two-stage or three-stage feedback, we need only look at the values in the diagonal of $M + M^2 + M^3$

$$
M + M^2 + M^3 = \begin{array}{c|ccc}
 & A & B & C \\
\hline
A & 4 & 5 & 5 \\
B & 5 & 4 & 5 \\
C & 5 & 5 & 4
\end{array}
$$

Thus, A, B and C have four ways to get feedback. The 5's in row i column j indicate the total number of ways for i to communicate with j using one stage, two stages or three stages.

● PROBLEM 6-39

The following figure shows the internal communication digraph as it exists in an organization headed by five officials, A, B, C, D and E:

a) Construct the incidence matrix corresponding to the digraph

b) Is the graph connected?

c) Which officials are liaison officials?

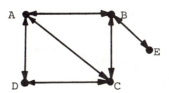

Solution: The graph in Fig. 1 is an example of a business communication model. This model obeys the following assumptions:

1) If official A communicates with official B then necessarily B communicates with A.

2) No official communicates with himself

3) Two officials may or may not communicate at all

4) There exists at most one line of direct communication between any two officials.

Assumptions 1) and 4) guarantee that the matrix of the digraph will be a symmetric incidence matrix.

a) The incidence matrix is

	A	B	C	D	E
A	0	1	1	1	0
B	1	0	1	0	1
C	1	1	0	1	0
D	1	0	1	0	0
E	0	1	0	0	0

b)   We first define the notion of a path. A path between two vertices v_1 and v_2 is a collection of edges and vertices of the form

$$v_1 \to v_3 \to v_4 \to \dots \to v_2$$

in which no vertex is repeated. For example, in Fig. 1,

$$A \to B \to C \to A \to D$$

is not a path but $A \to B \to C \to D$ is a path. Also, $E \to B \to C \to B \to A$ is not a path but $E \to B \to A$ is a path from E to A.

A graph is said to be connected if there is a path between every two officials. Observe from Fig. 1 that the given digraph is connected.

c)   A liaison official in a connected digraph is an official whose removal from the digraph results in a disconnected graph. In terms of organization theory, the removal of a liaison official will cause a break-down in some line of communication between two other officials.

The following theorem is useful in determining liaison officials:

A business communications graph is disconnected if and only if the corresponding incidence matrix M of the graph has the property that $M + M^2 + M^3 + \dots + M^{n-1}$ has one or more zero entries.

Intuitively, this means that a graph is disconnected when there is no path between two officials (with n officials, any path can have at most n-1 vertices). Recall that the kth power of an incidence matrix corresponds to the kth stage communication between officials.

To determine whether a person P from among n officials is a liaison official, delete from the incidence matrix M the column and the row corresponding to P. Call the deleted matrix N. This matrix represents a disconnected digraph if

$$N + N^2 + \dots + N^{n-2}$$

contains zero entries. In this case P is a liaison official.

Thus, to check whether A is a liaison official, first delete the first row and column from the incidence matrix to obtain

	B	C	D	E
B	0	1	0	1
C	1	0	1	0
D	0	1	0	0
E	1	0	0	0

Now find N^2 and N^3:

$$N^2 = \begin{bmatrix} 2 & 0 & 1 & 0 \\ 0 & 2 & 0 & 1 \\ 1 & 0 & 1 & 0 \\ 0 & 1 & 0 & 1 \end{bmatrix};$$

$$N^3 = \begin{bmatrix} 0 & 3 & 0 & 2 \\ 3 & 0 & 2 & 0 \\ 0 & 2 & 0 & 1 \\ 2 & 0 & 1 & 0 \end{bmatrix}$$

Then $N + N^2 + N^3$

$$= \begin{bmatrix} 2 & 4 & 1 & 3 \\ 4 & 2 & 3 & 1 \\ 1 & 3 & 1 & 1 \\ 3 & 1 & 1 & 1 \end{bmatrix}$$

A is not a liaison official.

Similarly, we find that C, D and E are not liaison officials and that only B is a liaison official. Examining the digraph, observe that if B is removed, no communication is possible between E and the other officials.

● PROBLEM 6-40

Develop a pseudocoded program that converts the nodes of the following undirected graph into an adjacency matrix.

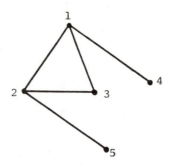

Solution: A graph is a collection of nodes and branches in which each branch links one node to another. The particular graph above is called an undirected graph because there is no particular direction associated with any of the branches. As you can see, the nodes are numbered. Two nodes that are connected by branches may be represented by pairs; for this graph, we notice the pairs:(1,2), (1,3), (1,4), (2,3), and (2,5).

One way of representing the graph is to use a list of the branches as defined by the pairs of nodes. We will look at the representation of the graph through an adjacency matrix. We define an adjacency matrix as one whose (i,j)th element is nonzero if nodes i and j are joined by a branch. Otherwise, the (i,j)th element is zero. For this graph, we have the adjacency matrix below:

$$\begin{pmatrix} 0 & 1 & 1 & 1 & 0 \\ 1 & 0 & 1 & 0 & 1 \\ 1 & 1 & 0 & 0 & 0 \\ 1 & 0 & 0 & 0 & 0 \\ 0 & 1 & 0 & 0 & 0 \end{pmatrix}$$

Notice that the nonzero elements are found at both the (i,j)th elements and the (j,i)th elements. This is because the graph is undirected: nodes may be accessed along branches from either side.

For this simple graph, we store nodes in a 5 by 5 array. Other data structures may be used to represent graphs; a chained list could be used if the nodes are to be changed often.

The pseudocode is presented below. After the array is filled with zeros, the (i,j)th elements and the (j,i)th elements are changed to ones.

```
C      TO CONVERT FROM PAIRS TO THE ADJACENCY MATRIX,
C      ASSUME THE BRANCHES ARE GIVEN AS PAIRS A(K), B(K)
           integer A,B,I,J,K,m(5,5)
           INPUT (A(1),B(1)),(A(2),B(2)), ..., (A(5),B(5))
           do for I←1 to 5
               do for J←1 to 5
                   m(I,J)←0
               end do for
           end do for
C      NOW CHANGE ELEMENTS TO NONZERO FOR PAIRS A(K),B(K)
C      FOR EACH PAIR
```

K indicates the branch number and A and B the nodes of that
branch.

```
       do for K←1 to 5
          m(A(K),B(K))←1
          m(B(K),A(K))←1
       end do for
       output m(K,K)
       end program
```

ISOMORPHIC AND HOMEOMORPHIC GRAPHS

● **PROBLEM** 6-41

Show that the following two graphs G and G' are isomorphic graphs.

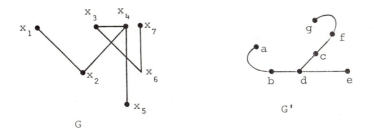

Solution: Let $G(V,E)$ and $G*(V*,E*)$ be graphs and $f : V \to V*$
be a one-to-one correspondence between the sets of vertices
(V and V*) defined by the following rule: $\{x,y\}$ is an edge
of G if and only if $\{f(x),f(y)\}$ is an edge of G*. Then f is
called an isomorphism between G and G*, and G and G* are said
to be isomorphic graphs.

In order to show that G and G' are isomorphic graphs,
one needs to verify that the number of edges that join any
pair of vertices v_i and v_j in G is the same as the number of
edges joining vertices $f(v_i) = v_i'$ and $f(v_j) = v_j'$ in G'.

315

For this problem, f is found to be defined as the following:

v	f(v')
x	a
x	b
x	c
x	d
x	e
x	f
x	g

and f is an isomorphism. Furthermore, from the following table it's observed that the number of edges that join any pair of vertices v_i and v_j in G is the same as the number of edges joining vertices v_i' and v_j' in G'.

Vertices in Graph G	Number of Joining Edges in G	Vertices in Graph G'	Number of Joining Edges in G'
x_1 and x_2	1	a and b	1
x_1 and x_3	0	a and c	0
x_1 and x_4	0	a and d	0
x_1 and x_5	0	a and e	0
x_1 and x_6	0	a and f	0
x_1 and x_7	0	a and g	0
x_2 and x_3	0	b and c	0
x_2 and x_4	1	b and d	1
x_2 and x_5	0	b and e	0
x_2 and x_6	0	b and g	0
x_2 and x_7	0	b and f	0
x_3 and x_4	1	c and d	1
x_3 and x_5	0	c and e	0
x_3 and x_6	1	c and f	1
x_3 and x_7	0	c and g	0
x_4 and x_5	1	d and e	1
x_4 and x_6	0	d and f	0
x_4 and x_7	0	d and g	0
x_5 and x_6	0	e and f	0
x_5 and x_7	0	e and g	0
x_6 and x_7	1	f and g	1

Therefore, G and G' are isomorphic graphs.

316

Determine whether the graphs G and G' in Fig. 1 are isomorphic.

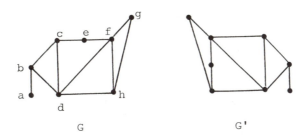

Fig. 1.

Solution: Observe that the graph G has only one 1-degree vertex, one beings by labeling the only 1-degree vertex in G' as shown in Fig. 2.

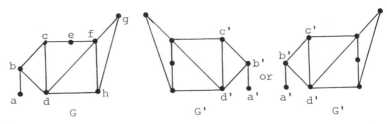

Fig. 2.

Vertex c' must be joined to one 2-degree vertex, one 3-degree vertex and one 4-degree vertex (since vertex c is joined to one 2-degree vertex, one 3-degree vertex and one 4-degree vertex) if G and G' are to be isomorphic. However, vertex c' is joined to one 3-degree vertex and two 4-degree vertices. Therefore, G and G' are not isomorphic.

Determine whether the graphs G and G' in Fig. 1 are isomorphic.

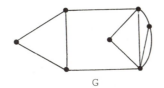

Fig. 1.

G   G'

<u>Solution</u>: First, label either of the given graphs arbitrarily since none of them are labeled (see Fig. 2).

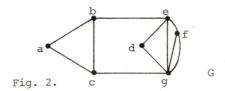

Fig. 2.

Note that vertex g in G is the only 3-degree vertex. Furthermore, if G and G' are to be isomorphic, their corresponding vertices must have the same degrees. Therefore, the only 5-degree vertex in G is labeled g'. By the same token, the only 2- and 4-degree vertices joint to g' are labeled d' and e', respectively (see Fig. 3).

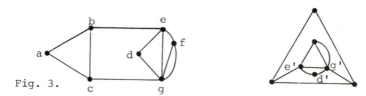

Fig. 3.

A choice needs to be made as how to label the two 3-degree vertices connected to g'. The possibilities are shown in Fig. 4.

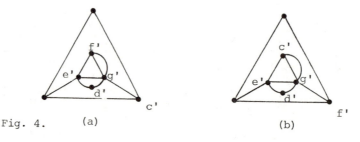

Fig. 4.   (a)   (b)

Note that c in G is joined to a 2-degree and a 3-degree vertices b and a, respectively. It's clear that the choice in Fig. 4(b) is incorrect. Continue the labeling by using Fig. 4(a), the result is shown in Fig. 5.

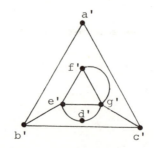

Fig. 5.   b'   c'

318

It's observed that G and G' are isomorphic. A table may be constructed to verify this final result:

Vertices in Graph G	Number of Joining Edges in G	Vertices in Graph G'	Number of Joining Edges in G'
a and b	1	a' and b'	1
a and c	1	a' and c'	1
a and d	0	a' and d'	0
a and e	0	a' and e'	0
a and f	0	a' and f'	0
a and g	0	a' and g'	0
b and c	1	b' and c'	1
b and d	0	b' and d'	0
⋮	⋮	⋮	⋮

It is shown in Fig. 6 that the two graphs are actually identical.

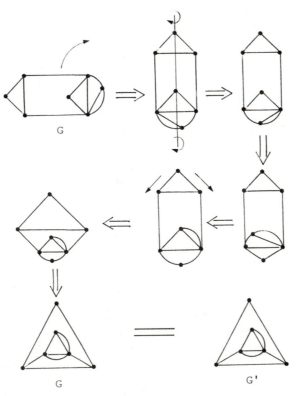

Fig. 6.

319

Indicate which of the following graphs are isomorphic.

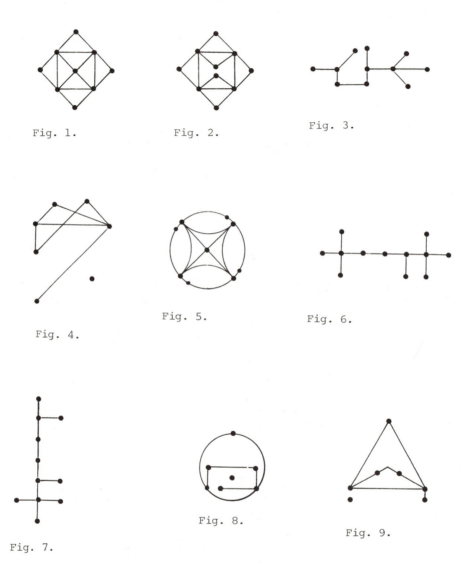

Fig. 1.

Fig. 2.

Fig. 3.

Fig. 4.

Fig. 5.

Fig. 6.

Fig. 7.

Fig. 8.

Fig. 9.

<u>Solution</u>: Isomorphic graphs are graphs in which there exists a direct, one-to-one relationship between the vertices of each graph. Each corresponding node will have the same characteristics. Isomorphic graphs may look different, but they are mathematically identical.

Therefore, the graphs in Figs. 3 and 7, Figs. 1 and 5, and the graphs in Figs. 4, 8, 9 are isomorphic graphs.

Indicate which of the following graphs are homeomorphic.

Fig. 1.

Fig. 2.

Fig. 3.

Fig. 4.

Fig. 5.

Solution: For any graph G one can obtain a new graph G' by dividing an edge of G with additional vertices; G and G' are said to be homeomorphic.

Therefore, graphs in Figs. 1, 3 and 5 are homeomorphic.

PLANAR GRAPHS AND COLORATIONS

● **PROBLEM** 6-46

Three families, a, b and c rely upon underground supply lines for their water from point x, their gas from point y, and their electricity from point z. Is it possible to arrange the three families and the three utility stations so that no supply lines cross one another except at their endpoints?

Solution: Assume it is possible to have the graph $K_{3,3}$ made planar. Then, one can apply Euler's Formula which states that for a planar graph $n - m + f = 2$ where n is number of vertices, m is number of edges and f is number of faces of a convex polyhedron. The graph for three families and three utilities has $m = 9$, $n = 6$ and $f = 2-n+m = 2-6+9 = 5$. Each

face has at least 4 edges in its contour because if a face F
had only 3 edges, then it would be bordered by 3 vertices of
which 2 must be in the same class (since there are only two
classes, families or utilities), however, two vertices of the
same class should not be adjacent. For the bipartite face-
edge incidence graph, the number of arcs A is such that
$4f \leq A \leq 2m$. However, $4f = 20 \nleq 2m = 18$ which is contradictory to the initial assumption. Hence, the graph for three
families and three utilities cannot be planar. It can be
shown that 8 supply lines can always be placed without cross-
ing one another but that the 9th supply line has to cross at
least one other supply line (see the figure).

$K_{3,3}$

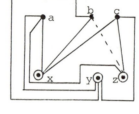

● **PROBLEM** 6-47

Draw a planar representation for each of the following
graphs:

(1)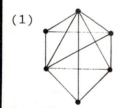

(2)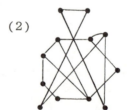

Solution: A planar graph or multigraph is a graph or multigraph which can be drawn in the plane so that its edges do
not cross.

The planar representation for the given graphs in (1)
and (2) are shown in Figs. 1 and 2, respectively.

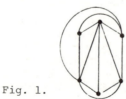

Fig. 1.

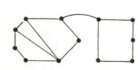

Fig. 2.

322

Draw a planar representation of the following graphs if
possible.

(1)   (2)

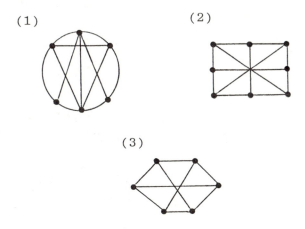

(3)

Solution: The graphs in (2) and (3) are non-planar graphs.

The planar representation of the graph in Fig. 1 is
shown in Fig. 4.

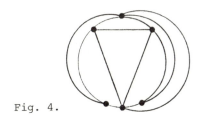

Fig. 4.

Find the dual of the following maps:

(1)   (2)

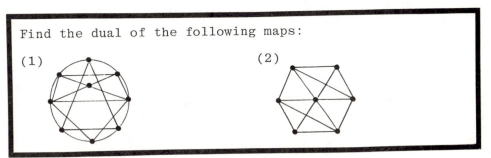

Solution: The dual of the two graphs are shown in Figs. 1
and 2.

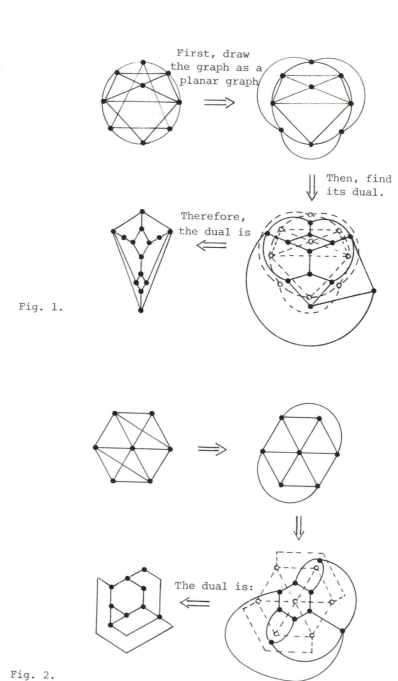

First, draw the graph as a planar graph

Then, find its dual.

Therefore, the dual is

Fig. 1.

The dual is:

Fig. 2.

● **PROBLEM** 6-50

Draw the map which is the dual of the following map:

Solution: A map is a planar representation of a finite
planar multigraph. Two regions of a map M are said to be
adjacent if they have an edge in common. To obtain the dual
of a map M, choose a point in each region of M and the corresponding points of two adjacent regions are connected by a
curve through the common edge of the two regions. The curves
can be drawn such that they are non-crossing. Therefore, we
obtain the dual of M, a new map, M*, such that each vertex of
M* corresponds to exactly one region of M.

The dual of the given map is shown in Fig. 1.

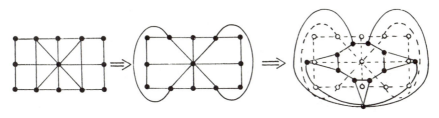

Fig. 1.

● PROBLEM 6-51

Can a complete graph K_5 be planar?

Solution: A graph G is complete if each vertex of G is connected to every other vertex of G. By Euler's formula for a
planar graph, one has $n - m + f = 2$, where n is number of vertices, m is number of edges and f is number of faces of a convex polyhedron. The graph K_5 is shown in the figure; if it
is to be planar, $m = 10$, $n = 5$, $f = 2 - n + m = 2 - 5 + 10 = 7$.
The contour of each face has at least 3 edges. For the bipartite face-edge incidence graph, the number of arcs A, is
such that $3f \leq A \leq 2m$. However, $3f = 21 \not\leq 2m = 20$ for K .

Hence, K_5 can't be planar.

● PROBLEM 6-52

Use the algorithm developed by Welch and Powell to paint
the following graphs, and find the chromatic number for
each graph.

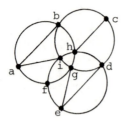

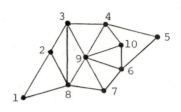

Fig. 1.   Fig. 2.

Solution: (1) First, order the vertices in decreasing degrees as follows:

d,g,b,h,i,e,a,f,c.

Note that the arrangement is not unique for a given graph, since some vertices may have the same degree. The graph is painted as shown in Fig. 3. The minimum number of colors needed to paint a graph is called the chromatic number. The chromatic number for this graph is 4.

Vertex: d, g, b, h, i, e, a, f, c

Color : 1, 2, 1, 3, 4, 3, 2, 1, 2

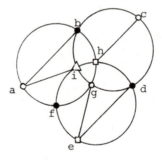

●: Color 1
○: Color 2
□: Color 3
△: Color 4

Fig. 3.

(2) The graph is painted as shown in Fig. 4. The chromatic number is 3.

Vertex: 9, 8, 6, 4, 3, 2, 7, 10, 1, 5

Colors used
to paint the
vertices
Color : A, B, B, B, C, A, C, C, C, A

●:Color A

○:Color B

□:Color C

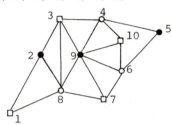

Fig. 4.

Find the minimum number of colors needed to paint the
region of the following maps.

Fig. 1.

Fig. 2.

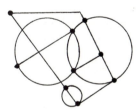

Fig. 3.

Solution: A map is a planar representation of a finite planar multigraph. By coloring, or painting, a map we mean an
assignment of a color to each region of the map such that
adjacent regions have different colors.

For simple maps like those in Fig. 1 and Fig. 2, one can paint the regions of the map simply by inspection.

For instance, the map in Fig. 1 is painted as shown in Fig. 4.

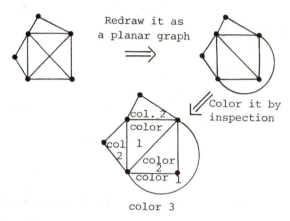

Redraw it as a planar graph

Color it by inspection

Fig. 4

Therefore, the minimum number of colors needed to paint the regions of the map in Fig. 1 is 3.

The minimum number of colors needed to paint the regions of the map in Fig. 2 is three. The graph is painted as shown in Fig. 5.

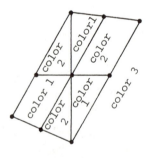

Fig. 5.

To color the regions of the map in Fig. 3, we first find its dual map which is a planar graph. We then use the Welch-Powell method to color the vertices of the dual map as shown in Fig. 6, which is equivilent to painting the regions of the original map.

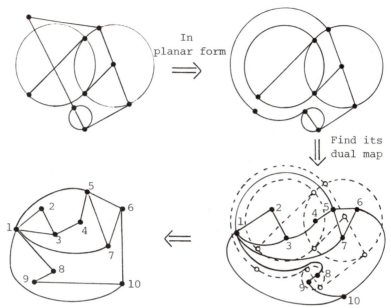

In
planar form

⟹

Find its
dual map

⟸

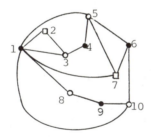

The graph is colored as shown below.

(vertices are ordered according to the decreasing degree)

Vertex : 1, 5, 3, 6, 7, 2, 4, 8, 9, 10
Color : A, B, B, A, C, C, A, B, A, B

Fig. 6.

The minimum number of colors needed to paint the regions of the map in Fig. 3 is three.

● **PROBLEM** 6-54

A zoo wishes to provide some large enclosures of a few acres or more in which a number of different species are allowed to roam freely and through which zoo visitors pass in enclosed or elevated vehicles for 11 species. Among these species some are natural enemies of each other; they must not allow to live in the same enclosure. These incompatibilities are indicated by the mark " " as shown in the table in Fig. 1. Find the minimum number of enclosures needed for these species.

	1	2	3	4	5	6	7	8	9	10	11
1			✓								
2				✓					✓		✓
3	✓										
4		✓				✓					
5							✓				
6				✓			✓				✓
7					✓	✓					
8									✓		
9		✓						✓		✓	
10									✓		
11		✓				✓					

Fig. 1.

Solution: If one represents the species by vertices in a graph and connects two vertices which represent two incompatible species. Then, the problem is converted to finding the minimum member of colors needed to color this graph. The graph is shown in Fig. 2.

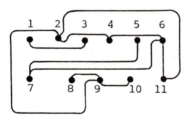

Fig. 2.

One of the algorithms which can be used to color a graph is the Welch-Powell algorithm. This algorithm is the following: "First order the vertices of the graph G to decreasing degrees. Then use the first color to paint the first vertex and to paint, in sequential order, each vertex which is not adjacent to a previously painted vertex (of the same color). Repeat the process using the second color and the subsequence of unpainted vertices. Continue the process with the third color, and so on until all vertices are colored. Note that the ordering of the vertices may not be unique since some vertices may have the same degree. Follow the Welch-Powell algorithm the graph in Fig. 2 is colored as shown in Fig. 3. (The vertices are ordered as 2, 6, 9, 4, 11, 7, 1, 3, 5, 8, 10.)

330

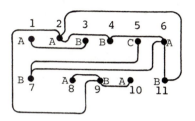

A, B, and C are three
different colors.

Fig. 3

The minimum number of colors needed to color the graph
is three.   Therefore, the minimum number of enclosures is
three.

● PROBLEM 6-55

Let G be a graph with every vertex having a degree $\leq n$.
Show that G can be colored with n + 1 colors.

Solution: Start with any vertex of G, say v_i.   Then find all
the other vertices connected to v_i by an edge.   The maximum
number of colors is needed when all the vertices of G are
connected to v_i and they are also connected to each other.

In this case, the maximum, or n + 1 colors, are needed.   Next
find any vertex which hasn't yet been colored.   This vertex
can be connected to at most n of the vertices which have already been colored, hence at most, we need n + 1 colors.
Repeat this argument until all vertices are colored.   It is
clear that in no case does one need more than n + 1 colors.

TREES

● PROBLEM 6-56

If one can go from city A to city B in 4 different ways,
a, b, c, and d, and from city B to city C in 2 different
ways, say x and y.   How many different ways can one go
from city A to city B and then to city C?   Illustrate
your answer by a tree.

Solution: By the rule of product, there are $4 \times 2 = 8$ different ways one can go to city C from city A via city B.
This is illustrated by a tree in Fig. 1.

From City A
to City B

From City A
to City C

From City B
to City C

Fig. 1.

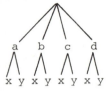

a   b   c   d

x y x y x y x y

Draw   (1)   all trees with 4 vertices.

(2)   all trees with 8 vertices.

Solution: A tree is a connected graph with no cycles.

(1) The trees with 4 vertices are shown in Fig. 1. There
are two such trees.

Fig. 1.

(2) There are 22 trees with 8 vertices as shown in Fig. 2.

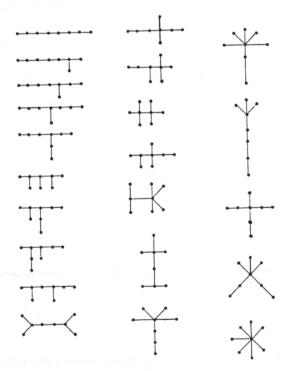

Fig. 2.

Prove the following theorem:

"A graph G = (V,E) has a partial graph that
is a tree if, and only if, G is connected."

Solution: The proof consists of two parts.

Part I: If a graph G is not a connected graph. Then, no
partial graph of G is connected. Hence, G does
not have any partial graph that is a tree.

Part II: If G is connected, then one of the following must
be true: (a) one can't find an edge whose removal
does not disconnect the graph. In this case, G is
a tree by virtue of the property of the connected
graphs (which states that a connected graph
G = (V,E) of order $|V|$ = n > 2 is a tree if the
removal of any edge of G disconnects G).

(b) One can find edges whose removal does not disconnect the graph. In this case, one can remove
such edges until no more edges can be removed without disconnecting G.

Therefore, the remaining graph (which is a partial
graph of G) is a tree.

Find all the spanning trees in the following graph.

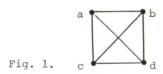

Fig. 1.

Solution: Let G = {V,E} be a simple graph, then the number
of spanning trees in G is equal to the minor (which is independent of the coefficients of the principle diagonal) of the
square matrix of order n, and the entry of the matrix a_{ij} is
defined as follows:

$$a_{ij} = \begin{cases} \text{degree of } x_i & \text{if } i = j \\ -1 & \text{if } i \neq j \text{ and } \{x_i,x_j\} \quad E \\ 0 & \text{if } i \neq j \text{ and } \{x_i,x_j\} \quad E \end{cases}$$

The matrix for the given graph is:

$$
M = \begin{array}{c} \\ a \\ b \\ c \\ d \end{array}
\begin{array}{c}
\begin{array}{cccc} a & b & c & d \end{array} \\
\left|\begin{array}{cccc}
3 & -1 & -1 & -1 \\
-1 & 3 & -1 & -1 \\
-1 & -1 & 3 & -1 \\
-1 & -1 & -1 & 3
\end{array}\right|
\end{array}
$$

The number of spanning trees in G is found by calculating

$$
\det \begin{vmatrix}
3 & -1 & -1 \\
-1 & 3 & -1 \\
-1 & -1 & 3
\end{vmatrix} = 16
$$

These 16 spanning trees are shown in Fig. 2.

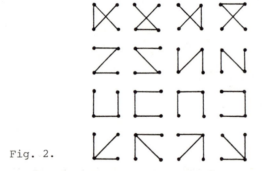

Fig. 2.

● **PROBLEM** 6-60

Draw all the possible spanning trees of the graph in
Fig. 1.

Fig. 1.

Solution: A subgraph of a graph G is called a spanning tree
of G if the subgraph includes all the vertices of G and is a
tree.

Therefore, the given graph has four spanning trees as shown in Fig. 2.

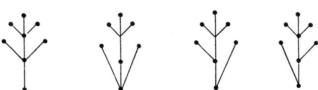

Fig. 2.

Draw all the spanning trees of the graph in Fig. 1.

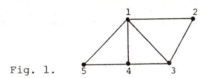

Fig. 1.

Solution: The spanning trees for the graph in Fig. 1 are shown in Fig. 2.

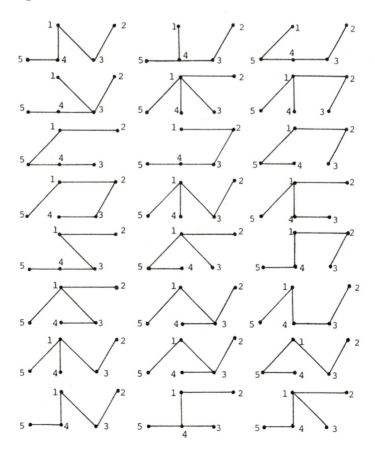

Fig. 2.

A tree of n vertices must have a spanning tree of n - 1 edges. Since the given graph has 5 vertices, the spanning trees must have 5 - 1 = 4 edges. Furthermore, the given graph has 7 edges, so that there are $_7C_3$ different ways to delete three edges from the graph in Fig. 1. However, among the

$$_7C_3 = \frac{7!}{3!(7-1)!} = 35 \text{ ways to delete 3 edges, 11 of}$$

them yield subgraphs which are not trees. Therefore, there must be 35 - 11 = 24 spanning trees. See Fig. 2.

● **PROBLEM** 6-62

Find a minimal spanning tree of the graph G in Fig. 1.

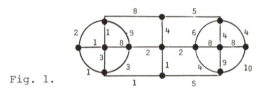

Fig. 1.

Solution: For graph G, whose edges with lengths are shown in Fig. 1, the minimal spanning tree of G is a spanning tree of G such that the sum of the lengths of the edges is minimal among all spanning trees of G.

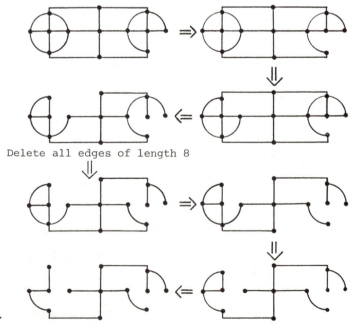

Delete all edges of length 8

Fig. 2.

One of the methods for finding a minimal spanning tree
is to repeatedly delete the edges with maximum length without
disconnecting the graph. The process of finding a minimal
tree for the given graph is shown in Fig. 2.

The minimal spanning tree we found is shown in Fig. 3.

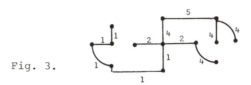

Fig. 3.

• **PROBLEM** 6-63

Find a minimal spanning tree for the graph in Fig. 1.

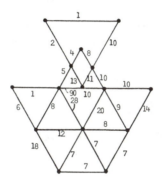

Fig. 1.

Solution: One of the methods to obtain a minimal tree for a
given graph is to place all the vertices first, then keep
adding edges of minimum length without forming a cycle until
a tree is obtained.

This process is shown in Fig. 2.

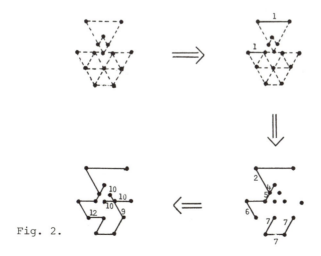

Fig. 2.

337

Find the shortest spanning tree in Figure 1, by applying the greedy algorithm.

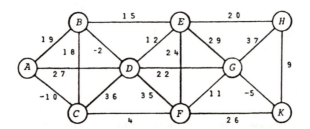

Fig. 1

<u>Solution</u>: The greedy algorithm: One begins with a sub-tree consisting of only one arbitrarily selected node, which will not affect the length of the final tree; this subtree is expanded by adding one arc at a time (with its incident node) until all p nodes have been connected by means of p - 1 arcs. At each stage the added arc is an arc of shortest length c_{ij} connecting a node i in the sub-tree with a node j not yet in the subtree. This produces a shortest spanning tree if $c_{ij} = c_{ji}$ and the arcs having finite c_{ij} form a connected graph.

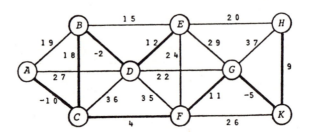

Fig. 2

The algorithm, applied to the problem produces the network in Figure 2 in which the heavy lines indicate the resulting shortest spanning tree of length 37. If one starts at node A one obtains the arc lengths

$$c_{AC} = -10 = \min(c_{AB} = 19, \ c_{AD} = 27, \ c_{AC} = -10)$$

$$c_{CF} = 4 = \min(19, 27, 18, 36, 4)$$

$$c_{FG} = 11 = \min(19, 27, 18, 36, 35, 24, 11, 26)$$

and so on to GK, KH, CB, BD and DE. The resulting undirected tree may be recorded compactly as:

EDBCFGKH.
A

The planned locations of computer terminals that are to be
installed in a multistory building are given in Figure 1.
Terminal A is the computer itself and phone cables must be
wired along some of the indicated branches in order that
there be a connected path from every terminal back to A.
The numbers along the arcs represent the costs (in hundreds
of dollars) of installing the lines between terminals.
Since operating costs are very low, the company would like
to find the branches that should be installed in order to
minimize total installation costs.

Solve this problem by applying the greedy (next-best)
rule.

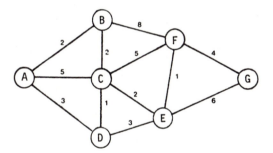

Fig. 1

<u>Solution</u>: Begin at any terminal and find the branch to
the nearest unconnected terminal. This branch is part of
the tree. From these two terminals find the branch to the
nearest unconnected terminal and add this branch to the
tree. Continue in this fashion until all terminals are
connected.

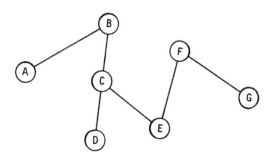

Fig. 2

339

Beginning arbitrarily at node C in Figure 1, the nearest terminal is D; CD is a branch on the tree. The nearest terminals to both C and D are B and E; arbitrarily
choose E. Next, find F closest to C, D, and E, and finally
choose B, then A and G. The spanning tree is as given in
Figure 2 and the total cost is $1200. This happens to be
the optimal solution even though the "next-best" rule has
been used to solve the problem. The next-best method always finds the minimum spanning tree.

● PROBLEM 6-66

Find the minimum spanning tree of the graph G(V,U) of
figure 1. Notice that it is an undirected graph.

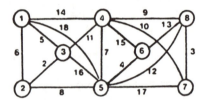

Fig. 1

Solution: Going through all the vertices v_1 to v_8 and
drawing the arc connecting each to its nearest neighbor,
yields the graph $G_1(V,U_1)$ of figure 2. The nearest
neighbor of v_i is v_3 , of v_2 is v_3 , of v_3 is v_2 , and
so on. The graph G_1 is not connected. It has three
components, A_1, A_2, A_3. Treat them as three 'vertices'.
The arcs of G connecting A_1 to A_2 are of lengths 14,18,
8,16,11, and so the distance between A_1 and A_2 is 8.
Similarly the distance between A_2 and A_3 is 9. Also
since there is no arc connecting A_1 and A_3, the distance
betweem them is ∞. The nearest neighbor of A_1 is A_2 and
of A_2 is A_1. Connect the two by arc (v_2,v_5), which
measures the distance between the two.

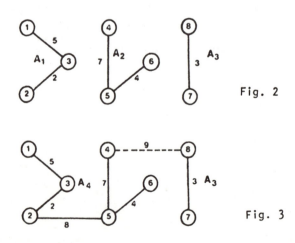

Fig. 2

Fig. 3

340

Thus results the graph $G_2(V,U_2)$ of figure 3 which has two components, A_4 and A_3. Since they are only two, each is the nearest neighbor of the other, and so connect them with the arc (v_4, v_8) (shown dotted) which measures the distance between them. Thus, there is a single connected graph which is the smallest spanning tree. The length of the tree is 38.

● **PROBLEM** 6-67

Represent the algebraic expression

$$((a-b) * c) + 7) * ((d + 4)/x)$$

by a tree, and determine the height of this tree,

Solution: The height of this tree is 4 (the length of the longest path).

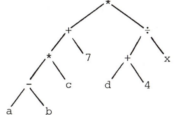

● **PROBLEM** 6-68

Find the prefix representation for the tree represented in Fig. 1.

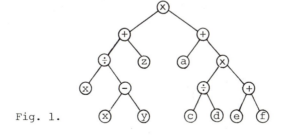

Fig. 1.

Solution: The prefix representation is found (as shown in Fig. 2) to be

$$\times + \div x - xyz + a \times \div cd + ef.$$

341

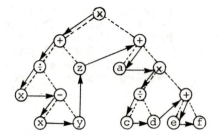

Fig. 2.

Another method that one may use to find the prefix representation for a tree is to order the tree in lexicographic order first, then list the symbol occurences in address order as illustrated in Fig. 3.

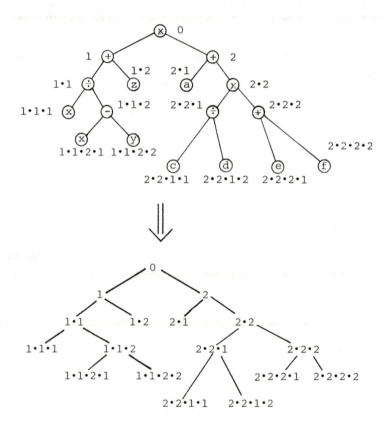

The lexicographic order is

$0 < 1 < 1 \cdot 1 < 1 \cdot 1 \cdot 1 < 1 \cdot 1 \cdot 2 < 1 \cdot 1 \cdot 2 \cdot 1 < 1 \cdot 1 \cdot 2 \cdot 2 < 1 \cdot 2 < 2 < 2 \cdot 1 < 2 \cdot 2 < 2 \cdot 2 \cdot 1 < 2 \cdot 2 \cdot 1 \cdot 1$

$< 2 \cdot 2 \cdot 1 \cdot 2 < 2 \cdot 2 \cdot 2 < 2 \cdot 2 \cdot 2 \cdot 1 < 2 \cdot 2 \cdot 2 \cdot 2$

The symbols are listed in the address order:

x + ÷ x - x y z + a x ÷ c d + e f

Fig. 3.

342

Define the following terms as they relate to Polish string notation:

a) infix

b) prefix

c) postfix

   Then, convert the following tree into the three types of expressions just defined.

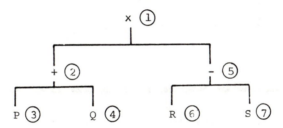

Solution: Infix notation refers to the way we normally write expressions, namely, with the operators between the two operands. This is also known as algebraic notation, which is different from Polish notation.

   Prefix notation refers to the fact that the operator precedes the expression in the string. So, if we have the string ÷ XY, it means that the division operator acts to divide X by Y.

   One inconvenience with prefix notation is that to understand the string, you must read from end to beginning. Postfix notation avoids this difficulty. For the string above, postfix notation would be XY ÷ . The meaning is the same as above, but the operator comes after the two operands.

   The infix translation of the tree diagram is then (P + Q) x (R - S). This is obvious if you follow the tree up from the terminal nodes to the root node.

   Prefix translation yields x + PQ - RS. By simply looking at the numbering of the individual nodes, you can see that they correspond with the order of the string. In other words, you can write the following to highlight the correspondence:

$$X + PQ - RS$$

①②③④⑤⑥⑦

Postfix notation looks somewhat different. For this tree, we get PQ + RS - x. This may be read from beginning to end as "add P and Q, then subtract S from R, then multiply the sum and the difference together."

Convert (a) - (c) from Infix to Prefix and Postfix expressions and (d) - (f) from Prefix to Infix and Postfix expressions.

(a)   A + B * C/D
(b)   A - C & D + B↑E
(c)   X↑Y↑Z
(d)   + * - ABCD
(e)   + mA * BmC
(f)   ↑X + * YZW

Solution: For conversion from Infix to Prefix and Postfix forms we need to recollect the hierarchy and associativity rule. The former says that ↑ has highest priority, then *, / and then +, - . The latter says that at the same level of priority associativity applies from left to right.

e.g.   A-B-C = (A-B) - C

(a)   A + B * C/D

Conversion to Prefix form: Scanning from left to right we observe the operators +, * and/ · As *, and/ have higher priority than + and * precedes/, BC becomes the operand followed by /D. Thus the whole second operand takes the form /*BCD. A is the first operand and + is the operator between them.

A + B * C / D   = + A/*BCD

Postfix form: — we have BC* then BC*D/ & finally ABC*D/+

A + B*C/D = ABC * D/+

(b) A-C * D + B↑E   scan left to right

First step   B↑E = ↑BE, BE↑ ↑has highest priority

Second step   C*D = *CD, CD*   * has 2nd highest priority

Third step   A–C*D = – A*CD, ACD+–   – and + are on the
             same level

Final step   A–C*D + B↑E = + – A*CD↑BE, ACD* – BE ↑ +

(c)   X↑Y↑Z   both ↑ on same level

                  Infix   Prefix   Postfix

First step   X↑Y   =   ↑XY   ,   XY↑

Final step   X↑Y↑Z   =   ↑↑XYZ   ,   XY↑Z↑

     Conversion from Prefix to Infix and Post fix expression.
     (d)   + * – A B C D

                  Prefix   Infix   Postfix

1st step   – AB   =   (A–B)   ,   AB –

2nd step   * –ABC   =   (A–B)*C   ,   AB–C*

Final step   + * – ABCD =   (A–B)*C+D ,   AB–C*D+

     (e)   + m A * B m C   m is unary operator acting
                               on immediate operand with
                               – sign & parenthesis.

                  Prefix   Infix   Postfix

1st step   mC   =   (–C)   ,   Cm

2nd step   &B mC   =   B&(–C)   ,   BCm*

Final step   +mA&BmC   =   (–A) + B&(–C), AmBCm*+

     (f)   ↑X + * YZW/

                  Prefix   Infix   Postfix

1st step   *Y Z   =   Y*Z   ,   YZ*

2nd step   +*YZW   =   Y*Z + W   ,   YZ*W+

Final step   ↑X+*YZW   =   X↑Y*Z+W   ,   XYZ*W+↑

● **PROBLEM** 6-71

Construct a tree which represents the Polish prefix expression +XA–B12÷C4. Remember that in Polish notation, all operators have the same precedence; only their position is significant to the order of operations.

<u>Solution</u>: Trees are useful to display the way in which the computer translates the algebraic form of an expression (the way in which you generally enter various formulae) into Polish notation during the process of compiling a program.

As a general rule, a tree will have operators (+,-,x, $\div$,$\uparrow$) in its nonterminal nodes and operands (constants or variables) in its terminal nodes. The root node is the exception; it may be thought of as representing the middle operator of an expression.

In Polish prefix notation, however, the middle operator actually comes at the beginning of the string. To begin constructing the tree, we number each element to correspond with the number of the node it will occupy. Hence,

+ × A - B 12 ÷ C 4

1 2 3 4 5 6 7 8 9

The first element (+) will be the root node. We see that the next element is (×). Remembering that trees are always constructed left to right, this element becomes the left subnode of the root. Since it is also an operator, we will neglect the right subnode for a moment in order to find the two operands which appear on either side of the (×). (See Figure 1.)

The following figures should illustrate the growth of the tree in a clearer fashion:

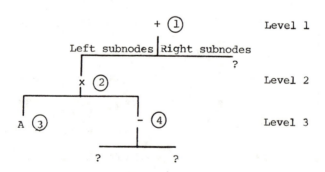

Fig. 1.

At this point, we need to find the operands which should appear on either side of the (-). B and 12, since they are the next two elements, will be those operators. Fig. 2 shows that the left side of the tree is complete:

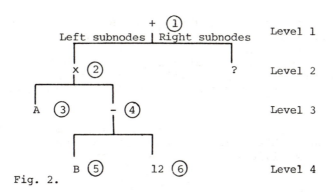

Fig. 2.

Finally, we notice that the next operator is (÷). The two operands, C and 4, follow directly, so we have the completed tree in Figure 3.

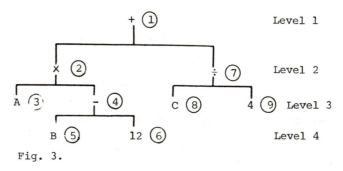

Fig. 3.

● **PROBLEM** 6-72

Evaluate the following tree expressions:

(a)

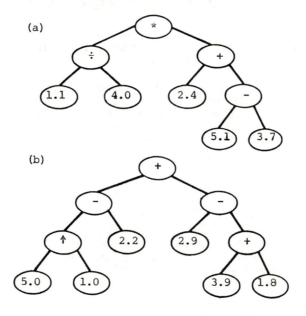

(b)

Solution: To begin evaluating trees, we look for terminal nodes, those which contain numerical values or variables at the end of the tree. Then, the rule becomes "left node, right node, operator." Starting at the leftmost node, we begin combining operands with their operators. The tree begins to lose its leaves as the process of combination continues.

a) Let us evaluate the first tree, combining nodes as we go along. Starting with the leftmost node, we see that /./ is to be divided by 4.0. After evaluation, we get the following tree:

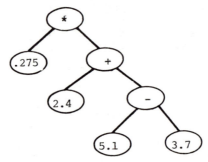

Note that the left side is now complete. Turning to the right side, we may combine 5.1 and 3.7 under the subtraction operator to obtain 1.4. A new node is created containing 1.4 because of this operation.

Moving up the tree from the right, we can combine 2.4 and 1.4 under the addition operator to obtain 3.8. The tree now has the following shape:

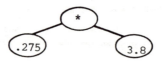

To complete it, we simply perform the multiplication to obtain the final answer 1.045.

b) This tree introduces the $\uparrow$ operator, which signifies the exponential operation. Starting once again from the left, we have 5.0 raised to the 1st power, which yields 5.0. Then, we move up the tree to find that the next task to be done is to subtract 2.2 from 5.0, which yields 2.8. At this stage, the tree is

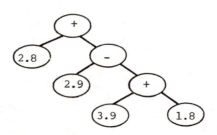

We now start with the right side to find that we must add 3.9 to 1.8, giving us the answer 5.7. We take this answer and move up to the next node, which indicates that we are to subtract 5.7 from 2.9. After this operation, we have

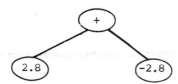

which evaluates to zero.

● **PROBLEM** 6-73

(1) Draw the ordered rooted tree for the algebraic expression

$$[4x^4 + 6y^2 + (6x+44z)^4 - 96(4+3x)] \div 45$$

(using ** for exponentiation).

(2) Rewrite the expression in

(a) postfix representation, and
(b) prefix representation.

Solution: (1)

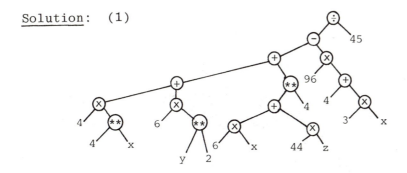

Fig. 1.

(2) (a) 44x ** × 6y2 ** × + 6x × 44z × +

4 ** + 9643 x × + × − 45 ÷

(b) ÷ − + + × 4 ** 4x × 6 ** y_2 + ** + × 6x × 44z

4 × 96 + 4 × 3x

Draw the ordered rooted tree for each of the following
algebraic expressions (using ** for exponentiation).

(1) $x^7 + (5 + 6x \div 4y) \times 14 - 24xyz + z^4$

(2) $(x^4 + y^8)^2 - z^2 + 100x - 50y$

Solution: (1)   An ordered rooted tree representation of an
algebraic expression is actually a graphic illustration of
the order of operations of that particular expression.   Each
edge represents a variable or a value, each vertice represents an algebraic operation.   Each node has two or more inputs and outputs a single value; the result of the operation.
This process continues through the tree until the final value
is determined.

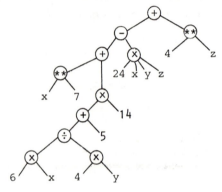

Fig. 1.

(2)

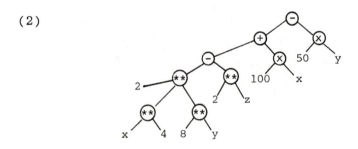

Fig. 2.

Arrange the addresses of the following system in the lexicographic order.

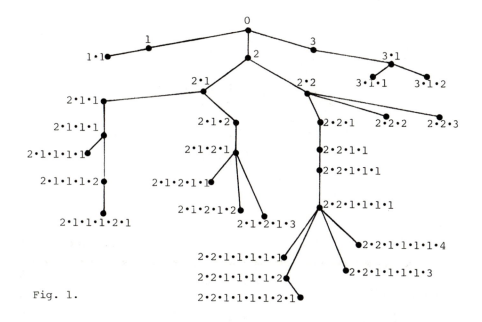

Fig. 1.

Solution: The lexicographic order can be easily obtained by moving down the leftmost branch of the ordered rooted tree, then the next branch to the right, then the second branch to the right, and so on. In short, we can describe this order as "root-left-right" or OLR (O for root). The arrangement for the given system is:

```
0         2           2.1.1.1.2
1         2.1         2.1.1.1.2.1
1.1       2.1.1
          2.1.1.1
          2.1.1.1.1


2.1.2     2.1.2.1.1   2.1.2.1.2   2.1.2.1.3
2.1.2.1


   2.2
   2.2.1
   2.2.1.1     2.2.1.1.1.1.1   2.2.1.1.1.1.2
   2.2.1.1.1                   2.2.1.1.1.1.2.1
   2.2.1.1.1.1


2.2.1.1.1.1.3   2.2.1.1.1.1.4   2.2.2   2.2.3   3      3.1.2
                                                3.1
                                                3.1.1
```

The process is also shown in Fig. 2.

351

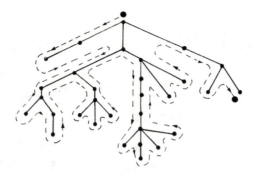

Fig. 2.

Use the universal address system to label the following
ordered rooted trees.

(1)

(2)

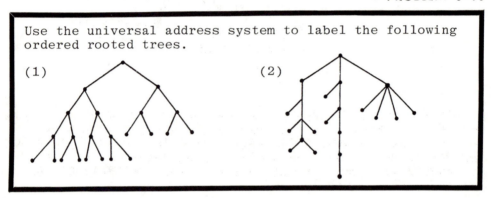

Solution: (1) This tree is labeled as shown in Fig. 3. We
first label the root of the tree by 0. Then we label the
vertices immediately following the root according to the manner in which the edges were ordered.

    For the vertices below this level we label them as fol-
lows: if x is the label of a vertex u then $x \cdot 1$, $x \cdot 2$, $x \cdot 3$
... are assigned to the vertices immediately following u according to the manner in which the edges were ordered.

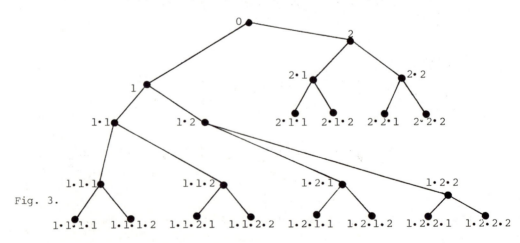

Fig. 3.

(2)

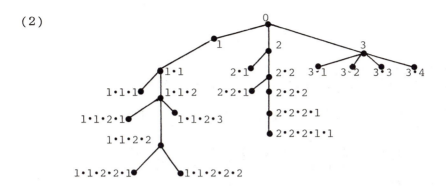

Fig. 4.

Determine what other addresses must be in a universal address system if address a = 4.4.4.4 is in the system.

Solution: In an ordered rooted tree labeled by the universal address system, in order to get the address a, one must pass through the addresses 0, 4, 4.4 and 4.4.4. The addresses in the same level as 4, and preceding 4, are 1, 2 and 3. The addresses in the same level as 4.4 and preceding 4.4 are 4.1, 4.2 and 4.3 The addresses preceding 4.4.4 and in the same level as 4.4.4 are 4.4.1, 4.4.2, and 4.4.3. Finally, the addresses preceding 4.4.4.4 and at the same level as 4.4.4.4 are: 4.4.4.1, 4.4.4.2 and 4.4.4.3. Therefore, addresses 0, 1, 2, 3, 4, 4.1, 4.2, 4.3, 4.4, 4.4.1, 4.4.2, 4.4.3, 4.4.4 4.4.4.1, 4.4.4.2, and 4.4.4.3 must be in the system if address a is in the system.

If address a = 1.1.1.2.4.7.7.1.2.2 is in a universal address system, what other addresses must also be in this system?

Solution: In order to get the address a from the root of the ordered rooted tree we must pass through the addresses 1, 1.1, 1.1.1, 1.1.1.2, 1.1.1.2.4, 1.1.1.2.4.7, 1.1.1.2.4.7.7, 1.1.1.2.4.7.7.1, 1.1.1.2.4.7.7.1.2.

Therefore, the addresses that are also in the system are:

0
1
1.1
1.1.1
1.1.1.1 and 1.1.1.2
1.1.1.2.1, 1.1.1.2.2, 1.1.1.2.3 and 1.1.1.2.4

1.1.1.2.4.1, 1.1.1.2.4.2, 1.1.1.2.4.3, 1.1.1.2.4.4,
1.1.1.2.4.5, 1.1.1.2.4.6 and 1.1.1.2.4.7

1.1.1.2.4.7.1, 1.1.1.2.4.7.2, 1.1.1.2.4.7.3,
1.1.1.2.4.7.4, 1.1.1.2.4.7.5, 1.1.1.2.4.7.6 and
1.1.1.2.4.7.7

1.1.1.2.4.7.7.1

1.1.1.2.4.7.7.1.1 and 1.1.1.2.4.7.7.1.2 and

1.1.1.2.4.7.7.1.2.1

● **PROBLEM** 6-79

What is the maximum number of entries that can be stored in
a binary tree if the longest path from the root to any node
does not exceed N? Find the general equation for the average search time for any entry in a tree with N levels.

Solution: If the longest path is N, the tree will have N
levels below the root node. At most, one node can be at
the root, two nodes at the next level, four nodes at the
next, etc. This expands geometrically, until the total
number of nodes possible is

$$1 + 2 + 4 + \ldots 2^{N-1} = 2^N - 1$$

If we want to search the nodes for a particular entry,
each time we descend to the next level, we must perform
another comparison. For example, if we have 3 levels, the
following distribution of search times is realized:

   1 node takes 1 comparison

   2 nodes take 2 comparisons

   4 nodes take 3 comparisons

The average over the seven nodes is given simply by multiplying nodes by comparisons, taking the sum, and dividing
by the total number of nodes. Hence,

   $(1 \times 1 + 2 \times 2 + 4 \times 3)/7 = 2.43$

We do not include the unit of time here, for it could be milli-, micro-, or nanoseconds, according to the machine doing the comparisons.

The general equation for N levels is given by

$$N-1 + (N/(2^N - 1))$$

• **PROBLEM** 6-80

Write a FORTRAN function subprogram that determines whether a particular item is contained in an ordered binary tree. All items in the tree are assumed to be integers.

Solution: A binary tree is said to be ordered if the following properties hold: For each node in the tree, all numbers contained in the left subtree are less than the number at the node being considered. All numbers contained in the right subtree of a node are greater than the number at that node.

When a search is begun, we start at the root, turning left if the number being searched is smaller, or turning right if the number is greater. Consider the tree below:

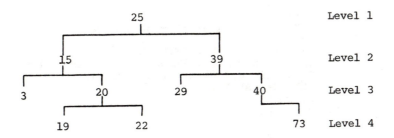

25	Level 1
15 39	Level 2
3 20 29 40	Level 3
19 22 73	Level 4

We must set up a linked list, complete with two pointers at each node, one for each branch. We also require a pointer for the root node. To implement this in FORTRAN, we use a two-dimensional array called TREE (100,3). We are searching the tree for an integer between 0 and 100, hence the size of the first parameter. The second parameter refers to the data for each node and the two pointers which check, respectively, the left and right subtrees. The following table shows the initial status of the array:

I	TREE(I,1)	TREE(I,2)	TREE(I,3)
1	3	0	0
2	15	1	4
3	19	0	0
4	20	3	5
5	22	0	0
6	25	2	8
7	29	0	0
8	39	7	9
9	40	0	10
10	73	0	0
Array Component	Node Data	Left Subtree	Right Subtree

The values of the pointers represent row numbers in the array. We take the root of the tree to be 6.

The strategy used involves logical variables. We declare the function to return a value of .TRUE. if the element is in the tree, and to return a value of .FALSE. if the element is not found. Nodes are checked in succession. If the value is not found at a particular node, the left branch is investigated first. Then the right branch is checked. If the node is terminal (i.e., having no branches), we exit from the program. The value of K is used to indicate the node under current consideration.

The steps are given as follows. Comments will illuminate the modular structrue of the program.

```
          LOGICAL FUNCTION SEARCH (TREE, ROOT, ITEM)
          INTEGER TREE (100,3), ROOT, ITEM, K
          K=ROOT
C         LOOP BEGINS HERE
   10     CONTINUE
C         IF NODE IS FOUND, SET SEARCH = .TRUE. AND EXIT
          IF (TREE (K,1).NE. ITEM) GO TO 20
          SEARCH = .TRUE.
          GO TO 99
   20     CONTINUE
C         TRY SEARCHING THE BRANCHES, STARTING WITH
C         THE LEFT
          IF (TREE(K,1).LT. ITEM) GO TO 30
C         IF THIS NODE IS TERMINAL,SET SEARCH = .FALSE. AND EXIT
          IF (TREE (K,2) . NE. 0) GO TO  40)
          SEARCH=.FALSE.
          GO TO 99
   40     CONTINUE
          K = TREE (K,2)
          GO TO 98
C         NOW TRY THE RIGHT BRANCH; IF NODE IS
C         TERMINAL, SET SEARCH=.FALSE. AND EXIT
          IF (TREE(K,3).NE.0) GO TO 50
          SEARCH=.FALSE.
          GO TO 99
   50     CONTINUE
          K=TREE (K,3)
   98     CONTINUE
          GO TO 10
C         END LOOP
   99     CONTINUE
          RETURN
          END
```

SHORTEST PATH(S)

Consider the network shown in Fig. 1. The numbers on the arcs give the distances d_{ij}. Find the shortest route from node 1 to each of the other nodes.

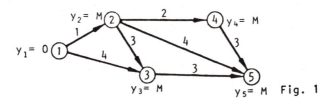

Fig. 1

Solution: Step 1: Set

$$y_1 = 0 \text{ and } y_2 = y_3 = y_4 = y_5 = M.$$

Node 1 is permanently labeled; the others are temporarily labeled. Set $i = 1$.

Step 2: From node 1, only nodes 2 and 3 can be reached. There are

$$y_2 = \min (M, 0 + 1) = 1 \text{ and } y_3 = \min (M, 0 + 4) = 4.$$

See Fig. 2.

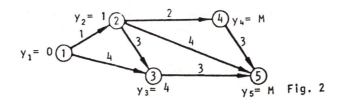

Fig. 2

Step 3: The smallest temporary label is $y_2 = 1$. Node 2 is now permanently labeled (with $y_2 = 1$). Set $i = 2$.

Step 4: Nodes 3, 4, and 5 still have temporary labels, so continue.

Step 2: From node 2, nodes 3, 4, and 5 can be reached. Thus,

$$y_3 = \min (4, 1 + 3) = 4$$

$$y_4 = \min (M, 1 + 2) = 3$$

$$y_5 = \min (M, 1 + 4) = 5$$

See Fig. 3.

357

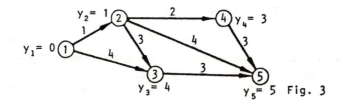

Fig. 3

Step 3: The smallest temporary label is $y_4 = 4$, so node 4 is now permanently labeled. Set $i = 4$.

Step 4: Nodes 3 and 5 are still temporarily labeled, so continue.

Step 2: From node 4, only node 5 can be reached and $y_5 = \min (5, 3 + 3) = 5$.

Step 3: The smallest temporary label is $y_3 = 4$, so permenently label node 3. Set $i = 3$.

Step 4: Node 5 still has a temporary label.

Step 2: From node 3, node 5 can be reached and $y_5 = \min (5, 4 + 3) = 5$.

Step 3: The only temporary label is $y_5 = 5$, so permanently label node 5.

Step 4: All nodes are permanently labeled.

The final (permanent) labels y_j give the length of the shortest path from node 1 to node j. Notice that once a node is permanently labeled, arcs are examined leading from it only once.

The answer is shown in Table 1.

Table 1

Node	Length of shortest route from node 1	Arcs in shortest route
2	1	(1,2)
3	4	(1,3) or (1,2),(2,3)
4	3	(1,2),(2,4)
5	5	(1,2),(2,5)

● **PROBLEM** 6-82

Use the algorithm developed by Dijkstra to find the shortest path between nodes 1 and 6.

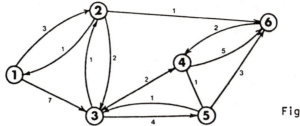

Fig. 1

Solution: The algorithm due to Dijkstra has the advantages that it: (i) requires $3n^3$ elementary operations, where each operation is either an addition or a comparison, and hence is more efficient than the other algorithms; (ii) can be applied in the case of non-symmetric distance matrices, with positive and negative arc lengths, hence it is quite general; and (iii) does not require storing all the data for all arcs simultaneously in the computer, irrespective of the size of the network, but only those for the arcs in sets I and II (described below), and this number is always less than n; hence it is quite economical in its demands on computer memory.

The algorithm follows. It capitalizes on the fact that if j is a node on the minimal path from s to t, knowledge of the latter implies knowledge of the minimal path from s to j. In the algorithm, the minimal paths from s to other nodes are constructed in order of increasing length until t is reached.

In the course of the solution the nodes are subdivided into three sets:

A. the nodes for which the path of minimum length from s is known; nodes will be added to this set in order of increasing minimum path length from node s.

B. the nodes from which the next node to be added to set A will be selected; this set comprises all those nodes that are connected to at least one node of set A but do not yet belong to A themselves;

C. the remaining nodes.

The arcs are also subdivided into three sets:

I. The arcs occurring in the minimal paths from node s to the nodes in set A;

II. The arcs from which the next arc to be placed in set I will be selected; one and only one arc of this set will lead to each node in set B;

III. The remaining arcs (rejected or not yet considered).

To start with, all nodes are in set C and all arcs are in set III. Now transfer node s to set A and from then onwards repeatedly perform the following steps.

Step 1. Consider all arcs as connecting the node just transferred to set A with nodes j in sets B or C. If node j belongs to set B, investigate whether the use of arc a gives rise to a shorter path from s to j than the known path that uses the corresponding arc in set II. If this is not so, arc a is rejected; if, however, use of arc a results in a shorter connection between s and j than hitherto obtained, it replaces the corresponding arc in set II and the latter is rejected. If the node j belongs to set C, it is added to set B and arc a is added to set II.

Table 1

	A[†]	B	C	I	II*	III*
Start	(1)		1,2,3, 4,5,6			(1,2) (2,1) (3,2) (1,3) (2,3) (3,4) (4,3) (2,6) (3,5) (4,5) (5,3) (6,4) (4,6) (5,6)
S1.1	(1)	2,3	4,5,6		(1,2), (1,3)	
S2.1	1(2)		4,5,6	(1,2)		
S1.2	1(2)	3,6	4,5	(1,2)	(2,3), (2,6)	
S2.2	1,2(6)	3	4,5	(1,2),(2,6)	(2,3)	
S1.3	1,2(6)	3,4	5	(1,2),(2,6)	(2,3),(6,4)	
S2.3	1,2(3) 6	4	5	(1,2),(2,3), (2,6)	(6,4)	
S1.4	1,2(3) 6	4	5	(1,2),(2,3), (2,6)	(6,4),(3,5)	
S2.4	1,2,3, (4)6		5	(1,2),(2,3), (2,6),(6,4)	(3,5)	
S1.5	1,2,3, (4)6	5		(1,2),(2,3), (2,6),(6,4)	(3,5),(4,5)	
S2.5	1,2,3, 4,(5)6			(1,2),(2,3), (2,6),(6,4), (4,5)	(4,5)	

† The circled node is the node 'just entered' in the set A

* A crossed arc is a 'rejected' arc. Arc (1,3) was rejected in step S1.2 and arc (3,4) was rejected in step S1.4

360

Step 2. Every node in set B can be connected to node s in only one way if restricted to arcs from set I and one from set II. In this sense each node in set B has a distance from node s: the node with minimum distance from s is transferred from set B to set A, and the corresponding arc is transerred from set II to set I. Then return to step 1 and repeat the process until node t is transferred to set A. Then the solution has been found.

Applying the algorithm to the problem, obtain the calculations in tableau form.

As can be seen, the result is obtained in five iterations and required only 10 comparisons.

● PROBLEM 6-83

Write down the flow conservation equations for the network in Figure 1.

Consider the partition

$$\vec{X} = \{1,3,5\} \ , \ \bar{X} = \{2,4,6\} \ ,$$

a cutset separating the source and the sink. What is the set of forward arcs, reverse arcs and the capacity of the cutset?

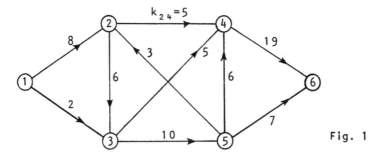

Network where k_{ij} is the capacity of the arc(i,j)

Solution: The flow conservation equations are as follows:

$$\sum_{j \in \vec{A}_i} f_{ij} - \sum_{j \in \vec{B}_i} f_{ji} = v \qquad \text{if i is source}$$

$$= - v \qquad \text{if i is sink}$$

$$= 0 \qquad \text{if i is an intermediate point}$$

where

$$\vec{A}_i = \{j: j \text{ such that } (i,j) \in T\},$$

$$\vec{B}_i = \{j: j \text{ such that } (j,i) \in T\},$$

where T stands for the set of all arcs and v is the net amount of material leaving the source (and arriving at the sink) is known as the value of the flow vector $f(=f_{ij})$. f specifies the amount of flow on each arc of the network. The equations require that the total amount of material reaching, should be equal to total amount of material leaving, at every intermediate point.

Hence:

Conservation equation for node	f_{12}	f_{13}	f_{23}	f_{24}	f_{52}	f_{34}	f_{35}	f_{54}	f_{46}	f_{56}	$-v=$	
1	1	1									1	0
2	-1		1	1	-1							0
3			-1	-1		1	1					0
4				-1		-1		-1	1			0
5					1		-1	1		1		0
6									-1	-1	-1	0

Letting G be a directed network, $\vec{X}$ a subset of points of G containing the source and not containing the sink, $\vec{\bar{X}}$ the set of all the points of G that are in $\vec{X}$, the partition $\vec{X},\vec{\bar{X}}$ generates "a cutset separating the source and the sink." The set of forward arcs of this cutset is

$$\{(i,j) : (i,j) \in T, i \in \vec{X}, j \in \vec{\bar{X}}\}.$$

The set of reverse arcs of this cutset is

$$\{(i,j) : (i,j) \in T, i \in \vec{\bar{X}}, j \in \vec{X}\}.$$

The cutset itself is denoted by $(\vec{X},\vec{\bar{X}})$.

The capacity of the cutset $(\vec{X},\vec{\bar{X}})$ separating the source and the sink is defined as

$$\sum_{(i,j)} k_{ij} - \sum_{(i,j)} \ell_{ij}$$

| (Forward arcs) | (Reverse arcs) |

In Figure 1 the set of forward arcs is

$$\{(1,2),(3,4),(5,2),(5,4),(5,6)\}$$

and the set of reverse arcs is $\{(2,3)\}$ and the capacity of the cutset is 17.

Consider an undirected network shown in Fig. 1, where numbers along the arcs (i,j) represent distances between nodes
i and j. Assume that the distance from i to j is the same
as from j to i (i.e., all arcs are two-way streets). Determine the shortest distance and the length of the shortest path from node 1 to node 6.

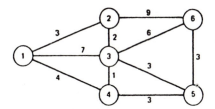

Fig. 1

Solution: Initially node 1 is labeled permanently as zero,
and all other nodes are given temporary labels equal to
their direct distance from node 1. Thus the node labels
at Step 0, denoted by L(0), are

$$L(0) = [0,3,7,4,\infty,\infty,].$$

(An asterisk indicates a permanent label.)

At Step 1 the smallest of the temporary labels is made permanent. Thus node 2 gets a permanent label equal to 3, and
it is the shortest distance from node 1 to node 2. To understand the logic behind this step, consider any other
path from node 1 to node 2 through an intermediate node
j = 3,4,5,6. The shortest distance from node 1 to node j
will be at least equal to 3 and d_{j2} is nonnegative since
all the distances are assumed to be nonnegative. Hence any
other path from node 1 to node 2 cannot have a distance less
than 3, and the shortest distance from node 1 to node 2 is 3.
Thus at Step 1 the node labels are

$$L(1) = [0,3,7,4,\infty,\infty].$$

For each of the remaining nodes j (j = 3,4,5,6), compute a
number which is the sum of the permanent label of node 2
and the direct distance from node 2 to node j. Compare
this number with the temporary label of node j, and the
smaller of the two values becomes the new tentative label
for node j. For example, the new temporary label for node
3 is given by

minimum of (3 + 2,7) = 5

363

Similarly, for nodes 4, 5, and 6, the new temporary labels are 4, ∞, and 12, respectively. Once again the minimum of the new temporary labels is made permanent. Thus at Step 2, node 4 gets a permanent label as shown below:

$$L(2) = [0,3,5,4,\infty,12].$$
$$* * *$$

Now using the permanent label of node 4, the new temporary labels of nodes 3, 5, and 6 are computed as 5, 7, and 12, respectively. Node 3 gets a permanent label and the node labels at Step 3 are

$$L(3) = [0,3,5,4,7,12].$$
$$* * * *$$

At each step, only the node which has been recently labeled permanent is used for further calculations. Thus at Step 4

the permanent label of node 3 is used to update the temporary labels of nodes 5 and 6 (if possible). Node 5 gets a permanent label and the node labels at Step 4 are

$$L(4) = [0,3,5,4,7,11].$$
$$* * * * *$$

Using the permanent label of node 5, the temporary label of node 6 is changed to 10 and is made permanent. The algorithm now terminates, and the shortest distance from node 1 to node 6 is 10. The shortest distance from node 1 to every other node in the network is obtained as

$$L(5) = [0,3,5,4,7,10].$$
$$* * * * * *$$

To determine the sequence of nodes in the shortest path from node 1 to node 6, work backwards from node 6. Node j (j = 1,2,3,4,5) precedes node 6 if the difference between the permanent labels of nodes 6 and j equals the length of the arc from j to 6. This gives node 5 as its immediate predecessor. Similarly node 4 precedes node 5, and the immediate predecessor of node 4 is node 1. Thus the shortest path from node 1 to node 6 is 1→4→5→6.

● **PROBLEM** 6-85

Find the minimum path from v_0 to v_7 in the graph G of figure 1. Notice that it has no circuit whose length is negative.

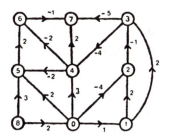

Fig. 1

Solution: An algorithm for arc lengths unrestricted in sign:

Let v_a, v_b be two vertices in the graph $G(V,U)$ whose arc lengths are real numbers, positive, negative or zero. Find the minimum path from v_a to v_b. Assume that there are no circuits in the graph whose arc lengths add up to a negative number. For, if there is any such circuit, one can go round and round it and decrease the length of the path without limit, getting an unbounded solution.

Construct an arborescence $A_1(V_1,U_1)$, $V_1 \subseteq V$, $U_1 \subseteq U$. with center v_a and V_1 containing all those vertices of V which can be reached from v_a along a path, and U_1 containing some arcs of U which are necessary to construct the arborescence. If V_1 contains v_b, a path connects v_a to v_b. In a particular arborescence this path is unique. There may be many arborescences and therefore many paths. A_1 is any one arborescence. If in any problem only one arborescence is possible, there is only one path from v_a to v_b, and that is the solution. If V_1 does not contain v_b, there is no path from v_a to v_b and the problem has no solution.

The method of construction of the arborescence is straightforward. Mark out the arcs going from v_a. From the vertices so reached mark out the arcs (not necessarily all of them) going out to the other vertices. No vertex should be reached by more than one arc, that is not more than one arc should be incident to any vertex. If there is a vertex to which no arc is incident, it cannot be reached from v_a and so is left out. No arc incident to v_a should be drawn.

Let f_j denote the length of the path from v_a to any vertex v_j in the arborescence. The arborescence determines f_j uniquely for each v_j in V_1, but f_j is not necessarily minimum. Let (v_k, v_j) be an arc in G but not in A_1. Consider the length $f_k + x_{kj}$ and compare it with f_j. If

$f_j \leq f_k + x_{kj}$, make no change. If $f_j > f_k + x_{kj}$, delete
the arc incident to v_j in A_1 and include instead the arc
(v_k, v_j). This modifies the arborescence from A_1 to A_2 and
reduces f_j to its new value $f_k + x_{kj}$, the reduction in the
value of f_j being $f_j - f_k - x_{kj}$. The lengths of the paths
to the vertices going through v_j are also reduced by the
same amount. These adjustments are made and thus the new

values of f_j for all v_j in A_2 are calculated.

Now repeat the operation in A_2, that is, select a vertex and see if any alternative arc gives a smaller path to
it. If yes, modify A_2 to A_3 and adjust f_j accordingly.
Ultimately an arborescence A_r is reached which cannot be
further changed by the above procedure. A_r marks out the
minimum path to each v_j from v_a, and f_b in this arborescence
is the minimum path to v_b. The proof is as follows.

Draw an arborescence A_1 (figure 2) with center v_0 consisting of all those vertices of the graph which can be
reached from v_0, (v_8 is thus excluded), and the necessary
number of arcs. Notice that there can be many such
arborescences. A_1 is one of them.

The lengths f_j of the paths from v_0 to different vertices v_j of A_1 are as follows.

$f_0 = 0$, $f_1 = 1$, $f_2 = -4$, $f_3 = 3$, $f_4 = 3$,

$f_5 = 2$, $f_6 = 4$, $f_7 = 5$.

Consider the vertex v_2. There is an arc (v_1, v_2) in G
which is not in A_1, such that

$$f_2 = -4 < f_1 + x_{12} = 1 + 2 = 3$$

A_1 is left unchanged.

Now consider the vertex v_3. There is an arc (v_2, v_3)
in G which is not in A_1 such that

$$f_3 = 3 > f_2 + x_{23} = -4 - 1 = -5.$$

Delete the arc (v_1, v_3) which is incident to v_3 in A_1 and
instead include the arc $\cdot(v_2, v_3)$. This gives a new
arborescence A_2 with $f_3 = -5$. Since no vertex is reached
in A_1 through v_3, all other f_j remain unchanged.

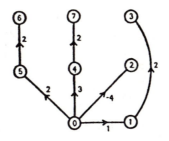

Fig. 2

Coming now to v_4 in A_2 (figure not drawn), arc (v_3, v_4) is in G but not in A_2 such that

$$f_4 = 3 > f_3 + x_{34} = -5 - 4 = -9$$

Delete the arc (v_0, v_4), include (v_3, v_4), get another arborescence A_3 with $f_4 = -9$ and consequently $f_7 = -7$.

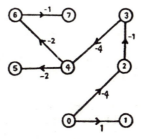

Fig. 3

Continuing like this, get the arborescence (figure 3) which cannot be further modified. No alternative arc decreases the length of the path from v_0 to any vertex. This is seen by testing for every vertex and every possible alternative arc. The minimum path from v_0 to v_7 is

$$(v_0, v_2, v_3, v_4, v_6, v_7)$$

with length -12.

● PROBLEM 6-86

Determine the shortest chain from the source to all other nodes of the network in Figure 1, where the distances associated with the arcs and edges are indicated.

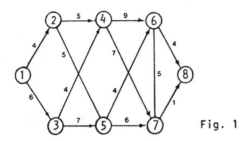

Fig. 1

Solution: The following is Dijkstra's algorithm for finding the shortest chain from the source to all other nodes of a network.

Step 1. Let $L'_{sk} = d_{sk}$. Initially, s (source) is the only node in the tree and $L_{ss} = 0$.

Step 2. $L_{sr} = \min_k L'_{sk} = L_{sj} + d_{jr}$. The k are neighbor nodes of the current tree.

Step 3. Make the arc (j,r) a tree arc.

Step 4. If the number of tree arcs is n - 1, terminate the algorithm. Otherwise, proceed to step 5.

Step 5. $L'_{sk} : = \min(L'_{sk}, L_{sr} + d_{rk})$, where: = means to be replaced by. Go to step 2.

The algorithm may be implemented by labeling the nodes. A label of the type (L,i) will be given to each node of the network. The L in the label is the value L'_{sk} or L_{sk}, and i refers to the last node on the shortest chain from the source to node k. There are two types of labels: temporary and permanent. A label is temporary if $L = L'_{sk}$, and it is permanent if $L = L_{sk}$.

Initially, there are no tree arcs and node 1 is the sole tree node. The neighbor nodes of 1 are 2 and 3. Since $L'_{1k} = d_{1k}$, attach the temporary labels (4,1) to node 2 and (6,1) to node 3. By step 1 of the algorithm, $L_{11} = 0$; thus $L_{1r} = \min \{4,6\} = 4$, and (4,1) is a permanent label of 2. In addition, the arc (1,2) is now a tree arc.

The neighbor nodes of the two-node tree are 3, 4, and 5:

$$L'_{13} : = \min(L'_{13}, L_{12} + d_{23}) = \min(6, 4 + \infty) = 6$$

$$L'_{14} : = \min(L'_{14}, L_{12} + d_{24}) = \min(\infty, 4 + 5) = 9$$

$$L'_{15} : = \min(L'_{15}, L_{12} + d_{25}) = \min(\infty, 4 + 5) = 9.$$

Since it is noted that the smallest of the L'_{1k} is L'_{13} ,
then (6,1) becomes a permanent label for node 3, and (1,3)
is a tree arc. The tree now consists of two tree arcs,
and there are only five more to determine. The neighbor
nodes of the current tree are 4 and 5. Then

$$L'_{14} \; : \; = \min (L'_{14} \; , L_{13} \; + \; d_{34} \;) = \min (9, 6 + 4) = 9$$

$$L'_{15} \; : \; = \min (L'_{15} \; , L_{13} \; + \; d_{35} \;) = \min (9, 6 + 7) = 9.$$

Since L'_{14} equals L'_{15} , the tie is broken arbitrarily, and
(9,2) is selected as a permanent label for node 4. Also,
(2,4) is a tree arc. The neighbor nodes of the current
tree are 5, 6, and 7. Thus

$$L'_{15} \; : \; = \min (L'_{15} \; , L_{14} \; + \; d_{45} \;) = \min (9, 9 + \infty) = 9$$

$$L'_{16} \; : \; = \min (L'_{16} \; , L_{14} \; + \; d_{46} \;) = \min (\infty, 9 + 9) = 18$$

$$L'_{17} \; : \; = \min (L'_{17} \; , L_{14} \; + \; d_{47} \;) = \min (\infty, 9 + 7) = 16.$$

Since the smallest of the L'_{1k} is L'_{15} , then (9,2) becomes
a permanent label for node 5, and (2,5) is a tree arc.
The neighbor nodes of the current tree are 6 and 7. Hence

$$L'_{16} \; : \; = \min (L'_{16} \; , L_{15} \; + \; d_{56} \;) = \min (18, 9 + 4) = 13$$

$$L'_{17} \; : \; = \min (L'_{17} \; , L_{15} \; + \; d_{57} \;) = \min (16, 9 + 6) = 15.$$

Since the smaller of the L'_{1k} is L'_{16} , then (13,5) becomes
a permanent label for node 6, and (5,6) is a tree arc. The
neighbor nodes of the current tree are 7 and 8. The tree
now consists of five tree arcs, and there are only two more
to determine. Continuing

$$L'_{17} \; : \; = \min (L'_{17} \; , L_{16} \; + \; d_{67} \;) = \min (15, 13 + 5) = 15$$

$$L'_{18} \; : \; = \min (L'_{18} \; , L_{16} \; + \; d_{68} \;) = \min (\infty, 13 + 14) = 17.$$

Since L'_{17} is the smaller, then (15,5) becomes a permanent
label for node 7, and (5,7) is a tree arc. Finally

$$L'_{18} \; : \; = \min (L'_{18} \; , L_{17} \; + \; d_{78} \;) = \min (17, 15 + 1) = 16.$$

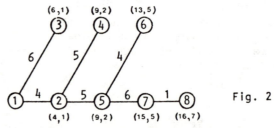

Fig. 2

Thus (7,8) becomes a tree arc, and node 8 receives the permanent label (16,7). There are now seven tree arcs, and the algorithm terminates. Figure 2 illustrates the complete tree where the permanent labels are next to each node.

MAXIMUM FLOW

● PROBLEM 6-87

In the graph of figure 1, numbers along arcs are values of c_i. Find the maximum flow in the graph.

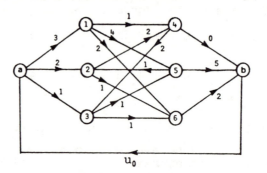

Fig. 1

Solution: Assuming the initial flow as zero in all arcs, let $W_1 = \{v_a\}$, $W_2 = \{$all other nodes$\}$. Now define a criterion (i): if (v_j, v_k) is an arc u and $x_u < c_u$, transfer v_k to W_1.

There is an arc (v_a, v_1) from v_a in W_1 to v_1 in W_2 in which the flow (zero) is less than its capacity 3. Therefore by the criterion (i) transfer v_1 to W_1. Now there is an arc (v_1, v_4) with $v_1 \in W_1$ and $v_4 \in W_2$ such that the flow in it is less than its capacity 1. So transfer v_4 to W_1. In the arc (v_4, v_b), $v_4 \in W_1$, $v_b \in W_2$, the flow is zero which is equal to its capacity, and therefore v_b cannot be transferred to W_1. But there is another arc (v_4, v_3), $v_4 \in W_1$,

370

$v_3 \in W_2$, which is such that v_3 is transferrable to W_1. Further, because the flow in arc (v_3,v_5), $v_3 \in W_1$, $v_5 \in W_2$ is below capacity, v_5 is transferred to W_1, and finally because the arc (v_5,v_b) satisfies the same criterion, v_b is transferred to W_1. Thus it is possible to transfer v_b to W_1 and so the flow is not optimal.

The chain $(v_a,v_1,v_4,v_3,v_5,v_b)$ has been examined. The least capacity in this chain is 1. So in each arc of this chain and also in the return arc (v_b,v_a) increase the flow to 1, keeping the flow as it was in all other arcs. The modified flow is feasible because in each arc it is less than or equal to its capacity, and also at every vertex the flow in equals the flow out.

The above reasoning is repeated with every modified feasible flow until it is not possible to transfer v_b to W_1. The iterations are shown in Table 1. In each feasible flow the bold numbers indicate the chain along which it is possible to proceed to bring v_b into W_1. The asterisk

TABLE 1

Arcs	Capacity c_i	Feasible flows					
		I	II	III	IV	V	VI
(a, 1)	3	0	1	3*	3*	3*	3*
(a, 2)	2	0	0	0	1	2*	2*
(a, 3)	1	0	0	0	0	0	1*
(1, 4)	1	0	1*	1*	1*	1*	0
(1, 5)	4	0	0	2	2	2	3
(1, 6)	2	0	0	0	0	0	0
(2, 4)	2	0	0	0	0	1	1
(2, 6)	1	0	0	0	1*	1*	1*
(3, 5)	1	0	1*	1*	1*	1*	1*
(3, 6)	1	0	0	0	0	1*	1*
(4, 3)	2	0	1	1	1	2	1
(4, b)	0	0*	0*	0*	0*	0*	0*
(5, 2)	1	0	0	0	0	0	0
(5, b)	5	0	1	3	3	3	4
(6, b)	2	0	0	0	1	2*	2*
(b, a)		0	1	3	4	5	6

(*) indicates that the flow in the corresponding arc is equal to its capacity and cannot be further increased.

The change from flow V to flow VI deserves to be followed carefully. The chain in V the flow through which has been modified is $(v_a,v_3,v_4,v_1,v_5,v_b)$. Starting with $W_1 = \{v_a\}$, v_3 can be transferred to W_1 by (i). There is no unsaturated arc going out from v_3, both (v_3,v_6) and (v_3,v_5) carrying capacity flows. But (v_4,v_3) is an arc such that $v_3 \in W_1$, $v_4 \in W_2$, and the flow in it is 2 which is greater than zero. Hence, by criterion (ii) if

(v_k,v_j)

371

is an arc u and $x_u > 0$, transfer v_k to W_1, v_4 is transferred to W_1. Again there is an arc (v_1, v_4) with $v_4 \in W_1$
and $v_1 \in W_2$ and with the flow in it greater than zero. So
v_1 is also transferred to W_1 by criterion (ii). This time
there is an arc (v_1, v_5) with $v_1 \in W_1$, $v_5 \in W_2$ with flow 2
in it which is less than its capacity 4. Consequently, by
criterion (i), v_5 is transferred to W_1, and finally, by
the same criterion, v_b is transferred to W_1. So the flow

is not optimal. In this chain arcs (v_4, v_3) and (v_1, v_4)

occur in reverse directions. Reduce flows in them by 1
and increase flows in other arcs of the chain by 1 thereby
saturating the arc (v_a, v_3).

   The iterations stop at this stage because v_b cannot

be brought into W_1. In fact one cannot even proceed one
step from the initial position of W_1 containing only one
point v_a. This is so because the arcs going out from v_a

are all saturated and so neither v_1 nor v_2 nor v_3 can be
brought in W_1. The maximum flow in the graph is 6.

● **PROBLEM** 6-88

Consider the network shown in Fig. 1. The problem is to
maximize the flow from node 1 to node 6 given the capacities
shown on the arcs. Solve by Ford and Fulkerson algorithm.

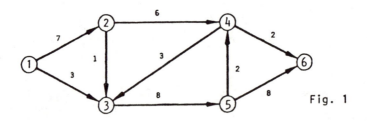

Fig. 1

Solution: An algorithm to find maximal flow is:

Step 1: Label node 1 with (+0,M). +0 indicates there
         is no predecessor node and M indicates an unlim-
         ited supply of additional flow. Node 1 is labeled
         and unscanned, all other nodes are unlabeled.
         The initial flow v can be any easily determined
         amount or zero.

372

Step 2: Select a labeled and unscanned node, say node i.
If all nodes are either labeled and scanned or un-
labeled, the current solution is optimal; stop.
Otherwise, go to Step 3.

Step 3: Suppose node i is labeled $(\pm k, y_i)$. Then for each

unlabeled node j for which an arc (i,j) exists
with $x_{i,j} < b_{i,j}$, where $b_{i,j}$ is the capacity of the

arc from i to j, $x_{i,j}$ is the initial flow between

i and j, assign the label $(+i, y_i)$ to node j, where

y_j is the minimum of $b_{i,j} - x_{i,j}$ and y_i; for each

unlabeled node j for which an arc (j,i) exists
with $x_{j,i} > 0$, assign the label $(-i, y_j)$ to node

j, where y_j is the minimum of $x_{j,i}$ and y_i.

Node i is now scanned; node j is labeled and
unscanned.

Step 4: If node n (the destination node) is labeled go to
step 5. If node n is unlabeled, go to step 2.

Step 5: Increase the value of the flow by y_n. This is
accomplished by using the first component of the
labels to go from node n back to node 1. For
each arc on the route traced, add y_n to the arc flow
when the first component (of the label) is positive and subtract y_n from the arc flow when the first
component is negative. Erase all labels and go to
step 1.

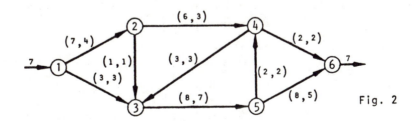

Fig. 2

Assume the initial flow shown in Fig. 2. The numbers on
the arcs are (capacity, flow). The value of the initial
flow is 7 units.

373

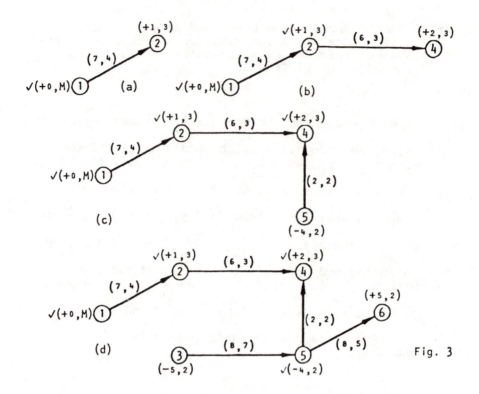

Fig. 3

Step 1: Label node 1 with (+0,M). Node 1 is unscanned, and all other nodes are unlabeled.

Steps 2 and 3: Select node 1. Since $x_{1,2} = 4 < b_{1,2} = 7$, one can label node 2 with (+1,3). See Fig. 3a. Use a check to indicate that node 1 is scanned.

Steps 4,2,3: Node 6 is not labeled, so select node 2. Since $x_{2,4} = 3 < b_{2,4} = 6$, one can label node 4 with (+2,3). Node 2 is now scanned. See Fig. 3b.

Steps 4,2,3: Node 6 is not labeled, so select node 4. Since $x_{5,4} = 2 > 0$, one can label node 5 with (-4,2) (here $x_{ji} = x_{5,4} = 2$ is less than $y_4 = 3$). Node 4 is now scanned. See Fig. 3c.

Steps 4,2,3: Node 6 is not labeled, so select node 5, Since $x_{3,5} = 7 > 0$, one can label node 3 with (-5,2) (here $x_{3,5} = 7$ is greater than $y_5 = 2$). Also, since $x_{5,6} = 5 < b_{5,6} = 8$, one can label node 6 with (+5,2). Node 5 is now scanned. See Fig. 3d.

Steps 4 and 5:   Node 6 is labeled, so breakthrough has occurred.   Since y_6 = 2, one can increase the flow through the

network by 2 units as follows.   Add 2 units to $x_{5,6}$,

subtract 2 units from $x_{5,4}$, add 2 units to $x_{2,4}$, and add 2

units to $x_{1,2}$.   The resulting flows are shown in Fig. 4.

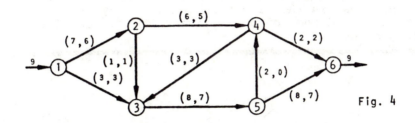

Fig. 4

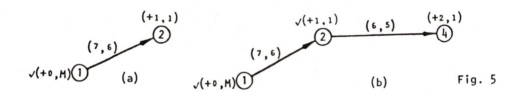

(a)

(b)

Fig. 5

Steps 1,2,3:   Starting again, label node 1 with (+0,M).
Scanning from node 1, see that $x_{1,2}$ = 6 < $b_{1,2}$ = 7, so one

can label node 2 with (+1,1).   Node 1 is scanned.   See Fig.
5a.

Steps 4,2,3:   Node 6 is not labeled, so select node 2.
$x_{2,4}$ = 5 < $b_{2,4}$ = 6, so one can label node 4 with (+2,1).

Node 2 is scanned.   See Fig. 5b.

Steps 4,2,3:   Node 6 is not labeled, so select node 4.   Note
that $x_{4,3}$ = 3 = $b_{4,3}$ and $x_{4,6}$ = 2 = $b_{4,6}$ and $x_{5,4}$ = 0, so

that one cannot label from node 4.   Node 4 is scanned.

Steps 4 and 2:   Node 6 is not labeled.   However, all nodes
are either labeled and scanned (nodes 1, 2, and 4) or unlabeled, so that current flows are optimal.   Thus, the flows
shown in Fig. 4 are optimal.   The value of the maximal flow
is 9 units.

The problem facing the Kughulu Park management during the peak season is to determine how to route the various tram trips from the park entrance (station O in Fig. 1) to the scenic wonder (station T) to maximize the number of trips per day. Strict upper limits have been imposed on the number of outgoing trips allowed in each direction on each individual road. These limits are shown in Fig. 1, where the number next to each station and road gives the limit for that road in the direction leading away from that station. Find the route maximizing the number of trips made per day.

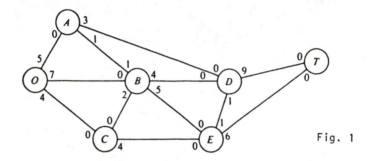

Fig. 1

Solution: Iteration 1:

Assign flow of 5 to 0→B→E→T. The resulting network is

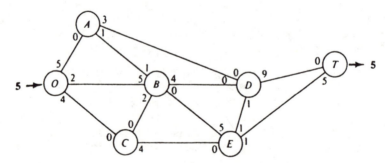

Iteration 2:

Assign flow of 3 to 0→A→D→T. The resulting network is

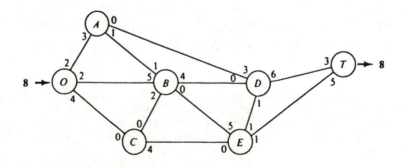

Iteration 3:

Assign flow of 1 to O→A→B→D→T.

Iteration 4:

Assign flow of 2 to O→B→D→T. The resulting network is

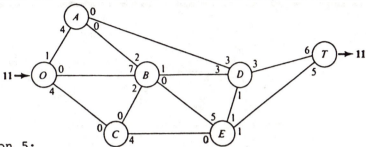

Iteration 5:

Assign flow of 1 to O→C→E→D→T.

Iteration 6:

Assign flow of 1 to O→C→E→T. The resulting network is

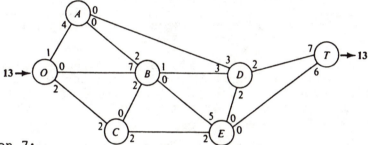

Iteration 7:

Assign flow of 1 to O→C→E→B→D→T. The resulting network is

No paths with strictly positive flow capacity remain.
The current flow pattern is optimal.

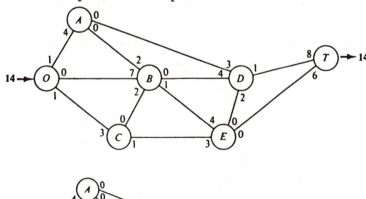

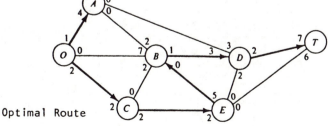

Optimal Route

Find the maximal flow between every pair of nodes in the network below, using the procedure of Gomory and Hu.

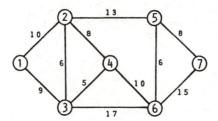

Fig. 1

Solution: The procedure according to Gomory and Hu requires the solution of only n-1 maximal flow problems. It rests on two fundamental, though essentially simple, ideas.

A spanning tree of a connected undirected network G is a connected subgraph of G which contains the same nodes as G but contains no loops. It is easy to prove, e.g., by induction, that a tree contains exactly n-1 arcs, where n is the number of nodes of G.

In general, a graph, connected or not, without cycles, is called a forest; each connected piece of a forest is clearly a tree. Finally, let $c(i,j)$ denote the capacity of the arc between i and j, and for any source-terminal pair (s,t), let $v(s,t)$ denote the value of the maximal flow between s and t. Such maximal flow can be determined by any well known approach.

The two fundamental concepts mentioned above are:

The first is a form of triangular inequality:

$$v(s,t) \geq \min [v(s,x), v(x,t)]; \quad s,x,t \in N. \tag{1}$$

The proof of this inequality relies on the maximum-flow minimum-cut theorem. For, in determining maximum flow between s and t a cut-set $C(X,\bar{X})$ must be obtained, with $s \in X$ and $t \in \bar{X}$, such that $c(X,\bar{X}) = v(s,t)$. Clearly, if $x \in X$, then $v(x,t)$ must be $< c(X,\bar{X})$; on the other hand, if $x \in \bar{X}$, then $v(s,x)$ must be $\leq c(X,\bar{X})$; and the inequality (1) follows. Simple as this inequality may seem, its consequences are far reaching. By simple induction one gets

$$v(s,t) \geq \min [v(s,x_1), v(x_1,x_2), \ldots, v(x_r,t)], \tag{2}$$

where

$$x_1, x_2, \ldots, x_r$$

are any sequence of nodes in N. Furthermore, if the net-
work consists only of three nodes, $\underline{s}$, x, and $\underline{t}$ and only
three arcs, then applying (1) to each 'side of the triangle'
shows that, among the three maximum flow values appearing
in (1), two must be equal and the third no smaller than their
common value. As a further consequence of (1) (or (2)) it
must be true that in a network of n nodes, the function v
can have at most n-1 numerically distinct values.

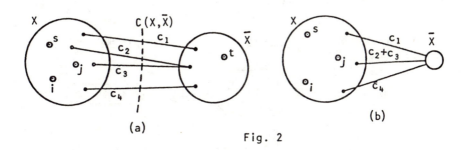

(a)

(b)

Fig. 2

The second fundamental concept is a 'condensation
property'. Suppose that with $\underline{s}$ as source and $\underline{t}$ as sink a
maximal flow problem has been solved, thereby locating a
minimal cut $C(X,\bar{X})$ with s $\in$ X and t $\in$ $\bar{X}$, see Fig. 2a. Suppose
now that it is desired to find $v(i,j)$ where both i and j
are on the same side of $C(X,\bar{X})$; say both are in X, as

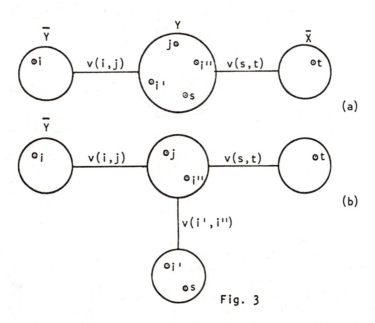

(a)

(b)

Fig. 3

shown in Fig. 2a. Then for this purpose, all nodes of $\bar{X}$
can be 'condensed' into a single node to which all the

arcs of the minimal cut are attached. The resulting network is a condensed network, as shown in Fig. 2b. In a
sense, one may think of the condensed network as having
accorded an infinite capacity to the arcs joining all pairs
of nodes of X̄.

The statement is certainly plausible. The formal proof
involves the demonstration that all the nodes of X̄ must lie
on one side of the cut-set between i and j. Consequently,
the condensation of all the nodes in X̄ cannot affect the
value of the maximum flow from i to j.

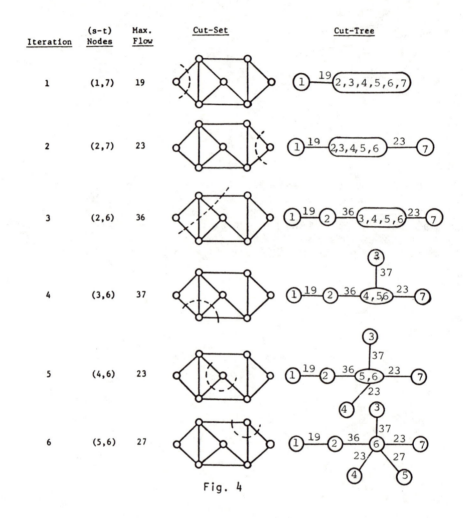

Iteration	(s-t) Nodes	Max. Flow	Cut-Set	Cut-Tree
1	(1,7)	19		
2	(2,7)	23		
3	(2,6)	36		
4	(3,6)	37		
5	(4,6)	23		
6	(5,6)	27		

Fig. 4

The Gomory-Hu construction exploits this second property to the extreme. Suppose that after choosing i, j ∈ X as
the new source-terminal pair (i or j may be s), the maximal
flow v(i,j) and the minimal cut-set C(Y,Ȳ) between i and j
are determined by the standard labeling procedure with all
the nodes of the subset X̄ condensed into one node. Notice
that X̄ is on one side of the new cut-set, see Fig. 3a.
Next choose two nodes i', i" ∈ Y, say; condense all the
nodes in X̄ and Ȳ to one node each and repeat the maximal

flow determination. Continue until each condensed subset contains only one node. Obviously, this requires exactly n-1 maximal flow calculations. Moreover, the resultant is a tree which is called the cut-tree R.

The steps of iteration, the minimal cut-set at each iteration, and the final cut-tree R for network in Fig. 1 are shown in Fig. 4, steps (1) to (6).

Notice that each arc of R represents a cut-set in G, and the number $v(\tau)$ attached to the τth arc of the cut-tree is the capacity of the corresponding cut-set in G.

The remarkable conclusion is that this cut-tree gives the maximal flow between any pair of nodes.

● **PROBLEM** 6-91

A building activity has been analyzed as follows. v_j stands for a job.

(i) v_1 and v_2 can start simultaneously, each one taking 10 days to finish.

(ii) v_3 can start after 5 days and v_4 after 4 days of starting v_1.

(iii) v_4 can start after 3 days of work on v_3 and 6 days of work on v_2.

(iv) v_5 can start after v_1 is finished and v_2 is half done.

(v) v_3, v_4 and v_5 take respectively 6, 8 and 12 days to finish. Find the critical path and the minimum time for completion.

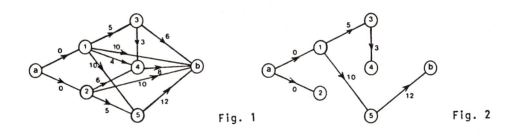

Fig. 1   Fig. 2

Solution: Figure 1 is the graph of the activity, vertices v_a and v_b representing the start and the finish, and the other

vertices the jobs to be done in between. The arc lengths denote the time between the start of two jobs.

The aborescence giving the maximum path is shown in figure 2.
The critical path is (v_a, v_1, v_5, v_b) of length 22 days

which is the minimum time of completion of the work.

CHAPTER 7

COUNTING AND BINOMIAL THEOREM

FACTORIAL NOTATION

● PROBLEM 7-1

Find the values of 6!, 10! and $\frac{11! \times 4!}{5!}$

Solution: In general, the symbol n! is defined for positive integral values of n to be the product of all the integers from 1 to n inclusive:

$$6! = 6 \times 5 \times 4 \times 3 \times 2 \times 1 = 720$$

$$10! = 10 \times 9 \times 8 \times 7 \times 6!$$

$$= 3628800$$

$$\frac{11! \times 4!}{5!} = \frac{11! \times 4!}{5 \times 4!} = \frac{11!}{5}$$

$$= 7983360$$

● PROBLEM 7-2

Find the value of $\frac{7!}{5!}$.

Solution: Apply the definition of factorial. If n is any positive integer, the symbol n! is the product of the integers from 1 up to and including n.

$$n! = 1 \cdot 2 \cdot 3 \ldots n.$$

$$\frac{7!}{5!} = \frac{(1)\,(\cancel{2})\,(\cancel{3})\,(\cancel{4})\,(\cancel{5})\,(6)\,(7)}{\cancel{1}\,(\cancel{2})\,(\cancel{3})\,(\cancel{4})\,(\cancel{5})}$$

$$= 6\,(7)$$

$$= 42$$

● **PROBLEM** 7-3

Simplify the following numbers.

(a) $\dfrac{8!}{11!}$ (b) $\dfrac{5! - 8!}{4! - 7!}$

Solution:

(a)   Note $n! = n \cdot (n-1) \cdot (n-2) \cdot (n-3) \cdot \ldots \cdot 1$;

      also $n! = n \cdot (n-1)!$ or $n \cdot (n-1) \cdot (n-2)!$,   etc.

     Thus, $\dfrac{8!}{11!} = \dfrac{\cancel{8!}}{11 \cdot 10 \cdot 9 \cdot \cancel{8!}} = \dfrac{1}{11 \cdot 10 \cdot 9} = \dfrac{1}{990}$.

(b)   Similarly,

$$\frac{5!-8!}{4!-7!} = \frac{5 \cdot 4! - 8 \cdot 7 \cdot 6 \cdot 5 \cdot 4!}{4! - 7 \cdot 6 \cdot 5 \cdot 4!}$$

Factoring 4!, $\quad = \dfrac{\cancel{4!}\,[5 - (8 \cdot 7 \cdot 6 \cdot 5)]}{\cancel{4!}\,[1 - (7 \cdot 6 \cdot 5)]}$

$$= \frac{5 - 1680}{1 - 210}$$

$$= \frac{-1675}{-209}$$

$$= \frac{1675}{209} \; .$$

● **PROBLEM** 7-4

Simplify the expression $\dfrac{5! - 4!}{6!}$.

Solution: Since $5! = 5 \cdot 4!$ and $6! = 6 \cdot 5 \cdot 4!$, the expression $\dfrac{5! - 4!}{6!}$
becomes:

$$\frac{5! - 4!}{6!} = \frac{(5 \cdot 4!) - 4!}{6 \cdot 5 \cdot 4!}$$

$$= \frac{(4! \cdot 5) - 4!}{4! \cdot 6 \cdot 5} \tag{1}$$

By the distributive property, which states that $ab + ac = a(b+c)$, equation (1) becomes:

$$\frac{5! - 4!}{6!} = \frac{\cancel{4!}(5-1)}{\cancel{4!} \cdot 6 \cdot 5}$$

$$= \frac{5 - 1}{6 \cdot 5} = \frac{4}{6 \cdot 5}$$

$$= \frac{4}{30} .$$

Thus,

$$\frac{5! - 4!}{6!} = \frac{2}{15} .$$

Find the value of $\dfrac{5!6!}{4!7!}$.

Solution: Apply the definition of factorial: If n is any positive integer, the symbol n! is the product of the integers from 1 up to and including n.

Also if r and n are both positive integers and r is less than n, then $n! = n \cdot (n - 1) \ldots (r + 2)(r + 1) \, r!$. Use these two ideas to expand each factorial.

$n! = 1 \cdot 2 \cdot 3 \ldots n$

$5! = (4!)(5)$ and $7! = (6!)(7)$ Substituting the values of 5! and 7! we have

$\dfrac{5!6!}{4!7!} = \dfrac{(4!)(5)(6!)}{(4!)(6!)(7)}$. Dividing the common factors

4! and 6!

$$= \frac{5}{7} .$$

If n and r are positive integers, and $r < n$, show that
$$n! = r!(r + 1)(r + 2) \cdot \ldots \cdot n.$$

Solution: By definition of factorial,
$$r! = 1 \cdot 2 \cdot \ldots \cdot r .$$

Then,

$$r!(r+1)(r+2) \cdot \ldots \cdot n = (1 \cdot 2 \cdot \ldots \cdot r)(r+1)(r+2) \cdot \ldots \cdot n$$
$$r!(r+1)(r+2) \cdot \ldots \cdot n = 1 \cdot 2 \cdot \ldots \cdot n \qquad (1)$$

385

Again, by definition of factorial,

$$n! = 1 \cdot 2 \cdot \ldots \cdot n .$$

Hence, equation (1) becomes:

$$r! (r+1)(r+2) \cdot \ldots \cdot n = n!$$

or

$$n! = r! (r+1)(r+2) \cdot \ldots \cdot n .$$

COUNTING PRINCIPLES

There are two roads between towns A and B. There are three roads between towns B and C. How many different routes may one travel between towns A and C.

Solution:

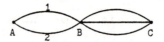

If we take road 1 from town A to town B and then any road from B to C there are three ways to travel from A to C. If we take road 2 from A to B and then any road from B to C there are again three ways to travel from A to C. These two possibilities are the only ones available to us. Thus there are 3 + 3 = 6 ways to travel from A to C.

This problem illustrates the fundamental principle of counting. This principle states that if an event can be divided into k components, and there are n_1 ways to carry out the first component, n_2 ways to carry out the second, n_i ways to carry out the ith, and n_k ways to carry out the kth then there are $n_1 \cdot n_2 \cdot n_3 \cdot \ldots \cdot n_k$ ways for the original event to take place.

How many ways can r different balls be placed in n different boxes? Consider the balls and boxes distinguishable.

Solution: If the balls are placed one at a time, then each ball may be placed in one of n (provided there is ample room) boxes. That is, there are n boxes in which to place the first ball, n boxes for the second, n boxes for the third and finally n boxes for the rth.

By the fundamental principle of counting there are

$$\underbrace{n \cdot n \cdot n \cdot \ldots \cdot n}_{r \text{ terms}} \text{ or } n^r \text{ ways to place r different balls}$$
in n boxes.

A man has 15 ties and 3 jackets. In how many different ways can he combine a jacket with a tie?

Solution: The ties can be chosen in 15 ways while the jackets can be chosen in 3 ways. Therefore, by the principle of counting, he has 15 x 3 = 45 ways to combine a jacket with a tie.

Calculate the number of ways in which a nonempty string of numbers from the set {1, 2, 3, 4, 5} can be selected such that the string lists each symbol at most once.

Solution: A nonempty string of numbers from the set {1, 2, 3, 4, 5} can have length either 1, 2, 3, 4, or 5. These sub-events are denoted by S_1, S_2, S_3, S_4, and S_5 respectively. By the Rule of Sum, if A and B are mutually exclusive events (i.e. A and B can't both occur at the same time) and A can happen in m different ways while B can happen in n different ways. Then the event (A or B) can happen in (m + n) ways. Therefore, if the number of ways in which S_1, S_2, S_3, S_4 and S_5 can occur are respectively m_1, m_2, m_3, m_4 and m_5; then, event (S_1 or S_2 or S_3 or S_4 or S_5) can happen in ($m_1 + m_2 + m_3 + m_4 + m_5$) different ways. A string of length 1 (event S_1) can happen in five different ways: 1, 2, 3, 4 and 5. Thus $m_1 = 5$. A string of length 2 can have its first number be chosen in 5 different ways and its second number in 4 different ways(since the number placed in the first position cannot be a possibility for the second position). Hence, by the Rule of Product (which states that if A and B are mutually exclusive events, and A can happen in m different ways, while B can happen in n different ways; then the event (A and B) can happen in m x n different ways), thus $m_2 = 5 \times 4 = 20$. Similarly, m_3, m_4 and m_5 can be found as the following, $m_3 = 5 \times 4 \times 3 = 60$

$m_4 = 5 \times 4 \times 3 \times 2 = 120$ and

$m_5 = 5 \times 4 \times 3 \times 2 \times 1 = 120$.

Therefore, strings of length ≤ 5 with nonrepeating symbols from the set $\{1, 2, 3, 4, 5\}$ can be chosen in $5 + 20 + 60 + 120 + 120 = 325$ different ways.

● **PROBLEM 7-11**

How many different 2 or 4 number sequences can be chosen from the set $\{1, 2, 3, 4, 5, 6, 7, 8, 9, 10\}$?

Solution: The number of possible sequences of two different numbers from the set$\{1, 2, 3, 4, 5, 6, 7, 8, 9, 10\}$ is found by calculating the number of permutations of 10 elements taken 2 at a time, i.e. $P(10,2)$.

$$P(n, r) = \frac{n!}{(n - r)!}$$

$$P(10,2) = \frac{10!}{(10 - 2)!} = \frac{10!}{8!} = 9 \times 10 = 90$$

By the same token, there are $P(10,4) = \frac{10!}{(10 - 4)!} = 5040$ different 4 number sequences which are possible from the given set of numbers.

Finally, by the Rule of Sum, the number of the different 2 or 4 number sequences that are possible from the set $\{1, 2, 3, 4, 5, 6, 7, 8, 9, 10\}$ is found to be $90 + 5040 = 5130$.

● **PROBLEM 7-12**

How many different numbers of 3 digits can be formed from the numbers 1, 2, 3, 4, 5 (a) If repetitions are allowed? (b) If repetitions are not allowed? How many of these numbers are even in either case?

Solution: (a) If repetitions are allowed, there are 5 choices for the first digit, 5 choices for the second and 5 choices for the third. By the Fundamental Principle of counting, there are $5 \times 5 \times 5 = 125$ possible three digit numbers. If the number is even, the final digit must be either 2 or 4, 2 choices. So there will be 5 choices for the first digit, 5 choices for the second and 2 for the third, $5 \times 5 \times 2 = 50$ such numbers.

     (b) If repetitions are not allowed, there are 5 choices for the first digit. After this has been picked, there will be 4 choices for the second digit, and 3 choices for the third. Hence, $5 \times 4 \times 3 = 60$ such numbers can be selected.

If the number must be even, then there are 2 choices for the final digits, 2 or 4. This leaves 4 choices for the next digit and 3 choices for the first digit. Hence there are 4 × 3 × 2 = 24 possible even numbers that can be selected in this way.

PERMUTATIONS

● **PROBLEM 7-13**

Find $_9P_4$.

Solution: Using the general formula for permutations of b different things taken a at a time, $_bP_a = \frac{b!}{(b-a)!}$, substitute 9 for b and 4 for a, $_9P_4 = \frac{9!}{(9-4)!} = \frac{9!}{5!}$. Evaluating our factorials, we obtain:

$$_9P_4 = \frac{9 \cdot 8 \cdot 7 \cdot 6 \cdot (5 \cdot 4 \cdot 3 \cdot 2 \cdot 1)}{(5 \cdot 4 \cdot 3 \cdot 2 \cdot 1)}$$

cancelling 5! in the numerator and denominator:

$$_9P_4 = 9 \cdot 8 \cdot 7 \cdot 6$$

$$= 3,024 \quad .$$

● **PROBLEM 7-14**

Calculate the number of permutations of the letters a,b,c,d taken four at a time.

Solution: The number of permutations of the four letters taken four at a time equals the number of ways the four letters can be arranged or ordered. Consider four places to be filled by the four letters. The first place can be filled in four ways choosing from the four letters. The second place may be filled in three ways selecting one of the three remaining letters. The third place may be filled in two ways with one of the two still remaining. The fourth place is filled one way with the last letter. By the fundamental principle, the total number of ways of ordering the letters equals the product of the number of ways of filling each ordered place, or 4 · 3 · 2 · 1 = 24 = P(4,4) = 4! (read 'four factorial').

In general, for n objects taken r at a time,

$$P(n,r) = n(n-1)(n-2)\ldots(n-r+1) = \frac{n!}{(n-r)!} \quad (r < n).$$

For the special case where $r = n$,

$$P(n,n) = n(n-1)(n-2)\ldots(3)(2)(1) = n!,$$

since $(n-r)! = 0!$ which $= 1$ by definition.

● **PROBLEM 7-15**

How many permutations of two letters each can be formed from the letters a,b,c,d,e? Actually write these permutations.

Solution: We recall the general formula for the number of permutations of n different things taken r at a time $_nP_r = n!/(n-r)!$. The number of permutations of 2 letters that can be formed from the 5 given letters is $_5P_2$.

$$_5P_2 = \frac{5!}{(5-2)!} = \frac{5!}{3!} = \frac{5\cdot4\cdot\cancel{3!}}{\cancel{3!}} = 20$$

Thus, the 20 permutations are:

ab	ac	ad	ae
ba	bc	bd	be
ca	cb	cd	ce
da	db	dc	de
ea	eb	ec	ed

● **PROBLEM 7-16**

Determine the number of permutations of three elements taken from a set of four elements {a, b, c, d}.

Solution:

Method A

In this example we can use the formula for permutations $P_a^b = \frac{b!}{(b-a)!}$:

hence, $P_3^4 = \frac{4!}{(4-3)!} = \frac{4!}{1!} = \frac{4\cdot3\cdot2\cdot1}{1} = 24$

Method B

If you do not recall the formula for permutations

you may determine the number of possible permutations of
3 elements taken from a set of four elements by recalling
the fundamental principle: **If an act can be performed in m**
ways and if, after this first act has been performed, a
second act can be performed in n ways then the number of
ways in which both acts can be performed, in the order
given is m × n ways.

Thus, there are 4 ways of filling our first box
   of 3 elements × 3 ways of filling our second box
                   × 2 ways of filling our third box

| 4
 (a or b or c or d) | × | 3
 (the 3 remaining <br> letters) | × | 2
 (the 2 re-
 maining <br> letters) | = 24. |

Method C

   We can also determine the number of permutations
using a tree diagram:

Hence our permutations are:

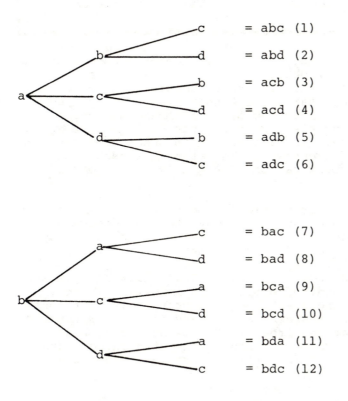

= abc (1)

= abd (2)

= acb (3)

= acd (4)

= adb (5)

= adc (6)

= bac (7)

= bad (8)

= bca (9)

= bcd (10)

= bda (11)

= bdc (12)

391

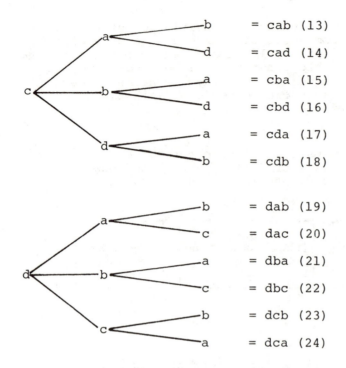

	b	= cab (13)
a	d	= cad (14)
	a	= cba (15)
c — b	d	= cbd (16)
	a	= cda (17)
d	b	= cdb (18)

	b	= dab (19)
a	c	= dac (20)
	a	= dba (21)
d — b	c	= dbc (22)
	b	= dcb (23)
c	a	= dca (24)

● **PROBLEM** 7-17

If a group of 26 members is to elect a president and a secretary, in how many ways could the 2 officers be elected?

Solution: The group consists of 26 members, anyone of the 26 can serve as president. After the president has been elected, there are still 25 other members that could be elected as the secretary. The Fundamental Principle of Counting states that, if one thing can be done in a different ways, and a second thing can be done in b different ways, then the two things in succession can be done in a • b different ways. Therefore the number of ways the two officers can be chosen is (26)(25) or 650 ways.

The fundamental principle can be extended to more than two events. The total number of ways the successive events could be performed is the product of the numbers of ways each of the events could be performed.

This can also be seen by using the following approach. Since the arrangement of officers is important (x serving as president and y serving as secretary is different than y serving as president and x serving as secretary), this is a permutations problem. Recalling the general formula for the number of permutations of n things taken r at a time, $_nP_r = n!/(n - r)!$, replace n by 26 and r by 2 to obtain

$$_{26}P_2 = \frac{26!}{(26 - 2)!} = \frac{26!}{24!} = \frac{26 \cdot 25 \cdot \cancel{24!}}{\cancel{24!}} = 650$$

Thus, once again we find there are 650 ways to elect a president and secretary from the 26 members.

● PROBLEM 7-18

How many telephone numbers of four different digits each can be made from the digits 0,1,2,3,4,5,6,7,8,9?

Solution: A different arrangement of the same four digits produces a different telephone number. Since we are concerned with the order in which the digits appear, we are dealing with permutations.

There are ten digits to choose from and four different ones are to be chosen at a time. The general formula for the number of permutations of n things taken r at a time is

$$P(n,r) = \frac{n!}{(n - r)!} \ .$$

Here n = 10, r = 4, and the desired number is

$$P(10,4) = \frac{10!}{(10 - 4)!} = \frac{10!}{6!} = \frac{10 \cdot 9 \cdot 8 \cdot 7 \cdot 6!}{6!}$$

$$= 5040$$

Thus 5040 telephone numbers of four digits each can be made from the 10 digits.

● PROBLEM 7-19

In how many ways can the letters in the word "Monday" be arranged?

Solution: The word Monday contains 6 different letters. Since different letter arrangements yield different "words", we seek the number of permutations of 6 different objects taken 6 at a time.

Recall the general formula for the number of permutations of n things taken r at a time:

$$_nP_r = \frac{n!}{(n - r)!} \ . \quad \text{Thus,} \quad _6P_6 = \frac{6!}{(6 - 6)!} = \frac{6!}{0!} \ .$$

Since 0! = 1 by definition,

$$_6P_6 = \frac{6!}{1} = 6 \cdot 5 \cdot 4 \cdot 3 \cdot 2 \cdot 1 = 720.$$

Thus, the letters in the word "Monday" may be

393

arranged in 720 ways. We can arrive at the same conclusion using the fundamental theorem of counting,
which states that for a given sequence of n events
E_1, E_2, ... E_n, if for each i, E_i can occur m_i ways,

then the total number of distinct ways the event may
take place is $m_1 \cdot m_2 \cdot m_3 \cdot \ldots \cdot m_n$.

Thus, the first of the 6 letters may be chosen 6 ways
      the second of the 6 letters may be chosen 5 ways
      the third of the 6 letters may be chosen 4 ways
      the fourth of the 6 letters may be chosen 3 ways
      the fifth of the 6 letters may be chosen 2 ways
      the sixth of the 6 letters may be chosen 1 way

Hence the total number of ways the letters may be
arranged is $6 \times 5 \times 4 \times 3 \times 2 \times 1 = 720$.

● **PROBLEM 7-20**

Find the number of permutations of the seven letters of the word
"algebra."

Solution: A permutation is an ordered arrangement of a set of objects.
For example, if you are given 4 letters a,b,c,d and you choose two
at a time, some permutations you can obtain are: ab, ac, ad, ba, bc,
bd, ca, cb.

For n things, we can arrange the first object in n different
ways, the second in n-1 different ways, the third can be done in
n-2 different ways, etc. Thus the n objects can be arranged in order
in

$$n! = n \cdot n-1 \cdot n-2 \ldots 1 \text{ ways}$$

Temporarily place subscripts, 1 and 2, on the a's to distinguish
them, so that we now have $7! = 5040$ possible permutations of the seven
distinct objects. Of these 5040 arrangements, half will contain the
a's in the order a_1, a_2 and the other half will contain them in the
order a_2, a_1. If we assume the two a's are indistinct, then we apply
the following theorem. The number P of distinct permutations of n
objects taken at a time, of which n_1 are alike, n_2 are alike of
another kind,. . . ,n_k are alike of still another kind, with $n_1+n_2+..+$
$n_k = n$ is $P = \dfrac{n!}{n_1! \, n_2! \, \ldots \, n_k!}$ Then, here in this example, the
2 a's are alike so

$$P = \frac{7!}{2!} = 2520 \text{ permutations of the letters of}$$

the word algebra, when the a's are indistinguishable.

In how many ways may a party of four women and four men be seated at a round table if the women and men are to occupy alternate seats?

Solution: If we consider the seats indistinguishable, then this is a problem in circular permutations, as opposed to linear permutations. In the standard linear permutation approach each chair is distinguishable from the others. Thus, if a woman is seated first, she may be chosen 4 ways, then a man seated next to her may be chosen 4 ways, the next woman can be chosen 3 ways and the man next to her can be chosen in 3 ways ... Our diagram to the linear approach shows the number of ways each seat can be occupied.

4	4	3	3	2	2	1	1

By the Fundamental Principle of Counting there are thus 4 · 4 · 3 · 3 · 2 · 2 · 1 · 1 = 576 ways to seat the people.

However, if the seats are indistinguishable then so long as each person has the same two people on each side, the seating arrangement is considered the same. Thus we may suppose one person, say a woman is seated in a particular place, and then arrange the remaining three women and four men relative to her. Because of the alternate seating scheme, there are three possible places for the remaining three women, so that there are 3! = 6 ways of seating them. There are four possible places for the four men, whence there are 4! = 24 ways in which the men may be seated. Hence the total number of arrangements is 6 · 24 = 144. In general, the formula for circular permutations of n things and n other things which are alternating is $(n - 1)!n!$. In our case we have

$$(4 - 1)!4! = 3!4! = 3 \cdot 2 \cdot 4 \cdot 3 \cdot 2 = 144.$$

In how many ways can 9 different chairs be arranged in a circle?

Solution: In arranging objects in a circle, one object must be set in place first in order to have a starting point. Hence, in arranging n objects in a circle, one has (n - 1)! arrangements. Note that the number of circular permutations of n elements taken n at a time is not n!. This is because in finding the number of circular permutations , as opposed to finding the number of linear permutations of n elements, the n different ways of choosing the starting point (or the first object) cannot be distinguished from each other. For example, if n = 4 the four "different" ways of choosing the starting point, give the following 4 arrangements.

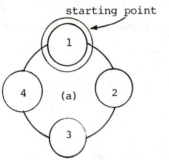

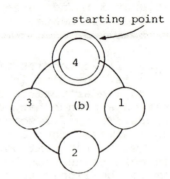

Figure 1

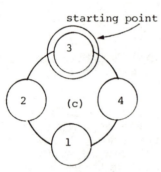

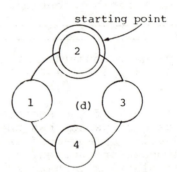

These four arrangements are actually the same since one can obtain (a) by merely rotating (clockwise) arrangements (b), (c) and (d) by 90°, 180° and 270°, respectively. In other words, there is only one way to choose the starting point. For this problem, n = 9 so the number of circular permutations is (n - 1)! = (9 - 1)! = 8! = 40320.

● **PROBLEM 7-23**

Prove this identity: P(n,n-1) = P(n,n).

Solution: The general formula for the number of permutations of x objects taken r at a time is

$$P(x,r) = \frac{x!}{(x - r)!}$$

Thus, evaluating the left side of the given identity, we obtain

$$P(n,n-1) = \frac{n!}{[n-(n-1)]!} = \frac{n!}{(n-n+1)!} = \frac{n!}{1!} = \frac{n!}{1} = n!$$

Evaluating the right side of the identity we obtain

$$P(n,n) = \frac{n!}{(n-n)!} = \frac{n!}{0!}$$

396

$0! = 1$, by definition, hence $\dfrac{n!}{0!} = \dfrac{n!}{1} = n!$

Thus $P(n,n-1) = n! = P(n,n)$. Therefore, by the transitive property (If $a = b$ and $b = c$, $a = c$),

$$P(n,n-1) = P(n,n).$$

COMBINATIONS

● **PROBLEM** 7-24

Find the value of $C(n,0)$.

Solution: Starting with the formula for combinations:

$$C(n,r) = \frac{n!}{(n-r)!\,r!}$$

and substituting $r = 0$, we have

$$C(n,0) = \frac{n!}{(n-0)!\,0!}$$

Hence, $\qquad C(n,0) = \dfrac{n!}{n!\,0!}$

Recall that $0! = 1$ by definition. Then

$$C(n,0) = \frac{n!}{n!}$$

$$= 1.$$

● **PROBLEM** 7-25

Find $_9C_4$.

Solution: By definition, combinations of b different things taken a at a time, $_bC_a = \dfrac{b!}{a!\,(b-a)!}$; hence by substitution, $_9C_4 = \dfrac{9!}{4!\,(9-4)!}$.

$$= \frac{9!}{4!\,5!}$$

$$= \frac{9 \cdot 8 \cdot 7 \cdot 6 \cdot (5 \cdot 4 \cdot 3 \cdot 2 \cdot 1)}{4 \cdot 3 \cdot 2 \cdot 1 \cdot (5 \cdot 4 \cdot 3 \cdot 2 \cdot 1)}$$

Cancelling $5! = (5 \cdot 4 \cdot 3 \cdot 2 \cdot 1)$ out of numerator and denominator, we multiply to obtain:

$$_9C_4 = \frac{3024}{24} = 126 \ .$$

Evaluate each of the following symbols:
(a) C(6,3)   (b) C(18,16)

Solution: Recalling the general formula for the number of combinations of n different things taken r at a time
$$C(n,r) = \frac{n!}{r!(n-r)!}:$$

(a) $C(6,3) = \frac{6!}{3!(6-3)!} = \frac{6!}{3!3!} = \frac{\overset{2}{\cancel{6}} \cdot 5 \cdot \cancel{4} \cdot \cancel{3!}}{\cancel{3} \cdot \cancel{2} \cdot 1 \cdot \cancel{3!}} = 20.$

(b) $C(18,16) = \frac{18!}{16!(18-16)!} = \frac{18!}{16!2!} = \frac{\overset{9}{\cancel{18}} \cdot 17 \cdot \cancel{16!}}{\cancel{2} \cdot 1 \cdot \cancel{16!}} = 153$

In how many different ways may a pair of dice fall?

Solution: The Fundamental Principle of Counting states that if one thing can be done in a different ways, and when it is done in any one of these ways, a second thing can be done in b different ways, then both things in succession can be done in a×b different ways. A die has 6 sides, thus it may land in any of six ways. Since each die may land in 6 ways, by the Fundamental Principle both die may fall in 6×6 = 36 ways. We can verify this result by enumerating all the possible ordered pairs of dice throws:

1,1	1,2	1,3	1,4	1,5	1,6
2,1	2,2	2,3	2,4	2,5	2,6
3,1	3,2	3,3	3,4	3,5	3,6
4,1	4,2	4,3	4,4	4,5	4,6
5,1	5,2	5,3	5,4	5,5	5,6
6,1	6,2	6,3	6,4	6,5	6,6

How many baseball teams of nine members can be chosen from among twelve boys, without regard to the position played by each member?

Solution: Since there is no regard to position, this is a combinations problem (if order or arrangement had been important it would have been a permutations problem). The general formula for the number of combinations of n things taken r at a time is

$$C(n,r) = \frac{n!}{r!(n-r)!} \quad .$$

We have to find the number of combinations of 12 things taken 9 at a time. Hence we have

$$C(12,9) = \frac{12!}{9!(12-9)!} = \frac{12!}{9!3!} = \frac{12 \cdot 11 \cdot 10 \cdot \cancel{9!}}{3 \cdot 2 \cdot 1 \cdot \cancel{9!}} = 220$$

Therefore, there are 220 possible teams.

● **PROBLEM** 7-29

A manufacturer produces 7 different items. He packages assortments of equal parts of 3 different items. How many different assortments can be packaged?

Solution: Since we are not concerned with the order of the items, we are dealing with combinations. Thus the number of assortments is the number of combinations of 7 items taken 3 at a time. Recall the general formula for the number of combinations of n items taken r at a time,

$$C(n,r) = \frac{n!}{r!(n-r)!}$$

$$C(7,3) = \frac{7!}{3!(7-3)!}$$

$$= \frac{7!}{3!4!}$$

$$= \frac{7 \cdot \cancel{6} \cdot 5 \cdot \cancel{4!}}{\cancel{3} \cdot \cancel{2} \cdot \cancel{4!}}$$

$$= 35$$

Thus, 35 different assortments can be packaged.

● **PROBLEM** 7-30

A man and his wife decide to entertain 24 friends by giving 4 dinners with 6 guests each. In how many ways can the first group be chosen?

Solution: In the first group we are considering one dinner and there are 6 people out of 24 friends to be invited. We must find the number of ways to choose 6 out of 24. We are dealing with combinations. To select r things out of n objects, we use the definition of combinations:

$$\binom{n}{r} = \frac{n!}{r!(n-r)!} = c(n,r)$$

$$c(24,6) = \binom{24}{6} = \frac{24!}{6!18!} = \frac{24 \cdot 23 \cdot 22 \cdot 21 \cdot 20 \cdot 19 \cdot 18!}{6! \; 18!}$$

$$= \frac{24 \cdot 23 \cdot 22 \cdot 21 \cdot 20 \cdot 19}{6 \cdot 5 \cdot 4 \cdot 3 \cdot 2 \cdot 1}$$

$$= 134,596$$

● **PROBLEM 7-31**

A Sunday school class of 12 members is to be seated on seven chairs and a bench that accommodates five persons. In how many ways can the bench be occupied?

Solution: If we are concerned with the order of people on the bench (so that we consider the same five people sitting in different arrangements as distinct ways), then this is a permutations problem. Recalling the general formula for the number of permutations of n elements taken r at a time

$$p(n,r) = \frac{n!}{(n-r)!}$$

we find the number of permutations of 12 elements taken 5 at a time, or p(12,5). Thus

$$p(12,5) = \frac{12!}{(12-5)!} = \frac{12!}{7!} = 12 \cdot 11 \cdot 10 \cdot 9 \cdot 8 = 95,040$$

If we are not concerned with the order of the people on the bench this becomes a combinations problem. Recalling the general formula for the number of combinations of n elements taken r at a time

$$c(n,r) = \frac{n!}{r!(n-r)!}$$

we find the number of combinations of 12 elements taken 5 at a time, or c(12,5).

$$c(12,5) = \frac{12!}{5!(12-5)!} = \frac{12!}{5!7!} = \frac{12 \cdot 11 \cdot 10 \cdot 9 \cdot 8 \cdot 7!}{5 \cdot 4 \cdot 3 \cdot 2 \cdot 7!} = 792.$$

● **PROBLEM 7-32**

How many different sums of money can be obtained by choosing two coins from a box containing a penny, a nickel, a dime, a quarter, and a half dollar?

Solution: The order makes no difference here, since a selection of a penny and a dime is the same as a selection of a dime and a penny, insofar as a sum of money is concerned. This is a case of combinations, then, rather than permutations. Then the number of combinations of n different objects taken r at a time is equal to:

$$\frac{n(n-1)\ldots(n-r+1)}{1\cdot 2\cdots r}\quad.$$

In this example, n = 5, r = 2, therefore

$$C(5,2) = \frac{5\cdot 4}{1\cdot 2} = 10.$$

As in the problem of selecting four committee members from a group of seven people, a distinct two coins can be selected from five coins in

$$\frac{5\cdot 4}{1\cdot 2} = 10 \text{ ways (applying the fundamental principle).}$$

● **PROBLEM** 7-33

From 12 books in how many ways can a selection of 5 be made, (1) when one specified book is always included, (2) when one specified book is always excluded?

Solution: Here the formula for combinations is appropriate: the number of combinations of   n   things taken   r   at a time:

$$C(n,r) = {}_nC_r = \frac{n!}{r!(n-r)!}$$

where   n = 11, and   r = 4.

(1)   Since the specified book is to be included in every selection, we have only to choose 4 out of the remaining 11.
Hence the number of ways = ${}^{11}C_4$

$$11^{C_4} = \frac{11!}{4!(11-4)!}$$

$$= \frac{11!}{4!7!}$$

$$= \frac{11\cdot 10\cdot 9\cdot 8\cdot 7!}{4\cdot 3\cdot 2\cdot 1\cdot 7!}$$

$$= \frac{11 \times 10 \times 9 \times 8}{1 \times 2 \times 3 \times 4}$$

$$= 330.$$

(2)   Since the specified book is always to be excluded, we have to select the 5 books out of the remaining   11.

Hence the number of ways = ${}^{11}C_5$

$$11^{C_5} = \frac{11!}{5!(11-5)!}$$

$$= \frac{11!}{5!6!}$$

$$= \frac{11 \cdot 10 \cdot 9 \cdot 8 \cdot 7 \cdot \cancel{6!}}{5 \cdot 4 \cdot 3 \cdot 2 \cdot 1 \cdot \cancel{6!}}$$

$$= \frac{11 \times 10 \times 9 \times 8 \times 7}{1 \times 2 \times 3 \times 4 \times 5}$$

$$= 462$$

● **PROBLEM** 7-34

How many groups can be formed from ten objects taking at least three at a time?

Solution: The number of combinations of n objects taken r at a time is $C(n,r) = n!/r!(n - r)!$. Thus, the number of groups that can be formed from 10 objects taking three at a time is $C(10, 3)$,

from 10 objects taking 4 at a time is $C(10,4)$,

from 10 objects taking 5 at a time is $C(10,5)$,

from 10 objects taking 6 at a time is $C(10,6)$,

.
.
.

from 10 objects taking 10 at a time is $C(10, 10)$.

Therefore, the number of groups that can be formed from 10 objects taking at least three at a time is $C(10, 3) + C(10, 4) + C(10, 5) + C(10, 6) + C(10, 7) + C(10, 8) + C(10, 9) + C(10, 10)$,

$$\frac{10!}{3!7!} + \frac{10!}{4!6!} + \frac{10!}{5!5!} + \frac{10!}{6!4!} + \frac{10!}{7!3!} + \frac{10!}{8!2!} + \frac{10!}{9!1!} + \frac{10!}{10!0!} =$$

$$\frac{10 \cdot 9 \cdot 8 \cdot 7!}{\cancel{3} \cdot \cancel{2} \cdot 7!}^{3 \cdot 4} + \frac{10 \cdot 9 \cdot 8 \cdot 7 \cdot 6!}{4 \cdot 3 \cdot 2 \cdot 6!}^{5 \cdot 3 \cdot 2} + \frac{10 \cdot 9 \cdot 8 \cdot 7 \cdot 6 \cdot 5!}{5 \cdot 4 \cdot 3 \cdot 2 \cdot 5!}^{2 \cdot 3 \cdot 2 \; \cdot \; 3} + \frac{10 \cdot 9 \cdot 8 \cdot 7 \cdot 6!}{4 \cdot 3 \cdot 2 \cdot 6!}$$

$$+ \frac{10 \cdot 9 \cdot 8 \cdot 7!}{3 \cdot 2 \cdot 7!} + \frac{10 \cdot 9 \cdot 8!}{2 \cdot 8!} + \frac{10 \cdot 9!}{1 \cdot 9!} + \frac{10!}{10! \cdot 1}$$

$$= 120 + 210 + 252 + 210 + 120 + 45 + 10 + 1$$

$$= 968.$$

● **PROBLEM** 7-35

A man has 6 friends; in how many ways may he invite one or more of them to dinner?

<u>Solution:</u> He has to select some or all of his 6 friends. Here the appropriate procedure is to utilize the formula for combinations in which no particular order is important. Since the question specifies some or all, every combination must be investigated and the sum will give the total number of ways which are possible. The formula for the number of combinations of n things taken r at a time is

$$_nC_r = \frac{n!}{r!(n-r)!} \qquad \text{(where } 0! = 1)$$

The six people can be selected one at a time, two at a time, three at a time etc., so r will vary between 1 and 6 inclusive; therefore the number of selections

$$= {}^6C_1 + {}^6C_2 + {}^6C_3 + {}^6C_4 + {}^6C_5 + {}^6C_6$$

$$= \frac{6!}{1!(6-1)!} + \frac{6!}{2!(6-2)!} + \frac{6!}{3!(6-3)!} + \frac{6!}{4!(6-4)!}$$

$$+ \frac{6!}{5!(6-5)!} + \frac{6!}{6!(6-6)!}$$

$$= \frac{6!}{1!5!} + \frac{6!}{2!4!} + \frac{6!}{3!3!} + \frac{6!}{4!2!} + \frac{6!}{5!1!} + \frac{6!}{6!0!}$$

$$= \frac{6 \cdot 5!}{1 \cdot 5!} + \frac{6 \cdot 5 \cdot 4!}{2 \cdot 1 \cdot 4!} + \frac{6 \cdot 5 \cdot 4 \cdot 3!}{3 \cdot 2 \cdot 1 \cdot 3!} + \frac{6 \cdot 5 \cdot 4!}{2 \cdot 4!} + \frac{6 \cdot 5!}{5! \cdot 1} + \frac{6! \cdot 1}{6! \cdot 1}$$

$$= 6 + 15 + 20 + 15 + 6 + 1$$

$$= 63$$

● **PROBLEM 7-36**

A boy has in his pocket a penny, a nickel, a dime, and a quarter. How many different sums of money can he take out if he removes one or more coins?

<u>Solution:</u> In this problem, we are not considering order; that is, we are not concerned whether we choose a penny first and a nickel second or vice versa. (It is still the same arrangement.) Thus, we are considering combinations, not permutations. We consider the following cases to solve this problem:

a)   the boy removes one coin
b)   the boy removes two coins
c)   the boy removes three coins
d)   the boy removes four coins

Now, a combination of n things taken r at a time is:

$$C(n, r) = \frac{n!}{r!(n-r)!} \; .$$

Thus if for a) the boy removes one coin, then we want to find the number of combinations of 4 coins taken one at a time. Similarly for b), c), and d). The total number of combinations of 4 things taken 1, 2, 3, or 4 at a time is

$$C(4,1) + C(4,2) + C(4,3) + C(4,4) = \frac{4!}{1!3!} + \frac{4!}{2!2!} + \frac{4!}{3!1!} + \frac{4!}{4!0!}$$

$$= \frac{4 \cdot 3!}{1 \cdot 3!} + \frac{4 \cdot 3 \cdot 2!}{2 \cdot 1 \cdot 2!} + \frac{4 \cdot 3!}{3! \cdot 1} + \frac{4!}{4! \cdot 1}$$

$$= 4 + \frac{12}{2} + 4 + 1$$

$$= 4 + 6 + 4 + 1$$

$$= 15.$$

● **PROBLEM** 7-37

From 10 men and 6 women, how many committees of 5 people can be chosen:

(a) If each committee is to have exactly 3 men?
(b) If each committee is to have at least 3 men?

Solution:
(a) The order in which the people on the committee are chosen is unimportant, thus this is a problem involving combinations. The general formula for the number of combinations of n different things taken r at a time is $C(n,r) = n!/r!(n-r)!$. Thus, the number of ways to choose 3 men from 10 men is $C(10,3)$

$$= \frac{10!}{3!(10-3)!} = \frac{10!}{3!7!} = \frac{10 \cdot 9 \cdot 8 \cdot 7!}{3 \cdot 2 \cdot 1 \cdot 7!} = 120.$$

The number of ways to choose 2 women from 6 women is $C(6,2)$

$$= \frac{6!}{2!(6-2)!} = \frac{6!}{2!4!} = \frac{6 \cdot 5 \cdot 4!}{2 \cdot 1 \cdot 4!} = 15.$$

The Fundamental Principle of Counting states that if the first of two independent acts can be performed in a ways, and if the second act can be performed in b ways, then the number of ways of performing the two acts in the order stated is ab. Thus by the fundamental principle, the number of ways to choose the committee is $C(10,3) \cdot C(6,2) = 120 \cdot 5 = 1,800$.

(b) If the committee is to contain at least 3 men, the possibilities are 3 men and 2 women, 4 men and 1 woman, 5 men and no women.

We have just shown that the number of committees consisting of 3 men and 2 women is 1,800. The number of committees containing 4 men and 1 woman is

$$C(10,4) \cdot C(6,1) = \frac{10!}{4!(10-4)!} \cdot \frac{6!}{1!(6-1)!} = \frac{10!}{4!6!} \cdot \frac{6!}{1!5!}$$

$$= \frac{10 \cdot 9 \cdot 8 \cdot 7 \cdot 6!}{4 \cdot 3 \cdot 2 \cdot 1 \cdot 6!} \cdot \frac{6 \cdot 5!}{1 \cdot 5!} = 210.6 = 1,260 .$$

The number of committees consisting of 5 men is

404

$$C(10,5) = \frac{10!}{5!(10-5)!} = \frac{10!}{5!5!} = \frac{\overset{2}{\cancel{10}}\cdot\overset{3}{\cancel{9}}\cdot\overset{2}{\cancel{8}}\cdot7\cdot\overset{3}{\cancel{6}}\cdot\cancel{5}!}{\cancel{5}\cdot\cancel{4}\cdot\cancel{3}\cdot\cancel{2}\cdot1\cdot\cancel{5}!} = 252$$

The probability that any of several mutually exclusive events will occur is the sum of the probabilities of the separate events.

Hence the number of committees containing at least 3 men is

$$1,800 + 1,260 + 252 = 3,312.$$

● PROBLEM 7-38

Two ordinary dice are rolled. In how many different ways can they fall? How many of these ways will give a sum of nine?

Solution: The first die can fall in any one of six positions, and for each of these positions the second die can also fall in six positions; so there are 6 x 6 = 36 ways that the dice can fall. This is shown in the accompanying construction of possible pairs for the faces of the two dice.

```
1,1   1,2   1,3   1,4   1,5   1,6

2,1   2,2   2,3   2,4   2,5   2,6

3,1   3,2   3,3   3,4   3,5   3,6

4,1   4,2   4,3   4,4   4,5   4,6

5,1   5,2   5,3   5,4   5,5   5,6

6,1   6,2   6,3   6,4   6,5   6,6
```

Of these 36 ways, there are four ways of obtaining a sum of nine: (6,3), (5,4),(4,5), (3,6) (circled in the figure).

● PROBLEM 7-39

If in a series of license plates the letters I and O are not used and if four successive zeroes can not be used as the four digits, how many different license plates can be made?

Solution: A license plate consists of two letters and four digits. Of the 26 letters of the alphabet, since I and O are not used, there are now 24 choices for each of the two letters. Since four successive zeroes can not be

used as the four digits, the digits would form the numbers
from 0001 to 9999; so there are 9999 choices for the four
digits. The fundamental principle states that if one
thing can be done in m different ways and, when it is done
in any one of these ways, a second thing can be done in n
different ways, and if a third thing can then be done in
p ways, ... then the successive things can be done in
mnp... different ways. Therefore,

$$\text{Number of license plates} = (24)(24)(9999)$$

$$= 5,759,424$$

● PROBLEM 7-40

A baseball manager after determining his starting players
now must determine his batting order. If the pitcher is
to bat last, how many different ways can the manager turn
in his batting order?

Solution: Since the pitcher is to bat last only the order
of the other eight players must be determined. The fundamental principle states that if one thing can be done in
m different ways and, when it is done in any one of these
ways, a second thing can be done in n different ways, and
if after it has been done in any one of these ways, a third
thing can be done in p different ways, . . .,the several
things can be done in m·n·p·...different ways. Using the
fundamental principle, he has 8 choices for the lead-off man.
Once this choice is made, he has 7 choices for the second
batter, and so on.

$$\text{Number of batting orders} = 8(7)(6)(5)(4)(3)(2)(1)$$

$$= 8!$$

$$= 40,320$$

● PROBLEM 7-41

Two cards are to be drawn in order from a pack of 4 cards
(say, an ace, king, queen, and jack), the drawn card not
being replaced before the second card is drawn. How many
different drawings are possible?

Solution: The first card can be drawn in four ways, and
then the second card in three ways. By the fundamental
principle of counting, which states that if the first of
two independent acts can be performed in x ways, and if the
second act can be performed in y ways, then the number of
ways of performing the two acts, in the order stated, is
xy, there are 4 · 3 = 12 different drawings. (We regard

here ace first, king second as a different drawing from king first, ace second.)

If the first card were to be replaced before the second is drawn in this example, then the answer would be $4 \cdot 4 = 16$ different drawings.

THE BINOMIAL THEOREM

(a)   Find the 9th term of $(x + y)^{11}$.

(b)   Find the first three terms of $(a + 4b)^{16}$.

Solution:

(a)   The binomial theorem may be expressed in the form

$$(x + y)^n = {_nC_0}x^n + {_nC_1}x^{n-1}y + {_nC_2}x^{n-2}y^2 + \ldots + {_nC_n}y^n$$

It's observed that there are $n + 1$ terms in the expression above and the $(r + 1)$th term is ${_nC_r}x^{n-r}y^r$. Hence, for $(x + y)^{11}$, there are $(n + 1) = (11 + 1) = 12$ terms in the expansion and $9 = r + 1$, $r = 8$. The 9th term of $(x + y)^{11}$ is ${_{11}C_8}x^{11-8}y^8$, that is, ${_{11}C_8}x^3 y^8$.

$$_{11}C_8 = C(n, r) = \frac{n!}{r!(n - r)!} = \frac{11!}{8!(11 - 8)!}$$

$$= \frac{8! \times 9 \times 10 \times 11}{8! \times 3!} = 165$$

So the 9th term of $(x + y)^{11}$ is $165\ x^3 y^8$.

(b)   From the binomial theorem, one obtains

$$(a + 4b)^{16} = a^{16} + 16a^{15}(4b) + \frac{16 \times 15}{1 \times 2} a^{14}(4b)^2$$

$$+ \frac{16 \times 15 \times 14}{1 \times 2 \times 3} a^{13} (4b)^3 + \ldots\ldots$$

$$= a^{16} + 64a^{15}b + 1920a^{14}b^2 + \ldots\ldots$$

Therefore, the first three terms of $(a + 4b)^{16}$ are a^{16}, $64a^{15}b$ and $1920a^{14}b^2$.

● PROBLEM 7-43

Find the term of the expansion of $(a + b + c + d)^6$ that involves ac^2d^3.

Solution: For any multinomial, the expansion of $(x_1 + x_2 + \ldots x_m)^n$ is equal to the sum of all possible terms of the form $\dfrac{n!}{n_1! n_2! \ldots n_m!} x_1^{n_1} x_2^{n_2} \ldots x_m^{n_m}$ where $n_1 + n_2 + \ldots + n_m = n$. Thus, the term of expansion of $(a + b + c + d)^6$ that involves ac^2d^3 is given by

$$\frac{6!}{(1!)(2!)(3!)} ac^2d^3 = 60ac^2d^3$$

● PROBLEM 7-44

Find the binomial expansion of $(2x - 5)^4$.

Solution: The generalized form of the binomial expansion is
$$(a+b)^n = a^n + {}_nC_1 a^{n-1}b + {}_nC_2 a^{n-2}b^2 + {}_nC_3 a^{n-3}b^3 + \ldots + {}_nC_{n-1} a^1 b^{n-1}$$
$$+ {}_nC_n a^0 b^n$$

Here we take $a = 2x$, $b = -5$, and $n = 4$.
$$(2x-5)^4 = (2x)^4 + {}_4C_1 (2x)^3(-5)^1 + {}_4C_2 (2x)^2(-5)^2 + {}_4C_3 (2x)^1(-5)^3$$
$$+ {}_4C_4 (2x)^0(-5)^4$$

$$= 16x^4 + {}_4C_1 \, 8x^3 \cdot (-5) + {}_4C_2 \, 4x^2(25) + {}_4C_3 \, 2x(-125) + {}_4C_4 1(-5)^4$$

Note that ${}_nC_r = \dfrac{n!}{r!(n-r)!}$. Therefore

$$(2x-5)^4 = 16x^4 + \frac{4!}{1!3!} 8x^3(-5) + \frac{4!}{2!2!}\left(4x^2\right)(25) + \frac{4!}{3!1!}(2x)(-125)$$

$$+ \frac{4!}{4!0!}(1)(-5)^4$$

$$= 16x^4 + \frac{4 \cdot \cancel{3!}}{\cancel{3!}} 8x^3(-5) + \frac{4 \cdot 3 \cdot \cancel{2!}}{\cancel{2!} \cdot 1 \cdot \cancel{2!}} \left(100x^2\right) + \frac{4 \cdot \cancel{3!}}{\cancel{3!}}(-250x) + 625$$

$$= 16x^4 - 160x^3 + 600x^2 - 1{,}000x + 625.$$

Find the expansion of $(x + y)^6$.

Solution: Use the Binomial Theorem which states that

$$(a+b)^n = \frac{1}{0!} a^n + \frac{n}{1!} a^{n-1}b + \frac{n(n-1)}{2!} a^{n-2}b^2 + \ldots + nab^{n-1} + b^n .$$

Replacing a by x and b by y:

$$(x+y)^6 = \frac{1}{0!} x^6 + \frac{6}{1!} x^5 y + \frac{6 \cdot 5}{2!} x^4 y^2 + \frac{6 \cdot 5 \cdot 4}{3!} x^3 y^3 + \frac{6 \cdot 5 \cdot 4 \cdot 3}{4!} x^2 y^4$$

$$+ \frac{6 \cdot 5 \cdot 4 \cdot 3 \cdot 2}{5!} x^1 y^5 + \frac{6 \cdot 5 \cdot 4 \cdot 3 \cdot 2 \cdot 1}{6!} x^0 y^6$$

$$= \frac{1}{1} x^6 + \frac{6}{1} x^5 y + \frac{6 \cdot 5}{2 \cdot 1} x^4 y^2 + \frac{6 \cdot 5 \cdot 4}{3 \cdot 2 \cdot 1} x^3 y^3 + \frac{6 \cdot 5 \cdot 4 \cdot 3}{4 \cdot 3 \cdot 2 \cdot 1} x^2 y^4$$

$$+ \frac{6 \cdot 5 \cdot 4 \cdot 3 \cdot 2}{5 \cdot 4 \cdot 3 \cdot 2 \cdot 1} xy^5 + \frac{6 \cdot 5 \cdot 4 \cdot 3 \cdot 2 \cdot 1}{6 \cdot 5 \cdot 4 \cdot 3 \cdot 2 \cdot 1} y^6$$

$$(x+y)^6 = x^6 + 6x^5 y + 15x^4 y^2 + 20x^3 y^3 + 15x^2 y^4 + 6xy^5 + y^6 .$$

Expand $(x + 2y)^5$.

Solution: Apply the binomial theorem. If n is a positive integer, then

$$(a + b)^n = \binom{n}{0} a^n b^0 + \binom{n}{1} a^{n-1}b + \binom{n}{2} a^{n-2}b^2 + \ldots + \binom{n}{r} a^{n-r}b^r$$

$$+ \ldots + \binom{n}{n} b^n .$$

Note that $\binom{n}{r} = \frac{n!}{r!(n-r)!}$ and that $0! = 1$. Then, we obtain:

$$(x + 2y)^5 = \binom{5}{0} x^5 (2y)^0 + \binom{5}{1} x^4 (2y)^1 + \binom{5}{2} x^3 (2y)^2$$

$$+ \binom{5}{3} x^2 (2y)^3 + \binom{5}{4} x^1 (2y)^4 + \binom{5}{5} x^0 (2y)^5$$

$$= \frac{5!}{0!5!} x^5 + \frac{5!}{1!4!} x^4 2y + \frac{5!}{2!3!} x^3 \left(4y^2\right)$$

$$+ \frac{5!}{3!2!} x^2 \left(8y^3\right) + \frac{5!}{4!1!} x \left(16y^4\right) + \frac{5!}{5!0!} 1 \left(32y^5\right)$$

$$= x^5 + \frac{5 \cdot 4! }{4!}x^4 2y + \frac{5 \cdot 4 \cdot 3!}{2 \cdot 1 \cdot 3!}x^3 \left(4y^2\right)$$

$$+ \frac{5 \cdot 4 \cdot 3!}{3! \cdot 2 \cdot 1}x^2 \left(8y^3\right) + \frac{5 \cdot 4!}{4!1!}x \left(16y^4\right) + \frac{5!}{5!0!}\left(32y^5\right)$$

$$= x^5 + 10x^4 y + 40x^3 y^2 + 80x^2 y^3 + 80xy^4 + 32y^5.$$

● **PROBLEM** 7-47

Expand $\left(3p^2 - 2q^{\frac{1}{2}}\right)^4$ by means of the binomial theorem.

<u>Solution</u>: We apply the binomial theorem: If n is a positive integer, then

$$(a + b)^n = \binom{n}{0} a^n b^0 + \binom{n}{1} a^{n-1}b + \binom{n}{2} a^{n-2}b^2 + \ldots + \binom{n}{r} a^{n-r}b^r + \ldots + \binom{n}{n} a^0 b^n$$

where $\binom{n}{r} = \frac{n!}{r!(n-r)!}$.

We identify $3p^2$ with a, $-2q^{\frac{1}{2}}$ with b, and n with 4. The binomial theorem then gives us

$$\left(3p^2 - 2q^{\frac{1}{2}}\right)^4 = \binom{4}{0}\left(3p^2\right)^4\left(-2q^{\frac{1}{2}}\right)^0 + \binom{4}{1}\left(3p^2\right)^3\left(-2q^{\frac{1}{2}}\right) + \binom{4}{2}\left(3p^2\right)^2\left(-2q^{\frac{1}{2}}\right)^2$$

$$+ \binom{4}{3}\left(3p^2\right)^1\left(-2q^{\frac{1}{2}}\right)^3 + \binom{4}{4}\left(3p^2\right)^0\left(-2q^{\frac{1}{2}}\right)^4$$

$$= \frac{4!}{0!(4)!} 3^4 p^8 + \frac{4!}{1!3!} 3^3 p^6 \left(-2q^{\frac{1}{2}}\right) + \frac{4!}{2!2!} 3^2 p^4 (-2)^2 q$$

$$+ \frac{4!}{3!1!} 3p^2 (-2)^3 q^{3/2} + \frac{4!}{4!0!}(-2)^4 q^2$$

$$= 81p^8 + 4 \cdot \frac{3!}{3!} 27p^6 \left(-2q^{\frac{1}{2}}\right) + \frac{4 \cdot 3 \cdot 2!}{2! 2!} 9p^4 4q$$

$$+ \frac{4 \cdot 3!}{3! \cdot 1!} 3p^2 (-8)q^{3/2} + 16q^2$$

$$= 81p^8 - 216p^6 q^{\frac{1}{2}} + 216p^4 q - 96p^2 q^{3/2} + 16q^2 .$$

● **PROBLEM** 7-48

Give the expansion of $\left(r^2 - \frac{1}{s}\right)^5$.

<u>Solution:</u> Write the given expression as the sum of two terms raised to the 5th power:

$$\left(r^2 - \frac{1}{s}\right)^5 = \left[r^2 + \left(-\frac{1}{s}\right)\right]^5 \tag{1}$$

The Binomial Theorem can be used to expand the expression on the right side of equation (1). The Binomial Theorem is stated as:

$$(a+b)^n = a^n + na^{n-1}b + \frac{n(n-1)}{1 \cdot 2}a^{n-2}b^2 + \frac{n(n-1)(n-2)}{1 \cdot 2 \cdot 3}a^{n-3}b^3$$

$$+ \ldots + nab^{n-1} + b^n, \text{ where a and b are any two}$$

numbers.

Let $a = r^2$, $b = -\frac{1}{s}$, and $n = 5$. Then, using the Binomial Theorem:

$$\left(r^2 - \frac{1}{s}\right)^5 = \left[r^2 + \left(-\frac{1}{s}\right)\right]^5$$

$$= \left(r^2\right)^5 + 5\left(r^2\right)^{5-1}\left(-\frac{1}{s}\right) + \frac{5(5-1)}{1 \cdot 2}\left(r^2\right)^{5-2}\left(-\frac{1}{s}\right)^2$$

$$+ \frac{5(5-1)(5-2)}{1 \cdot 2 \cdot 3}\left(r^2\right)^{5-3}\left(-\frac{1}{s}\right)^3$$

$$+ \frac{5(5-1)(5-2)(5-3)}{1 \cdot 2 \cdot 3 \cdot 4}\left(r^2\right)^{5-4}\left(-\frac{1}{s}\right)^4$$

$$+ \frac{5(5-1)(5-2)(5-3)(5-4)}{1 \cdot 2 \cdot 3 \cdot 4 \cdot 5}\left(r^2\right)^{5-5}\left(-\frac{1}{s}\right)^5$$

$$= r^{10} - \frac{5\left(r^2\right)^4}{s} + \frac{5(\cancel{4})}{1 \cdot \cancel{2}}\left(r^2\right)^3\left(\frac{1}{s^2}\right) - \frac{5(\cancel{4})(\cancel{3})}{1 \cdot 2 \cdot \cancel{3}}\left(r^2\right)^2\left(\frac{1}{s^3}\right)$$

$$+ \frac{5(\cancel{4})(\cancel{3})(\cancel{2})}{1 \cdot \cancel{2} \cdot \cancel{3} \cdot \cancel{4}}\left(r^2\right)^1\left(\frac{1}{s^4}\right) - \frac{\cancel{5}(\cancel{4})(\cancel{3})(\cancel{2})(\cancel{1})}{\cancel{1} \cdot \cancel{2} \cdot \cancel{3} \cdot \cancel{4} \cdot \cancel{5}}\left(r^2\right)^0\left(\frac{1}{s^5}\right)$$

$$= r^{10} - \frac{5r^8}{s} + \frac{10r^6}{s^2} - \frac{10r^4}{s^3} + \frac{5r^2}{s^4} - (1)(1)\left(\frac{1}{s^5}\right)$$

$$\left(r^2 - \frac{1}{s}\right)^5 = r^{10} - \frac{5r^8}{s} + \frac{10r^6}{s^2} - \frac{10r^4}{s^3} + \frac{5r^2}{s^4} - \frac{1}{s^5}$$

411

Find the first five terms of the expansion of $(1 + x)^{-2}$.

Solution: The Binomial Theorem states that:

$$(a+b)^n = \frac{1}{0!} a^n + \frac{n}{1!} a^{n-1}b + \frac{n(n-1)}{2!} a^{n-2}b^2 + \ldots + nab^{n-1} + b^n .$$

This theorem can be used to find the first five terms of the expansion of $(1 + x)^{-2}$. Replacing a by 1 and b by x, the expression $(1 + x)^{-2}$ becomes:

$$(1+x)^{-2} = \frac{1}{0!} 1^{-2} + \frac{-2}{1!} 1^{-3}x + \frac{(-2)(-3)}{2!} 1^{-4}x^2 + \frac{(-2)(-3)(-4)}{3!} 1^{-5}x^3$$

$$+ \frac{(-2)(-3)(-4)(-5)}{4!} 1^{-6}x^4 + \ldots + (-2)1x^{-3} + x^{-2} .$$

Writing only the first five terms of this expansion:

$$(1+x)^{-2} = \frac{1}{0!}1^{-2} + \frac{-2}{1!}1^{-3}x + \frac{(-2)(-3)}{2!}1^{-4}x^2 + \frac{(-2)(-3)(-4)}{3!}1^{-5}x^3$$

$$+ \frac{(-2)(-3)(-4)(-5)}{4!}1^{-6}x^4 + \ldots$$

$$= \frac{1}{1}\left(\frac{1}{1^2}\right) - 2x\left(\frac{1}{1^3}\right) + \frac{6x^2}{2 \cdot 1}\left(\frac{1}{1^4}\right) + \frac{(-24)x^3}{3 \cdot 2 \cdot 1}\left(\frac{1}{1^5}\right) + \frac{\overset{30}{120}x^4}{\underset{1}{4} \cdot 3 \cdot 2 \cdot 1}\left(\frac{1}{1^6}\right) + \ldots$$

$$(1+x)^{-2} = 1 - 2x + 3x^2 - 4x^3 + 5x^4 + \ldots \qquad (1)$$

Hence, the right side of equation (1) represents the first five terms of the expansion of $(1 + x)^{-2}$.

Find the first four terms of the expansion of $(2 - 1)^{\frac{1}{2}}$.

Solution: The Binomial Theorem states that:

$$(a+b)^n = \frac{1}{0!}a^n + \frac{n}{1!}a^{n-1}b + \frac{n(n-1)}{2!}a^{n-2}b^2 + \ldots + nab^{n-1} + b^n .$$

This theorem can be used to expand the expression $(2 - 1)^{1/2}$. Replacing a by 2 and b by -1, the expression $(2 - 1)^{1/2}$ becomes:

$$(2-1)^{1/2} = [2+(-1)]^{1/2} = \frac{1}{0!}2^{1/2} + \frac{1/2}{1!} 2^{\frac{1}{2}-1}(-1) + \frac{\frac{1}{2}(\frac{1}{2}-1)}{2!}(2)^{\frac{1}{2}-2}(-1)^2$$

$$+ \frac{\frac{1}{2}(\frac{1}{2}-1)(\frac{1}{2}-2)}{3!}(2)^{\frac{1}{2}-3}(-1)^3 + \ldots + \frac{1}{2}(2)(-1)^{\frac{1}{2}-1} + (-1)^{\frac{1}{2}}$$

$$= \frac{1}{1}\sqrt{2} + \frac{1}{2}(2)^{-\frac{1}{2}}(-1) + \frac{\frac{1}{2}(-\frac{1}{2})}{2 \cdot 1}(2)^{-\frac{3}{2}} \qquad (1)$$

$$+ \frac{\frac{1}{2}(-\frac{1}{2}) \ -\frac{3}{2}}{3 \cdot 2 \cdot 1}(2)^{-\frac{5}{2}}(-1) + \ldots + 1(-1)^{-\frac{1}{2}} + \sqrt{-1}$$

$$= \sqrt{2} - \frac{1}{2}\left(\frac{1}{2^{\frac{1}{2}}}\right) - \frac{1}{8}\left(\frac{1}{2^{\frac{3}{2}}}\right) - \frac{3}{48}\left(\frac{1}{2^{\frac{5}{2}}}\right) + \ldots + 1\left(\frac{1}{\sqrt{-1}}\right) + \sqrt{-1} \ .$$

The last result is true because of the law of exponents which states that

$$(N)^{-a/b} = \frac{1}{(N)^{a/b}}$$

where   a   and   b   are any positive integers.   Writing only the first four terms of the expansion:

$$(2-1)^{\frac{1}{2}} = \sqrt{2} - \frac{1}{2}\left(\frac{1}{2^{\frac{1}{2}}}\right) - \frac{1}{8}\left(\frac{1}{2^{\frac{3}{2}}}\right) - \frac{3}{48}\left(\frac{1}{2^{\frac{5}{2}}}\right) + \ldots$$

$$= \sqrt{2} - \frac{1}{2}\left(\frac{1}{\sqrt{2}}\right) - \frac{1}{8}\left(\frac{1}{\sqrt{2}^3}\right) - \frac{3}{48}\left(\frac{1}{\sqrt{2}^5}\right) + \ldots$$

$$= \sqrt{2} - \frac{1}{2\sqrt{2}} - \frac{1}{8}\left(\frac{1}{\sqrt{2}^2 \sqrt{2}}\right) - \frac{3}{48}\left(\frac{1}{\sqrt{2}^2 \sqrt{2}^2 \sqrt{2}}\right) + \ldots$$

$$= \sqrt{2} - \frac{1}{2\sqrt{2}} - \frac{1}{8}\left(\frac{1}{2\sqrt{2}}\right) - \frac{3}{48}\left(\frac{1}{(2)(2)\sqrt{2}}\right) + \ldots$$

$$= \sqrt{2} - \frac{1}{2\sqrt{2}} - \frac{1}{16\sqrt{2}} - \frac{3}{48}\left(\frac{1}{4\sqrt{2}}\right) + \ldots$$

$$= \sqrt{2} - \frac{1}{2\sqrt{2}} - \frac{1}{16\sqrt{2}} - \frac{3}{192}\frac{1}{\sqrt{2}} + \ldots$$

$$= \sqrt{2} - \frac{1}{2\sqrt{2}} - \frac{1}{16\sqrt{2}} - \frac{1}{64}\frac{1}{\sqrt{2}} + \ldots$$

$$= \sqrt{2} - \frac{1}{2\sqrt{2}} - \frac{1}{16\sqrt{2}} - \frac{1}{64\sqrt{2}} + \ldots$$

Since $\sqrt{2} = 1.414$,

$$(2-1)^{\frac{1}{2}} \doteq 1.414 - \frac{1}{2(1.414)} - \frac{1}{16(1.414)} - \frac{1}{64(1.414)}$$

$$= 1.414 - \frac{1}{2.828} - \frac{1}{22.624} - \frac{1}{90.496}$$

$$(2-1)^{\frac{1}{2}} = 1.414 - 0.354 - 0.044 - 0.011 \tag{1}$$

Hence, the right side of equation (1) represents the first four terms of the expansion of $(2 - 1)^{\frac{1}{2}}$.

Note that the sum of these four terms is 1.005.

● PROBLEM 7-51

Find the fifth term of $\left(2 + 2x^3\right)^{17}$.

<u>Solution:</u> Use the Binomial Theorem which states that

$$(c+b)^n = \frac{1}{0!} c^n + \frac{n}{1!} c^{n-1}b + \frac{n(n-1)}{2!} c^{n-2}b^2 + \ldots + ncb^{n-1} + b^n .$$

Replacing c by a and b by $2x^3$:

$$\left(a+2x^3\right)^{17} = \frac{1}{0!} a^{17} + \frac{17}{1!} a^{16}\left(2x^3\right)^1 + \frac{17\cdot 16}{2!} a^{15}\left(2x^3\right)^2$$

$$+ \frac{17\cdot 16\cdot 15}{3!} a^{14}\left(2x^3\right)^3 + \frac{17\cdot 16\cdot 15\cdot 14}{4!} a^{13}\left(2x^3\right)^4$$

$$+ \ldots$$

The fifth term of this expansion is:

$$\frac{17\cdot 16\cdot 15\cdot 14}{4!} a^{13}\left(2x^3\right)^4 = \frac{17\cdot \overset{4}{\cancel{16}}\cdot \overset{5}{\cancel{15}}\cdot \overset{7}{\cancel{14}}}{\underset{1}{\cancel{4}}\cdot \underset{1}{\cancel{3}}\cdot \underset{1}{\cancel{2}}\cdot 1} a^{13}\left(2^4\right)\left(x^3\right)^4$$

$$= \frac{17\cdot 4\cdot 5\cdot 7}{1} a^{13}\ 16x^{12}$$

$$= 38{,}080\ a^{13}\ x^{12}$$

● **PROBLEM** 7-52

Find the 5th term of the expansion of $(3y - 4w)^8$.

<u>Solution:</u> Use the binomial formula:

$$(u+v)^n = u^n + nu^{n-1}v + \frac{n(n-1)}{2} u^{n-2}v^2 + \frac{n(n-1)(n-2)}{2\cdot 3} u^{n-2}v^3 + \ldots + v^n .$$

Applying the formula for the rth term where $r \leq n+1$; the rth term is:

$$\frac{n(n-1)(n-2)\ldots(n-r+2)}{1\cdot 2\cdot 3\ \ldots(r-1)}\ u^{n-r+1}v^{r-1} .$$ Let u = 3y, v = −4w, n = 8, r = 5.

Then $n - r + 1 = 8 - 5 + 1 = 4$

$$r - 1 = 5 - 1 = 4$$

Therefore $u^{n-r+1}v^{r-1} = u^4v^4$

$$r - 1 = 4$$

$$n - r + 2 = 8 - 5 + 2 = 5$$

So: $n(n-1)(n-2)\ldots(5) = 8\cdot 7\cdot 6\cdot 5$

Thus the coefficient is:

$$\frac{8\cdot 7\cdot 6\cdot 5}{1\cdot 2\cdot 3\cdot 4}$$

$$u^4 = (3y)^4$$

$$v^4 = (-4w)^4$$

Therefore: $(3y - 4w)^8 = \frac{8\cdot 7\cdot 6\cdot 5}{4\cdot 3\cdot 2\cdot 1} (3y)^4(-4w)^4$

$$= \frac{\overset{2}{\cancel{8}}(7)(\cancel{6})(5)}{\cancel{4}(\cancel{3})(\cancel{2})} \left(81y^4\right)\left(256w^4\right)$$

$$= 70\left(81y^4\right)\left(256w^4\right)$$

$$= 1{,}451{,}520 y^4 w^4$$

Find the term involving y^5 in the expansion of $(2x^2 + y)^{10}$.

Solution: The formula for the binomial expansion is:

$$(a + b)^n = a^n + na^{n-1}b + \frac{n(n-1)}{1 \cdot 2} a^{n-2}b^2 + \frac{n(n-1)(n-2)}{1 \cdot 2 \cdot 3} a^{n-3}b^3 + \ldots + nab^{n-1} + b^n.$$

The rth term of the expansion of $(a + b)^n$ is

$$\text{rth term} = \frac{n(n-1)(n-2) \ldots (n-r+2)}{(r-1)!} a^{n-r+1}b^{r-1}$$

In this example,

$$b^{r-1} = y^5$$
$$r-1 = 5$$
$$r = 6 \quad \text{and} \quad n = 10$$

Thus,

$$\text{6th term} = \frac{10 \cdot 9 \cdot 8 \cdot 7 \cdot 6}{5!} (2x^2)^5 y^5$$

$$= \frac{\overset{2}{\cancel{10}} \cdot \overset{3}{\cancel{9}} \cdot \overset{\cancel{8}1}{\cancel{8}} \cdot 7 \cdot 6}{\underset{1}{\cancel{5}} \cdot \underset{1}{\cancel{4}} \cdot \underset{1}{\cancel{3}} \cdot \underset{1}{\cancel{2}} \cdot 1} \quad 32x^{10}y^5$$

$$= 8064 \ x^{10} y^5$$

Find the constant term in the expansion of $\left(2x^2 + \frac{1}{x}\right)^9$.

Solution: The rth term in the expansion of $(a+b)^n$ is given by:

$$\frac{n(n-1)(n-2)\ldots(n-r+2)}{1 \cdot 2 \cdot 3 \cdot \ldots \cdot (r-1)} a^{n-r+1} b^{r-1}$$

Replacing a by $2x^2$ and b by $1/x$ in this formula, the rth term in the expansion of $(2x^2 + 1/x)^9$ is given by:

$$\frac{9(8)(7)\ldots(9-r+2)}{(1)(2)(3)\ldots(r-1)}(2x^2)^{9-r+1}\left(\frac{1}{x}\right)^{r-1} = \frac{9(8)(7)\ldots(11-r)}{(1)(2)(3)\ldots(r-1)}(2x^2)^{10-r}\left(\frac{1}{x}\right)^{r-1}$$

Then the rth term in the expansion will contain the factors $(2x^2)^{10-r}$ and $\left(\frac{1}{x}\right)^{r-1}$. Hence, as far as powers of x are concerned, the rth term will involve

$$\left(x^2\right)^{10-r}\left(\frac{1}{x}\right)^{r-1} \quad \text{or} \quad \left(x^2\right)^{10-r}\frac{(1)^{r-1}}{x^{r-1}} \quad \text{or} \quad \frac{x^{20-2r}}{x^{r-1}}$$

$$\text{or} \quad x^{20-2r-(r-1)}$$

$$\text{or} \quad x^{20-2r-r+1}$$

$$\text{or} \quad x^{21-3r}$$

The desired constant term is free of x; that is, the constant term has a factor of x^0 since $kx^0 = k(1) = k$, where k is the constant. Hence,

$$21 - 3r = 0$$
$$21 = 3r$$
$$\frac{21}{3} = r$$

$$7 = r$$

The rth term or seventh term can be found by using the Binomial Theorem which states that:

$$(a+b)^n = \frac{1}{0!}a^n + \frac{n}{1!}a^{n-1}b + \frac{n(n-1)}{2!}a^{n-2}b^2 + \ldots + nab^{n-1} + b^n .$$

Replacing a by $2x^2$ and b by $1/x$, $\left(2x^2 + 1/x\right)^9$ can be expanded as:

$$\left(2x^2 + \frac{1}{x}\right)^9 = \frac{1}{0!}\left(2x^2\right)^9 + \frac{9}{1!}\left(2x^2\right)^8\left(\frac{1}{x}\right) + \frac{9(8)}{2!}\left(2x^2\right)^7\left(\frac{1}{x}\right)^2$$

$$+ \frac{9(8)(7)}{3!}\left(2x^2\right)^6\left(\frac{1}{x}\right)^3 + \frac{9(8)(7)(6)}{4!}\left(2x^2\right)^5\left(\frac{1}{x}\right)^4$$

$$+ \frac{9(8)(7)(6)(5)}{5!}\left(2x^2\right)^4\left(\frac{1}{x}\right)^5 + \frac{9(8)(7)(6)(5)(4)}{6!}\left(2x^2\right)^3\left(\frac{1}{x}\right)^6$$

$$+ \frac{9(8)(7)(6)(5)(4)(3)}{7!}\left(2x^2\right)^2\left(\frac{1}{x}\right)^7$$

$$+ \frac{9(8)(7)(6)(5)(4)(3)(2)}{8!}\left(2x^2\right)\left(\frac{1}{x}\right)^8$$

$$+ \frac{9(8)(7)(6)(5)(4)(3)(2)(1)}{9!}\left(2x^2\right)^0\left(\frac{1}{x}\right)^9$$

Hence, the rth term or 7th term of this expansion is:

$$\frac{9(8)(7)(6)(5)(4)}{6!}\left(2x^2\right)^3\left(\frac{1}{x}\right)^6$$

$$= \frac{\overset{3}{\cancel{9}}\cdot 8\cdot 7\cdot \overset{4}{\cancel{6}}\cdot \cancel{5}\cdot \cancel{4}}{\cancel{6}\cdot \cancel{5}\cdot \cancel{4}\cdot \cancel{3}\cdot 2\cdot \cancel{1}}\left(2x^2\right)^3\left(\frac{1}{x}\right)^6$$

$$= 3\cdot 4\cdot 7\left(2x^2\right)^3\left(\frac{1}{x}\right)^6$$

$$= 84\left(2^3\right)\left(x^2\right)^3\left(\frac{1^6}{x^6}\right)$$

$$= 84(8)\cancel{x}^6\frac{1}{\cancel{x}^6}$$

$$= 672 .$$

Hence, 672 is the desired constant term in the expansion of $\left(2x^2 + \frac{1}{x}\right)^9$.

Find the coefficient of $a_1^2 a_2 a_3$ in the expansion of $\left(a_1 + a_2 + a_3\right)^4$.

Solution: The binomial theorem states that, if n is a positive integer, then

$$(a + b)^n = a^n + na^{n-1}b + \frac{n(n-1)}{1 \cdot 2} a^{n-2}b^2 + \frac{n(n-1)(n-2)}{1 \cdot 2 \cdot 3} a^{n-3}b^3$$
$$+ \ldots + nab^{n-1} + b^n.$$

Use the binomial theorem, but for convenience, associate the terms $\left(a_2 + a_3\right)$, then expand the expression.

$$\left[a_1 + \left(a_2 + a_3\right)\right]^4 = a_1^4 + 4a_1^3\left(a_2 + a_3\right) + \frac{4 \cdot 3}{1 \cdot 2} a_1^2\left(a_2 + a_3\right)^2$$
$$+ \frac{4 \cdot 3 \cdot 2}{1 \cdot 2 \cdot 3} a_1\left(a_2 + a_3\right)^3 + \left(a_2 + a_3\right)^4.$$

Notice that the only term involving $a_1^2 a_2 a_3$ is the third term with coefficient $\frac{4 \cdot 3}{2}$ and, further, that $\left(a_2 + a_3\right)^2$ must be expanded also.

$$\left(a_2 + a_3\right)^2 = a_2^2 + 2a_2a_3 + a_3^2.$$

Therefore, the third term becomes:

$$\frac{4 \cdot 3}{1 \cdot 2} a_1^2\left(a_2 + a_3\right)^2 = \frac{4 \cdot 3}{1 \cdot 2} a_1^2\left(a_2^2 + 2a_2a_3 + a_3^2\right)$$
$$= 6a_1^2a_2^2 + 12a_1^2a_2a_3 + 6a_1^2a_3^2$$

Hence, the coefficient of $a_1^2 a_2 a_3$ is 12.

Find the term involving x^3yz^2 in the expansion $(x + 2y - 3z)^6$.

Solution: Use the binomial formula: $a^n + na^{n-1}b + \frac{n(n-1)}{1 \cdot 2}$ x $a^{n-2}b^2 + \frac{n(n-1)(n-2)}{1 \cdot 2 \cdot 3} a^{n-3}b^3 + \ldots + nab^{n-1} + b^n = (a+b)^n$

and associate 2y - 3z and substitute it for b in the formula, x for a, and 6 for n:

$$[x + (2y - 3z)]^6 = x^6 + 6x^5(2y - 3z)^1 + \frac{6 \cdot 5}{1 \cdot 2} x^4(2y - 3z)^2$$

$$+ \frac{6 \cdot 5 \cdot 4}{1 \cdot 2 \cdot 3} x^3(2y - 3z)^3 + \frac{6 \cdot 5 \cdot 4 \cdot 3}{1 \cdot 2 \cdot 3 \cdot 4} x^2(2y - 3z)^4$$

$$+ \frac{6 \cdot 5 \cdot 4 \cdot 3 \cdot 2}{1 \cdot 2 \cdot 3 \cdot 4 \cdot 5} x(2y - 3z)^5 + (2y - 3z)^6.$$

The term that involves x^3yz^2 is: $\frac{6 \cdot 5 \cdot 4}{1 \cdot 2 \cdot 3} x^3 (2y - 3z)^3$ in which $(2y - 3z)^3$ must be expanded.

$$(2y - 3z)^3 = (2y - 3z)(2y - 3z)^2 = (2y - 3z)(4y^2 - 12yz + 9z^2)$$
$$= 8y^3 - 36zy^2 + 54yz^2 - 27z^3$$

When $(2y - 3z)^3$ is multiplied by $\frac{6 \cdot 5 \cdot 4}{1 \cdot 2 \cdot 3} x^3$, the final term is

$$\frac{6 \cdot 5 \cdot 4}{1 \cdot 2 \cdot 3} x^3 \left[8y^3 - 36zy^2 + 54yz^2 - 27z^3 \right].$$

Distributing, notice that the term of interest involves

$$\frac{6 \cdot 5 \cdot 4}{1 \cdot 2 \cdot 3} x^3 \left(54yz^2 \right) = 1080x^3yz^2.$$

● **PROBLEM** 7-57

Find the value of $\left(a + \sqrt{a^2 - 1} \right)^7 + \left(a - \sqrt{a^2 - 1} \right)^7$.

Solution: Use the binomial theorem to expand the first term:

$$(a + b)^n = a^n + na^{n-1}b + \frac{n(n-1)}{1 \cdot 2} a^{n-2}b^2 + \frac{n(n-1)(n-2)}{1 \cdot 2 \cdot 3} a^{n-3}b^3$$
$$+ \ldots + b^n.$$

Now $\left(a + \sqrt{a^2 - 1} \right)^7 = a^7 + 7a^6 \left(\sqrt{a^2 - 1} \right)^1 + \frac{7 \cdot 6}{1 \cdot 2} (a)^5 \left(\sqrt{a^2 - 1} \right)^2$

$$+ \frac{7 \cdot 6 \cdot 5}{1 \cdot 2 \cdot 3} a^4 \left(\sqrt{a^2 - 1} \right)^3 + \frac{7 \cdot 6 \cdot 5 \cdot 4}{1 \cdot 2 \cdot 3 \cdot 4} a^3 \left(\sqrt{a^2 - 1} \right)^4$$

$$+ \frac{7 \cdot 6 \cdot 5 \cdot 4 \cdot 3}{1 \cdot 2 \cdot 3 \cdot 4 \cdot 5} a^2 \left(\sqrt{a^2 - 1} \right)^5 + 7a \left(\sqrt{a^2 - 1} \right)^6 + \left(\sqrt{a^2 - 1} \right)^7$$

Expand $\left(a - \sqrt{a^2 - 1} \right)^7$ in the same fashion.

$$\left(a - \sqrt{a^2 - 1} \right)^7 = a^7 + 7a^6 \left(-\sqrt{a^2 - 1} \right)^1 + \frac{7 \cdot 6}{1 \cdot 2} (a)^5 \left(-\sqrt{a^2 - 1} \right)^2$$

$$+ \frac{7 \cdot 6 \cdot 5}{1 \cdot 2 \cdot 3} a^4 \left(-\sqrt{a^2 - 1} \right)^3 + \frac{7 \cdot 6 \cdot 5 \cdot 4}{1 \cdot 2 \cdot 3 \cdot 4} a^3 \left(-\sqrt{a^2 - 1} \right)^4$$

$$+ \frac{7 \cdot 6 \cdot 5 \cdot 4 \cdot 3}{1 \cdot 2 \cdot 3 \cdot 4 \cdot 5} a^2 \left(-\sqrt{a^2 - 1} \right)^5 + 7a \left(-\sqrt{a^2 - 1} \right)^6 + \left(-\sqrt{a^2 - 1} \right)^7.$$

In this expansion, the odd powers of $-\sqrt{a^2 - 1}$ will cancel the corresponding terms in the previous expansion.

$$\left(a + \sqrt{a^2 - 1}\right)^7 + \left(a - \sqrt{a^2 - 1}\right)^7 = a^7 + 7a^6\left(\sqrt{a^2 - 1}\right)^1$$

$$+ \frac{7 \cdot \cancel{6}^3}{1 \cdot \cancel{2}_1} \, (a)^5 \left(\sqrt{a^2 - 1}\right)^2 + \frac{7 \cdot 6 \cdot 5}{1 \cdot 2 \cdot 3} \, a^4 \left(\sqrt{a^2 - 1}\right)^3$$

$$+ \frac{7 \cdot \cancel{6} \cdot 5 \cdot \cancel{4}}{1 \cdot \cancel{2} \cdot \cancel{3} \cdot \cancel{4}} \, a^3 \left(\sqrt{a^2 - 1}\right)^4 + \frac{7 \cdot 6 \cdot 5 \cdot 4 \cdot 3}{1 \cdot 2 \cdot 3 \cdot 4 \cdot 5} \, a^2 \left(\sqrt{a^2 - 1}\right)^5$$

$$+ 7a\left(\sqrt{a^2 - 1}\right)^6 + \left(\sqrt{a^2 - 1}\right)^7 + a^7 - 7a^6\left(\sqrt{a^2 - 1}\right)$$

$$+ \frac{7 \cdot \cancel{6}^3}{1 \cdot \cancel{2}_1} \, (a)^5 \left(\sqrt{a^2 - 1}\right)^2 - \frac{7 \cdot 6 \cdot 5}{1 \cdot 2 \cdot 3} \, a^4 \left(\sqrt{a^2 - 1}\right)^3 + \frac{7 \cdot \cancel{6}^2 \cdot 5 \cdot 4}{1 \cdot \cancel{2} \cdot 3 \cdot 4_1} \, a^3 \left(\sqrt{a^2 - 1}\right)^4$$

$$- \frac{7 \cdot 6 \cdot 5 \cdot 4 \cdot 3}{1 \cdot 2 \cdot 3 \cdot 4 \cdot 5} \, a^2 \left(\sqrt{a^2 - 1}\right)^5 + 7a\left(\sqrt{a^2 - 1}\right)^6 - \left(\sqrt{a^2 - 1}\right)^7$$

$$= a^7 + 21a^5(a^2 - 1) + 35a^3(a^2 - 1)^2 + 7a(a^2 - 1)^3 + a^7 + 21a^5(a^2 - 1)$$

$$+ 35a^3(a^2 - 1)^2 + 7a(a^2 - 1)^3$$

$$= 2a^7 + 42a^5(a^2 - 1) + 70a^3(a^2 - 1)^2 + 14a(a^2 - 1)^3$$

$$= 2\left[a^7 + 21a^5(a^2 - 1) + 35a^3(a^2 - 1)^2 + 7a(a^2 - 1)^3\right]$$

$$= 2\left[a^7 + 21a^7 - 21a^5 + 35a^3(a^4 - 2a^2 + 1) + 7a(a^2 - 1)^2(a^2 - 1)\right]$$

$$= 2\left[22a^7 - 21a^5 + 35a^7 - 70a^5 + 35a^3 + 7a(a^4 - 2a^2 + 1)(a^2 - 1)\right]$$

$$= 2\left[57a^7 - 91a^5 + 35a^3 + 7a(a^6 - 2a^4 + a^2 - a^4 + 2a^2 - 1)\right]$$

$$= 2a\left[57a^6 - 91a^4 + 35a^2 + 7(a^6 - 2a^4 + a^2 - a^4 + 2a^2 - 1)\right]$$

$$= 2a\left[57a^6 - 91a^4 + 35a^2 + 7a^6 - 14a^4 + 7a^2 - 7a^4 + 14a^2 - 7\right]$$

$$= 2a\left[64a^6 - 112a^4 + 56a^2 - 7\right]$$

● PROBLEM 7-58

Compute the approximate value of $(1.01)^5$

Solution: The Binomial Theorem can be used to find the approximate value of $(1.01)^5$. The Binomial Theorem is stated as:

$$(a + b)^n = a^n + na^{n-1}b + \frac{n(n-1)}{1 \cdot 2} a^{n-2}b^2 + \frac{n(n-1)(n-2)}{1 \cdot 2 \cdot 3} \times$$

$$a^{n-3}b^3 + \cdots + nab^{n-1} + b^n,$$

where a and b are any two numbers. In order to use this theorem, express $(1.01)^5$ as the sum of two numbers raised to the fifth power. Then,

$$(1.01)^5 = (1 + .01)^5$$

Now, let $a = 1$, $b = .01$, and $n = 5$ in the Binomial Theorem. Calculating the first four terms of this theorem with these substitutions:

$$(1.01)^5 = (1 + .01)^5$$

$$= (1)^5 + 5(1)^{5-1}(.01) + \frac{5(5-1)}{1 \cdot 2} \times$$

$$(1)^{5-2}(.01)^2 + \frac{5(5-1)(5-2)}{1 \cdot 2 \cdot 3} \times$$

$$(1)^{5-3}(.01)^3 + \cdots$$

$$= 1 + 5(1)^4(.01) + \frac{5(\overset{2}{\cancel{4}})}{\underset{1}{\cancel{2}}}(1)^3(.01)^2$$

$$+ \frac{5(4)(\overset{1}{\cancel{3}})}{\underset{2}{\cancel{6}}}(1)^2(.01)^3 + \cdots$$

$$= 1 + 5(1)(.01) + 10(1)(.0001) + 10(1)$$

$$(.000001) + \cdots$$

$$= 1 + 0.05 + 0.001 + 0.00001$$

$$= 1.05101$$

● PROBLEM 7-59

Use the binomial formula with $n = 1/3$ to find an approximation to $\sqrt[3]{28}$.

Solution: To apply the binomial formula, try to express $\sqrt[3]{28}$ as the sum of two numbers raised to a power. Note that the formula is simplified if one of the numbers is one.

$$\sqrt[3]{28} = (28)^{(1/3)} = (27+1)^{1/3} .$$

We can write the expansion of $(x+y)^{1/3}$ to four terms and later substitute for x and y. We write out the binomial expansion to four terms when $n = 1/3$.

$$(x+y)^{1/3} = x^{1/3} + \frac{1}{3} x^{(1/3)-1} y + \frac{\left(\frac{1}{3}\right)\left(\frac{1}{3}-1\right)}{1\cdot 2} x^{(1/3)-2} y^2$$

$$+ \frac{\left(\frac{1}{3}\right)\left(\frac{1}{3}-1\right)\left(\frac{1}{3}-2\right)}{1\cdot 2\cdot 3} x^{(1/3)-3} y^3 + \dots$$

$$= x^{1/3} + \frac{1}{3} x^{-2/3} y + \frac{\left(\frac{1}{3}\right)\left(\frac{-2}{3}\right)}{2} x^{-5/3} y^2$$

$$+ \frac{\left(\frac{1}{3}\right)\left(\frac{-2}{3}\right)\left(\frac{-5}{3}\right)}{1\cdot 2\cdot 3} x^{-8/3} y^3 + \dots$$

$$= x^{1/3} + \frac{1}{3} x^{-2/3} y + \left(\frac{-2}{9}\right)\left(\frac{1}{2}\right) x^{-5/3} y^2$$

$$+ \frac{\overset{5}{\cancel{10}}\cdot 1}{3\cdot 3\cdot 3\cdot 1\cdot\underset{1}{\cancel{2}}\cdot 3} x^{-8/3} y^3 + \dots$$

$$(x+y)^{1/3} = x^{1/3} + \frac{1}{3} x^{-2/3} y - \frac{1}{9} x^{-5/3} y^2$$

$$+ \frac{5}{81} x^{-8/3} y^3 + \dots \qquad\qquad (1)$$

In this case, n is fractional. We obtain an infinite series and we can expand it for the first few terms if $|x| < |y|$. $x = 27$ and $y = 1$, and $|y| < |x|$, i.e., $|1| < |27|$. Therefore, using equation (1) with $x = 27$, $y = 1$ and $n = 1/3$ (writing only the first four terms):

$$\sqrt[3]{28} = (28)^{1/3}$$

$$= (27 + 1)^{1/3}$$

$$= 27^{1/3} + 1/3\left(27^{-2/3}\right)(1) - \frac{1}{9}\left(27^{-5/3}\right)\left(1^2\right)$$

$$+ \frac{5}{81}\left(27^{-8/3}\right)\left(1^3\right)$$

$$= 3 + \frac{1}{3}\left(\frac{1}{9}\right) - \frac{1}{9}\left(\frac{1}{243}\right) + \frac{5}{81}\left(\frac{1}{6,561}\right)$$

$$= 3 + 0.037037 - 0.000457 + 0.000009$$

$$= 3.036589$$

CHAPTER 8

PROBABILITY

PROBABILITY

● PROBLEM 8-1

A deck of playing cards is thoroughly shuffled and a card is drawn from the deck. What is the probability that the card drawn is the ace of diamonds?

<u>Solution:</u> The probaility of an event occurring is

$$\frac{\text{the number of ways the event can occur}}{\text{the number of possible outcomes}}$$

In our case there is one way the event can occur, for there is only one ace of diamonds and there are fifty two possible outcomes (for there are 52 cards in the deck).Hence the probability that the card drawn is the ace of diamonds is 1/52.

● PROBLEM 8-2

There are 23 white balls and 2 blue balls in a box. If three balls are drawn at random, what is the probability that none of the three balls are blue?

<u>Solution:</u> The total number of ways of drawing 3 balls out of 25 is

$$_{25}C_3 = \frac{25!}{3!(25 - 3)!} = 2300.$$

The total number of ways in which three white balls may be drawn is given by

422

$$23C_3 = \frac{23!}{3!(23-3)!} = 1771$$

Hence, the probability that all three balls drawn are white is $\frac{1771}{2300} = \frac{77}{100}$.

• PROBLEM 8-3

A box contains 7 red, 5 white, and 4 black balls. What is the probability of your drawing at random one red ball? One black ball?

Solution: There are 7 + 5 + 4 = 16 balls in the box. The probability of drawing one red ball,

$$P(R) = \frac{\text{number of possible ways of drawing a red ball}}{\text{number of ways of drawing any ball}}$$

$$P(R) = \frac{7}{16}.$$

Similarly, the probability of drawing one black ball

$$P(B) = \frac{\text{number of possible ways of drawing a black ball}}{\text{number of ways of drawing any ball}}$$

Thus,

$$P(B) = \frac{4}{16} = \frac{1}{4}$$

• PROBLEM 8-4

If two dice are thrown, what is the probability that the sum of the dice will be either 9 or 10?

Solution: The yielding of a sum of 9 or 10 by throwing two dice are two mutually exclusive events. The probability that either one of two mutually exclusive events will occur is the sum of the probabilities of the separate events. Since a die has 6 faces, the total number of ways in which the two dice may turn up is 6 x 6 = 36. The two dice will yield a sum of 9 in the following ways:

4+5, 5+4, 3+6, 6+3, i.e., 4 ways

Hence, the probability of obtaining a sum of 9 is $\frac{4}{36} = \frac{1}{9}$.
Similarly, the probability of obtaining a sum of 10 is
$\frac{3}{36} = \frac{1}{12}$. (There are three ways to obtain a sum of 10; 5+5,

6+4 and 4+6). Thus, the probability of obtaining either a
9 or 10 is the sum of the individual probabilities, i.e.,

$$\frac{1}{9} + \frac{1}{12} = \frac{7}{36} \ .$$

In a single throw of a single die, find the probability of
obtaining either a 2 or a 5.

Solution: In a single throw, the die may land in any of
6 ways:

    1 2 3 4 5 6.

The probability of obtaining a 2,

$$P(2) = \frac{\text{number of ways of obtaining a 2}}{\text{numbers of ways the die may land}}$$

$$P(2) = \frac{1}{6}.$$

Similarly, the probability of obtaining a 5,

$$P(5) = \frac{\text{number of ways of obtaining a 5}}{\text{number of ways the die may land}}$$

$$P(5) = \frac{1}{6}.$$

The probability that either one of two mutually exclusive
events will occur is the sum of the probabilities of the
separate events. Thus the probability of obtaining either
a 2 or a 5, P(2) or P(5), is

$$P(2) + P(5) = \frac{1}{6} + \frac{1}{6} = \frac{2}{6} = \frac{1}{3}$$

If a card is drawn from a deck of playing cards, what is
the probability that it will be a jack or a ten?

Solution: The probability that an event A or B occurs, but
not both at the same time, is $P(A \cup B) = P(A) + P(B)$. Here
the symbol "$\cup$" stands for "or."
    In this particular example, we only select one card
at a time. Thus, we either choose a jack "or" a ten.

P(a jack or a ten) = P(a jack) + P(a ten).

$$P(a \text{ jack}) = \frac{\text{number of ways to select a jack}}{\text{number of ways to choose a card}} = \frac{4}{52} = \frac{1}{13}.$$

$$P(a \text{ ten}) = \frac{\text{number of ways to choose a ten}}{\text{number of ways to choose a card}} = \frac{4}{52} = \frac{1}{13}.$$

$$P(a \text{ jack or a ten}) = P(a \text{ jack}) + P(a \text{ ten}) = \frac{1}{13} + \frac{1}{13} = \frac{2}{13}.$$

● **PROBLEM 8-7**

An urn contains 6 white, 4 black, and 2 red balls. In a single draw, find the probability of drawing: (a) a red ball; (b) a black ball; (c) either a white or a black ball. Assume all outcomes equally likely.

<u>Solution:</u> The urn contains 6 white balls, 4 black balls, and 2 red balls, or a total of 12 balls.

(a) The probability of drawing a red ball,

$$P(R) = \frac{\text{number of ways of drawing a red ball}}{\text{number of ways of selecting a ball}}$$

$$P(R) = \frac{2}{12} = \frac{1}{6}.$$

(b) The probability of drawing a black ball,

$$P(B) = \frac{\text{number of ways of drawing a black ball}}{\text{number of ways of selecting a ball}}$$

$$P(B) = \frac{4}{12} = \frac{1}{3}.$$

(c) The probability that either one of two mutually exclusive events will occur is the sum of the probabilities of the separate events. Thus the probability of drawing either a white [P(W)] or a black ball [P(B)] is P(W) + P(B).

$$P(W) = \frac{\text{number of ways of drawing a white ball}}{\text{number of ways of selecting a ball}}$$

$$= \frac{6}{12} = \frac{1}{2}.$$

$$P(B) = \frac{1}{3} \text{ [shown in part (b)].}$$

Thus, $P(W \text{ or } B) = P(W) + P(B) = \frac{6}{12} + \frac{4}{12}$

$$= \frac{10}{12}$$

$$= \frac{5}{6}.$$

Determine the probability of getting 6 or 7 in a toss of two dice.

Solution: Let A = the event that a 6 is obtained in a toss of two dice

B = the event that a 7 is obtained in a toss of two dice.

Then, the probability of getting 6 or 7 in a toss of two dice is

$$P(A \text{ or } B) = P(A \cup B).$$

The union symbol "∪" means that A and/or B can occur. Now P(A ∪ B) = P(A) + P(B) if A and B are mutually exclusive. Two or more events are said to be mutually exclusive if the occurrence of any one of them excludes the occurrence of the others. In this case, we cannot obtain a six and a seven in a single toss of two dice. Thus, A and B are mutually exclusive.
To calculate P(A) and P(B), use the following table.

Note: There are 36 different tosses of two dice.

A = a 6 is obtained in a toss of two dice

$$= \{(1,5), (2,4), (3,3), (4,2), (5,1)\}$$

B = a 7 is obtained in a toss of two dice

$$= \{(1,6), (2,5), (3,4), (4,3), (5,2), (6,1)\}.$$

$$P(A) = \frac{\text{number of ways to obtain a 6 in a toss of two dice}}{\text{number of ways to toss two dice}}$$

$$= \frac{5}{36}$$

$$P(B) = \frac{\text{number of ways to obtain a 7 in a toss of two dice}}{\text{number of ways to toss two dice}}$$

$$= \frac{6}{36} = \frac{1}{6}.$$

Therefore, $P(A \cup B) = P(A) + P(B) = \frac{5}{36} + \frac{6}{36} = \frac{11}{36}.$

On an examination, a student answers a series of 5 multiple-choice questions each of which has 3 choices. This student, unfortunately, doesn't know how to answer any of the questions. He decides to answer each question by choosing an answer at random. Find the probability that he answers all 5 questions correctly.

Solution: In this problem, the occurrence of one event (choosing an answer for a problem) has no effect upon the occurrence or non-occurrence of the other; hence the events are said to be independent. If two events are independent and the probability of occurrence of the first event is x, and the probability of occurrence of the second event is y, then the probability that both will happen in the order stated is xy.

The probability of answering all five questions correctly is

$(\frac{1}{3})^5 = \frac{1}{243}$ since the probability of answering each question

correctly is $\frac{1}{3}$.

A box contains 5 blue balls and 8 green balls. If 2 balls are drawn in succession without replacing any ball after it has been drawn, what is the probability of drawing 2 green balls?

Solution: The event of successively drawing balls from the box is said to be dependent since the probability of drawing 2 green balls is affected by the fact that each successive drawing changes the probability of drawing a green ball.

The probability of drawing a green ball on the first draw is $\frac{8}{13}$.

The probability of drawing a green ball on the second draw is
$\frac{8-1}{13-1} = \frac{7}{12}$

(Since after the first draw there is a total of 13-1 = 12 balls left and among them 7 balls are green.) Therefore, he probability of drawing 2 green balls in succession is

$$\frac{8}{13} \times \frac{7}{12} = \frac{56}{156} = \frac{14}{39} .$$

A box contains 10 white balls, 10 green balls, 10 yellow balls and 10 blue balls. Two balls are drawn from the box. Find the probability that both balls are yellow.

Solution: There are 40 ways to draw the first ball. For each of these, the second ball may be drawn in 39 ways. By the principle of counting, the two balls can be drawn in 40 x 39 ways; two yellow balls may be drawn in 10 x 9 ways, there are 10 ways to draw the first yellow ball and 9 ways to draw the second.

Therefore, the probability of drawing two yellow balls is

$$\frac{10 \times 9}{40 \times 39} = \frac{3}{52} \ .$$

The probability that Mr. Smith will be elected Governor of State A is $\frac{3}{10}$. The probability that Mr. Brown will be elected Mayor of City B is $\frac{4}{5}$.

(1) What is the probability that both will be elected?

(2) What is the probability that both will be defeated?

(3) What is the probability that Mr. Smith will win and Mr. Brown will lose?

Solution: (1) We assume the elections of Governor of State A and Mayor of City B to be two independent events. (Since the candidates for mayor can't at the same time be the candidates for governor.) Hence, the probability that both will be elected is

$$\left(\frac{3}{10}\right) \times \left(\frac{4}{5}\right) = \frac{6}{25} \ .$$

(2) If the probability that an event will occur is P then the probability that this event will not happen is 1 - P.

Therefore, the probability that both will be defeated is

$$\left(1 - \frac{3}{10}\right) \left(1 - \frac{4}{5}\right) = \frac{7}{50} \ .$$

(3) The probability that Mr. Smith will be elected and Mr. Brown defeated is:

$$(\frac{3}{10}) \times (1 - \frac{4}{5}) = (\frac{3}{10}) \times (\frac{1}{5}) = \frac{3}{50} \ .$$

● **PROBLEM** 8-13

Find the probability of throwing two sixes in one toss of a pair of dice.

Solution: To find the probability of throwing two sixes in one toss of a pair of dice, first we express it symbolically.

P(throwing two sixes in one toss of a pair of dice) =

P(throwing a six in one toss of a die) × P(throwing a six in one toss of a die).
This is true because the event of tossing a die is independent of tossing another die. That is, the occurrence of one event has no effect upon the occurrence or non-occurrence of the other. Now,

P(throwing a six in one toss) =

$$\frac{\text{number of ways to obtain a six}}{\text{number of ways to obtain any face value of a die}} = \frac{1}{6} \ .$$

Hence, the probability of obtaining two sixes is $(\frac{1}{6})(\frac{1}{6}) = \frac{1}{36} \ .$

● **PROBLEM** 8-14

If a pair of dice is tossed twice, find the probability of obtaining 5 on both tosses.

Solution: We obtain 5 in one toss of the two dice if they fall with either 3 and 2 or 4 and 1 uppermost, and each of these combinations can appear in two ways. The ways to obtain 5 in one toss of the two dice are:

$$(1,4),(4,1),(3,2), \text{ and } (2,3).$$

Hence we can throw 5 in one toss in four ways. Each die has six faces and there are six ways for a die to fall. Then the pair of dice can fall in 6·6 = 36 ways. The probability of throwing 5 in one toss is:

$$\frac{\text{the number of ways to throw a 5 in one toss}}{\text{the number of ways that a pair of dice can fall}} = \frac{4}{36} = \frac{1}{9} \ .$$

Now the probability of throwing a 5 on both tosses is:

P(throwing five on first toss and throwing five on second toss).

"And" implies multiplication if events are independent, thus
p(throwing 5 on first toss and throwing 5 on second toss)

= p(throwing 5 on first toss) × p(throwing 5 on second toss)

Since the results of the two tosses are independent. Consequently, the

429

probability of obtaining 5 on both tosses is

$$\left(\frac{1}{9}\right)\left(\frac{1}{9}\right) = \frac{1}{81} \ .$$

A bag contains 4 black and 5 blue marbles. A marble is drawn and then replaced, after which a second marble is drawn. What is the probability that the first is black and second blue?

Solution: Let C = event that the first marble drawn is black.

D = event that the second marble drawn is blue.

The probability that the first is black and the second is blue can be expressed symbolically:

$$P(C \text{ and } D) = P(CD).$$

We can apply the following theorem. If two events A and B, are independent, then the probability that A and B will occur is,

$$P(A \text{ and } B) = P(AB) = P(A) \cdot P(B).$$

Note that two or more events are said to be independent if the occurrence of one event has no effect upon the occurrence or non-occurrence of the other. In this case the occurrence of choosing a black marble has no effect on the selection of a blue marble and vice versa; since, when a marble is drawn it is then replaced before the next marble is drawn. Therefore, C and D are two independent events.

$$P(CD) = P(C) \cdot P(D)$$

$$P(C) = \frac{\text{number of ways to choose a black marble}}{\text{number of ways to choose a marble}}$$

$$= \frac{4}{9}.$$

$$P(D) = \frac{\text{number of ways to choose a blue marble}}{\text{number of ways to choose a marble}}$$

$$= \frac{5}{9}.$$

$$P(CD) = P(C) \cdot P(D) = \frac{4}{9} \cdot \frac{5}{9} = \frac{20}{81}.$$

430

A box contains 4 black marbles, 3 red marbles, and 2 white marbles. What is the probability that a black marble, then a red marble, then a white marble is drawn without replacement?

Solution: Here we have three dependent events. There is a total of 9 marbles from which to draw. We assume on the first draw we will get a black marble. Since the probability of drawing a black marble is the

$$\frac{\text{number of ways of drawing a black marble}}{\text{number of ways of drawing 1 out of } (4+3+2) \text{ marbles}},$$

$$P(A) = \frac{4}{4+3+2} = \frac{4}{9}$$

There are now 8 marbles left in the box.

On the second draw we get a red marble. Since the probability of drawing a red marble is

$$\frac{\text{number of ways of drawing a red marble}}{\text{number of ways of drawing 1 out of the 8 remaining marbles}},$$

$$P(B) = \frac{3}{8}$$

There are now 7 marbles remaining in the box.

On the last draw we get a white marble. Since the probability of drawing a white marble is

$$\frac{\text{number of ways of drawing a white marble}}{\text{number of ways of drawing 1 out of the 7 remaining marbles}},$$

$$P(C) = \frac{2}{7}$$

When dealing with two or more dependent events, if P_1 is the probability of a first event, P_2 the probability that, after the first has happened, the second will occur, P_3 the probability that, after the first and second have happened, the third will occur, etc., then the probability that all events will happen in the given order is the product $P_1 \cdot P_2 \cdot P_3 \ldots$

Thus, $P(A \cap B \cap C) = P(A) \cdot P(B) \cdot P(C)$

$$= \frac{4}{9} \cdot \frac{3}{8} \cdot \frac{2}{7}$$

$$= \frac{1}{21}.$$

There is a box containing 5 white balls, 4 black balls, and 7 red balls. If two balls are drawn one at a time from the box and neither is replaced, find the probability that
(a)   both balls will be white.
(b)   the first ball will be white and the second red.
(c)   if a third ball is drawn, find the probability that the three balls will be drawn in the order white, black, red.

Solution: This problem involves dependent events. Two or more events are said to be dependent if the occurrence of one event has an effect upon the occurrence or non-occurrence of the other. If you are drawing objects without replacement, the second draw is dependent on the occurrence of the first draw. We apply the following theorem for this type of problem. If the probability of occurrence of one event is p and the probability of the occurrence of a second event is q, then the probability that both events will happen in the order stated is pq.

(a) To find the probability that both balls will be white, we express it symbolically.
p (both balls will be white) =
p (first ball will be white and the second ball will be white) =
p (first ball will be white) p(second ball will be white) =

$$= \left(\frac{\text{number of ways to choose a white ball}}{\text{number of ways to choose a ball}} \right) \left(\frac{\text{number of ways to choose a second white ball after removal of the first white ball}}{\text{number of ways to choose a ball after removal of the first ball}} \right)$$

$$= \frac{\overset{1}{\cancel{5}}}{\underset{4}{\cancel{16}}} \cdot \frac{\overset{1}{\cancel{4}}}{\underset{3}{\cancel{15}}} = \frac{1}{12}$$

(b) p (first ball will be white and the second red)
   = p (first ball will be white) p(the second ball will be red)

$$= \left(\frac{\text{number of ways to choose a white ball}}{\text{number of ways to choose a ball}} \right) \left(\frac{\text{number of ways to choose a red ball}}{\text{number of ways to choose a ball after the removal of the first}} \right)$$

$$= \frac{\overset{1}{\cancel{5}}}{16} \cdot \frac{7}{\underset{3}{\cancel{15}}} = \frac{7}{48}$$

(c) p (three balls drawn in the order white, black, red)
=   p (first ball is white) p(second ball is black) p(third ball is red)

$$= \left(\frac{\text{number of ways to choose that the first ball is white}}{\text{number of ways to choose the first ball}}\right)\left(\frac{\text{number of ways to choose that second one is black}}{\text{number of ways to choose the second one}}\right)$$

$$\left(\frac{\text{number of ways to choose that the third one is red}}{\text{number of ways to choose the third one}}\right)$$

$$= \frac{\cancel{5}^{\,1}}{\cancel{16}_{\,4}} \quad \frac{\cancel{4}^{\,1}}{\cancel{16}_{\,3}} \quad \frac{\cancel{7}^{\,1}}{\cancel{14}_{\,2}} = \frac{1}{24}$$

● PROBLEM 8-18

A box contains 4 black marbles, 3 red marbles, and 2 white marbles. What is the probability that a black marble, then a red marble, then a white marble is drawn without replacement?

__Solution:__ Here we have three dependent events. There is a total of 9 marbles from which to draw. We assume on the first draw we will get a black marble. Since the probability of drawing a black marble is the

$$\frac{\text{number of ways of drawing a black marble}}{\text{number of ways of drawing 1 out of (4+3+2) marbles}},$$

$$P(A) = \frac{4}{4 + 3 + 2} = \frac{4}{9}.$$

There are now 8 marbles left in the box.

On the second draw we get a red marble. Since the probability of drawing a red marble is

$$\frac{\text{number of ways of drawing a red marble}}{\text{number of ways of drawing 1 out of the 8 remaining marbles}},$$

$$P(B) = \frac{3}{8}$$

There are now 7 marbles remaining in the box.

On the last draw we get a white marble. Since the probability of drawing a white marble is

$$\frac{\text{number of ways of drawing a white marble}}{\text{number of ways of drawing 1 out of the 7 remaining marbles}},$$

$$P(C) = \frac{2}{7}$$

When dealing with two or more dependent events, if P_1 is the probability of a first event, P_2 the probability that, after the first has happened, the second will occur, P_3 the probability that, after the first and second have happened, the third will occur, etc., then the probability that all events will happen in the given order is the product $P_1 \cdot P_2 \cdot P_3 \ldots$

Thus, $P(A \cap B \cap C) = P(A) \cdot P(B) \cdot P(C)$

$$= \frac{4}{9} \cdot \frac{3}{8} \cdot \frac{2}{7} = \frac{1}{21} .$$

● **PROBLEM** 8-19

A traffic count at a highway junction revealed that out of 5,000 cars that passed through the junction in 1 week, 3,000 turned to the right. Find the probability that a car will turn (a) to the right, and (b) to the left.

<u>Solution:</u> (a) If an event can happen in s ways and fail to happen in f ways, and if all these ways (s + f) are assumed to be equally likely, then the probability (p) that the event will happen is

$$p = \frac{s}{s + f} = \frac{(successful\ ways)}{(total\ ways)}$$

In this case s = 3,000 and s + f = 5,000. Hence, $p = \frac{3,000}{5,000} = \frac{3}{5} .$

(b) If the probability that an event will happen is $\frac{a}{b}$ then the probability that this event will not happen is $1 - \frac{a}{b}$. Thus the probability that a car will not turn right, but left, is $1 - \frac{3}{5} = \frac{2}{5} .$ This same conclusion can also be arrived at using the following reasoning: Since 3,000 cars turned to the right, 5,000 - 3,000 = 2,000 cars turned to the left. Hence, the probability that a car will turn to the left is

$$\frac{2,000}{5,000} = \frac{2}{5} .$$

● **PROBLEM** 8-20

10 white balls and 19 blue balls are in a box. If a man draws a ball from the box at random, what are the odds in favor of him drawing a blue ball?

<u>Solution:</u> If an event can happen in p ways and fail to happen in q ways. Then, if p > q the odds are p to q in favor of the event happening.

If p < q then the odds are q to p against the event happening.

434

In this case,

p = 19 and q = 10, p > q.   Thus, the odds in favor of the event of drawing a blue ball are 19:10.

From 20 tickets marked with the first 20 numerals, one is drawn at random: find the chance that it is a multiple of 3 or of 7.

Solution: If an event can happen in s ways and fail to happen in f ways, and if all these ways (s + f) are assumed to be equally likely, then the probability (p) that the event will happen is

$$p = \frac{s}{s + f} = \frac{\text{(successful ways)}}{\text{(total ways)}}$$

There are 6 multiples of 3 in the first 20 numerals (3,6,9,12,15,18), and there are 2 multiples of 7 (7,14).   Thus

$$p(\text{multiple of 3}) = \frac{6}{20}$$

and

$$p(\text{multiple of 7}) = \frac{2}{20} .$$

Since the probability that either of two mutually exclusive events will occur is the sum of the probabilities of the separate events, the chance that it is a multiple of 3 or of 7 is

$$\frac{6}{20} + \frac{2}{20} = \frac{8}{20} = \frac{2}{5} .$$

But if the question had been: find the chance that the number is a multiple of 3 or of 5, it would have been incorrect to reason as follows: The chance that the number is a multiple of 3 is $\frac{6}{20}$, and the chance that the number is a multiple of 5 is $\frac{4}{20}$, therefore the chance that it is a multiple of 3 or 5 is $\frac{6}{20} + \frac{4}{20}$, or $\frac{1}{2}$.   This is erroneous for the number on the ticket might be a multiple both of 3 and of 5, so that the two events considered are not mutually exclusive.

What is the probability that the sum 11 will appear in a single throw of 2 dice?

Solution: There are 6 ways the first die may be tossed and 6 ways the second die may be tossed.   The Fundamental Principle of Counting states that if the first of two independent acts can be performed in x ways, and if the second act can be performed in y ways, then the number of ways of performing the two acts, in the order stated, is xy.   Thus there are   6 × 6 = 36 ways that two dice can be thrown (see accompanying figure).

1,1	1,2	1,3	1,4	1,5	1,6
2,1	2,2	2,3	2,4	2,5	2,6
3,1	3,2	3,3	3,4	3,5	3,6
4,1	4,2	4,3	4,4	4,5	4,6
5,1	5,2	5,3	5,4	5,5	5,6
6,1	6,2	6,3	6,4	6,5	6,6

The number of possible ways that an 11 will appear are circled in the figure. Let us call this set A. Thus,

$$A = \{(5,6),(6,5)\}.$$

The probability that an 11 will appear,

$$p(11) = \frac{\text{number of possible ways of obtaining an 11}}{\text{number of ways that 2 dice can be thrown}}$$

Therefore

$$p(11) = \frac{2}{36} = \frac{1}{18}.$$

● **PROBLEM** 8-23

What is the probability of making a 7 in one throw of a pair of dice?

Solution: There are 6 X 6 = 36 ways that two dice can be thrown, as shown in the accompanying figure.

1,1	1,2	1,3	1,4	1,5	1,6
2,1	2,2	2,3	2,4	2,5	2,6
3,1	3,2	3,3	3,4	3,5	3,6
4,1	4,2	4,3	4,4	4,5	4,6
5,1	5,2	5,3	5,4	5,5	5,6
6,1	6,2	6,3	6,4	6,5	6,6

The number of possible ways that a 7 will appear are circled in the figure. Let us call this set B. Thus,

$$B = \{(1,6),(2,5),(3,4),(4,3),(5,2),(6,1)\}.$$

The probability that a 7 will appear,

$$p(7) = \frac{\text{number of possible ways of obtaining a 7}}{\text{number of ways that 2 dice can be thrown}}$$

$$p(7) = \frac{6}{36} = \frac{1}{6}.$$

436

Find the probability that when a pair of dice are thrown, the sum of the two up faces is greater than 7 or the same number appears on each face.

Solution: The sample space consists of 36 equally likely outcomes as shown in the accompanying figure. Those out-comes that give a sum greater than 7 are

$$G = \{(6,2),(6,3),(6,4),(6,5),(6,6),(5,3),(5,4),(5,5),$$
$$(5,6),(4,4),(4,5),(4,6),(3,5),(3,6),(2,6)\}$$

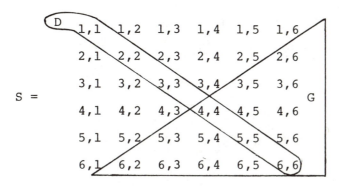

Then P(G) =

$$= \frac{\text{number of ways the two faces will be greater than 7}}{\text{total number of ways that the dice may fall}}$$

$$= \frac{15}{36} .$$

Those outcomes where each die is the same are

$$D = \{(1,1),(2,2),(3,3),(4,4),(5,5),(6,6)\}$$

Then P(D) =

$$= \frac{\text{number of ways that the same number can appear on each face}}{\text{total number of ways that the dice may fall}}$$

$$= \frac{6}{36} .$$

The probability of G or D is P(G∪D). Recall that

$$P(G∪D) = P(G) + P(D) - P(G∩D).$$

But G ∩ D = {(4,4),(5,5),(6,6)}.

Hence, $\quad P(G ∩ D) = \frac{3}{36}$

and $P(G ∪ D) = \frac{15}{36} + \frac{6}{36} - \frac{3}{36} = \frac{18}{36} = \frac{1}{2} .$

If two dice are cast, what is the probability the sum will be less than 5?

Solution: If A, B, and C are mutually exclusive events, that is, their intersection is the null set, then $P(A \cup B \cup C) = P(A) + P(B) + P(C)$. Since the obtaining of sums of 2, 3, and 4 are mutually exclusive events, the probability of obtaining a sum less than 5 is the sum of the probabilities of obtaining a sum of 2,3, and 4. To obtain the sum of 2 with 2 die, we have the following possibilities: (1,1).

Similarly for the sum of 3, we have: (1,2) and (2,1).

For the sum of 4, we obtain: (1,3), (3,1), and (2,2).

Thus P_1 = probability of obtaining a sum of 2

$$= \frac{\text{number of ways to obtain a sum of 2}}{\text{number of ways to throw 2 dice}}$$

$$= \frac{1}{36}$$

P_2 = probability of obtaining a sum of 3

$$= \frac{\text{number of ways to obtain a sum of 3}}{\text{number of ways to throw 2 dice}}$$

$$= \frac{2}{36} = \frac{1}{18}.$$

P_3 = probability of obtaining a sum of 4

$$= \frac{\text{number of ways to obtain a sum of 4}}{\text{number of ways to throw 2 dice}}$$

$$= \frac{3}{36} = \frac{1}{12}.$$

The probability of obtaining a sum less than 5 is

$$P_1 + P_2 + P_3 = \frac{1}{36} + \frac{1}{18} + \frac{1}{12}.$$

$$= \frac{1}{36} + \frac{2}{36} + \frac{3}{36} = \frac{6}{36} = \frac{1}{6}$$

A lottery has a prize of $6,000,000. If 5,000,000 tickets are sold and John buys 25 tickets, what is the probability that he will win? What is his expectation?

Solution: The probability that John will win is

$$\frac{25}{5 \times 10^6} = \frac{1}{2 \times 10^5} = \frac{1}{200,000}.$$

If the probability of acquiring an amount of money M is p then the expectation is the product of M and p. Therefore, John's expectation is

$$E = \frac{1}{2 \times 10^5} \times 6 \times 10^6 = \$30.$$

● PROBLEM 8-27

The probability that A wins a certain game is $\frac{2}{3}$. If A plays 5 games, what is the probability that A will win (a) exactly 3 games? (b) at least 3 games?

Solution: We shall apply the following theorem. If P is the probability that an event will happen in a single trial and q is the probability that this event will fail in this trial, then $_nC_r \, p^r q^{n-r}$ is the probability that this event will happen exactly r times in n trials. $_nC_r$, the number of combinations of n different objects taken r at a time, is

$$_nC_r = \frac{n!}{r!\,(n-r)!}.$$

Note that p + q = 1.

(a) We are given the probability of a success, p, which is winning a game: $p = \frac{2}{3}$. Therefore from p + q = 1, $q = 1 - p = 1 - \frac{2}{3} = \frac{1}{3}$. The number of ways of winning 3 games out of 5 is

$$_5C_3 = \binom{5}{3} = \frac{5!}{3!\,2!} = \frac{5 \cdot \cancel{4}^2 \cdot \cancel{3!}}{\cancel{3!} \; \cancel{2} \cdot 1} = 10.$$

Thus, the probability of A winning 3 games is

$$_nC_r \, p^r q^{n-r} = {}_5C_3\left(\frac{2}{3}\right)^3\left(\frac{1}{3}\right)^2 = 10 \; \frac{2}{3} \cdot \frac{2}{3} \cdot \frac{2}{3} \cdot \frac{1}{3} \cdot \frac{1}{3} = \frac{80}{243}$$

(b) To win at least 3 games A must win either exactly 3 or exactly 4 or all 5 games. In order that A will win at least 3 games, we must calculate the probability that A will win three games, four games, and five games.

$$P = {}_5C_3\left(\frac{2}{3}\right)^3\left(\frac{1}{3}\right)^2 + {}_5C_4\left(\frac{2}{3}\right)^4\left(\frac{1}{3}\right)^1 + {}_5C_5\left(\frac{2}{3}\right)^5\left(\frac{1}{3}\right)^0.$$

$$= \frac{5!}{3!\,2!}\left(\frac{2}{3}\right)^3\left(\frac{1}{3}\right)^2 + \frac{5!}{4!\,1!}\left(\frac{2}{3}\right)^4\left(\frac{1}{3}\right)^1 + \frac{5!}{5!\,0!}\left(\frac{2}{3}\right)^5$$

439

$$= \frac{5 \cdot 4^2 \cdot 3}{3! \cdot 2 \cdot 1} \frac{8}{243} + \frac{5 \cdot 4}{4! \cdot 1!} \frac{16}{243} + \frac{32}{243}$$

$$= 10 \cdot \frac{8}{243} + 5 \cdot \frac{16}{243} + \frac{32}{243} = \frac{192}{243} = \frac{64}{81}.$$

Find the chance of throwing at least one ace in a single throw with two dice.

$$S = \begin{matrix} 1,1 & 1,2 & 1,3 & 1,4 & 1,5 & 1,6 \\ 2,1 & 2,2 & 2,3 & 2,4 & 2,5 & 2,6 \\ 3,1 & 3,2 & 3,3 & 3,4 & 3,5 & 3,6 \\ 4,1 & 4,2 & 4,3 & 4,4 & 4,5 & 4,6 \\ 5,1 & 5,2 & 5,3 & 5,4 & 5,5 & 5,6 \\ 6,1 & 6,2 & 6,3 & 6,4 & 6,5 & 6,6 \end{matrix}$$

Solution: The chance of throwing at least one ace is the chance of throwing an ace on either the first die, or the second die, or both.

The sample space, S, consists of $6 \times 6 = 36$ equally likely outcomes (see figure). Those outcomes that give an ace on the first die are shown in the first row of the figure. Let us call this row A. Thus,

A = {(1,1),(1,2),(1,3),(1,4),(1,5),(1,6)}.

Then P(A) = $\dfrac{\text{number of ways of getting ace on first die}}{\text{total number of ways that the dice may fall}}$

$$= \frac{6}{36}$$

Those outcomes that give an ace on the second die are shown in the first column of the figure. Let us call this column B. Thus,

B = {(1,1),(2,1),(3,1),(4,1),(5,1),(6,1)}.

Then P(B) = $\dfrac{\text{number of ways of getting ace on second die}}{\text{total number of ways that the dice may fall}}$

$$= \frac{6}{36}$$

The probability of an ace on the first die or on the second die is P(A) or P(B), P(A∪B). Recall that

$$P(A \cup B) = P(A) + P(B) - P(A \cap B).$$

Now, $A \cap B = (1,1);$
then $P(A \cap B)$

$$= \frac{\text{number of ways of getting ace on both dice}}{\text{total number of ways that the dice may fall}} = \frac{1}{36}$$

Therefore, $P(A \cup B) = \frac{6}{36} + \frac{6}{36} - \frac{1}{36} = \frac{11}{36}$.

Thus, the chance of throwing at least one ace in a single throw of 2 dice is 11/36.

● **PROBLEM** 8-29

Find the probability of drawing three consecutive face cards on three consecutive draws (with replacement) from a deck of cards.

Let: Event A: face card on first draw,
           Event B: face card on second draw, and
           Event C: face card on third draw.

Solution: This problem illustrates sampling with replacement. After each draw, the card drawn is returned to the deck. The deck, or "population" from which we draw the second card is identical to the original deck. Thus the drawing of a face card on the second draw is independent of the first draw. Similarly, the result of the third drawing is independent of the first or second drawings. This sampling without replacement implies that

$$P(ABC) = P(A) \cdot P(B) \cdot P(C).$$

But we know that the probability of drawing a face card on any draw is

$$\frac{\text{number of face cards in deck}}{\text{number of cards in deck}} \quad \text{or} \quad \frac{12}{52}.$$

Therefore $P(ABC) = P(A)P(B)P(C)$

$$= \frac{12}{52} \cdot \frac{12}{52} \cdot \frac{12}{52} = \frac{27}{2197} = .012.$$

● **PROBLEM** 8-30

If 4 different balls are placed at random in 3 different cells, find the probability that no cell is empty. Assume that there is ample room in each cell for all 4 balls.

Solution: There are 3 ways to place each of the 4 balls into a cell. Thus, $3 \times 3 \times 3 \times 3 = 3^4$ ways to put the 4 balls into three cells. If each arrangement is equally likely, then any 1 arrangement will occur with the probability

$$\frac{1}{3^4} = \frac{1}{81} \cdot$$

We now must count the number of ways the balls can be placed in the cells so that none of the cells are empty.

First, we know that one cell will have two balls in it. Choose these two balls from the four; there are $\binom{4}{2}$ ways to do this. Now place these two balls in a cell; there are three ways to do this. There are 2 ways and 1 way respectively to place the two remaining balls in the two remaining cells to insure that all the cells are filled. Together, by the Fundamental Counting Principle, there are $\binom{4}{2} \cdot 3 \cdot 2 \cdot 1$ or 36 arrangements. Thus the probability of observing an arrangement with 0 cells empty if the balls are dropped

in at random is $\frac{36}{81} = \frac{4}{9} \cdot$

● **PROBLEM** 8-31

A die is tossed five times. What is the probability that an ace will appear: (a) at least twice; (b) at least once?

Solution: This is a problem involving repeated trials of an experiment. The experiment is "tossing a die five times". Apply the following theorem: If p is the probability that an event will happen in a single trial and q is the probability that this event will fail in this trial, then

$${}_nC_r p^r q^{n-r}$$

is the probability that this event will happen exactly r times in n trials.
(a) To find the probability that an ace will occur at least twice, find the probability that it will occur twice, or three times, or four times, or five times. The sum (the word "or" implies addition in set notation) of these probabilities will be the probability that an ace will happen at least twice. p = probability that an ace will occur in a given trial

$$= \frac{\text{number of ways to obtain an ace}}{\text{number of ways to obtain any face of a die}}$$

$$= \frac{1}{6}$$

An experiment can only succeed or fail, hence the probability of success, p, plus the probability of failure, q, is one; p+q = 1. Then q = 1-p = 1 - 1/6 = 5/6. Therefore, using ${}_nC_r p^r q^{n-r}$, p (at least two aces) =

$$_5C_2 \, (1/6)^2(5/6)^3 + {}_5C_3(1/6)^3(5/6)^2$$

$$+ \; _5C_4 \, (1/6)^4(5/6)^1 + {}_5C_5 \, (1/6)^5(5/6)^0$$

$_nC_r$ is a symbol for a combination of n things, r at a time, where r objects are chosen from n objects.

$$_nC_r = \frac{n!}{r! \; (n-r)!}$$

Apply this formula. Then,

$$_5C_2(1/6)^2(5/6)^3 + {}_5C_3(1/6)^3(5/6)^2$$

$$+ \; _5C_4(1/6)^4(5/6)^1 + {}_5C_5(1/6)^5(5/6)^0$$

$$= \frac{5!}{2!3!}\left(\frac{125}{6^5}\right) + \frac{5!}{2!3!}\left(\frac{25}{6^5}\right) + \frac{5!}{4!1!}\left(\frac{5}{6^5}\right) + \frac{5!}{5!0!}\left(\frac{1}{6^5}\right)$$

$$= \frac{5 \cdot \cancel{4} \cdot \cancel{3!}}{\cancel{2} \cdot 1 \cdot \cancel{3!}}\left(\frac{125}{6^5}\right) + \frac{5 \cdot \cancel{4} \cdot \cancel{3!}}{\cancel{2} \cdot 1 \cdot \cancel{3!}}\left(\frac{25}{6^5}\right) + \frac{5 \cdot \cancel{4!}}{\cancel{4!}1!}\left(\frac{5}{6^5}\right) + \frac{1}{6^5}$$

$$= 10\left(\frac{125}{6^5}\right) + 10\left(\frac{25}{6^5}\right) + 5\left(\frac{5}{6^5}\right) + \frac{1}{6^5}$$

$$= \frac{1250 + 250 + 25 + 1}{6^5} = \frac{1526}{7776} = \frac{763}{3888}$$

Therefore, the probability that an ace will appear at least twice is

$$\frac{763}{3888}.$$

(b) An ace can be obtained at least once by tossing one ace, 2 aces, 3 aces,..., or 5 aces. Hence, the probability of obtaining at least one ace is the sum of the individual probabilities of obtaining one, two, three,..., up to five aces. Apply the same method as in part (a).

$$p(\text{at least one ace}) = {}_5C_1 \, (1/6)^1(5/6)^4 + {}_5C_2(1/6)^2(5/6)^3$$

$$+ \; _5C^3 \, (1/6)^3(5/6)^2 + {}_5C_4(1/6)^4(5/6)^1$$

$$+ \; _5C_5 \, (1/6)^5(5/6)^0$$

$$= \frac{5!}{1!4!}\left(\frac{625}{6^5}\right) + \frac{5!}{2!3!}\left(\frac{125}{6^5}\right) + \frac{5!}{3!2!}\left(\frac{25}{6^5}\right) + \frac{5!}{4!1!}\left(\frac{5}{6^5}\right)$$

$$+ \; \frac{5!}{5!0!}\left(\frac{1}{6^5}\right) = \frac{5 \cdot \cancel{4!}}{1 \cdot \cancel{4!}}\left(\frac{625}{6^5}\right) + \frac{5 \cdot \cancel{4} \cdot \cancel{3!}}{\cancel{2} \cdot 1 \cdot \cancel{3!}}\left(\frac{125}{6^5}\right) + \frac{5 \cdot \cancel{4} \cdot \cancel{3!}}{\cancel{3!} \cdot \cancel{2}}\left(\frac{25}{6^5}\right)$$

$$+ \; \frac{5 \cdot \cancel{4!}}{\cancel{4!} \cdot 1}\left(\frac{5}{6^5}\right) + \left(\frac{1}{6^5}\right) = 5\left(\frac{625}{6^5}\right) + 10\left(\frac{125}{6^5}\right) + 10\left(\frac{25}{6^5}\right)$$

$$+ \; 5\left(\frac{5}{6^5}\right) + \frac{1}{6^5}$$

$$= \frac{3125 + 1250 + 250 + 25 + 1}{6^5} = \frac{4651}{7776}$$

An alternate, shorter method, is to calculate the probability of failure, (obtaining no aces) and subtract this from one. This is true because $q+p = 1$, hence $q = 1-p$.

$$p \text{ (at least one ace)} = 1 - p \text{ (no aces)}$$

$$p \text{ (no aces)} = {}_5C_0(1/6)^0(5/6)^5 = \frac{5!}{0!\,5!}\left(\frac{5^5}{6^5}\right)$$

$$= \frac{3125}{7776}$$

Thus,

$$p \text{ (at least one ace)} = 1 - p \text{ (no aces)}$$

$$= 1 - \frac{3125}{7776} = \frac{4651}{7776}$$

Therefore, the probability that an ace appears at least once is

$$\frac{4651}{7776} \, .$$

If a die is thrown 10 times in succession, what is the probability that 6 will be thrown at least 6 times?

Solution: The probability that an event will occur at least r times in n trials is

$$p^n + {}_nC_1p^{n-1}q + {}_nC_2p^{n-2}q^2 + \ldots + {}_nC_rp^rq^{n-r}$$

where p is the probability that this event will happen in a single trial and q is the probability that this event will fail in this trial. Observe that this expression is the sum of the first n−r+1 terms of the expansion of the binomial $(p+q)^n$.

In this problem, $p = \frac{1}{6}$, $q = \frac{5}{6}$, $n = 10$, and $r = 6$. Hence, n−r+1 = 10−6 + 1 = 5. The sum of the first 5 terms of the expansion of $(\frac{1}{6} + \frac{5}{6})^{10}$ are:

$$\left(\frac{1}{6}\right)^{10} + {}_{10}C_1 \left(\frac{1}{6}\right)^9 \left(\frac{5}{6}\right) + {}_{10}C_2 \left(\frac{1}{6}\right)^8 \left(\frac{5}{6}\right)^2 + {}_{10}C_3 \left(\frac{1}{6}\right)^7 \left(\frac{5}{6}\right)^3$$

$$+ {}_{10}C_4 \left(\frac{1}{6}\right)^6 \left(\frac{5}{6}\right)^4$$

Therefore, the probability that 6 will be thrown at least 6 times is computed as:

$$\left(\frac{1}{6}\right)^{10} + {}_{10}C_1 \left(\frac{1}{6}\right)^9 \left(\frac{5}{6}\right) + {}_{10}C_2 \left(\frac{1}{6}\right)^8 \left(\frac{5}{6}\right)^2 + {}_{10}C_3 \left(\frac{1}{6}\right)^7 \left(\frac{5}{6}\right)^3$$

$$+ {}_{10}C_4 \left(\frac{1}{6}\right)^6 \left(\frac{5}{6}\right)^4 = \frac{156,926}{60,466,176} = \frac{26,071}{10,077,696} = .00258 = 0.258\%$$

CONDITIONAL PROBABILITY AND BAYES' THEOREM

● **PROBLEM 8-33**

Find the probability that a face card is drawn on the first draw and an ace on the second in two consecutive draws, without replacement, from a standard deck of cards.

Solution: This problem illustrates the notion of conditional probability. The conditional probability of an event, say event B, given the occurrence of a previous event, say event A, is written P(B|A). This is the conditional probability of B given A.

P(B|A) is defined to be $\frac{P(AB)}{P(A)}$, where P(AB) =

Probability of the joint occurrence of events A and B.

Let A = event that a face card is drawn on the first draw

B = event that an ace is drawn on the second draw.

We wish to find the probability of the joint occurrence of these events, P(AB).

We know that P(AB) = P(A) · P(B|A).

P(A) = probability that a face card is drawn on the

first draw = $\frac{12}{52} = \frac{3}{13}$.

P(B|A) = probability that an ace is drawn on the second draw given that a face card is drawn on the first

= number of ways an ace can be drawn on the second draw given a face card is drawn on the first divided by the total number of possible outcomes of the second draw.

= $\frac{4}{51}$; remember there will be only 51 cards left in the deck after the face card is drawn.

Thus P(AB) = $\frac{3}{13}$ · $\frac{4}{51}$ = $\frac{4}{13 \times 17}$ = $\frac{4}{221}$.

445

A coin is tossed 3 times, and 2 heads and 1 tail fall. What is the probability that the first toss was heads?

Solution: This problem is one of conditional probability. Given two events, P_1 and P_2, the probability that event P_2 will occur on the condition that we have event P_1 is

$$P\left(P_2/P_1\right) = \frac{P\left(P_1 \text{ and } P_2\right)}{P\left(P_1\right)} = \frac{P\left(P_1 P_2\right)}{P\left(P_1\right)}$$

Define

$$P_1: \quad \text{2 heads and 1 tail fall,}$$
$$P_2: \quad \text{the first toss is heads.}$$

$$P\left(P_1\right) = \frac{\text{number of ways to obtain 2 heads and 1 tail}}{\text{number of possibilities resulting from 3 tosses}}$$

$$= \Big(\{H,H,T\},\{H,T,H\},\{T,H,H\}\Big)/\Big(\{H,H,H\},\{H,H,T\},\{H,T,T\},\{H,T,H\},$$
$$\{T,T,H\},\{T,H,T\},\{T,H,H\},\{T,T,T\}\Big)$$

$$= 3/8$$

$$P\left(P_1 P_2\right) = P(\text{2 heads and 1 tail and the first toss is heads})$$

$$= \frac{\text{number of ways to obtain } P_1 \text{ and } P_2}{\text{number of possibilities resulting from 3 tosses}}$$

$$= \frac{\Big(\{H,H,T\},\{H,T,H\}\Big)}{8} = 2/8 = 1/4$$

$$P\left(P_2/P_1\right) = \frac{P\left(P_1 P_2\right)}{P\left(P_1\right)} = \frac{1/4}{3/8} = 2/3$$

A coin is tossed 3 times. Find the probability that all 3 are heads,
    (a) if it is known that the first is heads,
    (b) if it is known that the first 2 are heads,
    (c) if it is known that 2 of them are heads.

Solution: This problem is one of conditional probability. If we have two events, A and B, the probability of event A given that event B has occurred is

$$P(A/B) = \frac{P(AB)}{P(B)}.$$

(a) We are asked to find the probability that all

three tosses are heads given that the first toss is heads. The first event is A and the second is B.

P(AB) = probability that all three tosses are heads given that the first toss is heads

$$= \frac{\text{the number of ways that all three tosses are heads given that the first toss is a head}}{\text{the number of possibilities resulting from 3 tosses}}$$

$$= \frac{\{H,H,H\}}{\{\{H,H,H\}, \{H,H,T\}, \{H,T,H\}, \{H,T,T\}, \{T,T,T\}, \{T,T,H\}, \{T,H,T\}, \{T,H,H\}\}}$$

$$= \frac{1}{8}.$$

P(B) = P(first toss is a head)

$$= \frac{\text{the number of ways to obtain a head on the first toss}}{\text{the number of ways to obtain a head or a tail on the first of 3 tosses}}$$

$$= \frac{\{H,H,H\}, \{H,H,T\}, \{H,T,H\}, \{H,T,T\}}{8}$$

$$= \frac{4}{8}$$

$$= \frac{1}{2}.$$

$$P(A/B) = \frac{P(AB)}{P(B)} = \frac{\frac{1}{8}}{\frac{1}{2}} = \frac{1}{8} \quad \frac{2}{1} = \frac{1}{4}.$$

To see what happens, in detail, we note that if the first toss is heads, the logical possibilities are HHH, HHT, HTH, HTT. There is only one of these for which the second and third are heads. Hence,

$$P(A/B) = \frac{1}{4}.$$

(b) The problem here is to find the probability that all 3 tosses are heads given that the first two tosses are heads.

P(AB) = the probability that all three tosees are heads given that the first two are heads

$$= \frac{\text{the number of ways to obtain 3 heads given that the first two tosses are heads}}{\text{the number of possibilities resulting from 3 tosses}}$$

$$= \frac{1}{8}.$$

P(B) = the probability that the first two are heads

$$= \frac{\text{number of ways to obtain heads on the first two tosses}}{\text{number of possibilities resulting from three tosses}}$$

$$= \frac{\{H,H,H\},\ \{H,H,T\}}{8} = \frac{2}{8} = \frac{1}{4}.$$

$$P(A/B) = \frac{P(AB)}{P(B)} = \frac{\frac{1}{8}}{\frac{1}{4}} = \frac{4}{8} = \frac{1}{2}.$$

(c) In this last part, we are asked to find the probability that all 3 are heads on the condition that any 2 of them are heads.

Define:

A = the event that all three are heads

B = the event that two of them are heads

P(AB) = the probability that all three tosses are heads knowing that two of them are heads

$$= \frac{1}{8}.$$

P(B) = the probability that two tosses are heads

$$= \frac{\text{number of ways to obtain at least two heads out of three tosses}}{\text{number of possibilities resulting from 3 tosses}}$$

$$= \frac{\{H,H,T\},\ \{H,H,H\},\ \{H,T,H\},\ \{T,H,H\}}{8}$$

$$= \frac{4}{8}$$

$$= \frac{1}{2}.$$

$$P(A/B) = \frac{P(AB)}{P(B)} = \frac{\frac{1}{8}}{\frac{1}{2}} = \frac{2}{8} = \frac{1}{4}.$$

A survey was made of 100 customers in a department store. Sixty of the 100 indicated they visited the store because of a newspaper advertisement. The remainder had not seen the ad. A total of 40 customers made purchases; of these customers, 30 had seen the ad. What is the probability that a person who did not see the ad made a purchase? What is the probability that a person who saw the ad made a purchase?

Solution: In these two questions we have to deal with conditional probability, the probability that an event occurred given that another event occurred. In symbols, $P(A|B)$ means "the probability of A given B". This is defined as the probability of A and B, divided by the probability of B. Symbolically,

$$P(A|B) = \frac{P(A \cap B)}{P(B)} .$$

In the problem, we are told that only 40 customers made purchases. Of these 40, only 30 had seen the ad. Thus, 10 of 100 customers made purchases without seeing the ad. The probability of selecting such a customer at random is

$$\frac{10}{100} = \frac{1}{10} .$$

Let A represent the event of "a purchase", B the event of "having seen the ad", and $\overline{B}$ the event of "not having seen the ad."

Symbolically, $P(A \cap \overline{B}) = \frac{1}{10}$. We are told that 40 of the customers did not see the ad. Thus $P(\overline{B}) = \frac{40}{100} = \frac{4}{10}$.

Dividing, we obtain $\frac{1/10}{4/10} = \frac{1}{4}$, and, by definition of conditional probability, $P(A|\overline{B}) = \frac{1}{4}$. Thus the probability that a customer purchased given they did not see the ad is $\frac{1}{4}$.

E₂ Purchases ... μ = 100 ... 10 30 30 ... 30 ... E₁ Ad Viewers

To solve the second problem, note that 30 purchasers saw the ad. The probability that a randomly selected customer saw the ad and made a purchase is $\frac{30}{100} = \frac{3}{10}$. Since 60 of the 100 customers saw the ad, the probability that a randomly-picked customer saw the ad is $\frac{60}{100} = \frac{6}{10}$.

Dividing we obtain

$$P(A|B) = \frac{P(A \cap B)}{P(B)} = \frac{\frac{3}{10}}{\frac{6}{10}} = \frac{3}{6} = \frac{1}{2}.$$

● **PROBLEM 8-37**

A committee is composed of six Democrats and five Republicans. Three of the Democrats are men, and three of the Republicans are men. If a man is chosen for chairman, what is the probability that he is a Republican?

Solution: Let E_1 be the event that a man is chosen, and E_2 the event that the man is a Republican.

We are looking for $P(E_2|E_1)$. From the definition of conditional probability $P(E_2|E_1) = \frac{P(E_1 \cap E_2)}{P(E_1)}$.

Of the eleven committee members, 3 are both male and Republican, hence

$$P(E_1 \cap E_2) = \frac{\text{number of male Republicans}}{\text{number of committee members}} = \frac{3}{11}.$$

Of all the members, 6 are men (3 Democrats and 3 Republicans), therefore

$$P(E_1) = \frac{6}{11}.$$

Furthermore, $P(E_2|E_1) = \frac{P(E_1 \cap E_2)}{P(E_1)} = \frac{3/11}{6/11} = \frac{3}{6} = \frac{1}{2}.$

● **PROBLEM 8-38**

A hand of five cards is to be dealt at random and without replacement from an ordinary deck of 52 playing cards. Find the conditional probability of an all spade hand given that there will be at least 4 spades in the hand.

Solution: Let C_1 be the event that there are at least 4 spades in the hand and C_2 that there are five. We want $P(C_2|C_1)$.

$C_1 \cap C_2$ is the intersection of the events that there are at least 4 and there are five spades. Since C_2 is contained in C_1, $C_1 \cap C_2 = C_2$. Therefore

$$P(C_2|C_1) = \frac{P(C_1 \cap C_2)}{P(C_1)} = \frac{P(C_2)}{P(C_1)};$$

450

$$P(C_2) = P \text{ (5 spades)} = \frac{\text{number of possible 5 spade hands}}{\text{number of total hands}}.$$

The denominator is $\binom{52}{5}$ since we can choose any 5 out of 52 cards. For the numerator we can have only spades, of which there are 13. We must choose 5, hence we have $\binom{13}{5}$ and $P(C_2) = \binom{13}{5} / \binom{52}{5}$

$$P(C_1) = P(4 \text{ or } 5 \text{ spades}) = \frac{\text{\# of possible 4 or 5 spades}}{\text{\# of total hands}}$$

The denominator is still $\binom{52}{5}$. The numerator is $\binom{13}{5}$ + (number of 4 spade hands). To obtain a hand with 4 spades we can choose any 4 of the 13, $\binom{13}{4}$. We must also choose one of the 39 other cards, $\binom{39}{1}$. By the Fundamental Principle of Counting, the number of four spade hands is $\binom{13}{4}\binom{39}{1}$. Hence the numerator is $\binom{13}{5} + \binom{13}{4}\binom{39}{1}$ and

$$P(C_1) = \frac{\binom{13}{5} + \binom{13}{4}\binom{39}{1}}{\binom{52}{5}}. \text{ Thus}$$

$$P(C_2 \mid C_1) = \frac{P(C_2)}{P(C_1)} = \frac{\dfrac{\binom{13}{5}}{\binom{52}{5}}}{\dfrac{\binom{13}{5} + \binom{13}{4}\binom{39}{1}}{\binom{52}{5}}}$$

$$= \frac{\binom{13}{5}}{\binom{13}{5} + \binom{13}{4}\binom{39}{1}} = .044.$$

● **PROBLEM 8-39**

Find the probability that Event A, drawing a spade on a single draw from a deck of cards, and Event B, rolling a total of 7 on a single roll of a pair of dice, will both occur.

<u>Solution</u>:   Pr (Event A) $= \dfrac{13}{52}$, We have previously that

Pr (rolling 7) = $\frac{1}{6}$, Pr(Event B) = $\frac{1}{6}$. = $\frac{1}{6}$.

We must now somehow combine these two probabilities to compute the joint probability of the two events A and B. To do this we assume that the conditional probability of event A given B, $P(A|B) = P(A)$.

Because drawing a spade is physically unconnected to rolling a seven, the probabilities of these two events should be unrelated. This is reflected in the statement that $P(A|B) = P(A)$. By our rule for conditional probability this implies that $P(AB) = P(B)P(A|B) = P(B) \times P(A)$.

Two events with this property are called independent and in general the probability of the joint occurrence of independent events is equal to the product of the probability that the events occur in isolation.

In our example, $P(AB) = P(A)P(B)$

$$= \frac{13}{52} \cdot \frac{1}{6}$$

$$= \frac{1}{4} \cdot \frac{1}{6} = \frac{1}{24} .$$

● **PROBLEM 8-40**

A bowl contains eight chips. Three of the chips are red and the remaining five are blue. If two chips are drawn successively, at random and without replacement, what is the probability that the first chip drawn is red and the second drawn is blue?

Solution: The probability that the first chip drawn is red is denoted $P(R_1)$. Since sampling is performed at random and without replacement, the classical probability model is applicable. Thus,

$$Pr(R_1) = \frac{\text{number of red chips}}{\text{total number of chips}} = \frac{3}{8} .$$

We now wish to calculate the conditional probability that a blue chip is drawn on the second draw given a red chip was drawn on the first. Denote this by $P(B_2|R_1)$. The second chip is sampled without replacement. Thus,

$$P(B_2|R_1) = \frac{\text{Number of blue chips}}{\text{total of chips after 1 red chip is drawn}}$$

$$= \frac{5}{8 - 1} = \frac{5}{7} ,$$

The probability we wish to find is $P(R_1 \text{ and } B_2)$.

By the multiplication rule,

$$P(R_1 \text{ and } B_2) = P(R_1)P(B_2 | R_1)$$

$$= \left(\frac{3}{8}\right)\left(\frac{5}{7}\right) = \frac{15}{56} .$$

Thus the probability that a red chip and than a blue chip are respectively drawn is $\frac{15}{56}$.

• **PROBLEM** 8-41

From an ordinary deck of playing cards, cards are drawn successively at random and without replacement. Compute the probability that the third spade appears on the sixth draw.

Solution: Recall the following form of the multiplication rule: $P(C_1 \cap C_2) = P(C_1) \ P(C_2 | C_1)$.

Let C_1 be the event of 2 spades in the first five draws and let C_2 be the event of a spade on the sixth draw. Thus the probability that we wish to compute is $P(C_1 \cap C_2)$.

After 5 cards have been picked there are $52 - 5 = 47$ cards left. We also have $13 - 2 = 11$ spades left after 2 spades have been picked in the first 5 cards. Thus, by the classical model of probability,

$$P(C_2 | C_1) = \frac{\text{favorable outcomes}}{\text{total possibilities}} = \frac{11}{47} .$$

To compute $P(C_1)$, use the classical model of probability. $P(C_1) = \frac{\text{ways of drawing 2 spades in 5}}{\text{All ways of drawing 5}}$.

The number of ways to choose 5 cards from 52 is $\binom{52}{5}$.

Now count how many ways one can select two spades in five draws. We can take any 2 of 13 spades, $\binom{13}{2}$. The other 3 cards can be chosen from any of the 39 non-spades, there are $\binom{39}{3}$ ways to choose 3 from 39.

To determine the total number of ways of drawing 2 spades and 3 non-spades we invoke the basic principle of counting and obtain $\binom{13}{2}\binom{39}{3}$. Hence

453

$$P(C_1) = \frac{\binom{13}{2}\binom{39}{3}}{\binom{52}{5}} \ .$$

$$P(C_1 \cap C_2) = P(C_1)P(C_2|C_1) = \frac{\binom{13}{2}\binom{39}{3}}{\binom{52}{5}} = \frac{11}{47} \ .$$

More generally, suppose X is the number of draws required to produce the 3rd spade. Let C_1 be the event of 2 spades in the first X - 1 draws and let C_2 be the event that a spade is drawn on the Xth draw. Again we want to compute the probability $P(C_1 \cap C_2)$. To find $P(C_2|C_1)$ note that after X-1 cards have been picked, 2 of which were spades, 11 of the remaining 52-(X-1) cards are spades. The classical model of probability gives

$$P(C_2|C_1) = \frac{11}{52 - (x - 1)} \ .$$

Again by the classical model,

$$P(C_1) = \frac{\text{ways of 2 spaced in X-1}}{\text{All ways of X-1 cards}} \ .$$ The denominator

is the number of ways to choose X-1 from 52 or $\binom{52}{X-1}$. Now determine the number of ways of choosing 2 spades in X-1 cards.

There are still only 13 spades in the deck, 2 of which we must choose. Hence we still have a $\binom{13}{2}$ term. The other (X-1) - 2 = X - 3 cards must be non-spades. Thus we must choose X - 3 out of 39 possibilities. This is $\binom{39}{X-3}$. The basic principle of counting says that to get the number of ways of choosing 2 spades and X - 3 non-spades we must multiply the two terms, $\binom{13}{2} \times \binom{39}{X-3}$.

Therefore, $P(C_1) = \dfrac{\binom{13}{2}\binom{39}{X-3}}{\binom{52}{X-1}}$ and the probability of

drawing the third spade on the Xth card is

$$P(C_1 \cap C_2) = P(C_1) \times P(C_2|C_1) = \frac{\binom{13}{2}\binom{39}{X-3}}{\binom{52}{X-1}} \times \frac{11}{52-(X-1)}$$

Find the probability that three successive face cards are drawn in three successive draws (without replacement) from a deck of cards.

Define Events A, B, and C as follows:

Event A: a face card is drawn on the first draw,
Event B: a face card is drawn on the second draw,
Event C: a face card is drawn on the third draw.

Solution: Let ABC = the event that three successive face cards are drawn on three successive draws.

Let D = AB = the event that two successive face cards are drawn on the first two draws.

Then $P(ABC) = P(CD) = P(D) P(C|D)$ by the properties of conditional probability. But

$$P(D) = P(AB) = P(A)P(B|A).$$

We have shown that $P(ABC) = P(A)P(B|A)P(C|AB)$. Now that the event is broken down into these component parts we can solve this problem.

$$P(A) = \frac{\text{number of face cards}}{\text{total number of cards}} = \frac{12}{52} .$$

$$P(B|A) = \frac{\text{number of face cards} - 1}{\text{total number of cards in the deck} - 1} = \frac{11}{51} .$$

$$P(C|AB) = \frac{12 - 2}{52 - 2} = \frac{10}{50}$$

and $P(ABC) = \dfrac{12}{52} \cdot \dfrac{11}{51} \cdot \dfrac{10}{50} = \dfrac{11}{1105} = .010.$

It is important to note that this is an example of sampling without replacement.

If 4 cards are drawn at random and without replacement from a deck of 52 playing cards, what is the chance of drawing the 4 aces as the first 4 cards?

Solution: We will do this problem in two ways. First we will use the classical model of probability which tells us

Probability = $\dfrac{\text{Number of favorable outcomes}}{\text{All possible outcomes}}$, assuming all

outcomes are equally likely.

There are four aces we can draw first. Once that is gone any one of 3 can be taken second. We have 2 choices for third and only one for fourth. Using the Fundamental Principle of Counting we see that there are 4 × 3 × 2 × 1 possible favorable outcomes. Also we can choose any one of 52 cards first. There are 51 possibilitites for second, etc. The Fundamental Principle of Counting tells us that there are 52 × 51 × 50 × 49 possible outcomes in the drawing of four cards. Thus,

$$\text{Probability} = \frac{4 \times 3 \times 2 \times 1}{52 \times 51 \times 50 \times 49} = \frac{1}{270,725} = .0000037.$$

Our second method of solution involves the multiplication rule and shows some insights into its origin and its relation to conditional probability.

The formula for conditional probability $P(A|B) = \dfrac{P(A \cap B)}{P(B)}$ can be extended as follows:

$$P(A|B \cap C \cap D) = \frac{P(A \cap B \cap C \cap D)}{P(B \cap C \cap D)} \; ; \qquad \text{thus}$$

$$P(A \cap B \cap C \cap D) = P(A|B \cap C \cap D)\ P(B \cap C \cap D)\ \text{but}$$

$$P(B \cap C \cap D) = P(B|C \cap D)\ P(C \cap D) \qquad \text{therefore}$$

$$P(A \cap B \cap C \cap D) = P(A|B \cap C \cap D)\ P(B|C \cap D)\ P(C \cap D) \qquad \text{but}$$

$$P(C \cap D) = P(C|D)\ P(D) \qquad \text{hence}$$

$$P(A \cap B \cap C \cap D) = P(A|B \cap C \cap D)\ P(B|C \cap D)\ P(C|D)\ P(D) .$$

Let event D = drawing an ace on the first card

C = drawing an ace on second card

B = ace on third draw

A = ace on fourth card .

Our conditional probability extension becomes

P (4 aces) = P (on 4th|first 3) × P(3rd|first 2) ×

P(2nd|on first) × P(on first).

Assuming all outcomes are equally likely;

P (on 1st draw) = $\dfrac{4}{52}$. There are 4 ways of success in

52 possibilities. Once we pick an ace there are 51 remain-

ing cards, 3 of which are aces. This leaves a probability
of $\frac{3}{51}$ for picking a second ace once we have chosen
the first. Once we have 2 aces there are 50 remaining
cards, 2 of which are aces, thus P (on 3rd|first 2) =
$\frac{2}{50}$. Similarly P (4th ace|first 3) = $\frac{1}{49}$. According
to our formula above

$$P(4 \text{ aces}) = \frac{1}{49} \times \frac{2}{50} \times \frac{3}{51} \times \frac{4}{52} = .000037.$$

● **PROBLEM** 8-44

Four cards are to be dealt successively, at random and
without replacement, from an ordinary deck of playing
cards. Find the probability of receiving a spade, a
heart, a diamond, and a club, in that order.

Solution: Let the events of drawing a spade, heart,
diamond, or club be denoted by S, H, D, or C. We wish to
find P (S, H, D, C) where the order of the symbols indicates the order in which the cards are drawn. This can be
rewritten as

$$P(S, H, D, C) = P (S,H,D,) P (C/S,H,D)$$

by the multiplication rule.

Continuing to apply the multiplication rule yields

$$P(S, H, D, C) = P(S) P(H|S) P(D|S, H) P(C|S, H, D).$$

The product of these conditional probabilites will
yield the joint probability.

Because each card is drawn at random, the classical
model is an apt one.

Pr (drawing a spade on the first draw)

$$= \frac{\text{number of spades}}{\text{number of cards in deck}} .$$

$$P(S) = \frac{13}{52} .$$

$$Pr(H|S) = \frac{\text{number of hearts}}{\text{number of cards after spade is drawn}}$$

$$= \frac{13}{52 - 1} = \frac{13}{51} .$$

$$Pr(D|S,H) = \frac{\text{number of diamonds}}{\text{number of cards after a heart and spade are drawn}}$$

$$= \frac{13}{52 - 2} = \frac{13}{50} \ .$$

$\Pr(C|S, H, D)$

$$= \frac{\text{number of clubs}}{\text{number of cards after heart, spade and diamond are drawn}}$$

$$= \frac{13}{52 - 3} = \frac{13}{49} \ .$$

Thus, $\Pr(S, H, D, C) = \frac{13}{52} \cdot \frac{13}{51} \cdot \frac{13}{50} \cdot \frac{13}{49} = 0.0044$

● **PROBLEM** 8-45

Find the probability that on a single draw from a deck of playing cards we draw a spade or a face card or both. Define Events A and B as follows:

    Event A: drawing a spade,
    Event B: drawing a face card.

Solution: We wish to find the probability of drawing a spade or a face card or both.

Let $A \cup B$ = the event of drawing a spade or face card or both.

$$P(A \cup B) = \frac{\text{number of ways a spade can occur}}{\text{total number of possible outcomes}} \ +$$

$$+ \ \frac{\text{number of ways a face card can occur}}{\text{total number of possible outcomes}} \ .$$

    But we have counted too much. Some cards are spades and face cards so we must subtract from the above expression

the $\dfrac{\text{number of ways a spade and face card can occur}}{\text{total number of possible outcomes}}$.

    This can be rewritten as

$$P(A \cup B) = P(A) + P(B) - P(AB).$$

$$P(AB) = \frac{13}{52} \ , \qquad P(B) = \frac{12}{52}$$

$$P(AB) = P(B)P(A|B).$$

$P(A|B)$ = Probability that a spade is drawn given
              that a face card is drawn.

$$= \frac{\text{number of spades that are face cards}}{\text{total number of face cards which could be drawn}}$$

$$= \frac{3}{12} \quad .$$

$$P(AB) = \frac{12}{52} \cdot \frac{3}{12} = \frac{3}{52} \quad .$$

We could have found P(AB) directly by counting the number of spades that are face cards and then dividing by the total possibilities.

Thus $P(A \cup B) = P(A) + P(B) - P(AB)$

$$= \frac{13}{52} + \frac{12}{52} - \frac{3}{52} = \frac{22}{52} = \frac{11}{26} \quad .$$

We could have found the answer more directly in the following way.

$$P(A \cup B) = \frac{\text{number of spades or face cards or both}}{\text{total number of cards}}$$

$$= \frac{22}{52} = \frac{11}{26} \quad .$$

● **PROBLEM** 8-46

Your company uses a pre-employment test to screen applicants for the job of repairman. The test is passed by 60% of the applicants. Among those who pass the test 80% complete training successfully. In an experiment, a random sample of applicants who do not pass the test is also employed. Training is successfully completed by only 50% of this group. If no pre-employment test is used, what percentage of applicants would you expect to complete training successfully?

<u>Solution</u>:   This is an exercise in conditional probability. We can make a tree diagram:

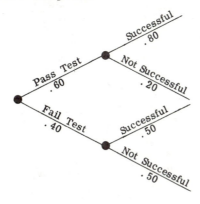

From the definition of conditional probability and

the multiplication rule:

P(Successful ∩ Pass) = P(Successful|Pass) P(pass)

$$= (.80)(.60) = .48.$$

Similarly, P(Successful∩Fail) = P(Successful|Fail) P(Fail)

$$= (.50)(.40) = .20.$$

    The event "an applicant is successful" is composed of two mutually exclusive events. These events are; "an applicant passed the test and was successful" (denoted by S∩P) and "an applicant failed the test and was successful" (denoted S∩F). Thus

Pr (Success) = P(S∩P) + P(S∩F)

$$= .48 + .20 = .68.$$

Thus we would expect 68 percent of the applicants to successfully complete the training.

● **PROBLEM** 8-47

An electronic device contains two easily removed sub-assemblies, A and B. If the device fails, the probability that it will be necessary to replace A is 0.50. Some failures of A will damage B. If A must be replaced, the probability that B will also have to be replaced is 0.70. If it is not necessary to replace A, the probability that B will have to be replaced is only 0.10. What percentage of all failures will you require to replace both A and B?

Solution: This situation may be pictured by the following tree diagram. Each "branch" of the tree denotes a possible event which might occur if device A fails.

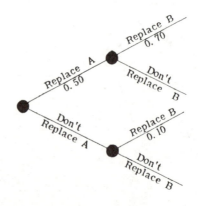

If device A fails, A will be replaced or not re-placed. These first two outcomes are represented by the first two branches of the tree diagram. The branches are labeled with their respective probabilities.

Given that A is replaced, the behavior of B is described by the two secondary branches emanating from the primary branch denoting replacement of A.

If A is not replaced, B's possible behavior is described by the secondary branches emanating from the branch denoting non-replacement of A.

The tree diagram, each branch labeled by its re-spective probability, is thus:

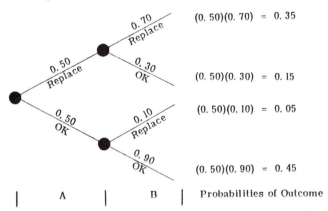

The Probability that both A and B must be replaced is 0.35.

Three probabilities are given but there are six branches to the tree. The probabilities of the remaining branches can be found. If an experiment has only two outcomes, denoted R and DR, and these outcomes are mutually exclusive and exhaustive, the sum

P(R) + P(DR) equals 1.

If P(R) or P(DR) is known, then the other unknown probability may be found.

P(R) = 1 - P(DR).

Let P(R) = probability that component A is replaced. Then P(DR) = probability that A is not replaced = 1 - P(DR). Substituting for P(R) gives

P(DR) = 1 - P(R) = 1 - .5 = .5.

The conditional probabilities that B is not replaced given A is replaced and that B is not replaced given A is not replaced can also be found in this way.

Denote these probabilities by P(B´|A) and P(B´|A´). Also let

461

$P(B|A)$ = probability B is replaced given A is replaced and

$P(B|A')$ = probability B is replaced given A is not replaced.

If A is replaced, then B can be replaced or not replaced. These events are mutually exclusive and exhaustive; thus

$P(B'|A) + P(B|A) = 1$

$P(B|A) = .7$

thus $P(B'|A) = 1 - P(B|A) = 1 - .7 = .3.$

Similarly, if A is not replaced, B may be replaced or not replaced. Given that A is not replaced, these events are mutually exclusive and exhaustive, thus

$P(B'|A') + P(B|A') = 1$

But $P(B|A') = .1$;

thus $P(B'|A') = 1 - .1 = .9.$

The problem asks for the probability that both A and B are replaced. Using the multiplication rule,

P(A and B are replaced)

= P(A is replaced) P(B is replaced | A replaced)

= P(A is replaced) $P(B|A)$

= (.5)(.7) = .35.

The probability that both A and B are replaced is .35

● **PROBLEM** 8-48

A bag contains 1 white ball and 2 red balls. A ball is drawn at random. If the ball is white then it is put back in the bag along with another white ball. If the ball is red then it is put back in the bag with two extra red balls. Find the probability that the second ball drawn is red. If the second ball drawn is red, what is the probability that the first ball drawn was red?

Solution: Let W_i or R_i = the event that the ball chosen on the ith draw is white or red.

Assuming that each ball is chosen at random, a tree diagram of this problem can be drawn showing the possible outcomes.

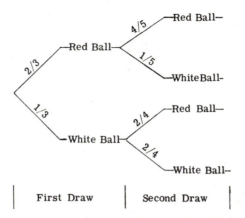

| | First Draw | Second Draw | |

The probabilities of different outcomes are labeled on the "branches" of this tree. These probabilities depend on the number of balls of each color in the bag at the time of the draw. For example, before the first draw there are 2 red balls and 1 white ball. Thus 2 of 3 balls on the average will result in a red ball being chosen or $P(R_1) = 2/3$.

If a red ball is chosen on the first draw then 2 more red balls are added to the bag. There are now 4 red balls and 1 white ball in the bag so on the average 4 of 5 balls chosen will result in a red ball. Thus Pr(R_2 given a red ball on the first draw) = $Pr(R_2|R_1) = 4/5$. The other probabilitites are computed in a similar fashion.

We wish to compute $Pr(R_2)$, the probability that the second ball drawn is red. As we see from the tree diagram there are two ways this can happen; thus, $Pr(R_2)$ = Pr(R_1 and R_2 or W_1 and R_2). The events R_1 and R_2 and W_1 and R_2 are mutually exclusive.

Thus $Pr(R_2) = Pr(R_1$ and $R_2) + Pr(W_1$ and $R_2)$.

But $Pr(R_1$ and $R_2) = Pr(R_2|R_1) \times Pr(R_1)$

and $Pr(W_1$ and $R_2) = Pr(R_2|W_1) \times Pr(W_1)$

by the definition of conditional probability.

From our diagram,

$Pr(R_2|R_1) = 4/5$ and $Pr(R_2|W_1) = 2/4$

also $Pr(R_1) = 2/3$ and $Pr(W_1) = 1/3$.

Thus $Pr(R_2) = Pr(R_2|R_1)Pr(R_1) + Pr(R_2|W_1)Pr(W_1)$

$$= \frac{4}{5} \cdot \frac{2}{3} + \frac{2}{4} \cdot \frac{1}{3}$$

$$= \frac{8}{15} + \frac{1}{6} = \frac{16 + 5}{30} = \frac{21}{30} = \frac{7}{10} \; .$$

Now we wish to find $Pr(R_1|R_2)$.

From the definition of conditional probability

$$Pr(R_1|R_2) = \frac{Pr(R_1 \text{ and } R_2)}{Pr(R_2)} \ .$$

We also know that

$$Pr(R_2) = Pr(R_2|R_1)Pr(R_1) + Pr(R_2|W_1)Pr(W_1) \quad \text{by the}$$

previous problem.

We also know that

$$Pr(R_1 \text{ and } R_2) = Pr(R_2|R_1) \times Pr(R_1).$$

Putting these together we see that

$$Pr(R_1|R_2) = \frac{Pr(R_2|R_1) \ Pr(R_1)}{Pr(R_2|R_1)Pr(R_1) + Pr(R_2|W_1)Pr(W_1)}$$

$$= \frac{\dfrac{4}{5} \cdot \dfrac{2}{3}}{\dfrac{4}{5} \cdot \dfrac{2}{3} + \dfrac{2}{4} \cdot \dfrac{1}{3}}$$

$$= \frac{\dfrac{8}{15}}{\dfrac{7}{10}} = \frac{10 \cdot 8}{7 \cdot 15} = \frac{16}{21} \ .$$

Note that in order to compute these probabilities we needed to know the number of red and white balls in the bag at the beginning.

● PROBLEM 8-49

Twenty percent of the employees of a company are college graduates. Of these, 75% are in supervisory position. Of those who did not attend college, 20% are in supervisory positions. What is the probability that a randomly selected supervisor is a college graduate?

Solution: Let the events be as followed:

      E : The person selected is a supervisor

      E_1 : The person is a college graduate.

      E_2 : The person is not a college graduate.

We are searching for $P(E_1|E)$.

By the definition of conditional probability

$$P(E_1|E) = \frac{P(E_1 \cap E)}{P(E)} \ .$$

But also by conditional probability $P(E_1 \cap E) = P(E|E_1)$, $P(E_1)$. Since, E is composed of mutually exclusive events, E_1 and E_2, $P(E) = P(E_1 \cap E) + P(E_2 \cap E)$. Furthermore, $P(E_2 \cap E) = P(E|E_2) P(E_2)$, by conditional probability. Inserting these expressions into $\dfrac{P(E_1 \cap E)}{P(E)}$, we obtain

$$P(E_1|E) = \frac{P(E_1)P(E|E_1)}{P(E_1)P(E|E_1) + P(E_2)P(E|E_2)} .$$

This formula is a special case of the well-known Bayes' Theorem. The general formula is

$$P(E_1|E) = \frac{P(E_1) \; P(E|E_1)}{\sum\limits_{1}^{n} P(E_n)P(E|E_n)} .$$

In our problem,

$P(E_1) = P(\text{College graduate}) = 20\% = .20$

$P(E_2) = P(\text{Not graduate}) = 1 - P(\text{Graduate}) = 1 - .2 = .80$

$P(E|E_1) = P(\text{Supervisor}|\text{Graduate}) = 75\% = .75.$

$P(E|E_2) = P(\text{Supervisor}|\text{Not a graduate}) = 20\% = .20.$

Substituting,

$$P(E_1|E) = \frac{(.20)(.75)}{(.20)(.75) + (.80)(.20)} = \frac{.15}{.15 + .16}$$

$$= \frac{15}{31} .$$

● **PROBLEM** 8-50

In a factory four machines produce the same product. Machine A produces 10% of the output, machine B, 20%, machine C, 30%, and machine D, 40%. The proportion of defective items produced by these follows: Machine A: .001; Machine B: .0005; Machine C: .005; Machine D: .002. An item selected at random is found to be defective. What is the probability that the item was produced by A? by B?. by C? by D?

Solution: Each question requires us to find the probability that a defective item was produced by a particular machine. Bayes' Rule allows us to calculate this using known (given) probabilities. First we define the necessary symbols: M_1 means the item was produced at A, M_2 means it was produced at B, and M_3 and M_4 refer to machines C and D, respectively. Let M mean that an item is defective. Using Bayes' Rule,

$$P(M_1 \mid M) = \frac{P(M_1)\ P(M \mid M_1)}{P(M_1)P(M \mid M_1)+P(M_2)P(M \mid M_2)+P(M_3)P(M \mid M_3)+P(M_4)P(M \mid M_4)}$$

we substitute the given proportions as follows:

$$P(M_1 \mid M) = \frac{(.1)(.001)}{(.1)(.001)+(.2)(.0005)+(.3)(.005)+(.4)(.002)}$$

$$= \frac{.0001}{.0001 + .0001 + .0015 + .0008} = \frac{.0001}{.0025} = \frac{1}{25}$$

To compute $P(M_2 \mid M)$ we need only change the numerator to $P(M_2)P(M \mid M_2)$. Substituting given proportions, we have $(.20)(.0005) = .0001$. We see that $P(M_2 \mid M) = \frac{1}{25} = P(M_1 \mid M)$. By the same procedure we find that $P(M_3 \mid M) = \frac{3}{5}$ and $P(M_4 \mid M) = \frac{8}{25}$.

To check our work, note that a defective item can be produced by any one of the 4 machines and that the four events "produced by machine i and defective" (i=1,2,3,4) are mutually exclusive. Thus

$$P(M) = \sum_{i=1}^{4} P(M \text{ and } M_i) \quad \text{or} \quad 1 = \sum_{i=1}^{4} \frac{P(M \text{ and } M_i)}{P(M)} \ ;$$

but $\quad \dfrac{P(M \text{ and } M_i)}{P(M)} = P(M_i \mid M).$

Thus $\quad \displaystyle\sum_{i=1}^{4} P(M_i \mid M) = 1.$ Adding we see that

$$\frac{1}{25} + \frac{1}{25} + \frac{15}{25} + \frac{8}{25} = \frac{25}{25} = 1.$$

● PROBLEM 8-51

In the St. Petersburg Community College, 30% of the men and 20% of the women are studying mathematics. Further, 45% of the students are women. If a student selected at random is studying mathematics, what is the probability that the student is a woman?

Solution: This problem involves conditional probabilities. The first two percentages given can be thought of as conditional probabilites; "30% of the men are studying mathematics" means that the probability that a male student selected at random is studying mathematics, is .3. Bayes' formula allows us to use the probabilities we know to

compute the probability that a mathematics student is a
woman. Using the symbols M (the student is studying mathematics); W (the sudent is a woman); and N (the student is
not a woman), we write:

$$P(W|M) = \frac{P(W) \ P(M|W)}{P(W) \ P(M|W) + P(N) \ P(M|N)} \text{ , substituting}$$

$$= \frac{(.45)(0.2)}{(.45)(0.2) + (.55)(0.3)} = \frac{.09}{.09 + 0.165}$$

$$= \frac{.09}{.255} = \frac{6}{17} .$$

Thus, the probability that a randomly selected
math student is a woman equals

$$\frac{6}{17} = .353.$$

CHAPTER 9

STATISTICS

DESCRIPTIVE STATISTICS

● **PROBLEM** 9-1

Discuss and distinguish between discrete and continuous
values.

Solution: The kinds of numbers that can taken on any
fractional or integer value between specified limits are
categorized as continuous, whereas values that are usually
restricted to whole-number values are called discrete.
Thus, if we identify the number of people who use each of
several brands of toothpaste, the data generated must be
discrete. If we determine the heights and weights of a
group of college men, the data generated is continuous.

However, in certain situations, fractional values
are also integers. For example, stock prices are generally
quoted to the one-eighth of a dollar. Since other fractional
values between, say, 24.5 and 24.37 cannot occur, these
values can be considered discrete. However, the discrete
values that we consider are usually integers.

● **PROBLEM** 9-2

Twenty students are enrolled in the foreign language
department, and their major fields are as follows:
Spanish, Spanish, French, Italian, French, Spanish,
German, German, Russian, Russian, French, German,
German, German, Spanish, Russian, German, Italian,
German, Spanish.
(a) Make a frequency distribution table.
(b) Make a frequency histogram.

Solution: (a) The frequency distribution table is constructed by writing down the major field and next to it
the number of students.

Major Field	Number of Students
German	7
Russian	3
Spanish	5
French	3
Italian	2
Total	20

(b) A histogram follows:

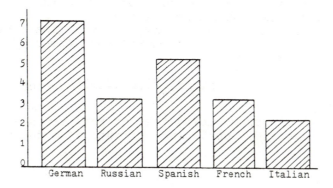

In the histogram, the fields are listed and spaced
evenly along the horizontal axis. Each specific field
is represented by a rectangle, and all have the same
width. The height of each, identified by a number on the
vertical axis, corresponds to the frequency of that field.

● **PROBLEM 9-3**

The IQ scores for a sample of 24 students who are entering
their first year of high school are:

115	119	119	134
121	128	128	152
97	108	98	130
108	110	111	122
106	142	143	140
141	151	125	126

(a) Make a cumulative percentage graph using classes of
seven points starting with 96 - 102.
(b) What scores are below the 25th percentile?
(C) What scores are above the 75th?
(d) What is the median score?

Solution:

Interval	Interval Midpoint	Frequency	Cumulative Frequency	Cumulative Percentage
96-102	99	2	2	8.34
103-109	106	3	5	20.83
110-116	113	3	8	33.33
117-123	120	4	12	50.00
124-130	127	5	17	71.00
131-137	134	1	18	75.00
138-144	141	4	22	91.33
145-151	148	1	23	96.00
152-158	155	1	24	100.00

The frequency is the number of students in that interval. The cumulative frequency is the number of students in intervals up to and including that interval. The cumulative percentage is the percentage of students whose IQ's are at that level or below.

$$\text{Cumulative Percentage} = \frac{\text{Cumulative Frequency}}{24} \times 100 \text{ \%} .$$

One can plot the graph using the interval midpoint as the x coordinate and the cumulative percentage as the y coordinate.

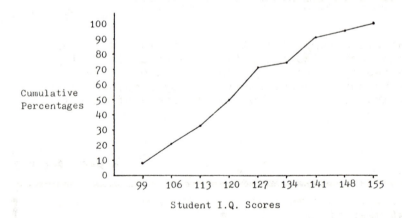

Student I.Q. Scores

(b)   The 25th percentile is defined to be a number that is exactly greater than the lowest 25% of the scores.   We want to know the score that is at least greater than (.25) 24=6 other students.   The 6 lowest are 97, 98, 106, 108, 108, 110. One cannot use 111 as a 25th percentile since another student has that score so 110.5 is used.

(c) The 75th percentile is the score which exceeds the lowest 75% of the population but is less than the top 25% of scores. We want the score below 6 students and above 18.   The 6 highest scores are 152, 151, 143, 142, 141, 140. The next highest is 134. As the 75% percentile one can take any value between 134 and 140. The average of 134 and 140, 137 is taken.

(d) The median is the the value which half of the values of the population excede and half do not, There are 12 values < 123 and 12 values > 124. Therefore we take as our median 123.5, the average of these two values. The median is the 50th percentile.

The following data is a sample of the accounts receivable of a small merchandising firm.

37	42	44	47	46	50	48	52	90
54	56	55	53	58	59	60	62	92
60	61	62	63	67	64	64	68	
67	65	66	68	69	66	70	72	
73	75	74	72	71	76	81	80	
79	80	78	82	83	85	86	88	

Using a class interval of 5, i.e. 35 - 39,

(a) Make a frequency distribution table.
(b) Construct a histogram.
(c) Draw a frequency polygon.
(d) Make a cumulative frequency distribution.
(e) Construct a cumulative percentage ogive.

Class Interval	Class Boundaries	Tally	Interval Median	Frequency
35 - 39	34.5 - 39.5	/	37	1
40 - 44	39.5 - 44.5	//	42	2
45 - 49	44.5 - 49.5	///	47	3
50 - 54	49.5 - 54.5	////	52	4
55 - 59	54.5 - 59.5	////	57	4
60 - 64	59.5 - 64.5	ЖЖ ///	62	8
65 - 69	64.5 - 69.5	ЖЖ ///	67	8
70 - 74	69.5 - 74.5	ЖЖ /	72	6
75 - 79	74.5 - 79.5	////	77	4
80 - 84	79.5 - 84.5	ЖЖ	82	5
85 - 89	84.5 - 89.5	///	87	3
90 - 94	89.5 - 94.5	//	92	2

Use fractional class boundaries. One reason for this is that one cannot break up the horizontal axis of the histogram into only integral values. One must do something with the fractional parts. The usual thing to do is to assign all values to the closest integer. Hence the above class boundaries. The appropriate histogram follows.

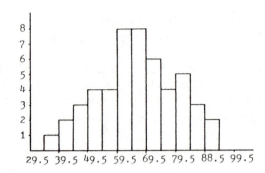

Construct a frequency polygon as follows:

Plot points (x_i, f_i), where x_i is the interval median and f_i, the class frequency. Connect the points by successive line segments.

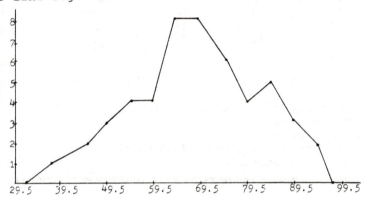

Accounts Receivable

Interval	Interval Median	Frequency (f_i)	Cumulative Frequency	Cumulative Percentage
35 - 39	37	1	1	2
40 - 44	42	2	3	6
45 - 49	47	3	6	12
50 - 54	52	4	10	20
55 - 59	57	4	14	28
60 - 64	62	8	22	44
65 - 69	67	8	30	60
70 - 74	72	6	36	72
75 - 79	77	4	40	80
80 - 84	82	5	45	90
85 - 89	87	3	48	96
90 - 94	92	2	50	100

The cumulative frequency is the number of values in all classes up to and including that class. It is obtained by addition. For example, the cumulative frequency for 65 - 69 is 1 + 2 + 3 + 4 + 4 + 8 + 8 = 30. The cumulative percentage is the percent of all observed values found in that class or below. We can use the formula -

$$\text{cumulative percentage} = \frac{\text{cumulative frequency}}{\text{total observations}} \times 100 \text{ \%.}$$

For example, Cum. per. (65-69) = $\frac{30}{50} \times 100 \text{ \%} = 60 \text{ \%}$.

We construct the cumulative percentage given by plotting points (x_i, f_i) where x_i is the interval median and f_i is the cumulative frequency. Finally we connect the points with successive line segments.

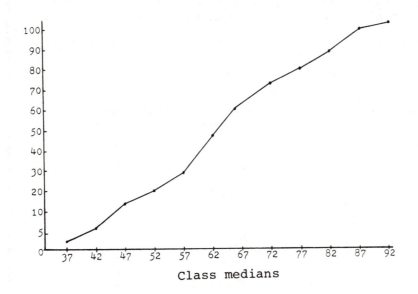

Class medians

● **PROBLEM 9-5**

Find the median age in a sample of 6 children with ages 3, 6, 4, 7, 9, 8.

Solution: Arranging the observations in order, one has 3, 4, 6, 7, 8, 9. Since there are six observations in the sample the median will be the average of the third and fourth observations, 6 and 7. The median is

$$\frac{6 + 7}{2} = 6 \frac{1}{2}.$$

Find the mode of the sample 14, 19, 16, 21, 18, 19, 24, 15 and 19.

Solution: The mode is another measure of central tendency in a data set. It is defined as the observation or observations that occur with the greatest frequency. The number 19 is observed three times in this sample and no other observation appears as frequently.   The mode of this sample is 19.

In the following chart, make 2 additional columns and fill in the cumulative frequencies and cumulative percentages. Also draw a histogram and a cumulative frequency diagram.

Relative Frequency Distribution of 100 Sixth-Grade Students and their Weights

Class	Frequency (f_i)	Relative Frequency	Percentage Distribution
59-61	4	4/100	4
62-64	8	8/100	8
65-67	12	12/100	12
68-70	13	13/100	13
71-73	21	21/100	21
74-76	15	15/100	15
77-79	12	12/100	12
80-82	9	9/100	9
83-85	4	4/100	4
86-88	2	2/100	2
	100		100%

Solution: The relative frequency of a class is found by dividing the class frequency by the total number of observations in the sample. The results, when multiplied by 100, form a percentage distribution. The class relative frequencies and the percentage distribution of the weights of 100 sixth grade students are given in the above table.

The relative frequency of a class is the empirical probability that a random observation from the population will fall into that class. For example, the relative frequency of the class 59-61 in the table is 4/100, and therefore, the empirical probability that a random observation falling in this interval is 4/100.

The table allows us to determine the percentage of the observations in a sample that lie in a particular class. When we want to know the percentage of observations that is above or below a specified interval, the cumulative frequency distribution can be used to advantage. The cumulative frequency distribution is obtained by adding the frequencies in all classes less than or equal to the class with which we are concerned. To find the percentage in each class just divide the frequency by the total number of observations and multiply by 100 %. In this example, $\frac{x}{100} \times 100\% = x\%$. Now we can find the cumulative percentages by taking the cumulative frequencies.

$$\text{Cum. percentage} = \frac{y \text{ cumulative frequency}}{\text{total observations}} \times 100\% =$$

$$= \frac{y}{100} \times 100\% = y\%.$$

Cumulative Frequency and Cumulative Percentage Distribution of the 100 Sixth-Grade Students

Class	Frequency (f_i)	Cumulative Frequency	Cumulative Percentage
59-61	4	4	4
62-64	8	12	12
65-67	12	24	24
68-70	13	37	37
71-73	21	58	58
74-76	15	73	73
77-79	12	85	85
80-82	9	94	94
83-85	4	98	98
86-89	2	100	100 %
	100		

The data in a frequency distribution may be represented graphically by a histogram. The histogram is constructed by marking off the class boundaries along a horizontal axis and drawing a rectangle to represent each class. The base of the rectangle corresponds to the class width, and the height to that class' frequency. See the accompanying histogram depicting the data on the table in the beginning of the problem. Note that the areas above the various classes are proportional to the frequency of those classes.

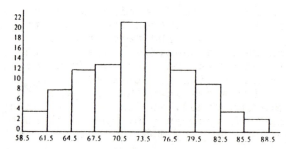

Frequency Histogram of the Weights of the 100
Sixth-Grade Students

Often a frequency polygon is used instead of a
histogram. In constructing a frequency polygon, the
points (x_i, f_i) are plotted on horizontal and vertical
axes. The polygon is completed by adding a class mark with
zero   frequency to each end of the distribution and
joining all the points with line segments. The frequency
diagram for the data in this problem follows:

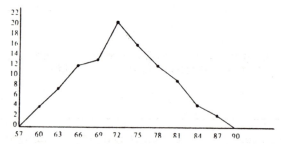

Frequency Polygon of the Weights of 100 Sixth-Grade
Students

A frequency polygon may also be constructed by
connecting the midpoints of the bars in a frequency
histogram by a series of line segments. The main advantage of the frequency polygon compared to the frequency
histogram is that it indicates that the observations in
the interval are not all the same. Also, when several
sets of data are to be shown on the same graph, it is
clearer to superimpose frequency polygons than histograms,
especially if class boundaries coincide.

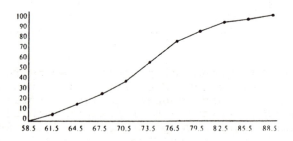

Class Boundaries

Cumulative Frequency Graph of the Weights of 100 Sixth-
Grade Students

It is often advantageous and desirable to make a graph showing the cumulative frequency within a sample. The data for such a graph depicting the cumulative frequency of the weights of the 100 sixth grade students are found in column three of the table in the solution. The graph, called an ogive, is illustrated below. To avoid the confusion of less than or greater than, the class boundaries are plotted on the horizontal axis rather than the interval medians.

A cumulative frequency graph makes it easy to read such items as the percentage of students whose weights are less than or greater than a specified weight. If the cumulative percentage had been plotted, the graph would appear the same as above but would be called a percentage ogive.

● **PROBLEM 9-8**

A motor car traveled 3 consecutive miles, the first mile at x_1 = 35 miles per hour (mph), the second at x_2 = 48 mph, and the third at x_3 = 40 mph. Find the average speed of the car in miles per hour.

<u>Solution:</u>   Distance = Rate × Time. Therefore,

$$\text{Time} = \frac{\text{Distance}}{\text{Rate}} \ .$$

For the first mile, Time = $\frac{1 \text{ mile}}{35 \text{ miles/hour}} = \frac{1}{35}$ hour.

For the second mile, Time = $\frac{1 \text{ mile}}{48 \text{ miles/hour}} = \frac{1}{48}$ hour.

For the third mile, Time = $\frac{1 \text{ mile}}{40 \text{ miles/hour}} = \frac{1}{40}$ hour.

Total time = $T_1 + T_2 + T_3 = \frac{1}{35} + \frac{1}{48} + \frac{1}{40}$.

Converting to decimals, Time_{tot} = .0286+.0208+.025

$$= .0744 \text{ hours} .$$

The average speed can be computed by the following formula:

$$\text{Average Speed} = \frac{\text{Total distance}}{\text{Total time}} = \frac{3 \text{ miles}}{.0744 \text{ hours}} = 40.32 \text{ mph.}$$

Average speed is an example of a harmonic mean. The harmonic is,

$$H = \frac{3}{\frac{1}{T_1} + \frac{1}{T_2} + \frac{1}{T_3}} \ .$$

Find the median of the sample 34, 29, 26, 37, 31.

Solution: The median, a measure of central tendency, is the middle number. The number of observations that lie above the median is the same as the number of observations that lie below it.

Arranged in order we have 26, 29, 31, 34 and 37. The number of observations is odd and thus the median is 31. Note that there are two numbers in the sample above 31 and two below 31.

Arrange the values 7, 8, 5, 10, 3 in ascending order and identify the value of the median.

Solution: In ascending order the numbers are

$$3, 5, 7, 8, 10.$$

The median is the middle value. In the case of 5 observations, it will be the third, 7.

Nine rats run through a maze. The time each rat took to traverse the maze is recorded and these times are listed below.

1 min., 2.5 min., 3 min., 1.5 min., 2 min., 1.25 min., 1 min., .9 min., 30 min.

Which of the three measures of central tendency would be the most appropriate in this case?

Solution: We calculate the three measures of central tendency and then compare them to determine which would be the most appropriate in describing these data.

The mean, $\overline{X}$, is the sum of observations divided by the number of observations. In this case,

$$\overline{X} = \frac{1 + 2.5 + 3 + 1.5 + 2 + 1.25 + 1 + .9 + 30}{9}$$

$$= \frac{43.15}{9} = 4.79.$$

The median is the "middle number" in an array of the observations from the lowest to the highest.

0.9, 1.0, 1.0, 1.25, 1.5, 2.0, 2.5, 3.0, 30.0

The median is the fifth observation in this array or 1.5. There are four observations larger than 1.5 and four observations smaller than 1.5.

The mode is the most frequently occurring observation in the sample. In this data set the mode is 1.0.

mean, $\overline{X}$ = 4.79

median = 1.5

mode = 1.0

The mean is not appropiate here. Only one rat took more than 4.79 minutes to run the maze and this rat took 30 minutes. Observe that the mean has been distorted by this one large observation.

The median or mode seem to describe this data set better and would thus be more appropiate to use.

● **PROBLEM 9-12**

Given the following set of ungrouped measurements

3, 5, 6, 6, 7, and 9,

determine the mean, median, and mode.

Solution: The mean is the average value of the measurements,

$$\overline{X} = \frac{\sum_i x_i}{n} = \frac{3 + 5 + 6 + 6 + 7 + 9}{6} = \frac{36}{6} = 6.$$

The median is the middle value. Since we have an even number of measurements, we take as the median the value halfway between the 2 middle values. In this case the 2 middle values are both 6 and hence the median is

$$\frac{6 + 6}{2} = 6.$$

The mode is the most common value. Therefore, it is 6, the same as the mean and the median.

The staff of a small company sign a timesheet indicating the time they leave the office.

These times for a randomly chosen day are given below,

5:15	4:50	1:50
5:30	2:45	5:15
5:00	5:30	5:30
5:30	4:55	4:20
5:20	5:30	5:20
5:25	5:00	

How do the three measures of central tendency describe the data? How variable is the data? What is the variance and standard deviation?

Solution: The following table, lists the departure times in ascending order. It also aids in the computation of the moments. In order to compute averages, we must first convert each departure time to a score which can be meaningfully added and multiplied. We will then convert back to a time after the computations. Let 12:00 noon be zero and convert each departure time to the number of hours from 12 noon. Thus 5:00 PM would be converted to 5. 5:20 PM would be converted to $5 \frac{1}{3}$ hours or 5.33, etc.

The median is the middle observation, in this case the 9th observation. Thus the median is 5.25 = 5 1/4 or 5:15 PM.

The mode is the most frequent observation. For our sample, the mode is 5:30 PM which appears 5 times.

The variance and standard deviation describe the variation or dispersion in a sample. The formula for the sample variance is,

$$s^2 = \frac{\sum (X_i - \overline{X})^2}{n} = \frac{\sum X_i^2 - n\overline{X}^2}{n} = \frac{\sum X_i^2}{n} - \overline{X}^2$$

Converted Departure Time X_i	X_i^2
$1+\frac{50}{60}=1.83$	3.35
2.75	7.56
4.34	18.83
4.83	23.32
4.92	24.21
5.00	25.00
5.00	25.00
5.25	27.56
5.25	27.56
5.34	28.52
5.34	28.52
5.42	29.38
5.5	30.25
5.5	30.25
5.5	30.25
5.5	30.25
5.5	30.25

One can now compute the three measures of central tendency:

$$\bar{X} = \frac{\Sigma X_i}{n}$$

$$= \frac{\text{sum of observations}}{\text{number of observations}}$$

$$= \frac{82.77}{17} = 4.87$$

$$= 4\frac{87}{100} = 4\frac{52}{60} \text{ PM}$$

$$= 4:52 \text{ PM}$$

$\Sigma X_i = 82.77$ $\Sigma X_i^2 = 420.06$.   Thus,

$$s^2 = \frac{420.06}{17} - (4.87)^2$$

$$= .9925 \text{ (hours)}^2 = 3573 \text{ (min.)}^2 .$$

The standard deviation is

$$s = \sqrt{s^2} = \sqrt{\frac{\Sigma X_i^2}{n} - \bar{X}^2} = \sqrt{.9925} = .9962 \text{ hours}$$

$$= 59.7 \text{ minutes} .$$

Note that the standard deviation is slightly preferable because it is expressed in meaningful units, minutes. The variance is expressed in (minutes)^2.

● **PROBLEM 9-14**

A family had eight children. The ages were 9, 11, 8, 15, 14, 12, 17, 14.

(a) Find the measures of central tendency for the data.
(b) Find the range of the data.

Solution: (a) The mean is the average age.

$$\bar{X} = \frac{\Sigma x_i}{n} = \frac{9 + 11 + 8 + 15 + 14 + 12 + 17 + 14}{8}$$

481

$$= \frac{100}{8} = 12.5 \text{ years.}$$

The median is the middle value. To find it, first we must arrange our data in ascending order;

$$8, \ 9, \ 11, \ 12, \ 14, \ 14, \ 15, \ 17.$$

We have an even number of measurements, eight.

The median will be the midway point between the fourth and fifth observations, 12 and 14. The median is 13.

The mode is the most common age. Only one age, 14 appears more than once. 14 must be the mode.

(b) Often we want to know how spread out the observations of data were. The range of the sample is a quantity which measures dispersion. We define the range to be the difference between the largest and smallest observations in our sample. In this case, R = 17 - 8 = 9.

● **PROBLEM** 9-15

From the sample of data 5, 8, 2, 1, compute the standard deviation of the sample.

Solution: The degree to which numerical data tends to spread about an average value is usually called dispersion or variation of data. One way to measure the degree of dispersion is with the standard deviation. It is defined as

$$s = \sqrt{\frac{\sum_{i} (X_i - \overline{X})^2}{n}} \ .$$

It gives a feeling for how far away from the mean one can expect an observation to be. Sometimes the standard deviation for the data of a sample is defined with n - 1 replacing n in for the denominator of the expression. The resulting value represents a "better" estimate of the true standard of deviation of the entire population. For large n (n > 30) there is practically no difference between the 2 values. The mean of the given sample is:

$$\overline{X} = \frac{\sum X_i}{n} = \frac{5 + 8 + 2 + 1}{n} = \frac{16}{4} = 4 \ .$$

X_i	$X_i - \bar{X}$	$(X_i - \bar{X})^2$
5	$5 - 4 = 1$	$1^2 = 1$
8	$8 - 4 = 4$	$4^2 = 16$
2	$2 - 4 = -2$	$(-2)^2 = 4$
1	$1 - 4 = -3$	$(-3)^2 = 9$

$$\Sigma (X_i - \bar{X})^2 = 1 + 16 + 4 + 9 = 30$$

$$n = 4$$

$$s = \sqrt{\frac{\Sigma (X_i - \bar{X})^2}{n}} = \sqrt{\frac{30}{4}} = \sqrt{\frac{15}{2}} = \sqrt{7.5} = 2.74.$$

● **PROBLEM** 9-16

Find the midrange of this sample of SAT-Verbal scores. The sample had the smallest observation of 426 and the largest at 740.

Solution: The midrange is the number halfway between the largest and smallest observations. In this case;

$$\text{midrange} = \frac{426 + 740}{2} = \frac{1166}{2} = 583.$$

● **PROBLEM** 9-17

Given the values 4, 4, 6, 7, 9 give the deviation of each from the mean.

Solution: First find the mean.

$$\bar{X} = \frac{\Sigma x_i}{n} = \frac{4 + 4 + 6 + 7 + 9}{5} = \frac{30}{5} = 6.$$

$X - \bar{X} = $ the deviation from the mean.

We will provide the deviations in tabular form.

X	$X - \bar{X}$
4	$4 - 6 = -2$
4	$4 - 6 = -2$
6	$6 - 6 = 0$
7	$7 - 6 = +1$
9	$9 - 6 = +3$

Find the variance of the sample of observations 2, 5, 7, 9, 12.

Solution: The variance of the sample is defined as

$$s^2 = \frac{\Sigma(X_i - \overline{X})^2}{n}$$. This is an average of the squared

deviations from the sample mean, $\overline{X}$.

$$\overline{X} = \frac{\Sigma X_i}{n} = \frac{2 + 5 + 7 + 9 + 12}{5} = \frac{35}{5} = 7$$

and $$s^2 = \frac{(2-7)^2 + (5-7)^2 + (7-7)^2 + (9-7)^2 + (12-7)^2}{5}$$

$$= \frac{25 + 4 + 0 + 4 + 25}{5} = \frac{58}{5} = 11.6.$$

A survey asking for the number of times toast is burned during one week was distributed to eight randomly selected households. The survey yielded the following results:

2, 3, 0, 3, 4, 1, 3, 0.

What is the range, variance and standard deviation for this data set?

Solution: The range is the difference between the largest and smallest observations is 4 - 0 = 4.

The variance is the mean   or average squared deviation from $\overline{X}$. To compute the variance of this sample

we use the formula
$$s^2 = \frac{\Sigma X^2 - n\overline{X}^2}{n}$$.

To facilitate the computation we use the following table,

X	X^2
2	4
3	9
0	0
3	9
4	16
1	1
3	9
0	0
$\Sigma X = 16$	$\Sigma X^2 = 48$

Thus $\bar{X} = \dfrac{\Sigma X}{n} = \dfrac{16}{8} = 2$ and

$$s^2 = \frac{48 - 8(2)^2}{8} = \frac{48 - 8(4)}{8} = \frac{16}{8} = 2 .$$

The standard deviation is

$$s = \sqrt{s^2} = \sqrt{2} = 1.414.$$

● **PROBLEM** 9-20

What can be said about a sample of observations whose standard deviation is zero?

Solution: The standard deviation is a measure of dispersion in the sample. If the standard deviation is zero we expect little or no variation in the sample. In fact, a standard deviation of zero is the most extreme example of lack of variation in a sample possible.

$$s = \sqrt{\frac{\sum_{i=1}^{n} (X_i - \bar{X})^2}{n}} \qquad \text{by definition.}$$

If $s = 0$, then $\sqrt{\dfrac{\sum_{i=1}^{n} (X_i - \bar{X})^2}{n}} = 0$

squaring both sides and multiplying by n we see that,

$$\sum_{i=1}^{n} (X_i - \bar{X})^2 = 0.$$

Any number squared is positive or zero. Thus, in order for the sum of squared numbers to be zero each term in the sum must be zero. Thus,

$$(X_i - \bar{X})^2 = 0 \qquad \text{for } i = 1, \ldots n$$

485

squaring and adding $\overline{X}$ to each of these n equations we see that

$$X_i = \overline{X} \qquad \text{for } i = 1, \ldots n.$$

The n observations are identical and equal to the sample mean. The observations are all the same, the most extreme example of lack of dispersion in a sample.

In an office, the employer notices his employees spend more time drinking coffee than working. He counts the number of coffee breaks each of his seven employees takes in the course of a day. The data are

1 , 1 , 2 , 2 , 3 , 5 , and 7 .

Find the mean, variance, standard deviation and the median number of coffee breaks a day.

Solution: To aid in these computations use the following table,

X_i	$X_i - \overline{X}$	$(X_i - \overline{X})^2$
1	- 2	4
1	- 2	4
2	- 1	1
2	- 1	1
3	0	0
5	2	4
7	4	16

$$\overline{X} = \frac{\Sigma X_i}{n} = \frac{21}{7} = 3.$$

The median is the fourth observation, 2, the observation such that 3 other observations are higher and 3 are lower. We compute the variance using:

$$s^2 = \frac{\Sigma (X_i - \overline{X})^2}{n} = \frac{4 + 4 + 1 + 1 + 4 + 16}{7}$$

$$= \frac{30}{7} = 4.29.$$

Thus the variance is 4.29 and the standard deviation is,

$$s = \sqrt{s^2} = \sqrt{4.29} = 2.07.$$

486

The following measurements were taken by an antique dealer as he weighed to the nearest pound his prized collection of anvils. The weights were,

84,　92,　37,　50,　50,　84,　40,　98.

What was the mean weight of the anvils?

Solution:　　The average or mean weight of the anvils is

$$\overline{X} = \frac{\text{sum of observations}}{\text{number of observations}}$$

$$= \frac{84 + 92 + 37 + 50 + 50 + 84 + 40 + 98}{8}$$

$$= \frac{535}{8} = 66.88 \cong 67 \text{ pounds.}$$

An alternate way to compute the sample mean is to rearrange the terms in the numerator, grouping the numbers that are the same. Thus,

$$\overline{X} = \frac{(84 + 84) + (50 + 50) + 37 + 40 + 92 + 98}{8} .$$

One can express the mean in terms of the frequency of observations. The frequency of an observation is the number of times a number appears in a sample.

$$\overline{X} = \frac{2 (84) + 2 (50) + 37 + 40 + 92 + 98}{8} .$$

The observations 84 and 50 appear in the sample twice and thus each observation has frequency 2.

In more general terms, the mean can be expressed as,

$$\overline{X} = \frac{\Sigma \ f_i \ X_i}{\text{number of observations}} ,$$

where X_i is the ith observation and f_i is the frequency of the ith observations.

The sum of the frequencies is equal to the total number of observations, $\Sigma f_i = n$.

Thus,　　　　$$\overline{X} = \frac{\Sigma f_i \ X_i}{\Sigma f_i} .$$

Compute the arithmetic mean for the following grouped data.

Class Limits	Class Mark X_i	f_i
6 – 8	7	4
9 – 11	10	6
12 – 14	13	7
15 – 17	16	4
18 – 20	19	3

Solution: We know that the arithmetic mean is

$$\overline{X} = \frac{\Sigma f_i \, X_i}{\Sigma f_i} \qquad .$$

$$\Sigma f_i = 4 + 6 + 7 + 4 + 3 = 24$$

and $\Sigma f_i \, X_i = (4)(7) + (6)(10) + (13)(7) + (16)(4) + (19)(3)$

$$= 300$$

Thus, $\overline{X} = \dfrac{300}{24} = 12.5$.

Find the mean of the data given in the frequency distribution below.

Class	Class Mark X_i'	Frequency f_i	$X_i \, f_i$
1	46	4	184
2	51	1	51
3	56	2	112
4	61	2	122
5	66	2	132
6	71	9	639
7	76	5	380
8	81	10	810
9	86	4	344
10	91	8	728
11	96	3	288
Totals:		50	3790

Solution: The sample mean or average of the data is

$$\bar{X} = \frac{1}{n} \sum_{i=1}^{K} X_i' \, f_i \qquad \text{where}$$

X_i = class mark or midpoint of class

f_i = frequency or number of observations in a particular class

$$n = \text{total observations} = \sum_{i=1}^{K} f_i$$

K = number of classes .

We wish to obtain some measure of central tendency in the data. However we do not have all the observations available to us. Instead we are given the class marks and frequencies.

One possibility would be a straight average of the class marks or $\frac{1}{11} \sum_{i=1}^{11} X_i'$. The drawback of this method is that if one class has a greater frequency, more observations, than another class this difference will not be reflected. Each class mark will be weighted the same.

A better way would be to weight each class mark by its relative frequency, $\frac{f_i}{n}$. This weighting system will cause the classes with more observations to receive greater weight.

Computing the sample mean gives:

$$\bar{X} = X_1' \cdot \frac{f_1}{n} + X_2' \cdot \frac{f_2}{n} + \ldots X_K \cdot \frac{f_K}{n} = \sum_{i=1}^{K} \frac{X_i' \cdot f_i}{n}$$

$$= \frac{1}{n} \sum_{i=1}^{K} X_i' \, f_i .$$

Referring to the given table, $X_i' f_i$ = 3790 and n = 50,

$$\bar{X} = \frac{3790}{50} = 75.8 .$$

Class	Class Boundaries	Frequencies
1	49.5 - 99.5	17
2	99.5 - 149.5	38
3	149.5 - 199.5	61
4	199.5 - 249.5	73
5	249.5 - 299.5	56
6	299.5 - 349.5	29
7	349.5 - 399.5	16
8	399.5 - 449.5	10
		300

Consider this Table. What is the median class? Estimate the median by linear interpolation.

Solution: The Table does not supply enough information to allow us to read off the median. Rather than observing the individual data, the observations have been grouped into classes. We are given the boundaries of each class, the range of numbers which are included in each class, and the frequencies or number of observations in each class.

The median class is the class which contains the median of the entire data set. In this case there are 300 total observations, 116 observations in the first three classes and 189 observations in the first four classes. The median will be the 150th observation and will be in the fourth class. Thus, the median class is the fourth class.

To estimate the median by linear interpolation we must assume that the observations are distributed uniformly throughout the fourth class. To see what this means, consider the following; If the interval from 199.5 to 249.5 is marked off as on a scale, the scale will be 50 units long. The observations will be distributed uniformly if they are spread out evenly over the 50 units.

In this case, there are 73 observations and by assuming that the observations are distributed uniformly, we are assuming that the observations are $\frac{50}{73}$ units apart. That is, if the 50 units were divided into 73 equal intervals, each interval would contain one observation.

If we make this assumption, then we can estimate the median. There are 116 observations in the first three classes and we wish to find the 150th observation. To do this we divide the fourth class into 73 intervals, by our assumption we know that each observation is in an interval $\frac{50}{73}$ units long. The 34th observation in the fourth class will be the median and this observation will be $34 \cdot \left(\frac{50}{73}\right)$ units from the lower boundary of the fourth class.

The median = 199.5 (the lower boundary of the fourth

class + 34 $\cdot$ $\left(\dfrac{50}{73}\right)$ = 199.5 + 23.3 = 222.8.

What is the value of the median for the data in the following frequency distribution?

Class Limits	Class Boundaries	Class Mark X_i	Frequency f_i
1 - 2	.5 - 2.5	1.5	2
3 - 4	2.5 - 4.5	3.5	5
5 - 6	4.5 - 6.5	5.5	15
7 - 8	6.5 - 8.5	7.5	10
9 - 10	8.5 - 10.5	9.5	5

Solution:

The median is the 50th-percentile or the number such that 50% of the cumulative frequency lies below it. This corresponds to $\dfrac{37}{2}$ = 18.5.

There are 7 observations in the first two classes and 22 observations in the first three classes thus the median will be the 18.5 - 7 = 11.5th "observation" in the third class.

Through linear interpolation we assume that each observation in the third class lies in an interval that is

$$\dfrac{\text{length of class}}{\text{number of observations in class}} = \dfrac{2}{15} \quad \text{units.} \qquad \text{The}$$

11.5th "observation" in the class lies $(11.5)\left(\dfrac{2}{15}\right)$ units from the lower boundary of the third class.

The median is

$$4.5 + (11.5) \left(\dfrac{2}{15}\right) = 6.0.$$

Find the average deviation for the grouped data given below:

Class	Frequency f_i	Class Mark X_i	$\overline{X}$	$|X_i-\overline{X}|$	$f_i|X_i - \overline{X}|$
49 - 54	6	51.5	66.5	15	90
55 - 60	15	57.5	66.75	9	135
61 - 66	24	63.5	66.5	3	72
67 - 72	33	69.5	66.5	3	99
73 - 78	22	75.5	66.5	9	198

Solution: In the above table the frequency represents the number of observations in a particular class. The class mark is the midpoint of each class. The sample mean $\overline{X}$ is defined as

$$\overline{X} = \frac{\Sigma f_i X_i}{\Sigma f_i} = [6(51.5) + 15(57.5) + 24(63.5) + 33(69.5)$$

$$+ 22(75.5)] \div 100$$

$$= \frac{6650}{100} = 66.5.$$

$X_i - \overline{X}$ is the absolute value of the difference between the class mark and the sample mean for each class i.

We wish to compute the measure of dispersion known as the average deviation. This measure gives an indication of the spread of the observations around the sample mean. The average deviation is defined to be,

$$\text{A.D.} = \frac{\Sigma f_i |X_i - \overline{X}|}{\Sigma f_i}$$

$$= \frac{90 + 135 + 72 + 99 + 198}{100} = \frac{594}{100} = 5.94.$$

A history test was taken by 51 students. The scores ranged from 50 to 95 and were classified into 8 classes of width 6 units. The resulting frequency distribution appears below. Find s^2 by applying the definition for s^2. Then find s.

Class i	Class Mark X_i	Frequency f_i	$X_i f_i$
1	51	2	102
2	57	3	171
3	63	5	315
4	69	8	552
5	75	10	750
6	81	12	972
7	87	10	870
8	93	1	93
		51	$3825 = \sum_{i=1}^{8} X_i' f_i$

Solution: The problem asks us to find s^2, the measure of dispersion of the observations about the sample mean for this classified data.

In the table one is given:

X_i' = the midpoint (class mark) of the ith class

f_i = the number of observations in the ith class

n = the toal number of observations = $\sum_{i=1}^{n} f_i$

By definition, the sample variance for classified data is

$$s^2 = \frac{\sum_{i=1}^{K} (X_i' - \bar{X})^2 f_i}{n}$$

where K is tne number of classes. First find $\bar{X}$,

$$\bar{X} = \frac{\sum_{i=1}^{K} X_i f_i}{n} = \frac{3825}{51} = 75.$$

The computations used in finding s^2 are displayed in the following table.

Class	X_i'	f_i	$X_i' - \bar{X}$	$(X_i' - \bar{X})^2$	$(X_i' - \bar{X})^2 f_i$
1	51	2	− 24	576	1152
2	57	3	− 18	324	972
3	63	5	− 12	144	720
4	69	8	− 6	36	288
5	75	10	0	0	0
6	81	12	6	36	432
7	87	10	12	144	1440
8	93	1	18	324	324

$$\sum_{i=1}^{8} f_i = 51, \qquad \sum_{i=1}^{8} (X_i' - \overline{X}) f_i = 5328.$$

$$s^2 = \frac{1}{n} \sum_{i=1}^{8} (X_i' - \overline{X})^2 f_i = \frac{1}{51} (5328) = 104.47.$$

and the standard deviation,

$$s = \sqrt{s^2} = 10.22.$$

The word "moment" is used quite often in a statistical context. It refers to the sum of the deviations from the mean in respect to sample size. The first moment is defined then as $\dfrac{\sum (X_i - \overline{X})}{n}$ $\left(\dfrac{\sum f_i (X_i - \overline{X})}{\sum f_i} \text{ for grouped data} \right)$.

What numerical value does the first moment always have?

Solution: Let us perform some algebraic manipulations on the definition of the first moment:

$$m_1 = \frac{\sum\limits^{n} (X_i - \overline{X})}{n} .$$ Substitute $\dfrac{\sum\limits^{n} X_i}{n}$ for $\overline{X}$ and obtain

$$\frac{\sum\limits^{n} \left(X_i - \dfrac{\sum\limits^{n} X_i}{n} \right)}{n} .$$ We now examine the numerator,

$$\sum^{n} \left(X_i - \frac{\sum\limits^{n} X_i}{n} \right).$$ The second term is a constant so we subtract it once for each of the n terms in the summation. Consequently,

$$m_1 = \frac{\left(X_1 - \dfrac{\sum X_i}{n} \right) + \left(X_2 - \dfrac{\sum X_i}{n} \right) + \left(X_3 - \dfrac{\sum X_i}{n} \right) + \dots + \left(X_n - \dfrac{\sum X_i}{n} \right)}{n}$$

$$= \frac{(X_1 + X_2 + \dots + X_n) - \dfrac{\sum X_i}{n} - \dfrac{\sum X_i}{n} - \dots - \dfrac{\sum X_i}{n}}{n}$$

$$= \frac{\sum X_i - n \dfrac{\sum X_i}{n}}{n}$$

$$= \frac{\sum X_i - \sum X_i}{n} = \frac{0}{n} = 0.$$

What is the relative measure of skewness for the data
listed below? This data represents the waist measurements
of six randomly selected chocolate rabbits.

> 3 inches,   2   inches,   3.7 inches,   5 inches,
>   2.7 inches,   3 inches.

Solution: The relative measure of symmetry is defined to

be $a_3 = \frac{m_3}{s^3}$ where s^3 is the standard deviation cubed and
m_3 is the third moment.

The third moment is defined as;

$$m_3 = \frac{\Sigma(X_i - \bar{X})^3}{n} .$$

The first moment is

$$m_1 = \frac{\Sigma(X_i - \bar{X})^1}{n} .$$

Observe that this moment has only one value.

$$m_1 = \frac{\Sigma(X_i - \bar{X})^1}{n} = \frac{\Sigma X_i}{n} - \frac{n\bar{X}}{n} = \frac{\Sigma X_i}{n} - \bar{X}$$

but $\bar{X} = \frac{\Sigma X_i}{n}$ thus $m_1 = 0.$

The second moment is $\frac{\Sigma(X_i - \bar{X})^2}{n}$ or the sample
variance.

The fourth moment is defined as

$$m_4 = \frac{\Sigma(X_i - \bar{X})^4}{n} .$$

The measure of symmetry has the following interpre-
tation, if $a_3 = \frac{m_3}{s^3}$ is equal to zero, the distribution is
symmetrical. If $a_3 < 0$ then the distribution is negatively
skewed. If $a_3 > 0$ the distribution is positively skewed.

To calculate the measure of symmetry use the table
below:

X_i	$\bar{X}$	$(X_i - \bar{X})$	$(X_i - \bar{X})^2$	$(X_i - \bar{X})^3$
3	3.23	$-$ .23	.053	$-$ .012
2	3.23	$-$ 1.23	1.51	$-$ 1.86
3.7	3.23	.47	.22	.103
5	3.23	1.77	3.13	5.54
2.7	3.23	$-$ .53	.28	$-$ .148
3	3.23	$-$ .23	.053	$-$ .012

$$\Sigma(X_i - \bar{X})^2 = 5.246 \qquad \Sigma(X_i - \bar{X})^3 = 3.611$$

$$s^2 = \frac{\Sigma(X_i - \bar{X})^2}{n} = \frac{5.246}{6} = .8743$$

$$s = \sqrt{s^2} = .9351$$

$$s^3 = .817$$

$$m_3 = \frac{\Sigma(X_i - \bar{X})^3}{n} = \frac{3.611}{6} = .6018$$

and $\quad a_3 = \frac{m_3}{s^3} = \frac{.6018}{.817} = .73659$.

As a result of the above calculation we observe that the distribution of the chocolate rabbits' waist measurements is skewed to the right or positively skewed.

● PROBLEM 9-31

What are two ways to describe the form of a frequency distribution? How would the following distributions be described?

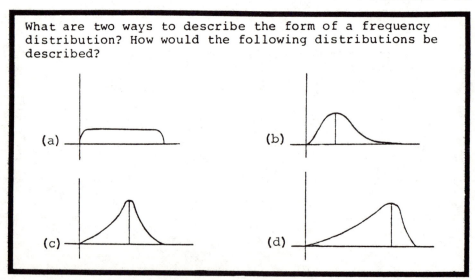

(a)

(b)

(c)

(d)

Solution: The form of a frequency distribution can be described by its departure from symmetry or skewness and its degree of peakedness or kurtosis.

If the few extreme values are higher than most of the others, we say that the distribution is "positively skewed" or "skewed" to the right.

If the few extreme values are lower than most of the others, we say that the distribution is "negatively skewed" or "skewed" to the left.

A distribution that is very flat is called "platykurtic" and a distribution that has a high peak is called "leptokurtic".

(a) This curve is quite flat indicating a wide dispersion of measurements. Thus it is platykurtic.

(b) This distribution has extreme values in the upper half of the curve and is skewed to the right or positively skewed.

(c) This frequency distribution is very peaked and is   leptokurtic.

(d) The extreme values of this distribution are in the lower half of the curve. Thus the distribution is negatively skewed or skewed to the left.

PROBABILITY DISTRIBUTIONS

● **PROBLEM 9-32**

Let X be the random variable denoting the result of the single toss of a fair coin. If the toss is heads, X = 1. If the toss results in tails, X = 0.
What is the probability distribution of X?

Solution: The probability distribution of X is a function which assigns probabilities to the values X may assume.

This function will have the following properties if it defines a proper probability distribution.

Let $f(x) = Pr(X = x)$. Then $\Sigma Pr(X = x) = 1$ and $Pr(X = x) \geq 0$ for all x.

We have assumed that X is a discrete random variable. That is, X takes on discrete values.

The variable X in this problem is discrete as it only takes on the values 0 and 1.

To find the probability distribution of X, we must find $Pr(X = 0)$ and $Pr(X = 1)$.

Let $p_0 = Pr(X = 0)$ and $Pr(X = 1) = p_1$. If the coin is fair, the events X = 0 and X = 1 are equally likely. Thus $p_0 = p_1 = p$. We must have $p_0 > 0$ and $p_1 > 0$.

In addition,

$$Pr(X = 0) + Pr(X = 1) = 1$$

or

$$P_0 + P_1 = p + p = 1$$

or

$$2p = 1$$

and

$$P_0 = P_1 = p = \tfrac{1}{2} \, ,$$

thus the probability distribution of X is f(x): where

$$f(0) = Pr(X = 0) = \tfrac{1}{2} \text{ and }$$

$$f(1) = Pr(X = 1) = \tfrac{1}{2}$$

f (anything else) = Pr(X = anything else) = 0 . We see that this is a proper probability distribution for our variable X.

$$\Sigma \ f(x) = 1 \quad \text{and}$$

$$f(x) \geq 0 \ .$$

● **PROBLEM** 9-33

If f(x) = 1/4, x = 0,1,2,3 is a probability mass function, find F(t), the cumulative distribution function and sketch its graph.

Solution: $F(t) = \sum\limits_{x=0}^{t} f(x) = Pr(X \leq t).$ F(t) changes for integer

values of t. We have:

$$F(t) = 0 \qquad\qquad t < 0$$

$$F(t) = f(0) = 1/4, \quad 0 \leq t < 1$$

$$F(t) = f(0) + f(1) = 1/4 + 1/4 = 1/2 \, ,$$

$$1 \leq t < 2$$

$$F(t) = f(0) + f(1) + f(2) \qquad 2 \leq t < 3$$

$$= \tfrac{1}{4} + \tfrac{1}{4} + \tfrac{1}{4} = 3/4 \ .$$

$$F(t) = \sum\limits_{x=0}^{t} f(x) = 1 \qquad\qquad 3 \leq t \ .$$

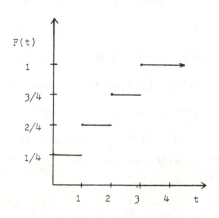

An absent-minded professor has 5 keys. One of the keys opens the door to his apartment. One night he arrives at his building, reaches into his pocket and selects a key at random from those on his chain. He tries it in the lock. If it doesn't work, he replaces the key and again selects at random from the 5 keys. He continues this process until he finally finds his key, then stops. Let X = the number of attempts the professor makes. What is the probability distribution of X?

Solution: X is a special type of binomial random variable. Each time the professor reaches into his pocket can be considered a trial. The professor always replaces the key so we assume that the trials are independent except if the correct key is chosen.

There are 5 keys and he is selecting at random. The probability of "success" on a given trial is thus 1/5 and the probability that he selects the wrong key is 4/5.

We are not sure how many attempts he will make, but we know that if he selects the correct key he will stop the process.

Let's compute some of the probability distribution of X,

$Pr(X = 1)$ = probability that he picks the right key on the first draw = 1/5.

$Pr(X = 2)$ = probability that he picks a wrong key first and the right key second

$$= (4/5)(1/5).$$

$Pr(X = 3)$ = probability that he picks two wrong keys first and the right key third

$$= (4/5)(4/5)(1/5) .$$

In general, $Pr(X = k)$ = probability that k-1 wrong keys are tried and the kth key selected is the correct one.

$$Pr(X = k) = (4/5)^{k-1}(1/5) .$$

This distribution is known as the _geometric distribution_ and is a special case of the binomial distribution.

Defects occur along the length of a cable at an average of 6 defects per 4000 feet. Assume that the probability of k defects in t feet of cable is given by the probability mass function:

$$Pr(k \text{ defects}) = \frac{e^{-\frac{6t}{4000}}\left(\frac{6t}{4000}\right)^k}{k!}$$

for k = 0,1,2,... . Find the probability that a 3000-foot cable will have at most two defects.

<u>Solution</u>: The probability of exactly k defects in 3000 feet is determined by the given discrete probability distribution as Pr(k defects in 3000 ft.)

$$= \frac{e^{-\frac{6(3000)}{4000}}\left(\frac{6(3000)}{4000}\right)^k}{k!}$$

$$= \frac{e^{-4.5}(4.5)^k}{k!} \ , \quad k = 0,1,2,\ldots .$$

We use the probability distribution to find the probability of at most two defects.

Pr(at most two defects) = Pr(0,1 or 2 defects).

The events "0 defects", "1 defect" and "2 defects" are all mutually exclusive, thus,

Pr(at most two defects) = Pr(0 defects) + Pr(1 defect) + Pr(2 defects)

$$= \frac{e^{-4.5}(4.5)^0}{0!} + \frac{e^{-4.5}(4.5)^1}{1!} + \frac{e^{-4.5}(4.5)^2}{2!}$$

$$= e^{-4.5}\left(1 + 4.5 + \frac{(4.5)^2}{2!}\right)$$

$$= .1736 .$$

● **PROBLEM** 9-36

Let X be a continuous random variable. We wish to find probabilities concerning X. These probabilities are determined by a density function. Find a density function such that the probability that X falls in an interval (a,b) (0 < a < b < 1) is proportional to the length of the interval (a,b). Check that this is a proper probability density function.

<u>Solution</u>: The probabilities of a continuous random variable are computed from a continuous function called a density function in the following way. If f(x) is graphed and is

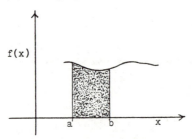

then, Pr(a ≤ X ≤ b) = the area under the curve f(x) from a to b.

With this definition some conditions on f(x) must be imposed.

f(x) must be positive and the total area between f(x) and the x-axis must be equal to 1.

We also see that if probability is defined in terms of area under a curve, the probability that a continuous random variable is equal to a particular value, Pr(X = a) is the area under f(x) at the point a. The area of a line is 0, thus Pr(X = a) = 0. Therefore

$$Pr(a < X < b) = Pr(a \le X \le b).$$

To find a density function for $0 < X < 1$, such that Pr(a < X < b) is proportional to the length of (a,b), we look for a function f(x) that is positive and the area under f(x) between 0 and 1 is equal to 1. It is reasonable to expect that the larger the interval the larger the probability that x is in the interval.

A density function that satisfies these criteria is

$$f(x) = \begin{cases} 1 & 0 < x < 1 \\ 0 & \text{otherwise} . \end{cases}$$

A graph of this density function is

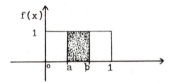

The probability that X is between a and b is the area of the shaded region. This is the area of a rectangle. The area of a rectangle is base $\times$ height.

Thus $Pr(a \le X \le b) = (b - a) \times 1 = b - a$. Similarly

$$Pr(X \le k) = (k - 0) \times 1 = k \quad \text{for} \quad 0 < k < 1.$$

Often the density function is more complicated and integration must be used to calculate the area under the density function.

To check that this is a proper probability density function, we must check that the total area under f(x) is 1. The total area under this density function is $(1 - 0) \times 1 = 1$.

● **PROBLEM 9-37**

Given that the continuous random variable X has distribution function F(x) = 0 when x < 1 and $F(x) = 1 - 1/x^2$ when $x \ge 1$, graph F(x), find the density function f(x) of X, and show how F(x) can be obtained from f(x).

Solution: To graph F(x), observe that as x approaches 1 from the right side, $1 - 1/x^2$ approaches 0. As x approaches $+\infty$, $1 - 1/x^2$ approaches 1 because $\lim_{x \to \infty} 1/x^2 = 0$.

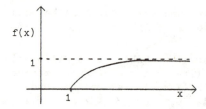

The curve is said to be asymptotic to the line $F(x) = 1$ because it comes closer and closer to it, as x increases without limit, but never touches it. $F(x)$ is a continuous function for all real numbers x because it satisfies 2 conditions: (1) $F(x)$ is defined for all values of x; (2) the function's value at any point c equals the left and right limits:

$$\text{limit of } F(x) = F(c) = \text{limit of } F(x)$$
$$\text{(as } x \rightarrow c \text{ from the left)} \qquad \text{(as } x \rightarrow c \text{ from the right)}.$$

Differentiating $F(x) = 1 - 1/x^2$ $(1 \leq x)$ yields

$$F'(x) = 0 - (-2x^{-3}) = 2/x^3 .$$

When $x < 1$, $F'(x) = d(0)/dx = 0$. The derivative does not exist at x = 1 because

$$\text{limit of } F'(x) = 0 \neq 2 = \frac{2}{1^3} = \text{limit of } F'(x)$$
$$\text{(as } x \rightarrow 1 \text{ from left)} \qquad \text{(}x \rightarrow 1 \text{ from right)}.$$

$$F'(x) = f(x) = \frac{2}{x^3} \text{ when } 1 \leq x < \infty = 0 \text{ when } x < 1$$

is the density function of **X**.

We can obtain $F(x)$ by integrating $f(x)$ from 1 to x when $x \geq 1$, and from $-\infty$ to x when $x < 1$. When $x \geq 1$,

$$F(x) = \int_1^x \frac{2}{t^3} dt = 2 \int_1^x t^{-3} dt = 2 \frac{t^{-2}}{-2} \Big]_1^x$$

$$= \frac{-1}{t^2} \Big]_1^x = -\frac{1}{x^2} - (-1) = -\frac{1}{x^2} + 1 = 1 - \frac{1}{x^2} = F(x).$$

When $x < 1$, $F(x) = \int_{-\infty}^x 0 \, dt = 0$.

● **PROBLEM** 9-38

The cumulative distribution functions of the   latitude angle $\theta(w)$ and the longitude angle $\phi(w)$ of the random orientation on the earth's surface are

$$F_\theta(\lambda) = \begin{cases} 0 & ; \ \lambda < 0 \\ 1-\cos\lambda & ; \ 0 \leq \lambda \leq \pi/2 \\ 1 & ; \ \lambda > \pi/2 \end{cases}$$

$$F_\phi(\lambda) = \begin{cases} 0 & ; \ \lambda < 0 \\ \lambda/2\pi & ; \ 0 \leq \lambda \leq 2\pi \\ 1 & ; \ \lambda > 2\pi \end{cases}$$

Find the corresponding density functions.

Solution: We know that $F(x) = \int_{-\infty}^{x} f(t)\, dt$. But by the Fundamental Theorem of Integral Calculus: $\dfrac{dF(x)}{dx} = f(x)$.

Hence $F_\theta(\lambda) = \dfrac{dF_\theta(\lambda)}{d\lambda}$

$$= \begin{cases} \dfrac{d}{d\lambda}(0) & ; \quad \lambda < 0 \\[2mm] \dfrac{d}{d\lambda}(1-\cos \lambda) & ; \quad 0 \le \lambda \le 2\pi \\[2mm] \dfrac{d}{d\lambda}(1) & ; \quad \lambda > 2\pi \end{cases}$$

$$= \begin{cases} 0 & ; \quad \lambda < 0 \\[2mm] \sin \lambda & ; \quad 0 \le \lambda \le 2\pi \\[2mm] 0 & ; \quad \lambda > 2\pi \;. \end{cases}$$

Also $F_\emptyset(\lambda) = \dfrac{dF(\lambda)}{d\lambda}$

$$= \begin{cases} \dfrac{d(0)}{d\lambda} & ; \quad \lambda < 0 \\[2mm] \dfrac{d}{d\lambda}\left(\dfrac{\lambda}{2\pi}\right) & ; \quad 0 \le \lambda \le 2\pi \\[2mm] \dfrac{d}{d\lambda}(1) & ; \quad \lambda > 2\pi \end{cases}$$

$$= \begin{cases} 0 & ; \quad \lambda < 0 \\[2mm] \dfrac{1}{2\pi} & ; \quad 0 \le \lambda \le 2\pi \\[2mm] 0 & ; \quad \lambda > 2\pi \end{cases}$$

● **PROBLEM 9-39**

X is a continuous random variable with probability density function

$$f(x) = \begin{cases} 1/k & 0 \le x \le k \\ 0 & \text{otherwise}. \end{cases}$$

Show that this is a proper density function and find $\Pr(a \le X \le b)$ for $0 \le a < b \le k$.

Solution: To show that $f(x)$ is a proper probability density function we must show that $f(x) \ge 0$ for all x and that $\int_{-\infty}^{\infty} f(x)dx = 1$.

From the way $f(x)$ is defined, for any value of $k > 0$, $f(x) = 1/k \ge 0$.

Furthermore, $\displaystyle\int_{-\infty}^{\infty} f(x)\, dx = \int_{0}^{k} f(x)\, dx$

because $f(x) = 0$ for $x \ge k$ or $x \le 0$.

$\displaystyle\int_{0}^{k} f(x)\, dx = \int_{0}^{k} 1/k\, dx = 1/k \cdot x\Big]_{0}^{k} = 1/k[k - 0] = 1$.

$\Pr(a \le X \le b)$ for $0 \le a < b \le k$ is defined to be $\displaystyle\int_{a}^{b} f(x)\, dx$.

Thus $\Pr(a \le X \le b) = \int_a^b f(x)\ dx$

$$= \int_a^b 1/k\ dx = 1/k\ x\ \big]_a^b$$

$$= \frac{b - a}{k}\ .$$

A graph of $f(x)$ is shown below:

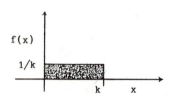

Suppose the length of time an electric bulb lasts, X, is a random variable with cumulative distribution

$$F(x) = \Pr(X \le x) = \begin{cases} 0 & x < 0 \\ 1 - e^{-x/500} & x \ge 0\ . \end{cases}$$

Find the probability that the bulb lasts (a) between 100 and 200 hours (b) beyond 300 hours.

Solution: (a) $\Pr(100 \le X \le 200) = 1 - [\Pr(X \ge 200) + \Pr(X \le 100)]$

$$= 1 - \Pr(X \ge 200) - \Pr(X \le 100)\ ;$$

but $\qquad 1 - \Pr(X \ge 200) = \Pr(X \le 200)$

thus, $\qquad \Pr(100 \le X \le 200)\ = \Pr(X \le 200) - \Pr(X \le 100)$

$$= F(200) - F(100)$$

$$= 1 - e^{-200/500} - (1 - e^{-100/500})$$

$$= e^{-1/5} - e^{-2/5}$$

$$= .1484\ .$$

Similarly, $\qquad \Pr(X \ge 300) = \Pr(\text{bulb lasts longer than 300 hours})$

$$= 1 - \Pr(X < 300)$$

$$= 1 - F(300)$$

$$= 1 - (1 - e^{-300/500})$$

$$= e^{-3/5} = .5488\ .$$

THE BINOMIAL AND JOINT DISTRIBUTIONS

Find the probabilities that X, the number of "successes" of a binomial experiment with 4 independent trials and π = probability of "success" = $\frac{1}{3}$, equals 0, 1, 2, 3, or 4. Sketch a histogram for the distribution.

Solution: A binomial experiment is an experiment made up of a certain number of independent trials. Independence of the trials implies that the results of the later trials are not influenced in any way by the results of the earlier trials. Each of these trials has two possible outcomes, denoted "success" and "failure." The probability of "success" in any particular trial is denoted by the Greek letter π.

We are often interested in the probability of a certain number of "successes" throughout the course of a particular experiment. The number of successes is denoted by the capital letter 'X'. X can take on the values from 0 to n, where n is the total number of trials performed in a binomial experiment.

What is the probability distribution of X? That is, what is Pr(X=0), Pr(X=1), ..., Pr(X=n)? Consider first Pr(X=0), the probability that there are no "successes" in n trials and the probability of success equals π. Each trial can be represented by the following tree diagram,

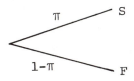

Here S represents the outcome "success" and will occur with probability π, F represents the outcome failure and will occur with probability $1 - \pi$. The only way that X will be equal to zero is if there is a failure on each of the n trials. Since the trials are independent, Pr (failure on each trial)= Pr (failure on the first trial) × Pr (failure on the second) × Pr (failure on the third) ... × Pr (failure on the nth)

$$= \underbrace{(1 - \pi) \cdot (1 - \pi) \cdot \ldots \cdot (1 - \pi)}_{\text{n terms}} = (1 - \pi)^n$$

The probability of exactly 1 "success" in n trials Pr (X=1), is found in a similar way.

A "success" in the first trial and "failures" there-after will occur with probability

$$\underbrace{(1 - \pi) \cdot \ldots \cdot (1 - \pi)}_{n - 1 \text{ terms}} = \pi (1 - \pi)^{n-1}.$$

Another possibility for X = 1 will be if there is a "fail-ure" on the first trial, a "success" on the second and "failures" thereafter. This will occur with probability

$$\underbrace{(1 - \pi) \cdot \pi \cdot (1 - \pi) \cdot \ldots \cdot (1 - \pi)}_{n - 2 \text{ terms}} = \pi (1 - \pi)^{n-1}.$$

There are n possible trials on which the single "success" might take place. Each of these events is mutual-ly exclusive. Therefore Pr (X = 1) = Pr (the single success is on the first trial) + Pr (success is on second) + ... +

Pr (success is on the nth trial) $= \pi (1 - \pi)^{n-1} +$

$\pi (1 - \pi)^{n-1} + \ldots + \pi (1 - \pi)^{n-1} = n \pi (1 - \pi)^{n-1}.$

In order to find the probability that x = 2 select the two trials on which successes will take place. This is "the number of ways 2 objects can be chosen from n" or $\binom{n}{2}$.

The probability of two "successes" and n - 2 "failures" in n independent trials is thus

$$\binom{n}{2} \pi \cdot \pi \cdot \underbrace{(1 - \pi) \cdot \ldots \cdot (1 - \pi)}_{n - 2 \quad \text{terms}}$$

$$\Pr (X = 2) = \binom{n}{2} \pi^2 (1 - \pi)^{n-2}, \text{ in general}$$

$$\Pr (X = j) = \binom{n}{j} \pi^j (1 - \pi)^{n-j} \quad j = 0, 1, 2, \ldots n.$$

In this problem n = 4 and $\pi = \frac{1}{3}$ and

$$\Pr (X=0) = \binom{4}{0} \left(\frac{1}{3}\right)^0 \left(\frac{2}{3}\right)^4 = 1 \cdot 1 \cdot \frac{16}{81} = \frac{16}{81}$$

$$\Pr (X=1) = \binom{4}{1} \left(\frac{1}{3}\right)^1 \left(\frac{2}{3}\right)^3 = 4 \cdot \frac{1}{3} \cdot \frac{8}{27} = \frac{32}{81}$$

$$\Pr (X=2) = \binom{4}{2} \left(\frac{1}{3}\right)^2 \left(\frac{2}{3}\right)^2 = 6 \cdot \frac{1}{9} \cdot \frac{4}{9} = \frac{24}{81}$$

$$\Pr (X=3) = \binom{4}{3} \left(\frac{1}{3}\right)^3 \left(\frac{2}{3}\right)^1 = 4 \cdot \frac{1}{27} \cdot \frac{2}{3} = \frac{8}{81}$$

$$\Pr \ (X=4) \ = \ \binom{4}{4}\left(\frac{1}{3}\right)^4\left(\frac{2}{3}\right)^0 \ = \ 1 \cdot \frac{1}{81} \cdot 1 \ = \ \frac{1}{81} \ .$$

Note that $\sum\limits_{j=0}^{4} \Pr \ (X = j) \ = \ \frac{16}{81} + \frac{32}{81} + \frac{24}{81} + \frac{8}{81} + \frac{1}{81}$

$$= \ \frac{81}{81} = 1 \quad .$$

as it must in order for this to be a proper probability distribution.

Histogram

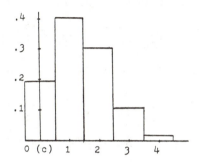

● **PROBLEM** 9-42

Calculate the binomial probability distribution for n = 10 and $p = \frac{1}{8}$. Compare this distribution with the distribution for n = 10 and $p = \frac{1}{2}$. Do they differ in degree of dispersion? Position of peak? Maximum height of peak? Degree of symmetry?

Solution: The binomial distribution formula is $P(X = k) = \binom{n}{k} p^k (1 - p)^{n-k}$ where n is the number of trials, k the number of "successes", and p the probability of a "success."

For n = 10 $p = \frac{1}{8}$,

$$P(X = k) \ = \ \binom{10}{k}\left(\frac{1}{8}\right)^k\left(1 - \frac{1}{8}\right)^{10-k} \ = \ \binom{10}{k}\left(\frac{1}{8}\right)^k\left(\frac{7}{8}\right)^{10-k}$$

$$P(X = 0) \ = \ \binom{10}{0}\left(\frac{1}{8}\right)^0\left(\frac{7}{8}\right)^{10} \ = \ \frac{10!}{10!0!} \times \left(\frac{7}{8}\right)^{10} \ = \ \left(\frac{7}{8}\right)^{10} \ ,$$

since by definition 0! and any quantity to the zero power is one.

507

$$P(X = 1) = \binom{10}{1}\left(\frac{1}{8}\right)^1\left(\frac{7}{8}\right)^9 = \frac{10!}{9!1!}\frac{1}{8}\left(\frac{7}{8}\right)^9 = 10\left(\frac{1}{8}\right)\left(\frac{7}{8}\right)^9$$

$$= 10\ (.125)(.3007) = .376$$

$$P(X = 2) = \binom{10}{2}\left(\frac{1}{8}\right)^2\left(\frac{7}{8}\right)^8 = \frac{10!}{8!2!}\left(\frac{1}{8}\right)^2\left(\frac{7}{8}\right)^8$$

$$= 90\ \left(\frac{1}{8}\right)^2\left(\frac{7}{8}\right)^8 = 45\ (.016)(.3436) = .247$$

$$P(X = 3) = \binom{10}{3}\left(\frac{1}{8}\right)^3\left(\frac{7}{8}\right)^7 = \frac{10!}{7!3!}\left(\frac{1}{8}\right)^3\left(\frac{7}{8}\right)^7$$

$$= 120\ (.002)(.3927) = .094$$

$$P(X = 4) = \binom{10}{4}\left(\frac{1}{8}\right)^4\left(\frac{7}{8}\right)^6 = \frac{10!}{6!4!}\left(\frac{1}{8}\right)^4\left(\frac{7}{8}\right)^6$$

$$= 210\ (.00024)(.4488) = .023$$

$$P(X = 5) = \binom{10}{5}\left(\frac{1}{8}\right)^5\left(\frac{7}{8}\right)^5 = \frac{10!}{5!5!}\left(\frac{1}{8}\right)^5\left(\frac{7}{8}\right)^5$$

$$= 252\ (.000016) = .004$$

$$P(X = 6) = \binom{10}{6}\left(\frac{1}{8}\right)^6\left(\frac{7}{8}\right)^4 = \frac{10!}{6!4!}\left(\frac{1}{8}\right)^6\left(\frac{7}{8}\right)^4$$

$$= 210\ (.000004)(.5862) = .0005$$

$$P(X = 7) = \binom{10}{7}\left(\frac{1}{8}\right)^7\left(\frac{7}{8}\right)^3 = \frac{10!}{7!3!}\left(\frac{1}{8}\right)^7\left(\frac{7}{8}\right)^3 = .00004$$

$$P(X = 8) = \binom{10}{8}\left(\frac{1}{8}\right)^8\left(\frac{7}{8}\right)^2 = \frac{10!}{8!2!}\left(\frac{1}{8}\right)^8\left(\frac{7}{8}\right)^2$$

$$= .000002$$

$$P(X = 9) = \binom{10}{9}\left(\frac{1}{8}\right)^9\left(\frac{7}{8}\right)^1 = \frac{10!}{9!1!}\left(\frac{1}{8}\right)^9\left(\frac{7}{8}\right)^1$$

$$= .00000006$$

$$P(X = 10) = \binom{10}{10}\left(\frac{1}{8}\right)^{10}\left(\frac{7}{8}\right)^0 = \frac{10!}{10!0!}\left(\frac{1}{8}\right)^{10}\left(\frac{7}{8}\right)^0$$

$$= .000000001\ .$$

The probabilities total 1.007542061. The discrepancy from 1 is due to approximations and rounding off.

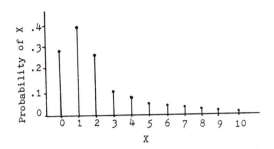

This graph shows the previous distribution. Now consider the binomial distribution for $n = 10$ $p = \frac{7}{8}$.

$$P\left(X_{\frac{7}{8}} = r\right) = \binom{10}{r}\left(\frac{7}{8}\right)^{r}\left(\frac{1}{8}\right)^{10-r}.$$

Recall that $\binom{n}{r} = \frac{n!}{(n-r)!\,r!} = \frac{n!}{r!\,(n-r)!} = \binom{n}{n-r}$ so that

$$P\left(X_{\frac{7}{8}} = r\right) = \binom{10}{r}\left(\frac{7}{8}\right)^{r}\left(\frac{1}{8}\right)^{10-r} = \binom{10}{10-r}\left(\frac{1}{8}\right)^{10-r}\left(\frac{7}{8}\right)^{r} =$$

$P\left(X_{\frac{1}{8}} = 10 - r\right).$ Therefore

$P_{\frac{1}{8}}\ (X = 0) = P_{\frac{7}{8}}\ (X = 10)$ $\qquad$ $P_{\frac{1}{8}}\ (X = 5) = P_{\frac{7}{8}}\ (X = 5)$

$P_{\frac{1}{8}}\ (X = 1) = P_{\frac{7}{8}}\ (X = 9)$ $\qquad$ $P_{\frac{1}{8}}\ (X = 6) = P_{\frac{7}{8}}\ (X = 4)$

$P_{\frac{1}{8}}\ (X = 2) = P_{\frac{7}{8}}\ (X = 8)$ $\qquad$ $P_{\frac{1}{8}}\ (X = 7) = P_{\frac{7}{8}}\ (X = 3)$

$P_{\frac{1}{8}}\ (X = 3) = P_{\frac{7}{8}}\ (X = 7)$ $\qquad$ $P_{\frac{1}{8}}\ (X = 8) = P_{\frac{7}{8}}\ (X = 2)$

$P_{\frac{1}{8}}\ (X = 4) = P_{\frac{7}{8}}\ (X = 6)$ $\qquad$ $P_{\frac{1}{8}}\ (X = 9) = P_{\frac{7}{8}}\ (X = 1)$

$\qquad\qquad\qquad\qquad\qquad$ $P_{\frac{1}{8}}\ (X = 10) = P_{\frac{7}{8}}\ (X = 0).$

Hence the probability density function for $n = 10$ and
$p = \frac{7}{8}$ is the mirror image of $n = 10$ and $p = \frac{1}{8}$, as shown
below:

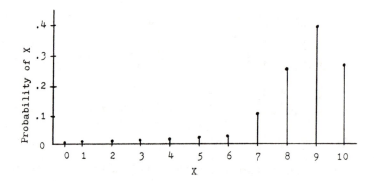

Let us proceed to calculate the binomial density function for $n = 10$, $p = \frac{1}{2}$.

$$P(X=k) = \binom{n}{k} p^k (1 - p)^{n-k} = \binom{10}{k} \left(\frac{1}{2}\right)^k \left(1 - \frac{1}{2}\right)^{10-k}$$

$$= \binom{10}{k} \left(\frac{1}{2}\right)^k \left(\frac{1}{2}\right)^{10-k} = \binom{10}{k} \left(\frac{1}{2}\right)^{10} = \binom{10}{k} \frac{1}{1024} .$$

A few representative calculations;

$$P(X = 0) = \binom{10}{0} \frac{1}{1024} = \binom{10}{10} \frac{1}{1024} = P(X = 10) = .0010$$

$$P(X = 1) = \binom{10}{1} \frac{1}{1024} = \frac{10!}{9!1!} \frac{1}{1024} = \binom{10}{9} \frac{1}{1024}$$

$$= P(X = 9) = .0098$$

$$P(X = 2) = \binom{10}{2} \frac{1}{1024} = \frac{10!}{8!2! \cdot 1024} = \binom{10}{8} \frac{1}{1024}$$

$$= P(X = 8) = .0439$$

$$P(X = 3) = \binom{10}{3} \frac{1}{1024} = \frac{10!}{7!3! \cdot 1024} = \binom{10}{7} \frac{1}{1024}$$

$$= P(X = 7) = .1172$$

$$P(X = 4) = \binom{10}{4} \frac{1}{1024} = \frac{10!}{6!4! \cdot 1024} = \binom{10}{6} \frac{1}{1024}$$

$$= P(X = 6) = .2051$$

$$P(X = 5) = \binom{10}{5} \frac{1}{1024} = \frac{10!}{5!5! 1024} = .2460.$$

The graph follows:

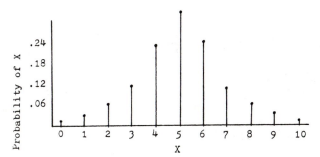

One may observe that distributions for $p = \frac{1}{8}$ and $p = \frac{7}{8}$ are mirror images. They are skewed with peaks at $X = 1$ and $X = 9$ respectively. The peaks are relatively high (.376). The functions are concentrated near the peaks and are asymmetric.

On the other hand, the distribution for $p = \frac{1}{2}$ is completely symmetric. The peak is lower (.2460) and the function is more widely dispersed.

● **PROBLEM 9-43**

Give the expansion of $\left(r^2 - \frac{1}{s} \right)^5$.

Solution: Write the given expression as the sum of two terms raised to the 5th power:

$$\left(r^2 - \frac{1}{s} \right)^5 = \left[r^2 + \left(-\frac{1}{s} \right) \right]^5 . \tag{1}$$

The Binomial Theorem can be used to expand the expression on the right side of equation (1). The Binomial Theorem is stated as:

$$(a+b)^n = a^n + na^{n-1}b + \frac{n(n-1)}{1 \cdot 2}a^{n-2}b^2 + \frac{n(n-1)(n-2)}{1 \cdot 2 \cdot 3}a^{n-3}b^3$$

$$+ \ldots + nab^{n-1} + b^n, \text{ where a and b are any two}$$

numbers.

In order to apply the Binomial Theorem let $a = r^2$, $b = -\frac{1}{s}$, and $n = 5$.

$$\left(r^2 - \frac{1}{s}\right)^5 = \left[r^2 + \left(-\frac{1}{s}\right)\right]^5$$

$$= \left(r^2\right)^5 + 5\left(r^2\right)^{5-1}\left(-\frac{1}{s}\right) + \frac{5(5-1)}{1 \cdot 2}\left(r^2\right)^{5-2}\left(-\frac{1}{s}\right)^2$$

$$+ \frac{5(5-1)(5-2)}{1 \cdot 2 \cdot 3}\left(r^2\right)^{5-3}\left(-\frac{1}{s}\right)^3$$

$$+ \frac{5(5-1)(5-2)(5-3)}{1 \cdot 2 \cdot 3 \cdot 4}\left(r^2\right)^{5-4}\left(-\frac{1}{s}\right)^4$$

$$+ \frac{5(5-1)(5-2)(5-3)(5-4)}{1 \cdot 2 \cdot 3 \cdot 4 \cdot 5}\left(r^2\right)^{5-5}\left(-\frac{1}{s}\right)^5$$

$$= r^{10} - \frac{5\left(r^2\right)^4}{s} + \frac{5(\cancel{4})}{1 \cdot \cancel{2}}\left(r^2\right)^3\left(\frac{1}{s^2}\right) - \frac{5(\cancel{4})(\cancel{3})}{1 \cdot \cancel{2} \cdot \cancel{3}}\left(r^2\right)^2\left(\frac{1}{s^3}\right)$$

$$+ \frac{5(\cancel{4})(\cancel{3})(\cancel{2})}{1 \cdot \cancel{2} \cdot \cancel{3} \cdot \cancel{4}}\left(r^2\right)^1\left(\frac{1}{s^4}\right) - \frac{\cancel{5}(\cancel{4})(\cancel{3})(\cancel{2})(\cancel{1})}{\cancel{1} \cdot \cancel{2} \cdot \cancel{3} \cdot \cancel{4} \cdot \cancel{5}}\left(r^2\right)^0\left(\frac{1}{s^5}\right)$$

$$= r^{10} - \frac{5r^8}{s} + \frac{10r^6}{s^2} - \frac{10r^4}{s^3} + \frac{5r^2}{s^4} - (1)(1)\left(\frac{1}{s^5}\right)$$

$$\left(r^2 - \frac{1}{s}\right)^5 = r^{10} - \frac{5r^8}{s} + \frac{10r^6}{s^2} - \frac{10r^4}{s^3} + \frac{5r^2}{s^4} - \frac{1}{s^5}$$

● **PROBLEM 9-44**

What is the probability of getting exactly 3 heads in 5 flips of a balanced coin?

Solution: The situation here is often refered to as a Bernoulli trial. There are two possible outcomes, head or tail, each with a finite probability. Each flip is independent. This is the type of situation to which the binomial distribution,

$$P(X = k) = \binom{n}{k} p^k (1 - p)^{n-k},$$

applies. The a priori probability of tossing a head is $p = \frac{1}{2}$. the probability of a tail is $q = 1 - p = 1 - \frac{1}{2} = \frac{1}{2}$. Also $n = 5$ and $k = 3$ (number of heads required).

Applying the above formula one obtains:

$$P(X = 3) = \binom{5}{3} \left(\frac{1}{2}\right)^3 \left(1 - \frac{1}{2}\right)^{\underset{2}{2}} = \frac{5!}{3!2!} \left(\frac{1}{2}\right)^3 \left(\frac{1}{2}\right)^2$$

$$= \frac{5 \cdot \cancel{4} \cdot \cancel{3} \cdot \cancel{2} \cdot \cancel{1}}{\cancel{3} \cdot \cancel{2} \cdot \cancel{1} \cdot \cancel{2} \cdot \cancel{1}} \left(\frac{1}{2}\right)^5 = \frac{10}{2^5} = \frac{10}{32} = \frac{5}{16} .$$

● PROBLEM 9-45

On three successive flips of a fair coin, what is the probability of observing 3 heads? 3 tails?

Solution: The three successive flips of the coin are three independent events. Since the coin is fair, the probability of throwing a head on any particular toss is $p = \frac{1}{2}$.

Let X = the number of heads observed in three tosses of the coin. We wish to find $Pr(X = 3)$. By our assumptions, X is binomially distributed with parameters n = 3 and p = $\frac{1}{2}$.

Thus, $Pr(X = 3) = \binom{3}{3} \left(\frac{1}{2}\right)^3 \left(\frac{1}{2}\right)^0 = 1 \left(\frac{1}{2}\right)^3 \left(\frac{1}{2}\right)^0 = \frac{1}{8} .$

Similarly, let T = the number of tails observed in three successive flips of a fair coin. T is distributed binomially with parameters n = 3 and p = $\frac{1}{2}$ = the probability that a tail is observed on a particular toss of the coin.

Thus $Pr(T = 3) = \binom{3}{3} \left(\frac{1}{2}\right)^3 \left(\frac{1}{2}\right)^0 = 1 \cdot \frac{1}{8} \cdot 1 = \frac{1}{8} .$

● PROBLEM 9-46

In a family of 4 children, what is the probability that there will be exactly two boys?

Solution: The case of the sex of a born child   may be classically described by the binomial distribution. There are 2 possible outcomes, boy or girl. The probability of giving birth to a boy is $p = \frac{1}{2}$. The probability of having a girl is $q = 1 - p = 1 - \frac{1}{2} = \frac{1}{2}$. Also n = 4 (number of children) and k = 2 (number of boys). Furthermore,

$$P(X = k) = \binom{n}{k} p^k (1 - p)^{n-k}$$

$$P(X = 2) = \binom{4}{2} \left(\frac{1}{2}\right)^2 \left(\frac{1}{2}\right)^{4-2} = \frac{4!}{2!2!} \left(\frac{1}{2}\right)^2 \left(\frac{1}{2}\right)^2$$

$$= \frac{4 \cdot 3 \cdot 2 \cdot 1}{2 \cdot 1 \cdot 2 \cdot 1} \left(\frac{1}{2}\right)^4 = \frac{6}{2^4}$$

$$= \frac{3 \cdot 2}{2^3 \cdot 2} = \frac{3}{2^3} = \frac{3}{8} .$$

● **PROBLEM 9-47**

The probability that a basketball player makes at least one of six free throws is equal to 0.999936. Find: (a) the probability function of X, the number of times he scores; (b) the probability that he makes at least three baskets.

Solution: This problem involves the binomial distribution for three reasons: (1) there six indipendent trials (the outcome of each throw is indipendent of the others); (2) each throw has only 2 possible outcomes - score or no score; (3) since the player is shooting " free throws" (from a standard position) one can assume that the probability of a score remains the same from throw to throw.

In order to determine the probability function of the number of scores, one needs to know the probability that any free throw will score. The event that no free throw scores is the complement of the event that one or more throws score. By the binomial distribution,

$$P(\underline{no} \text{ scores}) = \binom{6}{0} p^0 (1 - p)^{6-0} = 1 \cdot 1 \cdot (1 - p)^6$$

$$= (1 - p)^6 ,$$

since $\binom{6}{0} = \frac{6!}{(0!)(6-0)!} = \frac{6!}{1 \cdot 6!} = 1$ and

$p^0 = 1.$

P(at least one score) + P(no scores) = 1 since the events are complementary. Substituting $(1 - p)^6$ for P(no scores), the equation becomes

P(at least one score) + $(1 - p)^6$ = 1. Using the given information, this becomes p (at least one score) = .999936 = $1 - (1-p)^6$

$$= (1 - p)^6 = 0.000064, \text{ so}$$

$1 - p = \sqrt[6]{0.000064} = .2$ and so p = .8

Adding p to both sides and subtracting .2 from both sides, one has $1 - .2 = .8 = p$.

Again using the binomial distribution, where r is the number of scores in 6 throws,

$$P(r) = \binom{6}{r} (.8)^r (.2)^{6-r}.$$

Substitution of any integer from 1 to 6 inclusive for r will yield the probability of that number of scores in 6 free throws.

Part (b) asks for the probability of at least three scores. This includes 4 possibilities: exactly 3 scores; exactly 4 scores; exactly 5 scores; or exactly 6 scores in 6 throws. Since only one of these events can occur, they are mutually exclusive. It follows that their probabilities can be added to give the probability that any one event will occur. Using summation notation,

P (3 or more scores in 6 throws) =

$$= \sum_{r=3}^{6} \binom{6}{r} (.8)^r (.2)^{6-r}.$$

By the table, P (3 scores) = .082, P(4) = .246, P(5) = .393, and P(6) = .262. Then

P(3 or more) = .082 + .246 + .393 + .262 = .983.

Many binomial tables only give values for $p \leq .5$. Since in this problem $p = .8$, one would have to convert the binomial distribution to an equivalent form in the following way:

Let $\Pr(X=k) = \binom{n}{k} p^k (1 - p)^{n-k}$ with $p \geq .5$. If $p \geq .5$ then $1 - p \leq .5$.

Find a probability for a new random variable Y with probability of success, $1 - p \leq .5$.

$$\Pr(X=k) = \binom{n}{k} p^k (1 - p)^{n-k}$$

$$= \binom{n}{n-k} p^k (1 - p)^{n-k} = \Pr(Y = n-k).$$

Remember that $\binom{n}{k} = \dfrac{n!}{(n-k)!k!} = \binom{n}{n-k}$.

One can thus use the table to find $\Pr(Y=n-k)$ where the probability of success is $1 - p$ and one can thus find the $\Pr(X=k)$ when the probability of success is p.

Over a period of some years, a car manufacturing firm finds that 18% of their cars develop body squeaks within the guarantee period. In a randomly selected shipment, 20 cars reach the end of the guarantee period and none develop squeaks. What is the probability of this?

Solution: The car can either squeak or not squeak. The probability of a car squeaking is 18% or .18. The probability of not squeaking is $q = 1 - p = 1 - .18 = .82$. The situation here is a dichotomy of the type that fits the binomial distrobution. There are indipendent trials with two possible outcomes each with a finite, constant probability.

Using the binomial distrobution we obtain:

$$P(X=k) = \binom{n}{k} p^k (1 - p)^{n-k}.$$

Here, $p = .18$, $1 - p = .82$, $n = 20$, and $k = 0$. Hence,

$$P(X = 0) = \binom{20}{0} (.18)^0 (.82)^{20} = \frac{20!}{20!0!} (.18)^0 (.82)^{20}$$

$$= (.82)^{20} = .019.$$ (since any quantity to the zero power and 0! are both defined to be one.)

A proportion p of a large number of items in a batch is defective. A sample of n items is drawn and if it contains no defective items the batch is accepted while if it contains more than two defective items the batch is rejected. If, on the other hand, it contains one or two defectives, an independent sample of m is drawn, and if the combined number of defectives in the samples does not exceed two, the batch is accepted. Calculate the possibility of accepting this batch.

Solution: The batch will be accepted only if

(1) the first sample contains no defectives.

(2) The first sample contains 1 defective and the second sample contains 0 or 1 defective.

(3) The first sample contains 2 defectives and the second sample contains 0 defectives.

These three probabilities are mutually exclusive. If one occurs then none of the others can occur. Thus if we compute the probability of each of these events, the sum of the three will be the probability of acceptance.

Let: X = the number of defectives in the first sample.

Y = the number of defectives in the second sample.

If the sampling is done with replacement, X will be binomially distributed with the parameter n equal to the number of trials (or size of sample) and p = probability of selecting a defective on 1 trial. Similarly, Y is binomially distributed with parameters m and p.

Again by the addition law, Pr(acceptance) = Pr(0 defectives in first batch) + Pr(1 in first and 0 or 1 in second) + Pr(2 in first and 0 in second) = $Pr(X=0)$ + $Pr(X=1, Y=0 \text{ or } 1)$ + $Pr(X = 2, Y=0)$.

$$Pr(X=0) = \binom{n}{0} p^0 (1 - p)^{n-0}.$$

$$Pr(X=1, Y=0 \text{ or } 1) = Pr(X=1, Y=0) + Pr(X=1, Y=1)$$

by the addition rule, since we are dealing with mutually exclusive events. $Pr(X=1, Y=0) = Pr(X=1) \cdot Pr(Y=0)$ by the multiplication law. Hence

$$Pr(X=1, Y=0) = \binom{n}{1} p^1 (1 - p)^{n-1} \cdot \binom{m}{0} p^0 (1 - p)^m.$$

Similarly, $Pr(X=1, Y=1) = Pr(X=1) \cdot Pr(Y=1)$

$$= \binom{n}{1} p^1 (1 - p)^{n-1} \cdot \binom{m}{1} p^1 (1 - p)^{m-1},$$

x	π= 01	02	03	04	05	06	07	08	09	10	
1	0394	0776	1147	1507	1855	2193	2519	2836	3143	3439	3
2	0006	0023	0052	0091	0140	0199	0267	0344	0430	0523	2
3			0001	0002	0005	0008	0013	0019	0027	0037	1
4									0001	0001	0
x	99	98	97	96	95	94	93	92	91	90 =π	x

n = 4

x	π= 11	12	13	14	15	16	17	18	19	20	
1	3726	4003	4271	4530	4780	5021	5254	5479	5695	5904	3
2	0624	0732	0847	0968	1095	1228	1366	1509	1656	1808	2
3	0049	0063	0079	0098	0120	0144	0171	0202	0235	0272	1
4	0001	0002	0003	0004	0005	0007	0008	0010	0013	0016	0
x	89	88	87	86	85	84	83	82	81	80 =π	x

n = 4

x	π= 21	22	23	24	25	26	27	28	29	30	
1	6105	6298	6485	6664	6836	7001	7160	7313	7459	7599	3
2	1963	2122	2285	2450	2617	2787	2959	3132	3307	3483	2
3	0312	0356	0403	0453	0508	0566	0628	0694	0763	0837	1
4	0019	0023	0028	0033	0039	0046	0053	0061	0071	0081	0
x	79	78	77	76	75	74	73	72	71	70 =	x

n = 4

x	π= 31	32	33	34	35	36	37	38	39	40	
1	7733	7862	7985	8103	8215	8322	8425	8522	8615	8704	3
2	3660	3837	4015	4193	4370	4547	4724	4900	5075	5248	2
3	0915	0996	1082	1171	1265	1362	1464	1596	1679	1792	1
4	0092	0105	0119	0134	0150	0168	0187	0209	0231	0256	0
x	69	68	67	66	65	64	63	62	61	60 =π	x

n = 4

x	π= 41	42	43	44	45	46	47	48	49	50	
1	8788	8868	8944	9017	9085	9150	9211	9269	9323	9375	3
2	5420	5590	5759	5926	6090	6252	6412	6569	6724	6875	2
3	1909	2030	2155	2283	2415	2550	2689	2831	2977	3125	1
4	0283	0311	0342	0375	0410	0448	0488	0531	0576	0625	0
x	59	58	57	56	55	54	53	52	51	50 =π	x

For similar reasons, $\Pr(X=2, Y=0) = \Pr(X=2) \cdot \Pr(Y=0) = \binom{n}{2} p^2 (1-p)^{n-2} \binom{m}{0} p^0 (1-p)^m$.

Hence $\Pr(\text{acceptance}) =$

$$= \binom{n}{0} p^0 (1-p)^{n-0} + \binom{n}{1} p^1 (1-p)^{n-1} \binom{m}{0} p^0 (1-p)^m + \binom{n}{1} p^1 (1-p)^{n-1}$$

$$\binom{m}{1} p^1 (1-p)^{m-1} + \binom{n}{2} p^2 (1-p)^{n-2} \binom{m}{0} p^0 (1-p)^m$$

518

$$=(1-p)^n + np(1-p)^{m+n-1} + nmp^2(1-p)^{m+n-2}$$
$$+\frac{1}{2} n(n=1) \ p^2(1-p)^{m=n-2}.$$

Letting $1-p=q$, we can write this more concisely as

$$pr(\text{acceptance}) = q^n \ 1 = npq^{m-1} + mnp^2 q^{m-2}$$
$$+ \frac{1}{2} n(n-1) \ p^2 q^{m-2} \ .$$

Let X be a binomially distributed random variable with parameters n and π. where n = the number of independent trials and π is the probability of success on a particular trial.

Use the table above to find $Pr(X \geq 2)$ and $Pr(X=2)$ if n = 4 and π = .23.

Solution: $Pr(X<2) = Pr(X = 2 \text{ or } 3 \text{ or } 4)$. In order to find this possibility, resort to the table of cumulative binomial probabilities.

First find π =.23 in the body of this table. Then read down the left side of the table until x = 2. The number in the row of x = 2 and π = 23 is the $Pr(X>2)$. Observe that $Pr(X>2) = .2285$.

To find the $Pr(X=2)$ from the cumulative binomial table one must first express an exact probability in terms of a cumulative probability.

$$Pr(X=2) = Pr(X=2) + Pr(X=3) + Pr(X=4) - [Pr(X=3 + Pr(X=4)]$$
$$= Pr(X \geq 2) - Pr(X \geq 3).$$

Now, find the two cumulative probabilities, $Pr(X>2)$ and $Pr(X>3)$ from the table. Reading down the column headed by π = 23 and from the left across the row labeled x = 3 $Pr(X>3)$ is found to be 0.0403.

Thus, $Pr(X=2) = Pr(X \geq 2) - Pr(X \geq 3)$
$$= .2285 - .0403 = .1882.$$

The probability of hitting a target on a shot is $\frac{2}{3}$. If a person fires 8 shots at a target, Let X denote the number of times he hits the target, and find:

(a) $P(X = 3)$ (b) $P(1 < X \leq 6)$ (c) $P(X > 3)$.

Solution: If one assumes that each shot is independent of any other shot then X is a binomial distributed random variable with parameters $n = 8$ and $\pi = \frac{2}{3}$. (π equals the probability of hitting the target on any particular shot and n = the number of shots.)

Thus, $Pr(X=3) = \binom{8}{3}\left(\frac{2}{3}\right)^3\left(\frac{1}{8}\right)^{8-3} = \frac{8!}{3!5!}\left(\frac{2}{3}\right)^3\left(\frac{1}{3}\right)^5$

$= \frac{8 \cdot 7 \cdot 6}{3 \cdot 2 \cdot 1}\left(\frac{8}{27}\right)\left(\frac{1}{243}\right) = \frac{448}{6561} = .06828$

$Pr(1 < X \leq 6) = Pr(X = 2, 3, 4, 5 \text{ or } 6)$.

Each of these events is mutually exclusive and thus,

$Pr(X = 2, 3, 4, 5, \text{ or } 6) = Pr(X=2) + Pr(X=3) + Pr(X=4)$

$+ Pr(X=5) + Pr(X=6)$

$= \sum_{n=2}^{6} Pr(X=n)$.

$Pr(1 < X \leq 6) = \sum_{n=2}^{6} \binom{8}{n}\left(\frac{2}{3}\right)^n\left(\frac{1}{3}\right)^{8-n}$.

One may use the tables of cumulative probabilities and the fact that $Pr(1 < X \leq 6) = Pr(X \leq 6) - Pr(X \leq 1)$, or calculate single probabilities and add, in order to observe that $Pr(1 < X \leq 6) = .8023$.

$Pr(X > 3) = Pr(X = 4, 5, 6, 7 \text{ or } 8)$

$= Pr(X=4) + Pr(X=5) + Pr(X=6) + Pr(X=7)$

$+ Pr(X=8)$

$= \sum_{n=4}^{8} Pr(X = n)$

$$= \sum_{n=4}^{8} \binom{8}{n} \left(\frac{2}{3}\right)^n \left(\frac{1}{3}\right)^{8-n}.$$

Again, using a table of cumulative probabilities or calculating each single probability, one obtains $\Pr(X > 3) = .912$.

If a bag contains three white, two black, and four red balls and four balls are drawn at random with replacement, calculate the probabilities that

(a) The sample contains just one white ball.
(b) The sample contains just one white ball given that it contains just one red ball.

Solution: Since there are nine balls and one is sampling with replacement and choosing the balls at random, on each draw

$$\Pr(\text{white ball}) = \frac{3}{9} = \frac{1}{3}.$$

$$\Pr(\text{black ball}) = \frac{2}{9}.$$

$$\Pr(\text{red ball}) = \frac{4}{9}.$$

(a) On each draw, $\Pr(\text{white}) + \Pr(\text{black or red}) = 1$. Let X = number of white balls. Then X is distributed binomially with $n = 4$ trials and $\Pr(\text{white ball}) = \frac{1}{3}$. Thus one obtains

$$\Pr(\text{just one white}) = \Pr(X=1) = \binom{4}{1}\left(\frac{1}{3}\right)^1 \left(1 - \frac{1}{3}\right)^{4-1}$$

$$= 4 \left(\frac{1}{3}\right)\left(\frac{2}{3}\right)^3 = \frac{32}{81}.$$

(b) $\Pr(\text{just 1 white} \mid \text{just 1 red})$

$$= \frac{\Pr(\text{just 1 white and just 1 red})}{\Pr(\text{just 1 red})}.$$

If Y = number of red balls then Y is distributed binomially with parameters $n = 4$ and $p = \frac{4}{9}$.

Thus $\Pr(\text{just 1 red}) = \Pr(Y = 1) = \binom{4}{1}\left(\frac{4}{9}\right)^1 \left(1 - \frac{4}{9}\right)^{4-1}$

$$= 4 \binom{4}{9}\left(\frac{5}{9}\right)^3 .$$

Pr(just 1 white and just 1 red)

= Pr(1 white, 1 red and 2 blacks).

Any particular sequence of outcomes in which 1 white ball is chosen, 1 red ball is chosen and 2 black balls are chosen has probability $\left(\frac{3}{9}\right)^1 \left(\frac{2}{9}\right)^2 \left(\frac{4}{9}\right)^1$. One now must find the number of such distinguishable arrangements. There are $\binom{4}{1}$ ways to select the position of the white ball. There are now three positions available to select the position of the red ball and $\binom{3}{1}$ ways to do this. The position of the black balls are now fixed. There are thus

$$\binom{4}{1}\binom{3}{1} = \frac{4!}{1!3!} \frac{3!}{1!2!} = \frac{4!}{1!2!1!} \quad \text{distinguishable arrange-}$$

ments.

Thus the Pr(1 red ball, 1 white ball and 2 black balls)

$$= \frac{4!}{1!2!1!} \cdot \left(\frac{3}{9}\right)\left(\frac{2}{9}\right)^2\left(\frac{4}{9}\right)^1 = \frac{4 \cdot 3 \cdot 3 \cdot 4 \cdot 4}{9^4}$$

$$\text{Pr(just 1 white | just 1 red)} = \frac{\dfrac{4 \cdot 3 \cdot 3 \cdot 4 \cdot 4}{9^4}}{4 \left(\dfrac{4}{9}\right)\left(\dfrac{5}{9}\right)^3}$$

$$= \frac{4 \cdot 3 \cdot 3 \cdot 4 \cdot 4}{4 \cdot 4 \cdot 5 \cdot 5 \cdot 5} = \frac{36}{125}.$$

● **PROBLEM** 9-53

A package in the mail can either be lost, delivered or damaged while being delivered. If the probability of loss is .2, the probability of damage is .1 and the probability of delivery is .7 and 10 packages are sent to Galveston, Texas, what is the probability that 6 arrive safely 2 are lost and 2 are damaged?

Solution: If each package being sent can be considered an independent trial with three outcomes, one can assume the event of 6 safe arrivals, 2 losses and 2 smashed packages to have a multinomial probability. Thus,

$$\Pr(6, \ 2 \ \text{and} \ 2) \ = \ \begin{pmatrix} 10 \\ 6, \ 2, \ 2 \end{pmatrix} \ (.7)^6 \ (.2)^2 \ (.1)^2$$

$$= \ \frac{10!}{6!2!2!} \ (.7)^6 \ (.2)^2 \ (.1)^2 \ = \ .059.$$

The probability of 6 safe arrivals, 2 losses and 2 damaged packages is .059.

Consider the joint distribution of X and Y given in the form of a table below. The cell (i,j) corresponds to the joint probability that X = i, Y = j, for i = 1,2,3, j = 1,2,3.

Y \ X	1	2	3
1	0	1/6	1/6
2	1/6	0	1/6
3	1/6	1/6	0

Check that this is a proper probability distribution. What is the marginal distribution of X? What is the marginal distribution of Y?

Solution: A joint probability mass function gives the probabilities of events. These events are composed of the results of two (or more) experiments. An example might be the toss of two dice. In this case, each event or outcome has two numbers associated with it. The numbers are the outcomes from the toss of each die. The probability distribution of the pair (X,Y) is

$$\Pr(X = i, \ X = j) \ = \ \frac{1}{36} \qquad\qquad \begin{aligned} i &= 1,2,3,4,5,6 \\ j &= 1,2,3,4,5,6 \ . \end{aligned}$$

Another example is the toss of two dice where X = number observed on first die ; Y = the larger of the two numbers.

In order for f(x,y) = Pr(X = x, Y = y) to be a proper joint probability, the sum of Pr(X = x, Y = y) over all (x,y), over all points in the sample space must equal 1.

In the case of the pair of tossed dice,

$$\sum_x \ \sum_y \ \Pr(X = x, Y = y) \ = \ \sum_{i=1}^{6} \ \sum_{j=1}^{6} \ \frac{1}{36} \ = \ \sum_{i=1}^{6} \ \frac{6}{36} \ = \ \frac{6 \cdot 6}{36} \ = \ 1 \ .$$

Thus, this is a proper probability distribution.

In the original example,

$$\sum_{i=1}^{3} \ \sum_{j=1}^{3} \ \Pr(X = i, \ Y = j) \ = \ \sum_{i=1}^{3} \ [\Pr(X{=}i,Y{=}1) \ + \ \Pr(X{=}i,Y{=}2) \ + \ \Pr(X{=}i,Y{=}3)]$$

$$= \sum_{i=1}^{3} \Pr(X=i, Y=1) + \sum_{i=1}^{3} \Pr(X=i, Y=2) + \sum_{i=1}^{3} \Pr(X=i, Y=3)$$

$$= (0 + 1/6 + 1/6) + (1/6 + 0 + 1/6) + (1/6 + 1/6 + 0)$$

$$= 1/3 + 1/3 + 1/3 = 1 .$$

Thus, the probability distribution specified in the table is a proper distribution.

One can compute the individual probability distributions of X and Y. These are called the marginal distributions of X and Y and are calculated in the following way.
Find the probability that X = 1,2,3,

$$\Pr(X = 1) = \Pr(X = 1, Y = 1, 2, \text{ or } 3) .$$

Because the events "X = 1, Y = 1", "X = 1, Y = 2" ,"X = 1, Y = 3" are mutually exclusive,

$$\Pr(X=1) = \Pr(X=1,Y=1) + \Pr(X=1,Y=2) + \Pr(X=1,Y=3)$$

$$= \sum_{i=1}^{3} \Pr(X=1,Y=i) .$$

Thus, $$\Pr(X=1) = 0 + 1/6 + 1/6 = 1/3 .$$

Similarly, $$\Pr(X=2) = \sum_{i=1}^{3} \Pr(X=2,Y=i) = 1/6 + 0 + 1/6 = 1/3$$

and $$\Pr(X=3) = \sum_{i=1}^{3} \Pr(X=3,Y=i) = 1/6 + 1/6 + 0 = 1/3 .$$

Computing the marginal probabilities of Y in a similar way.

$$\Pr(Y=1) = \Pr(X=1,Y=1) + \Pr(X=2,Y=1) + \Pr(X=3,Y=1)$$

$$= 0 + 1/6 + 1/6 = 2/6 = 1/3$$

$$\Pr(Y=2) = \sum_{j=1}^{3} \Pr(X=j,Y=2) = 1/6 + 0 + 1/6 = 1/3 .$$

$$\Pr(Y=3) = \sum_{j=1}^{3} \Pr(X=j,Y=3) = 1/6 + 1/6 + 0 .$$

To see why these are called marginal probabilities, examine the way they were computed.
The marginal probabilities of X were found by summing along the rows of the table of the joint distribution. The marginal probabilities of Y were found by summing along the columns of the table of the joint distribution.
The probabilities resulting from these summations are often placed in the margins, as in the table below, hence the name marginal probabilities.

Show, by altering the joint density of X and Y in the previous problem, that it is not always possible to construct a unique joint distribution from a pair of given marginal distributions.

Solution: The joint density of X and Y with its marginal distributions is given by the table below:

Y \ X	1	2	3	4
1	1/16	0	0	0
2	1/16	2/16	0	0
3	1/16	1/16	3/16	0
4	1/16	1/16	1/16	4/16
Pr(X = x)	4/16	4/16	4/16	4/16

Imagine one is given the marginal distributions of X and Y above and asked to construct the joint distribution of X and Y. There are an infinite number of possibilities for this distribution as seen by the table below.

For any ε , 0 < ε < 1/16, this joint distribution will yield the given marginal distributions. Thus, these marginal distributions do not specify a unique joint distribution.

Use

$$f(x,y) = \begin{cases} e^{-x}\,e^{-y} & \begin{array}{l} x > 0 \\ y > 0 \end{array} \\ 0 & \text{otherwise} \end{cases}$$

to find the probability that $\{1 < X < 2 \text{ and } 0 < Y < 2\}$.

Solution: Pr(1 < X < 2 and 0 < y < 2) is the volume indicated by the shaded rectangle:

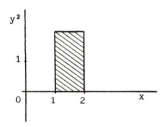

This volume over the rectangle and under f(x,y) is pictured below:

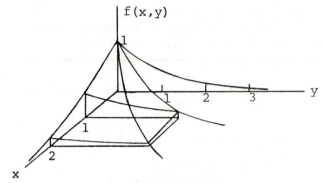

To find this volume, intergrating X from 1 to 2 and Y from 0 to 2, one has

$$Pr(1 < X < 2 \quad \text{and} \quad 0 < Y < 2) = \int_0^2 \int_1^2 f(x,y) \; dx \; dy$$

$$= \int_0^2 e^{-y} \left(\int_1^2 e^{-x} \; dx \right) dy$$

$$= \int_0^2 e^{-y} dy \left[-e^{-x} \Big|_1^2 \right] = \int_0^2 e^{-y} dy \, (e^{-1} - e^{-2})$$

$$= -e^{-y} \Big]_0^2 \, (e^{-1} - e^{-2})$$

$$= (e^0 - e^{-2})(e^{-1} - e^{-2}) = (1 - e^{-2})(e^{-1} - e^{-2})$$

$$= (.865)(.233) = .20 \; .$$

● **PROBLEM** 9-57

Two individuals agree to meet at a certain spot sometime between 5:00 and 6:00 P.M. They will each wait 10 minutes starting from when they arrive. If the other person does not show up, they will leave. Assume the arrival times of the two individuals are independent and uniformly distributed over the hour-long interval, find the probability that the two will actually meet.

Solution: Let X = arrival time of the first individual and
          Y = arrival time of the second individual.
     X and Y have uniform distributions over any hour-long period, thus in minutes the densities are:

$$f(x) = \frac{1}{60} \qquad 0 < x < 60$$

$$g(y) = \frac{1}{60} \qquad 0 < y < 60 \; .$$

Furthermore X and Y are independent. Thus the joint density of X and Y will be the product of the individual density functions.

$$h(x,y) = f(x)g(y) = \begin{cases} (\frac{1}{60})(\frac{1}{60}) = \frac{1}{3600} & \begin{array}{l} 0 < x < 60 \\ 0 < y < 60 \end{array} \\ \\ 0 & \text{otherwise.} \end{cases}$$

One can now try to formulate the event "a meeting takes place" in terms of X and Y.

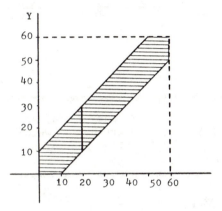

Consider the shaded region above. If the point (x,y) lies within this shaded region a meeting will take place. To see that this is true, we arbitrarily test the point X = 20. If X = 20, the first individual arrives at 5:20. If the second individual arrived at any time between 5:10 and 5:30 there will be a meeting. Thus Y may take on a value between 10 and 30. This region is described mathematically by $|X-Y| < 10$. The absolute value signs reflect the fact that the order of arrival is unimportant in assuring a meeting, only the proximity or closeness of the arrival times is important. Thus,

$$\begin{aligned} \text{Pr(a meeting)} &= \text{Pr}(|X-Y| < 10) \\ &= \text{volume over the shaded region in the x-y plane} \\ &\quad \text{under } f(x,y). \end{aligned}$$

One may divide this volume into three regions, A_1, A_2, A_3 .

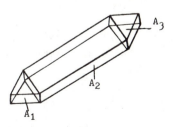

The volume of A_2 is the volume of a rectangular parellelapiped, (box-shaped region) and A_1 and A_3 are right prisms of equal volume.

527

The volume of A_2 is length $\times$ width $\times$ height $= (10\sqrt{2})(50\sqrt{2})(\frac{1}{3600})$

$$= 2(500)(\frac{1}{3600}) \ .$$

And the volume of A_3 and A_1 each is (Area of base) $\times$ height

$$= \tfrac{1}{2}(10)(10) \cdot \frac{1}{3600}$$

$$= \frac{50}{3600} \ .$$

$$Pr(|X-Y| < 10) = \text{Volume of } A_2 + \text{Volume of } A_1 + \text{Volume of } A_3$$

$$= \frac{1000}{3600} + \frac{50}{3600} + \frac{50}{3600} \quad = \frac{1100}{3600} = \frac{11}{36} \ .$$

● **PROBLEM** 9-58

Two continuous random variables X and Y may also be jointly distributed. Suppose (X,Y) has a distribution which is uniform over a unit circle centered at (0,0). Find the joint density of (X,Y) and the marginal densities of X and Y. Are X and Y independent?

<u>Solution</u>: The pairs of points (X,Y) lie in the unit circle with center at (0,0).

The probability that a random point (X,Y) lies in a particular region of this circle is given by the volume over the region, A, and under a joint density function f(x,y).

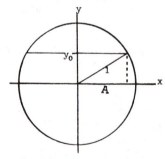

In the case of the uniform joint density function , the density function is a constant such that the total volume over some area in the (x,y) plane and under f(x,y) = c > 0 is 1.

The total area of the unit circle is πr^2 where r = 1. Thus the area is π . In order for the total volume to equal 1,

$$c\pi = 1 \quad \text{or} \quad c = 1/\pi \ .$$

Thus, $\qquad f(x,y) = \begin{cases} 1/\pi & \text{for} \quad x^2 + y^2 < 1 \\ 0 & \text{otherwise} \ . \end{cases}$

The marginal distributions are found by "summing" over all values of the other variable after one variable is fixed. Because the

528

variables are continuous, the "summing" must be performed by integration. Thus the marginal distributions of X and Y are respectively:

$$g(x) = \int_{\text{all } y} f(x,y)\,dy \qquad x^2 + y^2 < 1$$

and

$$h(y) = \int_{\text{all } x} f(x,y)\,dx \qquad x^2 + y^2 < 1 .$$

In our problem, let x be fixed, $x = x_0$, then $y^2 < 1-x_0^2$ or $y < \sqrt{1-x_0^2}$ or $y < \sqrt{1-x_0^2}$ and $-y > \sqrt{1-x_0^2}$. Thus

$$g(x_0) = \int_{-\sqrt{1-x_0^2}}^{\sqrt{1-x_0^2}} f(x,y)\,dy = \int_{-\sqrt{1-x_0^2}}^{\sqrt{1-x_0^2}} \frac{1}{\pi}\,dy$$

$$= \frac{2\sqrt{1-x_0^2}}{\pi} , \qquad 1 < x_0 < -1 .$$

Similarly, for fixed $y = y_0$, $-\sqrt{1-y_0^2} < x < \sqrt{1-y_0^2}$

$$h(y_0) = \int_{-\sqrt{1-y_0^2}}^{\sqrt{1-y_0^2}} f(x,y)\,dx = \int_{-\sqrt{1-y_0^2}}^{-\sqrt{1-y_0^2}} \frac{1}{\pi}\,dx = \frac{x}{\pi}\Bigg]_{-\sqrt{1-y_0^2}}^{\sqrt{1-y_0^2}}$$

$$= \frac{2\sqrt{1-y_0^2}}{\pi} \qquad -1 < y < 1 .$$

X and Y will be independent if and only if the joint density is the product of the marginal densities or $f(x,y) = g(x)h(y)$.

In this problem, we see that

$$g(x_0)h(y_0) = \frac{2\sqrt{1-x_0^2}}{\pi} \cdot \frac{2\sqrt{1-y_0^2}}{\pi}$$

$$= \frac{4}{\pi^2} \sqrt{1-x_0^2}\, \sqrt{1-y_0^2}$$

$$\neq \frac{1}{\pi} = f(x,y) .$$

Thus X and Y are dependent.

● **PROBLEM** 9-59

Let X and Y be jointly distributed with density function

$$f(x,y) = \begin{cases} 1 & \begin{array}{l} 0 < x < 1 \\ 0 < y < 1 \end{array} \\ 0 & \text{otherwise} . \end{cases}$$

Find

$$F(\lambda \,|\, X > Y) = \Pr(X \leq \lambda \,|\, X > Y).$$

529

<u>Solution</u>: By the definition of conditional probability,

$$Pr(A|B) = \frac{Pr(A \text{ and } B)}{Pr(B)} \;.$$

Thus,

$$F(\lambda|X > Y) = Pr(X \le \lambda | X > Y)$$

$$= \frac{Pr(X \le \lambda \text{ and } X > Y)}{Pr(X > Y)} \;.$$

The shaded region represents the area where $X \le \lambda$ and $X > Y$. Thus $Pr(X \le \lambda$ and $X > y)$ is the volume over the shaded area under the curve $f(x,y) = 1$.

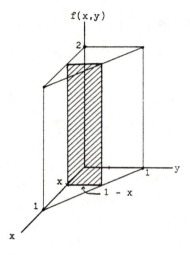

This volume is that of a right prism. Thus $Pr(X \le \lambda$ and $X > Y)$ = volume of right prism whose base is the shaded region in the figure and whose height is 1.

$Pr(X \le \lambda$ and $X > Y)$ = (Area of base) x height = $\frac{1}{2}\lambda$ x λ x 1 = $\frac{\lambda^2}{2}$.

$Pr(X > Y)$ = the volume over the triangle with vertices $(0,0)$, $(1,1)$ and $(1,0)$ and under $f(x,y) = 1$.

This volume is an the shape of a right prism with a base of area $\frac{1}{2}$ and a height of 1.

The volume of this region is thus $\frac{1}{2} \cdot 1 = \frac{1}{2}$ and $Pr(X > Y) = \frac{1}{2}$.

Thus $\quad \dfrac{Pr(X \le \lambda \text{ and } X > Y)}{Pr(X > Y)} = \dfrac{\lambda^2/2}{\frac{1}{2}} = \lambda^2$.

The conditional cumulative distribution function for X given

$X > Y$ is therefore $\quad Pr(X \le \lambda | X > Y) = \begin{cases} \lambda^2 & 0 < \lambda < 1 \\ 0 & \lambda < 0 \\ 1 & \lambda > 1 \;. \end{cases}$

530

Let

$$f(x,y) = \begin{cases} 2 - x - y & \begin{array}{l} 0 < x < 1 \\ 0 < y < 1 \end{array} \\ 0 & \text{otherwise .} \end{cases}$$

Find the conditional distribution of Y given X.

Solution: The conditional distribution of Y given X = x is defined to be

$$f(y \mid X = x) = \frac{f(x,y)}{g(x)} ,$$

where g(x) is the marginal distribution of X and f(x,y) is the joint distribution of X and Y.

The marginal distribution of X is

$$g(x) = \int_{-\infty}^{\infty} f(x,y)dy = \int_{0}^{1} (2-x-y)dy$$

$$= (2-x)y - \frac{y^2}{2} \Big]_{0}^{1} = 2 - x - \tfrac{1}{2} = \frac{3}{2} - x \quad \text{for} \quad 0 < x < 1 .$$

Thus the conditional density is

$$f(y \mid x) = \frac{2-y-x}{3/2 - x} , \quad 0 < y < 1 \quad \text{and} \quad x \quad \text{fixed.}$$

FUNCTIONS OF RANDOM VARIABLES

X is a discrete random variable with probability mass function, $f(x) = 1/n$, $x = 1,2,3,\ldots,n; = 0$ otherwise.

If $Y = X^2$, find the probability mass function of Y.

Solution: Since $Y = X^2$ as x takes the values $1,2,3,\ldots,n$, y takes the values $1,4,9,\ldots,n^2$. Thus one obtains

$$\Pr(Y = r^2) = \Pr(X = r) = 1/n, \quad r = 1,2,\ldots,n$$

and so

$$h(y) \begin{cases} = 1/n , \quad y = 1,4,\ldots,n^2 \\ = 0 \quad\quad\quad \text{otherwise.} \end{cases}$$

> Given that the random variable X has probability density function $f(x) = (\frac{1}{2})^X$ when x is a positive integer, find the density function of the random variable
> $$Y = \sin(\pi \frac{X}{2}).$$

Solution: Determine the possible values of $Y = \sin(\pi \frac{X}{2})$. When $x = 1$, $Y = \sin(\pi \cdot \frac{1}{2}) = 1$. When $x = 2$, $Y = \sin(\pi) = 0$. When $x = 3$, $Y = \sin(\frac{3\pi}{2}) = -1$. When $x = 4$, $Y = \sin(2\pi) = 0$. When $x = 5$, $Y = \sin(\frac{5}{2}\pi) = 1$. The same values $(1, 0, -1)$ will be generated as x increases by 1, because the sine function is periodic with period 2π.

$Y = 0$ if and only if $\pi(\frac{X}{2}) = k\pi$, where k is a positive integer. Then $\frac{X}{2} = k$ and $x = 2k$.

$Y = 1$ if and only if $\pi(\frac{X}{2}) = \frac{1}{2}\pi$ or $\pi(\frac{X}{2}) = \frac{5}{2}\pi$ or $\pi(\frac{X}{2}) = \frac{4k+1}{2}\pi$ where $k = 0, 1, 2, \ldots$. Dividing each side of this equation by $\pi/2$, one obtains $x = 4k + 1$.

$Y = -1$ if and only if $\pi(\frac{X}{2}) = \frac{3}{2}\pi$ or $\pi(\frac{X}{2}) = \frac{7}{2}\pi$ or $\pi(\frac{X}{2}) = \frac{11}{2}\pi$ or $\pi(\frac{X}{2}) = (\frac{4k+3}{2})\pi$ where $k = 0, 1, 2, 3, \ldots$. Again dividing each side by $\pi/2$, one finds that $X = 4k + 3$, $X = 0, 1, 2, 3 \ldots$.

Now evaluate the probability that Y takes each of its possible values $(1, 0, -1)$.

$P(X = x) = (\frac{1}{2})^X$ when $x = 1, 2, 3, \ldots$ is given. The event $X = 2k$ for some k is exclusive of any other event $x = 2j$, $R \neq j$. Therefore one can add the probabilities of their occurrence to find

$$P(Y = 0) = \sum_{k=1}^{\infty} P(X = 2k) = \sum_{k=1}^{\infty} (\tfrac{1}{2})^{2k}$$

(by substituting $2k$ for x in the given probability function) and

$$\sum_{k=1}^{\infty} (\tfrac{1}{2})^{2k} = \sum_{k=1}^{\infty} \frac{(1)^{2k}}{2^{2k}} = \sum_{k=1}^{\infty} \frac{1}{4^k} = \sum_{k=1}^{\infty} (\tfrac{1}{4})^k .$$

Factoring $\frac{1}{4}$ from this series, one has

$$\sum_{k=1}^{\infty} (\tfrac{1}{4})^k = \tfrac{1}{4} \sum_{k=0}^{\infty} (\tfrac{1}{4})^k .$$

The series from $k = 0$ is geometric with sum (limit of sequence of partial sums)

$$\frac{1}{1-\frac{1}{4}} = \frac{1}{3/4} = \frac{4}{3} .$$

Thus,

$$P(Y = 0) = \tfrac{1}{4} \sum_{k=0}^{\infty} (\tfrac{1}{4})^k = \tfrac{1}{4} \cdot \frac{4}{3} = \frac{1}{3} .$$

The events $X = 4k + 1$ are mutually exclusive for different K. Thus one can add their probabilities to find

$$P(Y = 1) = \sum_{k=0}^{\infty} P(X = 4k+1) = \sum_{k=0}^{\infty} (\tfrac{1}{2})^{4k+1} = \sum_{k=0}^{\infty} (\tfrac{1}{2})^{4k} (\tfrac{1}{2})^{1}$$

$$= \tfrac{1}{2} \sum_{k=0}^{\infty} \left(\frac{1}{2^4}\right)^k = \tfrac{1}{2} \sum_{k=0}^{\infty} \left(\frac{1}{16}\right)^k$$

by the same reasoning one used to find $P(Y = 0)$. $\sum_{k=0}^{\infty} (\frac{1}{16})^k$ is a geometric series and so $\sum_{k=0}^{\infty} (\frac{1}{16})^k = \frac{1}{1-1/16} = \frac{1}{15/16} = 16/15.$ Therefore $P(Y = 1) = \frac{1}{2} (\frac{16}{15}) = \frac{8}{15}.$

Exactly the same method is used to find

$$P(Y = -1) = \sum_{k=0}^{\infty} P(X = 4k+3) = \sum_{k=0}^{\infty} (\tfrac{1}{2})^{4k+3}$$

$$= \sum_{k=0}^{\infty} (\tfrac{1}{2})^{4k} (\tfrac{1}{2})^3 = \tfrac{1}{8} \sum_{k=0}^{\infty} \left(\frac{1}{2^4}\right)^k = \tfrac{1}{8} \sum_{k=0}^{\infty} \left(\frac{1}{16}\right)^k$$

$$= \tfrac{1}{8} \left(\frac{16}{15}\right) = \frac{2}{15}.$$

Therefore, the distribution of Y is;

$$h(y) = \begin{cases} \frac{2}{15}, & y = -1 \\ \frac{1}{3}, & y = 0 \\ \frac{8}{15}, & y = 1 \\ 0, & \text{otherwise.} \end{cases}$$

Checking we see that

$$Pr(Y = 0) + Pr(Y = 1) + Pr(Y = -1)$$

$$= 1/3 + 8/15 + 2/15 = \frac{5+8+2}{15} = \frac{15}{15} = 1,$$

as it should to have a proper probability distribution.

● **PROBLEM** 9-63

Suppose X takes on the values 0,1,2,3,4,5 with probabilities P_0, P_1, P_2, P_3, P_4 and P_5. If $Y = g(X) = (X-2)^2$, what is the distribution of Y?

Solution: First, examine the values that the random variable Y may assume. If $X = 0$ or 4,

$$Y = (0-2)^2 = (4-2)^2 = 2^2 = 4.$$

If $X = 1$ or 3,

$$Y = (1-2)^2 = (3-2)^2 = 1^2 = 1.$$

If $X = 2$,

$$Y = (2-2)^2 = 0^2 = 0$$

and if $X = 5$,

$$Y = (5-2)^2 = 3^2 = 9.$$

Thus, Y assumes the values $0, 1, 4$ and 9. We now find the probability distribution of Y.

$$Pr(Y = 0) = Pr((X-2)^2 = 0)$$
$$= Pr(X = 2) = p_2$$

$$Pr(Y = 1) = Pr((X-2)^2 = 1) \qquad = Pr(X = 1 \text{ or } X = 3)$$

$$= Pr(X = 1) + Pr(X = 3) \qquad = p_1 + p_3$$

$$Pr(Y = 4) = Pr((X-2)^2 = 4) \qquad = Pr(X = 0 \text{ or } X = 4)$$

$$= Pr(X = 0) + Pr(X = 4) \qquad = p_0 + p_4$$

and

$$Pr(Y = 9) = Pr((X-2)^2 = 9)$$
$$= Pr(X = 5) = p_5 .$$

Therefore, the distribution of Y can be written

$$h(y) = \begin{cases} p_2 & , y = 0 \\ p_1 + p_3, & y = 1 \\ p_0 + p_4, & y = 4 \\ p_5 & , y = 9 \\ 0 & , \text{otherwise.} \end{cases}$$

● **PROBLEM** 9-64

Suppose $X_1, \ldots, X_n$ are independent Bernoulli random variables, that is, $Pr(X_i = 0) = 1 - p$ and $Pr(X_i = 1) = p$ for $i = 1, \ldots, n$. What is the distribution of

$$Y = \sum_{i=1}^{n} X_i ?$$

Solution: One has a sum of indipendent and identically distributed random variables. The moment generating technique is useful here. Find the moment generating funtion of Y and use the one to one correspondence between moment generating functions and distribution functions to find the distribution of Y.

$$M_Y(t) = E[e^{Yt}] \qquad = E[e^{(X_1 + \ldots + X_n)t}]$$

$$= E[e^{X_1 t}] \, E[e^{X_2 t}] \cdot \, \ldots \, \cdot \, E[e^{X_n t}] \; .$$

But for each i,

$$E[e^{X_i t}] = e^{0 \cdot t} Pr(X_i = 0) + e^{1 \cdot t} Pr(X_i = 1)$$

$$= Pr(X_i = 0) + e^t Pr(X_i = 1)$$

$$= 1 - p + pe^t \, .$$

Thus

$$M_Y(t) = (1 - p + pe^t)^n$$

which is the moment generating function of a binomially distributed random variable with parameters n and p.

● **PROBLEM** 9-65

Let T be distributed with density function

$$f(t) = \begin{cases} \lambda e^{-\lambda t} & \text{for } t > 0 \\ 0 & \text{otherwise} \end{cases}$$

If S is a new random variable defined as S = ln T, find the density function of S.

Solution: The technique used here to find the density function of S is known as the distribution function technique. If G(t) = Pr(T < t) is known, one can sometimes use it to find F(x) = Pr(s < x). Differentiating F(s) will then yield the density function of the random variable S.

$$Pr(S \leq x) = Pr(\ln T \leq x)$$

because ln T = S but the natural logarithm function of a variable is one-to-one. That is, if ln y = x, then given y one can find a unique x such that y = $\ln^{-1} x$.

$\ln^{-1}(x)$ is the invers of the natural log function and is known to be $e^{(x)}$. Thus ln y = x if and only if y = e^x.

Similarly, ln T $\leq$ x if and only T $\leq e^x$. Thus,

$$Pr(S \leq x) = Pr(\ln T \leq x)$$

$$= Pr(T \leq e^x) \qquad x > 0.$$

But $Pr(T \leq e^x) = G(e^x)$,   the cumulative distribution function of T evaluated at e^x .

$$\Pr(T \le e^x) = \int_{-\infty}^{e^x} f(t)dt \qquad x > 0$$

$$= \int_{-\infty}^{e^x} \lambda e^{-\lambda t} \, dt \qquad x > 0$$

$$= \int_{0}^{e^x} \lambda e^{-\lambda t} \, dt \qquad x > 0$$

$$= -e^{-\lambda t}\Big]_{0}^{e^x} \qquad x > 0$$

$$= -[e^{-\lambda e^x} - 1] \ .$$

Thus,

$$\Pr(S \le x) = 1 - e^{-\lambda e^x} \qquad x > 0 \ .$$

Differentiating one obtains

$$f(s) = \begin{cases} (\lambda e^x)e^{-\lambda e^x} & x > 0 \\ 0 & \text{otherwise.} \end{cases}$$

Let X have the probability distribution defined by

$$F(x) = \begin{cases} 1 - e^{-x} & \text{for } x \ge 0 \\ 0 & \text{for } x < 0 \ . \end{cases}$$

Let $Y = \sqrt{X}$ be a new random variable. Find $G(y)$, the distribution function of Y, using the cumulative distribution function technique.

Solution: We find $G(y) = \Pr(Y \le y)$. $Y = \sqrt{X}$ thus

$$\Pr(Y \le y) = \Pr(\sqrt{X} \le y) \ .$$

Because X takes on only positive values or zero, $\sqrt{X}$ will make sense. $\sqrt{X}$ will always be positive; thus

$$G(Y) = \Pr(Y \le y) = \begin{cases} \Pr(\sqrt{X} \le y) & y \ge 0 \\ 0 & y < 0 \end{cases}$$

$$= \begin{cases} \Pr(X \le y^2) & y \ge 0 \\ 0 & y < 0 \ . \end{cases}$$

But for $y \ge 0$, $\Pr(X \le y^2) = F(y^2)$, the distribution function of X evaluated at y^2. Thus,

$$\Pr(X \le y^2) = 1 - e^{-y^2}$$

and so

$$G(Y) = \Pr(Y \le y) = \begin{cases} 1 - e^{-y^2} & y \ge 0 \\ 0 & y < 0 \ . \end{cases}$$

The density function of Y is found by differentiating $G(y)$

with respect to y. Thus,

$$g(y) = \frac{dG(y)}{dy} = \begin{cases} 2ye^{-y^2} & y \geq 0 \\ 0 & \text{otherwise} \end{cases}$$

We have shown that the square root of an exponentially distributed random variable is distributed as a Weibull random variable
with parameters a = 1, b = 2 .
The general density of a Weibull distribution is

$$f(x) = \begin{cases} abx^{b-1}e^{-ax^b} & \infty > x > 0 \\ 0 & \text{otherwise} \end{cases}$$

● PROBLEM 9-67

Let $X_1, \ldots, X_n$ represent the incomes of n randomly selected tax-
payers.
Assume that each observation is independent and has probability
density function

$$f(x) = \begin{cases} \dfrac{\theta(x_0)^\theta}{x^{\theta+1}} & \text{for } x > x_0 \\ 0 & \text{otherwise} \end{cases}$$

This is the Pareto distribution. Assume θ = 100 and
x_0 = \$4,000. Find the density function of the minimum of the n
observations.

Solution: We first find $Pr(Y_1 \leq y)$ where Y_1 = minimum of
$\{X_1, X_2, \ldots, X_n\}$. The minimum of the n observations will be the
smallest observation in the sample.

$$Pr(Y_1 \leq y) = \begin{cases} 1 - Pr(Y_1 \geq y) & y > x_0 \\ 0 & y < x_0 \end{cases}$$

But $Y_1 \geq y$ only if $X_1 \geq y$, $X_2 \geq y, \ldots, X_n \geq y$. Thus

$$Pr(Y_1 \geq y) = Pr(X_1 \geq y, X_2 \geq y, \ldots, X_n \geq y) .$$

By the independence of the X_i ,

$$Pr(X_1 \geq y, X_2 \geq y, \ldots, X_n \geq y)$$

$$= Pr(X_1 \geq y) Pr(X_2 \geq y), \ldots, Pr(X_n \geq y) .$$

And for i = 1,2,3,\ldots,n ,

$$Pr(X_i \geq y) = 1 - Pr(X_i \leq y)$$

$$Pr(X_i \leq y) = \int_{x_0}^{y} f(x)dx \qquad y > x_0$$

$$= \int_{x_0}^{y} \frac{\theta x_0^{\theta}}{x^{\theta+1}} dx \qquad = \theta x_0^{\theta} \int_{x_0}^{y} x^{-(\theta+1)} dx \qquad y > x_0$$

$$= \frac{\theta x_0^{\theta}}{-(\theta+1)+1} x^{-(\theta+1)+1} \Big]_{x_0}^{y} \qquad = \frac{\theta x_0^{\theta}}{-\theta} \left[y^{-\theta} - x_0^{-\theta} \right]$$

$$= x_0^{\theta} \left[x_0^{-\theta} - y^{-\theta} \right] \qquad = 1 - \left(\frac{x_0}{y} \right)^{\theta}$$

is the cumulative distribution function of each of the observations. And

$$Pr(X_i \geq y) = 1 - Pr(X_i \leq y)$$

$$= 1 - \left[1 - \left(\frac{x_0}{y} \right)^{\theta} \right]$$

$$= \left(\frac{x_0}{y} \right)^{\theta} \; .$$

And

$$Pr(Y_1 \geq y) = Pr(X_1 \geq y)Pr(X_2 \geq y) \ldots Pr(X_n \geq y) \quad \text{for} \quad y > x_0$$

$$= \underbrace{\left(\frac{x_0}{y} \right)^{\theta} \cdot \left(\frac{x_0}{y} \right)^{\theta} \cdot \ldots \cdot \left(\frac{x_0}{y} \right)^{\theta}}_{n \text{ terms}}$$

or

$$Pr(Y_1 \leq y) = 1 - Pr(Y_1 \geq y)$$

$$= 1 - \left(\frac{x_0}{y} \right)^{n\theta} \qquad y > x_0$$

The density function is found by differentiating $Pr(Y_1 \leq y)$ with respect to y.

Thus,

$$f(y_1) = \begin{cases} \dfrac{d}{dy} \left[1 - \left(\dfrac{x_0}{y} \right)^{n\theta} \right] & y > x_0 \\ 0 & \text{otherwise} \end{cases}$$

$$= \begin{cases} -n\theta \left(\dfrac{x_0}{y} \right)^{n\theta-1} \left(\dfrac{-x_0}{y^2} \right) & y > x_0 \\ 0 & \text{otherwise} \end{cases}$$

$$= \begin{cases} \dfrac{n\theta \, x_0^{n\theta}}{y^{n\theta+1}} & y > x_0 \\ 0 & \text{otherwise} \, . \end{cases}$$

Substituting our values of θ and x_0 we see that

$$f(y_1) = \begin{cases} \dfrac{(100n)(4,000)^{100n}}{(y_1)^{100n+1}} & y_1 > 4,000 \\\\ 0 & \text{otherwise .} \end{cases}$$

● **PROBLEM** 9-68

Let X and Y be independent, standard normal random variables, and let $R = \sqrt{X^2 + Y^2}$ be the distance of (X,Y) from $(0,0)$. Find the distribution of R.

Solution: We shall find the distribution of R by quoting some of the previous results from problems in this chapter.

The random variables X^2 and Y^2 are independent and both are distributed with a Chi-square distribution with 1 degree of freedom. The density function of X^2 and Y^2 is

$$f(x) = \frac{1}{2^{\frac{1}{2}}\Gamma(\frac{1}{2})} x^{\frac{1}{2}-1} e^{-x/2} \qquad x > 0 .$$

$X^2 + Y^2$ must then be Chi-square distributed with two degrees of freedom. Thus $X^2 + Y^2$ has density,

$$f(x) = \frac{1}{\Gamma(1)}(\tfrac{1}{2})^{2/2} w^{2/2-1} e^{-\frac{1}{2}x} \qquad x > 0$$

$$= \tfrac{1}{2} e^{-\frac{1}{2}x} \qquad x > 0 .$$

An exponential density function, $X^2 + Y^2$ is exponentially distributed with parameter $\beta = \tfrac{1}{2}$.

$R = \sqrt{X^2 + Y^2}$ is the square root of an exponentially distributed random variable. We have shown that such a random variable is distributed with a Weibull distribution with parameters $a = 1, b = 2$. Thus R has density

$$h(r) = 2re^{-r^2} \qquad r > 0 .$$

● **PROBLEM** 9-69

Let X and Y be jointly distributed continuous random variables with density $f(x,y)$. If $Z = X + Y$, find the density function of Z.

Solution: We use the cumulative distribution function technique and integrate to find

539

$$\Pr(Z \leq z) = \Pr(X+Y \leq z)$$

$$= \iint\limits_{x+y<z} f(x,y)dx\ dy.$$

The range of integration is over the shaded region in the x,y plane, i.e., all x,y such that x + y < z.

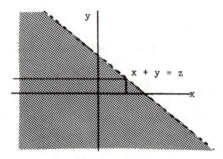

To represent this integral as an iterated integral, we fix y and then integrate with respect to x from z-y to -∞ . Next integrate with respect to y from -∞ to ∞. Thus

$$\Pr(Z \leq z) = \Pr(X+Y \leq z)$$

$$= \iint\limits_{x+y<z} f(x,y)dx\ dy$$

$$= \int_{-\infty}^{\infty} \left[\int_{-\infty}^{z-y} f(x,y)dx \right] dy .$$

Let u = x+y; for fixed y, du = dx . If y is fixed then when x = z-y, u = z. Thus

$$= \int_{-\infty}^{\infty} \left[\int_{-\infty}^{z} f(u-y,y)du \right] dy$$

Assume that f(x,y) has continuouse partial derivatives. Therefore one can exchange the order of integration and

$$F(z) = \Pr(Z \leq z) = \int_{-\infty}^{z} \left[\int_{-\infty}^{\infty} f(u-y,y)dy \right] du$$

Differentiating with respect to z will yield the density function of the random variable Z. Hence,

$$\frac{dF}{dz} = \frac{d}{dz} \left[\int_{-\infty}^{z} \left(\int_{-\infty}^{\infty} f(u-y,y)dy \right) du \right]$$

$$= \int_{-\infty}^{\infty} f(z-y,y)\ dy$$

540

This integral is called a convolution. Thus the density function of z is

$$\int_{-\infty}^{\infty} f(z-y,y) \, dy \ .$$

Suppose X and Y are independent random variables with densities f(x) and g(y). If Z = XY, what is the density of Z?

Solution: We will use the cumulative distribution function technique. Let

$$F(z) = Pr(Z \le z) = Pr(XY \le z)$$

$$= \int \int_{xy \le z} f(x)g(y)dx \, dy$$

The range of integration, that is, the region of the x,y-plane where xy ≤ z is the shaded region below.

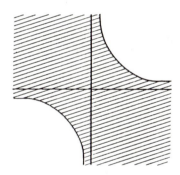

Dividing this region along the y-axis we compute our integral as the sum of two integrals.

$$Pr(Z \le z) = \int \int_{0<xy<z} f(x,y)dy \, dx + \int \int_{xy<0<z} f(x,y)dy \, dx \ .$$

$$\int \int_{0<xy<z} f(x,y)dy \, dx = \int_0^\infty \left[\int_{-\infty}^{z/x} f(x)g(y)dy \right] dx$$

and

$$\int \int_{xy<0<z} f(x,y)dy \, dx = \int_{-\infty}^0 \left[\int_{z/x}^\infty f(x)g(y)dy \right] dx \ .$$

Let $z_1 = xy$ if x > 0, y > 0; then for fixed x, y = z_1/x , dy = dz_1/x . When y = z/x , z_1 = z; when y = ∞ , z_1 = ∞. If x < 0, y < 0 then for fixed x , y = z_1/x , dy = dz_1/x. When y = z/x, z_1 = z and when y = ∞, z_1 = -∞ .

The last statement is true because x < 0.
Substituting yields

$$Pr(Z \le z) = \int_0^\infty \left[\int_{-\infty}^z f(x)g\left(\frac{z_1}{x}\right) \frac{dz_1}{x} \right] dx$$

$$+ \int_{-\infty}^0 \left[\int_z^{-\infty} f(x)g\left(\frac{z_1}{x}\right) \frac{dz_1}{x} \right] dx \quad = \int_{-\infty}^z \left[\int_0^\infty f(x)g\left(\frac{z_1}{x}\right) \frac{dx}{x} \right] dz_1$$

$$+ \int_z^{-\infty} \left[\int_{-\infty}^0 f(x)g\left(\frac{z_1}{x}\right) \frac{dx}{x} \right] dz_1 \quad = \int_{-\infty}^z \left[\int_0^\infty f(x)g\left(\frac{z_1}{x}\right) \frac{dx}{x} \right] dz_1$$

$$+ \int_{-\infty}^z \left[\int_{-\infty}^0 f(x)g\left(\frac{z_1}{x}\right) \left(-\frac{dx}{x}\right) \right] dz_1$$

Differentiating with respect to z and invoking the fundamental theorem, the density of Z is

$$h(z) = \int_0^\infty f(x)g\left(\frac{z}{x}\right) \frac{dx}{x} + \int_{-\infty}^0 f(x)g\left(\frac{z}{x}\right) \frac{dx}{-x} \; .$$

But we see when $x > 0$ the integrand is $f(x)g(z/x)\, dx/x$ and when $x < 0$ it is $f(x)g(z/x)\, dx/-x$. This recalls the absolute value function which is defined as

$$|x| = \begin{cases} x & x > 0 \\ -x & x < 0 \; . \end{cases}$$

Substituting,
$$h(z) = \int_{-\infty}^\infty \frac{1}{|x|} f(x)\ g\left(\frac{z}{x}\right) dx \; .$$

● **PROBLEM** 9-71

Consider a probability distribution for random orientations in which the probability of an observation in a region on the surface of the unit hemisphere is proportional to the area of that region. Two angles, u and v, will determine the position of an observation.

It can be shown that the position of an observation is jointly distributed with density function

$$f(u,v) = \frac{\sin u}{2\pi} \qquad \begin{array}{l} 0 < v < 2\pi \\ 0 < u < \pi/2 \; . \end{array}$$

Two new variables, X and Y are defined, where

$$X = \sin u \cos v$$
$$Y = \sin u \sin v \; .$$

Find the joint density function of X and Y.

Solution: In the region $0 < u < \pi/2$ and $0 < v < 2\pi$ this transformation is one-to-one and the mapping is pictured below.

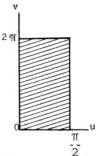 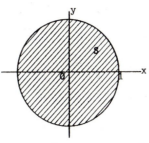

We now find u and v in terms of x and y. Let

$$x^2 = \sin^2 u \cos^2 v$$
$$y^2 = \sin^2 u \sin^2 v ;$$

adding we see that

$$x^2 + y^2 = \sin^2 u (\cos^2 v + \sin^2 v)$$
$$= \sin^2 u$$

Thus $\sin u = \sqrt{x^2 + y^2}$ and $u = \sin^{-1}\left[\sqrt{x^2 + y^2}\right]$.

Also $\dfrac{y}{x} = \dfrac{\sin u \sin v}{\sin u \cos v} = \sin v / \cos v = \tan v$.

Hence

$$v = \tan^{-1} \frac{y}{x} .$$

The Jacobian of the original transformation is

$$\frac{\partial(u,v)}{\partial(x,y)} = \frac{1}{\dfrac{\partial(x,y)}{\partial(u,v)}} .$$

Thus we find,

$$\frac{\partial(x,y)}{\partial(u,v)} = \begin{bmatrix} \dfrac{\partial x}{\partial u} & \dfrac{\partial x}{\partial v} \\[2ex] \dfrac{\partial y}{\partial u} & \dfrac{\partial y}{\partial v} \end{bmatrix}$$

$$= \left| \det \begin{pmatrix} \cos u \cos v & -\sin u \sin v \\ \sin v \cos u & \cos v \sin u \end{pmatrix} \right|$$

$$= \left| \cos^2 v (\cos u)(\sin u) + \sin^2 v (\cos u)(\sin u) \right|$$

$$= \left| (\cos u)(\sin u) \right|$$

Thus

$$\frac{\partial(u,v)}{\partial(x,y)} = \left[(\cos u)(\sin u) \right]^{-1}.$$

The joint density of X and Y is

$$g(x,y) = f\left(\sin^{-1}(\sqrt{x^2+y^2}), \tan^{-1}\frac{y}{x} \right) \cdot \frac{\partial(u,v)}{\partial(x,y)}$$

$$= \frac{\sin\left[\sin^{-1}\sqrt{x^2+y^2} \right]}{2\pi} \cdot \frac{1}{(\cos u)(\sin u)}$$

$$= \frac{\sqrt{x^2+y^2}}{2\pi} \cdot \frac{1}{(\cos u)(\sin u)}$$

But $\sin^2 u = X^2 + Y^2$; thus

$$1 - \cos^2 u = X^2 + Y^2$$

543

or

$$\cos u = \sqrt{1 - (x^2 + y^2)}$$

and

$$\sin u = \sqrt{x^2 + y^2} .$$

Thus

$$g(x,y) = \frac{1}{2\pi\sqrt{1-(x^2+y^2)}} \text{for } x^2 + y^2 < 1$$

is the joint density of X and Y.

EXPECTED VALUE

● **PROBLEM** 9-72

Let X be a random variable whose value is determined by the flip of a fair coin. If the coin lands heads up X = 1, if tails then X = 0. Find the expected value of X.

Solution: The expected value of X, written E(X), is the theoretical average of X. If the coin were flipped many, many times and the random variable X was observed each time, the average of X would be considered the expected value.

The expected value of a discrete variable such as X is defined to be

$$E(X) = x_1 \Pr(X = x_1) + x_2 \Pr(X = x_2) \ldots +$$

$$x_n \Pr(X = x_n)$$

where $x_1, x_2, x_3, \ldots x_n$, are the values X may take on and $\Pr(X = x_j)$ is the probability that X actually equals the value x_j.

For this problem, the random variable X takes on only two values, 0 and 1. X assumes these values with

$$\Pr(X = 1) = \Pr(X = 0) = \frac{1}{2} .$$

Thus, according to the definition,

$$E(X) = 0 \cdot \Pr(X = 0) + 1 \cdot \Pr(X = 1)$$

$$= 0 \cdot \frac{1}{2} + 1 \cdot \frac{1}{2} = 0 + \frac{1}{2} = \frac{1}{2} .$$

Let X be the random variable defined as the number of dots observed on the upturned face of a fair die after a single toss. Find the expected value of X.

Solution: X can take on the values 1, 2, 3, 4, 5, or 6. Since the die is fair one can assume that each value is observed with equal probability. Thus,

$$Pr(X = 1) = Pr(X = 2) = \ldots = Pr(X = 6)$$

$$= \frac{1}{6} .$$

The expected value of X is by definition:

$$E(X) = \Sigma x Pr(X = x) . \qquad \text{Hence}$$

$$E(X) = 1 \cdot \frac{1}{6} + 2 \cdot \frac{1}{6} + 3 \cdot \frac{1}{6} + 4 \cdot \frac{1}{6} + 5 \cdot \frac{1}{6} + 6 \cdot \frac{1}{6}$$

$$= \frac{1}{6} (1 + 2 + 3 + 4 + 5 + 6)$$

$$= \frac{21}{6} = 3 \frac{1}{2} .$$

The State of New Hampshire conducts an annual lottery to raise funds for the school districts in the state. Assume a million tickets are sold. One ticket is the winning ticket and the winner receives $10,000. If each ticket costs $.25, find the expected value of a randomly purchased ticket and the revenue that the lottery generates for the school districts in the state.

Solution: Let X be the value of a randomly purchased lottery ticket.

$$X = - \$.25 \text{ with probability } \frac{999,999}{1,000,000} .$$

The reason for assigning this value to x is that 999,999 of 1,000,000 lottery tickets have no value and the buyer of these tickets lose the $.25 price.

However,

$$X = \$10,000 - \$.25 \text{ with probability } \frac{1}{1,000,000} .$$

This reflects the fact that one of the million tickets wins $10,000 minus the purchase price of the ticket, thus the winner receives $10,000 - $.25.

The expected value of the random variable X is the expected value of a randomly purchased lottery ticket. By the definition of expected value,

$$E(X) = \$[10,000 - .25] \cdot \frac{1}{1,000,000}$$

$$+ [- \$.25] \cdot \frac{999,999}{1,000,000} \cdot$$

Rearranging terms we see that

$$E(X) = \$10,000 \cdot \left[\frac{1}{1,000,000} \right]$$

$$+ \left[\frac{- \$.25 - (\$.25)(999,999)}{1,000,000} \right]$$

$$= \$10,000 \left[\frac{1}{1,000,000} \right] - \$.25 \left[\frac{1,000,000}{1,000,000} \right]$$

$$= \$ \frac{1}{100} - \$.25$$

$$= \$.01 - \$.25 = - \$.24 \ .$$

Thus, the expected value of an average lottery ticket is - $.24. Each buyer loses an average of 24 cents on a lottery ticket.

The total revenue is the number of tickets sold times the price of each ticket or $(.25)(1,000,000) = $250,000. The net revenue, after the prize is paid is

$250,000 - 10,000 = $240,000.

Thus the school districts receive $240,000.

● **PROBLEM** 9-75

Let the random variable X represent the number of defective radios in a shipment of four radios to a local appliance store. Assume that each radio is equally likely to be defective or non-defective, hence the probability that a radio is defective is $p = \frac{1}{2}$. Also assume whether or not each radio is defective or non-defective is indipendent of the status of the other radios. Find the expected number of defective radios.

Solution: First find the probability distribution of x, the number of defective radios in the shipment of four. X can assume 5 values, 0, 1, 2, 3, or 4.

If X is 0, then 0 radios are defective. This can only take place if each is non-defective. By the independence assumption

$$Pr(X = 0) = \left(\frac{1}{2}\right)\left(\frac{1}{2}\right)\left(\frac{1}{2}\right)\left(\frac{1}{2}\right)$$

$$= \frac{1}{2^4} = \frac{1}{16} \ .$$

Similarly, $Pr(X = 1) = Pr(1 \text{ radio is defective, } 3 \text{ are not})$

$$= \frac{\text{number of favorable outcomes}}{\text{number of possible outcomes}}$$

$$= \frac{4}{2^4} = \frac{4}{16} = \frac{1}{4} \ .$$

$Pr(X = 2) = Pr(2 \text{ radios are defective})$

$$= \frac{\text{number of ways two can be chosen from four}}{\text{number of ways to choose 4 radios}}$$

$$= \binom{4}{2}\left(\frac{1}{2}\right)^4 = \frac{6}{16} \ .$$

By symmetry,

$$Pr(X = 1) = Pr(X = 3) = \frac{4}{16} = \frac{1}{4} \qquad \text{and}$$

$$Pr(X = 0) = Pr(X = 4) = \frac{1}{16} \ .$$

The expected number of defective radios is thus:

$$E(X) = 0 \cdot \frac{1}{16} + 1 \cdot \frac{4}{16} + 2 \cdot \frac{6}{16} + 3 \cdot \frac{4}{16} + 4 \cdot \frac{1}{16}$$

$$= \frac{4}{16} + \frac{12}{16} + \frac{12}{16} + \frac{4}{16} = \frac{32}{16} = 2 \ .$$

● **PROBLEM** 9-76

Let Y = the Rockwell hardness of a particular alloy of steel. Assume that Y is a continuous random variable that can take on any value between 50 and 70 with equal probability. Find the expected Rockwell hardness.

Solution: The random variable Y has a density function that is sketched below.

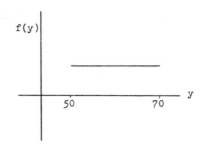

In order for Y to have a proper probability density function, the area under the density function must be 1. The area under the density function of Y has the shape of a rectangle with length 20. Thus the height of the rectangle must be

$$f(y)(20) = 1 \qquad\qquad 50 < y < 70,$$

Where the probability density function $f(y)$ represents the width of this rectangle. Solving for $f(y)$, the probability density function is found to be

$$f(y) = \frac{1}{20} \qquad\qquad 50 < y < 70 \ .$$

To find the expected value of a continuous random variable use the technique of integration:

$$E(Y) = \int_{50}^{70} yf(y)\ dy$$

$$E(Y) = \int_{50}^{70} y\ \frac{1}{20}\ dy \qquad = \frac{1}{20}\left(\frac{y^2}{2}\right) \Bigg|_{50}^{70}$$

$$= \frac{1}{20}\left(\frac{70^2 - 50^2}{2}\right) = \frac{(70 + 50)(70 - 50)}{40}$$

$$= \frac{70 + 50}{2} = \frac{120}{2} = 60 \ .$$

Thus, the expected Rockwell hardness of this alloy is 60.

Let X be a random variable denoting the hours of life in an electric light bulb. Suppose X is distributed with density function

$$f(x) = \frac{1}{1,000} \; e^{-x/1000} \qquad \text{for } x > 0$$

Find the expected lifetime of such a bulb.

Solution: The expected value of a continuous random variable is the "sum" of all the values of the random variable multiplied by their probabilities. In the continuous case, this "summing" necessitates integration. Thus by defintion:

$$E(X) = \int_{\text{all } x} x \, f(x) \, dx \, .$$

In this problem, x can take all positive values; thus:

$$E(X) = \int_0^\infty x \, f(x) \, dx$$

$$= \int_0^\infty \frac{x}{1000} \; e^{-\frac{x}{1000}} \, dx \, .$$

Integrating by parts, let u = x and

$$dv = \frac{1}{1000} \, e^{-\frac{x}{1000}} \, dx; \text{ then } du = dx \text{ and } v = - \, e^{-\frac{x}{1000}}$$

Thus one solves for the expected value by integrating as follows

$$E(X) = uv \Big|_0^\infty - \int_0^\infty v \, du$$

$$E(X) = - \, xe^{-\frac{x}{1000}} \Big|_0^\infty - \int_0^\infty - e^{-\frac{x}{1000}} \, dx$$

$$= 0 - 1000 \int_0^\infty - \frac{1}{1000} \, e^{-\frac{x}{1000}} \, dx$$

$$= - 1000 \cdot e^{-\frac{x}{1000}} \Big]_0^\infty$$

$$= - 1000 \, [0 - 1] = 1000 .$$

Thus, the expected lifetime of the bulb is 1000 hours.

● **PROBLEM** 9-78

Find E(X) for the continuous random variables with probability density functions;

a) $f(x) = 2x, \quad 0 < x < 1.$

b) $f(x) = \frac{1}{(2\sqrt{x})}, \quad 0 < x < 1.$

c) $f(x) = 6x(1 - x), \quad 0 < x < 1.$

d) $f(x) = \frac{1}{2}x^2 e^{-x}, \quad 0 < x < \infty.$

e) $f(x) = \frac{1}{x^2}, \quad 1 \le x < \infty .$

f) $f(x) = 1 - |1 - x|, \quad 0 \le x \le 2.$

<u>Solution:</u>   For a continuous random variable, X,

$$E(X) = \int_{-\infty}^{\infty} x \, f(x) \, dx.$$

It is possible that $f(x) = 0$ for large portions of the real line reducing E(X) to a proper integral.

(a) $E(X) = \int_0^1 x \cdot 2x \, dx = \int_0^1 2x^2 \, dx = \frac{2}{3} x^3 \Big|_0^1$

$$= \frac{2}{3} [1 - 0] = \frac{2}{3} .$$

(b) $E(X) = \int_0^1 x \frac{1}{2 \sqrt{x}} \, dx = \frac{1}{2} \int_0^1 \sqrt{x} \, dx$

550

$$= \frac{1}{2} \frac{x^{\frac{1}{2}+1}}{1+\frac{1}{2}} \Bigg]_0^1 \qquad \frac{1}{2} \cdot \frac{2}{3} x^{\frac{3}{2}} \Bigg]_0^1 \qquad = \frac{1}{3} \cdot$$

(c) $E(X) = \displaystyle\int_0^1 x(6x(1-x))\,dx = 6\int_0^1 (x^2 - x^3)\,dx$

$$= 6 \left[\frac{x^3}{3} - \frac{x^4}{4} \right]_0^1 = 6\left(\frac{1}{3} - \frac{1}{4}\right) = \frac{6}{12} = \frac{1}{2}\cdot$$

(d) $E(X) = \displaystyle\int_0^\infty x \cdot f(x)\,dx = \int_0^\infty \frac{1}{2} x^3 e^{-x}\,dx .$

Using integration by parts,

let $\qquad u = x^3 \qquad\qquad$ thus $\qquad du = 3x^2\,dx$

$\qquad\qquad dv = e^{-x}\,dx \qquad\qquad\qquad v = -e^{-x}$

and we see that

$$E(X) = \frac{1}{2}\left[-x^3 e^{-x} \Bigg|_0^\infty - \int_0^\infty -e^{-x} 3x^2\,dx \right]$$

$$= \frac{1}{2}\left[0 + 0 + 3 \int_0^\infty x^2 e^{-x}\,dx \right]$$

$$= \frac{3}{2}\int_0^\infty x^2 e^{-x}\,dx \qquad = 3\int_0^\infty \frac{1}{2} x^2 e^{-x}\,dx ,$$

but the integrand is $f(x) = \frac{1}{2} x^2 e^{-x}$ our original density function and by definition a density function is a positive-valued function $f(x)$ such that

$$\int_0^\infty f(x)\,dx = 1; \qquad\qquad \text{thus}$$

$$E(X) = 3 \int_0^\infty f(x)\,dx = 3.$$

(e) $E(X) = \displaystyle\int_1^\infty x \cdot \frac{1}{x^2}\, dx = \int_1^\infty \frac{1}{x}\, dx = \lim_{b\to\infty} \int_1^b \frac{dx}{x}$

$= \displaystyle\lim_{b\to\infty} [\log b - \log 1]$

$= \displaystyle\lim_{b\to\infty} \log b = \infty;$

thus the expected value of x does not exist.

(f) $E(X) = \displaystyle\int_0^2 x\, f(x)\, dx$

$= \displaystyle\int_0^2 x(1 - |1 - x|)\, dx$

$= \displaystyle\int_0^2 [x - x\,|1 - x|]\, dx$

$= \dfrac{1}{2} x^2 \Big|_0^2 - \displaystyle\int_0^2 x\,|1 - x|\, dx$

$= 2 - \displaystyle\int_0^2 x\,|1 - x|\, dx;$

but $x\,|1 - x| = \begin{cases} x(1 - x) & \text{for } 0 \le x \le 1 \\ x(x - 1) & \text{for } 1 \le x \le 2 \end{cases}.$

Thus

$E(X) = 2 - \left[\displaystyle\int_0^1 x(1 - x)\, dx + \int_1^2 x(x - 1)\, dx \right]$

$= 2 - \displaystyle\int_0^1 x\,dx + \int_0^1 x^2\, dx - \int_1^2 x^2 + \int_1^2 x\, dx$

$= 2 - \dfrac{1}{2} x^2 \Big|_0^1 + \dfrac{1}{3} x^3 \Big|_0^1 - \dfrac{1}{3} x^3 \Big|_1^2 + \dfrac{1}{2} x^2 \Big|_1^2$

$= 2 - \dfrac{1}{2} + \dfrac{1}{3} - \dfrac{8}{3} + \dfrac{1}{3} + \dfrac{4}{2} - \dfrac{1}{2}$

$= 1.$

Find the expected value of the random variable X if X is distributed with probability density function

$$f(x) = \lambda e^{-\lambda x} \qquad \text{for} \quad 0 < X < \infty .$$

Solution: To find this expected value we will use another method.

This new method computes the expected value from $F(x) = \Pr(X \le x)$. For our random variable,

$$\Pr(X \le x) = \int_0^x f(t) \, dt = \int_0^x \lambda e^{-\lambda t} \, dt$$

$$= -\int_0^x (-\lambda) e^{-\lambda t} \, dt$$

$$= -e^{-\lambda t} \Big|_0^x = -e^{-\lambda x} - \left(-e^{-\lambda \cdot 0}\right)$$

$$= 1 - e^{-\lambda x}$$

We have defined $E(X) = \int_0^\infty x \, f(x) \, dx = \int_0^\infty x \, \lambda e^{-\lambda x} \, dx;$

but $x = \int_0^x dt.$ Thus substituting,

$$E(X) = \int_0^\infty f(x) \left[\int_0^x dt \right] dx .$$

This is an iterated integration over the shaded region,

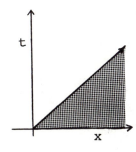

Reversing the order of integration, we integrate with respect to x first. The variable x is integrated from t to ∞ and then t is integrated from 0 to ∞. Thus,

$$\int_0^\infty f(x) \left[\int_0^x dt \right] dx = \int_0^\infty \left[\int_t^\infty f(x) \ dx \right] dt .$$

But $\int_t^\infty f(x) \ dx = \Pr(X \geq t) = 1 - \Pr(X < t) = 1 - F(t)$

or $\quad E(X) = \int_0^\infty [1 - F(t)] \ dt .$

Thus, $\quad E(X) = \int_0^\infty \left[1 - \left(1 - e^{-\lambda t} \right) \right] dt$

$$= \int_0^\infty e^{-\lambda t} dt = -\frac{1}{\lambda} e^{-\lambda t} \Big|_0^\infty = -\frac{1}{\lambda} [0 - 1] = \frac{1}{\lambda} .$$

● **PROBLEM** 9-80

Find the expected value of the random variable Y = f(X), when X is a discrete random variable with probability mass function g(x). Let f(X) = X² + X + 1

and $\Pr(X = x) = g(x) = \begin{cases} \frac{1}{2} & x = 1 \\ \frac{1}{3} & x = 2 \\ \frac{1}{6} & x = 3 . \end{cases}$

Solution: To find the expected value of a function of a random variable, we define

$$E(Y) = E(f(X)) = \sum_x f(X) \ g(x) = \sum_x f(X) \ \Pr(X = x).$$

As an example, we consider the above problem.

$$E(Y) = f(1) \Pr(X = 1) + f(2) \Pr(X = 2) + f(3) \Pr(X = 3).$$

But $f(1) = 1^2 + 1 + 1 = 3$

$f(2) = 2^2 + 2 + 1 = 7$

$f(3) = 3^2 + 3 + 1 = 13$.

Substituting we see that,

$E(Y) = 3 \Pr(X = 1) + 7 \Pr(X = 2) + 13 \Pr(X = 3)$

$= 3 \cdot \frac{1}{2} + 7 \cdot \frac{1}{3} + 13 \cdot \frac{1}{6}$

$= \frac{3}{2} + \frac{7}{3} + \frac{13}{6} = \frac{9}{6} + \frac{14}{6} + \frac{13}{6}$

$= \frac{36}{6} = 6$.

● **PROBLEM** 9-81

A factory has to decide the amount y of lengths of a certain cloth to produce. The demand for X lengths is uniformly distributed over the interval (a, b). For each length sold a profit of m dollars is made, while for each length not sold a loss of n dollars is incurred. Find the expected profit and maximize this with respect to y.

Solution: The form of the profit is related to the value of y, the amount of lengths of cloth produced. If the demand is greater than the amount produced we will sell everything made and P(X) = profit when demand is X = my.

If we make too many lengths and the demand is less than y then we will sell X for a profit of mX but will lose n dollars on each of the extra y - X lengths for a total loss of n(y - X). The net profit will be mX - n(y - X).

More succinctly, if $X \geq y$ $P(X) = my$,

if $X < y$ $P(X) = mX - n(y - X)$.

The expected profit is $E(P(X)) = \int_a^b P(X) \ f(x) \ dx$

where f(x) is the p.d.f. of X.

X is uniformly distributed over (a, b) thus

$f(x) = \dfrac{1}{b - a}$ $a \leq x \leq b$

$= 0$ otherwise.

Hence $E(P(X)) = \frac{1}{b-a} \int_a^b P(x)\ dx$.

If $y \le a$ then always $X \ge a \ge y$ and

$$E(P(X)) = \frac{1}{b-a} \int_a^b my\ dx = \frac{my}{b-a}\ x\ \Big|_a^b = my$$

which will increase until $y = a$.

If $y \ge b$ then always $X \le b \le y$ and

$$E(P(X)) = \frac{1}{b-a} \int_a^b [mx - n(y-x)]\ dx$$

$$= \frac{1}{b-a} \int_a^b [(m+n)\ x - ny]\ dx$$

$$= \frac{m+n}{2(b-a)}\ x^2\ \Big|_a^b - \frac{ny}{(b-a)}\ x\ \Big|_a^b$$

$$= \frac{m+n}{2(b-a)}\ [b^2 - a^2] - ny$$

$$= \frac{m+n}{2(b-a)}\ (b-a)(b+a) - ny$$

$$= \left(\frac{m+n}{2}\right)(b+a) - ny$$

which is strictly decreasing from $y = b$.

A graph of our $E(\text{Profit}\ (X))$ with respect to y so far is

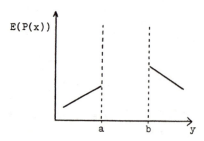

We do not know anything about the expected profit when y is in the interval (a, b) but we can see what the maximum profit with respect to y will be for $a \le y \le b$.

When $a \le y \le b$

$$E(P(X)) = \frac{1}{b - a} \left[\int_a^b P(x)\ dx \right]$$

$$= \frac{1}{b - a} \left[\int_a^y P(x)\ dx + \int_y^b P(x)\ dx \right]$$

$$= \frac{1}{b - a} \left[\int_a^y [mx - n(y - x)\ dx + \int_y^b my\ dx \right]$$

$$= \frac{1}{b - a} \left[\int_a^y [(m + n)x - ny]dx + myx \Big|_y^b \right]$$

$$= \frac{1}{b - a} \left[\frac{m + n}{2} x^2 \Big|_a^y - nyx \Big|_a^y + my(b - y) \right]$$

$$= \frac{1}{b - a} \left[\frac{m + n}{2} (y^2 - a^2) - ny(y - a) + mby - my^2 \right]$$

$$E(P(X)) = \frac{1}{b - a} \left[\frac{-(m + n)}{2} y^2 + (na + mb)y + (\frac{m + n}{2}) a^2 \right].$$

We see that the expected profit is a polynomial in y.

To maximize $E(P(X))$ we differentiate and then

solve for y when $\dfrac{d\ E(P(X))}{dy} = 0$.

$$\frac{d\ E(P(X))}{dy} = \frac{1}{b - a} [-(m + n)y + (na + mb)] = 0$$

implies that $y = \dfrac{na + mb}{m + n}$ is the amount of cloth lengths

that will maximize the expected profit.

Find the expected values of the random variables X and Y

if $\quad \Pr(X = 0) = \frac{1}{2}\quad$ and $\Pr(X = 1) = \frac{1}{2}$

and $\quad \Pr(Y = 1) = \frac{1}{4}\quad$ and $\Pr(Y = 2) = \frac{3}{4}$.

Compare the sum of $E(X) + E(Y)$ with $E(X + Y)$ if $\Pr(X = x, Y = y) = \Pr(X = x)\Pr(Y = y)$.

<u>Solution:</u>   The expected value of X is

$$E(X) = 0 \cdot \Pr(X = 0) + 1 \cdot \Pr(X = 1) = (0)\left(\frac{1}{2}\right) + (1)\left(\frac{1}{2}\right) =$$

The expected value of Y is

$$E(Y) = 1 \cdot \Pr(Y = 1) + 2 \cdot \Pr(Y = 2)\quad = \frac{1}{4} + \frac{6}{4} = \frac{7}{4}\ .$$

Thus, $E(X) + E(Y) = \frac{1}{2} + \frac{7}{4} = \frac{9}{4}$.

    To find the expected value of the random variable $(X + Y)$ we need the joint distribution of X and Y. This has been given to be

$$\Pr(X = x, Y = y) = \Pr(X = x)\ \Pr(Y = y).$$

The distribution for X and Y is

$$\Pr(X = 0,\ Y = 1) = \frac{1}{2} \cdot \frac{1}{4} = \frac{1}{8}$$

$$\Pr(X = 0,\ Y = 2) = \frac{1}{2} \cdot \frac{3}{4} = \frac{3}{8}$$

$$\Pr(X = 1,\ Y = 1) = \frac{1}{2} \cdot \frac{1}{4} = \frac{1}{8}$$

$$\Pr(X = 1,\ Y = 2) = \frac{1}{2} \cdot \frac{3}{4} = \frac{3}{8}$$

$$E(X + Y) = \sum_x \sum_y (x + y)\ \Pr(X = x,\ Y = y).$$

    That is, the expected value of the random variable X + Y is the sum of the possible values that X + Y can assume times the probability that X + Y will assume these values.

Thus, $E(X + Y) = (0 + 1) \, Pr(X = 0, \, Y = 1)$ +

$$(0 + 2) \, Pr(X = 0, \, Y = 2) \quad +$$

$$(1 + 1) \, Pr(X = 1, \, Y = 1) \quad +$$

$$(1 + 2) \, Pr(X = 1, \, Y = 2) \, ;$$

$$E(X + Y) = 1 \cdot \frac{1}{8} + 2 \cdot \frac{3}{8} + 2 \cdot \frac{1}{8} + 3 \cdot \frac{3}{8}$$

$$= \frac{1}{8} + \frac{6}{8} + \frac{2}{8} + \frac{9}{8} = \frac{18}{8} = \frac{9}{4} \, .$$

Thus $E(X + Y) = E(X) + E(Y)$.

● **PROBLEM** 9-83

Show that the sum of the expected value of two discrete random variables with joint density $f(x, y)$ is equal to the expected value of the sum of these two random variables. That is

$$E(X + Y) = E(X) + E(Y).$$

Solution: The expected value of X is $E(X) = \sum_{x} x \, g(x)$

where $g(x) = Pr(X = x) =$ the marginal distribution of X. The marginal distribution of X is $\sum_{y} f(x, y)$.

The expected value of Y is $E(Y) = \sum_{y} y \, h(y)$

where $h(y)$ is the marginal distribution of Y or $\sum_{x} f(x, y)$.

The expected value of X + Y is

$$E(X + Y) = \sum_{x} \sum_{y} (x + y) \, f(x, y)$$

but rearranging the terms in this sum we see that

$$E(X + Y) = \sum_{x} \sum_{y} x \, f(x, y) + \sum_{x} \sum_{y} y \, f(x, y)$$

$$= \sum_{x} x \left[\sum_{y} f(x, y) \right] + \sum_{y} \sum_{x} y \, f(x, y)$$

$$= \sum_x x \left[\sum_y f(x, y) \right] + \sum_y y \left[\sum_x f(x, y) \right].$$

But $\sum_y f(x, y) = g(x)$ and $\sum_x f(x, y) = h(y)$.

Thus, $E(X + Y) = \sum_x x \, g(x) + \sum_y y \, h(y) = E(X) + E(Y)$.

This result is still valid if X and Y are continuous random variables. The summation signs are replaced by integral signs throughout the proof.

● PROBLEM 9-84

Find the expected value of the product of two independent random variables, X and Y.

Solution: Let f(x, y) be the joint density of the variables X and Y. Then in the discrete case,

$$E(X \cdot Y) = \sum_x \sum_y xy \, f(x, y).$$

If X and Y are independent, their joint density function can be factored into the marginal densities of X and Y. Thus if g(x) and h(y) are the densities of X and Y respectively $f(x, y) = g(x) h(y)$.

Substituting we see that: $E(X \cdot Y) = \sum_x \sum_y xy \, g(x) h(y)$.

Factoring we see that: $E(X \cdot Y) = \sum_x x \, g(x) \left[\sum_y y \, h(y) \right]$

but $\sum_y y \, h(y) = E(y)$ and $\sum_x x \, g(x) = E(X)$.

Thus, substituting again we derive: $E(X \cdot Y) = E(X) \cdot E(Y)$,

This result is also true if X and Y are continuous random variables and the summation signs are replaced by integral signs.

Suppose the random vector (X, Y) is distributed with probability density,

$$f(x, y) = \begin{cases} x + y & \begin{array}{l} 0 < x < 1 \\ 0 < y < 1 \end{array} \\ 0 & \text{otherwise.} \end{cases}$$

Find $E[XY]$, $E[X + Y]$ and $E(X)$.

Solution: By definition,

$$E(g(x, y)) = \int\int_{(x, y)} g(x, y) \; f(x, y) \; dx \; dy$$

Thus, if $g(x,y) = xy$, we have: $E(xy) = \int_0^1 \int_0^1 xy \; (x + y) \; dx \; dy$

$$= \int_0^1 \left[\int_0^1 (x^2 y + xy^2) \; dx \right] dy \quad = \quad \int_0^1 \left[\frac{x^3 y}{3} + \frac{x^2 y^2}{2} \right]_0^1 dy$$

$$= \int_0^1 \left(\frac{y}{3} + \frac{y^2}{2} \right) \; dy \quad = \left[\frac{y^2}{6} + \frac{y^3}{6} \right]_0^1 = \frac{2}{6} = \frac{1}{3} \; .$$

$$E(X + Y) = \int_0^1 \int_0^1 (x + y)(x + y) \; dx \; dy$$

$$= \int_0^1 \int_0^1 (x^2 + 2xy + y^2) \; dx \; dy$$

$$= \int_0^1 \left[\frac{x^3}{3} + \frac{2 \; x^2 y}{2} + y^2 x \right]_0^1 dy$$

$$= \int_0^1 \left[\frac{1}{3} + y + y^2 \right] dy$$

$$= \left[\frac{y}{3} + \frac{y^2}{2} + \frac{y^3}{3} \right]_0^1 = \frac{1}{3} + \frac{1}{2} + \frac{1}{3} = \frac{7}{6} \; .$$

$$E(X) = \int_0^1 \int_0^1 x(x + y) \, dx \, dy = \int_0^1 \left[\frac{x^3}{3} + \frac{x^2 y}{2} \right]_0^1 \, dy$$

$$= \int_0^1 \left[\frac{1}{3} + \frac{y}{2} \right] \, dy = \left[\frac{y}{3} + \frac{y^2}{4} \right]_0^1$$

$$= \frac{1}{3} + \frac{1}{4} = \frac{7}{12} \, .$$

● **PROBLEM** 9-86

Compute the conditional distribution of Y given X if X and Y are jointly distributed with density

$$f(x, y) = \begin{cases} x + y & \begin{array}{l} 0 < x < 1 \\ 0 < y < 1 \end{array} \\ \\ 0 & \text{otherwise} \, . \end{cases}$$

What is the conditional expectation of Y?

<u>Solution:</u>   The conditional distribution of Y given X is defined by analogy with conditional probability to be:

$$f(y|x) = \frac{f(x, y)}{f(x)}$$

where f(x, y) is the joint density of x and y and f(x) is the marginal distribution of x.

In our example, $f(x) = \int_0^1 f(x, y) \, dy$

$$= \int_0^1 (x + y) \, dy = \left[xy + \frac{y^2}{2} \right]_0^1$$

$$= x + \frac{1}{2} \qquad\qquad 0 < x < 1 \, .$$

Thus $f(y|x) = \dfrac{f(x, y)}{f(x)} = \dfrac{x + y}{x + \frac{1}{2}} \qquad \begin{array}{l} 0 < y < 1 \\ \\ 0 < x < 1 \end{array} \, .$

To see that f(y|x) is a proper density function,

$$\int_0^1 f(y|x) \, dy = \int_0^1 \frac{x + y}{x + \frac{1}{2}} \, dy = \left(\frac{1}{x + \frac{1}{2}} \right) \int_0^1 (x + y) \, dy$$

562

$$= \left(\frac{1}{x + \frac{1}{2}} \right) \left(xy + \frac{y^2}{2} \right)_0^1 \qquad = \frac{x + \frac{1}{2}}{x + \frac{1}{2}} = 1 \ .$$

The conditional expectation of Y given X is the expectation of y against the conditional density $f(y|x)$.

Thus, $\qquad E(Y/X = x) = \int_{\text{all } y} y \, f(y/x) \, dy$.

For our example,

$$E(Y|X = x) = \int_0^1 y \left(\frac{x + y}{x + \frac{1}{2}} \right) dy = \left(\frac{1}{x + \frac{1}{2}} \right) \int_0^1 (xy + y^2) \ dy$$

$$= \frac{1}{x + \frac{1}{2}} \left[\frac{xy^2}{2} + \frac{y^3}{3} \right]_0^1 = \frac{\frac{x}{2} + \frac{1}{3}}{x + \frac{1}{2}}$$

$$= \frac{3x + 2}{3(2x + 1)} \qquad 0 < x < 1 \ .$$

● **PROBLEM** 9-87

Given the probability distribution of the random variable X in the table below, compute E(X) and Var (X).

x_i	$Pr(X = x_i)$
0	$\frac{8}{27}$
1	$\frac{12}{27}$
2	$\frac{6}{27}$
3	$\frac{1}{27}$

Solution:

$$E(X) = \sum_i x_i \; Pr(X = x_i) \quad \text{and} \quad Var \; X = E\,[(X - E(X))^2]\,.$$

Thus,
$$E(X) = (0)\;Pr(X = 0) + (1)\;Pr(X = 1)$$
$$+ (2)\;Pr(X = 2) + (3)\;Pr(X = 3)$$
$$= (0)\;\frac{8}{27} + (1)\;\frac{12}{27} + (2)\;\frac{6}{27} + 3\;\left(\frac{1}{27}\right)$$
$$= 0 + \frac{12}{27} + \frac{12}{27} + \frac{3}{27} = \frac{27}{27} = 1.$$

$$Var \; X = (0 - 1)^2 \; Pr(X = 0) + (1 - 1)^2 \; Pr(X = 1)$$
$$+ (2 - 1)^2 \; Pr(X = 2) + (3 - 1)^2 \; Pr(X = 3)$$
$$= (1^2)\;\frac{8}{27} + (0^2)\;\frac{12}{27} + (1^2)\;\frac{6}{27} + (2^2)\;\frac{1}{27}$$
$$= \frac{8}{27} + \frac{6}{27} + \frac{4}{27} = \frac{18}{27} = \frac{2}{3}\,.$$

● **PROBLEM** 9-88

Suppose that 75% of the students taking statistics pass the course. In a class of 40 students, what is the expected number who will pass. Find the variance and standard deviation.

Solution: Let X be a random variable denoting the number of students in the class of 40 who will pass the course.

If 75% of the students pass the course it is reasonable to assume that a randomly chosen student will pass the course with probability .75 and fail the course with probability .25. It is also reasonable to assume that a student passes or fails the course independently of what other students do.

With these two assumptions, it can be shown that X is a binomially distributed random variable with parameters p = .75 and n = 40. The parameter p indicates the probability of a student passing the course and n represents the number of students in the class.

In the previous problem, we have shown that the expected value of such a random variable is

$$E(X) = np\,.\qquad\text{In this case,}$$

$$E(X) = (40)(.75) = 30.$$

It has also been shown that the variance of X is

$\sigma^2 = \text{Var } X = np(1 - p)$ and substituting we see that:

$\sigma^2 = \text{Var } X \quad (40)(.75)(1 - .75) \quad = 7.5$.

The standard deviation is defined as $\sigma = \sqrt{\text{Var } X}$;

thus $\sigma = \sqrt{np(1 - p)} = \sqrt{7.5} = 2.74$.

● **PROBLEM** 9-89

Find the variance of the random variable X + b where X has variance, Var X and b is a constant.

Solution:
$$\text{Var } (X + b) = E[(X + b)^2] - [E(X + b)]^2$$
$$= E[X^2 + 2bX + b^2] - [E(X) + b]^2$$

$$= E(X^2) + 2bE(X) + b^2 - [E(X)]^2 - 2E(X)b - b^2,$$
thus $\text{Var } (X + b) = E(X^2) - [E(X)]^2 = \text{Var } X$.

● **PROBLEM** 9-90

Find the expected value and variance of a random variable,

$$Y = a_1 X_1 + a_2 X_2 + \ldots + a_n X_n$$

where the X_i are independent and each have mean μ and variance σ^2. The a_i are constants.

Solution: By a generalization of the property that the expected value of a sum is the sum of the expected values,

$$E(Y) = E\left(a_1 X_1 + \ldots + a_n X_n\right) = E(a_1 X_1) + \ldots + E\left(a_n X_n\right)$$

Also the expected value of a constant multiplied by

a random variable is the constant multiplied by the random variable or $E(ax) = a E(X)$. Thus,

$$E\left(a_i X_i\right) = a_i E\left(X_i\right) \qquad \text{and}$$

565

$$E(Y) = a_1 \, E(X_1) + a_2 \, E(X_2) + \ldots + a_n \, E\!\left(X_n\right).$$

But $E(X_1) = E(X_2) = E(X_3) \ldots = E\!\left(X_n\right) = \mu \, ;$

hence $E(Y) = a_1\mu + a_2\mu + a_3\mu + \ldots + a_n\mu$

$$= \mu \left[a_1 + a_2 + a_3 + \ldots + a_n \right] .$$

To find the variance of Y we generalize the properties of variance. Remember that if two variables, X_1 and X_2, are independent then the variance of $X_1 + X_2$ is the variance of X_1 + the variance of X_2.

Because X_1, X_2, $\ldots$ X_n are independent,

$$\text{Var } Y = \text{Var} \left[a_1 X_1 + \ldots + a_n X_n \right]$$

$$= \text{Var } (a_1 X_1) + \text{Var } (a_2 X_2) + \ldots + \text{Var} \left[a_n X_n \right].$$

Also, the variance of a constant multiplied with a random variable is the constant squared times the variance of the random variable. Equivalently,

$$\text{Var } (aX) = a^2 \, \text{Var } X.$$

Thus, $\text{Var } a_i X_i = a_i{}^2 \, \text{Var } X_i$.

But $\text{Var } X_i = \sigma^2$ for all i, hence

$$\text{Var } Y = a_1{}^2 \, \text{Var } X_1 + a_2{}^2 \, \text{Var } X_2 + \ldots + a_n{}^2 \, \text{Var } X_n$$

$$= a_1{}^2\sigma^2 + a_2{}^2\sigma^2 + \ldots + a_n{}^2\sigma^2$$

$$= \sigma^2 \left[a_1{}^2 + a_2{}^2 + \ldots + a_n{}^2 \right]$$

Find the variance of the random variable, $Z = X + Y$
if X and Y are not independent.

Solution:

$$\text{Var } Z = \text{Var } (X + Y) \quad = E[((X + Y) - E(X + Y))^2]$$

$$= E[(X - E(X) + Y - E(Y))^2]$$

$$(\text{because } E(X + Y) = E(X) + E(Y)),$$

$$= E[(X - E(X))^2 + 2(X - E(X))(Y - E(Y)) + (Y - E(Y))^2],$$

and by the properties of expectation,

$$= E[(X - E(X))^2] + 2E[(X - E(X))(Y - E(Y))] + E[Y - E(Y))^2]$$

Thus Var $Z = $ Var $X + $ Var $Y + 2E[(X - E(X))(Y - E(Y))]$.

If X and Y are independent, $E[(X - E(X))(Y - E(Y))] = 0$
but since X and Y are not independent, we may not assume
that this cross product is zero.

$$E[(X - E(X))(Y - E(Y))]$$

is called the covariance of X and Y and is a measure of
the linear relation between X and Y. It is a measure in
the   sense that , if X is greater than E(X) at the same
time that Y is greater than E(Y) with high probability, then
the covariance of X and Y will be positive. If X is below
E(X) at the same time Y is above E(Y) with high probability,
the covariance of X and Y will be negative.

Related to the covariance is the correlation
coefficient defined as;

$$\rho = \frac{\text{Cov } (X, Y)}{\sqrt{\text{Var } X} \sqrt{\text{Var } Y}} \quad .$$

The correlation coefficient gives a clearer picture
of the linear relation between X and Y because it takes
account of the variation in the individual variables X
and Y.

Other properties of covariance are: Cov $(X, Y) = 0$,
if X and Y are independent. The converse is not true.

$$\text{Cov } (X, Y) = \text{Cov } (Y, X).$$

Let Y be distributed with a Pareto distribution with parameters X_0 and θ. The density function of such a random variable is:

$$f(y) = \begin{array}{l} \dfrac{\theta\,X_0^{\theta}}{y^{\theta+1}} \qquad y > X_0 \\[2ex] \qquad\qquad \text{with } X_0,\ \theta > 0 \\[2ex] \qquad\qquad \text{otherwise}. \end{array}$$

What is the variance of Y?

Solution: We first find E(Y), the expected value of Y. By definition,

$$E(Y) = \int_{-\infty}^{\infty} y\,f(y)\,dy = \int_{X_0}^{\infty} \frac{y\,\theta\,X_0^{\theta}}{y^{\theta+1}}\,dy$$

$$= \int_{0}^{\infty} \frac{\theta X_0^{\theta}}{y^{\theta}}\,dy = \theta X_0^{\theta} \int_{0}^{\infty} \frac{dy}{y^{\theta}}$$

$$= \theta X_0^{\theta} \left[\frac{y^{-\theta+1}}{-\theta+1} \right]_{X_0}^{\infty} = \theta X_0^{\theta}\ \frac{X_0^{-\theta+1}}{\theta-1}$$

$$= \frac{\theta X_0}{\theta-1} \qquad\qquad \text{for} \qquad \theta > 1.$$

Similarly, $E(Y^2) = \displaystyle\int_{X_0}^{\infty} y^2\,\frac{\theta X_0^{\theta}}{y^{\theta+1}}\,dy = \int_{X_0}^{\infty} \frac{\theta X_0^{\theta}}{y^{\theta-1}}\,dy$

$$= \theta X_0^{\theta} \int_{X_0}^{\infty} y^{-\theta+1}\,dy = \theta X_0^{\theta} \left[\frac{y^{-\theta+2}}{-\theta+2} \right]_{X_0}^{\infty}$$

$$= \frac{\theta X_0\,\theta\,X_0^{-\theta+2}}{(\theta-2)} = \frac{\theta X_0^2}{\theta-2} \qquad \text{for} \qquad \theta > 2.$$

By definition, the variance of Y is

Var Y = E(Y²) - [E(Y)]²

$$= \frac{\theta X_0^2}{\theta-2} - \frac{\theta^2\,X_0^2}{(\theta-1)^2}$$

$$= \frac{[(\theta - 1)^2\theta - \theta^2(\theta - 2)]X_0{}^2}{(\theta - 2)(\theta - 1)^2}$$

$$= \frac{[\theta^3 - 2\theta^2 + \theta - \theta^3 + 2\theta^2]\ X_0{}^2}{(\theta - 2)(\theta - 1)^2}$$

$$= \frac{\theta\ X_0{}^2}{(\theta - 2)(\theta - 1)^2} \qquad \text{for} \qquad \theta > 2.$$

MOMENT GENERATING FUNCTION

● **PROBLEM** 9-93

Consider a simple random variable X having just two possible values, $Pr(X = 1) = p$ and $Pr(X = 0) = 1-p$. Find the moment generating function of X and $E(X^k)$ for all $k = 1, 2, 3, \ldots$

Solution: The moment generating function, $M(t)$, is defined to be

$$M(t) = E(e^{tX})$$

where t is a constant and e is the base of the natural logarithm.

In the case of a discrete, integer-valued random variable, the

expected value of e^{tX} is, $M(t) = \sum_{x=0}^{\infty} e^{tx} Pr(X = x)$.

For this problem, X has only two values, 0 and 1, hence

$$M(t) = e^{t \cdot 0}Pr(X = 0) + e^{t \cdot 1}Pr(X = 1)$$

$$= Pr(X = 0) + e^t Pr(X = 1)$$

$$= (1-p) + e^t p.$$

The moment-generating function has the interesting and useful

property that $\qquad \dfrac{d^n M(0)}{dt^n} = E(X^n)$.

That is, the nth derivative of the moment-generating function is equal to the expected value of X^n.

To see that this is true,

$$M(t) = \sum_{x=0}^{\infty} e^{tx} \Pr(X = x)$$

$$\frac{d^n M(t)}{dt^n} = \frac{d^n}{dt^n} \left(\sum_{x=0}^{\infty} e^{tx} \Pr(X = x) \right)$$

$$= \sum_{x=0}^{\infty} \frac{d^n}{dt^n} [e^{tx} \Pr(X = x)],$$

if

$$\sum_{x=0}^{\infty} \frac{d^n}{dt^n} [e^{tx} \Pr(X = x)] \quad \text{converges}$$

uniformly. But,

$$\frac{d^n}{dt^n} [e^{tx} \Pr(X = x)] = \Pr(X = x)[x^n e^{tx}] .$$

Thus,

$$\frac{d^n M(t)}{dt^n} = \sum_{x=0}^{\infty} \Pr(X = x) x^n e^{tx}$$

and

$$\frac{d^n M(0)}{dt^n} = \sum_{x=0}^{\infty} \Pr(X = x) x^n e^{x \cdot 0}$$

$$= \sum_{x=0}^{\infty} x^n \Pr(X = x) = E(X^n) .$$

In our example,

$$\frac{d^k M(t)}{dt^k} = \frac{d^k}{dt^k} [(1-p) + pe^t]$$

$$= p \frac{d}{dt^k} [e^t] = pe^t$$

and thus

$$\frac{d^k M(0)}{dt^k} = pe^0 = p$$

or

$$E(X^k) = p \quad \text{for} \quad k = 1, 2, \ldots$$

Consider the distribution   defined   by the following distribution function:

$$F\ (x) = \begin{cases} 0 & \text{if} \quad x < 0 \\ 1\text{-}pe^{-x} & \text{if} \quad x \geq 0 \text{ for } 0 < p < 1. \end{cases}$$

This distribution is partly discrete and partly continuous. Find the moment generating function of X and use it to find the mean and variance of X.

<u>Solution</u>: The density function of X is

$$Pr\,(X = 0) = 1 - p$$

and

$$f(x) = \frac{dF(x)}{dx} = pe^{-x} \qquad x > 0 \ .$$

Thus, the moment generating function of X is

$$M(t) = E(e^{Xt}) = \int_{-\infty}^{\infty} e^{tx} \ dF(x)$$

where this integral is a Riemann-Stieljes integral reflecting both the discrete and continuous nature of X.

$$E(e^{Xt}) = e^{t \cdot 0} Pr\,(X = 0) + \int_{0}^{\infty} e^{tx} \ pe^{-x} \ dx$$

$$= (1\text{-}p) + \int_{0}^{\infty} pe^{-x(1-t)} \ dx \ ;$$

but

$$\int_{0}^{\infty} pe^{-x(1-t)} dx = \frac{-p}{1-t} \ e^{-x(1-t)} \Big]_{0}^{\infty} \ ; \text{ for } \ |t| < 1$$

this equals

$$\frac{-p}{1-t} \ [0 - 1] = \frac{p}{1-t} \ .$$

Thus,

$$M(t) = (1\text{-}p) + \frac{p}{1-t} \quad \text{for} \quad |t| < 1; \ \frac{1}{1-t} \quad \text{may be expanded}$$

in a geometric series

$$\frac{1}{1-t} = 1 + t + t^2 + \dots \ .$$

Hence

$$M(t) = 1\text{-}p + p(1 + t + t^2 + \dots)$$

$$= 1 + pt + pt^2 + \dots$$

$$= 1 + \sum_{n=1}^{\infty} pt^n \ ,$$

$$E(X) = \frac{dM(0)}{dt} = 0 + p \sum_{n=1}^{\infty} n \, t^{n-1} \Big]_{t=0}$$

$$= 0 + p + p \sum_{n=1}^{\infty} (n+1) t^{n} \Big]_{t=0}$$

$$= p + p \sum_{n=1}^{\infty} (n+1)(0)^n = p .$$

Thus the mean of X is $E(X) = p$.

The variance of X is $\operatorname{Var} X = E(X^2) - [E(X)]^2$

$$E(X^2) = \frac{d^2 M(0)}{dt^2} = \frac{d^2}{dt^2}\left(1 + p \sum_{n=1}^{\infty} t^n\right)\Big|_{t=0}$$

$$= p \frac{d^2}{dt^2}(t + t^2 + t^3 + \dots)\Big|_{t=0}$$

$$= p \frac{d}{dt}(1 + 2t + 3t^2 + \dots)\Big|_{t=0}$$

$$= p(0 + 2 + 6t + \dots)\Big|_{t=0}$$

$$= 2p .$$

Thus,

$$\operatorname{Var} X = E(X^2) - [E(X)]^2$$

$$= 2p - p^2 = p(2 - p).$$

SPECIAL DISCRETE DISTRIBUTIONS

● **PROBLEM** 9-95

Find the uniform distribution for the subsets of months of size 3.

Solution: Since there are 12 possible months, we may choose 3 at random in

$$\binom{12}{3} = 220 \text{ ways. Numbering these subsets from 1}$$

to 220 the probability distribution is given by

$$F(x) = \frac{1}{220}, \ x = 1, 2, \dots, 220,$$

since any choice is equally likely. Thus the probability of choosing any given subset is:

$$F(92) = \frac{1}{220} .$$

Find the probability that a person flipping a balanced coin requires four tosses to get a head.

Solution: Suppose one performs a series of repeated Bernoulli trials until a success is observed, and then stops. The total number of trials is random and equal to the number of failures + 1 for the success. The probability that there will be K trials is equal to the probability that there are K - 1 failures followed by a success. If the probability of a success is p,

$$\Pr(X = K) = \underbrace{(1 - p)(1 - p) \ldots (1 - p)}_{K - 1 \text{ times.}} p$$

In other words $F(x) = (1 - p)^{x - 1} p; \ x = 1, 2, 3, \ldots$

This is called the geometric distribution. Our problem is one of this type. We have $p = 1 - p = \dfrac{1}{2}$.

We want $F(4) = \left(\dfrac{1}{2}\right)^{4-1} \dfrac{1}{2} = \left(\dfrac{1}{2}\right)^{4} = \dfrac{1}{16}$.

Suppose that flaws in plywood occur at random with an average of one flaw per 50 square feet. What is the probability that a 4 foot × 8 foot sheet will have no flaws? At most one flaw? To get a solution assume that the number of flaws per unit area is Poisson distributed.

Solution: This problem will serve to introduce the Poisson distribution. A random variable X is defined to have a Poisson distribution if the density of X is given

by $\Pr(X = K) = \dfrac{e^{-\lambda} \lambda^{K}}{K!}$ for $K = 0, 1, 2, \ldots$

where $\lambda > 0$.

The Poisson distribution has the unique property that the expectation equals the variance and they equal the value of the parameter λ. We will prove this by using the moment generating function.

$$m_x(t) = E(e^{tx}) = \sum_{x=0}^{\infty} e^{tx} \, P(X=x) = \sum_{x=0}^{\infty} e^{tx} \, \frac{e^{-\lambda} \lambda^x}{x!}$$

$$= e^{-\lambda} \sum_{x=0}^{\infty} e^{tx} \, \frac{\lambda^x}{x!} = e^{-\lambda} \sum_{x=0}^{\infty} \frac{(\lambda e^t)^x}{x!}$$

$$= e^{-\lambda} e^{\lambda e^t} = e^{\lambda(e^t - 1)}.$$

$$m'_x(t) = e^{\lambda(e^t-1)} \, \frac{d}{dt} (\lambda e^t - \lambda) = e^{\lambda(e^t-1)} \, \lambda e^t.$$

By the properties of the moment generating function:

$$E(x) = m'_x(0) = e^{\lambda(e^0-1)} \, \lambda e^0 = e^{\lambda(1-1)} \lambda e^0 = e^0 \lambda e^0 = \lambda.$$

Also $Var(x) = E(x^2) - (E(x))^2 = m''_x(0) - (m'_x(0))^2$

$$m''_x(t) = \lambda \left[e^{\lambda(e^t-1)} \, e^t + e^t \left(e^{\lambda(e^t-1)} \, \lambda e^t \right) \right]$$

$$m''(0) = \lambda \left[e^{\lambda(e^0-1)} \, e^0 + e^0 \left(e^{\lambda(e^0-1)} \, \lambda \, e^0 \right) \right]$$

$$= \lambda(1 + \lambda) = \lambda + \lambda^2.$$

So $Var(x) = (\lambda + \lambda^2) - \lambda^2 = \lambda.$

For our problem, we can calculate the expected value and use that as λ. We expect 1 flaw per 50 square feet. Hence we expect $\frac{1}{50}$ flaw per square foot. We have $4 \times 8 = 32$ sq. ft. We expect $\lambda = \frac{32}{50}$ flaws.

$$Pr(\text{no flaws}) = Pr(x=0) = \frac{e^{-\frac{32}{50}} \left(\frac{32}{50}\right)^0}{0!} = e^{-\frac{32}{50}} = e^{-.64}.$$

$$Pr(\text{at most one flaw}) = Pr(\text{no flaws}) + Pr(\text{1 flaw})$$

$$= e^{-.64} \frac{e^{-\frac{32}{50}} \left(\frac{32}{50}\right)^1}{1!}$$

$$= e^{-.64} + .64 \, e^{-.64}.$$

Consider the Poisson distribution $\dfrac{e^{-\lambda} \lambda^k}{k!}$.

Prove $\quad \dfrac{e^{-\lambda} \lambda^{k-1}}{(k-1)!} < \dfrac{e^{-\lambda} \lambda^k}{k!} \qquad$ for $k < \lambda$,

$\qquad \dfrac{e^{-\lambda} \lambda^{k-1}}{(k-1)!} > \dfrac{e^{-\lambda} \lambda^k}{k!} \qquad$ for $k > \lambda$,

$\qquad \dfrac{e^{-\lambda} \lambda^{k-1}}{(k-1)!} = \dfrac{e^{-\lambda} \lambda^k}{k!} \qquad$ if λ is an integer and

$k = \lambda$.

Solution: Consider the following ratio:

$$R = \dfrac{\dfrac{e^{-\lambda} \lambda^{k-1}}{(k-1)!}}{\dfrac{e^{-\lambda} \lambda^k}{k!}}$$

If $R > 1$ then $\qquad \dfrac{e^{-\lambda} \lambda^{k-1}}{(k-1)!} > \dfrac{e^{-\lambda} \lambda^k}{k!}$.

If $R < 1$ then $\qquad \dfrac{e^{-\lambda} \lambda^{k-1}}{(k-1)!} < \dfrac{e^{-\lambda} \lambda^k}{k!}$.

If $R = 1$ then $\qquad \dfrac{e^{-\lambda} \lambda^{k-1}}{(k-1)!} = \dfrac{e^{-\lambda} \lambda^k}{k!}$.

$$R = \dfrac{\dfrac{e^{-\lambda} \lambda^{k-1}}{(k-1)!}}{\dfrac{e^{-\lambda} \lambda^k}{k!}} = \dfrac{\dfrac{\lambda^{k-1}}{(k-1)!}}{\dfrac{\lambda^k}{k!}} = \dfrac{\lambda^{k-1}}{(k-1)!} \dfrac{k!}{\lambda^k}$$

$$= \dfrac{k}{\lambda} .$$

Hence if $k < \lambda$, $R < 1$; if $k > \lambda$ $R > 1$; and if $k = \lambda$, $R = 1$. The result follows immediately. Note that λ must be an integer for $\lambda = k$, since k is an integer.

A lot consisting of 100 fuses, is inspected by the following procedure. Five of these fuses are chosen at random and tested; if all 5 "blow" at the correct amperage, the lot is accepted. Find the probability distribution of the number of defectives in a sample of 5 assuming there are 20 in the lot.

Solution: We want to find $\Pr(X = x)$. We will use the classical model of probability;

$$\Pr(X = x) = \frac{\text{no. of ways of getting x defectives in 5 draws}}{\text{total possible draws of 5}}$$

We can draw 5 fuses out of a lot of 100 in $\binom{100}{5}$ ways.

Also we can select x defectives out of the 20 in $\binom{20}{x}$ ways. In addition we must choose $5 - x$ out of $100 - 20 = 80$ nondefectives. We can do this in $\binom{80}{5-x}$ ways. By the multiplication law we can get x defectives out of a lot of 100 in $\binom{20}{x}\binom{80}{5-x}$ ways.

Hence $F(x) = \Pr(X = x) = \dfrac{\binom{20}{x}\binom{80}{5-x}}{\binom{100}{5}}$

for $x = 0, 1, 2, 3, 4, 5$.

This is an example of the hypergeometric distribution.

NORMAL DISTRIBUTIONS

If Z is a standard normal variable, use the table of standard normal probabilities to find:

    (a) $\Pr(Z < 0)$
    (b) $\Pr(-1 < Z < 1)$
    (c) $\Pr(Z > 2.54)$.

Solution: The normal distribution is the familiar

"bell-shaped" curve. It is a continuous probability
distribution that it widely used to describe the
distribution of heights, weights, and other characteristics.

The density function of the standard normal
distribution is

$$f(x) = \frac{1}{\sqrt{2\pi}} \exp \left(\frac{-x^2}{2} \right) \qquad -\infty < x < \infty \ .$$

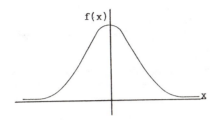

is a graph of this density. The probability of a standard
normal variable being found in a particular interval can be
found with the help of tables found in the backs of most
statistics text books.

(a) To find the probability $Pr(Z < 0)$ we can take
advantage of the fact that the normal distribution is
symmetric about its mean of zero. Thus

$Pr(Z > 0) = Pr(Z < 0)$.   We know that

$Pr(Z > 0) + Pr(Z < 0) = 1$

because $Z > 0$ and $Z < 0$ are exhaustive events.   Thus

$2Pr(Z < 0) = 1$ or $Pr(Z < 0) = \frac{1}{2}$.

(b) To find the $Pr(-1 < Z < 1)$ we use the tables
of the standard normal distribution.

$Pr(-1 < Z < 1) = Pr(Z < 1) - Pr(Z < -1)$.

Reading across the row headed by 1 and down the
column labeled .00 we see that $Pr(Z < 1.0) = .8413$.

$Pr(Z < -1) = Pr(Z > 1)$ by the symmetry of the
normal distribution. We also know that

$Pr(Z > 1) = 1 - Pr(Z < 1)$.

Substituting we see,

$Pr(-1 < Z < 1) = Pr(Z < 1) - [1 - Pr(Z < 1)]$

$= 2Pr(Z < 1) - 1 \quad = 2(.8413) - 1 = .6826.$

(c) $Pr(Z > 2.54) = 1 - Pr(Z < 2.54)$ and reading across the row labeled 2.5 and down the column labeled .04 we see that $Pr(Z < 2.54) = .9945$.

Substituting,

$Pr(Z > 2.54) = 1 - .9945 = .0055$.

Find $\Phi(- .45)$.

Solution: $\quad \Phi(- .45) = Pr(Z \leq - .45),$

where Z is distributed normally with mean 0 and variance 1.

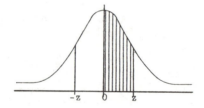

Let $A(Z) =$ the area under the curve from 0 to Z. From our table we find $A(.45) = .17364$ and by the symmetry of the normal distribution,

$A(- .45) = A(.45) = .17364.$

We wish to find $\Phi(- .45)$, the shaded area below.

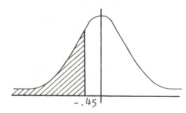

We know that $\Phi(0) = .5000$ and from the diagram below we know that $\Phi(0) - A(- .45) = \Phi(- .45)$.

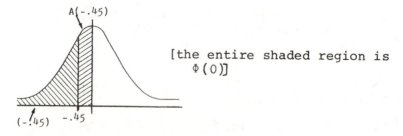

[the entire shaded region is $\Phi(0)$]

578

Substituting,

$$\phi(0) - A(-.45) = .5000 - .17364 = \phi(-.45)$$

and $\phi(-.45) = .32636.$

In a normal distribution, what is the Z-score equivalent
of the median? What is the Z score above which only
16 percent of the distribution lies? What percentage of
the scores lie below a Z score of +2.0?

Solution: The median is the number such that 1/2 of a
probability distribution lies above or below it. Equi-

valently the median is a number $\tilde{m}$ such that a random
observation X from a distribution is equally likely to be
above or below it. Thus

$$Pr(X \geq \tilde{m}) = Pr(X \leq \tilde{m}) = \frac{1}{2} .$$

To find the Z-score equivalent of the median we
wish to find some number $\tilde{m}$ such that

$$Pr(Z \geq \tilde{m}) = Pr(Z \leq \tilde{m}) = \frac{1}{2}$$

where Z is a normally distributed random variable with
mean 0 and variance 1.

From the tables or from the fact that the normal
distribution is symmetrical about its mean we have

$$Pr(Z \geq 0) = Pr(Z \leq 0) = \frac{1}{2} .$$

Thus the median is $\tilde{m} = 0.$

To find the Z-score above which 16 percent of the
distribution lies we find a constant C such that

$$Pr(Z \geq C) = 16\% = .160 \quad \text{or equivalently}$$

$$Pr(Z \leq C) = 1 - .160 = .840.$$

Searching for .8400 in the body of the table and

then reading up the appropriate row and column we find that

Pr(Z < 1) = .84, thus C = 1.

To find the percentage of scores that lie below a Z-score of 2, we wish to find Pr(Z < 2.00).

Reading across the column labeled 2.0 and then down the row headed by .00, we find

Pr(Z < 2.00) = .9772,

but .9772 is 97.72% of 1; thus 97.72% of the Z-scores lie below 2.00.

● **PROBLEM** 9-103

Given that x has a normal distribution with mean 10 and standard deviation 4, find P(x < 15).

Solution: x is a normal random variable with a mean or location parameter of 10 and a standard deviation or scale parameter of 4. A graph of its density function might look like this:

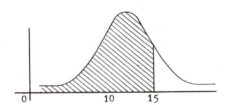

We wish to find the Pr(x < 15) or the area of the shaded area. It would be possible to construct tables which would supply such probabilities for many different values of the mean and standard deviation. Luckily this is not necessary. We may shift and contort our density function in such a way so that only one table is needed. How is such a change accomplished?

First the mean is subtracted from x giving a new random variable, x - 10. This new random variable is normally distributed but E(x - 10), the mean of x - 10, is E(x) - 10 = 0. We have shifted our distribution so that it is centered at 0.

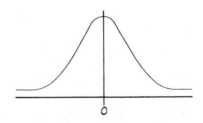

We can contort our new random variable by dividing
by the standard deviation creating a new random variable
$Z = \frac{x - 10}{4}$; the variance of Z is

$$\text{Var} \left(\frac{x - 10}{4}\right) = \frac{\text{Var } x}{16} = \frac{(\text{standard deviation of x})^2}{16} = 1.$$

Fortunately, after all this twisting and shifting, our
new random variable Z is still normally distributed and
has mean 0 and variance 1. This new random variable is
referred to as a Z-score or standard random variable and
tables for its probabilities are widespread.

To solve our problem we first convert an x-score to
a Z-score and then consult the appropriate table.

$$\text{Pr}(x < 15) = \text{Pr} \left(\frac{x - 10}{4} < \frac{15 - 10}{4}\right)$$

$$= \text{Pr} \left(Z < \frac{5}{4}\right) = \Phi(1.25) = .5000 + A(1.25)$$

$$= .5000 + .39439 = .89439.$$

● **PROBLEM** 9-104

Given a normal population with $\mu = 25$ and $\sigma = 5$, find
the probability that an assumed value of the variable
will fall in the interval 20 to 30.

<u>Solution</u>:   We wish to find the probability that a
normally distributed random variable, X, mean $\mu = 25$
and standard deviation $\sigma = 5$, will lie in the interval
(20, 30), $\text{Pr}(20 < X < 30)$. We convert X from a normally
distributed variable with mean 25 and standard deviation
5 to Z, a normally distributed random variable with mean
0 and standard deviation 1. The formula for conversion
is $Z = \frac{X - \mu}{\sigma}$. Thus,

$$\text{Pr}(20 < X < 30) = \text{Pr} \left(\frac{20 - \mu}{\sigma} < \frac{X - \mu}{\sigma} < \frac{30 - \mu}{\sigma}\right)$$

$$= \text{Pr} \left(\frac{20 - 25}{5} < Z < \frac{30 - 25}{5}\right) = \text{Pr}(-1 < Z < 1).$$

To find this   probability involving the random
variable Z, we resort to prepared tables which are
usually found in statistics texts. There are a variety of
such tables and we find $\text{Pr}(-1 < Z < 1)$ three different
ways to illustrate the various types of tables.

(1) This type of table gives $Pr(0 < Z < a)$. To find the $Pr(-1 < Z < 1)$, we first find $Pr(0 < Z < 1)$. This is $Pr(0 < Z < 1) = .341$. Next use the fact that the standard normal distribution is symmetrical about 0, hence $Pr(0 < Z < a) = Pr(0 > Z > -a)$.

From this fact,

$$Pr(0 > Z > -1) = Pr(0 < Z < 1) = .341$$

$$Pr(-1 < Z < 1) = Pr(0 > Z > -1) + Pr(0 < Z < 1)$$

$$= 2(.341) = .682 .$$

(2) Another type of table gives $Pr(Z \leq a)$ for various values of a. To use this table we note that;

$$Pr(-1 \leq Z \leq 1) = Pr(Z \leq 1) - Pr(Z \leq -1);$$

from the table we see that

$$Pr(Z \leq 1) = .841 \qquad \text{and} \qquad Pr(Z \leq -1) = .159 .$$

Thus, $Pr(-1 \leq Z \leq 1) = .841 - .159 = .682.$

(3) This type of table gives $Pr(Z \geq a)$ for certain values of a.

$$Pr(-1 \leq Z \leq 1) = Pr(-1 \leq Z) - Pr(1 \leq Z)$$

But $Pr(-1 \leq Z) = Pr(1 \geq Z)$ by symmetry ;

$$= 1 - Pr(1 \leq Z)$$

Thus,

$$Pr(-1 \leq Z \leq 1) = [1 - Pr(1 \leq Z)] - Pr(1 \leq Z)$$

$$= 1 - 2Pr(1 \leq Z) = 1 - 2(.1587) = .682.$$

● **PROBLEM** 9-105

The average grade on a mathematics test is 82, with a standard deviation of 5. If the instructor assigns A's to the highest 12% and the grades follow a normal distribution, what is the lowest grade that will be assigned A?

Solution: We relate the given information to the normal curve by thinking of the highest 12% of the grades as 12% of the area under the right side of the curve. Then the lowest grade assigned A is that point on the X-axis for which the area under the curve to its right is 12% of the total area.

582

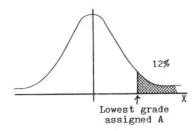

Lowest grade
assigned A

The standard normal curve is symmetric about the Y-axis; this means that if we take the additive inverse of any Z-score, its Y value will be unchanged. For example, the area to the right of Z = 2 is equal to the area to the left of Z = - 2; $Pr(Z \geq 2) = Pr(Z \leq - 2)$. It follows that the area under the curve to the right of Z = 0 is exactly half the total area under the curve. Therefore the area between the Y-axis and the desired X-score is 50% - 12% = 38% of the total area, which is 1. If we can find a K for which $P(0 < Z < K) = .38$, K can be converted to an X-score which is the lowest grade assigned A.

Using the table of areas under the standard normal curve, we locate .380 and see that its Z-score is 1.175. This means that $P(0 < Z < 1.175) = .380$. This is equivalent to $P(Z > 1.175) = .500 - .380 = .120$.

Now we can convert Z = 1.175 to an X-score by solving the equation

$$1.175 = \frac{X - \mu}{\sigma}$$ for X. We are told that the average

grade on the test is 82 = μ (the mean) and the standard deviation is 5. Substituting we have

$$1.175 = \frac{X - 82}{5}$$ and $5(1.175) = X - 82$, so that

$$X = 5(1.175) + 82 = 87.875.$$ This means that

$$P(Z > 1.175) = P(X > 87.875) = .120.$$

Since all grades are integers, the integer just above 87.875 is 88, which is the lowest grade assigned an A.

Let X be a normally distributed random variable representing the hourly wage in a certain craft. The mean of the hourly wage is \$4.25 and the standard deviation is \$.75.

    (a) What percentage of workers receive hourly wages between \$3.50 and \$4.90?

    (b) What hourly wage represents the 95th percentile?

Solution: (a) We seek,

$$Pr(3.50 \leq X \leq 4.90).$$

Converting to Z-scores we see that

$$Pr(3.50 \leq X \leq 4.90) = Pr\left(\frac{3.50 - \mu}{\sigma} \leq Z \leq \frac{4.90 - \mu}{\sigma}\right)$$

$$= Pr\left(\frac{3.50 - 4.25}{.75} \leq Z \leq \frac{4.90 - 4.25}{.75}\right)$$

$$= Pr\left(\frac{-.75}{.75} \leq Z \leq \frac{.65}{.75}\right) \quad = Pr(-1 \leq Z \leq .87)$$

$$= Pr(Z \leq .87) - Pr(Z \leq -1) = Pr(Z \leq .87) - (1 - Pr(Z \geq -1))$$

$$= Pr(Z \leq .87) - [1 - Pr(Z \leq 1)] = .809 - [1 - .841]$$
$$= .650 .$$

    Thus 65% of the hourly wages are between \$3.50 and \$4.90.

(b) The 95th percentile is that number Z_α such that

$$Pr(X \leq K) = .95.$$

To find Z_α, we first convert to Z-scores. Thus

$$Pr(X \leq Z_\alpha) = Pr\left(\frac{X - 4.25}{.75} \leq \frac{K - 4.25}{.75}\right) = .95$$

$$= Pr\left(Z \leq \frac{K - 4.25}{.75}\right) = .95 .$$

But $Pr(Z \leq 1.645) = .95$ thus $\dfrac{K - 4.25}{.75} = 1.645,$

$K = 4.25 + (.75)(1.645) \quad = 5.48.$

    Thus 95% of the craftsmen have hourly wages less than \$5.48.

A pair of dice *is* thrown 120 times. What is the approximate probability of throwing at least 15 sevens? Assume that the rolls are independent and remember that the probability of rolling a seven on a single roll is $\frac{6}{36} = \frac{1}{6}$.

Solution: The answer to this problem is a binomial probability. If X = number of sevens rolled, n = 120, then

$$Pr(X \geq 15) = \sum_{j=15}^{120} \binom{120}{j}\left(\frac{1}{6}\right)^{j}\left(\frac{5}{6}\right)^{120-j} .$$

This sum is quite difficult to calculate. There is an easier way. If n is large, $Pr_B(X \geq 15)$ can be approximated by $Pr_N(X \geq 14.5)$ where X is normally distributed with the same mean and variance as the binomial random variable. Remember that the mean of a binomially distributed random variable is np; n is the number of trials and p is the probability of "success" in a single trial. The variance of a binomially distributed random variable is np(1 - p) and the standard deviation is $\sqrt{np(1 - p)}$.

Because of this fact $\dfrac{X - np}{\sqrt{np(1 - p)}}$ is normally distributed with mean 0 and variance 1.

$$np = (120)\left(\frac{1}{6}\right) = 20 \qquad\qquad \text{and}$$

$$\sqrt{np(1 - p)} = \sqrt{120\left(\frac{1}{6}\right)\left(\frac{5}{6}\right)} = \sqrt{\frac{50}{3}} = 4.08248 .$$

Thus $Pr(X \geq 15) = Pr\left(\dfrac{X - 20}{4.08248} > \dfrac{14.5 - 20}{4.08248}\right)$

$= Pr(Z > -1.35) = 1 - Pr(Z < -1.35)$

$= 1 - \Phi(-1.35)$ $= 1 - .0885$
$$= .9115 .$$

$$Pr_B(X > 15) \stackrel{\sim}{\sim} Pr_N(X \geq 14.5) .$$

The reason 15 has become 14.5 is that a discrete random variable is being approximated by a continuous random variable.

Consider the example below:

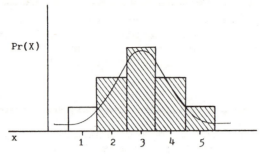

$Pr_B(2 \leq X \leq 5)$ = sum of the areas of the shaded

rectangles. In approximating this area with a curve we must start at the edge of the first shaded rectangle and move to the edge of the last shaded rectangle. This implies

$$Pr_B(2 \leq X \leq 5) \overset{\sim}{\,} Pr_N(1.5 \leq X \leq 5.5).$$

SPECIAL CONTINUOUS DISTRIBUTIONS

● **PROBLEM** 9-108

The simplest continuous random variable is the one whose distribution is constant over some interval (a, b) and zero elsewhere. This is the uniform distribution.

$$f(x) = \begin{cases} \dfrac{1}{b - a}, & a \leq X \leq b \\[2mm] 0, & \text{elsewhere} \end{cases}$$

Find the mean and variance of this distribution.

Solution: By definition,

$$E(x) = \int_{-\infty}^{\infty} x\, f(x)\, dx = \int_{a}^{b} \frac{1}{b - a}\, x\, dx$$

$$= \frac{1}{b - a} \int_{a}^{b} x\, dx = \frac{1}{b - a}\, \frac{x^2}{2} \bigg|_{a}^{b}$$

$$= \frac{b^2 - a^2}{2} \left(\frac{1}{b - a}\right) = \frac{a + b}{2}\ .$$

For the variance we must first find $E(X^2)$. By definition

$$E(X^2) = \int_{-\infty}^{\infty} x^2 \, f(x) \, dx$$

$$= \int_{a}^{b} x^2 \, \frac{1}{b-a} \, dx$$

$$= \frac{1}{b-a} \int_{a}^{b} x^2 \, dx$$

$$= \frac{1}{b-a} \left. \frac{x^3}{3} \right|_{a}^{b}$$

$$= \frac{b^3 - a^3}{3(b-a)} \quad .$$

But Var $(X) = E(X^2) - (E(x))^2$

$$= \frac{b^3 - a^3}{3(b-a)} - \left(\frac{b+a}{2}\right)^2$$

$$= \frac{b^3 - a^3}{3(b-a)} - \frac{(a^2 + 2ab + b^2)}{4}$$

$$= \frac{(b^2 + ab + a^2)(b-a)}{3(b-a)} - \frac{(a^2 + 2ab + b^2)}{4}$$

$$= \frac{(b^2 + ab + a^2)}{3} - \frac{(a^2 + 2ab + b^2)}{4}$$

$$= \frac{(4b^2 + 4ab + 4a^2)}{12} - \frac{(3a^2 + 6ab + 3b^2)}{12}$$

$$= \frac{b^2 - 2ab + a^2}{12} = \frac{(b-a)^2}{12} \quad .$$

587

Consider the exponential distribution $f(x) = \lambda e^{-\lambda x}$ for $x > 0$. Find the moment generating function and from it, the mean and variance of the exponential distribution.

<u>Solution:</u>   By definition $M_x(t) = E(e^{tx})$

$$= \int_{-\infty}^{\infty} e^{tx} f(x) \, dx$$

$$= \int_{x=0}^{\infty} e^{tx} \lambda e^{-\lambda x} \, dx$$

$$= \int_{0}^{\infty} \lambda e^{(t - \lambda)x} \, dx = \lambda \int_{0}^{\infty} e^{(t - \lambda)x} \, dx$$

$$= \lambda \left[\frac{-1}{t - \lambda} \ e^{(t - \lambda)x} \right]_{0}^{\infty}$$

$$= \frac{\lambda}{\lambda - t} \left[e^{(t - \lambda)x} \right]_{0}^{\infty} .$$

Consider $t < \lambda$. Then $\lambda - t > 0$ and $t - \lambda < 0$. Hence $e^{(t - \lambda)x} = e^{-kx}$ and $M_x(t) = \frac{\lambda}{\lambda - t} (0 - (-1)) = \frac{\lambda}{\lambda - t}$

for $t < \lambda$.

The mean is

$$E(x) = M_x'(t) \Big|_{t=0}$$

$$M_x'(t) = \frac{d}{dt} \left[\frac{\lambda}{\lambda - t} \right] = \lambda \left[\frac{d}{dt} \ \frac{1}{\lambda - t} \right]$$

$$= \lambda \left[\frac{-1}{(\lambda - t)^2} \right] \frac{d}{dt} (\lambda - t) = \frac{\lambda}{(\lambda - t)^2}$$

$$M_x'(0) = E(x) = \frac{\lambda}{(\lambda - 0)^2} = \frac{\lambda}{\lambda^2} = \frac{1}{\lambda} .$$

Also by the moment generating function's properties

$$E(x^2) = M_x''(t) \Big|_{t=0}$$

$$M_x''(t) = \frac{d}{dt} \frac{\lambda}{(\lambda - t)^2} = \lambda \frac{d}{dt} \frac{1}{(\lambda - t)^2} .$$

$$= \lambda \frac{-2}{(\lambda - t)^3} \frac{d}{dt} (\lambda - t) = \frac{2\lambda}{(\lambda - t)^3}$$

$$M_x''(0) = E(x^2) = \frac{2\lambda}{\lambda^3} = \frac{2}{\lambda^2} \quad ,$$

Now $Var(X) = E(x^2) - (E(x))^2$

$$= \frac{2}{\lambda^2} - \left(\frac{1}{\lambda}\right)^2$$

$$= \frac{1}{\lambda^2} \quad .$$

● **PROBLEM** 9-110

The following density function defines the gamma distribution:

$$f(x) = \frac{\lambda^\alpha}{\Gamma(\alpha)} x^{\alpha-1} e^{-\lambda x}, \text{ for } X > 0,$$

where α and λ are positive parameters. $\Gamma(t)$, the gamma function, is defined by

$$\Gamma(t) = \int_0^\infty x^{t-1} e^{-x} dx \quad \text{for } t > 0.$$

Integration by parts yields the following recursion relations: $\Gamma(t + 1) = t \Gamma(t)$, and if n is an integer $\Gamma(n + 1) = n!$. Find the moment generating function, mean, and variance of the gamma distribution.

Solution: First find the moment generating function.

$$M_x(t) = E(e^{tx}) = \int_0^\infty e^{tx} f(x) dx$$

$$= \int_0^\infty e^{tx} \frac{\lambda^\alpha}{\Gamma(\alpha)} x^{\alpha-1} e^{-\lambda x} dx$$

$$= \lambda^\alpha \int_0^\infty \frac{e^{(t-\lambda)x}}{\Gamma(\alpha)} x^{\alpha-1} dx$$

$$= \left(\frac{\lambda}{\lambda - t}\right)^\alpha \int_0^\infty \frac{(\lambda - t)^\alpha}{\Gamma(\alpha)} x^{\alpha-1} e^{-(\lambda-t)x} dx,$$
$$(t < \lambda).$$

The integral is that of a gamma distribution with parameters α and $\lambda - t$ and hence must be 1. Therefore

$$M_x(t) = \left(\frac{\lambda}{\lambda - t}\right)^\alpha .$$

By the properties of the moment generating function,

$$E(x) = M_x'(t) \text{ at } t = 0.$$

$$M_x'(t) = \frac{d}{dt}\left(\frac{\lambda}{\lambda - t}\right)^\alpha = \lambda^\alpha \frac{d}{dt} \frac{1}{(\lambda - t)^\alpha}$$

$$= \lambda^\alpha \frac{-\alpha}{(\lambda - t)^{\alpha+1}} (-1) = \frac{\alpha\,\lambda^\alpha}{(\lambda - t)^{\alpha+1}}$$

$$E(x) = M_x'(0) = \frac{\alpha\,\lambda^\alpha}{\lambda^{\alpha+1}} = \frac{\alpha}{\lambda} .$$

We also know

$$E(x^2) = M_x''(t) \Big|_{t=0} .$$

$$M_x''(t) = \frac{d}{dt} \frac{\alpha\,\lambda^\alpha}{(\lambda - t)^{\alpha+1}} = \alpha\lambda^\alpha \frac{d}{dt} \frac{1}{(\lambda - t)^{\alpha+1}}$$

$$= \alpha\lambda^\alpha \frac{-(\alpha + 1)}{(\lambda - t)^{\alpha+2}} (-1) = \frac{\alpha\,(\alpha + 1)\,\lambda^\alpha}{(\lambda - t)^{\alpha+2}}$$

$$E(x^2) = M_x''(0) = \frac{\alpha\,(\alpha + 1)\lambda^\alpha}{\lambda^{\alpha+2}} = \frac{\alpha\,(\alpha + 1)}{\lambda^2}$$

Finally, $\text{Var}(X) = E(x^2) - (E(x))^2$

$$= \frac{\alpha\,(\alpha + 1)}{\lambda^2} - \frac{\alpha^2}{\lambda^2}$$

$$= \frac{\alpha^2 + \alpha - \alpha^2}{\lambda^2} = \frac{\alpha}{\lambda^2} .$$

SAMPLING THEORY

● **PROBLEM** 9-111

> Suppose that a sociologist desires to study the religious habits of 20-year-old males in the United States. He draws a sample from the 20-year-old males of a large city to make his study. Describe the sampled and target populations. What problems arise in drawing conclusions from this data?

Solution: The sampled population consists of the 20-year-old males in the city which the sociologist samples.

The target population consists of the 20-year-old males in the United States.

In drawing conclusions about the target population, the researcher must be careful. His data may not reflect religious habits of 20-year-old males in the United States but rather the religious habits of 20-year-old males in the city that was surveyed. The reliability of the extrapolation cannot be measured in probabilistic terms.

● **PROBLEM** 9-112

> It helps to remember that the total of all probabilities in a sampling distribution is always 1.   If the probability of a sample mean between 19 and 21 is 0.9544 (i.e., 95% of the time), what is the probability of a sample mean that is not between 19 and 21 (either less than 19 or more than 21)?

Solution: Let $\bar{x}$ be the random variable denoting the sample mean. We are given that

$$\Pr(19 < \bar{x} < 21) = .9544$$

and are asked to find

$$\Pr(\bar{x} \text{ is not between 19 and 21}) .$$

If $\bar{x}$ is not between 19 and 21 then either $\bar{x} < 19$ or $\bar{x} > 21$. These two events are mutually exclusive. That is, if $\bar{x} < 19$ then $\bar{x}$ cannot possibly be greater than 21 and if $\bar{x} > 21$ then $\bar{x}$ cannot be less than 19. Thus,

$$\Pr(\bar{x} \text{ is not between 19 and 21})$$
$$= \Pr(\bar{x} < 19 \text{ or } \bar{x} > 21)$$
$$= \Pr(\bar{x} < 19) + \Pr(\bar{x} > 21) .$$

We also know that the events $19 < \bar{x} < 21$ and $\bar{x}$ not between 19 and 21 are exhaustive, i.e., they 'exhaust' all the alternatives for the values of $\bar{x}$. Thus,

591

$$\Pr(19 < \bar{x} < 21) + \Pr(\bar{x} \text{ not between 19 and 21}) = 1 \ .$$

Thus,

$$\Pr(\bar{x} \text{ not between 19 and 21})$$

$$= \Pr(\bar{x} < 19 \ \text{ or } \ \bar{x} > 21) = 1 - \Pr(19 < \bar{x} < 21)$$

$$= 1 - .9544$$

$$= .0456 \ .$$

Suppose that some distribution with an unknown mean has variance equal to 1. How large a sample must be taken in order that the probability will be at least .95 that the sample mean $\bar{X}_n$ will lie within .5 of the population mean?

Solution: This problem involves the Weak Law of Large Numbers. In order to establish the law we must first verify the following lemma.

Lemma: Let X be a random variable and $g(X)$ a non-negative function with the real line as a domain; then

$$\Pr[g(X) \geq K] \leq \frac{E[g(X)]}{K} \quad \text{for all } K > 0 \ .$$

Proof of Lemma: Assume that X is a continuous random variable.(a similar proof will hold for X discrete) with probability density $f(x)$; then by definition

$$E(g(x)) = \int_{-\infty}^{\infty} g(x) \ f(x) dx$$

$$= \int_{\{x:g(x) \geq K\}} g(x)f(x)dx \ + \int_{\{x:g(x) < K\}} g(x)f(x)dx$$

$$\geq \int_{\{x:g(x) \geq K\}} g(x)f(x)dx \ .$$

Since $g(x) \geq h(x) = K$ in the domain of integration, this is greater than or equal to

$$\int_{\{x:g(x) \geq K} Kf(x)dx = K \int_{\{x:g(x) \geq K} f(x)dx$$

$$= K \Pr(g(x) \geq K) \ .$$

Now divide by K and obtain

$$\frac{E(g(x))}{K} \geq \Pr(g(x) \geq K) \ .$$

An equivalent result to this lemma is
$$\Pr(g(x) < K) \geq 1 - \frac{E(g(x))}{K}$$

because
$$\Pr(g(x) < K) = 1 - \Pr(g(x) \geq K) .$$

Now let $g(x) = (\bar{X}_n - \mu)^2$ and $K = \epsilon^2$; then

$$\Pr((\bar{X}_n - \mu)^2 < \epsilon^2) \geq 1 - \frac{1}{\epsilon^2} E((\bar{X}_n - \mu)^2) .$$

Equivalently

$$\Pr(-\epsilon < \bar{X}_n - \mu < \epsilon) \geq 1 - \frac{E((\bar{X}_n - \mu)^2)}{\epsilon^2}$$

or

$$\Pr(|\bar{X}_n - \mu| < \epsilon) \geq 1 - \frac{E((\bar{X}_n - \mu)^2)}{\epsilon^2} .$$

But $E((\bar{X}_n - \mu)^2) = \mathrm{Var}(\bar{X}) = \mathrm{Var}(\frac{\Sigma x}{n}) = \mathrm{Var}[\frac{1}{n}(X_1 + X_2 + \ldots + X_n)]$

$$= \frac{1}{n^2} \mathrm{Var}(X_1 + \ldots + X_n)$$

since $\mathrm{Var}(aX) = a^2 \mathrm{Var}(X)$. But the X_i will be assumed independent.

Therefore

$$\frac{1}{n^2} \mathrm{Var}(X_1 + \ldots + X_n) = \frac{1}{n^2} [\mathrm{Var}(X_1) + \mathrm{Var}(X_2) + \ldots + \mathrm{Var}(X_n)]$$

$$= \frac{1}{n^2} n \mathrm{Var}(X) = \frac{\sigma^2}{n} .$$

Thus,
$$\Pr(|\bar{X}_n - \mu| < \epsilon) \geq 1 - \frac{\sigma^2}{n\epsilon^2} .$$

This is the Weak Law of Large Numbers.

In the present problem, we want

$$\Pr(|\bar{X}_n - \mu| < .5) \geq .95 = 1 - .05.$$

Comparing this with the general statement of the Weak Law of Large Numbers we set $\epsilon = .5$, $\sigma^2/n\epsilon^2 = .05$. Multiplying through by $n/.05$ we see that we would need an $n \geq \sigma^2/.05\epsilon^2$. Here then

$$n \geq \frac{1}{(.05)(.5)^2} = \frac{1}{.05(.25)} = \frac{1}{.0125}$$

$$n \geq 80.$$

We must therefore choose a sample of at least 80.

Briefly discuss the Central Limit Theorem.

Solution: The theorem has to do with the means of large greater than (30) samples. As the sample size increases, the distribution of the sample mean, $\bar{X}$, has a distribution which is approximately normal. This distribution has a mean equal to the population mean and a standard deviation equal to the population standard deviation divided by the square root of the sample size.

Since $\bar{X}$ is approximately normal, $\dfrac{\bar{X} - E(\bar{X})}{\sigma_{\bar{X}}} = \dfrac{\bar{X} - \mu}{\sigma/\sqrt{n}} = \dfrac{\sqrt{n}(\bar{X} - \mu)}{\sigma}$

will have a standard normal distribution.

For a large sample, the distribution of $\bar{X}$ is always approximately normal. Find the probability that a random sample mean lies within
a) one standard error of the mean.
b) two standard errors.

Solution: a) The question asks what is $\Pr(\mu_{\bar{X}} - \sigma_{\bar{X}} < \bar{X} < \mu_{\bar{X}} + \sigma_{\bar{X}})$?

If $\bar{X}$ is approximately normally distributed,

$$\frac{\bar{X} - \mu_{\bar{X}}}{\sigma_{\bar{X}}}$$

will be standard normal. Let $\mu_{\bar{X}} = \mu$, $\sigma_{\bar{X}} = \sigma$.

Note $\Pr(\mu - \sigma < \bar{x} < \mu + \sigma)$

$$= \Pr\left(\frac{(\mu - \sigma) - \mu}{\sigma} < \frac{\bar{x} - \mu}{\sigma} < \frac{(\mu - \sigma) - \mu}{\sigma}\right)$$

$$= \Pr(-1 < \text{Standard Normal Quantity} < 1)$$

$$= .6826 \text{ from the tables.}$$

b) We now want

$$\Pr(\mu_{\bar{X}} - 2\sigma_{\bar{X}} < \bar{X} < \mu_{\bar{X}} + 2\sigma_{\bar{X}}).$$

Again assuming $\bar{X}$ is approximately normally distributed, $\dfrac{\bar{X} - \mu_{\bar{X}}}{\sigma_{\bar{X}}}$

will be standard normal. Again, for ease of reading, let $\mu_{\bar{X}} = \mu, \sigma_{\bar{X}} = \sigma$.

Using a similar procedure,

$$Pr(\mu_{\bar{X}} - 2\sigma_{\bar{X}} < \bar{X} < \mu_{\bar{X}} + 2\sigma_{\bar{X}}) =$$

$$Pr\left(\frac{\mu - 26 - \mu}{\sigma} < \frac{\bar{x} - \mu}{\sigma} < \frac{\mu + 26 - \mu}{\sigma}\right)$$

$= Pr(-2 < \text{Standard Normal Quantity} < 2)$

$= .9544$ from the table.

These problems can be completely generalized. Any normal distribution has 68.26 % of the probability within one standard deviation of the mean and 95.44% within two.

● **PROBLEM** 9-116

A research worker wishes to estimate the mean of a population using a sample large enough that the probability will be .95 that the sample mean will not differ from the population mean by more than 25 percent of the standard deviation. How large a sample should he take?

Solution: 25 percent of the standard deviation is ¼ of it, $\sigma/4$. We want

$$Pr(|\bar{X} - \mu| < \sigma/4) = .95.$$

Equivalently,

$$Pr\left(-\frac{\sigma}{4} < \bar{X} - \mu < \frac{\sigma}{4}\right) = .95 .$$

Divide through by $\sigma/\sqrt{n}$:

$$Pr\left(\frac{\bar{X} - \mu}{\sigma/\sqrt{n}} < \frac{\sigma/4}{\sigma/\sqrt{n}}\right) = .95$$

or

$$Pr\left(\frac{\bar{X} - \mu}{\sigma/\sqrt{n}} < \frac{\sqrt{n}}{4}\right) = .95 .$$

By the Central Limit Theorem, $\dfrac{\bar{X} - \mu}{\sigma/\sqrt{n}}$ has a standard normal distribution.

595

Therefore $\qquad$ Pr(Standard Normal Quantity $< \sqrt{n}/4) = .95$.

From the standard normal tables we know,

$$\Pr(-1.96 < \text{Standard Normal} < 1.96) = .95.$$

Now set

$$1.96 = \frac{\sqrt{n}}{4}$$

$$4(1.96) = \sqrt{n}$$

$$\sqrt{n} = 7.84$$

$$n = 61.4656.$$

The research worker would have to take a sample of at least 62 observations.

● **PROBLEM** 9-117

Find the expected value of the random variable

$$S_*^2 = \frac{1}{n} \sum_{i=1}^{n} (X_i - \bar{X})^2 \text{ , where } \bar{X} = \sum_{i=1}^{n} X_i/n \text{ and the } X_i \text{ are}$$

independent and identically distributed with $E(X_i) = \mu$, $\text{Var } X_i = \sigma^2$ for $i = 1, 2, \ldots n$.

<u>Solution</u>: $E(S_*^2) = E\left[\dfrac{1}{n} \sum_{i=1}^{n} (X_i - \bar{X})^2\right]$ by definition. We will use the common mathematical trick of adding and subtracting the same quantity, thereby leaving everything unchanged. Hence,

$$E(S_*^2) = E\left[\frac{1}{n} \sum_{i=1}^{n} ((X_i - \mu) - (\bar{X} - \mu))^2\right] = E\left[\frac{1}{n} \sum_{i=1}^{n} ((X_i - \mu)^2 - 2(\bar{X} - \mu)(X_i - \mu) + (\bar{X} - \mu)^2))\right]$$

$$= E\left[\frac{1}{n} \sum_{i=1}^{n} (X_i - \mu)^2 - \frac{2}{n} \sum_{i=1}^{n} (\bar{X} - \mu)(X_i - \mu) + \frac{1}{n} \sum_{i=1}^{n} (\bar{X} - \mu)^2\right]$$

Since $(\bar{X} - \mu)$ is constant with respect to i, we have

$$E(S_*^2) = E\left[\frac{1}{n} \sum_{i=1}^{n} (X_i - \mu)^2 - \frac{2}{n}(\bar{X} - \mu) \sum_{i=1}^{n} (X_i - \mu) + \frac{1}{n} (\bar{X} - \mu)^2 \sum_{i=1}^{n} 1\right].$$

Since $\sum X_i = n\bar{X}$, we have

$$E(S_*^2) = E\left[\frac{1}{n} \sum_{i=1}^{n} (X_i - \mu)^2 - \frac{2}{n}(\bar{X} - \mu)(n\bar{X} - n\mu) + \frac{1}{n} (\bar{X} - \mu)^2 \cdot n\right]$$

$$= E\left[\frac{1}{n} \sum_{i=n}^{n} (X_i-\mu)^2 - 2(\bar{X}-\mu)^2 + (\bar{X}-\mu)^2\right]$$

$$= E\left[\frac{1}{n} \sum_{i=1}^{n} (X_i-\mu)^2 - (\bar{X}-\mu)^2\right]$$

$$= \frac{1}{n} \sum_{i=1}^{n} E(X_i-\mu)^2 - E(\bar{X}-\mu)^2 ,$$

by the linearity properties of the expectation operator; then,

$$E(S_*^2) = \frac{1}{n} \sum_{i=1}^{n} \sigma^2 - \sigma_{\bar{X}}^2 \; ; \; \sigma_{\bar{X}} = \text{standard deviation of the sample mean},$$

$$= \frac{1}{n}(n\sigma^2) - \frac{\sigma^2}{n} = \sigma^2 - \frac{\sigma^2}{n} = \frac{n-1}{n} \sigma^2 .$$

If we estimate σ^2 by $\frac{1}{n} \sum_{i=1}^{n} (X_i-\bar{X})^2$, we see that $E(S_*^2) \neq \sigma^2$. The

word given to this type of estimator is biased.

● **PROBLEM** 9-118

Z is a standard normal random variable. U is chi-square with k degrees of freedom. Assume Z and U are independent. Using the change of variable technique, find the distribution of

$$X = \frac{Z}{\sqrt{u/k}} .$$

Solution: Note that

$$g(z) = \frac{1}{\sqrt{2\pi}} e^{-Z^2/2} \qquad -\infty < Z < \infty$$

$$h(u) = \frac{1}{\Gamma(k/2)} \left(\frac{1}{2}\right)^{k/2} u^{k/2-1} e^{-(\frac{1}{2})u} \qquad u > 0 .$$

Since Z and U are independent, the joint density $f(z,u) = g(z)h(u)$.

Therefore

$$f(z,u) = \frac{1}{\sqrt{2\pi}} \frac{1}{\Gamma(k/2)} \left(\frac{1}{2}\right)^{k/2} u^{(k/2)-1} e^{-\frac{1}{2}u} e^{-\frac{1}{2}z^2} \quad \text{for} \quad u > 0 .$$

We make the change of variables $X = Z/\sqrt{u/k}$ and $Y = U$. The change of variables technique tells us

$$f_{x,y}(x,y) = f_{u,v}(x,y)\,|J|\ ,$$

where J is the Jacobian determinant of the transformation,

$$J = \det \begin{bmatrix} \dfrac{\partial x}{\partial z} & \dfrac{\partial x}{\partial u} \\[2mm] \dfrac{\partial y}{\partial z} & \dfrac{\partial y}{\partial u} \end{bmatrix} = \det \begin{bmatrix} \dfrac{1}{\sqrt{u/k}} & -\dfrac{z}{2}\sqrt{\dfrac{k}{u^3}} \\[2mm] 0 & 1 \end{bmatrix}$$

$$= \sqrt{k/u} \quad \text{but} \quad y = u \quad \text{so} \quad |J| = J = \sqrt{k/y}\ .$$

Hence

$$f(x,y) = \sqrt{y/k}\ \frac{1}{\sqrt{2\pi}}\ \frac{1}{\Gamma(k/2)}\ (\tfrac{1}{2})^{k/2}\ y^{(k/2)-1} e^{-\frac{1}{2}y} e^{(-x^2/2)y}$$

$$y > 0\ .$$

$$f(x) = \int_{-\infty}^{\infty} f(x,y)\,dy$$

$$= \int_{0}^{\infty} \sqrt{y/k}\ \frac{1}{\sqrt{2\pi}}\ \frac{1}{\Gamma(k/2)}\ (\tfrac{1}{2})^{k/2}\ y^{(k/2)-1}\ e^{-(y/2)}\ e^{-x^2 y/2}\ dy$$

$$= \frac{1}{\sqrt{2k\pi}}\ \frac{1}{\Gamma(k/2)}\ (\tfrac{1}{2})^{k/2} \int_{0}^{\infty} y^{(k/2)-1+\frac{1}{2}}\ e^{-\frac{1}{2}(1+x^2/k)y}\ dy$$

$$= \frac{1}{\sqrt{k\pi}}\ \frac{1}{\Gamma(k/2)}\ (\tfrac{1}{2})^{(k+1)/2} \int_{0}^{\infty} y^{k-1/2}\ e^{-\frac{1}{2}(1+x^2/k)y}\ dy\ .$$

Let

$$r = \tfrac{1}{2}(1+x^2/k)\,y \quad \text{then}$$

$$dr = \tfrac{1}{2}(1+x^2/k)\,dy \quad \text{and}$$

$$y = \frac{r}{\tfrac{1}{2}(1+x^2/k)}\quad .$$

We now have

$$\frac{1}{\sqrt{k\pi}}\ \frac{1}{\Gamma(k/2)}(\tfrac{1}{2})^{k+1/2}\ \frac{1}{(\tfrac{1}{2}(1+\frac{x^2}{k}))^{k-1/2}\ \cdot\ \tfrac{1}{2}(1+x^2/k)} \int_{0}^{\infty} r^{\frac{k-1}{2}}\ e^{-r}\ dr$$

$$= \frac{1}{\sqrt{k\pi}}\ \frac{1}{\Gamma(k/2)}\ \frac{(\tfrac{1}{2})^{k+1/2}}{(\tfrac{1}{2})^{k+1/2}\ (1+x^2/k)^{k+1/2}}\ \Gamma(\tfrac{k+1}{2})\ ,$$

$$= \frac{\Gamma(k+\frac{1}{2})}{\Gamma(k/2)}\ \frac{1}{\sqrt{k\pi}\ (1+x^2/k)^{k+1/2}}\quad.$$

This is the density function of what is usually called the **Student's t** distribution.

Let U be a chi-square random variable with m degrees of freedom. Let V be a chi-square random variable with n degrees of freedom. Consider the quantity

$$X = \frac{U/m}{V/n} , \quad \text{where} \quad U \text{ and } V \text{ are independent.}$$

Using the change of variable technique, find the probability density of X.

Solution: We start with

$$g(U) = \frac{1}{\Gamma(m/2)} (\tfrac{1}{2})^{m/2} U^{m/2-1} e^{-u/2} \quad U > 0$$

$$h(V) = \frac{1}{\Gamma(n/2)} (\tfrac{1}{2})^{n/2} V^{n/2-1} e^{-v/2} \quad V > 0 .$$

Since U and V are independent, the joint density $f(U,V)$ equals the product $g(u)h(v)$. Hence

$$f(u,v) = \frac{1}{\Gamma(m/2)\Gamma(n/2)2^{m+n/2}} U^{m-2/2} V^{n-2/2} e^{-\frac{1}{2}(u+v)} .$$

Consider the transformation $X = \frac{U/m}{V/n}$, $Y = V$. By the change of variable formula, $f_{x,y}(x,y) = f_{u,v}(\bar{x},y)|J|$ where J is the Jacobian determinant of the transformation,

$$J = \det \begin{bmatrix} \dfrac{\partial x}{\partial u} & \dfrac{\partial x}{\partial v} \\[2mm] \dfrac{\partial y}{\partial u} & \dfrac{\partial y}{\partial v} \end{bmatrix} = \det \begin{bmatrix} \dfrac{mv}{n} & \dfrac{-nu}{mv^2} \\[2mm] 0 & 1 \end{bmatrix}$$

$$= \frac{mv}{n} = \frac{my}{n} \quad \text{since} \quad y = v. \quad \text{Therefore}$$

$$f(x,y) = \frac{my}{n} \frac{1}{\Gamma(m/2)\Gamma(n/2)2^{m+n/2}} \left(\frac{mxy}{n}\right)^{m-2/2} y^{n-2/2} e^{-\frac{1}{2}(\frac{m}{n} xy + y)} .$$

Now $f(x) = \int_{-\infty}^{\infty} f(x,y)dy = \int_{0}^{\infty} \frac{my}{n} \frac{1}{\Gamma(m/2)\Gamma(n/2)2^{m+n/2}} \left(\frac{mxy}{n}\right)^{m-2/2} y^{n-2/2}$

$$e^{-\frac{1}{2}(\frac{m}{n} xy + y)} \quad dy$$

599

$$= \frac{1}{\Gamma(m/2)\Gamma(n/2)2^{m+n/2}} \left(\frac{m}{n}\right)^{m/2} x^{m-2/2} \int_0^\infty y^{m+n-2/2} e^{-\frac{1}{2}\left(\frac{m}{n}xy + y\right)} dy .$$

Let $r = \frac{1}{2}\left(\frac{m}{n}xy + y\right)$. Then $y = \dfrac{2r}{\frac{m}{n}x + 1}$ and $dy = \dfrac{2}{\frac{m}{n}x + 1} dr$.

We now have

$$\frac{1}{\Gamma(m/2)\Gamma(n/2)2^{m+n/2}} \left(\frac{m}{n}\right)^{m/2} x^{m-2/2} \left(\frac{2}{\frac{m}{n}x + 1}\right)^{m+n/2} \int_0^\infty r^{m+n-2/2} e^{-r} dr$$

$$= \frac{1}{\Gamma(m/2)\Gamma(n/2)2^{m+n/2}} \left(\frac{m}{n}\right)^{\frac{m}{2}} x^{m-2/2} \left(\frac{2}{\frac{m}{n}x + 1}\right)^{m+n/2} \Gamma\left(\frac{m+n}{2}\right)$$

$$= \frac{\Gamma((m+n)/2)}{\Gamma(m/2)\Gamma(n/2)} \left(\frac{m}{n}\right)^{m/2} \frac{x^{(m-2)/2}}{[1 + (m/n)x]^{m+n/2}} \qquad X > 0 .$$

The above is called an F distribution with m and n degrees of freedom.

● PROBLEM 9-120

Let $X_1, \ldots, X_n$ be a random sample from $N(0,1)$. Define

$$\bar{X}_k = \frac{1}{k} \sum_1^k X_i \quad \text{and} \quad \bar{X}_{n-k} = \frac{1}{n-k} \sum_{k+1}^n X_i .$$

(a) What is the distribution of $\frac{1}{2}(\bar{X}_k + \bar{X}_{n-k})$?

(b) What is the distribution of $k\bar{X}_k^2 + (n-k)\bar{X}_{n-k}^2$?

(c) What is the distribution of X_1^2/X_2^2 ?

Solution: a) $\bar{X}_k$ and $\bar{X}_{n-k}$ are sample means from a normal distribu-

tion. As such they are normally distributed with means equal to the original distribution mean, 0. Their variances are

$$\frac{\sigma^2 \text{ orig.}}{\text{sample size}} = \frac{1}{k} \text{ and } \frac{1}{n-k} \text{ respectively.}$$

$\bar{X}_k + \bar{X}_{n-k}$ is normally distributed, being the sum of two normal distributions. By the linearity properties of expectation

$$E(\bar{X}_k + \bar{X}_{n-k}) = E(\bar{X}_k) + E(\bar{X}_{n-k}) = 0 + 0 = 0 .$$

$\bar{X}_k$ and $\bar{X}_{n-k}$ are concerned with different observations; hence they are independent. In light of this

600

$$Var(\bar{X}_k + \bar{X}_{n-k}) = Var(\bar{X}_k) + Var(\bar{X}_{n-k}) = \frac{1}{k} + \frac{1}{n-k} \ .$$

We now see $Y = \bar{X}_k + \bar{X}_{n-k}$ is distributed $N(0, \frac{1}{k} + \frac{1}{n-k})$. $\frac{1}{2}Y$ will still be normally distributed. By linearity properties

$$E(\tfrac{1}{2}Y) = \tfrac{1}{2}E(Y) = \tfrac{1}{2}\cdot 0 = 0 \ .$$

Since $Var(a\ y) = a^2 Var(y)$, $Var(\tfrac{1}{2}\ y) = \tfrac{1}{4}Var\ y = \tfrac{1}{4}(\frac{1}{k} + \frac{1}{n-k}) = \frac{1}{4k} + \frac{1}{4n-4k}$.

Therefore, $\frac{1}{2}(\bar{X}_k + \bar{X}_{n-k})$ has a distribution which is $N(0, \frac{1}{4k} + \frac{1}{4n-4k})$.

b) Here we concern ourselves with $k\bar{X}_k^2 + (n-k)\bar{X}_{n-k}^2$. Note that $\bar{X}_k$ is distributed normally with mean 0 and variance $1/k$. $\sqrt{k}\ \bar{X}_k$ will still be normally distributed with mean zero but the variance will be 1.

$1(Var(a\bar{X}) = a^2 Var(X); \ Var(\sqrt{k}\ \bar{X}_k) = k\ Var(\bar{X}_k) = k\ \frac{1}{k} = 1)$. $\sqrt{k}\ \bar{X}_k$ is distributed $N(0,1)$. An identical argument applies to $\sqrt{n-k}\ \bar{X}_{n-k}$.

Note that $k\bar{X}_k^2 + (n-k)\bar{X}_{n-k}^2 = (\sqrt{k}\ \bar{X}_k)^2 + (\sqrt{n-k}\ \bar{X}_{n-k})^2$ which is distributed as $[N(0,1)]^2 + [N(0,1)]^2$. Since $[N(0,1)]^2$ is $\chi^2(1)$, we have $\chi^2(1) + \chi^2(1)$. We know that $\chi^2(m) + \chi^2(n) = \chi^2(m+n)$, so our final answer is $\chi^2(1+1) = \chi^2(2)$.

c) X_1 is a standard normal random variable as is X_2. Hence X_1^2/X_2^2 is the quotient of the squares of 2 standard normal quantities. Equivalently it is the quotient of 2 chi-square random variables, each with 1 degree of freedom. But

$$\frac{\chi^2(1)}{\chi^2(1)} = \frac{\chi^2(1)/1}{\chi^2(1)/1}$$

which by definition is distributed $F(1,1)$.

● **PROBLEM** 9-121

Suppose that the life of a certain light bulb is exponentially distributed with mean 100 hours. If 10 such light bulbs are installed simultaneously, what is the distribution of the life of the light bulb that fails first, and what is its expected life? Let X_i denote the life of the ith light bulb; then $Y_1 = \min[X_1,\ldots,X_{10}]$ is the life of the light bulb that fails first. Assume that the X_i's are independent.

<u>Solution</u>: We discuss the case in complete generality first. Let $X_1, \ldots, X_n$ be n given random variables. Define $Y_1 = \min (X_1, \ldots, X_n)$ and $Y_n = \max(X_1, \ldots, X_n)$. Y_1 and Y_n are called the first and nth order statistics.

In order to find $f_{y_1}(y)$, we first obtain $F_{y_1}(y)$ and then differentiate to find $F_{y_1}'(y)$.

$$F_{y_1}(y) = \Pr(Y_1 \leq y) = 1 - \Pr(Y_1 > y) = 1 - \Pr(X_1 > y; \ldots; X_n > y)$$

since Y_1 is greater than y if and only if every $X_i > y$. And if $X_1, \ldots, X_n$ are independent, then

$$1 - \Pr(X_1 > y; \ldots; X_n > y) = 1 - \Pr(X_1 > y) \ldots \Pr(X_n > y)$$
$$= 1 - \prod_{i=1}^{n} \Pr(X_i > y) = 1 - \prod_{i=1}^{n} [1 - F_{x_i}(y)].$$

We further assume that $X_1, \ldots, X_n$ are identically distributed with common cumulative distribution function $F_x(\cdot)$; then,

$$1 - \prod_{i=1}^{n} [1 - F_{x_i}(y)] = 1 - [1 - F_x(y)]^n .$$

To find $f_{y_1}(y)$, we note

$$f_{y_1}(y) = \frac{d}{dy} F_{y_1}(y) = \frac{d}{dy}[1 - [1 - F_x(y)]^n]$$
$$= -n[1 - F_x(y)]^{n-1} \frac{d}{dy}[1 - F_x(y)]$$
$$= n[1 - F_x(y)]^{n-1} f_x(y).$$

We now apply this to our example. The mean of an exponential distribution is $1/\lambda$ where λ is the parameter of the distribution. We know then that $1/\lambda = 100$ or $\lambda = 1/100$. The distribution of a certain light bulb is exponential with parameter $1/100$;

$$f(y) = \frac{1}{100} e^{-(1/100)y} , \quad y \geq 0 .$$

Also by definition

$$F(y) = \int_{-\infty}^{y} f(t)dt = \int_{0}^{y} \frac{1}{100} e^{-(1/100)t} dt$$

$$= -e^{-\frac{1}{100}t}\Big|_{0}^{y} = 1 - e^{-y/100} .$$

In our problem, $n = 10$; therefore,

$$f_{y_1}(y) = n[1 - F(y)]^{n-1} f(y)$$

$$= 10[1 - (1 - e^{-y/100})]^{9} \frac{1}{100} e^{-y/100}$$

$$= \frac{10}{100}(e^{-y/100})^{10} = \frac{1}{10} e^{-y/10} .$$

The first order statistic, y_1 is exponentially distributed with parameter $1/10$. Hence $E(y_1) = \frac{1}{1/10} = 10$.

CONFIDENCE INTERVALS

● **PROBLEM** 9-122

Let X be $\chi^2(16)$. What is the probability that the random interval $(X, 3.3X)$ contains the point $x = 26.3$? and what is the expected length of the interval?

Solution: We will begin by trying to transform the interval $X < 26.3 < 3.3X$ into an equivalent event with which we can more easily deal. It is clear that X must be less than 26.3. Examine now the right hand side of the inequality $26.3 < 3.3X$. This is equivalent to $X > \frac{26.3}{3.3}$ or $X > 7.97$.

Now we see $\Pr(X < 26.3 < 3.3X) = \Pr(7.97 < X < 26.3)$. Recall that X is $\chi^2(16)$, therefore:

$$\Pr(7.97 < X < 26.3) = \Pr(\chi^2(16) < 26.3) - \Pr(\chi^2(16) < 7.97).$$

From the table of the Chi-square distribution, this equals $.95 - .05 = .90$.

The length of the interval is $3.3X - X = 2.3X$.

$$E(\text{Length}) = E(2.3X) = 2.3 E(X)$$

by the linearity properties of expectation. Since X is $\chi^2(16)$, $E(X) = 16$ $[E(\chi^2(n)) = n.]$.

$$E(\text{Length}) = 2.3(16) = 36.8.$$

Find a 95 per cent confidence interval for μ, the true mean of a normal population which has variance $\sigma^2 = 100$. Consider a sample of size 25 with a mean of 67.53.

<u>Solution</u>:   We have a sample mean $\overline{X} = 67.53$. We want to transform that into a standard normal quantity, for we know from the standard normal tables that

$$\Pr\,(-\,1.96 < \text{Standard Normal Quantity} < 1.96) = .95\;.$$

$\dfrac{\overline{X} - E(\overline{X})}{\sqrt{\text{Var}\;(\overline{X})}}$ is a standard normal quantity.

Recall now that the expectation of a sample mean is μ, the true mean of a population. Also recall that the variance of a sample mean is $\dfrac{\sigma^2}{n}$ where σ^2 is the true variance of the population and n is the size of our sample. Applying this to our case, $E(\overline{X}) = \mu$ and $\sqrt{\text{Var}(\overline{X})} = \sqrt{\dfrac{\sigma^2}{n}} = \sqrt{\dfrac{100}{25}} = 2$.

For our sample: $\Pr\left(-\,1.96 < \dfrac{\overline{X} - \mu}{2} < 1.96\right) = .95\;.$

Multiplying by 2: $\Pr(-\,3.92 < \overline{X} - \mu < 3.92) = .95.$

Transposing: $\Pr(\overline{X} - 3.92 < \mu < \overline{X} + 3.92) = .95\;.$

$\overline{X} - 3.92 < \mu < \overline{X} + 3.92$ is our required confidence interval. If we insert our given sample mean, we come up with $67.53 - 3.92 < \mu < 67.53 + 3.92$

or $63.61 < \mu < 71.45.$

Barnard College is a private institution for women located in New York City. A random sample of 50 girls was taken. The sample mean of grade point averages was 3.0. At neighboring Columbia College a sample of 100 men had an average gpa of 2.5. Assume all sampling is normal and Barnard's standard deviation is .2, while Columbia's is .5. Place a 99% confidence interval on $\mu_{Barnard} - \mu_{Columbia}$.

Solution: The main idea behind all of these problems is the same. We want to find a standard normal pivotal quantity involving $\mu_B - \mu_C$ and use the fact, obtainable from the standard normal tables, that

$$\Pr(-2.58 < \text{Standard Normal Quantity} < 2.58) = .99.$$

We want to find a quantity $\overline{B} - \overline{C}$, where $\overline{B}$ = Barnard's average and $\overline{C}$ = Columbia's average.

The sample mean from a normal population is normally distributed. $\overline{B} - \overline{C}$ is a difference in normal distributions

and is thus also normal. Hence $\dfrac{(\overline{B} - \overline{C}) - E(\overline{B} - \overline{C})}{\text{S.D. } (\overline{B} - \overline{C})}$ is

standard normal.

Recall that the expectation of a sample mean is μ, the expectation of the original distribution. Thus, $E(\overline{B}) = \mu_B$ and $E(\overline{C}) = \mu_C$.

$$E(\overline{B} - \overline{C}) = E(\overline{B} + (-\overline{C})) = E(\overline{B}) + E(-\overline{C}) = E(\overline{B}) - E(\overline{C})$$

by the linearity properties of expectation.

Hence $E(\overline{B} - \overline{C}) = E(\overline{B}) - E(\overline{C}) = \mu_B - \mu_C$.

Also, $\sigma \quad = \sqrt{\text{Var}(\overline{B} - \overline{C})} = \sqrt{\text{Var}(\overline{B}) + \text{Var}(\overline{C})}$ since

$\text{Var}(ax + by) = a^2 \text{Var}(X) + b^2 \text{Var}(Y)$. Furthermore,

$$\sigma \quad = \sqrt{\frac{\sigma_B^2}{n_B} + \frac{\sigma_C^2}{n_C}} \quad \text{since the standard deviation}$$

of a sample mean is $\dfrac{\sigma}{\sqrt{n}}$.

Now we see when substituting that

$$\Pr\left[-2.58 < \frac{(B - C) - (\mu_B - \mu_C)}{\sqrt{\dfrac{\sigma^2_B}{n_B} + \dfrac{\sigma^2_C}{n_C}}} < 2.58\right] = .99 \, .$$

We are given

$\bar{B} = 3.0; \ \bar{C} = 2.5; \ \sigma_B = .2; \ \sigma_C = .5; \ n_B = 50; \ n_C = 100.$

Inserting these values into the inequality, we obtain:

$$-2.58 < \frac{(3.0 - 2.5) - (\mu_B - \mu_C)}{\sqrt{\dfrac{(.2)^2}{50} + \dfrac{(.5)^2}{100}}} < 2.58.$$

Combining: $\quad -2.58 < \dfrac{.5 - (\mu_B - \mu_C)}{\sqrt{.0033}} < 2.58.$

Multiplying by $\sqrt{.0033}$: $\quad -.148 < .5 - (\mu_B - \mu_C) < .148.$

Subtracting .5: $\quad -.648 < -(\mu_B - \mu_C) < -.352$.

Multiplying by -1: $\quad .352 < \mu_B - \mu_C < .648$

Our required interval is

$.352 < \mu_B - \mu_C < .648.$

● **PROBLEM 9-125**

The Harvard class of 1927 had a reunion.which 36 attended. Among them they discovered they had been married an average of 2.6 times apiece. From the Harvard Alumni Register Dean Epps learned that the standard deviation for the 1927 alumni was 0.3 marriages. Help Dean Epps construct a 99 per cent confidence interval for the marriage rate of all Harvard alumni.

Solution: We are not given a standard deviation for the entire population here, but we are given a sample standard deviation. Our sample of 36 is fairly large thus the Laws of Large Numbers tell us that we can use the sample deviation as a good approximation to this real standard deviation. Again we search for a standard normal pivotal quantity upon which we can construct our confidence interval.

Recall the Central Limit Theorem: Let X be a random variable with mean μ and standard deviation σ; then the random variable $Z = \dfrac{(\overline{X} - \mu)\sqrt{n}}{\sigma}$ has a distribution that approaches standard normal as n gets large. We have n = 36. (That is large enough). We know from the tables that

$$Pr(-\,2.58 < \text{Standard Normal} < 2.58) = .99 \; .$$

Inserting our standard normal random variable:

$$Pr\left(-\,2.58 < \frac{(\overline{X} - \mu)\sqrt{n}}{\sigma} < 2.58\right) = .99 \; .$$

We know $\overline{X} = 2.6$, $\sqrt{n} = \sqrt{36} = 6$, and $\sigma = 0.3$.

Substituting these values, the inequality becomes:

$$-\,2.58 < \frac{(2.6 - \mu)\,6}{0.3} < 2.58 \; .$$

Multiplying through by $\dfrac{0.3}{6}$ $-\,0.129 < 2.6 - \mu < 0.129$.

Subtracting 2.6: $-\,2.729 < -\,\mu < 2.471$.

Multiplying through by -1: $2.471 < \mu < 2.729$.

Tell Dean Epps that he can be 99% sure that Harvard men on the average marry between 2.471 and 2.729 times.

● **PROBLEM** 9-126

What size sample is required to establish a .95 confidence interval for the grade point average of students attending Ponoma State Teachers College if a random sample of 100 students had a mean grade point average of 2.8 with a standard deviation of .4 and if the length of the interval is .1?

Solution: We are dealing with a large sample, 100. We can approximate σ, the true standard deviation, by S, the sample standard deviation   . Also the Central Limit Theorem will apply and therefore $\dfrac{(\overline{X} - \mu)\sqrt{n}}{\sigma}$ can be considered standard normal. In light of this, we have

607

$$Pr\left(-1.96 < \frac{(\overline{X} - \mu)\sqrt{n}}{\sigma} < 1.96\right) = .95.$$

We need only concern ourselves with the interval

$$-1.96 < \frac{(\overline{X} - \mu)\sqrt{n}}{\sigma} < 1.96.$$

We want the smallest n such that the above inequality will hold uniformly under the conditions of the problem.

Set $\frac{(\overline{X} - \mu)\sqrt{n}}{\sigma} = 1.96$. Multply through by σ.

$(\overline{X} - \mu)\sqrt{n} = 1.96\ \sigma$. Next divide by $(\overline{X} - \mu)$ and square

both sides. The result, $n = \left(\frac{1.96\ \sigma}{(\overline{X} - \mu)}\right)^2$. σ is

given as .4, but to resolve $(\overline{X} - \mu)$ is trickier. The length of $(X - \mu)$ will be at most .1. We can minimize $|\overline{X} - \mu|$ by assuming μ is the center of the interval. $|\overline{X} - \mu|$ now must $< .05$. $(.05 = (\frac{1}{2})\ (.1))$. If μ were not in the center it would have to be further then .5 from one end and $|\overline{X} - \mu|$ could be greater than .05. By minimizing $|\overline{X} - \mu|$, we minimize $(\overline{X} - \mu)^2$ and increase our value of n thereby obtaining a safer estimate for n. With our numbers

$$n = \left(\frac{(1.96)\ (.4)}{(.05)}\right)^2 = 245.86.$$

We can be confident that a random sample of size 246 will provide a 95% interval estimate of μ with length no more than .1 units.

● **PROBLEM** 9-127

During the 1976-77 season Coach Jerry Tarkanian outfitted his University of Nevada at Las Vegas basketball team with new sneakers. The 16 member team had an average size of 14.5 and a standard deviation of 5. Find a 90 percent confidence interval for the mean sneaker size of all collegiate basketball players. Assume the population is normal and the variance is not known.

Solution: In confidence interval problems the approch is to find a pivotal quantity, i.e. a function of the observations which has a distribution independent of the parameter we are trying to estimate. For example, the

quantity $\frac{(\overline{X} - \mu)\sqrt{n}}{\sigma}$ is normal with mean 0 and standard deviation 1, provided that X is normal. This is true regardless of the value of μ. We can now use facts about the standard normal distribution to rearrange our expressionand obtain a confidence interval for μ. In a case such as our present example where σ is unknown

$$\frac{(\overline{X} - \mu)\sqrt{n}}{\sigma}$$ does us no good in searching for μ. There is

no obvious way to rid ourselves of σ. We look for a pivotal quantity involving only μ.

We know that $\dfrac{\overline{X} - \mu}{\sigma/\sqrt{n}}$ is standard normal.

Recall also that $\dfrac{\Sigma(X_i - \overline{X})^2}{\sigma^2}$ has a Chi-Square dis-

tribution with $n - 1$ degrees of freedom. We also know that

if Z is standard normal and U is $\chi^2_{(k)}$, $\dfrac{Z}{\sqrt{\dfrac{U}{k}}}$ follows a t

distribution with k degrees of freedom. In our case then

$$\frac{(\overline{X} - \mu)/(\sigma/\sqrt{n})}{\sqrt{\Sigma(X_i - \overline{X})^2/(n - 1)\,\sigma^2}} = \frac{\text{Normal }(0,1)}{\sqrt{\chi^2_{(n-1)}}/n - 1}$$

has a t distribution with $n - 1$ degrees of freedom. Further-

more, $\dfrac{(\overline{X} - \mu)/(\sigma/\sqrt{n})}{\sqrt{\Sigma(X_i - \overline{X})^2/(n - 1)\,\sigma^2}} = \dfrac{\overline{X} - \mu}{S/\sqrt{n}}$ where S is the

sample standard deviation, $\sqrt{\dfrac{\Sigma(X_i - \overline{X})}{n - 1}}$. Note that

some books define the sample standard deviation as

$\sqrt{\dfrac{\Sigma(X_i - \overline{X})}{n}}$. In that case $\dfrac{(\overline{X} - \mu)/(\sigma/\sqrt{n})}{\sqrt{\Sigma(X_i - \overline{X})^2/(n - 1)\,\sigma^2}} = \dfrac{\overline{X} - \mu}{S/\sqrt{n - 1}}$

and $\dfrac{\overline{X} - \mu}{S/\sqrt{n - 1}}$ not $\dfrac{\overline{X} - \mu}{S/\sqrt{n}}$ would have a t distribution with n - 1 degrees of freedom. Notice that in the statistic $\frac{\overline{X} - \mu}{S/\sqrt{n}}$ there is no mention of σ. The fact that σ is un-known need no longer bother us. A look at the t tables tells us that $\Pr(-1.753 < t_{(15)} < 1.753) = .90$.

We chose 15 degrees of freedom since our sample is of size 16 and $U = \frac{(n - 1)\, S^2}{\sigma^2}$ is thus χ^2 with n - 1 degrees of freedom. Returning to our problem:

$$\Pr\left(-1.753 < \frac{\overline{X} - \mu}{S/\sqrt{n}} < 1.753\right) = .90 .$$

We know $\overline{X} = 14.5$, $\sqrt{n} = \sqrt{16} = 4$, and $S = 5$.

Inserting these values, the inequality becomes

$$-1.753 < \frac{14.5 - \mu}{5/4} < 1.753$$

The inequality $-1.753 < \frac{14.5 - \mu}{5/4} < 1.753$ will give us our confidence interval.

Multiplying by $\frac{5}{4}$: $-2.19 < 14.5 - \mu < 2.19$.

Subtracting 14.5: $-16.69 < -\mu < -12.31$.

Multiplying by -1: $12.31 < \mu < 16.69$.

$$12.31 < \mu < 16.69 .$$

● **PROBLEM 9-128**

In the Idaho State Home for Runaway Girls, 25 residents were polled as to what age they ran away from home. The sample mean was 16 years old with a standard deviation of 1.8 years. Establish a 95% confidence interval for μ, the mean age at which runaway girls leave home in Idaho.

Solution: We do not know precisely the population standard deviation and our sample is not large, 25. We might be all right in using the sample deviation to approximate the real one, but it is better to be safe and use a t-statistic. We know $\frac{\overline{X} - \mu}{\sigma/\sqrt{n}}$ is a standard normal random variable. Also note that $\sqrt{\dfrac{\Sigma(X_i - \overline{X})^2}{\sigma^2(n-1)}}$ is the square

root of a Chi-square random variable with n- 1 degrees of freedom divided by n - 1. The quotient of these two

quantities $\dfrac{(\overline{X} - \mu)/(\sigma/\sqrt{n})}{\sqrt{\Sigma(X_i - \overline{X})^2/\sigma^2(n-1)}}$ is a t random

variable with n - 1 degrees of freedom. Factoring

$$\frac{(\overline{X} - \mu)/(\sigma/\sqrt{n})}{\sqrt{\Sigma(X_i - \overline{X})^2/\sigma^2(n-1)}} = \frac{\dfrac{1}{\sigma}\dfrac{(\overline{X} - \mu)}{1/\sqrt{n}}}{\dfrac{1}{\sigma}\sqrt{\Sigma(X_i - \overline{X})^2/(n-1)}} \ .$$

The denominator is now the sample standard deviation, S. A convenient form for our t-statistic (n - 1 d.o.f.) is $\dfrac{\overline{X} - \mu}{S/\sqrt{n}}$.

In our case n - 1 = 25 - 1 = 24. From t-tables we see that $\Pr(-2.064 < t_{(24)} < 2.064) = .95$.

Inserting our t-statistic: $\Pr\left(-2.064 < \dfrac{\overline{X} - \mu}{S/\sqrt{n}} < 2.064\right) = .95$.

It is the interval, $-2.064 < \dfrac{\overline{X} - \mu}{S/\sqrt{n}} < 2.064$, with

which we are concerned. Substituting our values $\overline{X} = 16$, $S = 1.8$, $\sqrt{n} = \sqrt{25} = 5$, we see the result is $-2.064 < \dfrac{16 - \mu}{1.8/5} < 2.064$.

Multiplying through by $\dfrac{1.8}{5}$: $-.743 < 16 - \mu < .743$.

Subtracting 16: $-16.743 < -\mu < -15.267$.

Multiplying by -1: $15.267 < \mu < 16.743$.

Thus a 95% confidence interval for the true mean age at which Idaho girls run away from home is (15.267, 16.743).

Assume you have two populations $N(\mu_1, \sigma^2)$ and $N(\mu_2, \sigma^2)$. The distributions have the same, but unknown, variance σ^2. Derive a method for determining a confidence interval for $\mu_1 - \mu_2$.

Solution: Let $X_1, X_2, \ldots, X_n$ and $Y_1, Y_2, \ldots Y_m$ denote different random samples from the two independent distributions. We shall denote the sample means by $\bar{X}$ and $\bar{Y}$, and the sample variances by S_1^2 and S_2^2 $\left(S_1^2 = \frac{\Sigma(X - \bar{X})^2}{n - 1}\right)$.

These four statistics are all mutually stochastically independent. Therefore $\bar{X}$ and $\bar{Y}$ are normally distributed and independent with means μ_1 and μ_2 and variances $\frac{\sigma^2}{n}$ and $\frac{\sigma^2}{m}$.

The difference, $\bar{X} - \bar{Y}$, is normally distributed with mean $\mu_1 - \mu_2$ and variance $\frac{\sigma^2}{n} + \frac{\sigma^2}{m}$. The random variable

$$\frac{(\bar{X} - \bar{Y}) - (\mu_X - \mu_Y)}{\sqrt{\frac{\sigma^2}{n} + \frac{\sigma^2}{m}}}$$

will be standard normal and thus may serve as the numerator of a t random variable.

Also note that

$$\frac{(n - 1)S_1^2}{\sigma^2} = \frac{\Sigma(X - \bar{X})^2}{\sigma^2} \quad \text{and} \quad \frac{(m - 1)S_2^2}{\sigma^2} = \frac{\Sigma(Y - \bar{Y})^2}{\sigma^2} \quad \text{are}$$

stochastically independent chi-square random variables with $n - 1$ and $m - 1$ degrees of freedom respectively. Their

sum $\frac{(n - 1)S_1^2}{\sigma^2} + \frac{(m - 1)S_2^2}{\sigma^2}$ will then be chi-square with

$m + n - 2$ degrees of freedom, provided $m + n - 2 > 0$

$$\sqrt{\frac{U}{m + n - 2}} = \sqrt{\frac{\frac{(n - 1)S_1^2}{\sigma^2} + \frac{(m - 1)S_2^2}{\sigma^2}}{m + n - 2}}$$

is the square root of a chi-square random variable divided by its degrees of freedom. Thus it can serve as the denominator of a t random variable. Thus

$$T = \frac{(\bar{X} - \bar{Y}) - (\mu_1 - \mu_2)}{\sqrt{\dfrac{\sigma^2}{n} + \dfrac{\sigma^2}{m}}} \div \sqrt{\frac{U}{m + n - 2}}$$

$$= \frac{\dfrac{(\bar{X} - \bar{Y}) - (\mu_1 - \mu_2)}{\sigma \sqrt{\dfrac{1}{m} + \dfrac{1}{n}}}}{\dfrac{1}{\sigma} \sqrt{\dfrac{(n - 1)S_1^2 + (m - 1)S_2^2}{(m + n - 2)}}}$$

$$= \frac{(\bar{X} - \bar{Y}) - (\mu_1 - \mu_2)}{\sqrt{\dfrac{(n - 1)S_1^2 + (m - 1)S_2^2}{m + n - 2}\left[\dfrac{1}{n} + \dfrac{1}{m}\right]}}$$

$$= \frac{(\bar{X} - \bar{Y}) - (\mu_1 - \mu_2)}{\sqrt{W}}$$

$$\left(\text{where } \sqrt{W} \text{ is } \sqrt{\frac{(n-1)S + (m-1)S}{m+n-2}\left[\frac{1}{n} + \frac{1}{m}\right]}\right).$$

T is a t random variable with $m + n - 2$ degrees of freedom.

Say we wanted a $100(1 - \alpha)\%$ confidence interval. We can examine the tables to find a value b such that

$$\Pr(- b < t_{(m+n-2)} < b) = 1 - \alpha .$$

We have

$$\Pr\left(- b < \frac{(\bar{X} - \bar{Y}) - (\mu_1 - \mu_2)}{\sqrt{W}} < b\right) = 1 - \alpha .$$

Multiplying the inequality within the parentheses through by the denominator we obtain

$$- b \sqrt{W} < (\bar{X} - \bar{Y}) - (\mu_1 - \mu_2) < b \sqrt{W} .$$

We subtract $(\bar{X} - \bar{Y})$ and multiply by $- 1$ to obtain the final confidence interval.

$$(\bar{X} - \bar{Y}) - b \sqrt{\frac{(n - 1)S_1^2 + (m - 1)S_2^2}{n + m - 2}\left[\frac{1}{n} + \frac{1}{m}\right]} ,$$

$$(\bar{X} - \bar{Y}) + b \sqrt{\frac{(n - 1)S_1^2 + (m - 1)S_2^2}{n + m - 2}\left[\frac{1}{n} + \frac{1}{m}\right]} .$$

This is our $100(1 - \alpha)\%$ confidence interval.

The seven dwarfs challenged the Harlem Globetrotters to a basketball game. Besides the obvious difference in height, we are interested in constructing a 95% confidence in ages between dwarfs and basketball players. The respective ages are,

Dwarfs		Globetrotters	
Sneezy	20	Meadowlark	43
Grumpy	39	Curley	37
Dopey	23	Marques	45
Doc	41	Bobby Joe	25
Sleepy	35	Theodis	34
Happy	29		
Bashful	31		

Can you construct the interval? Assume the variance in age is the same for dwarfs and Globetrotters.

Solution: We have small samples with a common unknown variance. We have previously solved a problem dealing with such confidence intervals in general. We obtain a t statistic by taking the quotient of a standard normal random variable and the square root of a chi-square random variable with $n + m - 2$ degrees of freedom divided by $n + m - 2$. The resulting confidence interval was

$$\left[(\bar{X} - \bar{Y}) - b \sqrt{\frac{(n - 1)S_1^2 + (m - 1)S_2^2}{n + m - 2} \left[\frac{1}{n} + \frac{1}{m}\right]} \, , \right.$$

$$\left. (\bar{X} - \bar{Y}) + b \sqrt{\frac{(n - 1)S_1^2 + (m - 1)S_2^2}{n + m - 2} \left[\frac{1}{n} + \frac{1}{m}\right]} \right].$$

We need to compute $\bar{X}$, $\bar{Y}$, S_1^2, S_2^2, given $n = 7$, $m = 5$.

$$\bar{X} = \frac{\Sigma X}{n} = \frac{20 + 39 + 23 + 41 + 35 + 29 + 31}{7} = \frac{218}{7} = 31.1 \, .$$

$$\bar{Y} = \frac{\Sigma Y}{m} = \frac{43 + 37 + 45 + 25 + 34}{5} = \frac{184}{5} = 36.8 \, .$$

$$S_1^2 = \frac{\Sigma (X - \bar{X})^2}{n - 1} \quad . \text{ Use the following table:}$$

X	X - $\bar{X}$	(X - $\bar{X}$)^2
20	- 11.1	123.21
39	7.9	62.41
23	- 8.1	65.61
41	9.9	98.01
35	3.9	15.21
29	- 2.1	4.41
31	- .1	.01

$$\Sigma (X - \bar{X})^2 = 368.87$$

In the problem we are concerned with

$$(n - 1)S_1^2 = (n - 1) \frac{\Sigma(X - \overline{X})^2}{n - 1} = \Sigma(X - \overline{X})^2 = 368.87.$$

Similarly, $(m - 1)S_2^2 = \Sigma(Y - \overline{Y})^2$.

Y	$Y - \overline{Y}$	$(Y - \overline{Y})^2$
43	6.2	38.44
37	0.2	.04
45	8.2	67.24
25	- 11.8	139.24
34	- 2.8	7.84

$$\Sigma(Y - \overline{Y})^2 = 252.80$$
$$= (m - 1)S_2^2.$$

Now
$$\sqrt{\frac{(n - 1)S_1^2 + (m - 1)S_2^2}{n + m - 2} \left[\frac{1}{n} + \frac{1}{m}\right]}$$

$$= \sqrt{\frac{368.87 + 252.80}{7 + 5 - 2} \left[\frac{1}{7} + \frac{1}{5}\right]}$$

$$= \sqrt{\frac{621.67}{10} \left[\frac{12}{35}\right]} = \sqrt{21.314} = 4.62 .$$

$n + m - 2 = 7 + 5 - 2 = 10$. Our t statistic has 10 degrees of freedom and b is the value such that

$$Pr(- b < t(10) < b) = .95.$$

The t table tells us b = 2.228. Our confidence interval is now

$$((\overline{X} - \overline{Y}) - 2.228 (4.62), (\overline{X} - \overline{Y}) + 2.228 (4.62)).$$

But $\overline{X} - \overline{Y} = 31.1 - 36.8 = - 5.7$.

The interval is

$$(- 5.7 - 2.228 (4.62), - 5.7 + 2.228 (4.62)),$$

or $(- 15.99, 4.59)$

is a 95% confidence interval for $\mu_1 - \mu_2$, the difference between the dwarfs and Globetrotters' ages.

Consider a distribution $N(\mu, \sigma^2)$ where μ is known but σ^2 is not. Devise a method of producing a confidence interval for σ^2.

Solution: Let $X_1, X_2, \ldots, X_n$ denote a random sample of a size n from $N(\mu, \sigma^2)$, where μ is known. The random variable

$$Y = \frac{\sum_{1}^{n}(X_i - \mu)^2}{\sigma^2}$$

is a chi-square with n degrees of freedom. This is not to be confused with

$$\frac{\sum_{1}^{n}(X_i - \bar{X})^2}{\sigma^2}$$

which is $\chi^2(n-1)$. We select a probability, $1 - \alpha$, and for the constant n, determine values a and b, with $a < b$ such that

$$\Pr(a < Y < b) = 1 - \alpha.$$

Thus $\quad P\left(a < \dfrac{\sum_{1}^{n}(X_i - \mu)^2}{\sigma^2} < b\right) = 1-\alpha.$

We will concern ourselves with the central inequality. Taking reciprocals, we obtain

$$\frac{1}{b} < \frac{\sigma^2}{\sum_{1}^{n}(X_i - \mu)^2} < \frac{1}{a}.$$

Multiply through by $\sum_{1}^{n}(X_i - \mu)^2$ and the interval is

$$\frac{\sum_{1}^{n}(X_i - \mu)^2}{b} < \sigma^2 < \frac{\sum_{1}^{n}(X_i - \mu)^2}{a}.$$

The interval $\left(\dfrac{\sum_{1}^{n}(X_i - \mu)^2}{b}, \dfrac{\sum_{1}^{n}(X_i - \mu)^2}{a}\right)$ is a

random interval having probability $1 - \alpha$ of including the unknown fixed point (parameter) σ^2. Once we perform the experiment and find that $X_1 = x_1, X_2 = x_2, \ldots, X_n = x_n$, then the particular interval we calculate is a $1 - \alpha$ confidence interval for σ^2.

You should observe that there are no unique numbers a and b, a < b, such that $\Pr(a < Y < b) = 1 - \alpha$. A common convention, one which we will follow, is to find a and b such that $\Pr(Y < a) = \frac{\alpha}{2}$ and $\Pr(Y > b) = \frac{\alpha}{2}$. That way

$$\Pr(a < Y < b) = 1 - \frac{\alpha}{2} - \frac{\alpha}{2} = 1 - \alpha.$$

● **PROBLEM 9-132**

Eight scholars are working on a book. They are scheduled for an eight hour day, but no one works exactly eight hours. Yesterday the totals were 7.9, 7.8, 8.0, 8.1, 8.2, 7.9, 7.7, and 8.3 hours. Find a .95 confidence interval estimate for the variance of all 8 hour days these scholars will put in before their work is published.

Solution: We have a small sample with an unknown mean. We want a confidence interval for σ^2. The statistic we resort to is $\frac{(n-1)S^2}{\sigma^2}$ which is $\chi^2_{(n-1)}$ or $\chi^2_{(7)}$ here.

We must find $(n-1)S^2$. Since $S^2 = \frac{\Sigma(X - \bar{X})^2}{n-1}$,

$$(n-1)S^2 = \Sigma(X - \bar{X})^2.$$

We must find $\bar{X}$.

$$\bar{X} = \frac{\Sigma X}{n} = \frac{7.9+7.8+8+8.1+8.2+7.9+7.7+8.3}{8} = \frac{63.9}{8} = 7.99.$$

The following table will help us find $\Sigma(X - \bar{X})^2$.

X	$X - \bar{X}$	$(X - \bar{X})^2$
7.9	− .09	.0081
7.8	− .19	.0361
8.0	.01	.0001
8.1	.11	.0121
8.2	.21	.0441
7.9	− .09	.0081
7.7	− .29	.0841
8.3	.31	.0961

$$\Sigma(X - \bar{X})^2 = 0.2888 = (n-1)S^2$$

We know that there exists a and b so that

$$\Pr\left(a < \frac{(n-1)S^2}{\sigma^2} < b\right). \text{ Since } \frac{(n-1)S^2}{\sigma^2} \text{ is } \chi^2_{(7)}, \text{ we}$$

617

choose a and b such that

$$P(\chi^2_{(7)} < a) = .025 \quad \text{and} \quad P(\chi^2_{(7)} > b) = .025.$$

The chi-square tables tell us a = 1.69 and b = 16.0.

We will construct our confidence interval from the

inequality $\quad a < \dfrac{(n-1)S^2}{\sigma^2} < b \quad$ which equals

$1.69 < \dfrac{.2888}{\sigma^2} < 16$.

Dividing by .2888 yields: $\quad \dfrac{1.69}{.2888} < \dfrac{1}{\sigma^2} < \dfrac{16}{.2888}$.

Taking the reciprocal produces the result:

$\dfrac{.2888}{16} < \sigma^2 < \dfrac{.2888}{1.69} \quad$ or $\quad .018 < \sigma^2 < .171.$

● **PROBLEM** 9-133

Consider the following situation: A normal distribution of a random variable, X, has a variance σ_1^2, where σ_1^2 is unknown. It is found however that experimental values of X have a wide dispersion indicating that σ_1^2 must be quite large. A certain modification in the experiment is made to reduce the variance. Let the post-modification random variable be denoted Y, and let Y have a normal distribution with variance σ_2^2. Find a completely general method of determining confidence intervals for ratios of variances, $\dfrac{\sigma_1^2}{\sigma_2^2}$.

<u>Solution</u>: Consider a random sample $X_1, X_2, \ldots, X_n$ of

size n ≥ 2 from the distribution of X and a sample Y_1, $Y_2, \ldots, Y_m$ of size m ≥ 2. The X's are independent of the Y's and m and n may or may not be equal. Let the two means be denoted $\bar{X}$ and $\bar{Y}$, and the sample variances

$$S_1^2 = \frac{\Sigma(X_i - \bar{X})^2}{n-1} \quad \text{and} \quad S_2^2 = \frac{\Sigma(Y_i - \bar{Y})^2}{m-1} .$$

The independent random variables $\dfrac{(n-1)S_1^2}{\sigma_1^2}$ and

$\dfrac{(m-1)S_2^2}{\sigma_2^2}$ have chi-square distributions with n - 1 and

618

m - 1 degrees of freedom, respectively. The quotient of 2 chi-square random variables each divided by their degrees of freedom is called an F random variable. Then

$$\frac{\dfrac{(n-1)S_1^2}{\sigma_1^2}/n - 1}{\dfrac{(m-1)S_2^2}{\sigma_2^2}/m - 1} = \frac{S_1^2/\sigma_1^2}{S_2^2/\sigma_2^2} \text{ has an F distribution with}$$

n - 1 degrees of freedom in the numerator and m - 1 degrees of freedom in the denominator.

For given values of n and m, a specified probability 1 - α, and the use of F tables, we can determine a such that

$$\Pr\left(F_{(n-1,\ m-1)} < a\right) = \frac{\alpha}{2}$$

and b such that $\Pr\left(F_{(n-1,\ m-1)} > b\right) = \frac{\alpha}{2}$. With

these values $\Pr\left(a < F_{(n-1,\ m-1)} < b\right) = 1 - \frac{\alpha}{2} - \frac{\alpha}{2}$

$= 1 - \alpha$.

We know therefore that $\Pr\left(a < \dfrac{S_1^2/\sigma_1^2}{S_2^2/\sigma_2^2} < b\right) = 1 - \alpha$.

Our confidence interval can be constructed from the

inequality $a < \dfrac{S_1^2/\sigma_2^2}{S_2^2/\sigma_1^2} < b$. Inverting the fraction,

$a < \dfrac{S_1^2\sigma_2^2}{S_2^2\sigma_1^2} < b$. Multiply through by $\dfrac{S_2^2}{S_1^2}$ and the interval is

$a\ \dfrac{S_2^2}{S_1^2} < \dfrac{\sigma_2^2}{\sigma_1^2} < b\ \dfrac{S_2^2}{S_1^2}$.

It is seen that the interval has a probability 1 - α

of including the fixed but unknown point $\dfrac{\sigma_2^2}{\sigma_1^2}$.

Out of a group of 10,000 degree candidates of The University
of North Carolina at Chapel Hill, a random sample of 400
showed that 20 per cent of the students have an earning
potential exceeding $30,000 annually. Establish a .95 con-
fidence-interval estimate of the number of students with
a $30,000 plus earning potential.

Solution: We solve this problem in two steps. First we
will establish a 0.95 confidence-interval estimate for the
proportion of students with a chance at the $30,000 bracket
(a 95% confidence interval estimate of p). Secondly, we
will multiply the range established for p by 10,000, the
number of degree candidates, to obtain a range for the
number of students about to exceed $30,000.

We can assume that there is a population of 10,000
and we are looking for p, the probability of a "success".
This is a problem in binomial probabilities. Recall the
following theorem relating the normal distribution to the
binomial:

Theorem: If X represents the number of successes in n
independent trials of an event for which p is the probability
of success in a single trial, then the variable

$$\frac{(x - np)}{\sqrt{npq}}$$ has a distribution that approaches the standard

normal distribution as the number of trials, n, approaches
infinity.

If we divide both numerator and denominator by n we

obtain $\dfrac{\frac{x}{n} - p}{\sqrt{pq/n}}$ which is still standard normal. Therefore

$$Pr\left[- 1.96 < \frac{\frac{x}{n} - p}{\sqrt{pq/n}} < 1.96\right] = .95 .$$

Since we have a large sample, 400, the Law of Large

Numbers will allow us to estimate p and q in the denominator

by the proportions in the population, $\hat{p} = .20$ and $\hat{q} = 1 -$

.20 = .80. Also $\frac{x}{n}$ = 20% = .20 and n = 400.

Substitution yields:

$$\Pr\left[-1.96 < \frac{.20 - p}{\sqrt{(.20)(.80)/400}} < 1.96\right] = .95.$$

Manipulate the central inequality as follows:

$$-1.96 \sqrt{\frac{(.20)(.80)}{400}} < .20 - p < 1.96 \sqrt{\frac{(.20)(.80)}{400}}.$$

Simplification gives us:

$$-1.96 \, (.02) < .20 - p < 1.96 \, (.02)$$

or $-.0392 < .20 - p < .0392$.

Subtract .20 from all terms: $-.2392 < -p < -.1608$.

For our final confidence interval, multiply by -1:

$.1608 < p < .2392$.

Since we have 10,000 degree candidates, the .95 confidence interval estimate of the number of students with a \$30,000 earning potential is

$$10,000(0.1608 \leq p \leq 0.2392);$$

or between 1608 and 2392 students.

● **PROBLEM** 9-135

Harvey of Brooklyn surveyed a random sample of 625 students at SUNY-Stony Brook. Being a pre-medical student, he hoped that most students would major in the social sciences rather than the natural sciences, thus provide him with less competition. To Harvey's dismay, 60% of the students he surveyed were majoring in the natural sciences. Construct a 95% confidence interval for p, the population proportion of students majoring in the natural sciences.

Solution: Let $\hat{p}$ = .60, the sample proportion. $\hat{q}$ = 1 - $\hat{p}$ = .40. Since the sample size is large, n = 625, we can use the normal approximation to the binomial. $\dfrac{x - np}{\sqrt{npq}} = \dfrac{\dfrac{x}{n} - p}{\sqrt{\dfrac{pq}{n}}}$ can be considered a standard normal random variable.

Therefore: $\Pr\left(- 1.96 < \dfrac{\dfrac{x}{n} - p}{\sqrt{\dfrac{pq}{n}}} < 1.96\right) = .95$.

We have trouble with p and q however. Since we have a large sample we can estimate them by $\hat{p}$ and $\hat{q}$ for the purpose of computing $\sqrt{\dfrac{pq}{n}}$. Thus

$$\Pr\left(- 1.96 < \dfrac{\dfrac{x}{n} - p}{\sqrt{\dfrac{\hat{p}\hat{q}}{n}}} < 1.96\right) = .95 .$$

We multiply the central inequality by $\sqrt{\dfrac{\hat{p}\hat{q}}{n}}$:

$$- 1.96 \sqrt{\dfrac{\hat{p}\hat{q}}{n}} < \dfrac{x}{n} - p < 1.96 \sqrt{\dfrac{\hat{p}\hat{q}}{n}} .$$

Subtract $\dfrac{x}{n}$: $- \dfrac{x}{n} - 1.96 \sqrt{\dfrac{\hat{p}\hat{q}}{n}} < - p < - \dfrac{x}{n} + 1.96 \sqrt{\dfrac{\hat{p}\hat{q}}{n}}$.

Multiply by - 1: $\dfrac{x}{n} - 1.96 \sqrt{\dfrac{\hat{p}\hat{q}}{n}} < p < \dfrac{x}{n} + 1.96 \sqrt{\dfrac{\hat{p}\hat{q}}{n}}$.

Substitute our given values:

$$.60 - 1.96 \sqrt{\dfrac{(.60)(.40)}{625}} < p < .60 + 1.96 \sqrt{\dfrac{(.60(.40)}{625}} .$$

Simplifying yields .60 - .038 < p < .60 + .038

or   .562 < p < .638.

Determine a method for constructing a confidence interval for $p_1 - p_2$, the difference of two population proportions.

Solution: Consider Y_1 and Y_2 to be two independent random variables with binomial distributions $b(n_1, p_1)$ and $b(n_2, p_2)$. Examine the random variables $\frac{Y_1}{n_1}$ and $\frac{Y_2}{n_2}$. Assume n_1 and n_2 are known.

The expected values of $\frac{Y_1}{n_1}$ and $\frac{Y_2}{n_2}$ are p_1 and p_2 respectively. Since $\text{Var}\left(Y_1\right) = n_1 p_1 q_1$,

$$\text{Var}\left(\frac{1}{n_1}\ Y_1\right) = \left(\frac{1}{n_1}\right)^2 \text{Var}(Y_1) = \frac{1}{n_1^2}\ n_1 p_1 q_1 = \frac{p_1 q_1}{n_1}\ .$$

Similarly for Y_2.

The variances of $\frac{Y_1}{n_1}$ and $\frac{Y_2}{n_2}$ are $\frac{p_1 q_1}{n_1}$ and $\frac{p_2 q_2}{n_2}$ or equivalently $\frac{p_1(1 - p_1)}{n_1}$ and $\frac{p_2(1 - p_2)}{n_2}$.

The mean and variance of $\frac{Y_1}{n_1} - \frac{Y_2}{n_2}$, are $p_1 - p_2$, by the linearity properties of expectation, and $\frac{p_1(1 - p_1)}{n_1} + \frac{p_2(1 - p_2)}{n_2}$ by the additive properties of variance.

We will assume that n_1 and n_2 are large and apply the Central Limit Theorem. We will therefore consider

$$\frac{\left(\frac{Y_1}{n_1} - \frac{Y_2}{n_2}\right) - E\left(\frac{Y_1}{n_1} - \frac{Y_2}{n_2}\right)}{\sqrt{\text{Var}\left(\frac{Y_1}{n_1} - \frac{Y_2}{n_2}\right)}}$$

as a standard normal random variable. Hence we find $Z_{\alpha/2}$ such that

$$\text{Pr}\left[-Z_{\alpha/2} < \frac{\left(\frac{Y_1}{n_1} - \frac{Y_2}{n_2}\right) - E\left(\frac{Y_1}{n_1} - \frac{Y_2}{n_2}\right)}{\sqrt{\text{Var}\left(\frac{Y_1}{n_1} - \frac{Y_2}{n_2}\right)}} < Z_{\alpha/2}\right] = 1 - \alpha.$$

Substituting our known expressions, we obtain

$$\Pr\left[-Z_{\alpha/2} < \frac{\left(\frac{Y_1}{n_1} - \frac{Y_2}{n_2}\right) - (p_1 - p_2)}{\sqrt{\dfrac{p_1(1 - p_1)}{n_1} + \dfrac{(p_2(1 - p_2)}{n_2}}} < Z_{\alpha/2}\right] = 1 - \alpha.$$

We do not know p_1 and p_2 precisely but since our samples are large we can estimate them accurately by

$$\frac{\frac{Y_1}{n_1}\left(1 - \frac{Y_1}{n_1}\right)}{n_1} \quad \text{and} \quad \frac{\frac{Y_2}{n_2}\left(1 - \frac{Y_2}{n_2}\right)}{n_2}$$

for the purposes of calculating the square root of the variance. Therefore

$$\Pr\left[-Z_{\alpha/2} < \frac{\left(\frac{Y_1}{n_1} - \frac{Y_2}{n_2}\right) - (p_1 - p_2)}{\sqrt{\dfrac{\frac{Y_1}{n_1}\left(1 - \frac{Y_1}{n_1}\right)}{n_1} + \dfrac{\frac{Y_2}{n_2}\left(1 - \frac{Y_2}{n_2}\right)}{n_2}}} < Z_{\alpha/2}\right] = 1 - \alpha.$$

Multiplying through by the denominator:

$$-Z_{\alpha/2}\sqrt{\dfrac{\frac{Y_1}{n_1}\left(1 - \frac{Y_1}{n_1}\right)}{n_1} + \dfrac{\frac{Y_2}{n_2}\left(1 - \frac{Y_2}{n_2}\right)}{n_2}} < \left(\frac{Y_1}{n_1} - \frac{Y_2}{n_2}\right) -$$

$$(p_1 - p_2) < Z_{\alpha/2}\sqrt{\dfrac{\frac{Y_1}{n_1}\left(1 - \frac{Y_1}{n_1}\right)}{n_1} + \dfrac{\frac{Y_2}{n_2}\left(1 - \frac{Y_2}{n_2}\right)}{n_2}}.$$

Subtracting $\left(\frac{Y_1}{n_1} - \frac{Y_2}{n_2}\right)$:

$$-\left(\frac{Y_1}{n_1} - \frac{Y_2}{n_2}\right) - Z_{\alpha/2}\sqrt{\dfrac{\frac{Y_1}{n_1}\left(1 - \frac{Y_1}{n_1}\right)}{n_1} + \dfrac{\frac{Y_2}{n_2}\left(1 - \frac{Y_2}{n_2}\right)}{n_2}}$$

$$< - (p_1 - p_2) < - \left[\frac{Y_1}{n_1} - \frac{Y_2}{n_2}\right]$$

$$+ \; Z_{\alpha/2} \; \sqrt{\frac{\frac{Y_1}{n_1}\left(1 - \frac{Y_1}{n_1}\right)}{n_1} + \frac{\frac{Y_2}{n_2}\left(1 - \frac{Y_2}{n_2}\right)}{n_2}}$$

Multiply by -1:

$$\left[\frac{Y_1}{n_1} - \frac{Y_2}{n_2}\right] - Z_{\alpha/2} \; \sqrt{\frac{\frac{Y_1}{n_1}\left(1 - \frac{Y_1}{n_1}\right)}{n_1} + \frac{\frac{Y_2}{n_2}\left(1 - \frac{Y_2}{n_2}\right)}{n_2}}$$

$$< p_1 - p_2 < \left[\frac{Y_1}{n_1} - \frac{Y_2}{n_2}\right]$$

$$+ \; Z_{\alpha/2} \; \sqrt{\frac{\frac{Y_1}{n_1}\left(1 - \frac{Y_1}{n_1}\right)}{n_1} + \frac{\frac{Y_2}{n_2}\left(1 - \frac{Y_2}{n_2}\right)}{n_2}}$$

Our $1 - \alpha$ confidence interval is

$$\left[\left(\frac{Y_1}{n_1} - \frac{Y_2}{n_2}\right) - Z_{\alpha/2} \; \sqrt{\frac{\frac{Y_1}{n_1}\left(1 - \frac{Y_1}{n_1}\right)}{n_1} + \frac{\frac{Y_2}{n_2}\left(1 - \frac{Y_2}{n_2}\right)}{n_2}} \right.$$

$$\left. \left(\frac{Y_1}{n_1} - \frac{Y_2}{n_2}\right) + Z_{\alpha/2} \; \sqrt{\frac{\frac{Y_1}{n_1}\left(1 - \frac{Y_1}{n_1}\right)}{n_1} + \frac{\frac{Y_2}{n_2}\left(1 - \frac{Y_2}{n_2}\right)}{n_2}} \right]$$

or

$$\left[(\hat{p}_1 - \hat{p}_2) - Z_{\alpha/2} \; \sqrt{\frac{\hat{p}_1(1 - \hat{p}_1)}{n_1} + \frac{\hat{p}_2(1 - \hat{p}_2)}{n_2}} \; , \right.$$

$$\left. (\hat{p}_1 - \hat{p}_2) + Z_{\alpha/2} \; \sqrt{\frac{\hat{p}_1(1 - \hat{p}_1)}{n_1} + \frac{\hat{p}_2(1 - \hat{p}_2)}{n_2}} \; \right] \; .$$

POINT ESTIMATION

● **PROBLEM** 9-137

A physchologist wishes to determine the variation in I.Q.s of the population in his city. He takes many random samples of size 64. The standard error of the mean is found to be equal to 2. What is the population standard deviation?

Solution: The standard error of the mean is defined to be

$$\sigma_{\bar{x}} = \frac{\sigma}{\sqrt{n}} \tag{1}$$

where σ is the positive square root of the population variance and n is the size of the sample. Formula (1) is valid when sampling occurs with replacement or when the population is infinite.

We are given n = 64 and $\sigma_{\bar{x}}$ = 2.

Substituting into (1),

$$2 = \frac{\sigma}{\sqrt{64}} \quad \text{or,} \quad \sigma = 16. \tag{2}$$

Thus, the standard deviation of the distribution of I.Q.s in the city is 16.

If we assume that I.Q.s are normally distributed with mean 100, then a standard deviation of 16 tells us that approximately 68% of the population have I.Q.s between 84 and 116.

● **PROBLEM** 9-138

Let $X_1, \ldots, X_n$ be a random sample from a normal distribution with mean μ and variance σ^2. Let $(0_1, 0_2) = (\mu, \sigma)$. Estimate the parameters μ and σ by the method of moments.

Solution: Let x be a random variable having the probability density function $f(x; Q_1, \ldots, Q_k)$ where $Q_1, \ldots, Q_k$ are parameters that characterize the distribution. The rth moment about the origin, $\mu_r'(Q_1, \ldots, Q_k)$ is defined as

$$\mu_r'(Q_1, \ldots, Q_k) = E[x^r].$$

For example, if the distribution is continuous,

$$\mu_1' = E[x] = \int_{-\infty}^{\infty} xf(x) \, dx,$$

the population expected value, and

$$\mu_2' = E[x^2] = \int_{-\infty}^{\infty} x^2 f(x) \, dx.$$

Next, consider a random sample, $x_1, \ldots, x_n$ from the distribution having density function $f(x_1; Q_1, Q_2, \ldots Q_k)$. Then the jth sample moment is

$$M_j' = \frac{1}{n} \sum_{i=1}^{n} x_i^{\,j} \qquad (j = 1, \ldots, k). \tag{1}$$

These moments are statistics (i.e., functions of the random sample) and hence may be used to estimate the parameters $Q_1, \ldots, Q_k$. Thus,

$$M_j' = \mu_j (Q_1, \ldots, Q_k); \qquad j = 1, \ldots, k \tag{2}$$

in the k variables $Q_1, \ldots, Q_k$. A solution of (2), say $(\hat{Q}_1, \ldots, \hat{Q}_k)$ is called the method of moments estimator of $(Q_1, \ldots, Q_k)$.

In the present problem, we are asked to find the method of moments estimators of the two parameters μ and σ in the normal distribution. Since $\mu = \mu_1'$ and $\sigma^2 = \mu_2 - (\mu_1)^2$, the method of moments equations are

$$M_1' = \frac{1}{n} \sum_{i=1}^{n} x_i = \mu_1(\mu, \sigma) = \bar{x} \tag{3}$$

and $$M_2' = \frac{1}{n} \sum_{i=1}^{n} x_i^2 = \mu_2(\mu, \sigma) = \sigma^2 + \mu^2. \tag{4}$$

From (3), $\bar{x}$ is the method of moments estimator of μ while from (4)

$$\hat{\sigma} = \sqrt{M_2' - (\mu)^2} = \sqrt{\frac{1}{n} \sum_{i=1}^{n} x_i^2 - (\bar{x})^2}$$

$$= \sqrt{\frac{\sum_{i=1}^{n} (x_i - \bar{x})^2}{n}} \, .$$

Note that the estimator of σ^2 is biased.

An urn contains a number of black and a number of white balls, the ratio of the numbers being 3 : 1. It is not known, however, which color ball is more numerous. From a random sample of three elements drawn with replacement from the urn, estimate the probability of drawing a black ball.

Solution: Let p be the probability of drawing a black ball, and n the number of balls drawn. Then p is either $\frac{1}{4}$ or $\frac{3}{4}$. Since a drawn ball is either black or white, the number of black balls is given by the binomial distribution

$$f(x; p) = \binom{n}{x} p^x (1 - p)^{n-x} ; \quad x = 0, 1, \ldots, n. \quad (1)$$

Letting $p = \frac{1}{4}$ and then $p = \frac{3}{4}$ we obtain the following table from (1).

Outcome: x	0	1	2	3
f(x; 1/4)	27/64	27/64	9/64	1/64
f(x; 3/4)	1/64	9/64	27/64	27/64

Now assume that we draw a sample and find x = 2. Then, it is more likely that black balls are more numerous in the urn. If, on the other hand, no black balls were drawn, i.e., x = 0, then it is more likely that the white balls are three times more numerous than the black balls.

In general,

$$\hat{p} = \hat{p}(x) = \begin{array}{ll} .25 & \text{for } x = 0, 1 \\ .75 & \text{for } x = 2, 3. \end{array}$$

is the estimator of the parameter p. For given sample outcomes it yields the most likely values of the parameter.

Let $x_1, \ldots, x_n$ be a random sample from a distribution having mean μ. Show that the sample mean,

$$\bar{x} = \sum_{i=1}^{n} x_i,$$

is an unbiased estimator of the population mean.

Solution: Let the random variable x have the density function $f(x, \theta)$, where θ is an unknown parameter and let $x_1, \ldots, x_n$ be a random sample from the distribution having the above density function. Then $\hat{\theta} = \hat{\theta}(x_1, \ldots, x_n)$ is defined to be an unbiased estimator of θ if

$$E(\hat{\theta}) = \theta, \tag{1}$$

where $E(z)$ is the expected value of z. The expectation operator E has the properties,

$$E[cz] = cE(z) \tag{2}$$

and $E[z_1 + \ldots + z_n] = E[z_1] + \ldots + E[z_n].$ (3)

In the present problem,

$$\bar{x} = \frac{1}{n} \sum_{i=1}^{n} x_i = \frac{1}{n} (x_1 + x_2 + \ldots + x_n).$$

Hence, $E(\bar{x}) = E\left[\frac{1}{n} (x_1 + x_2 + \ldots + x_n)\right]$

$$= \frac{1}{n} E[x_1 + x_2 + \ldots + x_n],$$

$$= \frac{1}{n} [E(x_1) + \ldots + E(x_n)], \tag{4}$$

using (2) and (3). But the expected value of a random variable is the mean,

$$E(x) = \mu.$$

Hence, (4) becomes

$$E(\bar{x}) = \frac{1}{n} [n\mu] = \mu. \tag{5}$$

Comparing (5) and (1), $\bar{x}$, the sample mean, is an unbiased estimator of μ, the population mean.

Give an example of a sufficient statistic.

Solution: We shall derive a sufficient statistic, T,
for the parameter p of the family of Bernoulli distributions.   The probability density function of the Bernoulli
distribution is

$$f(x) = p^x (1 - p)^{1-x} \qquad\qquad x = 0, 1, \qquad\qquad (1)$$

where p is the probability of success.

Consider now the performance of three Bernoulli
trials.   There are then eight distinct samples of size
three and the probability function of (x_1, x_2, x_3) is

$$P(X = x) = P(X_1 = x_1, X_2 = x_2, X_3 = x_3)$$

$$= p^k (1 - p)^{3-k} \qquad\qquad (2)$$

where $x_1 + x_2 + x_3 = k$, is the number of 1's among the
values (x_1, x_2, x_3).   The number of 1's or successes in
three trials is a random variable, (say)

$$T = X_1 + X_2 + X_3,$$

with the probability density function,

$$P(T = k) = P(k \text{ successes in 3 trials})$$

$$= \binom{3}{k} p^k (1 - p)^{3-k} . \qquad\qquad (3)$$

We now ask, what is the conditional probability of
obtaining the point (x_1, x_2, x_3) given $T = k$?

$$P(x/T = k) = P\left(X = x \,\middle|\, T = k\right)$$

$$= \frac{P(X = x \text{ and } T = k)}{P(T = k)} , \qquad\qquad (4)$$

using the definition of conditional probability,

$$P(z_2/z_1) = \frac{P(z_2 \cap z_1)}{P(z_1)} , \quad \text{and letting}$$

$$x = x_1 + x_2 + x_3 .$$

The conditional probability density function of (4) is given by

$$P(x/T = k) = \begin{cases} 0, & \text{if } x_1 + x_2 + x_3 \neq k \\ \dfrac{p^k (1 - p)^{3-k}}{\binom{3}{k} p^k (1 - p)^{3-k}} = \dfrac{1}{\binom{3}{k}} & \\ & \text{if } x_1 + x_2 + x_3 = k \end{cases} \quad (5)$$

from (2) and (3).

Using (5) we may form the table:

Sample	T	Prob.	Prob. given $T = 0$	Prob. given $T = 1$	Prob. given $T = 2$	Prob. given $T = 3$
(0, 0, 0)	0	$(1 - p)^3$	1	0	0	0
(0, 0, 1)	1	$p(1 - p)^2$	0	1/3	0	0
(0, 1, 0)	1	$p(1 - p)^2$	0	1/3	0	0
(1, 0, 0)	1	$p(1 - p)^2$	0	1/3	0	0
(0, 1, 1)	2	$p^2(1 - p)$	0	0	1/3	0
(1, 0, 1)	2	$p^2(1 - p)$	0	0	1/3	0
(1, 1, 0)	2	$p^2(1 - p)$	0	0	1/3	0
(1, 1, 1)	3	p^3	0	0	0	1/3

Examining the table we see that the distribution of (X_1, X_2, X_3) given any particular value of T, does not involve the parameter p. That is, we cannot make any inferences about p using the conditional distribution of (x_1, x_2, x_3) given $T = X_1 + X_2 + X_3 = k$. It follows that the statistic T is sufficient for knowledge of the parameter p.

The reader should be aware that the general definition of a sufficient statistic involves the concepts of functional independence and the theory of transformations of a variable. The more modest purpose of this problem was to exhibit a concrete example of a sufficient statistic and how the concept of sufficiency arises.

● **PROBLEM** 9-142

Consider a probability distribution having mean μ and variance σ^2. Show that

$$\bar{x}_n = \frac{1}{n} \sum_{i=1}^{n} x_i \quad \text{and} \quad s_n^2 = \frac{1}{n - 1} \sum_{i=1}^{n} (x_i - \bar{x}_n)^2$$

are consistent sequences of estimators of σ^2.

<u>Solution</u>: The sample mean and the sample estimate of the distribution variance are unbiased estimates of the population mean μ and population variance σ^2. Recall that an estimate $\hat{\theta}$ of a parameter θ is consistent if

$$\lim_{n \to \infty} E(\hat{\theta}_n) = \theta \quad \text{and} \quad \lim_{n \to \infty} \text{Var} \ (\hat{\theta}_n) = 0$$

where $\hat{\theta}_n$ represents a sequence of estimators that depend on the sample size n.

Hence, we must show that $\text{Var} \ [\bar{x}_n]$ and $\text{Var} \ [s_n^2]$ both tend to zero as $n \to \infty$.

The variance of a random variable is defined as

$$\text{Var} \ (\theta) = E \ [(\theta - E(\theta))^2] \tag{1}$$

From this definition it follows that, for independent random variables x_1, x_2 and constants a, b,

$\text{Var} \ (aX_1 + bX_2) = a^2 \ \text{Var} \ (x_1) + b^2 \ \text{Var} \ (x_2).$

In general,

$$\text{Var} \left[\sum_{i=1}^{n} a_i x_i \right] = \sum_{i=1}^{n} a_i^2 \ \text{Var} \ (x_i)$$

Now, $\text{Var} \ (\bar{x}) = \text{Var} \left[\dfrac{1}{n} \sum_{i=1}^{n} x_i \right] = \dfrac{1}{n^2} \text{Var} \left[\sum_{i=1}^{n} x_i \right]. \tag{2}$

But since the sample is randomly chosen, the x_i are independent. That is,

$$\text{Var} \ [x_1 + x_2 + \ldots + x_n] = \text{Var} \left[\sum_{i=1}^{n} x_i \right]$$

$$= \sum_{i=1}^{n} \text{Var} \ (x_i).$$

Hence, (2) becomes

$$\frac{1}{n^2} \sum_{i=1}^{n} \text{Var} (x_i) = \frac{1}{n^2} n\sigma^2 = \frac{\sigma^2}{n} .$$

Hence, $\lim_{n \to \infty} \text{Var} (\bar{x}) = \lim_{n \to \infty} \frac{\sigma^2}{n} = 0$

and $\bar{x}$ is a consistent estimator of μ.

We now consider $\text{Var} (s_n^2)$. Using (1)

$$\text{Var} (s_n^2) = E [(s_n^2 - \sigma^2)^2]$$

$$= \frac{1}{n} \left[\mu_4 - \frac{n - 3}{n - 1} \sigma^4 \right] \qquad (3)$$

where μ_4 denotes the fourth moment about the origin. Letting $n \to \infty$ in (3) we find

$$\lim_{n \to \infty} \text{Var} (s_n^2) = \lim_{n \to \infty} \frac{1}{n} \left[\mu_4 - \frac{n - 3}{n - 1} \sigma^4 \right] = 0 .$$

Hence s_n^2 is a consistent estimator of σ^2.

● **PROBLEM** 9-143

Consider the exponential density $f(x; \theta) = \theta e^{-\theta x}$. Let $x_1, \ldots, x_n$ denote a random sample from this density. Show that the sample mean, $\bar{x}$, is a minimum variance estimator of the mean of the distribution $\frac{1}{\theta}$.

Solution: We use the Cramér-Rao inequality to solve the problem. Let T be an unbiased estimator of the parameter $\frac{1}{\theta}$. Then, by the Cramér-Rao inequality,

$$\text{Var} [T] \geq \frac{\left[\left(\frac{1}{\theta} \right)' \right]^2}{nE \left[\left[\frac{\partial}{\partial \theta} \ln f(x; \theta) \right]^2 \right]} \qquad (1)$$

where $\left(\frac{1}{\theta} \right)' = \frac{d}{d\theta} \left(\frac{1}{\theta} \right)$ and $\ln z \equiv \log_e z$.

Verbally, (1) states that the variance of any unbiased estimator of a parameter of a distribution is always greater than the expression on the right side of (1). Equality prevails only when there exists a function $K(\theta, n)$ such that

$$\sum_{i=1}^{n} \frac{\partial}{\partial \theta} \ln f(x_i; \theta)$$

$$= K(\theta, n) \left[u_1(x_1, \ldots, x_n) - \left(\frac{1}{\theta}\right) \right] \tag{2}$$

When (2) is possible, T is called a minimum variance unbiased estimator.

We first find $\left[\frac{\partial}{\partial \theta} \ln f(x; \theta) \right]^2$. Thus,

$$f(x; \theta) = \theta e^{-\theta x}$$

$$\ln f(x; \theta) = \ln \theta - \theta x$$

$$\frac{\partial}{\partial \theta} \ln f(x; \theta) = \frac{1}{\theta} - x. \qquad \text{Hence,}$$

$$E \left[\left[\frac{\partial}{\partial \theta} \ln f(x; \theta) \right]^2 \right] = E \left[\left(\frac{1}{\theta} - x\right)^2 \right]$$

$$= E \left[\left(x - \frac{1}{\theta}\right)^2 \right] = \text{Var} [x] = \frac{1}{\theta^2}, \tag{3}$$

since the variance of a random variable which has the negative exponential density is $\frac{1}{\theta^2}$.

Next, we find $\left[\left(\frac{1}{\theta}\right)' \right]^2$. $\frac{d}{d\theta} \left(\frac{1}{\theta}\right) = -\frac{1}{\theta^2}$.

$$\tag{4}$$

Hence, $\left[\left(\frac{1}{\theta}\right)' \right]^2 = \frac{1}{\theta^4}$.

Substituting (3) and (4) into (1)

$$\text{Var} [T] \geq \frac{\frac{1}{\theta^4}}{n \frac{1}{\theta^2}} = \frac{1}{n \, \theta^2} \quad . \tag{5}$$

Changing the inequality in (5) to equality,

$$\text{Var} [T] = \frac{1}{n \, \theta^2} \quad .$$

Let us try to put the exponential density into the form (2). Thus

$$\sum_{i=1}^{n} \frac{\partial}{\partial \theta} \ln f(x; \theta) = \sum_{i=1}^{n} \frac{\partial}{\partial \theta} \ln \theta - \theta x_i$$

$$= \sum_{i=1}^{n} \frac{1}{\theta} - x_i = -n \left(\overline{x} - \frac{1}{\theta} \right).$$

Letting $K(\theta, n) = -n$ and $u_1(x_1, \ldots, x_n) = \overline{x}$, we see that (2) is indeed possible for the negative exponential density.

This is a sufficient condition to show that $\overline{x}$ is an unbiased estimator of $\left(\frac{1}{\theta} \right)$ with minimum variance (equal to the lower bound).

HYPOTHESIS TESTING

• PROBLEM 9-144

In testing a hypothesis concerned with the value of a population mean, first the level of significance to be used in the test is specified and then the regions of acceptance and rejection for evaluating the obtained sample mean are determined. If the 1 percent level of significance is used, indicate the percentages of sample means in each of the areas of the normal curve, assuming that the population hypothesis is correct, and the test is two-tailed.

Solution: A level of significance of 1% signifies that when the population mean is correct as specified, the sample mean will fall in the critial areas of rejection only 1% of the time. Referring to the figure below, .005 or .5% of the sample means will fall in each area of rejection and 99% of the sample means will fall in the region of acceptance.

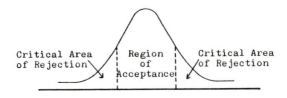

Critical Area of Rejection   Region of Acceptance   Critical Area of Rejection

A sample of size 49 yielded the values $\bar{x} = 87.3$ and $s^2 = 162$. Test the hypothesis that $\mu = 95$ versus the alternative that it is less. Let $\alpha = .01$.

<u>Solution</u>: The null and alternative hypotheses are given respectively by

$$H_0 : \mu = 95; \quad H_1 : \mu < 95 .$$

$\alpha = .01$ is the given level of significance.

Because the sample size is quite large (≥ 30), we can assume that the distribution of $\bar{X}$ is approximately normal. We are using the sample variance s^2 as an estimate of the true but unknown population variance and if the sample were not as large we would use a t-test.

The critical region consists of all z-scores that are less than $z_{.01} = -2.33$. The observed z-score is

$$z = \frac{\bar{X} - \mu}{\sqrt{s^2 / n}} = \frac{87.3 - 95}{\sqrt{162/49}} = \frac{(-7.7)(7)}{\sqrt{162}} = -4.23 .$$

This observed score is in the critial region; thus we reject the null hypothesis and accept the alternative that $\mu < 95$.

In investigating several complaints concerning the weight of the "NET WT. 12 OZ." jar of a local brand of peanut butter, the Better Business Bureau selected a sample of 36 jars. The sample showed an average net weight of 11.92 ounces and a standard deviation of .3 ounce. Using a .01 level of significance, what would the Bureau conclude about the operation of the local firm?

<u>Solution</u>: We use a one-tailed test because we are concerned with whether the actual population mean is 12 ounces or whether it is less than 12 ounces. Therefore we have

$$H_0 : \mu = 12; \quad H_1 : \mu < 12.$$

The figure below depicts this problem. Since the sample size, n, is large (≥ 30), the test statistic $z = (\bar{x} - \mu)/s_{\bar{x}}$ is normally dis-

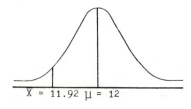

$\bar{X} = 11.92 \quad \mu = 12$

tributed with a mean of 0 and a standard deviation of 1. We must calculate

$$z = \frac{\bar{x} - \mu}{s_{\bar{x}}} \qquad \text{where} \qquad s_{\bar{x}} = \frac{s}{\sqrt{n}}$$

and compare the value of z obtained to a critical value. Our critical value of z for this problem is -2.33 since 1% of scores of the standard normal distribution have a z-value below -2.33 and we want to reject H_0 for values of $\bar{x}$ less than μ.

Therefore, we have for our decision rule: reject H_0 if z < -2.33; accept H_0 if z ≥ -2.33.
For the data of this problem

$$s_{\bar{x}} = \frac{.3}{\sqrt{36}} = \frac{.3}{6} = .05$$

and

$$z = \frac{11.92 - 12}{.05} = \frac{-.08}{.05} = -1.60$$

Since -1.60 ≥ - 2.33, one accepts H_0 and concludes that at a 1% level of significance the actual population mean of the local brand of peanut butter is 12 ounces.

● **PROBLEM** 9-147

Suppose it is required that the mean operating life of size "D" batteries be 22 hours. Suppose also that the operating life of the batteries is normally distributed. It is known that the standard deviation of the operating life of all such batteries produced is 3.0 hours. If a sample of 9 batteries has a mean operating life of 20 hours, can we conclude that the mean operating life of size "D" batteries is not 22 hours? Then suppose the standard deviation of the operating life of all such batteries is not known but that for the sample of 9 batteries the standard deviation is 3.0. What conclusion would we then reach?

Solution: Since the operating life of "D" batteries is normally distributed, and the standard deviation for all batteries is known, the sample mean will have a Z or normal distribution regardless of the sample size.
Hence for the first part of this problem, we will calculate

$$Z = \frac{\bar{X} - \mu}{\sigma_{\bar{X}}} \qquad \text{where} \qquad \sigma_{\bar{X}} = \frac{\sigma}{\sqrt{n}} .$$

The diagram for this problem is given below.

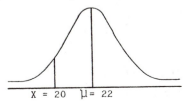

X = 20   μ = 22

The null hypothesis and alternate hypothesis are

$$H_0 : \mu = 22 \; ; \qquad H_1 : \mu \neq 22 .$$

The decision rule at $\alpha = .05$ is as follows: Reject H_0 if $Z > 1.96$ or $Z < -1.96$; accept H_0 if $-1.96 \leq Z \leq 1.96$. The value of 1.96 is chosen as the critial value because for this problem we have $\alpha = .05$ and a two-tailed test, and for the standard normal distribution, 2.5% of scores will have a Z-value greater than 1.96 and 2.5% of scores will have a Z-value less than -1.96.

For this set of data

$$\sigma_{\bar{X}} = \frac{3.0}{\sqrt{9}} = 1 \quad \text{and}$$

$$Z = \frac{20 - 22}{1} = -2.0 .$$

Therefore we will reject H_0 and conclude that the mean operating life of size "D" batteries is not 22 hours.

When the population standard deviation is not known the sample mean has a t-distribution with $n-1 = 8$ degrees of freedom. We calculate

$$t = \frac{\bar{X} - \mu}{S_{\bar{X}}}$$

and use the decision rule to reject H_0 if $t > 2.306$ or $t < -2.306$ (the critical t for 8 df's $\alpha = .05$, and a two-tailed test is 2.306) and to accept H_0 if $-2.306 \leq t \leq 2.306$. For this data

$$t = \frac{20 - 22}{1} = -2.0 .$$

So in this case we will accept H_0 that the mean operating life of size "D" batteries is 22 hours.

Given the eight sample observations 31, 29, 26, 33, 40, 28, 30, and 25, test the null hypothesis that the mean equals 35 versus the alternative that it does not. Let $\alpha = .01$.

Solution: $H_0 : \mu = 35$; $H_1 : \mu \neq 35$.

The level of significance α is given to be .01. This is a two-tailed test. Since the variance is unknown and must be estimated by the sample variance the t-test is appropriate.

$$\bar{X} = \frac{\sum_{i=1}^{n} x_i}{n} = 30.25 \; ; \quad n = 8 \qquad s^2 = \frac{1}{n-1} \left[\sum_{i=1}^{n} x_i^2 - \frac{(\sum_{i=1}^{n} x_i)^2}{n} \right]$$

$$= \frac{1}{7} \left[7476 - \frac{(242)^2}{8} \right] = \frac{155.5}{7} = 22.21 \; ; \qquad s = \sqrt{22.2} = 4.71.$$

The t-statistic is

$$t = \frac{\bar{x} - \mu}{s/\sqrt{n}} = \frac{30.25 - 35}{4.71/\sqrt{8}} = \frac{-4.7\sqrt{8}}{4.71} = -2.85.$$

The critical region will be that in which $Pr(c_1 < t < c_2) = 1 - .01$ or $Pr(t < c_1) = .005$ and $Pr(t > c_2) = .005$. Thus,

$$c_1 = -t_{.005}(7) = -3.499$$

$$c_2 = t_{.005}(7) = 3.499 .$$

Thus if our t-statistic lies between -3.499 and 3.499 we will accept H_0 and otherwise reject. Our calculated t-statistic is -2.85 which leads us to accept the null hypothesis that the mean equals 35 and reject the alternative hypothesis.

Suppose that you want to decide which of two equally-priced brands of light bulbs lasts longer. You choose a random sample of 100 bulbs of each brand and find that brand A has sample mean of 1180 hours and sample standard deviation of 120 hours, and that brand B has sample mean of 1160 hours and sample standard deviation of 40 hours. What decision should you make at the 5% significance level?

Solution: Arrange the data into a table:

	n	$\bar{X}$	s
Brand A	100	1180	120
Brand B	100	1160	40

Establish two hypotheses: H_0 asserts that A and B last the same, on the average, and H_1 asserts that A and B have different average lifespans. Thus:

$$H_0 : \mu_A = \mu_B , \text{ or, } \quad H_1 : \mu_A \neq \mu_B ;$$

equivalently, $\quad H_0 : \mu_A - \mu_B = 0; \quad H_1 : \mu_A - \mu_B \neq 0.$

Now define the acceptance region and rejection region for this test. We can use the standard normal curve to determine these regions because of the theorem that if two populations from which two independent random samples are taken are normally distributed or if $n_1 + n_2 > 30$, then the sampling distribution of the difference between the sample means is normal or approximately normal, and its standard error is

$$\sqrt{\frac{\sigma_1^2}{n_1} + \frac{\sigma_2^2}{n_2}}$$

when σ_1^2 and σ_2^2 are the variances of populations 1 and 2 respectively. In this problem, $n_1 + n_2 = 200 > 30$, so that the sampling distribution of $\overline{X}_1 - \overline{X}_2$ is approximately normal. The acceptance region is the interval which lies under 95% of the area under the standard normal curve, because your decision will be made at the 5% level of significance. The acceptance region is therefore $|Z| \leq 1.96$ and the rejection region is $|Z| > 1.96$.

$$\text{Now } Z = \frac{(\overline{X}_A - \overline{X}_B) - (\mu_A - \mu_B)}{\sqrt{\frac{\sigma_A^2}{n_A} + \frac{\sigma_B^2}{n_B}}} = \frac{(\overline{X}_A - \overline{X}_B)}{\sqrt{\frac{\sigma_A^2}{n_A} + \frac{\sigma_B^2}{n_B}}} \text{ by } H_0.$$

$(S_A)^2 = (120)^2$ and $(S_B)^2 = (40)^2$ can be used as estimates for σ_A^2 and σ_B^2. Substituting known values in the formula for Z, we have

$$Z = \frac{1180 - 1160}{\sqrt{\frac{14400}{100} + \frac{1600}{100}}} = \frac{20}{\sqrt{144 + 16}} = \frac{20}{\sqrt{160}} = \frac{20}{\sqrt{12.65}} = 1.58 .$$

Since $- 1.96 < (Z = 1.58) < 1.96$, Z is in the acceptance region. Therefore, we accept the hypothesis that there is no difference between the average lifespans of the two brands, at the 5% significance level.

A reading test is given to an elementary school class that consists of 12 Anglo-American children and 10 Mexican-American children. The results of the test are - Anglo-American children: $\overline{X}_1 = 74$, $S_1 = 8$; Mexican-American children: $\overline{X}_2 = 70$, $S_2 = 10$. Is the difference between the means of the two groups significant at the .05 level of significance?

Solution: Assuming the test scores are normally distributed, we may use the t-test with $n_1 + n_2 - 2$ degrees of freedom to test the significance of the difference between the means because the statistic

$$\frac{(\overline{X}_2 - \overline{X}_1) - (\mu_2 - \mu_1)}{S_{\overline{X}_2 - \overline{X}_1}}$$

has a t-distribution when $n_1 + n_2 \leq 30$. The figure below depicts this problem.

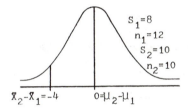

We have for our null and alternate hypotheses

$H_0: \mu_2 - \mu_1 = 0$ $H_1: \mu_2 - \mu_1 \neq 0$.

We must calculate $t = \dfrac{(\overline{X}_2 - \overline{X}_1) - 0}{S_{\overline{X}_2 - \overline{X}_1}}$ where

$$S_{\overline{X}_2 - \overline{X}_1} = \sqrt{\frac{(n_1 - 1)S_1^2 + (n_2 - 1)S_2^2}{n_1 + n_2 - 2}} \sqrt{\frac{1}{n_1} + \frac{1}{n_2}} .$$

The critical t for $n_1 + n_2 - 2 = 20$ df's and a two-tailed test at $\alpha = .05$ is 2.09 because for the t-distribution with mean of 0 and standard deviation of 1, 2.5% scores will have a t-value greater than 2.09 and 2.5% of scores will have a t-value less than - 2.09. Therefore, our decision rule is: reject H_0 if $t > 2.09$ or $t < - 2.09$; accept H_0 if $- 2.09 \leq t \leq 2.09$.

For the data of this problem,

$$S_{\overline{X}_2 - \overline{X}_1} = \sqrt{\frac{11(8)^2 + 9(10)^2}{12 + 10 - 2}} \sqrt{\frac{1}{12} + \frac{1}{10}}$$

$$= \sqrt{\frac{704 + 900}{20}} \sqrt{.0833 + .1}$$

$$= \sqrt{80.2} \sqrt{.1833} = 3.834 \qquad \text{and}$$

$$t = \frac{(70 - 74) - 0}{3.834} = \frac{-4}{3.834} = -1.04.$$

Since $-2.09 < -1.04 < 2.09$, we accept H_0 and conclude that the difference between the means of the two groups is not significant.

● **PROBLEM** 9-151

A sports magazine reports that the people who watch Monday night football games on television are evenly divided between men and women. Out of a random sample of 400 people who regularly watch the Monday night game, 220 are men. Using a .10 level of significance, can be conclude that the report is false?

<u>Solution</u>: We have for this problem as our hypotheses:

$$H_0: p = .50, \text{ where } p \text{ is the true population proportion.}$$

$$H_1: p \neq .50.$$

The following diagram depicts the data of this problem, where $\overline{p}$ is the sample proportion of men who watch Monday night football.

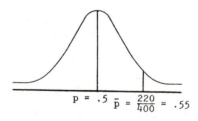

$$p = .5 \quad \overline{p} = \frac{220}{400} = .55$$

The statistic $(\overline{p} - p)/\sigma_p$ is approximately normally distributed with a mean of 0 and a standard deviation of 1. We calculate this value, which is called Z,

$$Z = \frac{\overline{p} - p}{\sigma_{\overline{p}}} \quad \text{where}$$

$$\sigma_{\overline{p}} = \sqrt{\frac{pq}{n}}$$

and compare the value of Z to a critical value. If Z lies beyond this critical value, we will reject H_0. For this problem, where we have $\alpha = 10\%$ and a two-tailed test, our critical value is 1.645, since for the normal distribution with mean of 0 and standard deviation of 1, 5% of scores will have a Z-value above 1.645 and 5% of scores will have a Z-value below -1.645. Therefore our decision rule is: reject H_0 if $|Z| > 1.645$, accept H_0 if $|Z| \leq 1.645$.

For the data of this problem,

$$\sigma_p = \sqrt{\frac{(.50)(.50)}{400}} \approx .025 \quad \text{and}$$

$$Z = \frac{.55 - .50}{.025} = \frac{.05}{.025} = 2.0 \ .$$

Since $2.0 > 1.645$, we reject H_0 and conclude that the report of the sports magazine is incorrect at a 10% level of significance.

● **PROBLEM** 9-152

In one income group, 45% of a random sample of people express approval of a product. In another income group, 55% of a random sample of people express approval. The standard errors for these percentages are .04 and .03 respectively. Test at the 10% level of significance the hypothesis that the percentage of people in the second income group expressing approval of the product exceeds that for the first income group.

<u>Solution:</u> This problem may be depicted by the following diagram.

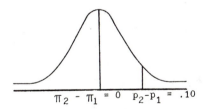

We would like to see whether $p_2 - p_1 = .10$ is far enough away from $p_2 - p_1 = 0$ to reject a null hypothesis that $\pi_2 - \pi_1 = 0$. Therefore our hypotheses are:

$$H_0 : \pi_2 - \pi_1 = 0 \ ,$$

where π_2 and π_1 are the population proportions for the second and first income group respectively.

$$H_1 : \pi_2 - \pi_1 > 0 .$$

The statistic

$$\frac{(P_2 - P_1) - (\pi_2 - \pi_1)}{\sigma_{P_2 - P_1}}$$

is approximately normally distributed with a mean of 0 and a standard deviation of 1.

We calculate this value, which is called Z:

$$Z = \frac{(P_2 - P_1) - (\pi_2 - \pi_1)}{\sigma_{P_2 - P_1}} \qquad \text{where} \qquad \sigma_{P_2 - P_1} = \sqrt{S_{P_1}^2 + S_{P_2}^2}$$

and compare the value of Z to a critical value. If Z lies beyond this critical value, we will reject H_0. For this problem where we have $\alpha = 10\%$ and a one-tailed test, our critical value is 1.28, since for the normal distribution with mean of 0 and standard deviation of 1 10% of scores will have a Z-value above 1.28.

For $\alpha = .10$ and a one-tailed test our decision rule is: reject H_0 if $Z > 1.28$, accept H_0 if $Z \le 1.28$.

We must calculate

$$Z = \frac{(P_2 - P_1) - (\pi_2 - \pi_1)}{\sigma_{P_2 - P_1}} \qquad \text{where} \qquad \sigma_{P_2 - P_1} = \sqrt{S_{P_1}^2 + S_{P_2}^2} .$$

For the data of this problem

$$\sigma_{P_2 - P_1} = \sqrt{(.04)^2 + (.03)^2} = \sqrt{.0016 + .0009}$$

$$= \sqrt{.0025} = .05 \quad \text{and}$$

$$Z = \frac{(.55 - .45) - 0}{.05} = 2.0 .$$

Since $2.0 > 1.28$, we reject H_0 and conclude that the percentage of people in the second income group expressing approval of the product exceeds that for the first income group at the 10% level of significance.

A new treatment plan for schizophrenia has been tried for 6 months with 54 randomly chosen patients. At the end of this time, 25 patients are recommended for release from the hospital; the usual proportion released in 6 months is $\frac{1}{3}$. Using the normal approximation to the binomial distribution, determine whether the new treatment plan has resulted in significantly more releases than the previous plan. Use a 0.05 level of significance.

Solution: The number of patients recommended for release from the hospital has a binomial distribution. We have p, the proportion of patients released from the hospital, equal to $\frac{1}{3}$, and q, the proportion of patients not released, equal to 1 - p or $\frac{2}{3}$. Since np = $(54)\left[\frac{1}{3}\right]$ = 18 and nq = $(54)\left[\frac{2}{3}\right]$ = 36, are both greater than or equal to 5, we may approximate this binomial distribution by the normal distribution in this case.

We use μ = np and σ = $\sqrt{npq}$ as the mean and standard deviation of this normal approximation.

Since n = 54, p = $\frac{1}{3}$ and q = $\frac{2}{3}$, we have μ = $54\left[\frac{1}{3}\right]$ = 18 and σ = $\sqrt{54\left[\frac{1}{3}\right]\left[\frac{2}{3}\right]}$ = 3.46.

The problem can now be depicted by the following diagram.

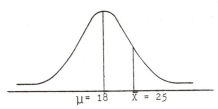

μ= 18 $\bar{X}$ = 25

We desire to see whether our sample mean of 25 exceeds 18 by enough for us to conclude that the actual population mean for the new treatment is greater than 18. Thus we choose as our alternate hypothesis

H_1: μ > 18.

Hence, our null hypothesis is

H_0: μ = 18.

The statistic $\frac{(x - \mu)}{\sigma}$ has a normal distribution with a mean of 0 and a standard deviation of 1.

We use a critical value of 1.645 for this problem, since we have $\alpha = .05$ and a one-tailed test, and for the normal distribution with mean of 0 and standard deviation of 1, 5% of scores will have a Z-value above 1.645. Therefore our decision rule is: reject H_0 if $Z > 1.645$; accept H_0 if $Z \leq 1.645$.

We must now calculate

$$Z = \frac{x - \mu}{\sigma} .$$

Since the binomial distribution is a discrete distribution and the normal distribution is continuous, we will use 24.5 instead of 25 as the value for x in the above formula.

We calculate Z.

$$Z = \frac{24.5 - 18}{3.46} = \frac{6.5}{3.46} = 1.88.$$

Since $1.88 > 1.645$, we reject H_0 and conclude that the population mean for the new treatment is greater than 18, and hence, it has resulted in significantly more releases than the previous plan.

● **PROBLEM** 9-154

The makers of a certain brand of car mufflers claim that the life of the mufflers has a variance of .8 year. A random sample of 16 of these mufflers showed a variance of 1 year. Using a 5% level of significance, test whether the variance of all the mufflers of this manufacturer exceeds .8 year.

Solution: Our hypotheses for this problem are:

$$H_0: \sigma^2 = .8 \qquad H_1: \sigma^2 > .8.$$

The statistic $\frac{(n - 1)s^2}{\sigma^2}$ has a χ^2 distribution with $n - 1$ degrees of freedom.

For $\alpha = .05$, a one-tailed test, and $n - 1 = 15$ degrees of freedom, we will have for our decision rule: reject H_0 if $\chi^2 > 24.996$; accept H_0 if $\chi^2 \leq 24.996$.

We calculate χ^2.

$$\chi^2 = \frac{(n - 1)s^2}{\sigma^2} = \frac{15(1)^2}{(.8)} = 18.75.$$

Since $18.75 < 24.996$, we accept H_0 and conclude that the variance of all the mufflers of this manufacturer does not exceed .8 year.

● **PROBLEM** 9-155

A random sample of 20 boys and 15 girls were given a standardized test. The average grade of the boys was 78 with a standard deviation of 6, while the girls made an average grade of 84 with a standard deviation of 8. Test the hypothesis that $\sigma_1^2 = \sigma_2^2$ against the alternate hypothesis $\sigma_1^2 < \sigma_2^2$ where σ_1^2 and σ_2^2 are the variances of the population of boys and girls. Use a .05 level of significance.

Solution: Our hypotheses for this problem are:

$H_0: \sigma_1^2 = \sigma_2^2$ $\qquad\qquad H_1: \sigma_2^2 > \sigma_1^2$.

The F random variable is defined as

$$F = V/r_2 \Big/ U/r_1$$

where U and V are chi-square random variables with r_1 and r_2 degrees of freedom.

Since $(n_1-1)S_1^2/\sigma_1^2$ and $(n_2-1)S_2^2/\sigma_2^2$ have chi-square distributions with n_1-1 and n_2-1 degrees of freedom, then this becomes,

$F = (n_2-1)S_2^2/(n_2-1)\sigma_2^2 \Big/ (n_1-1)S_1^2/(n_1-1)\sigma_1^2 = S_2^2/S_1^2$,

since under $H_0, \sigma_1^2 = \sigma_2^2$.

The statistic $\dfrac{S_2^2}{S_1^2}$ has an F distribution with $n_2 - 1$ degrees of freedom for the numerator (df_1) and $n_1 - 1$ degrees of freedom for the denominator (df_2).

To make a decision we must decide whether the calculated F given by $\dfrac{S_2^2}{S_1^2}$ is greater than $F_{.95}(df_1, df_2)$. Note that df_1 represents the degrees of freedom given by the sample size generating s_2^2. Similarly, the degrees of freedom in the denominator of the F-ratio, $d.f._2$, are found from the sample size n_1.

Since $F_{.95}(14, 19) = 2.26$ our decision rule is:
Reject H_0 if calculated $F > 2.26$; accept H_0 if calculated $F \leq 2.26$.

We calculate F.

$$F = \frac{8^2}{6^2} = \frac{64}{36} = 1.78$$

Since $1.78 \leq 2.26$, we accept H_0 and conclude that the variances of the populations of boys and girls are equal.

● **PROBLEM** 9-156

A plant manager claims that on the average no more than 5 service calls per hour are made by the plant's workers. Suppose in one particular hour, 9 service calls were required. At a 5% level of significance, could we now reject the plant manager's claim?

Solution: The number of accidents, claims, errors, or other such occurrences in a fixed time interval has a Poisson distribution.

We may use the Poisson distribution for this problem, where the variable in question is the number of service calls.

Our hypotheses in this problem are

H_0: $\lambda = 5$ H_1: $\lambda > 5$

where λ is the average number of service calls in the fixed time interval of one hour.

For the hypothesized value of $\lambda = 5$, the probability of obtaining 9 or more service calls in one hour is, using a table of Poisson probabilities, given by

$$.0363 + .0181 + .0082 + .0034 + .0013 + .0005 + .0002 = .068$$

which equals Pr(9 service calls) + Pr(10 service calls) + Pr(11 service calls) + Pr(12 service calls) + Pr(13 service calls) + Pr(14 service calls) + Pr(15 service calls). (Note: the probability of higher numbers of service calls is virtually 0).

Since this value of .068 is greater than .05, the probability of an hour occurring where 9 or more service calls are required is greater than 5%. Hence at a 5% level of significance, we would not reject H_0.

Two independent reports on the value of a tincture for treating a disease in camels were available. The first report made on a small pilot series showed the new tincture to be probably superior to the old treatment with a Yates' χ^2 of 3.84, df = 1, α = .05. The second report with a larger trial gave a "not significant" result with a Yates χ^2 = 2.71, df = 1, α = .10. Can the results of the 2 reports be combined to form a new conclusion?

Solution: The statistic χ^2 is additive, so we may combine the results by adding the values of χ^2 and the degrees of freedom. To make the combined result stronger and more accurate, we should recalculate each value of χ^2 from the 2 × 2 data tables of the two original studies without the Yates' correction. Since these tables are not available, we must be satisfied with the weaker result obtained by adding the Yates' χ^2 values. We obtain χ^2 = 6.55, df = 2, which is significant at α = .05. Thus the difference between the old and new treatments using the combined results of the 2 studies is significant.

Suppose an experimenter is about to analyze some data. In the data $\sigma_{\bar{X}_E - \bar{X}_C}$ = 1.0. The data require a two-tailed test, and the experimenter adopts the .05 level of significance.

Probability of Accepting H_0 and Rejecting H_0 for Different Sizes of $\mu_{\bar{E}}, \mu_C$.

$\mu_{\bar{E}}, \mu_C$ (true mean difference)	Probability of accepting H_0	An error to accept H_0?	Probability of rejecting H_0	An error to reject H_0?
4.0	.021	Type II Error	.979	Not an error: $\mu_0 \neq 0$
3.5	.062	"	.938	"
3.0	.149	"	.851	"
2.5	.295	"	.705	"
2.0	☐	"	.516	"
1.5	.677	"	.323	"
1.0	.830	"	.170	"
.5	.921	"	.079	"
0	.950	Not an error: $\mu_0 = 0$	.050	Type I Error
- .5	.921	"	.079	Not an error: $\mu_0 \neq 0$ "
-1.0	.830	"	.170	"
-1.5	.677	"	.323	"
-2.0	.484	"	.516	"
-2.5	.295	"	.705	"
-3.0	.149	"	.851	"
-3.5	.062	"	.938	"
-4.0	.021	"	.979	"

Let us consider differences of various sizes: Let $\mu_E - \mu_C$ range from -4.0 to +4.0. Then for each value, we compute the probability of a Type II error. **The table** reports all of the probabilities but one. Compute that missing value.

<u>Solution</u>: We will accept H_0 at a level of significance of .05 if $\bar{X}_E - \bar{X}_C$ lies within ± 1.96 standard deviations of 0, since $\mu_E - \mu_C = 0$ according to H_0. But the standard deviation of the differences between the means ($\sigma_{\bar{X}_E - \bar{X}_C}$) is 1.0. Therefore we will accept H_0 if $\bar{X}_E - \bar{X}_C$ lies between -1.96 and +1.96. But when $\mu_E - \mu_C = 2.0$, our normal curve will be as given in the figure below.

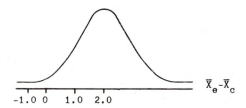

The probability of $\bar{X}_E - \bar{X}_C$ being less than -1.96 in this case is given by using the formula

$$Z = \frac{-1.96 - 2.00}{1.00} = \frac{-3.96}{1.00} = -3.96$$

since the statistic

$$\frac{(\bar{X}_E - \bar{X}_C) - (\mu_E - \mu_C)}{\sigma_{\bar{X}_E - \bar{X}_C}}$$

is normally distributed with a mean of 0 and standard deviation of 1.

Using the table for probabilities associated with the normal curve yields a probability of this occurring to be .00005.

The probability of $\bar{X}_E - \bar{X}_C$ being less than 1.96 is given by the formula

$$Z = \frac{1.96 - 2.00}{1.00} = \frac{-.04}{1.00} = -.04 \ .$$

Using the table for probabilities associated with the normal curve yields .4840 as the probability of $\bar{X}_E - \bar{X}_C$ being less than 1.96. Therefore the probability of $\bar{X}_E - \bar{X}_C$ lying between -1.96 and 1.96 when $\mu_E - \mu_C = 2.00$ is .4840 -.00005 or approximately .484.

Suppose we have a binomial distribution for which H_0 is $p = \frac{1}{2}$ where p is the probability of success on a single trial. Suppose the type I error, $\alpha = .05$ and $n = 100$. Calculate the power of this test for each of the following alternate hypotheses, H_1: $p = .55$, $p = .60$, $p = .65$, $p = .70$, and $p = .75$. Do the same when $\alpha = .01$.

<u>Solution</u>: Since $N(p)$ and $N(q)$ are both greater than 5, we may use the normal approximation to the binomial distribution for this problem. For $p = \frac{1}{2}$, the mean, μ, and standard deviation, σ, for this data are $p = \frac{1}{2}$, $\mu = np = 100(\frac{1}{2}) = 50$ and $\sigma = \sqrt{npq} = \sqrt{100(\frac{1}{2})(\frac{1}{2})} = 5$.

Since $\alpha = .05$, we will reject H_0 when $Z > 1.65$. Our formula for Z is

$$Z = \frac{X - \mu}{\sigma}.$$

Substituting the given values for Z, μ, and σ, we obtain

$$\frac{X - 50}{5} > 1.65.$$

Multiplying both sides of this equation by 5 yields

$$X - 50 > 8.25.$$

Adding 50 to both sides gives

$$X > 58.25.$$

So we will reject H_0 when $X > 58$.

The power of a test is given by the probability of accepting H_1 when H_1 is true. We must therefore calculate the probability of $X > 58$ for each of the specified H_1's.

We use the formula

$$Z = \frac{X - \mu}{\sigma}$$

where $X = 57.5$ ($58.5 - .5$ because the binomial distribution is discrete and the normal distribution is continuous). Also, for each case $\mu = np$ and $\sigma = \sqrt{npq}$.

The table below gives the values of μ, σ, Z, and the power of the test for each specified H_1.

H_1	$\mu = np$	$\sigma = \sqrt{npq}$	$\dfrac{X-\mu}{\sigma} = Z$	Power
p = .55	55	4.97	.5	.308
p = .60	60	4.90	-.51	.695
p = .65	65	4.77	-1.57	.942
p = .70	70	4.58	-2.73	.997
p = .75	75	4.33	-4.04	1.000

The power is obtained by using a table for the normal distribution and finding the probability that Z is greater than the value obtained in the prior column.

For $\alpha = .01$, we will reject H_0 when $Z > 2.33$. To find the value of X corresponding to $Z = 2.33$, we use

$$Z = \frac{X - \mu}{\sigma}$$

where $Z = 2.33$, $\mu = 50$, and $\sigma = 5$. Substituting $\dfrac{X - 50}{5} > 2.33$.

Multiplying both sides by 5 yields $X - 50 > 11.65$.

Adding 50 to each side yields $X > 61.65$.

Since 61.65 is in the interval 61.5 to 62.5, we will reject H_0 when X is greater than 61.5 for an $\alpha = .01$. Since the binomial distribution is discrete we use the value of X as 61.5, and we may now construct a table to give the power of the test for each specified H_1 when $\alpha = .01$.

H_1	$\mu = np$	$\sigma = \sqrt{npq}$	$\dfrac{X - \mu}{\sigma} = Z$	Power
p = .55	55	4.97	1.31	.095
p = .60	60	4.90	.31	.378
p = .65	65	4.77	-.73	.767
p = .70	70	4.58	-1.86	.967
p = .75	75	4.33	-3.12	.999

Note that a decrease in the size of the critical region from .05 to .01 uniformly reduced the power of the test for all H_1.

Let X possess a Poisson distribution with mean μ, i.e.

$$f(X,\mu) = e^{-\mu} \frac{\mu^X}{X!}$$

Suppose we want to test the null hypothesis $H_0: \mu = \mu_0$ against the alternative hypothesis, $H_1: \mu = \mu_1$, where $\mu_1 < \mu_0$.

Find the best critical region for this test.

<u>Solution</u>: We use a likelihood ratio test to find the best critical region. This is the method suggested by the Neyman-Pearson theorem.

$$L_1 = \prod_{i=1}^{n} e^{-\mu_1} \frac{\mu_1^{x_i}}{x_i!} \qquad L_0 = \prod_{i=1}^{n} e^{-\mu_0} \frac{\mu_0^{x_i}}{x_i!} \quad .$$

The best critical region is therefore given by the region in which

$$\frac{L_1}{L_0} = \frac{\displaystyle\prod_{i=1}^{n} e^{-\mu_1} \frac{\mu_1^{x_i}}{x_i!}}{\displaystyle\prod_{i=1}^{n} e^{-\mu_0} \frac{\mu_0^{x_i}}{x_i!}} = e^{n(\mu_0-\mu_1)}\left(\frac{\mu_1}{\mu_0}\right)^{\Sigma x_i} \geq k.$$

where k is a constant.

Taking logarithms of each side of this inequality yields

$$n(\mu_0 - \mu_1) + \Sigma x_i (\log \frac{\mu_1}{\mu_0}) \geq \log k. \tag{1}$$

Since $\log \dfrac{\mu_1}{\mu_0} = \log \mu_1 - \log \mu_0$, we have

$$n(\mu_0 - \mu_1) + \Sigma x_i ((\log \mu_1 - \log \mu_0) \geq k \quad . \tag{2}$$

Transposing $n(\mu_0 - \mu_1)$ to the right side of the equation gives

$$\Sigma x_i ((\log \mu_1 - \log \mu_0) \geq \log k + n(\mu_1 - \mu_0). \tag{3}$$

Dividing both sides of (3) by $\log \mu_1 - \log \mu_0$ yields

$$\Sigma x_i \leq \frac{\log k + n(\mu_1 - \mu_0)}{\log \mu_1 - \log \mu_0} \quad . \tag{4}$$

The inequality is reversed in (4) because $\log \mu_1 - \log \mu_0$ is negative under our stated assumption of $\mu_1 < \mu_0$.

Now x_i is a Poisson variable, and since the sum of independent Poisson variables is a Poisson variable with mean equal to the sum of the means, it follows that $Z = \Sigma x_i$ is a Poisson variable with mean $n\mu$. The critical region determined by (4) is therefore equivalent to a critical region of the type

$$Z \le \frac{\log k + n(\mu_1 - \mu_0)}{\log \mu_1 - \log \mu_0} \tag{5}$$

for the Poisson variable Z.

By choosing k properly the quantity on the right side of (5) can be made to have a value such that Z will be less than that value when H_0 is true 100α percent of the time for any specified α.

REGRESSION AND CORRELATION ANALYSIS

● **PROBLEM** 9-161

X	Fertilizer	.3	.6	.9	1.2	1.5	1.8	2.1	2.4
Y	Corn Yield	10	15	30	35	25	30	50	45

Plot the dependent variable against the independent variable. Find the least squares line for this data. What is the Y-intercept? If 3.0 units of fertilizer were used what would be a good guess as to the resultant corn yield?

Solution: The dependent variable in this problem is corn yield. The amount of fertilizer used will affect the corn yield that is observed, but the amount of fertilizer used will probably not depend on the corn yield. Fertilizer is used before the corn yield is even observed.

The vertical axis of a coordinate system is usually used to indicate the dependent variable Y and the horizontal axis for the independent variable X. Each point on the graph coincides with a pair of numbers, an X value and a Y value. In this case, the scatter plot, as such a graph is called is given below:

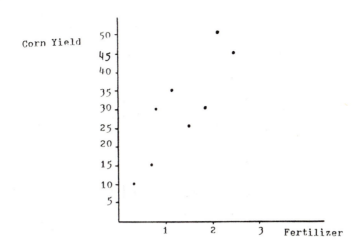

This graph seems to indicate a linear relationship between fertilizer and corn yield. As more fertilizer is used, more corn is produced.

The general equation of any line is Y = a + bX. Y and X represent the vertical and horizontal distances from the origin. Given 8 pairs of X and Y values we wish to estimate the Y-intercept a and the slope b. The Y-intercept, a, is the Y-value that corresponds to an X-value of zero. The point where the line Y = a + bX intersects the Y-axis is thus the Y-intercept. The slope b indicates the relative change of Y with respect to X. That is, b = amount Y changes for a unit increase in X.

The points in our scatter plot do not lie precisely on a line but appear to be scattered about a line. There are many lines that could be chosen to represent the true but unknown relationship between fertilizer and corn yield.

The most common choice of such a line is that line which minimizes the squared distance between the Y-values that lie on the chosen line and the observed Y-values.

Let the estimated values of a and b be denoted by $\hat{a}$ and $\hat{b}$. From $\hat{a}$ and $\hat{b}$ we can compute a new Y-value, $\hat{Y} = \hat{a} + \hat{b}X$. This equation denotes a line which is graph-ed below:

655

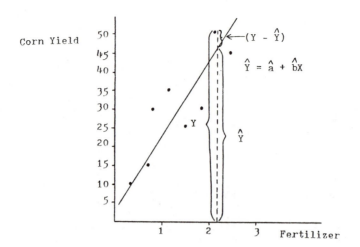

The vertical distances Y, $\hat{Y}$ and $(Y - \hat{Y})$ are indicated
on this graph. The least squares criterion says: estimate
the Y-intercept and slope in such a way that the quantity
$\Sigma\ (Y - \hat{Y})^2$ will be minimized. $\Sigma\ (Y - \hat{Y})^2$ is the sum of
squared deviation of each Y value from the estimated line.
If this condition is met, the line $\hat{Y} = \hat{a} + \hat{b}X$ will be the
least squares line and will best fit the data.

We now wish to find the values a and b which will
achieve this minimum. The derivation below is included
for completeness and may be omitted without loss of conti-
nuity.

let $\quad \Theta\ (\hat{a},\ \hat{b}) = \Sigma\ (Y - \hat{Y})^2$

$$= \Sigma\ \left[Y - (\hat{a} + \hat{b}X)\right]^2$$

Θ can be considered a function of two variables a
and b. To minimize $\Theta(\hat{a},\ \hat{b})$ we will find

$$\frac{\delta\Theta(\hat{a},\hat{b})}{\delta\hat{a}} \quad \text{and} \quad \frac{\delta\Theta(\hat{a},\hat{b})}{\delta\hat{b}}$$

and set these partial derivatives equal to zero.

$$\Theta(\hat{a}\ \hat{b}) = \Sigma\left[Y - (\hat{a} + \hat{b}X)\right]^2 ;$$

squaring inside the summation sign and factoring out constants where appropriate,

$$\Theta(\hat{a},\hat{b}) = \Sigma\left[Y^2 - 2Y(\hat{a} + \hat{b}X) + (\hat{a} + \hat{b}X)^2\right]$$

656

$$= \Sigma Y^2 - 2\Sigma\ Y(\hat{a} + \hat{b}X) + \Sigma(\hat{a} + \hat{b}X)^2$$

$$= \Sigma Y^2 - 2\hat{a}\Sigma Y - 2\hat{b}\Sigma XY$$

$$+ \Sigma(\hat{a}^2 + 2\hat{a}\hat{b}X + \hat{b}^2X^2)$$

$$= \Sigma Y^2 - 2\hat{a}\Sigma Y - 2\hat{b}\Sigma XY$$

$$+ \Sigma\hat{a}^2 + 2\hat{a}\hat{b}\Sigma X + \hat{b}^2\Sigma X^2.$$

Thus $(\hat{a},\hat{b}) = \Sigma Y^2 - 2\hat{a}\ \Sigma Y - 2\hat{b}\Sigma XY + n\hat{a}^2$

$$+ 2\hat{a}\hat{b}\Sigma X + \hat{b}^2\Sigma X^2.$$

Differentiating with respect to $\hat{a}$ and $\hat{b}$:

$$\frac{\partial\theta(a,b)}{\partial\hat{a}} = -2\Sigma y + 2\hat{a}n + 2\hat{b}\Sigma x$$

$$\frac{\partial\theta(a,b)}{\partial\hat{b}} = -2\Sigma xy + 2\hat{a}\Sigma x + 2\hat{b}\Sigma x^2.$$

Setting these equations equal to zero we get the so-called normal equations:

$$\Sigma y = n\hat{a} + \hat{b}\Sigma x \qquad (1)$$
$$\Sigma xy = \hat{a}\Sigma x + \hat{b}\Sigma x^2. \qquad (2)$$

We now proceed to solve (1) and (2) simultaneously for $\hat{a}$ and $\hat{b}$. Multiply (1) by Σx and (2) by n to obtain

$$(\Sigma x)(\Sigma y) = \hat{a}n\ \Sigma x + \hat{b}(\Sigma x)^2$$

$$n\Sigma xy = \hat{a}n\ \Sigma x + n\hat{b}\Sigma x^2.$$

Thus, $(\Sigma x)(\Sigma y) - n\Sigma xy = \hat{b}\ [(\Sigma x)^2 - n\Sigma x^2]$

and $\hat{b} = \dfrac{\Sigma x\Sigma y - n\Sigma xy}{(\Sigma x)^2 - n\Sigma x^2}$ \qquad (3)

Substituting (3) for $\hat{b}$ in (2),

$$\hat{a} = \frac{\Sigma y\ \Sigma x^2 - \Sigma x\Sigma xy}{n\Sigma x^2 - (\Sigma x)^2}. \qquad (4)$$

The two formulas (3) and (4) give the values of $\hat{a}$ and $\hat{b}$ and hence the line estimating the linear relationship between corn yield and fertilizer. We can use this line to predict or estimate the corn yield given certain amounts of fertilizer.

The first step in finding $\hat{a}$ and $\hat{b}$ is to find Σx, Σy, Σx^2, Σxy and n. This is done in the table below:

x	y	x^2	xy
0.3	10	.09	3.0
0.6	15	.36	9.0
0.9	30	.81	27.0
1.2	35	1.44	42.0
1.5	25	2.25	37.5
1.8	30	3.24	54.0
2.1	50	4.41	105.0
2.4	45	5.76	108.0

$$\Sigma = 10.8 \qquad 240 \qquad 18.36 \qquad 385.5 \qquad n = 8$$

Thus $\quad \hat{a} = \dfrac{(240)(18.36) - (10.8)(385.5)}{8(18.36) - (10.8)^2} = \dfrac{243}{30.24} = 8.03$

and $\quad \hat{b} = \dfrac{(10.8)(240) - 8(385.5)}{(10.8)^2 - 8(18.36)} = \dfrac{-492}{-30.24} = 16.27.$

Thus, $\quad \hat{y} = 8.03 + 16.27x \qquad\qquad\qquad (5)$

is the least squares line relating fertizilier to corn yield

To predict the corn yield when 3.0 units of fertilizer are used we let x = 3.0 and substitute into (5) giving

$$y = 8.03 + 16.27(3.0) = 56.84.$$

56.84 is the corn-yield we would predict based upon our least squares or regression line.

The table below lists the ranks assigned by two securities analysts to 12 investment opportunities in terms of the degree of investor risk involved.

Investment	Rank by analyst 1	Rank by analyst 2
A	7	6
B	8	4
C	2	1
D	1	3
E	9	11
F	3	2
G	12	12
H	11	10
I	4	5
J	10	9
K	6	7
L	5	8

Find the correlation between the two rankings. What is the relationship between the two rankings?

Solution: We compute r_s, the coefficient of rank correlation between the rankings given by Analyst 1 and Analyst 2.

Investment	Analyst 1	Analyst 2	$X_i - Y_i$	$(X_i - Y_i)^2$
A	7	6	1	1
B	8	4	4	16
C	2	1	1	1
D	1	3	-2	4
E	9	11	-2	4
F	3	2	1	1
G	12	12	0	0
H	11	10	1	1
I	4	5	-1	1
J	10	9	1	1
K	6	7	-1	1
L	5	8	-3	9

$$r_s = 1 - \frac{6 \sum_{i=1}^{12} (X_i - Y_i)^2}{n(n^2 - 1)}$$

$$\sum_{i=1}^{12} (X_i - Y_i)^2 = 40$$

$$n = 12 \ .$$

$$r_s = 1 - \frac{6(40)}{12(144 - 1)} = 1 - \frac{240}{1716}$$

$$= 1 - .14$$

$$r_s = .86 \cdot$$

The correlation between the rankings given by
Analyst 1 and Analyst 2 is .86. This is a very strong,
positive correlation between the two rankings. It
seems to imply that these two analysts have very similar
ideas about the degree of investor risk involved with
these twelve securities.

● **PROBLEM** 9-163

FISHER-z VALUES (z_f)

r	.00	.01	.02	.03	.04	.05	.06	.07	.08	.09
.0	.00000	.01000	.02000	.03001	.04002	.05004	.06007	.07012	.08017	.09024
.1	.10034	.11045	.12508	.13074	.14093	.15114	.16139	.17167	.18198	.19234
.2	.20273	.21317	.22366	.23419	.24477	.25541	.26611	.27686	.28768	.29857
.3	.30952.	.32055	.33165	.34283	.35409	.36544	.37689	.38842	.40006	.41180
.4	.42365	.43561	.44769	.45990	.47223	.48470	.49731	.51007	.52298	.53606
.5	.54931	.56273	.57634	.59014	.60415	.61838	.63283	.64752	.66246	.67767
.6	.69315	.70892	.72500	.74142	.75817	.77530	.79281	.81074	.82911	.84795
.7	.86730	.88718	.90764	.92873	.95048	.97295	.99621	1.02033	1.04537	1.07143
.8	1.09861	1.12703	1.15682	1.18813	1.22117	1.25615	1.29334	1.33308	1.37577	1.42192
.9	1.47222	1.52752	1.58902	1.65839	1.73805	1.83178	1.94591	2.09229	2.29756	2.64665

Using the table above, find z_f-values that correspond
to r = .48, r = .07, r = .55, r = .80.

Solution: In order to test hypotheses, construct
confidence intervals, or make any inferences about the
true but unknown correlation between two variables, we
must know something about the distribution of the
random variable r.

Unfortunately, the sample correlation coefficient

$$r = \frac{n\Sigma XY - (\Sigma X)(\Sigma Y)}{[\sqrt{n\Sigma X^2 - (\Sigma X)^2}][\sqrt{n\Sigma Y^2 - (\Sigma Y)^2}]}$$

has neither a normal distribution nor a distribution
that becomes approximately normal as the sample size
increases.

However we can transform this random variable r
to another random variable if we assume X and Y are
distributed with a bivariate normal distribution.

Let $z_f = (1.1513) \log \frac{1 + r}{1 - r}$, then z_f is approximately normal with mean $\mu_z = (1.1513) \log \frac{1 + \rho}{1 - \rho}$ where ρ is the true but unknown correlation coefficient.

To compute z_f-values from given values of r we can use the above table. To find the z_f-value that corresponds to r = .48, read down the first column of digits until .4 is reached. Then read across that row until the column headed by .08 is reached. Then read z_f. For r = .48, $z_f = .52298$.

If r = .55, then we read across the row labeled .5 until we reach the column headed by .05. We read $z_f = .61838$.

If r = .07, read across the row labled .00 until we reach .07. Read $z_f = .07012$.

If r = .80, we read down the first column of digits until we reach .80. Then read across this row until the column is headed by .00. We see that for r = .80, $z_f = 1.09861$.

● **PROBLEM** 9-164

Show that if (X, Y) has a bivariate normal distribution, then the marginal distributions of X and Y are univariate normal distributions; that is, X is normally distributed with mean μ_x and variance σ_x^2 and Y is normally distributed with mean μ_y and variance σ_y^2.

Solution: This problem involves some slightly advanced mathematical techniques and may be skipped by less advanced students.

The joint distribution of ordered pairs (X, Y) is

$$f(x, y) = \frac{1}{2\pi\sigma_x \sigma_y \sqrt{1 - \rho^2}}$$

$$\exp \left\{ -\frac{1}{2(1 - \rho)^2} \left[\left(\frac{x - \mu_x}{\sigma_x} \right)^2 - 2\rho \left(\frac{x - \mu_x}{\sigma_x} \right) \left(\frac{y - \mu_y}{\sigma_y} \right) \right. \right.$$

$$\left. \left. + \left(\frac{y - \mu_y}{\sigma_y^2} \right)^2 \right] \right\}$$

(where μ_x, σ_x are the mean and standard deviation of X, μ_x, σ_y are the mean and standard deviation of Y, ρ is the theoretical correlation coefficient between X and Y and exp{ } is $e^{\{\}}$, the exponential function).

The marginal density of one of the variables, say X, is by definition:

$$f_x(x) = \int_{-\infty}^{\infty} f(x, y) \, dy.$$

Returning to the expression $f(x, y)$ we make the following substitutions,

$$Z = \frac{x - \mu_x}{\sigma_x} \quad \text{and} \quad V = \frac{y - \mu_y}{\sigma_y}$$

These substitutions are convenient relabelings. Thus

$$\int_{-\infty}^{\infty} f(x, y) \, dy =$$

$$\int_{-\infty}^{\infty} \frac{1}{2\pi \, \sigma_x \, \sigma_y \, \sqrt{1 - \rho^2}} \, \exp\left\{ \frac{1}{2(1 - \rho)^2} [Z^2 - 2\rho ZV + V^2] \right\} dy$$

but $y = \sigma_y V + \mu_y$ \qquad hence \qquad $dy = \sigma_y \, dv$.

Thus, $\int_{\infty}^{\infty} f(x, y) \, dy$

$$= \int_{-\infty}^{\infty} \frac{\sigma y}{2\pi\sigma_x \, \sigma_y \, \sqrt{1 - \rho^2}} \, \exp\left\{ \frac{-1}{2(1 - \rho)^2} [Z^2 - 2\rho ZV + V^2] \right\} dv$$

Consider the term contained in the braces
We will complete the square on the variable V.
Thus

$$V^2 - 2\rho ZV + Z^2 = V^2 - 2\rho ZV + (-\rho Z)^2 - (-\rho Z)^2 + Z^2$$

$$= (V - \rho Z)^2 + Z^2 (1 - \rho^2).$$

Our integral becomes, (with cancellation of σ_y)

$$\int_{-\infty}^{\infty} f(x,y)dy$$

$$= \int_{-\infty}^{\infty} \frac{dv}{2\pi \sigma_x \sqrt{1 - \rho^2}} \exp \left\{ \frac{-1}{2(1-\rho^2)} [Z^2(1-\rho^2)+(V-\rho Z)^2] \right\}$$

$$= \int_{-\infty}^{\infty} \frac{dv}{2\pi \sigma_x \sqrt{1 - \rho^2}} \exp \left\{ \frac{-Z^2}{2} - \frac{1}{2(1-\rho^2)} (V-\rho Z)^2 \right\} .$$

We now carry out another relabeling. Let

$$w = \frac{V - \rho Z}{\sqrt{1 - \rho^2}} \quad . \text{ Then}$$

$$V = w\sqrt{1 - \rho^2} + \rho Z \quad \text{and}$$

$$dv = dw\sqrt{1 - \rho^2} \quad . \quad \text{Further note that}$$

$$\frac{1}{2(1-\rho^2)} (V - \rho Z)^2 = \frac{1}{2} \left[\frac{V - \rho Z}{\sqrt{1 - \rho^2}} \right]^2 = \frac{1}{2} w^2 .$$

Thus, $\int_{-\infty}^{\infty} f(x, y) \, dy$

$$= \int_{-\infty}^{\infty} \frac{dw \sqrt{1 - \rho^2}}{2\pi \sigma_x \sqrt{1 - \rho^2}} \exp \left\{ \frac{-Z^2}{2} - \frac{w^2}{2} \right\}$$

$$= \int_{-\infty}^{\infty} \frac{dw}{2\pi \sigma_x} \exp \left\{ \frac{-Z^2}{2} - \frac{w^2}{2} \right\} .$$

Because we are integrating with respect to w, we may factor out the terms which do not involve w.

$$= \frac{\exp\left\{\frac{-Z^2}{2}\right\}}{2\pi \quad \sigma_x} \int\limits_{-\infty}^{\infty} \exp\left\{\frac{-w^2}{2}\right\} \, dw .$$

The integrand should be familiar. It is the density function of a standard normal variable. We know that

$$\frac{1}{\sqrt{2\pi}} \int\limits_{-\infty}^{\infty} \exp\left\{\frac{-w^2}{2}\right\} \, dw = 1$$

or

$$\int\limits_{-\infty}^{\infty} \exp\left\{\frac{-w^2}{2}\right\} \, dw = \sqrt{2\pi} .$$

Thus

$$f_x(x) = \int\limits_{-\infty}^{\infty} f(x, y) \, dy = \frac{\exp\left\{\frac{-Z^2}{2}\right\}}{2\pi \quad \sigma_x} \cdot \sqrt{2\pi} .$$

Remember $Z = \dfrac{x - \mu_x}{x}$; resubstituting we see

that the marginal density function of a random variable X which is jointly distributed with Y in a bivariate normal distribution is:

$$f_x(x) = \frac{\exp\left\{-\frac{1}{2}\left[\frac{x - \mu_x}{\sigma_x}\right]^2\right\}}{\sigma_x \sqrt{2\pi}} .$$

This is the density of a normally distributed random variable with mean μ_x and variance σ_x^2.

The marginal density of y, $f_y(y) = \int\limits_{-\infty}^{\infty} f(x,y)\,dx$, is found

by the same argument to be

$$f_y(y) = \frac{\exp\left[-\frac{1}{2}\left(\frac{x-\mu}{\sigma}x\right)\right]^2}{\sigma_x\sqrt{2}\ \pi} .$$

The following is a two variable, one common factor model:

$$X_1 = .8F + .6U_1 \tag{1}$$
$$X_2 = .6F + .8U_2 \tag{2}$$

a) Draw a path model for the system.
b) Find the covariance and correlation between X_1, X_2 and

F, U_1, U_2 .

Solution: This is an example of a basic theoretical structure in Factor Analysis. The model (1)-(2) asserts that the observed variables X_1, X_2 are linear combinations of an unobserved common factor F and two unobserved unique factors, U_1 (uniquely affecting X_1) and U_2 (uniquely affecting X_2). The general form of the model is:

$$X_1 = b_1F + d_1U_1 \tag{3}$$
$$X_2 = b_2F + d_2U_2 \tag{4}$$

a) A path model for the system (3)-(4) consists of the observed and unobserved variables connected by lines showing the causal structure.

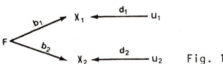

Fig. 1

Applying Fig. 1 to the given model (1)-(2) yields

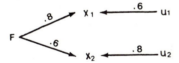

b) The covariance and correlation between the given variables is determined by the assumptions made involving their statistical nature. A convenient set of assumptions is the following:

i)   Cov (F, U_1) = Cov (F, U_2) = Cov (U_1, U_2) = 0 where

Cov $(Y, Z) = E[(Y-\bar{Y})(Z-\bar{Z})]$.

ii) All the variables are normally distributed with mean zero and variance one. In fact, they are standard normal variables.
Note that i) is implicit in the path diagram (Fig. 1) since there are no causal connections between F and U_1 or U_2 and between U_1 and U_2 . Using the general model (3)-(4):

$$\mathrm{Var}(X_1) = E(X_1 - \bar{X})^2 = E(X_1)^2$$

$$= E[b_1 F + d_1 U_1]^2 = E[b_1^2 F^2 + d_1^2 U_1^2 + 2b_1 d_1 F U_1]$$

$$= b_1^2 E[F^2] + d_1^2 E[U_1^2] + 2b_1 d_1 E[F U_1]$$

$$= b_1^2 \mathrm{Var}(F) + d_1^2 \mathrm{Var}(U_1) + 2b_1 d_1 \mathrm{Cov}(F, U_1)$$

$$\left(\text{since} \quad \mathrm{Var}(F) = E(F - \bar{F})^2 = E(F - 0)^2 = E(F^2) \quad \text{and} \quad \mathrm{Var}(U_1) = E(U_1 - \bar{U}_1)^2 \right.$$
$$\left. = E(U_1^2) \right)$$

$$= b_1^2 \mathrm{Var}(F) + d_1^2 \mathrm{Var}(U_1) \ .$$

(Since $\mathrm{Cov}(F U_1) = 0$ by assumption). (The property of the expectation operator $E[aX + bY] = aE(X) + bE(Y)$ was used in the above derivation).

But $\mathrm{Var}(F) = \mathrm{Var}(U_1) = 1$ (by assumption) and thus $\mathrm{Var}(X_1) = 1 = b_1^2 + d_1^2$. Similarly $\mathrm{Var}(X_2) = 1 = b_2^2 + d_2^2$.

Applying these results to the given model (1)-(2),
$$\mathrm{Var}(X_1) = (.8)^2 + (.6)^2 = 1$$
$$\mathrm{Var}(X_2) = (.6)^2 + (.8)^2 = 1 \ .$$

Thus the proportion of variance in X_1 determined by F is $.64$ and the proportion determined by U_1 is $.36$. Next, the **covariance** between a factor and an observed variable is given by

$$\mathrm{Cov}(F, X_1) = E[(F - \bar{F})(X_1 - \bar{X})] = E[(F - 0)(X_1 - 0)] = E(F X_1)$$

$$= E[F(b_1 F + d_1 U_1)] = b_1 E(F^2) + d_1 E(F U_1)$$

$$= b_1 \mathrm{Var}(F) + d_1 \mathrm{Cov}(F U_1) = b_1 \mathrm{Var}(F) \ .$$

But $\mathrm{Var}(F) = 1$ by assumption and thus $\mathrm{Cov}(F, X_1) = b_1$.

Similarly $\mathrm{Cov}(F, X_2) = b_2$. Now r_{xy} (the correlation coefficient)

$$= \frac{E(X - \bar{X})(Y - \bar{Y})}{\sigma_x \ \sigma_y} = \frac{\mathrm{Cov}(X, Y)}{\sigma_x \ \sigma_y} \ .$$

But here, $\sigma_x, \sigma_y = 1$ and thus $\text{Cov}(F,X_1) = b_1 = r_{F,X_1}$.

The covariance equals the correlation coefficient. For the given

model, $\text{Cov}(F,X_1) = .8$ and $\text{Cov}(F,X_2) = .6$.

Also, since $\text{Cov}(X_1,U_1) = r_{X_1 U_1} = d_1$ and $\text{Cov}(X_2,U_2) = r_{X_2 U_2} = d_2$,

$\text{Cov}(X_1,U_1) = .6$ and $\text{Cov}(X_2,U_2) = .8$.

Finally, the covariance between X_1 and X_2 is:

$$\text{Cov}(X_1,X_2) = E[(X_1-\bar{X}_1)(X_2-\bar{X}_2)]$$

$$= E[(b_1F + d_1U_1)(b_2F + d_2U_2)]$$

$$= E[b_1b_2F^2 + b_1d_2FU_2 + b_2d_1FU_1 + d_1d_2U_1U_2]$$

$$= b_1b_2\text{Var}(F) + b_1d_2\text{Cov}(FU_2) + b_2d_1\text{Cov}(F,U_1)$$

$$+ d_1d_2\text{Cov}(U_1,U_2)$$

$$= b_1b_2\text{Var}(F) = b_1b_2.$$

Thus the covariance between two observed variables sharing one common
factor is equivalent to the variance of the factor times the two
respective linear factors involved.

In the given model, $\text{Cov}(X_1,X_2) = (.8)(.6) = .48$. This shows that
the covariation between the observed variables is completely determined by the common factor; if F were removed, there would be no
correlation between X_1 and X_2.

667

ANALYSIS OF VARIANCE

An experimenter compared two groups, an experimental group and a control group. Each group contained 10 subjects. Do the two means of these groups differ significantly?

Control	Experimental
10	7
5	3
6	5
7	7
10	8
6	4
7	5
8	6
6	3
5	2

Solution: Let Y_{ij} be the ith observation in the jth population.   We can make the following assumptions about Y_{ij}.

$$Y_{ij} = \mu_j + \varepsilon_{ij} \qquad\qquad i = 1, \ldots n_j$$

$$j = 1, \ldots k$$

where $E(\varepsilon_{ij}) = 0$, $Var(\varepsilon_{ij}) = \sigma^2$.

In this problem, we have two populations and we wish to compare their means μ_1 and μ_2.

We have previously seen this problem as a test of equality of two means. Using this new notation, we have the following test:

$$H_0: \mu_1 = \mu_2$$

$$H_A: \mu_1 \neq \mu_2$$

and the test statistic for this test will be:

$$\frac{\left[\dfrac{\sum\limits_{i=1}^{n_1} Y_{i1}}{n_1} - \dfrac{\sum\limits_{i=1}^{n_2} Y_{i2}}{n_2}\right] - (\mu_1 - \mu_2)}{\sqrt{s_p^2 \left(\dfrac{1}{n_1} + \dfrac{1}{n_2}\right)}}$$

where $s_p{}^2$ is an estimate of the pooled population variance. $s_p{}^2$ is a weighted average of $s_1{}^2$ and $s_2{}^2$, the estimates of the individual population variances and it can be shown that

$$s_p{}^2 = \frac{(n_1 - 1)s_1{}^2 + (n_2 - 1)s_2{}^2}{n_1 + n_2 - 2} \; .$$

The following notation is often useful in the analysis of variance.

$$\overline{Y}._j = \frac{\sum\limits_{i=1}^{n_j} Y_{ij}}{n_j} \qquad \text{for } i = 1, \ldots k.$$

This is the average of the observations in the ith popula-tion. Let

$$\overline{Y}.. = \frac{\sum\limits_{j=1}^{k} \sum\limits_{i=1}^{n_j} Y_{ij}}{\sum\limits_{j=1}^{k} n_j} \; .$$

This is the grand average of all the observations.

With this notation,

$$s_1{}^2 = \frac{\sum\limits_{i=1}^{n_1} (Y_{i_1} - \overline{Y}._1)^2}{n_1 - 1} \; ;$$

$$s_2{}^2 = \frac{\sum\limits_{i=1}^{n_2} (Y_{i_2} - \overline{Y}._2)^2}{n_2 - 1} \; .$$

Under the null hypothesis that $\mu_1 = \mu_2$, the test statistic is;

$$\frac{\overline{Y}._1 - \overline{Y}._2 - (\mu_1 - \mu_2)}{\sqrt{s_p{}^2 \left(\frac{1}{n_1} + \frac{1}{n_2}\right)}} = \frac{\overline{Y}._1 - \overline{Y}._2}{s_p \sqrt{\frac{1}{n_1} + \frac{1}{n_2}}}$$

669

and will be approximately t-distributed with $n_1 + n_2 - 2$ degrees of freedom. (Ratio of normal and chi-square.)

For this problem, let the control group be population 1 and the experimental group be population 2. Then,

$$n_1 = n_2 = 10$$

$$\overline{Y}._1 = \frac{10+5+6+7+10+6+7+8+6+5}{10} = \frac{70}{10} = 7.$$

$$\overline{Y}._2 = \frac{7+3+5+7+8+4+5+6+3+2}{10} = \frac{50}{10} = 5.$$

Also, $$\sum_{i=1}^{10} Y_{i_1}{}^2 = 10^2 + 5^2 + 6^2 + \ldots + 6^2 + 5^2 = 520$$

$$\sum_{i=1}^{10} Y_{i_2}{}^2 = 7^2 + 3^2 + 5^2 + 7^2 + \ldots + 3^2 + 2^2 = 286.$$

We now must compute the pooled variance $s_p{}^2$.

First find $s_1{}^2$ and $s_2{}^2$. We have seen previously that for one sample, the estimate of σ^2 is

$$s^2 = \frac{\sum_{i=1}^{n} X_i{}^2 - n\overline{X}^2}{n - 1}.$$

Thus, $$s_1^2 = \frac{\sum_{i=1}^{n_1} Y_{i_1}{}^2 - n_1\overline{Y}._1^2}{n_1 - 1} \quad \text{and} \quad s_2{}^2 = \frac{\sum_{i=1}^{n_2} Y_{i_2}{}^2 - n_2\overline{Y}._2^2}{n_2 - 1}$$

Substituting we have,

$$s_1{}^2 = \frac{520 - 10 \cdot 7^2}{9} \qquad s_2{}^2 = \frac{286 - 10 \cdot 5^2}{9}$$

$$= 3.33 \qquad\qquad\qquad = 4$$

and thus,

$$s_p{}^2 = \frac{(n_1 - 1)s_1{}^2 + (n_2 - 1)s_2{}^2}{n_1 + n_2 - 2} = \frac{30 + 36}{18} = 3.66.$$

Our test statistic is thus,

$$t = \frac{\overline{Y}_1. - \overline{Y}_2.}{s_p \sqrt{\dfrac{1}{n_1} + \dfrac{1}{n_2}}} = \frac{7 - 5}{\sqrt{3.66} \sqrt{\dfrac{1}{10} + \dfrac{1}{10}}}$$

$$= \frac{2}{(1.91)(.447)} = 2.34 .$$

From a table of the t-distribution, we see that we will reject H_0 at level of significance .05 if $t > 2.101$ or $t < - 2.101$. The t-statistic is 2.34 which is greater than 2.101. Thus, we reject the null hypothesis that the mean is the same for both groups and accept that the means are different.

● **PROBLEM** 9-167

The following experiment was performed to determine the effect of two advertising campaigns on three kinds of cake mixes. Sales of each mix were recorded after the first advertising compaign and then after the second advertising campaign. This experiment was repeated 3 times for each advertising campaign with the following results:

	Campaign 1	Campaign 2
Mix 1	574, 564, 550	1092, 1086, 1065
Mix 2	524, 573, 551	1028, 1073, 998
Mix 3	576, 540, 592	1066, 1045, 1055

Set up an ANOVA table for this problem and find the appropriate sums of squares, degrees of freedom and mean squares.

Solution: The model describing this experiment is the following: let Y_{ijk} be the dollar sales observed in the kth repetition for the ith campaign and the jth cake mix.

Then, $$Y_{ijk} = \mu. + \alpha_i + \beta_j + \eta_{ij} + \varepsilon_{ijk} \qquad \text{for } \begin{array}{l} k = 1, 2, 3 \\ i = 1, 2 \\ j = 1, 2, 3 , \end{array}$$

671

where α_i is the effect on sales due to the ith campaign for some fixed j. β_j is the effect on sales due to the jth mix for some fixed i and η_{ij} is the effect on sales due to an interaction between mix and advertising campaign. The ε_{ijk} are random variables with mean 0 and variance σ^2.

The analysis of variance procedures are similar to those used for one-way ANOVA. The total variation of an observation about the grand mean,

$$\overline{Y}... = \frac{\sum\limits_{i=1}^{2} \sum\limits_{j=1}^{3} \sum\limits_{k=1}^{3} Y_{ijk}}{3 \times 3 \times 2}$$

is decomposed into variance due to mix, the advertising, interactions and the error. The error variance is the variation within each treatment combination.

The sums of squares associated with each of these variances are:

1) $$SSA = \sum\limits_{i=1}^{2} \sum\limits_{j=1}^{3} \sum\limits_{k=1}^{3} (\overline{Y}_i.. - \overline{Y}...)^2$$

$$= 3 \cdot 3 \cdot \sum\limits_{i=1}^{2} (\overline{Y}_i.. - \overline{Y}...)^2 \qquad \text{where}$$

$$\overline{Y}_i.. = \frac{\sum\limits_{j=1}^{3} \sum\limits_{k=1}^{3} Y_{ijk}}{3 \cdot 3} \quad .$$

This is the variation in sales due to the advertising campaign.

2) $$SSB = \sum\limits_{i=1}^{2} \sum\limits_{j=1}^{3} \sum\limits_{k=1}^{3} (\overline{Y}._j. - \overline{Y}...)^2$$

$$= 2 \cdot 3 \cdot \sum\limits_{j=1}^{3} (\overline{Y}._j. - \overline{Y}...)^2 \qquad \text{where}$$

$$\overline{Y}._j. = \frac{\sum\limits_{i=1}^{2} \sum\limits_{k=1}^{3} Y_{ijk}}{2 \cdot 3} \quad .$$

This is the variation in sales due to the type of mix.

3) $$SSAB = \sum_{i=1}^{2} \sum_{j=1}^{3} \sum_{k=1}^{3} (\overline{Y}_{ij\cdot} - \overline{Y}_{i\cdot\cdot} - \overline{Y}_{\cdot j\cdot} + \overline{Y}_{\cdots})^2$$

$$= 3 \cdot \sum_{i=1}^{2} \sum_{j=1}^{3} (\overline{Y}_{ij\cdot} - \overline{Y}_{i\cdot\cdot} - \overline{Y}_{\cdot j\cdot} + \overline{Y}_{\cdots})^2$$

where $$\overline{Y}_{ij\cdot} = \frac{\sum_{k=1}^{3} Y_{ijk}}{3} \quad .$$

This is the variation in sales due to an interaction between mix and advertising campaign.

4) $$SSE = \sum_{i=1}^{2} \sum_{j=1}^{3} \sum_{k=1}^{3} (Y_{ijk} - Y_{ij\cdot})^2 \quad .$$

This is the variation in sales within each treatment combination, which is the variation in sales due to error.

Computing the necessary summary statistics we have:

$$\sum_{j=1}^{3} \sum_{k=1}^{3} Y_{1jk} = 5044 \qquad \sum_{k=1}^{3} Y_{11k} = 1688$$

$$\sum_{j=1}^{3} \sum_{k=1}^{3} Y_{2jk} = 9508 \qquad \sum_{k=1}^{3} Y_{21k} = 3243$$

$$\sum_{i=1}^{2} \sum_{k=1}^{3} Y_{i1k} = 4931 \qquad \sum_{k=1}^{3} Y_{12K} = 1648$$

$$\sum_{i=1}^{2} \sum_{k=1}^{3} Y_{i2k} = 4747 \qquad \sum_{k=1}^{3} Y_{22k} = 3099$$

$$\sum_{i=1}^{2} \sum_{k=1}^{3} Y_{i3k} = 4874 \qquad \sum_{k=1}^{3} Y_{13k} = 1708$$

$$\sum_{k=1}^{3} Y_{23k} = 3166$$

$$\sum_{i=1}^{2} \sum_{j=1}^{3} \sum_{k=1}^{3} Y_{ijk}^{2} = 12,882,026.$$

$$\sum_{i=1}^{2} \sum_{j=1}^{3} \sum_{k=1}^{3} Y_{ijk} = 14,552 \quad .$$

We now compute the necessary sums of squares in the following manner:

1) $\quad SSA = \displaystyle\sum_{i=1}^{2} \sum_{j=1}^{3} \sum_{k=1}^{3} (\overline{Y}_i.. - \overline{Y}...)^2$

$\qquad = 9 \displaystyle\sum_{i=1}^{2} (\overline{Y}_i.. - \overline{Y}...)^2 \quad = 9 \displaystyle\sum_{i=1}^{2} \overline{Y}_i^2.. - 9 \cdot 2 \cdot \overline{Y}^2..$

$\qquad = 9 \left(\dfrac{5044}{9}\right)^2 + 9 \left(\dfrac{9508}{9}\right)^2 - 18 \left(\dfrac{14552}{18}\right)^2$

$\qquad = 12{,}871{,}553 - 11{,}764{,}484 \quad = 1{,}107{,}070.$

2) $\quad SSB = \displaystyle\sum_{i=1}^{2} \sum_{j=1}^{3} \sum_{k=1}^{3} (\overline{Y}._j. - \overline{Y}...)^2$

$\qquad = 6 \displaystyle\sum_{j=1}^{3} \overline{Y}^2._j. - 18 \overline{Y}^2..$

$\qquad = 6 \left[\left(\dfrac{4931}{6}\right)^2 + \left(\dfrac{4747}{6}\right)^2 + \left(\dfrac{4874}{6}\right)^2 \right] - 11{,}764{,}483$

$\qquad = 2957.$

3) $\quad SSAB = \displaystyle\sum_{i=1}^{2} \sum_{j=1}^{3} \sum_{k=1}^{3} (\overline{Y}_{ij}. - \overline{Y}_i.. - \overline{Y}._j. + \overline{Y}...)^2$

$\qquad = 3 \cdot \displaystyle\sum_{i=1}^{2} \sum_{j=1}^{3} (\overline{Y}_{ij}. - \overline{Y}_i.. - \overline{Y}._j. + \overline{Y}...)^2$

$\qquad = 1126.$

4) $\quad SSTO = \displaystyle\sum_{i=1}^{2} \sum_{j=1}^{3} \sum_{k=1}^{3} (Y_{ijk} - \overline{Y}...)^2$

$\qquad = \displaystyle\sum_{i=1}^{2} \sum_{j=1}^{3} \sum_{k=1}^{3} Y_{ijk}^2 - 18 \overline{Y}^2..$

$\qquad = 12{,}882{,}026 - 11{,}764{,}484 = 1{,}117{,}542.$

The total sum of squares:

$$SSTO = \sum_{i=1}^{2} \sum_{j=1}^{3} \sum_{k=1}^{3} (Y_{ijk} - \overline{Y}...)^2$$

674

may be decomposed as in one-way ANOVA. The decomposition is the following:

$$\sum_{i=1}^{2} \sum_{j=1}^{3} \sum_{k=1}^{3} (Y_{ijk} - \bar{Y}...)^2$$

$$= \sum_{i=1}^{2} \sum_{j=1}^{3} \sum_{k=1}^{3} (\bar{Y}_{i}.. - \bar{Y}...)^2 + \sum_{i=1}^{2} \sum_{j=1}^{3} \sum_{k=1}^{3} (\bar{Y}._{j}. - \bar{Y}...)^2$$

$$+ \sum_{i=1}^{2} \sum_{j=1}^{3} \sum_{k=1}^{3} (\bar{Y}_{ij}. - \bar{Y}_{i}.. - \bar{Y}._{j}. + \bar{Y}...)^2 + \sum_{i=1}^{2} \sum_{j=1}^{3} \sum_{k=1}^{3} (Y_{ijk} - \bar{Y}_{ij}.)^2$$

or   SSTO = SSA + SSB + SSAB + SSE.

Thus, SSE = SSTO - SSA - SSB - SSAB = 6389 .

The ANOVA table is

Source of Variation	Sum of Squares	Degrees of Freedom	Mean Squares
Advertising Campaign	1107070	2 - 1 = 1	1107070
Cake mix	2957	3 - 1 = 2	1478.5
Interaction	1126	(2 - 1)(3 - 1) = 2	563
Error	6389	3 · 2(3 - 1) = 12	532.42

The degrees of freedom are found in the usual way. To find the number of degrees of freedom for the interaction sum of squares we use the fact that the degrees of freedom are additive. That is

$$df_A + df_B + df_{AB} + df_E = df_{total} = n - 1 .$$

Thus, df_{AB} = 18 - 1 - (2 - 1) - (3 - 1) - [3 · 2 (3 - 1)]

$$= 3 \cdot 2 - (2 - 1) - (3 - 1) - 1$$

$$= (3 \cdot 2) - 2 - 3 + 1$$

$$= (3 - 1)(2 - 1) = 2 .$$

In general, $df_{AB} = (r - 1)(c - 1)$

where r = number of rows, c = number of columns.

675

A certain drug is thought to have an effect on the ability to perform mental arithmetic. The quantity of the drug used may vary from 0 to 100 milligrams.

One possible experiment would be to test all possible levels of this drug on groups of subjects and then use analysis of variance to detect differences. Because of limited funds, only six levels of the drug may be tested. Describe two methods of implementing and analyzing this experiment.

Solution: One possibility would be to systematically choose a set of levels covering the range of doses. For example, one might choose 0, 20, 40, 60, 80, and 100. The differences in mental arithmetic scores could then be analyzed using one-way analysis of variance. The model analyzed would be

$$Y_{ij} = \mu. + \alpha_i + \varepsilon_{ij} \qquad i = 1, 2, 3, 4, 5, 6$$
$$j = 1, 2, \ldots n$$

where Y_{ij} is the mental arithmetic test score, α_i is the fixed effect of dosage level i and ε_{ij} is a zero-mean normally distributed random variable.

Another possibility is to choose the six dose levels randomly from the set of numbers 1 to 100. If this experiment were repeated then different levels might be chosen. The ith dosage level changes and because of this the effect of the ith treatment is a random variable. The model to be analyzed becomes

$$Y_{ij} = \mu_1 + \alpha_i + \varepsilon_{ij} \qquad i = 1, \ldots 6$$
$$j = 1, \ldots n$$

where Y_{ij}, μ_1 and ε_{ij} are as before, but α_i (the effect due to the ith treatment) is a random variable. This is called the random effects model. The α_i are usually assumed to have a mean of zero and be normally distributed with variance σ_α^2.

NON-PARAMETRIC METHODS

● **PROBLEM** 9-169

Consider the following data obtained from testing the breaking strength of ceramic tile manufactured by a new cheaper process: 20, 42, 18, 21, 22, 35, 19, 18, 26, 20, 21, 32, 22, 20, 24.

Suppose that experience with the old process produced a median of 25. Then test the hypothesis $H_0 : M = 25$ against $H_1 : M < 25$.

Solution: We will use the Sign Test. M is the value

uniquely defined by $\int_{m}^{\infty} f(x)\ dx = \frac{1}{2}$, the median.

Let $X_1,\ \ldots,\ X_n$ be a random sample of size n. We will devise a method for testing

$$H_0 : M = M_0 \quad \text{against } H_1 : M < M_0.$$

It follows from the definition of the median that if H_0 is true

$$Pr(X - M_0 \geq 0) = \frac{1}{2} \qquad \text{and therefore}$$

$$Pr(X_i - M_0 \geq 0) = \frac{1}{2}, \quad i = 1, \ldots n.$$

$$\text{Let } Z_i = \begin{cases} 1 & \text{if } X_i - M_0 \geq 0 \\ 0 & \text{if } X_i - M_0 \leq 0. \end{cases}$$

The variable Z_i is a Bernoulli random variable corresponding to a single trial for which $p = \frac{1}{2}$.

$U = \sum_{i=1}^{n} Z_i$ will then be a binomial random variable corre-

sponding to n independent trials of an experiment for which $p = \frac{1}{2}$.

Under H_1, the X_i will tend to be larger than M_0 and U will exceed the value expected under H_0. Hence a good rejection region might be $U < K$. Under H_0, U is distributed binomially with parameters n and $p = \frac{1}{2}$. Thus

$$Pr(U < K) = \sum_{i=0}^{K} \binom{n}{i} \left(\frac{1}{2}\right)^n .$$

This is the probability that K or less of the $z_i = 1$. If $U = \Sigma z_i$ is less than K, it is likely that $M < M_0$, thus we reject the null hypothesis.

For very small samples it is necessary to calculate the binomial probabilities exactly until a total probability of roughly α has been computed to obtain the critical region for the test. Here $\alpha = Pr(\text{reject } H \text{ given } H \text{ is true})$.

For most problems it suffices to use normal approximation to the binomial. However, to use this approximation when the sample is below 30, it is best to employ a small correction of $\frac{1}{2}$ to correct for the approximation of a discrete random variable by a continuous one. The following is then approximately a standard normal random variable:

$$Z = \frac{U \pm \frac{1}{2} - np}{\sqrt{np(1 - p)}}$$

In our problem, $p = \frac{1}{2}$, thus $\quad Z = \dfrac{U \pm \frac{1}{2} - \frac{n}{2}}{\sqrt{\frac{n}{4}}}$

The correction $+ \frac{1}{2}$ is used for a $U < K$ type critical region and $- \frac{1}{2}$ is used for $U > K$.

Returning to our original problem, we subtract 25 from each observed value. $- 5, 17, - 7, - 4, - 3, 10, - 6, - 7, 1, - 5, - 4, 7, - 3, - 5, - 1$.

The following are the corresponding z_i:

$0, 1, 0, 0, 0, 1, 0, 0, 1, 0, 0, 1, 0, 0, 0.$

678

$$U = \sum_{i=1}^{n} Z_i = 4 .$$

Often it is customary to record the signs of the

values $X_i - M_0$ instead.

- + - - - + - - + - - + - - -.

The total number of + signs gives the value of U. Since $n = 15$ and $U = 4$,

$$Z = \frac{4 + \frac{1}{2} - \frac{15}{2}}{\sqrt{\frac{15}{4}}} = - 1.55 .$$

From standard normal tables, it will be seen that

$$\Pr(U < K) \underset{\sim}{\sim} \Pr(Z < - 1.645) = .05 .$$

Because $- 1.55$ is greater than $- 1.645$, we accept the null hypothesis that the median is $m = 25$ and reject that the median is less than 25.

● **PROBLEM** 9-170

For a sample of size 10, what is the confidence co-efficient of the confidence interval (Y_2, Y_q) which is an interval estimate of the population median?

Solution: We can obtain a confidence interval estimate of ξ_q, the qth quantile, by using two order statistics. We are interested in computing the confidence coefficient for a pair of order statistics, i.e.,

$$\Pr(Y_j \leq \xi_q \leq Y_i) .$$

The confidence coefficient is the probability that a random interval contains a parameter. Note that

$$\Pr(Y_j \leq \xi_q \leq Y_i) = \Pr(Y_j < \xi_q) - \Pr(Y_k < \xi_q) .   \text{But}$$

679

$$\Pr(Y_j < \xi_q) = \Pr(j^{th} \text{ order statistic} < \xi_q)$$

which is the same as

$$\Pr(j \text{ or more observations} < \xi_q).$$

This probability equals

$$\sum_{i=j}^{n} \Pr(\text{exactly } i \text{ observations} \le \xi_q).$$

We consider the event of having an observation $< \xi_q$ a success.

The probability of having an observation $< \xi_q$ is fixed. By definition it is $F(\xi_q)$. Putting this all together, the number of observations less than or equal to the qth quantile is a binomial random variable with parameters n and $F(\xi_q)$.

Hence $\Pr(Y < \xi_q)$ now becomes

$$\sum_{i=j}^{n} \binom{n}{i} F(\xi_q)^i (1 - F(\xi_q))^{n-1}.$$

By definition of ξ_q, ξ_q is the value such that $\Pr(X \le \xi_q) = q$. Therefore $F(\xi_q) = q$. Finally,

$$\Pr(Y < \xi_q) = \sum_{i=j}^{n} \binom{n}{i} q^i (1 - q)^{n - i}. \quad \text{And}$$

$$\Pr(Y_j < \xi_q < Y_k) = \Pr(Y_j < \xi_q) - \Pr(Y_k < \xi_q)$$

$$= \sum_{i=j}^{n} \binom{n}{i} q^i (1-q)^{n-i} - \sum_{i=k}^{n} \binom{n}{i} q^i (1-q)^{n-i}$$

$$= \sum_{i=j}^{n} \binom{n}{i} q^i (1-q)^{n-1}.$$

A table of the binomial distribution can be helpful in evaluating the confidence coefficient.

In our problem we are concerned with the median, $\xi_{\frac{1}{2}}$, so that $q = \frac{1}{2}$. We also want $\Pr(Y < \xi_{\frac{1}{2}} < Y_q)$

which tells us $j = 2$ and $k = 9$. Hence

$$\Pr(Y_2 < \xi_{\frac{1}{2}} < Y_a) = \sum_{i=2}^{9-1} \binom{n}{i} \left(\frac{1}{2}\right)^i \left(1 - \frac{1}{2}\right)^{n-i}$$

$$= \sum_{i=2}^{8} \binom{n}{i} \left(\frac{1}{2}\right)^n .$$

The sample is of size $n = 10$. We now want

$$\sum_{i=2}^{8} \binom{10}{i} \left(\frac{1}{2}\right)^{10} = \left(\frac{1}{2}\right)^{10} \sum_{i=2}^{8} \binom{10}{i}$$

$$= \frac{1}{1024} \left[\binom{10}{2} + \binom{10}{3} + \binom{10}{4} + \binom{10}{5} + \binom{10}{6} + \binom{10}{7} + \binom{10}{8} \right]$$

$$= \frac{1}{1024} (45 + 120 + 210 + 252 + 210 + 120 + 45)$$

$$= \frac{1002}{1024} = .9785.$$

Thus the confidence coefficient is

$$.9785 = \Pr(Y_2 < \xi_{\frac{1}{2}} < Y_q).$$

● **PROBLEM** 9-171

For a random sample of size 5, use the order statistics (Y_1, Y_5) as a tolerance interval for 75 per cent of the population. With what probability, γ, can we expect that 75% of the population falls in the interval.

Solution: We now concern ourselves with a nonpara-
metric method for estimating the variability of a random
variable. In general terms, let $f(\cdot)$ be a probability
density function, and on the basis of a sample of n
values we want to determine two numbers, L_1 and L_2
such that at least .75 of the area under $f(\cdot)$ is between
L_1 and L_2. On the basis of a sample we cannot be positive
that .75 of the area under $f(\cdot)$ is between L_1 and L_2, but
we can specify a probability to that effect. Our goal now
is to find two functions $L_1 = \ell_1(X_1, \ldots, X_n)$ and
$L_2 = \ell_2(X_2, \ldots, X_n)$ such that

$$\Pr\left(\int_{L_1}^{L_2} f(x)\ dx \geq .75 \right) = \gamma.$$

We now recall the general definition of Tolerance
Limits: Let $X_1, \ldots, X_n$ be a random sample from a continuous c.d.f $.F(\cdot)$ having a density function $F(\cdot)$. Let
$L_1 = \ell_1(X_1, \ldots, X_n) < L_2 = \ell_2(X_2, \ldots, X_n)$ be two
statistics satisfying:

(i) The distribution of $F(L_2) - F(L_1)$ does not depend
on $F(\cdot)$, and

(ii) $\Pr(F(L_2) - F(L_1) \geq 1 - \alpha) = \gamma$.

Then L_1 and L_2 will be defined to be $100(1 - \alpha)$
percent tolerance limits at probability γ.

For continuous random variables, order statistics
Y_j and Y_k ($j < k$) form tolerance limits. To obtain β
and γ, where $\Pr(F(L_2) - F(L_1) > \beta) \leq \gamma$, we need the
distribution of $F(L_2) - F(L_1)$. Recall that for the order
statistics Y_i and Y_j, the joint density is

$$f_{Y_j, Y_k}(Y_j, Y_k) = \frac{n!}{(j - 1)!(k - 1 - j)!(n - k)!} \times [F(Y_j)]^{j-1}$$

$$[F(Y_k) - F(Y_j)]^{k-1-j} [1 - F(Y_k)]^{n-k} f(Y_i) f(Y_k).$$

Now make the transformation $Z = F(Y_k) - F(Y_j)$ and
$Y = F(Y_j)$. Once we find the joint distribution of Y and Z,
we can integrate with respect to Y to get the marginal distribution of Z. We find that

$$f_Z(Z) = \frac{n!}{(k-1-j)!(n-k+j)!} \; Z^{k-1-j} (1-Z)^{n-k+j}$$

for $0 < Z < 1$.

This is a beta distribution with parameters $k - j$ and $n - k + j + 1$. Now we define

$$Pr(Z < \beta) = \int_0^\beta f_Z(z) \, dz = IB_\beta (k - j, \; n - k + j + 1),$$

the incomplete beta function, which is extensively tabulated. It can be shown that

$$IB_\beta (k - j, \; n - k + j + 1) = \sum_{k-j}^{n} \binom{n}{i} \beta^i (1-\beta)^{n-i}.$$

Thus for any β, we can calculate γ. In our problem, we are given $\beta = .75$ and we need γ. By definition:

$$\gamma = Pr[F(Y_5) - F(Y_1) \geq .75]$$

$$= 1 - Pr[F(Y_5) - F(Y_1) < .75]$$

$$= 1 - \int_0^\beta f_Z(z) \, dz$$

$$= 1 - IB(5 - 1, \; 5 - 5 + 1 + 1)$$

$$= 1 - \sum_{i=4}^{5} \binom{5}{i} (.75)^i (.25)^{5-i}$$

$$= 1 - \binom{5}{4} (.75)^4 (.25) - \binom{5}{5} (.75)^5 = .3672.$$

Use the Kolmogorov-Smirnov Statistic to find a 95%
confidence interval for F(x). F(x) is the cumulative
distribution function of a population from which the
following ordered samples was taken: 8.2, 10.4, 10.6,
11.5, 12.6, 12.9, 13.3, 13.3, 13.4, 13.4, 13.6, 13.8,
14.0, 14.0, 14.1, 14.2, 14.6, 14.7, 14.9, 15.0, 15.4,
15.6, 15.9, 16.0, 16.2, 16.3, 17.2, 17.4, 17.7, 18.1.

Solution: Let $X_1{}'$, $X_2{}'$, ..., $X_n{}'$ denote a random
sample from a population with distribution function $F(x)$
and let x_1, ..., x_n denote the ordered sample.

The Kolmogorov-Smirnov method uses the ordered
sample to construct an upper and lower step function such
that $F(x)$ will have a specified probability of lying
between them. Consider the sample cumulative distribution
given by

$$Sn(x) = \begin{cases} 0 & X < X_1 \\ \dfrac{K}{n} & X_K \leq X < X_{K+1} \\ 1 & X \geq X_n \end{cases}$$

A graph of this type of function is:

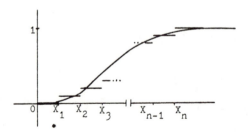

Superimposed on the above graph is a typical continuous
c.d.f.

If $F(x)$ were a known continuous function, we could
calculate $|F(x) - Sn(x)|$ for any X. From the above sketch
we can see that the function $|F(x) - Sn(x)|$ takes on its
largest value at the endpoint of the interval $[X_i, X_{i+1}]$.

Because $Sn(x)$ is constant over each interval $[X_i, X_{i+1}]$,

and $F(x)$ is nondecreasing and continuous, we can assume
$|F(x) - Sn(x)|$ attains its maximum at the left endpoint

since the right endpoint is not included in the interval.
What happens at the right of the interval is of concern
too. Therefore, instead of maxima, we study the function

$$\sup_{x} |F(x) - Sn(x)|.$$

This gives the maximum vertical distance possible
between the graphs of $F(x)$ and $Sn(x)$. It can be shown that
the distribution of $\sup_{x} |F(x) - Sn(x)|$ does <u>not</u> depend
upon $F(x)$. Hence, this quantity, which we shall call Dn
can be used as a non-parametric variable for constructing
a confidence band for $F(x)$.

Dn is a random variable. Certain critical values of
this distribution are given in the table below. Let
D_n^{α} denote such a critical value that satisfies the relation.

$$Pr(D \leq D_n^{\alpha}) = 1 - \alpha.$$

Critical Values for D_n in the Kolmogorov-Smirnov Test

n \ α	.20	.10	.05	.01
5	.45	.51	.56	.67
10	.32	.37	.41	.49
15	.27	.26	.34	.40
20	.23	.24	.29	.36
25	.21	.22	.27	.32
30	.19	.20	.24	.29
35	.18	.19	.23	.27
40	.17	.18	.21	.25
45	.16	.17	.20	.24
50	.15		.19	.23
Large values	$\dfrac{1.07}{\sqrt{n}}$	$\dfrac{1.22}{\sqrt{n}}$	$\dfrac{1.36}{\sqrt{n}}$	$\dfrac{1.63}{\sqrt{n}}$

We can now write the following equalities:

$$1 - \alpha = Pr\left(\sup_{x} |F(x) - Sn(x)| \leq D_n^{\alpha}\right)$$

$$= Pr(|F(x) - Sn(x)| \leq D_n^{\alpha}) \quad \text{for all } x$$

$$= Pr\left(Sn(x) - D_n^{\alpha} \leq F(x) \leq Sn(x) + D_n^{\alpha}\right).$$

The 2 step functions $Sn(x) - D_n^\alpha$ and $Sn(x) + D_n^\alpha$ yield a $1 - \alpha$ confidence interval for $F(x)$.

In our problem, $n = 30$ and $\alpha = .05$. $D_{30}^{.05} = .24$. Therefore .24 is the value that must be added to and subtracted from $S_{30}(x)$ to yield the desired confidence interval.

Observed Value	$S_{30}(x)$	Intervals * Lower Bound $(S_{30}(x) - .24)$	Intervals ** Upper Bound $(S_{30}(x) + .24)$
8.2	$\frac{1}{30}$	0	.273
10.4	$\frac{2}{30}$	0	.301
10.6	$\frac{3}{30}$	0	.340
11.5	$\frac{4}{30}$	0	.373
12.6	$\frac{5}{30}$	0	.401
12.9	$\frac{6}{30}$	0	.440
13.3	-	-	-
13.3	$\frac{8}{30}$	.027	.501
13.4	-	-	-
13.4	$\frac{10}{30}$	.093	.573
13.6	$\frac{11}{30}$	.127	.601
13.8	$\frac{12}{30}$	1.60	.640
14.0	-	-	-
14.0	$\frac{14}{30}$	.227	.701
14.1	$\frac{15}{30}$	.260	.740
14.2	$\frac{16}{30}$	.293	.773
14.6	$\frac{17}{30}$	.327	.801
14.7	$\frac{18}{30}$	.360	.840
14.9	$\frac{19}{30}$	.393	.873
15.0	$\frac{20}{30}$	.427	.901
15.4	$\frac{21}{30}$	.460	.940
15.6	$\frac{22}{30}$	.493	.973

15.9	$\dfrac{23}{30}$	.527	1
16.0	$\dfrac{24}{30}$	.560	1
16.2	$\dfrac{25}{30}$	.593	1
16.3	$\dfrac{26}{30}$	.627	1
17.2	$\dfrac{27}{30}$	.660	1
17.4	$\dfrac{28}{30}$	.693	1
17.7	$\dfrac{29}{30}$	.727	1
18.1	$\dfrac{30}{30}$	.760	1

* The reason for the presence of zeroes is that even though $S_{30}(x) - .24$ may be less than 0, we know $F(x)$ never is.

** The ones are there since $F(x)$ never exceeds one no matter what $S_{30}(x) + .24$ is.

The confidence interval is sketched below:

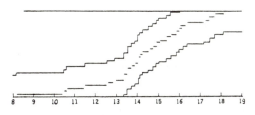

Confidence Band for a Distribution
Function

● **PROBLEM 9-173**

Random samples of 2 brands of almond candy bars were examined to see if one brand contained more almonds than the other. Altogether 18 bars of brand A and 22 of brand B were tested. The rank total of brand A, T_x, turned out to be 274. Does this indicate a 5% statistically significant difference between the two brands.

Solution: We have H_0: $F_A(\ ) = F_B(\ )$ and H_1: $F_A(\ ) \neq F_B(\ )$.

We will use the Mann-Whitney-Wilcoxon Rank Sum Test

and reject Ho if $\left| T_x - E(T_x) \right| \geq K.$

At the 5% level we will reject Ho if $\left| \dfrac{T_x - E(T_x)}{\text{Var } T_x} \right| \geq 1.96.$

We are given $T_x = 274.$

$E(T_x) = \dfrac{m \ (m + n + 1)}{2}$ where m = number of X's in sample and n = size of Y sample.

$E(T_x) = \dfrac{18 \ (18 + 22 + 1)}{2} = 369$

$\text{Var } (T_x) = \dfrac{mn \ (m + n + 1)}{12} = \dfrac{18 \cdot 22 \ (18 + 22 + 1)}{12} = 1353.$

$\sqrt{\text{Var } (T_x)} = \sqrt{1353} = 36.78$

$\left| \dfrac{T_x - E(T_x)}{\text{Var } (T_x)} \right| = \left| \dfrac{274 - 369}{36.78} \right| = 2.58.$

Since $2.58 \geq 1.96$ we reject H_0 and assume a statistically significant difference.

● **PROBLEM** 9-174

My wife wanted to know whether putting cut flowers into a certain chemical solution (we'll call it 'Flower-Life') would prolong their life, so we designed the following experiment. She bought 2 fresh blooms of 25 different kinds of flowers - 2 roses, 2 irises, 2 carnations, and so on We then put one of each pair in a vase of water, and their partners in a vase containing 'Flower-Life'. Both vases were put side by side in the same room, and the length of life of each flower was noted.

We then had 2 matched samples, so the results could be tested for significance by Wilcoxon's Signed Ranks Test. This revealed a smaller rank total of 50. Is there a statistical difference between 'Flower-Life' and plain water?

Solution: Wilcoxon's Signed Ranks Test compares 2 random samples of matched measurements.

The test depends on the fact that if no significant difference between 2 sets of paired measurements exist, then there should be equal numbers of plus and minus differences.

This test not only takes into account the direction of the difference but the size as well. If there is really no significant difference the total plus and minus rank values should be about equal.

First align all pairs from set A and set B. Subtract each member in set B from its partner in set A. Mark all appropriate minus signs. If any pairs are identical, we exclude them from our test on the grounds that they contribute nothing to our search for a significant difference. This unfortunately reduces our sample and the power of our test.

Rank all differences by absolute value and then ascribe a minus sign if appropriate. For example suppose the signed differences were

$$- \frac{1}{4} , \frac{1}{2} , \frac{3}{4} , \frac{11}{10} , - 2, - \frac{17}{18} , 3.$$

The signed ranks would be $- 1, 2, 3, 4, - 5, - 6, 7.$

Let $T_x = \Sigma$ positive signed ranks

and $T_y = \Sigma$ negative signed ranks.

In the above example $T_x = 2 + 3 + 4 + 7$ and $T_y = 1 + 5 + 6.$

In general, consider either one, T_+ or T_- . Call it T. If we assume that there is no significant difference,

$$E(T) = \frac{1}{2} \times \text{ the total sum of all rank absolute values}$$

$$= \frac{1}{2} \sum_{i=1}^{n} i = \frac{1}{2} \left(\frac{n (n + 1)}{2} \right) = \frac{n (n + 1)}{4} .$$

Var(T) can be calculated to be $\frac{n (n + 1)(2n + 1)}{24} .$

The computation is involved and will be omitted. It is important to note that we can use either T_+ or T_-.

For large values of n, the exact determination of T's distribution becomes tedious. Since all ranks do not have identical distributions we cannot apply the Central Limit Theorem. However, a more general theorem, due to Liapounov, states that if a random variable U_i has mean μ_i and σ_i^2 ($i = 1, 2, 3, \ldots, n$) and the U_i are mutually stochastically independent, if $E(|U_i - \mu_i|^3)$ is finite for all i, and if

$$\text{Lim}_{n \to \infty} \frac{\sum\limits_{i=1}^{n} E(|U_i - \mu_i|^3)}{\left(\sum\limits_{i=1}^{n} \sigma_i^2\right)^{\frac{3}{2}}} = 0$$

then
$$\frac{\sum\limits_{i=1}^{n} U_i - \sum\limits_{i=1}^{n} \mu_i}{\sqrt{\sum\limits_{i=1}^{n} \sigma_i^2}}$$
has a limiting distribution

which is standard normal. Let $U_i = R_i$, the ith signed rank, that is the appropriate rank, i, with the suitable + or - sign.

$\mu_i = E(R_i)$. Under the null hypothesis of equality of distributions, H_0, the positive rank sum would equal the negative rank sum and they are of opposite signs. Hence

$E(R_i) = 0$ since Σ signed ranks = 0.

$E(|R_i - \mu_i|)^3 = E(|R_i|)^3 = i^3$.

$$\sum\limits_{i=1}^{n} E(|R_i - \mu_i|)^3 = \sum\limits_{i=1}^{n} i^3 = \frac{n^2 (n+1)^2}{4}$$

by a known algebraic formula.

Now
$$\text{Lim}_{n \to \infty} \sum\limits_{i=1}^{n} \frac{E(|U_i - \mu_i|)^3}{\left(\sum\limits_{i=1}^{n} \sigma_i^2\right)^{\frac{3}{2}}}$$

$$= \frac{\sum\limits_{i=1}^{n} E(|U_i - \mu_i|)^3}{\left(\sum\limits_{i=1}^{n} \sigma_i^2\right)^{\frac{3}{2}}} = \frac{n^2 (n+1)^2/4}{(\text{Var } (T_+) + \text{Var } (T_-))^{\frac{3}{2}}}$$

$$= \frac{n^2(n+1)^2/4}{(2\,\mathrm{Var}(T))^{\frac{3}{2}}} = \frac{n^2(n+1)^2/4}{[n(n+1)(2n+6)/12]^{\frac{3}{2}}}$$

As $n \to \infty$, (1) becomes $\dfrac{n^4 + \text{other terms}}{n^{9/27} + \text{other terms}} \to 0$,

so Liapounov's Theorem holds

and $\quad \dfrac{\sum\limits_i R_i - 0}{\sum\limits_i \mathrm{Var}(R_i)}$ has a limiting standard normal

distribution. It can be now shown, although it won't be here, that since $(T_+) + (T_-) = \Sigma R_i$ (a constant),

$$\frac{(T_+) - \dfrac{n(n+1)}{4}}{\sqrt{\dfrac{n(n+1)(2n+1)}{24}}} \quad \text{and} \quad \frac{(T_-)\left[\dfrac{n(n+1)}{4}\right]}{\sqrt{\dfrac{n(n+1)(2n+1)}{24}}}$$

have approximate standard normal distributions. That we can use either T_+ or T_- comes from the fact that either one uniquely determines the other and they therefore have "mirror" distributions. Since the normal distribution is symmetric, this mirror quality is irrelevant.

Let us examine, in our problem,

$$K = \left| \frac{T - \dfrac{n(n+1)}{4}}{\sqrt{\dfrac{n(n+1)(2n+1)}{24}}} \right| = \left| \frac{50 - \dfrac{25(25+1)}{4}}{\sqrt{\dfrac{25(25+1)(2 \cdot 25+1)}{24}}} \right|$$

$= 3.03$.

Examination of the standard normal table shows this to indicate a random probability of .0002, if there is no difference between water and Flower-Life. We therefore reject the outside possibility.

Suppose we want to compare 2 treatments for curing acne (pimples). Suppose, too, that for practical reasons we are obliged to use a presenting sample of patients. We might then decide to alternate the 2 treatments strictly according to the order in which the patients arrive (A, B, A, B, and so on). Let us agree to measure the cure in terms of weeks to reach 90% improvement (this may prove more satisfactory than awaiting 100% cure, for some patients, may not be completely cured by either treatment, and many patients might not report back for review when they are completely cured).

This design would ordinarily call for Wilcoxon's Sum of Ranks Test, but there is one more thing to be considered: severity of the disease. For it could happen that a disproportionate number of mild cases might end up, purely by chance, in one of the treatment groups, which could bias the results in favor of this group, even if there was no difference between the 2 treatments. It would clearly be better to compare the 2 treatments on comparable cases, and this can be done by stratifying the samples. Suppose we decide to group all patients into one or other of 4 categories - mild, moderate, severe, and very severe. Then all the mild cases would be given the 2 treatments alternatively and likewise with the other groups.

Given the results tabulated below (in order of size, not of their actual occurrence), is the evidence sufficient to say that one treatment is better than the other?

Category	Treatment A Weeks	Treatment B Weeks
(I) Mild	2 3	2 4
(II) Moderate	3 5 6 10	4 6 7 9
(III) Severe	6 8 11	9 14 14
(IV) Very severe	8 10 11	12 14 15

Solution: Wilcoxon extended his sum of ranks to compare 2 independent stratified random samples of measurements which have comparable strata. Wilcoxon's Stratified Test serves to compare the effect of 2 treatments when both are tested at various levels or when both are applied to 2 or more independent sample groups.

The first step is to divide the sample into appropriate strata. In our case this has already been done. Next assign ranks to the observations in both samples combined, for each stratum separately, working from the smallest to the largest in each case. We give average rank values to identical measurements occurring in the same stratum. Add up the 2 sets of sample ranks. Call either one T (it should be easy to see that we can construct a test also using the other total).

It can be shown that as the strata groups increase in number

$$Z = \frac{T - \dfrac{n_I (2n_I + 1)}{2} - \dfrac{n_{II} (2n_{II} + 1)}{2} - \text{etc.}}{\sqrt{\dfrac{n_I^2 (2n_I + 1)}{12} + \dfrac{n_{II}^2 (2n_{II} + 1)}{12} + \text{etc.}}}$$

has a distribution which approximates standard normal (n_I, n_{II}, etc. are the number of observations in each stratum).

If each of the K strata contains n measurements, the above statistic reduces to

$$Z = \frac{T - \dfrac{kn (2n + 1)}{2}}{\sqrt{\dfrac{kn^2 (2n + 1)}{12}}}$$

The above formulae are only extensions of that given in the Wilcoxon Sum of Ranks Test.

The following extension of the problem's table should prove helpful.

Category	Treatment A Weeks	Rank	Treatment B Weeks	Rank
(I) Mild	2	$1,2 = 1\frac{1}{2}$	2	$1, 2 = 1\frac{1}{2}$
	3	3	4	4
(II) Moderate	3	1	4	2
	5	3	6	$4, 5 = 4\frac{1}{2}$
	6	$4,5 = 4\frac{1}{2}$	7	6
	10	8	9	7
(III) Severe	6	1	9	3
	8	2	14	$5, 6 = 5\frac{1}{2}$
	11	4	14	$5, 6 = 5\frac{1}{2}$
(IV) Very severe	8	1	12	4
	10	2	14	5
	11	3	15	6

Rank sum: 34   Rank sum: 54

Let T = 34

$n_I = 2$ so $n_I (2n_I + 1) = 2 \times 5 = 10$

$$n_I^2 (2n_I + 1) = 4 \times 5 = 20$$

$n_{II} = 4$ so $n_{II} (2n_{II} + 1) = 4 \times 9 = 36$

$$n_{II}^2 (2n_{II} + 1) = 16 \times 9 = 144$$

$n_{III} = 3$ so $n_{III} (2n_{III} + 1) = 3 \times 7 = 21$

$$n_{III}^2 (2n_{III} + 1) = 9 \times 7 = 63$$

$n_{IV} = 3$ so $n_{IV} (2n_{IV} + 1) = 3 \times 7 = 21$

$$n_{IV}^2 (2n_{IV} + 1) = 9 \times 7 = 63 \; .$$

694

Now our standard normal statistic Z

$$= \frac{T - \dfrac{n_I (2n_I + 1)}{2} - \dfrac{n_{II} (2n_{II} + 1)}{2} - \text{etc.}}{\sqrt{\dfrac{n_I^2 (2n_I + 1) + n_{II}^2 (2n_{II} + 1) + \text{etc.}}{12}}}$$

$$= \frac{34 - \dfrac{10}{2} - \dfrac{36}{2} - \dfrac{21}{2} - \dfrac{21}{2}}{\sqrt{\dfrac{20 + 144 + 63 + 63}{12}}} = \frac{-10}{4.92} = -2.03 \cdot$$

A standard normal statistic will produce a value with absolute value as large as 2.03 less than 5% of the time. There probably is a significant difference between the 2 treatments.

Note that ranks do not show the size of the differences and thus the sensitivity of the test is reduced. The test of Analysis of Variance will be more precise but much more laborious.

● **PROBLEM 9-176**

Mr. Smith was asked to judge the beauty contest. There were 8 lovely young ladies in the contest, and the results were as follows:

Contestant	Place
Amelia, aged 17	1
Betsy, aged 16	=2
Carolyn, aged 18	=2
Daisy, aged 20	4
Eve, aged 18	5
Freda, aged 18	6
Georgina, aged 20	7
Helen, aged 23	8

Their ages are also quoted, because I suspect that Mr. Smith shows a bias towards youth. Let us see if this apparent relationship between age and place is likely to be a mere chance correlation, or not. In other words, if Mr. Smith has shown no tendency to favor the young contestants, what is the probability of getting the observed results merely by chance?

<u>Solution:</u>   Spearman's Rank Correlation Test is suitable. Its purpose is to test for correlation between 2 measurable characteristics. When considering correlation between 2 things one must be on guard against the fallacy that one <u>causes</u> the other.

The test works on the following principle. Each of the two sets of measurements is given its own set of rank values. If the sets are perfectly correlated, there would be no difference in the ranking order of the 2 sets, so if we subtracted each rank value in one set from its partner the total of differences would be zero. On the other hand, if the 2 sets are perfectly inversely correlated the ranks will be reversed. You can convince yourself that in this case the total of the rank differences will be a maximum.

In testing for correlation we start by assuming the null hypothesis H_0; there is no significant correlation. We determine the probability, under H_0, that we can obtain our result by chance. If the probability is remote we usually reject the hypothesis of no correlation. We will use 5% as our dividing line.

The procedure for the test is a bit involved. First record all paired values of the 2 sets in question. Second assign ranks to each set independently. In case of ties, use average ranks.

Subtract to find the difference (d) between the rank values of each pair of observations. Square each difference value (d^2). Add up all the difference squares. Call this total D^2. If there are any ties, a correction factor (T) must be added to the value of D^2. This is needed because each set of ties involving x observations falsely lowers the value of D^2 by $\frac{x^3 - x}{12}$. Let tx = number of ties involving x elements. Then

$$T = \sum_{x} tx \left(\frac{x^3 - x}{12} \right).$$

If n = 5 to 30, we have provided a table giving the probabilities of observing certain values of $D^2 + T$ under the assumption of no correlation.

If n > 30

$$Z = \sqrt{n - 1} \left[1 - \frac{D^2 + T}{\frac{1}{6}(n^3 - n)} \right]$$

is approximately a standard normal statistic.

The following table rearranges the data for our problem.

Contestant	Rank of Place	Age Rank	d	d^2
Amelia	1	2	-1	1
Betsy	$2, 3 = 2\frac{1}{2}$	1	$1\frac{1}{2}$	$2\frac{1}{4}$
Carolyn	$2, 3 = 2\frac{1}{2}$	$3,4,5 = 4$	$-1\frac{1}{2}$	$2\frac{1}{4}$
Daisy	4	$6,7 = 6\frac{1}{2}$	$-2\frac{1}{2}$	$6\frac{1}{4}$
Eve	5	$3,4,5 = 4$	1	1
Freda	6	$3,4,5 = 4$	2	4
Georgina	7	$6,7 = 6\frac{1}{2}$	$\frac{1}{2}$	$\frac{1}{4}$
Helen	8	8	0	0

$$n = 8 \qquad D^2 = \Sigma d^2 = 17$$

We have 3 sets of ties

(1) Betsy and Carolyn, for second place.

(2) Daisy and Georgina are both aged 20.

(3) Carolyn, Eve, and Freda are all aged 18.

We have 2 sets of ties involving 2 observations each ($t_2 = 2$) and 1 set involving 3 ($t_3 = 1$). Therefore

$$T = 2\ \frac{2^3 - 2}{12} + 1\ \frac{3^3 - 3}{12} = 1 + 2 = 3.$$

$$D^2 + T = 20.$$

Use a $D^2 + T$ table. Look under $n = 8$ and find that we could expect a result like this less than 5% of the time. The correlation is probably significant.

In their article announcing their test, Kruskal and Wallis gave the following example.

Three machines were making bottle caps. Their output on days selected at random were -

Machine	Daily output				
A	340	345	330	342	338
B	339	333	344		
C	347	343	349	355	

Is there a significant difference between the output of these 3 machines?

Solution: The Kruskal and Wallis Test extends the range of Wilcoxon's Sum of Ranks Test to cases where there are more than two sets of measurements. As before, we make the tentative assumption, H_0, that there is no significant difference between the samples. Any differences are (tentatively) looked upon as being the result of chance variation. As in the Wilcoxon Rank Sum Test, we pool all the measurements and assign them rank values.

If H_0 is correct, extracting the rank totals for each subgroup from the total ranking should reveal that they are about equal, if the number in each group is equal. If each of these subtotals is squared, and the squares are summed, the result will be minimal when the rank totals are identical. For example, $15^2 + 15^2 + 15^2 = 675$, while $13^2 + 15^2 + 17^2 = 683$.

In this test the probability of getting any particular sum of squares, under Ho, is determined by converting the sum of squares into a value called chi-squared, by

$$\chi^2 = \frac{12}{n^2 + n} \left(\frac{R_1{}^2}{n_1} + \frac{R_2{}^2}{n_2} + \frac{R_3{}^2}{n_3} \text{, etc.} \right) - 3(n + 1)$$

where n is the size of the total samples and n_i is the size of the ith subgroup.

The higher the value of χ^2, the greater the likelihood that the observed differences are not just from chance but are due to genuine differences in the parent groups from which the samples have been drawn.

With up to five measurements in each sample, we can refer the value of χ^2 to Kruskal and Wallis' table below, which shows the values that χ^2 must reach so that the probability would be 5% or 1% that such a deviation was produced by chance alone.

With more than 5 measurements in each sample or with more than 3 samples of measurements, our new χ^2 will follow a χ^2 distribution with K-1 degrees of freedom where K is the number of samples being compared. We can obtain probabilities that way.

We use the following table to help with Kruskal and Wallis' initial problem.

Date Values	Tally			Rank Values	A ranks	B ranks	C ranks
330	A			1	1		
333		B		2		2	
338	A			3	3		
339		B		4		4	
340	A			5	5		
342	A			6	6		
343			C	7			7
344		B		8		8	
345	A			9	9		
347			C	10			10
349			C	11			11
355			C	12			12

$$n = 12 \qquad R_1 = 24 \qquad R_2 = 14 \qquad R_3 = 40$$

$$n_1 = 5 \qquad n_2 = 3 \qquad n_3 = 4$$

$$\chi^2 = \frac{12}{n^2 + n} \left(\frac{R_1^2}{n_1} + \frac{R_2^2}{n_2} + \frac{R_3^2}{n_3} \right) - 3(n - 1)$$

$$= \frac{12}{12^2 + 12} \left(\frac{24^2}{5} + \frac{14^2}{3} + \frac{40^2}{4} \right) - 3 \ (12 + 1)$$

$$= \frac{12}{156} \left(\frac{576}{5} + \frac{196}{3} + \frac{1600}{4} \right) - 3 \ (13)$$

$$= \frac{1}{13} \ (115.2 + 65.33 + 400) - 39$$

$$= \left(\frac{1}{13} \times 580.53 \right) - 39 = 44.66 - 39 = 5.66 \ .$$

All three samples contain 5 or less measurements, so we refer this value of χ^2 to Kruskal and Wallis' table. In the present instance $n_1 = 5$, $n_2 = 3$, and $n_3 = 4$, but for interpreting the results it does not matter which is which

- all that matters is that we have 3 samples sized 3, 4 and 5. Examine the line 3, 4, 5. This row shows that 5.66 has a probability of just less than 5% of occurring by chance. We conclude that the deviation is probably significant.

You may be interested to know that Analysis of Variance gives a probability of 5.1%. The correspondence is good.

● **PROBLEM** 9-178

A teacher wanted to find out the best way to demonstrate the proof of Sylow's Theorem to his class. There are three possible ways of proving this formula, but perhaps his students would not find them all equally easy to understand. He explained each of the proofs to his class of 18 students, and then asked each student to write down which of the proofs he had not understood. The results were:

Student			
1	A, B, C	10	-
2	B	11	-
3	B	12	B
4	B	13	B
5	B, C	14	B, C
6	-	15	B
7	B	16	A, B
8	A, B	17	A, B, C
9	A, B, C	18	B

   Does this represent a significant difference or might it be reasonably attributed to chance?

Solution: We want to see if there is a significant difference between three or more sets of matched observations. When the observations can be divided into two categories, Cochran's Test is usually used.

   The tentative assumption is made that there is no significant difference between the various sets of observations. On this basis, the probability of getting any observed deviation can be calculated.

   The ordinary χ^2 test is inapplicable in these circumstances, because the sample groups are <u>not</u> necessarily independent.

   All observations must fall into one of the two categories, positive or negative, good or bad, etc. Note in which category each observation falls. Choose one of the 2 classes and add the total number of observations of this class in each of observations. Call these totals, x_A, x_B, x_C, etc. Obtain $X = x_A^2 + x_B^2 + x_C^2$, etc.

Now for every matched triple or n-tuple of observations, note how many fall into the category in question. Call these values y_1, y_2, etc. Obtain $Y = \sum_n y_n$. Square each y_n and find $\sum_n y_n^2 = Z$. Finally calculate χ^2 from

Cochran's Formula
$$\chi^2 = \frac{(K - 1)(KX - Y^2)}{KY - Z},$$

where K is the number of sets being compared.

This value has a chi-square distribution with K - 1 degrees of freedom. The higher the value of χ^2, the less the likelihood of no significant difference between the three methods. The following table should aid in the application of Cochran's Test.

Student	Failed to understand A	B	C	y	y^2
1	✓	✓	✓	3	9
2		✓		1	1
3		✓		1	1
4		✓		1	1
5		✓	✓	2	4
6				0	0
7		✓		1	1
8	✓	✓		2	4
9	✓	✓	✓	3	9
10				0	0
11				0	0
12		✓		1	1
13		✓		1	1
14		✓	✓	2	4
15		✓		1	1
16	✓	✓		2	4
17	✓	✓	✓	3	9
18		✓		1	1
Column Totals	$x_A = 5$	$x_B = 15$	$x_C = 5$	Y=25	Z=51

$$X = x_A^2 + x_B^2 + x_C^2 = 25 + 225 + 25 = 275.$$

We also have $Y = 25$, $Y^2 = 625$, $Z = 51$, and $K = 3$. Hence,

$$\chi^2 = \frac{(K - 1)(KX - Y^2)}{KY - Z} = \frac{(3 - 1)(3 \cdot 275 - 625)}{3 \cdot 25 - 51}$$

$$= \frac{2(825 - 625)}{75 - 51} = \frac{2 \times 200}{24} = 16.67 .$$

The degrees of freedom are $K - 1$ which thus equal 2 in the present case. Reference to the χ^2 table shows that this value of χ^2 with 2 d.o.f. could be expected to occur by chance, if there was no significant difference between the 3 proofs, with a probability of less than .001. We conclude, then, that a very significant difference exists between the three methods.

● **PROBLEM 9-179**

A taxi driver kept a note of his monthly mileage for the period of a year. Do the following figures indicate a seasonal or any other periodic fluctuation, or might they reasonably be accounted to vary in a random manner?

Jan.	Feb.	Mar.	Apr.	May	June
4,690	4,910	3,520,	3,330	3,140,	2,850

July	Aug.	Sept.	Oct.	Nov.	Dec.
3,400	3,090,	3,480	4,650	3,830,	5,270

Solution: As before let $X_1, \ldots, X_m$ denote a random sample from $F_x(\cdot)$ and $Y_1, \ldots, Y_n$ a random sample from $F_y(\cdot)$. A test of $H_0 : F_x(z) = F_y(z)$ for all z is based on runs of values of X and values of Y. To understand the meaning of runs, combine the m x observations with the n y's and order the combined sample. For example, if m = 4 and n = 5, one might obtain

y x x y x y y y x.

A run is a sequence of letters of the same kind bounded by letters of another kind except for the first and last position. In the above sequence, the runs are y, xx, y, x, yyy, and x. If the two samples are from the same population, the x's and y's will be well mixed and the number of runs will be quite large. If the populations are so dissimilar that their ranges do not overlap, then the number of runs will be only 2 (e.g.: xxxx yyyyy). In general, population differences will reduce the number of runs.

Even if the two populations have the same mean and median, but one population is dispersed while the other is concentrated, the number of runs will be small (e.g.: yyyyy xxxxxx yyy).

We will perform a test by observing the total number of runs, Z, and reject H_0 if $Z <$ some specified z_0. Our task is now to find the distribution of Z.

If H_0 is true, we can see that all possible arrangements of the m x and n y values are equally likely.

There are $\binom{m + n}{m}$ such arrangements. To find $\Pr(Z = z)$, it is now sufficient to count all arrangements with z runs. Suppose z is even, say $2K$; then there must be K runs of x values and K runs of y values. To get K runs of x's, the m x's must be divided into K groups. These groups, or runs, can be formed by inserting $K - 1$ dividers into the $m - 1$ spaces between the m x values with no more than 1 divider per space. This can be done in $\binom{m - 1}{K - 1}$ ways. Similarly, we can construct the K runs of y values in $\binom{n - 1}{K - 1}$ ways. We can combine any particular x arrangement with any one of y. Now since the combined arrangement may begin with an x run or a y run there are a total of $2\binom{m - 1}{K - 1}\binom{n - 1}{K - 1}$ arrangements of $2K$ runs. Hence

$$\Pr(Z = z) = \Pr(Z = 2K) = \frac{2\binom{m - 1}{K - 1}\binom{n - 1}{K - 1}}{\binom{m + n}{m}} \; .$$

The only difference in determining $\Pr(Z = 2K + 1)$ is to note that there must be $K + 1$ runs of one type and K of the other. Using the same arguments we can obtain:

$$\Pr(Z = z) = \Pr(Z = 2k+1) = \frac{\binom{m - 1}{K}\binom{n - 1}{K - 1} + \binom{m - 1}{K - 1}\binom{n - 1}{K}}{\binom{m + n}{m}} \; .$$

To test H_0 at the α significance level, one finds the integer zo, so that as nearly as possible $\sum\limits_{Z=2}^{z_0} Pr(Z = z) = \alpha$.

Reject H_0 if $Z \leq z_0$.

The Runs Test can test for randomness. It can be used as a check to see if we can treat X_1, X_2, ..., X_s as a random sample from a continuous distribution.

For the sake of argument, take s to be even. We are given the s values X_1, X_2, ..., X_s in the order in which we observe them. There are $\frac{s}{2}$ of these values smaller than the remaining $\frac{s}{2}$. We have a "lower half" and an upper half. In the sequence X_1, X_2, ..., X_s, replace each value by either L or U, depending on whether X_i is in the lower or upper half.

For s = 10, an arrangement such as

LLLLULUUUU

may suggest a trend towards increasing values of x. We can make a test of randomness and reject if $Z < c$ using the probability distributions found earlier. On the other hand if we find a sequence such as

LULULULULU,

our suspicions will be aroused that there is a cyclic effect. Here we might want to use a rejection region of $Z > c_2$.

In our taxi driver problem the six lower half values are in order 2,850, 3,090, 3,140, 3,330, 3,400, and 3,480. These will be assigned the letter L. the upper half, or U, values are 3,520, 3,830, 4,650, 4,690, 4,910, 5,270.

The data in original order with the appropriate substitutions of L and U is

UUU LLLLLL UUU .

There are Z = 3 runs.

Let us determine $Pr(Z \leq 3)$. We have n = m = 6.

$Pr(Z \leq 3) = Pr(Z = 2) + Pr(Z = 3)$.

$Pr(Z = 2) = Pr(Z = 2 \cdot 1)$ so K = 1.

$$\Pr(z=2) = \frac{2 \binom{6-1}{1-1}\binom{6-1}{1-1}}{\binom{6+6}{6}} = \frac{2 \binom{5}{0}\binom{5}{0}}{\binom{12}{6}} = \frac{2}{\binom{12}{6}} = \frac{2}{924} \, .$$

$$\Pr(Z = 3) = \Pr(Z = 2 \cdot 1 + 1) \quad \text{so } K = 1$$

$$\frac{\binom{6-1}{1}\binom{6-1}{1-1} + \binom{6-1}{1-1}\binom{6-1}{1}}{\binom{6+6}{6}}$$

$$= \frac{\binom{5}{1}\binom{5}{0} + \binom{5}{0}\binom{5}{1}}{\binom{12}{6}} = \frac{10}{924} \, .$$

Hence $\quad \Pr(Z \le 3) = \dfrac{2}{924} + \dfrac{10}{924} = \dfrac{12}{924} = .013.$

The probability of as few as 3 runs in a random process is quite small. Hence we conclude that the process is not random.

CHI-SQUARE AND CONTINGENCY TABLES

● **PROBLEM** 9-180

The results of a survey show that the 518 respondents may be categorized as

Protestant - Republicans	126
Protestant - Democrats	71
Protestant - Independents	19
Catholic - Republicans	61
Catholic - Democrats	93
Catholic - Independents	14
Jewish - Republicans	38
Jewish - Democrats	69
Jewish - Independents	27

Given this data construct a contingency table.

Solution: A contingency table provides a way of simplifying the presentation of data. Each cell of such a table represents the number of observations belonging to a particular category or class of the data. The labels of this class are found in the column and row corresponding to the cell.

Contingency tables provide a useful method of comparing two variables. We are often interested in the possibility of relationship between two variables.

Also of interest are the degree or strength of relation between two variables and the significance of the relationship. Contingency tables are essential with nominal variables such as religion, political affiliation, or occupation.

The contingency table for the data in this problem is below:

Political Party	Protestants	Catholics	Jews	Row Totals
Republicans	126	61	38	225
Democrats	71	93	69	233
Independents	19	14	27	60
Column Totals	216	168	134	518

The row totals are formed by summing along the rows. Thus there are

225	Republicans	
233	Democrats	and
60	Independents	

in the sample.

The column totals are formed by summing down the columns. Thus there are

216	Protestants	
168	Catholics	and
134	Jews	in the sample.

The sum of the column totals equals the sum of the row totals and both are equal to the number of observations in the sample.

A die was tossed 120 times and the results are listed below.

Upturned face	1	2	3	4	5	6
Frequency	18	23	16	21	18	24

Compute the χ^2 statistic for this 1 by 6 contingency table under the hypothesis that the die was fair.

Solution: The χ^2 statistic provides a means of comparing observed frequencies with expected frequencies.

This statistic is defined to be

$$\chi^2 = \sum_{i=1}^{n} \frac{(O_i - E_i)^2}{E_i}$$

where O_i is the observed frequency in cell i, E_i is the expected frequency found from some underlying probability model which must be assumed, and n is the number of cells.

The chi-square statistic (χ^2) is approximately distributed with a chi-square distribution. This distribution is found tabulated in many statistics texts. The χ^2 statistic has a parameter called degrees of freedom associated with it.

In this problem, we assume that the die is fair. That is, any face will land upturned with probability $\frac{1}{6}$. If this model is valid we would expect equal numbers of each face to appear, or $\frac{120}{6} = 20$ occurrences of each face. The expected frequencies are

Upturned face	1	2	3	4	5	6
E_i, expected frequency if die is fair	20	20	20	20	20	20

Thus we can compute the χ^2 statistic from the observed and expected frequencies.

The degrees of freedom of a 1 by n contingency table is n - 1. In our problem n = 6 thus n - 1 = 6 - 1 = 5 degrees of freedom.

The chi-square statistic will give some idea of the difference between the hypothesized model and the observed frequencies. The χ^2 statistic, with its distribution, will allow us to test the significance of the hypothesized probability model.

$$\chi^2 = \sum_{i=1}^{6} \frac{(O_i - E_i)^2}{E_i}$$

$$= \frac{(18 - 20)^2}{20} + \frac{(23 - 20)^2}{20} + \frac{(16 - 20)^2}{20}$$

$$+ \frac{(21 - 20)^2}{20} + \frac{(18 - 20)^2}{20} + \frac{(24 - 20)^2}{20}$$

$$= \frac{1}{20} [(-2)^2 + 3^2 + (-4)^2 + 1^2 + (-2)^2 + 4^2]$$

$$= \frac{1}{20} [4 + 9 + 16 + 1 + 4 + 16] = \frac{50}{20}$$

$$= 2.5 \qquad \text{with 5 degrees of freedom.}$$

Often frequency data are tabulated according to two criteria, with a view toward testing whether the criteria are associated. Consider the following analysis of the 157 machine breakdowns during a given quarter.

Number of Breakdowns

	Machine				Total per Shift
	A	B	C	D	
Shift 1	10	6	13	13	41
Shift 2	10	12	19	21	62
Shift 3	13	10	13	18	54
Total per machine	33	28	44	52	157

We are interested in whether the same percentage of breakdown occurs on each machine during each shift or whether there is some difference due perhaps to untrained operators or other factors peculiar to a given shift.

Solution: If the number of breakdowns is independent of the shifts and machines, then the probability of a breakdown occurring in the first shift and in the first machine can be estimated by multiplying the proportion of first shift breakdowns by the proportion of machine A breakdowns.

If the attributes of shift and particular machine are independent,

Pr(Breakdown on Machine A during shift 1)

= Pr(breakdown on Machine A) ×

Pr(breakdown during shift 1)

where

Pr(Breakdown on Machine A) is estimated by

$$\frac{\text{number of breakdowns on A}}{\text{total number of breakdowns}} = \frac{33}{157}$$

709

and

Pr(Breakdown during shift 1)

$$= \frac{\text{number of breakdowns in shift 1}}{\text{total number of breakdowns}}$$

$$= \frac{41}{157} \quad .$$

Of the 157 breakdowns, given independence of machine and shift, we would expect

$$(157) \left[\frac{41}{157}\right]\left[\frac{33}{157}\right] = \frac{41 \cdot 33}{157}$$

breakdowns on machine A and during the first shift.

Similarly, for the third shift and second machine, we would expect

$$\left(\frac{54}{157}\right)\left(\frac{28}{157}\right) \cdot 157 = \frac{54 \cdot 28}{157} \quad \text{breakdowns.}$$

The expected breakdowns for different shifts and machines are

$$E_{11} = \frac{41 \times 33}{157} = 8.62$$

$$E_{12} = \frac{28 \times 41}{157} = 7.3$$

$$E_{13} = \frac{44 \times 41}{157} = 11.5$$

$$E_{14} = \frac{52 \times 41}{157} = 13.57 \cdot$$

Similarly, the other expected breakdowns given independence are:

	A	B	C	D
Shift 1	8.62	7.3	11.5	13.57
Shift 2	13.03	11.06	17.38	20.54
Shift 3	11.35	9.63	15.13	17.88

We will assume that the χ^2 test is applicable. There are $r = 3$ rows and $c = 4$ columns; thus, the chi-square statistic will have $(3 - 1)(4 - 1) = 6$ degrees of freedom. The level of significance of this hypothesis will be $\alpha = .05$. This is the probability of rejecting independence of attributes (the null hypothesis) given that the attributes are in fact independent.

If this statistic is greater than 12.6, then we will reject the hypothesis that the machine and shift are independent attributes in determining incidence of breakdown.

The chi-square statistic is

$$\chi_6^2 = \sum_{i=1}^{3} \sum_{j=1}^{4} \frac{(O_{ij} - E_{ij})^2}{E_{ij}} = 2.17.$$

We see that 2.17 < 12.6. Thus we accept that the attributes of machine and shift are independent in determining incidence of breakdown.

MISCELLANEOUS APPLICATIONS

● **PROBLEM** 9-183

XYZ Company is considering digging an oil well. The cost of the well is $50,000. If the well is successful XYZ will make a profit of $400,000, otherwise zero. The probability of the well being successful is 0.1. Is it worthwhile to dig the well?

Solution: We have the following choice :

We will choose the alternative with the higher expected profit. If we don't dig, our expected profit is zero. If we dig, we have the following uncertainty:

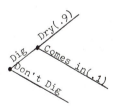

If there is oil, our profit is $400,000 - $50,000 = $350,000. If there isn't we loose the $50,000 cost of the well. Therefore our expected profit is $350,000·P (oil) + (-$50,000)·P (dry) = 350,000(.1) - 50,000(.9) = -$10,000. If we dig, our expected profit is less than zero. Our best bet is not to dig.

We can summarize the solution in a decision tree with the profit shown.

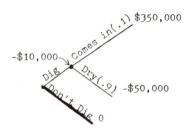

The optimal decision is indicated by the darkened line.

● **PROBLEM** 9-184

Flapjack Computers is interested in developing a new tape drive for a proposed new computer. Flapjack does not have research personnel available to develop the new drive itself and so is going to sub-contract the development to an independent research firm. Flapjack has set a fee of $250,000 for developing the new tape drive and has asked for bids from various research firms. The bid is to be awarded not on the basis of price (set at $250,000) but on the basis of both the technical plan shown in the bid and the firm's reputation.

Dyna Research Institute is considering submitting a proposal (i.e., a bid) to Flapjack to develop the new tape drive. Dyna Research Management estimated that it would cost about $50,000 to prepare a proposal; further they estimated that the chances were about 50-50 that they would be awarded the contract.

There was a major concern among Dyna Research engineers concerning exactly how they would develop the tape drive if awarded the contract. There were three alternative approaches that could be tried. One involved the use of certain electronic components. The engineers estimated that it would cost only $50,000 to develop a prototype of the tape drive using the electronic approach, but that there was only 50 percent chance that the prototype would be satisfactory. A second approach involved the use of certain magnetic apparatus. The cost of developing a prototype using this approach would cost $80,000 with 70 percent chance of success. Finally, there was a mechanical approach with cost of $120,000, but the engineers were certain of success.

Dyna Research could have sufficient time to try only two approaches. Thus, if either the magnetic or the electronic approach tried and failed, the second attempt would have to use the mechanical approach in order to guarantee a successful prototype.

The management of Dyna Research was uncertain how to take all this information into account in making the immediate decision-whether to spend $50,000 to develop a proposal for Flapjack. Can you help?

Solution: Since this decision seems complex, we will build the decision tree in steps. The first decision facing Dyna Research involves the actions "Prepare a Proposal" and "Do not Prepare a Proposal". If a proposal is developed and submitted to Flapjack, then either of the events "Contract Awarded to Dyna" or "Dyna Loses Contract". Each event has probability 0.5. These choices are shown below:

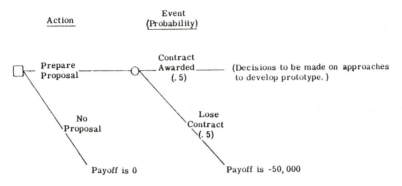

If Dyna Research decides not to prepare a bid, the net payoff is zero. If a bid is prepared but the contract is lost, Dyna Research loses the $50,000 cost of preparing the bid (i.e., the payoff is -$50,000). If the contract is awarded to Dyna, then the next decision, the choice between alternative methods of developing a successful tape drive, must be made.

In the second decision, Dyna Research must decide which of the three approaches - mechanical, electronic, or magnetic - to try first. This decision is shown below:

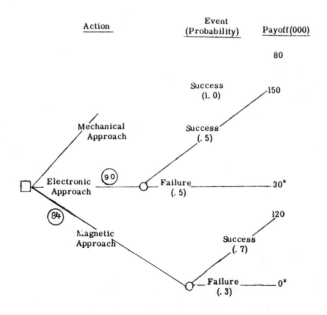

*Mechanical Approach must be used.

The payoffs in the above diagram are calculated as shown below:

713

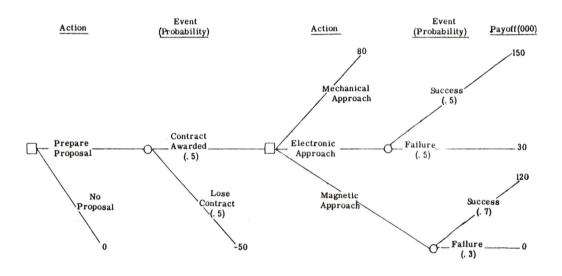

Payoff(thousands of dollars)

End of Branch	Fee	Cost of Proposal	Cost of Prototype Indicated	Cost of Mechanical Prototype	
Electronic Approach					
Success..................250 -		50	- 50		= 150
Failure..................250 -		50	- 50	- 120	= 30
Magnetic Approach					
Success..................250 -		50	- 80		= 120
Failure..................250 -		50	- 80	- 120	= 0

For the mechanical approach we take

Fee - Cost of Proposal - Cost of Mechanical Prototype

= 250 - 50 - 120 = 80.

The complete decision tree is shown as follows:

We proceed by working backwards. The expected values are calculated for each of the event forks in the right part of the tree. Thus the expected payoff associated with the electronic approach is $90,000 ((0.5 × 150) + (0.5 × 30) = 90) and for the magnetic approach, it is $84,000 ((0.7 × 120) + (0.3 × 0) = 84). These expected payoffs are inserted in circles beside the appropriate forks in the following figure:

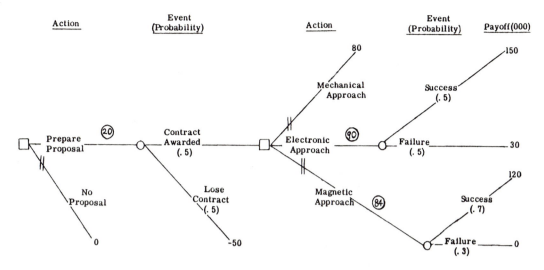

Moving left to the second decision point, we see that the electronic approach offers the highest expected payoff ($90,000) and is the best choice. The value $90,000 is written (circled) beside the decision point and the nonpreferred approaches are indicated by drawing ‖ on the branches.

The tree now has a payoff of +$90,000 if the contract is awarded and -$50,000 if not. The expected value of preparing a proposal is $20,000 ((0.5 x 90) + (0.5 x (-50)) = 20). This is written in a circle beside the event fork.

Finally, the choice must be made between the expected profit of $20,000 for preparing the proposal and zero if the proposal is not prepared. The first course is selected, and the mark "‖" drawn through the "No Proposal" branch.

In summary, Dyna Research should prepare the proposal. If the contract is awarded, the electronic approach should be tried first; but if this fails, the mechanical approach should be used.

● **PROBLEM** 9-185

A student was asked to write down a string of 50 digits that seemed to him to represent a random sequence. Reasoning that when digits are written down at random, then the same digit cannot be, repeated successively the student proceeded to write down the following digits:

```
5 8 3 1 9 4 6 7 9 2
6 3 0 8 7 5 1 3 6 2
1 9 5 4 8 0 3 7 1 4
6 0 4 3 8 2 1 3 9 8
5 6 1 8 7 0 3 5 2 5
```

On being questioned he broke down and confessed that he had also tried to keep successive digits at least two units apart. As penance he computed the chi-square statistic to test the hypothesis that the sequence was random. What did he find?

Solution: In a random sequence of numbers, the probability that a number will be the same as its predecessor is 1/10. The probability that two numbers next to each other differ by one is 2/10. Using these expected values we form the following table:

	Actual Frequency	Expected Frequency
Same	0	50(1/10) = 5
one away	8	50(2/10) = 10
other	42	50(7/10) = 35
Total	50	50

Before we compute the chi-square statistic for the data above we must first specify our level of significance. Let it be $\alpha = .05$. The degrees of freedom associated with the test is $(k-1) = 3-1 = 2$. Then, if the numbers are indeed a random sequence, we should obtain a value of χ^2 less than 5.99.

The computed chi-square statistic is

$$\frac{(0-5)^2}{5} + \frac{(8-10)^2}{10} + \frac{(42-35)^2}{35} = 6.80 > 5.99 .$$

Hence we reject the hypothesis that the sequence was random.

● PROBLEM 9-186

A highly specialized industry builds one device each month. The total monthly demand is a random variable with the following distribution.

Demand	0	1	2	3
P(D)	1/9	6/9	1/9	1/9

When the inventory level reaches 3, production is stopped until the inventory drops to 2. Let the states of the system be the inventory level. The transition matrix is found to be

$$P = \begin{array}{c|cccc} & 0 & 1 & 2 & 3 \\ 0 & 8/9 & 1/9 & 0 & 0 \\ 1 & 2/9 & 6/9 & 1/9 & 0 \\ 2 & 1/9 & 1/9 & 6/9 & 1/9 \\ 3 & 1/9 & 1/9 & 6/9 & 1/9 \end{array} \qquad (1)$$

Assuming the industry starts with zero inventory find the transition matrix as $n \to \infty$.

Solution: This problem involves stochastic processes. Since both the random variable and the time intervals are discrete and there is no carry-over effect, we consider a Markov chain.

The transition matrix (1) gives the probabilities of changing from one state to another in one step. For example, the probability of the inventory level changing from 1 to 2 is 1/9. The relationship between the initial transition matrix and the transition matrix after n steps is given by the matrix equation

$$P(n) = P^n .$$

For example after two steps, the transition matrix is given by
$P(2) = P^2$

$$= \begin{pmatrix} 8/9 & 1/9 & 0 & 0 \\ 2/9 & 6/9 & 1/9 & 0 \\ 1/9 & 1/9 & 6/9 & 1/9 \\ 1/9 & 1/9 & 6/9 & 1/9 \end{pmatrix} \begin{pmatrix} 8/9 & 1/9 & 0 & 0 \\ 2/9 & 6/9 & 1/9 & 0 \\ 1/9 & 1/9 & 6/9 & 1/9 \\ 1/9 & 1/9 & 6/9 & 1/9 \end{pmatrix}$$

$$= \frac{1}{81} \begin{pmatrix} 66 & 14 & 1 & 0 \\ 29 & 39 & 12 & 1 \\ 17 & 14 & 43 & 7 \\ 17 & 14 & 43 & 7 \end{pmatrix} . \tag{2}$$

The elements of (2) give the probabilities of proceeding from the initial state to another in exactly 2 steps. Similarly, after 3 steps
$P(3) = P^3$

$$= \begin{pmatrix} 8/9 & 1/9 & 0 & 0 \\ 2/9 & 6/9 & 1/9 & 0 \\ 1/9 & 1/9 & 6/9 & 1/9 \\ 1/9 & 1/9 & 6/9 & 1/9 \end{pmatrix}^3 = \frac{1}{729} \begin{pmatrix} 557 & 151 & 20 & 1 \\ 323 & 276 & 117 & 13 \\ 214 & 151 & 314 & 50 \\ 214 & 151 & 314 & 50 \end{pmatrix}$$

We are interested in finding the value of $P(n)$ as $n \to \infty$. First, however, we must include the initial inventory level in our calculations. The probabilities of reaching the various states in n steps are given by $p'(n) = p'(0)P(n)$. Thus,

$$p'(1) = (1,0,0,0) \begin{pmatrix} 8/9 & 1/9 & 0 & 0 \\ 2/9 & 6/9 & 1/9 & 0 \\ 1/9 & 1/9 & 6/9 & 1/9 \\ 1/9 & 1/9 & 6/9 & 1/9 \end{pmatrix}$$

$$= (8/9,\ 1/9,\ 0,0)$$
$$p'(2) = (66/81,\ 14/81,\ 1/81,\ 0/81)$$
$$p'(3) = (557/729,\ 151/729,\ 20/729,\ 1/729)$$

We note that the probability of zero inventory is decreasing. Furthermore, for $n > 3$, all the elements of P^n are greater than zero. For

Markov matrices having this property, P^n approaches a probability matrix A, where each row of A is the same probability vector α' (where α' denotes the transpose of the row vector α).
To find α, we solve the equations

$$\begin{pmatrix} 8/9 & 2/9 & 1/9 & 1/9 \\ 1/9 & 6/9 & 1/9 & 1/9 \\ 0 & 1/9 & 6/9 & 6/9 \\ 0 & 0 & 1/9 & 1/9 \end{pmatrix} \begin{pmatrix} \alpha_1 \\ \alpha_2 \\ \alpha_3 \\ \alpha_4 \end{pmatrix} = \begin{pmatrix} \alpha_1 \\ \alpha_2 \\ \alpha_3 \\ \alpha_4 \end{pmatrix} \tag{3}$$

subject to the requirement $\alpha_i \geq 0$, $i = 1,4$ and $\sum_{i=1}^{n} \alpha_i = 1$. Using either Cramer's rule or Gauss-Jordan elimination, the solution to (3) is found to be

$$\alpha' = (45/72, \; 18/72, \; 8/82, \; 1/72) \; .$$

The limit matrix is then

$$A = \begin{pmatrix} 45/72 & 18/72 & 8/72 & 1/72 \\ 45/72 & 18/72 & 8/72 & 1/72 \\ 45/72 & 18/72 & 8/72 & 1/72 \\ 45/72 & 18/72 & 8/72 & 1/72 \end{pmatrix} \; .$$

Thus, if the initial inventory level is zero, the system approaches the state where the probability of zero inventory is 45/72.

CHAPTER 10

BOOLEAN ALGEBRA

BOOLEAM ALGEBRA AND BOOLEAN FUNCTIONS

(a) Set up the truth table for each of the following Boolean expressions:

  (i) $\sim A$

 (ii) $A + B$ (inclusive or)

(iii) $A \cdot B$ (and)

(b) Show the equivalent switching circuits for (i) - (iii).

(c) Show the gate representations of (i)-(iii).

Solution: (a) A truth table is set up by specifying all the possible combinations of values of the Boolean variables, then evaluating each term according to the rule of relation of the operation. There are 2 possible values for each variable: 0 or 1. Then, if there are n variables, there will be 2^n possible combinations of the variables, and therefore, the truth table will have 2^n rows of entries.
(i) Since there is only one variable, the truth table has only $2^1 = 2$ rows of entries.

A	$\sim A$
0	1
1	0

The $\sim$ operation (called complement or invertion) changes 0 to 1 or 1 to 0.

(ii) Since there are 2 variables, the truth table has $2^2 = 4$ rows of entries:

A	B	$A + B$
0	0	0
0	1	1
1	0	1
1	1	1

719

The + operation (called disjunction) yields a value of 0 if and only if both variables have a value of 0. Otherwise, it yields a value of 1.

(iii)   Since there are 2 variables, the truth table has $2^2 = 4$ rows of entries.

A	B	A · B
0	0	0
0	1	0
1	0	0
1	1	1

The · operation (called conjunction) yields a value of 1 if and only if both variables have a value of 1. Otherwise, it yields a value of 0.

(b)   There is a one-to-one correspondence between Boolean expressions and switching circuits. An open switch or nonconductor can represent the 0-condition and a closed switch or conductor can represent the 1-condition.

   Boolean algebra can be used to determine what circuits are necessary to perform a specific logic function. The action of a two-position switch with an open position and a closed position can be represented by a Boolean variable.

(i)   If A represents a switch in the closed position, then ~A would represent the same switch in the open position, as shown in figure 1.

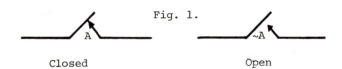

Fig. 1.

Closed   Open

(ii)   The + operation yields a value of 0 if and only if both variables have a value of 0. Therefore, the switching circuit needed to perform the + operation must be open (nonconducting) if and only if both switches are open. The parallel connection is the switching circuit realization of the + operation; as shown in figure 2.

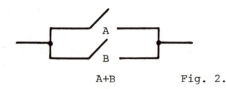

A+B   Fig. 2.

(iii)   The · operation yields a value of 1 if and only if both variables have a value of 1. The switching circuit needed to perform the · operation, therefore, must be closed (conducting) if and only if both switches are closed. The series connection is the switching circuit realization of the · operation, as shown in figure 3.

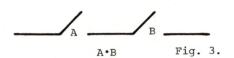

A·B   Fig. 3.

(c) Gates are switches that are sensitive to high and low voltages.
If 0 represents a low voltage and 1 represents a high voltage,
the output of a gate is either 0 or 1, depending on the type of
gate.
(i) The ~ operation changes 0 to 1 and 1 to 0. The gate
needed to perform the ~ operation, therefore, must output a 1 for
a 0 input and output a 0 for a 1 input. This is the operation
of a NOT gate shown in figure 4.

Fig. 4.

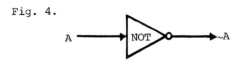

(ii) An OR-gate outputs a 0 if and only if all inputs are 0. Otherwise, it outputs a 1. This corresponds to the + operation and is
shown in figure 5.

Fig. 5.

Fig. 6.

(iii) An AND-gate outputs a 1 if and only if all inputs are 1.
Otherwise, it outputs a 0. This corresponds to the · operation;
shown in figure 6.

● **PROBLEM** 10-2

Translate the following Boolean expressions into Venn diagrams:
a) A + B   (Inclusive OR)
b) A · B   (AND)
c) ~ A   (NOT)

Solution: Venn diagrams offer a pictorial representation of logical
relations. When translating from Boolean algebra, sometimes the notation changes. We will give the diagrams and the alternate notations:
a) The expression A + B is an OR function, also known as the union
of two sets. Elements can be in A, or in B, or in both A and B.
In the Venn diagram of Fig. 1, the shaded area is the area of interest.

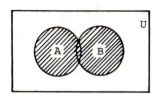

Fig. 1.

A + B can also be written as AUB, using set notation.

b) The AND function of Boolean algebra is given by A · B. This may also be considered to be the intersection of sets A and B, written as A∩B . By definition, an intersection is the set of all elements contained in both A and B simultaneously. Using a Venn diagram, we represent it as shown in Fig. 2.

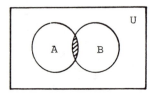

Fig. 2.

Again, the shaded area is the part described by the relation A · B (or A∩B).

c) The NOT function, ∼ A, is also known as the complement of a set A. The complement of a set A includes all elements in the universe that are not in A. We can write the complement as $\bar{A}$, and our Venn diagram is shown in Fig. 3. The shaded area represents $\bar{A}$.

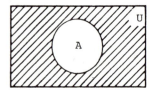

Fig. 3.

● **PROBLEM** 10-3

Use a Venn diagram to verify

   (a) $A + \bar{A}B = A + B$

   (b) $AB + \bar{A}C + BC = AB + \bar{A}C$

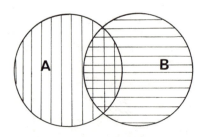

Fig. 1

Solution: (a) The Venn diagram for $A + \overline{AB}$ is shown in Fig. 1, where A is represented by the vertically shaded region and $\overline{AB}$ is the horizontally shaded area. The area which is in either A or in $\overline{AB}$ is then the entire shaded area. The Venn diagram for $A + B$ is given in Fig. 2. Observe that the shaded areas in Fig. 1 and Fig. 2 are identical, that is, $A + \overline{AB} = A + B$.

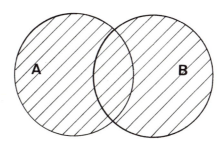

Fig. 2

(b) In Fig. 3 the vertical shading covers the region AB, i.e. the area that is in both A and B. The horizontal shading covers the area $\overline{AC}$. Note that BC (which is represented by the shaded area in the Venn diagram in Fig. 4) is encompassed by the total shaded area, $AB + \overline{AC}$, in Fig. 3. Therefore, $AB + \overline{AC}$ automatically includes BC, in other words, $AB + \overline{AC} = AB + \overline{AC} + BC$.

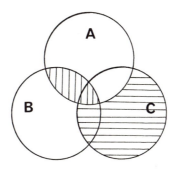

Fig. 3

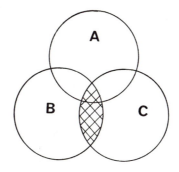

Fig. 4

• PROBLEM 10-4

Prove that $A+\overline{A}B = A+B$.

Solution: Construct a truth table as shown in Fig. 1. Since columns 5 and 6 are identical, $A+\overline{A}B = A+B$.

1	2	3	4	5	6
A	B	$\overline{A}$	$\overline{A}B$	$A+\overline{A}B$	A + B
0	0	1	0	0	0
0	1	1	1	1	1
1	0	0	0	1	1
1	1	0	0	1	1

Fig. 1

• PROBLEM 10-5

Verify that $(X+Y)(\overline{X}+Z)(Y+Z) = (X+Y)(\overline{X}+Z)$.

Solution: The proof of the given equation follows directly from a comparison of the truth table for $(X+Y)(\overline{X}+Z)(Y+Z)$ and $(X+Y)(\overline{X}+Z)$. The truth table is shown in Fig. 1. Observe that since columns 8 and 9 are identical, (that is, $(X+Y)(\overline{X}+Z)(Y+Z) = (X+Y)(\overline{X}+Z)$) for all the possible combinations of X, Y, and Z. The given equation is verified.

1	2	3	4	5	6	7	8	9
X	Y	Z	$\overline{X}$	X+Y	$\overline{X}$+Z	Y+Z	$(X+Y)(\overline{X}+Z)(Y+Z)$	$(X+Y)(\overline{X}+Z)$
0	0	0	1	0	1	0	0	0
0	0	1	1	0	1	1	0	0
0	1	0	1	1	1	1	1	1
0	1	1	1	1	1	1	1	1
1	0	0	0	1	0	0	0	0
1	0	1	0	1	1	1	1	1
1	1	0	0	1	0	1	0	0
1	1	1	0	1	1	1	1	1

Fig. 1

State the basic laws of Boolean algebra.

Solution: If A, B and C are Boolean variables, then the basic laws of Boolean algebras are (where + and • are binary operators representing OR (inclusive) and AND respectively):

1A: A + A = A [Idempotent law for +]
1B: A • A = A [Idempotent law for •]
2A: A + B = B + A [Commutative law for +]
2B: A • B = B • A [Commutative law for •]
3A: A + (B+C) = (A+B) + C [Associative law for +]
3B: A • (B•C) = (A•B) • C [Associative law for •]
4A: A • (B+C) = (A•B) + (A•C) [Distributive law for • over +]
4B: A + (B•C) = (A+B) • (A+C) [Distributive law for + over •]
5A: A + 1 = 1   (Law of Union)
5B: A • 0 = 0   (Law of Intersection)
6A: A • 1 = A [1 is the identity element for •]
6B: A + 0 = A [0 is the identity element for +]

The laws of $\sim$:

7 : $\sim(\sim A) = A$ [Double Negative Law or Involution Law]
8A: A + $\sim$A = 1 $\}$
8B: A • $\sim$ A = 0 $\}$ Law of Complement
9A: $\sim$(A+B) = $\sim$ A • $\sim$ B [deMorgan's law]
9B: $\sim$(A•B) = $\sim$ A + $\sim$ B [deMorgan's law]

10: $\tilde{1}$ = 0 and $\tilde{0}$ = 1

Any system obeying these laws is known as Boolean algebra. A set with its subsets and the subset operations union and intersect (+ and • , respectively) is a Boolean algebra. The one-to-one correspondence between Boolean expressions and switching circuits suggests that a switching algebra is a Boolean algebra. In fact, 1A - 3B obviously holds true for switching circuits. The switching analog of 4A claims that Fig. 1,

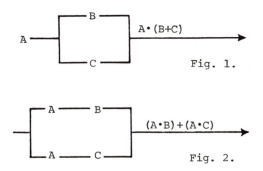

A — A•(B+C) →   Fig. 1.

(A•B)+(A•C) →   Fig. 2.

and Fig. 2 are equivalent circuits. Since both circuits conduct unless A is open or both B and C are open, they are, in fact, equivalent. Therefore, 4A holds true for switching circuits. And similarly, 4B - 10 can be shown to hold for switching circuits by exhibiting the switching circuits representing the left and right sides of the equation and showing that the two are equivalent. Once it is shown that all 18 laws hold for switching circuits, they can be used for simplifying circuits. For example, 4A and 4B reduce four-switch circuits to three-switch circuits. Actually all 18 laws are not a minimal set of Boolean algebra laws, since laws 1A, 1B, 3A, 3B, 5A, 5B, 7, 9B and 10 can be derived from the other laws. For example, once this fact is established, only 8 properties need be satisfied to determine a Boolean algebra, since the other ten follow immediately. Since the gate elements OR, AND and NOT are representations of the operations +, · and ~ respectively, it follows that a gating algebra is a Boolean algebra. This fact can be used for simplifying gating circuits. For instance 4A and 4B reduce 3-gate circuits to 2-gate circuits.

● **PROBLEM** 10-7

Write the duals of the following Boolean equations:

(1) (a+b)+c(1+d)+1 = 1

(2) $D \cdot (\overline{A+\overline{B}(C+\overline{D})}) = \overline{A} \, D(B+\overline{C})$

Solution: The dual of any statement in a Boolean algebra B is the statement obtained by interchanging the operations + and ·, and interchanging their identity elements 0 and 1 in the original statement. Hence, the duals of the given equations are:

(1) $ab(c + 0 \cdot d) \cdot 0 = 0$

(2) $D + \overline{A(\overline{B} + C\overline{D})} = \overline{A} + D + B\overline{C}$

● **PROBLEM** 10-8

Find the duals of the following Boolean expressions:

(1) $a + b(\overline{c}\overline{d} + (\overline{ef}+c))$

(2) $(\overline{x+y})z + w \cdot 0 + 1$

(3) $(a+b)c(\overline{d}+e)(1+a)$

Solution: The dual of any Boolean expression is the expression obtained by interchanging the operations + and ·, and interchanging their identity elements 0 and 1 in the original expression.

Therefore, the duals of the given expressions are:

(1) $a(b+(\overline{c}+\overline{d})(e+\overline{f})c)$

726

(2) $(\overline{xy}+z)(w+1) \cdot 0$

(3) $ab + c + \overline{d}e + 0 \cdot a$

● **PROBLEM** 10-9

Apply DeMorgan's theorem to the following equations:
a) $F = \overline{V + A + L}$
b) $F = \overline{A} + \overline{B} + \overline{C} + \overline{D}$
c) $F = \overline{WXYZ}$
d) $F = \overline{ABC} + D$

__Solution__: DeMorgan's theorem states that a logical expression can be complemented by complementing all its variables and exchanging AND ($\cdot$) operations with OR (+) operations. For example the complement of

$$F = AB$$

is

$$\overline{F} = \overline{A} + \overline{B} .$$

Another expression for F is found by complementing $\overline{F}$.

$$F = \overline{\overline{F}} = \overline{\overline{A} + \overline{B}}$$

Thus

$$\overline{\overline{A} + \overline{B}} = AB$$

a) Complementing variables V,A, and L and changing + to $\cdot$, we get

$$\overline{F} = \overline{V} \cdot \overline{A} \cdot \overline{L} \quad \text{(note AB = A} \cdot \text{B)}$$

Complement the entire expression to find F,

$$F' \; \overline{\overline{F}} = \overline{\overline{V} \; \overline{A} \; \overline{L}} = \overline{V} \; \overline{A} \; \overline{L}$$

Thus,

$$\overline{V + A + L} = \overline{V} \; \overline{A} \; \overline{L}$$

b) Again, complementing A,B,C, and D and changing + to $\cdot$ we get

$$\overline{F} = ABCD$$

$$F = \overline{\overline{F}} = \overline{ABCD}$$

c) This time exchanging $\cdot$ for +, we have

$$\overline{F} = \overline{W} + \overline{X} + \overline{Y} + \overline{Z}$$

$$F = \overline{\overline{F}} = \overline{\overline{W} + \overline{X} + \overline{Y} + \overline{Z}}$$

d) This time split the function into two parts

$$F = X + D$$

where $X = \overline{ABC}$. Apply DeMorgan's theorem to X.

$$\overline{X} = \overline{\overline{A} + \overline{B} + \overline{C}}$$

thus,
$$X = \overline{A} + \overline{B} + \overline{C}$$

$$F = \overline{A} + \overline{B} + \overline{C} + D$$

● PROBLEM 10-10

Simplify $Z = D(A + \overline{\overline{B}(C + \overline{D})})$.

<u>Solution:</u> $Z = D[\overline{A + \overline{B}(C + \overline{D})}]$

$\qquad = D[\overline{A}(\overline{\overline{B}(C + \overline{D})})]$ $\qquad$ DeMorgan's theorem

$\qquad = D[\overline{A}[B + \overline{(C + \overline{D})}]]$ $\qquad$ DeMorgan's theorem

$\qquad = D[\overline{A}[B + \overline{C}D]]$ $\qquad$ DeMorgan's theorem

$\qquad = \overline{A}BD + \overline{A}\,\overline{C}D \cdot D$ $\qquad$ $A(B+C) = AB+AC$

$\qquad = \overline{A}BD + \overline{A}\,\overline{C}D$ $\qquad$ $A \cdot A = A$

$\qquad = \overline{A}D(B + \overline{C})$ $\qquad$ $AB + AC = A(B+C)$

Note that sometimes the final result may not be unique. For instance, $\overline{A}D(B+\overline{C})$ may be written as $\overline{A}D(\overline{BC})$ (using DeMorgan's theorem and $\overline{\overline{A}} = A$), or we may just leave $\overline{A}BD + \overline{A}\,\overline{C}D$ as the final result.

● PROBLEM 10-11

Simplify $Y = \overline{a}[(a+b) + (\overline{\overline{c}+\overline{c}b})] + c(\overline{a}+b) + a + \overline{b}$.

<u>Solution:</u> We simplify using the laws of Boolean algebra:

$\quad Y = \overline{a}[(a+b) + (\overline{\overline{c}+\overline{c}b})] + c(\overline{a}+b) + a + \overline{b}$

$\qquad = \overline{a}(a+b) + \overline{a}(\overline{\overline{c}+\overline{c}b}) + c(\overline{a}+b) + a + \overline{b}$ $\qquad$ $x(y+z)=xy+xz$

$\qquad = \overline{a}\cdot a + \overline{a}\cdot b + \overline{a}(\overline{\overline{c}+\overline{c}b}) + c(\overline{a}+b) + a + \overline{b}$ $\qquad$ $x(y+z)=xy+xz$

$\qquad = \overline{a}b + \overline{a}(\overline{\overline{c}+\overline{c}b}) + c(\overline{a}+b) + a + \overline{b}$ $\qquad$ $x \cdot \overline{x} = 0$

$\qquad = \overline{a}b + \overline{a}\cdot(c(\overline{\overline{c}b})) + c(\overline{a}+b) + a + \overline{b}$ $\qquad$ DeMorgan's theorem

$$= \overline{a}b + \overline{a} \cdot c(c+\overline{b}) + c(\overline{a}+b) + a + \overline{b} \qquad \text{DeMorgan's theorem}$$

$$= \overline{a}b + \overline{a} \cdot c \cdot c + \overline{a}c\overline{b} + c(\overline{a}+b) + a + \overline{b} \qquad x(y+z)=xy+xz$$

$$* \quad = \overline{a}b + \overline{a}c + a\overline{b}c + c(\overline{a}+b) + a + \overline{b} \qquad x \cdot x = x$$

$$= \overline{a}b + c(\overline{a}+\overline{b}a) + c(\overline{a}+b) + a + \overline{b} \qquad xy+xz=x(y+z)$$

$$= \overline{a}b + c(\overline{a}+\overline{b}) + c(\overline{a}+b) + a + \overline{b} \qquad x+\overline{x}y = x+y$$

$$= \overline{a}b + c[(\overline{a}+\overline{b}) + (\overline{a}+b)] + a + \overline{b} \qquad xy+xz = x(y+z)$$

$$= \overline{a}b + c[(\overline{a}+\overline{a}) + (\overline{b}+b)] + a + \overline{b} \qquad \begin{array}{l} x+y=y+x \quad \text{and} \\ (x+y)+z=x+(y+z) \end{array}$$

$$= \overline{a}b + c[\overline{a} + (\overline{b}+b)] + a + \overline{b} \qquad x + x = x$$

$$= \overline{a}b + c[\overline{a} + 1] + a + \overline{b} \qquad x + \overline{x} = 1$$

$$= \overline{a}b + c[1] + a + \overline{b} \qquad x + 1 = 1$$

$$** \quad = \overline{a}b + c + a + \overline{b} \qquad x \cdot 1 = x$$

$$= \overline{a}b + c + (a+\overline{b}) \qquad (x+y)+z=x+(y+z)$$

$$= \overline{a}b + c + \overline{\overline{(a+\overline{b})}} \qquad x = \overline{\overline{x}}$$

$$= \overline{a}b + c + (\overline{\overline{a}b}) \qquad \text{DeMorgan's theorem}$$

$$= [(\overline{a}b) + \overline{(\overline{a}b)}] + c \qquad \begin{array}{l} x+y=y+x, (x+y)+z= \\ x+(y+z) \end{array}$$

$$= 1 + c = 1 \qquad x + \overline{x} = 1$$

$$= 1 \qquad 1 + x = 1$$

Note that, in general, the method used for simplifying an expression is not unique. For example, from step 7 of the above simplification (indicated by *), one may continue as the following:

Step 7

$$* \quad = \overline{a}b + \overline{a}c + a\overline{b}c + c(\overline{a}+b) + a + \overline{b}$$

$$= \overline{a}b + \overline{a}c + a(\overline{b}c+1) + c(\overline{a}+b) + \overline{b} \qquad \begin{array}{l} x+y=y+x, xy+xz= \\ x(y+z) \end{array}$$

$$= \overline{a}b + \overline{a}c + a \cdot 1 + c(\overline{a}+b) + \overline{b} \qquad x + 1 = 1$$

$$= \overline{a}b + \overline{a}c + a + c(\overline{a}+b) + \overline{b} \qquad x \cdot 1 = x$$

$$= (\overline{a}b+a) + \overline{a}c + c(\overline{a}+b) + \overline{b} \qquad \begin{array}{l} x+y=y+x, (x+y)+z = \\ x+(y+z) \end{array}$$

$$= b + a + \overline{a}c + c(\overline{a}+b) + \overline{b} \qquad x + \overline{x}y = x + y$$

729

$$= (b+\overline{b}) + a + \overline{a}c + c(\overline{a}+b) \qquad\qquad x+y=y+x,(x+y)+z=$$
$$\qquad\qquad\qquad\qquad\qquad\qquad\qquad\qquad x+(y+z)$$

$$= 1 + a + \overline{a}c + c(\overline{a}+b) \cdot \qquad\qquad x + \overline{x} = 1$$

$$= 1 \qquad\qquad\qquad\qquad\qquad\qquad\qquad 1 + x = 1$$

Also from step 15, indicated by "**", one can continue as the following:

$$** = \overline{a}b + c + a + \overline{b}$$

$$= (\overline{a}b+a) + c + \overline{b}$$

$$= a + b + c + \overline{b} = a + c + (b+\overline{b})$$

$$= a + c + 1 = 1$$

• **PROBLEM** 10-12

(1) Simplify $x(xy + \overline{x}zy + ypqx + \overline{x} + xpy) + x\overline{p}y$.

(2) Find an equivalent expression for
$xy + \overline{x}z + (x+p)q + (\overline{x}+p)r$ such that the expression has at most only one occurence of x and one occurrence of $\overline{x}$.

Solution: Theorem:

$$X \cdot f(X,\overline{X},Y...Z) = X \cdot f(1,0,Y...Z)$$

$$X + f(X,\overline{X},Y...Z) = X + f(0,1,Y...Z)$$

(1) The above theorem states that for any Boolean expression that can be expressed as a product of a variable X and a function f of variables: X, $\overline{X}$, Y...Z, one can replace all X's and $\overline{X}$'s in f by 1's and 0's respectively. Furthermore, if an expression can be expressed as a sum of a variable X and a function f of variables X, $\overline{X}$, Y...Z, all X's and $\overline{X}$'s in f can be replaced by 0's and 1's respectively. Hence, applying the theorem to the given expression, one obtains:

$$x(xy + \overline{x}zy + ypqx + \overline{x} + xpy) + xpyz$$

$$= x(1 \cdot y + 0 \cdot zy + ypq \cdot 1 + 0 + 1 \cdot py) + xpyz$$

$$= x(y + 0 + ypq + py) + xpyz \qquad\qquad 1 \cdot A=A,\ 0 \cdot A=0,\ 0+A=A$$

$$= xy(1 + pq + p) + xpyz \qquad\qquad AB+AC = A(B+C)$$

$$= xy + xypz \qquad\qquad\qquad\qquad 1 + A = 1 \quad A \cdot 1 = A$$

$$= xy(1 + pz) \qquad\qquad\qquad\qquad AB+AC = A(B+C)$$

$$= xy \cdot 1 \qquad\qquad\qquad\qquad\qquad 1 + A = 1$$

$$= xy \qquad\qquad\qquad 1 \cdot A = A$$

(2) Since $A + \overline{A} = 1$ and $1 \cdot A = A$ one can have

$$xy + \overline{x}z + (x+p)q + (\overline{x}+p)r$$

$$= (x+\overline{x}) \cdot (xy + \overline{x}z + (x+p)q + (\overline{x}+p)r)$$

$$= x(xy + \overline{x}z + (x+p)q + (\overline{x}+p)r) + \overline{x}(xy + \overline{x}z + (x+p)q + (\overline{x}+p)r)$$

$$\qquad\qquad\qquad\qquad\qquad A(B+C) = AB+AC$$

$$= x(1 \cdot y + 0 \cdot z + (1+p)q + (0+p)r)$$

$$\qquad + \overline{x}(0 \cdot y + 1 \cdot z + (0+p)q + (1+p)r)$$

$$\qquad\qquad\qquad\qquad\qquad X \cdot f(X,\overline{X},Y.,.Z) =$$

$$\qquad\qquad\qquad\qquad\qquad X \cdot f(1,0,Y..,Z)$$

$$= x(y + 1 \cdot q + pr) + \overline{x}(z + pq + 1 \cdot r)$$

$$= x(y + q + pr) + \overline{x}(z + pq + r)$$

Observe that this expression is equivalent to the original expression and it has only one occurence of x and $\overline{x}$.

● **PROBLEM** 10-13

Simplify the following expressions:

(1) $A = x\overline{y}z + y + yz + \overline{y}\overline{z} \cdot 0 + \overline{x}z$

(2) $B = x\overline{y}z + xy\overline{z} + \overline{y}z + (\overline{xy+\overline{x}z}) + \overline{y}\,\overline{z}$

Solution: (1) $A = x\overline{y}z + y + yz + \overline{y}\overline{z} \cdot 0 + \overline{x}z$

$$= x\overline{y}z + y + yz + \overline{x}z \qquad\qquad A \cdot 0 = 0, \ A + 0 = A$$

$$= x\overline{y}z + y + \overline{x}z \qquad\qquad\qquad A + AB = A$$

$$= xz + y + \overline{x}z \qquad\qquad\qquad A + \overline{A}B = A + B$$

$$= z(x + \overline{x}) + y \qquad\qquad\qquad AB + AC = A(B+C)$$

$$= z \cdot 1 + y \qquad\qquad\qquad\qquad \overline{A} + A = 1$$

$$= z + y \qquad\qquad\qquad\qquad\quad 1 \cdot A = A$$

(2) $B = x\overline{y}z + xy\overline{z} + \overline{y}z + (\overline{xy+\overline{x}z}) + \overline{y}\,\overline{z}$

$$= x\overline{y}z + xy\overline{z} + \overline{y}z + (\overline{x}+\overline{y}+\overline{x}+\overline{z}) + \overline{y}\,\overline{z} \qquad \text{DeMorgan's Theorem}$$

$$= x\overline{y}z + xy\overline{z} + \overline{y}z + (\overline{x}+\overline{y}+\overline{z}) + \overline{y}\,\overline{z} \qquad\qquad A + A = A$$

$$= x\overline{y}z + xy\overline{z} + \overline{y}z + xyz + \overline{y}\ \overline{z}$$

$$= xz(\overline{y}+y) + xy\overline{z} + \overline{y}z + \overline{y}\ \overline{z}$$

$$= xz + xy\overline{z} + \overline{y}z + \overline{y}\ \overline{z}$$

$$= x(z+y\overline{z}) + \overline{y}z + \overline{y}\ \overline{z}$$

$$= x(z+y) + \overline{y}z + \overline{y}\ \overline{z}$$

$$= xz + xy + \overline{y}z + \overline{y}\ \overline{z}$$

$$= xy + \overline{y}z + \overline{y}\ \overline{z}$$

$$= xy + \overline{y}(z+\overline{z})$$

$$= xy + \overline{y}\cdot 1$$

$$= xy + \overline{y}$$

$$= x + \overline{y}$$

DeMorgan's Theorem

$A+B=B+A$, $AB+AC = A(B+C)$

$A+\overline{A} = 1$, $A\cdot 1 = A$

$AB+AC = A(B+C)$

$A+\overline{A}B = A+B$

$A(B+C) = AB+AC$

$AB+\overline{A}C+BC = AB+\overline{A}C$

$AB+AC = A(B+C)$

$A + \overline{A} = 1$

$A \cdot 1 = A$

$A + \overline{A}B = A+B$

• **PROBLEM** 10-14

Simplify the following expressions

a) $A = ST + VW + RST$

b) $A = TUV + XY + Y$

c) $A = F(E + F + G)$

d) $A = \overline{(PQ + R + ST)}TS$

e) $A = \overline{D}\ \overline{D}\ E$

f) $A = Y(W + X + \overline{Y} + \overline{Z})Z$

g) $A = (BE + C + F)C$

Solution: We need the following laws to simplify the expressions:

Idempotent law : $\qquad AA = A$
$\qquad\qquad\qquad\qquad A + A = A$

Distributive law : $\qquad A(B+C) = AB + AC$
$\qquad\qquad\qquad\qquad A + BC = (A+B)(A+C)$

Law of Absorption : $\qquad A(A+B) = A$
$\qquad\qquad\qquad\qquad A + AB = A$

DeMorgan's law : $\qquad \overline{AB} = \overline{A} + \overline{B}$
$\qquad\qquad\qquad\qquad \overline{A+B} = \overline{A}\ \overline{B}$

a) $A = ST + VW + RST$ is equal to $ST + RST + VW$ by the associative law.
$A = ST(1 + R) + VW$ by distributive law but $1 + R = 1$ by law of
union. So,

$$A = ST + VW$$

b) $A = TUV + XY + Y$

Use the law of absorption on XY and Y.

$$A = TUV + Y$$

c) $A = F(E + F + G)$

Use the distributive law,

$$A = FE + FF + FG$$

Use the idempotent law on FF.

$$A = FE + F + FG$$

Use the law of absorption.

$$A = F$$

d) $A = (PQ + R + ST)TS$

Use the distributive law

$$A = TSPQ + TSR + STTS$$

Use the idempotent law on STTS.

$$A = TSPQ + TSR + TS$$

Use the law of absorption.

$$A = TS$$

e) $A = \overline{\bar{D} \bar{D} E}$

Use DeMorgan's law

$$A = D + D + \bar{E}$$

Use the idempotent law on D + D.

$$A = D + \bar{E}$$

f) $A = Y(W + X + \overline{\bar{Y} + \bar{Z}})Z$

Use DeMorgan's law on $\overline{\bar{Y} + \bar{Z}}$.

$A = Y(W + X + YZ)Z$

Use the distributive law

$$A = YZW + YZX + YZYZ$$

Use the idempotent law on YZYZ.

$$A = YZW + YZX + YZ$$

Use the law of absorption

$$A = YZ$$

g) $A = (BE + C + F)C$

Use the distributive law

$$A = CBE + CC + CF$$

Use the idempotent law

$$A = CBE + C + CF$$

Use the law of absorption

$$A = C$$

Although the law of absorption enables us to solve problems somewhat faster, it generally takes the new reader some time to familiarize himself with it. In such cases, even though it may take an extra step or two, you will certainly find that it is less confusing to use the Law of distribution.

● **PROBLEM** 10-15

Simplify the following Boolean expression:

$(a+b)(c+a+\overline{b})(\overline{b}+c)(acb+\overline{a}\,\overline{b}) + abcd + \overline{c} + \overline{b}$

Solution:

$(a+b)(c+a+\overline{b})(\overline{b}+c)(acb+\overline{a}\,\overline{b}) + abcd + \overline{c} + \overline{b}$

$= (a+b)(a+c)(\overline{b}+c)(acb+\overline{a}\,\overline{b})$ $\quad + abcd + \overline{c} + \overline{b}$	Theorem used: $(X+Y)(\overline{X}+Y+Z)$ $\quad = (X+Y)(Y+Z)$
$= (a+b)(\overline{b}+c)(acb+\overline{a}\,\overline{b})$ $\quad + abcd + \overline{c} + \overline{b}$	$(X+Y)(\overline{X}+Z)(Y+Z)$ $\quad = (X+Y)(\overline{X}+Z)$
$= (a+b)(\overline{b}+c)[(a+\overline{b})(\overline{a}+bc)]$ $\quad + abcd + \overline{c} + \overline{b}$	$XY + \overline{X}Z$ $\quad = (X+Z)(\overline{X}+Y)$
$= (a+b)(\overline{b}+c)(a+\overline{b})(\overline{a}+bc) + abcd + \overline{c} + \overline{b}$	
$= a(\overline{b}+c)(\overline{a}+bc) + abcd + \overline{c} + \overline{b}$	$(X+Y)(X+\overline{Y}) = X$
$= a(\overline{a}\,\overline{b} +\overline{b}bc+c\overline{a}+bc\cdot c) + abcd + \overline{c} + \overline{b}$	$X(Y+Z) = XY+XZ$
$= a(\overline{a}\,\overline{b} +0+c\overline{a}+bc) + abcd + \overline{c} + \overline{b}$	$X \cdot \overline{X} = 0$
$= a(\overline{a}\,\overline{b} +bc+\overline{a}c) + abcd + \overline{c} + \overline{b}$	$X + 0 = X$
$= a(\overline{a}\,\overline{b} +bc) + abcd + \overline{c} + \overline{b}$	$XY+\overline{X}Z+YZ = XY+\overline{X}Z$
$= a\overline{a}\,\overline{b} +abc + abcd + \overline{c} + \overline{b}$	$X(Y+Z) = XY+XZ$
$= 0 + abc + abcd + \overline{c} + \overline{b}$	$X \cdot \overline{X} = 0$ $0 \cdot X = 0$
$= abc + abcd + \overline{c} + \overline{b}$	$X + 0 = X$
$= abc + \overline{c} + \overline{b}$	$X+XY = X$
$= ab + \overline{c} + \overline{b}$	$X+\overline{X}Y = X+Y$
$= a + \overline{c} + \overline{b}$	$X+\overline{X}Y = X+Y$

Simplify the following expressions:

(1) $E_1 = (x+y)(x+\overline{z}+\overline{y})(xyz+\overline{x}\ \overline{y}+z)$

(2) $E_2 = (x+y+z)(\overline{x}+\overline{y})(z+y)x\overline{y}z$

Solution:

(1) $E_1 = (x+y)(x+\overline{z}+\overline{y})(xyz+\overline{x}\ \overline{y}+z)$

$= (x+y)(x+\overline{z}+\overline{y})(\overline{x}\ \overline{y}+z)$ $X+XY = X$

$= (x+y)(x+\overline{z})(\overline{x}\ \overline{y}+z)$ $(X+Y)(X+\overline{Y}+Z)$
$= (X+Y)(X+Z)$

$= (x+y)(x+\overline{z})(z+\overline{x})(z+\overline{y})$ $X+YZ = (X+Y)(X+Z)$

$= (x+y)(x+\overline{z})(z+\overline{x})(z+\overline{y})(x+z)$ $(X+Y)(\overline{X}+Z)$
$= (X+Y)(\overline{X}+Z)(Y+Z)$

$= (x+y)x(z+\overline{x})(z+\overline{y})$ $(X+Y)(X+\overline{Y}) = X$

$= x(z+\overline{x})(z+\overline{y})$ $X(X+Y) = X$

$= xz(z+\overline{y})$ $X(\overline{X}+Y) = XY$

$= x[z(z+\overline{y})]$ $XYZ = (XY)Z = X(YZ)$

$= x(z) = xz$ $X(X+Y) = X$

Note E_1, as most Boolean expressions, can be simplified in more than one way. For example, E_1 may also be simplified as follows:

$E_1 = (x+y)(x+\overline{z}+\overline{y})(xyz+\overline{x}\ \overline{y}+z)$

$= (x\cdot x+x\overline{z}+x\cdot\overline{y}+xy+\overline{z}\cdot y+\overline{y}\cdot y)(xyz+\overline{x}\ \overline{y}+z)$

$= (x+x\overline{z}+x\overline{y}+xy+y\overline{z}+0)(xyz+\overline{x}\ \overline{y}+z)$

$= (x+x\overline{y}+xy+y\overline{z})(xyz+\overline{x}\ \overline{y}+z)$ $X+0 = X, \ X+XY = X$

$= (x+xy+y\overline{z})(xyz+\overline{x}\ \overline{y}+z)$ $X+XY = X$

$= (x+y\overline{z})(xyz+\overline{x}\ \overline{y}+z)$ $X+XY = X$

$= (x+y\overline{z})(\overline{x}\ \overline{y}+z)$ $X+XY = X$

$= x\cdot\overline{x}y + xz + y\overline{z}\cdot\overline{x}\ \overline{y} + y\overline{z}\cdot z$

$= 0 + xz + 0 + 0$ $X\cdot\overline{X} = 0, \ 0\cdot X = 0$

$= xz$ $X+0 = X$

(2) $E_2 = (x+y+z)(\overline{x}+\overline{y})(z+y)x\overline{y}z$

$\qquad = (\overline{x}+\overline{y})(z+y)x\overline{y}z \qquad\qquad X(X+Y) = X$

$\qquad = (\overline{x}+\overline{y})x\overline{y}z \qquad\qquad\qquad X(X+Y) = X$

$\qquad = x\overline{y}z \qquad\qquad\qquad\qquad\quad X(X+Y) = X$

● **PROBLEM** 10-17

Simplify the following Boolean function:

$f(a,b,c,d) = (a+b+c) + \overline{ab} + cd + (\overline{abc+\overline{dc}}) + \overline{a+b} + abcd(a+\overline{a}\,\overline{b})$

Solution:

$f(a,b,c,d)$

$= (a+b+c) + \overline{ab} + cd + (\overline{abc+\overline{dc}}) + (\overline{a+b}) + abcd(a+\overline{a}\,\overline{b})$

$= a+b+c + \overline{ab} + cd + (\overline{abc+\overline{dc}}) + (\overline{a+b}) + abcd(a+\overline{b}\,\overline{a})$

$\qquad\qquad\qquad\qquad\qquad\qquad\qquad\qquad\qquad X+(Y+Z)=X+Y+Z$

$= a+b+c + cd + (\overline{abc+\overline{dc}}) + (\overline{a+b}) + abcd(a+\overline{b}\,\overline{a})$

$\qquad\qquad\qquad\qquad\qquad\qquad\qquad\qquad\qquad X+XY = X$

$= a+b+c + (\overline{abc+\overline{dc}}) + (\overline{a+b}) + abcd(a+\overline{b}\,\overline{a})$

$\qquad\qquad\qquad\qquad\qquad\qquad\qquad\qquad\qquad X+XY = X$

$= a+b+c + (\overline{abc+\overline{dc}}) + (\overline{a+b}) \qquad\qquad X+XY = X$

$\qquad\qquad\qquad\qquad\qquad\qquad\qquad$ treat a as X and
$\qquad\qquad\qquad\qquad\qquad\qquad\qquad$ bcd(a+$\overline{b}\,\overline{a}$) as Y

$= a+b+c + (\overline{abc})\cdot(\overline{\overline{dc}}) + \overline{a}\cdot\overline{b} \qquad$ DeMorgan's theorem

$= a+b+c + (\overline{a}+\overline{b}+\overline{c})dc + \overline{a}\cdot\overline{b} \qquad$ DeMorgan's theorem

$= a+b+c + \overline{a}\cdot\overline{b} \qquad\qquad\qquad\qquad X+XY = X$

$\qquad\qquad\qquad\qquad\qquad\qquad\qquad$ treat c as X
$\qquad\qquad\qquad\qquad\qquad\qquad\qquad$ ($\overline{a}+\overline{b}+\overline{c}$)d as Y

$= a+b+c+\overline{b} \qquad\qquad\qquad\qquad\qquad X+\overline{X}Y = X+Y$

$= a+c + b+\overline{b} \qquad\qquad\qquad\qquad\qquad X+Y = Y+X$

$= a + c + 1 \qquad\qquad\qquad\qquad\qquad\qquad X+\overline{X} = 1$

$= 1 \qquad\qquad\qquad\qquad\qquad\qquad\qquad\quad X+1 = 1$

Given the following Boolean functions:

(a) $f_1(a,b,c,d) = (a+c\overline{d})(c+a\overline{b})$

(b) $f_2(a,b,c,d) = a\overline{b}c + ab\overline{c}d + \overline{a}b\overline{c}d + \overline{a}\,\overline{d}$

Find the expanded product of sums for f_1 and f_2.

Solution: By definition the expanded product of sums (EPS) is an expression in which all the variables of the given Boolean function appear in every product term of the given function.

(a) If the given function is not expressed as a sum of products or a product of sums, rearrange the given function so that it is expressed as a sum of products or a product of sums (depending on which form is easier to obtain).

Thus, $f_1(a,b,c,d) = (a+c\overline{d})(c+a\overline{b})$

$$= (a+\overline{c}+d)(c+a\overline{b}) \qquad \text{DeMorgan's theorem}$$

$$= (a+\overline{c}+d)(c+a)(c+\overline{b}) \qquad X+YZ=(X+Y)(X+Z)$$

There are several methods by which one can obtain the expanded product of sums.

Method 1: Use the K-map. The K-map of four variables is shown in Fig. 1.

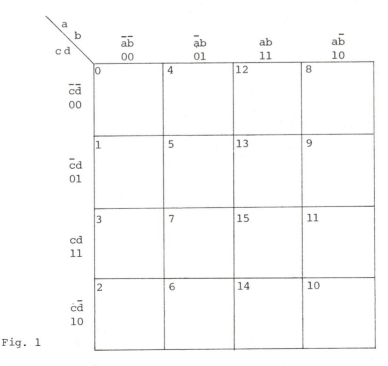

Fig. 1

In the K-map each box represents a sum or a product of four variables. If one wants a box to represent a sum, that box will be filled with a 0, otherwise it's filled with a 1. For example, box 4 can represent $\bar{a}b\bar{c}\bar{d}$ or $a+\bar{b}+c+d$. Note that $\overline{\bar{a}b\bar{c}\bar{d}} = \overline{a+\bar{b}+c+d}$ and $a+\bar{b}+c+d=\overline{\bar{a}b\bar{c}\bar{d}}$. This is true in general, i.e. a box can represent a sum of n variables or its complement and vice versa. Note also that $\bar{a}b\bar{c}\bar{d}$ may be expressed as 0100 which is the binary number for 4, and that is precisely how one numbers the boxes in a K-map.

The K-map for $f_1(a,b,c,d) = (a+\bar{c}+d)(a+c)(\bar{b}+c)$ is shown in Fig. 2. It is obtained by entering a 0 to every box that includes the terms in f. For instance, term $(a+c)$ is in f and it is included in $a+b+c+d$, $a+b+c+\bar{d}$, $a+\bar{b}+c+d$ and $a+\bar{b}+c+\bar{d}$ which are boxes 0, 1, 4 and 5 respectively.

The expanded product of sums of f is then the product of all sums represented by the boxes with zeros.

$f_1(a,b,c,d) = \Pi(0,1,2,4,5,6,12,13)$

$\qquad = (a+b+c+d)(a+b+c+\bar{d})(a+b+\bar{c}+d)$

$\qquad (a+\bar{b}+c+d)(a+\bar{b}+c+\bar{d})(a+\bar{b}+\bar{c}+d)$

$\qquad (\bar{a}+\bar{b}+c+d)(\bar{a}+\bar{b}+c+\bar{d})$

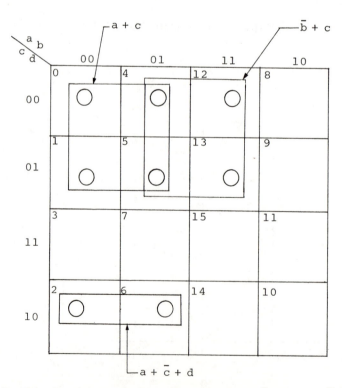

Fig. 2 This is the K-map for $f_1(a,b,c,d)=(a+\bar{c}+d)(a+c)(\bar{b}+c)$. It is also the K-map for the expanded product of sums of f_1.

Method 2:

$$f_1(a,b,c,d) = (a+\bar{c}+d)(c+a)(c+\bar{b})$$

$$= (a+\bar{c}+d+b\bar{b})(c+a)(c+\bar{b}) \qquad X\bar{X}=0,\ 0+X=X$$

$$= [(a+b+\bar{c}+d)(a+\bar{b}+\bar{c}+d)](c+a)(c+\bar{b})$$
$$\qquad\qquad\qquad\qquad\qquad\qquad X+YZ=(X+Y)(X+Z)$$

$$= (a+b+\bar{c}+d)(a+\bar{b}+\bar{c}+d)(a+b\bar{b}+c+d\bar{d})(a\bar{a}+\bar{b}+c+d\bar{d})$$

$$= (a+b+\bar{c}+d)(a+\bar{b}+\bar{c}+d)(a+b+c+d)(a+b+c+\bar{d})$$

$$\quad (a+\bar{b}+c+d)(a+\bar{b}+c+\bar{d})(a+\bar{b}+c+d)(\bar{a}+\bar{b}+c+d)$$

$$\quad (a+\bar{b}+c+\bar{d})(\bar{a}+\bar{b}+c+\bar{d})$$

$$= (a+b+c+d)(a+b+c+\bar{d})(a+\bar{b}+c+d)(a+\bar{b}+c+\bar{d})$$

$$\quad (a+b+\bar{c}+d)(\bar{a}+\bar{b}+c+d)(\bar{a}+\bar{b}+c+\bar{d})(a+\bar{b}+\bar{c}+d)$$

(b) Find the K-map of the given function; it is shown in
Fig. 3.

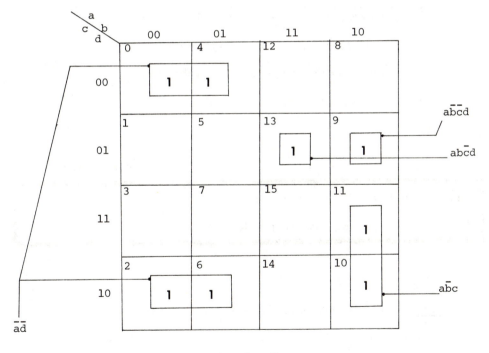

Fig. 3   The K-map for f_2.

Let the sum of the products of four variables that are represented by the empty boxes in the K-map in Fig. 3 be f_2^*

$$f_2^*\ (a,b,c,d) = \Sigma(1,3,5,7,8,12,14,15).$$

f_2^* is the complement of f_2.

739

A function of n variables can have 2^n different combinations of n variables. Let f_0 be a function of n variables and let f_0' be the expanded sum of products (or the expanded product of sums) of f_0^*. Then, if f_0' contains less that 2^n terms, the sum (or product, if f_0' is EPS) of all the missing combinations is f_0, which is the complement of f_0'.

The complement of $f_0^* = (\overline{f_0'}) = f_0' = f_0$ and is in the form of an expanded product of sums (or an expanded sum of products if f_0' is an EPS).

Therefore, for this problem,

$$\overline{f_2^* (a,b,c,d)} = \overline{\Sigma(1,3,5,7,8,12,14,15)}$$

$$= (\overline{\overline{f_2}}) = f_2$$

Since f_2^* is a sum of products, $(\overline{f_2^*})$ is a product of sums and is equal to f_2. Hence, the expanded product of sums of the given function is

$$\overline{f_2^* (a,b,c,d)} = \overline{\Sigma(1,3,5,7,8,12,14,15)}$$

$$= \overline{\overline{a}\,\overline{b}\,\overline{c}\,d + \overline{a}\,\overline{b}\,c\,d + \overline{a}\,b\,\overline{c}\,d + \overline{a}\,b\,c\,d + a\,\overline{b}\,\overline{c}\,\overline{d} + a\,b\,\overline{c}\,\overline{d} + a\,b\,c\,\overline{d} + a\,b\,c\,d}$$

$$= (a+b+c+\overline{d})(a+b+\overline{c}+\overline{d})(a+\overline{b}+c+\overline{d})(a+\overline{b}+\overline{c}+\overline{d})$$

$$(\overline{a}+b+c+d)(\overline{a}+\overline{b}+c+d)(\overline{a}+\overline{b}+\overline{c}+d)(\overline{a}+\overline{b}+\overline{c}+\overline{d})$$

(by DeMorgan's theorem)

● **PROBLEM** 10-19

Given $f(a,b,c,d) = (a+b\overline{c})(\overline{a}+c)$.

Find the expanded sum of products of f.

Solution:

Method 1: First express f as a sum of products.

$$f(a,b,c,d) = (a+b\overline{c})(\overline{a}+c)$$

$$= a\overline{a} + ac + \overline{a}b\overline{c} + b\overline{c}c \qquad\qquad X(Y+Z)=XY+XZ$$

$$= ac + \overline{a}b\overline{c} \qquad\qquad X\cdot0=0,\ X\cdot\overline{X}=0, X+0=X$$

Make up the missing variables in each term as follows:

$$f(a,b,c,d) = ac + \overline{a}b\overline{c}$$

$$= ac(b+\overline{b}) + \overline{a}b\overline{c}(d+\overline{d}) \qquad\qquad X+\overline{X}=1,\ 1\cdot X = X$$

740

$$= acb + ac\overline{b} + \overline{a}b\overline{c}d + \overline{a}b\overline{c}\,\overline{d} \qquad\qquad X(Y+Z)=XY+XZ$$

$$= acb(d+\overline{d}) + ac\overline{b}(d+\overline{d}) + \overline{a}b\overline{c}d + \overline{a}b\overline{c}\,\overline{d}$$

$$f(a,b,c,d) = acbd + acb\overline{d} + ac\overline{b}d + ac\overline{b}\,\overline{d} + \overline{a}b\overline{c}d + \overline{a}b\overline{c}\,\overline{d},$$

and this is the expanded sum of products of the given function.

<u>Method 2</u>: After expressing f as a sum of products, use a K-map to find the expanded sum of products of f.

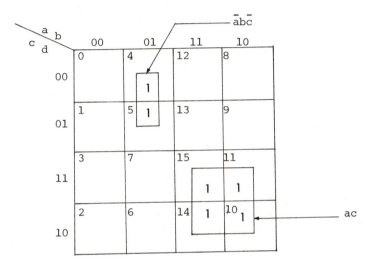

Fig. 1 $f(a,b,c,d)=ac+\overline{a}b\overline{c}$

    The K-map for f is given in Fig. 1; it is obtained by filling in a 1 for each box that includes the terms in f. For instance, term ac is included in terms abcd, $\overline{a}$bcd, abc$\overline{d}$ and a$\overline{b}$c$\overline{d}$, hence, boxes 10, 11, 14, and 15, which represent the above four terms, are filled with 1's.

    The expanded sum of products of $f(a,b,c,d) = ac + \overline{a}b\overline{c}$ is the sum of all the terms represented by the boxes with 1's.

That is, $f(a,b,c,d) = \Sigma(4,5,10,11,14,15)$

$$= \overline{a}b\,\overline{c}\,\overline{d} + \overline{a}b\,\overline{c}d + abcd + a\overline{b}cd + abc\overline{d} + a\overline{b}c\overline{d}$$

<u>Method 3</u>: Express f as a product of sums;

$$f(a,b,c,d)$$

$$= (a+b\overline{c})(\overline{a}+c)$$

$$= (a+b)(a+\overline{c})(\overline{a}+c) \qquad\qquad X+YZ=(X+Y)(X+Z)$$

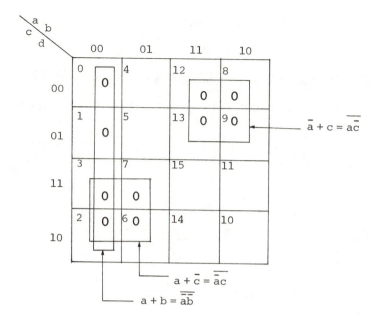

Fig. 2   K-map of $f(a,b,c,d)=(a+b)(a+\bar{c})(\bar{a}+c)$

Find the K-map of f as shown in Fig. 2.   Each box with a zero represents a sum of the four variables, which is the complement of the product represented by the same box. For example, box 1 represents $a\bar{b}\bar{c}d$ if it contains a 1; its complement is

$$\overline{(a\bar{b}\bar{c}d)} = a+b+c+\bar{d} \text{ (by DeMorgan's theorem).}$$

$a+b+c+\bar{d}$ is then represented by box 1 when it contains a 0.   Note that the product of all the sums that are represented by the boxes with zeros gives the expanded product of sums of f.

Let the product of sums which are represented by the empty boxes in the K-map in Fig. 2 be f*.

$$f*(a,b,c,d) = \Pi(4,5,10,11,14,15)$$

$$= (a+\bar{b}+c+d)(a+\bar{b}+c+\bar{d})(\bar{a}+b+\bar{c}+d)(\bar{a}+b+\bar{c}+\bar{d})$$

$$(\bar{a}+b+\bar{c}+d)(\bar{a}+b+\bar{c}+\bar{d})$$

$$(\bar{a}+\bar{b}+\bar{c}+d)(\bar{a}+\bar{b}+\bar{c}+\bar{d})$$

f* is the complement of f, i.e.   f* = $\bar{f}$, therefore,

$$\overline{(f*)} = \overline{(\bar{f})} = f.$$

That is,

$$f = \bar{\bar{f}} = \overline{f*(a,b,c,d)} = \overline{(a+\bar{b}+c+d)(a+\bar{b}+c+\bar{d})(\bar{a}+b+\bar{c}+d)(\bar{a}+b+\bar{c}+\bar{d})}$$

$$\overline{(\bar{a}+\bar{b}+\bar{c}+d)(\bar{a}+\bar{b}+\bar{c}+\bar{d})}$$

$$= \bar{a}\,b\,\bar{c}\,\bar{d} + \bar{a}\,b\,\bar{c}\,d + a\,\bar{b}\,c\,\bar{d} + a\,\bar{b}\,c\,d + a\,b\,c\,\bar{d} + a\,b\,c\,d$$

by DeMorgan's theorem.

Observe that this is the expanded sum of products of the given f.

Consider the function

$$f(x,y,z) = y + \bar{x}z.$$

Rewrite it so that it becomes an expression in which all three variables appear in each of the product terms.

Solution:

$$f(x,y,z) = y + \bar{x}z$$

$$= y(x+\bar{x}) + \bar{x}z(y+\bar{y}) \qquad A+\bar{A}=1, \ A \cdot 1 = A$$

$$= yx + y\bar{x} + \bar{x}yz + \bar{x}\,\bar{y}\,z$$

$$= xy(z+\bar{z}) + \bar{x}y(z+\bar{z}) + \bar{x}yz + \bar{x}\,\bar{y}\,z$$

$$= xyz + xy\bar{z} + \bar{x}yz + \bar{x}y\bar{z} + \bar{x}yz + \bar{x}\,\bar{y}\,z$$

$$= xyz + xy\bar{z} + \bar{x}yz + \bar{x}y\bar{z} + \bar{x}\,\bar{y}\,z$$

The function $f(x,y,z)$ is now in the standard sum-of-products form or the expanded sum-of-products form. Each of the products in $f(x,y,z)$ is called a minterm.

Given $f(x,y,z) = (\bar{x}+y)(\bar{z}+x+\bar{y})$ rewrite it so that it becomes an expression in which all three variables appear in each of the product terms.

Solution:

$$f(x,y,z) = (\bar{x}+y)(\bar{z}+x+\bar{y})$$

$$= (\bar{x}+y+z\bar{z})(\bar{z}+x+\bar{y}) \qquad A \cdot \bar{A} = 0, \ A+0 = A$$

$$= (\bar{x}+y+z)(\bar{x}+y+\bar{z})(\bar{z}+x+\bar{y}) \qquad A+BC = (A+B)(A+C)$$

The function $f(x,y,z)$ is now in the standard product-of-sums form or the expanded product-of-sums form in which each sum is called a maxterm.

Given the function $f(x,y,z)$ below, write $f(x,y,z)$ as a product of maxterms.

$$f(x,y,z) = (z + \bar{x})(y + \bar{z})(x + y + z)(\bar{x} + \bar{y})$$

Solution: If a switching function is specified in Product of Sum form, then it may be expanded to canonical PS form by repeatedly using the theorem given below.

THEOREM: $(a + b)(a + \bar{b}) = a$

Note: in the original function $f(x,y,z)$, the first, second, and the fourth terms can be changed using the theorem defined above. These changes will be;

$$(z + \bar{x}) = (\bar{x} + y + z)(\bar{x} + \bar{y} + z)$$
$$(y + \bar{z}) = (x + y + \bar{z})(\bar{x} + y + \bar{z})$$
$$(\bar{x} + \bar{y}) = (\bar{x} + \bar{y} + z)(\bar{x} + \bar{y} + \bar{z})$$

then the function $f(x,y,z)$ becomes;

$$f(x,y,z) = (\bar{x} + y + z)(\bar{x} + \bar{y} + \bar{z})(x + y + \bar{z})$$
$$\cdot (\bar{x} + y + \bar{z})(\bar{x} + \bar{y} + z)(\bar{x} + \bar{y} + \bar{z})$$
$$\cdot (x + y + z)$$

this is the desired final form of the function, now it can be translated into MAXTERM code by assigning 1's to false variables, and 0's to true variables. This procedure of coding MAXTERMS is illustrated below using the function $f(x,y,z)$;

$$f(x,y,z) = (\bar{x} + y + z)(\bar{x} + \bar{y} + \bar{z})(x + y + \bar{z})$$

$$1 \quad 0 \quad \ \ 0 \quad 1 \quad \ 1 \quad \ \ 1 \quad 0 \quad \ 0 \quad \ 1$$

$$(\bar{x} + y + \bar{z})(\bar{x} + \bar{y} + z)(\bar{x} + \bar{y} + \bar{z})$$

$$1 \quad \ \ 0 \quad \ 1 \quad \ \ 1 \quad \ \ 1 \quad 0 \quad \ 1 \quad \ 1 \quad \ 1$$

$$(x + y + z)$$

$$0 \quad \ 0 \quad \ 0$$

In decimal arithmetic each code group represents numbers such as 4,7,1,5,6,0, with the redundant 7 being used only once, and the solution can be written as;

$$f(x,y,z) = M_4 \ M_7 \ M_1 \ M_5 \ M_6 \ M_0$$

or

$$f(x,y,z) = \Pi \ M(0,1,4,5,6,7)$$

Given the function $f(x,y,z)$ below, write $f(x,y,z)$ as a sum of minterms.

$$f(x,y,z) = x\bar{z} + y\bar{z} + xyz$$

Solution: If a switching function is given in Sum of Products form, it may be expanded to the canonical Sum of Products through repeated use of the Theorem below.

THEOREM 1: $ab + a\bar{b} = a$

Therefore given the function $f(x,y,z) = x\bar{z} + y\bar{z} + xyz$, $f(x,y,z) = xy\bar{z} + x\bar{y}\bar{z} + y\bar{z} + xyz$ can be obtained, by using theorem 1, and saying $x\bar{z} = xy\bar{z} + x\bar{y}\bar{z}$. Using Theorem 1 a second time and making the equality $y\bar{z} = xy\bar{z} + \bar{x}y\bar{z}$, the function $f(x,y,z)$ becomes:

$$f(x,y,z) = xy\bar{z} + x\bar{y}\bar{z} + xy\bar{z} + \bar{x}y\bar{z} + xyz$$

At this point a second theorem may be used to reduce the number of terms present;

THEOREM 2: $a + a = a$

In the final form of the function it can easily be seen that the first and the third terms are in the same form, $xy\bar{z}$. Using Theorem 2 these terms can be reduced as follows;

$$xy\bar{z} + xy\bar{z} = xy\bar{z}$$

then the final form of the function is;

$$f(x,y,z) = xy\bar{z} + x\bar{y}\bar{z} + \bar{x}y\bar{z} + xyz .$$

After the desired reduced form is obtained where each term has the same number of variables, a special minterm code is used to CODE the function. Minterm code is basically giving a value of 1 for each true variable, and a value of 0 for each false variable. Taking this definition and applying it to the function gives;

$$xy\bar{z} + x\bar{y}\bar{z} + \bar{x}y\bar{z} + xyz$$

| 110 | 100 | 010 | 111 | (minterm code) |

In decimal arithmetic these terms represent the numbers 6,4,2, and 7. Therefore the desired Sum of Minterms form is written as;

$$f(x,y,z) = m_6 + m_4 + m_2 + m_7$$

or

$$f(x,y,z) = \Sigma m \ (2,4,6,7)$$

Given $f(x,y,z) = yz(\bar{x}+y) + (\bar{z}+x)(y+z)$; put f into

(1) the disjunctive normal form (dnf)

(2) the full disjunctive normal form

(3) the sum-of-products form

(4) the expanded sum-of-products.

Solution: (1) A Boolean expression is said to be in disjunctive normal form (dnf) if the expression is a fundamental product or the sum of two or more fundamental products of which none is included in another. A fundamental product is a product of two or more literals in which no two literals involve the same variable. Thus, applying the basic laws of Boolean algebra, the given function is put into the sum-of-products form as follows:

$$f(x,y,z) = yz(\bar{x}+y) + (\bar{z}+x)(y+z)$$

$$= yz\bar{x} + yzy + \bar{z}y + \bar{z}z + xy + xz$$

$$= yz\bar{x} + yz + y\bar{z} + 0 + xy + xz$$

$$= yz\bar{x} + yz + \bar{z}y + xy + xz$$

Next, using theorems A+A=A and A+AB=A, one obtains the dnf as follows:

$$f(x,y,z) = yz\bar{x} + yz + \bar{z}y + xy + xz$$

$$= yz + y\bar{z} + xy + xz$$

Any Boolean expression may be transformed into dnf by first writing it as a sum-of-products, and then eliminating all terms that can be eliminated by using theorems A+AB=A and A+A=A.

(2) The full disjunctive normal form is a dnf in which every fundamental product involves all the variables.

The full dnf is obtained from dnf by multiplying any fundamental product by $x + \bar{x}$ if x is missing from that product. Hence, the given f(x,y,z) is transformed into full dnf as follows:

$$f(x,y,z) = yz + y\bar{z} + xy + xz$$

$$= yz(x+\bar{x}) + y\bar{z}(x+\bar{x}) + xy(z+\bar{z}) + xz(y+\bar{y})$$

$$= xyz + y\bar{x}z + xy\bar{z} + \bar{x}y\bar{z} + xyz + xy\bar{z} + xyz + x\bar{y}z$$

Eliminating the repeated terms using A+A=A, one has the full dnf:

746

$$f(x,y,z) = xyz + xy\overline{z} + \overline{x}yz + \overline{x}y\overline{z} + x\overline{y}z$$

(3) The sum of product form of the given function can be easily obtained by using basic laws of Boolean algebra as follows:

$$f(x,y,z) = yz(\overline{x}+y) + (\overline{z}+x)(y+z)$$

$$= \overline{x}yz + yz + xy + xz + y\overline{z}$$

It can be further simplified to $f(x,y,z) = y+xz$.

(4) The expanded sum-of-products is the same as the full dnf for any Boolean expression.

Find the minimal dnf for the following Boolean function:

$$f(x,y,z) = x + \overline{xy}(z+x)(\overline{y}+\overline{z}) + (\overline{x+y+z})\overline{x} + yz$$

Solution:

$$f(x,y,z) = x + \overline{xy}(z+x)(\overline{y}+\overline{z}) + (\overline{x+y+z})\overline{x} + yz$$

$$= x + (\overline{x}+\overline{y})(z+x)(\overline{y}+\overline{z}) + \overline{x}\,\overline{y}\,\overline{z}\,\overline{x} + yz$$

$$= x + (\overline{x}x+\overline{x}z+\overline{y}z+\overline{y}x)(\overline{y}+\overline{z}) + \overline{x}\,\overline{y}\,\overline{z} + yz$$

$$= x + (0+\overline{x}z+\overline{y}z+\overline{y}x)(\overline{y}+\overline{z}) + \overline{x}\,\overline{y}\,\overline{z} + yz$$

$$= x + [\overline{x}z\overline{y}+\overline{y}z\overline{y}+\overline{y}x\overline{y}+\overline{z}\,\overline{x}\,z+\overline{y}z\cdot\overline{z}+\overline{y}x\overline{z}] + \overline{x}\,\overline{y}\,\overline{z} + yz$$

$$= x + [\overline{x}z\overline{y}+z\overline{y}+x\overline{y}+0+0+\overline{y}x\overline{z}] + \overline{x}\,\overline{y}\,\overline{z} + yz$$

$$= x + \overline{x}\,\overline{y}\,z + \overline{y}z + x\overline{y} + x\,\overline{y}\,\overline{z} + \overline{x}\,\overline{y}\,\overline{z} + yz$$

$$= x + \overline{y}z(\overline{x}+1) + x\overline{y}(1+\overline{z}) + \overline{x}\,\overline{y}\,\overline{z} + yz$$

$$= x + \overline{y}z\cdot 1 + x\overline{y}\cdot 1 + \overline{x}\,\overline{y}\,\overline{z} + yz$$

$$= x(1+\overline{y}) + \overline{y}z + yz + \overline{x}\,\overline{y}\,\overline{z}$$

$$= x\cdot 1 + z(\overline{y}+y) + \overline{x}\,\overline{y}\,\overline{z}$$

$$= x + z\cdot 1 + \overline{x}\,\overline{y}\,\overline{z}$$

$$= x + z + \overline{x}\,\overline{y}\,\overline{z}$$

$$= x + z + \overline{y}\,\overline{z}$$

$$= x + z + \overline{y}$$

It can be shown that the set of operations $S = \{ +, \cdot, \sim\}$ is functionally complete. That is, every Boolean function can be represented by a form $f(x_1,\ldots,x_n)$ in variables $x_1,\ldots,x_n$ and operations $+$, $\cdot$, $\sim$. Equivalently, the set of gates $S' = \{OR, AND, NOT\}$ is functionally complete. Show that:

(a) $S_1 = \{ +, \sim\}$

(b) $S_0 = \{ \cdot, \sim\}$

(c) $S_3 = \{\uparrow\}$ where $x_1 \uparrow x_2 = \sim(x_1 \cdot x_2)$ (1)

(d) $S_4 = \{\downarrow\}$ where $x_1 \downarrow x_2 = \sim(x_1 + x_2)$ (2)

are functionally complete.

<u>Solution</u>: (a) It is sufficient to show that we can construct the operation from S_1, since we know that $S = \{ +, \cdot, \sim\}$ is functionally complete. DeMorgan's law suggests itself here:

$$\sim(A \cdot B) = \sim A + \sim B \qquad\qquad 9_B$$

Applying the $\sim$ operator to both sides and using the involution law for the left side, we get:

$$A \cdot B = \sim [\sim (A \cdot B)] = \sim [\sim A + \sim B]$$

Hence, $S_1 = \{+, \sim\}$ is functionally complete. The gate representation of

$$A \cdot B = \sim [\sim A + \sim B]$$

is shown in figure 1

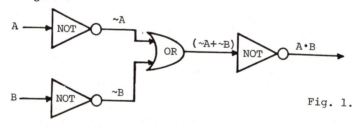

Fig. 1.

This means that every gating function can be implemented by a network of OR and NOT gates, by substituting the network of fig. 1 for any AND gate in the original OR-AND-NOT network.

(b) It is sufficient to show that we can construct the $+$ operation from S_2. But DeMorgan's law states that:

$$\sim(A + B) = \sim A \cdot \sim B$$

Applying the $\sim$ operator to both sides and using the involution law for the left side, we get:

$$A + B = \sim[\sim(A + B)] = \sim[\sim A \cdot \sim B]$$

Hence, $S_2 = \{ \cdot, \sim\}$ is functionally complete. The gate representation of

$$A + B = \sim[\sim A \cdot \sim B]$$

is shown in fig. 2 .

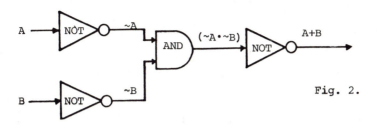

Fig. 2.

This means that every gating function can be implemented by a network of AND and NOT gates.

(c) The ↑ operator is called the Shaeffer stroke function and its associated gate is a NAND(for NOT-AND) gate. It is sufficient to show that we can construct $S_2 = \{ \cdot , \sim \}$ from the stroke function alone, since it has already been shown that S_2 is functionally complete. The idempotent law for $\cdot$ allows substitution of $x_1 \cdot x_1$ for x_1 so that:

$$\sim x_1 = \sim (x_1 \cdot x_1)$$

But from the definition of the stroke function, equation (1), we have:

$$\sim (x_1 \cdot x_1) = x_1 \uparrow x_1$$

Therefore, $\sim x_1 = x_1 \uparrow x_1$.   (3)

To obtain $\cdot$ from the stroke function, recall that the involution law allows us to write:

$$x_1 \cdot x_2 = \sim [\sim (x_1 \cdot x_2)]$$

But from the definition of the stroke function, equation (1), we have:

$$\sim [\sim (x_1 \cdot x_2)] = \sim [x_1 \uparrow x_2]$$

Using the result obtained in equation (3), we can write:

$$\sim [x_1 \uparrow x_2] = (x_1 \uparrow x_2) \uparrow (x_1 \uparrow x_2)$$

Therefore, $x_1 \cdot x_2 = (x_1 \uparrow x_2) \uparrow (x_1 \uparrow x_2)$ (4)

and S_3 is functionally complete. The gate representation of equation (3) is shown in fig. 3.

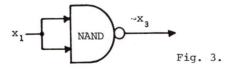

Fig. 3.

The gate representation of equation (4) is shown in fig. 4 .

x1
x2
NAND
x1↑x2
NAND
x1·x2

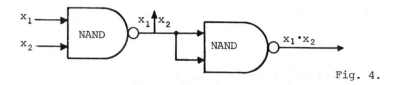

Fig. 4.

This means that every gating function can be implemented by a network of NAND gates only.

(d) The $\downarrow$ operator is called the NOR (from NOT-OR) function and its associated gate is a NOR gate. It is sufficient to show that we can construct the $S_1 = \{ + , \sim\}$ from the NOR function alone. Using the idempotent law for + and the definition of the NOR function, equation (2), we can write:

$$\begin{aligned} \sim x_1 &= \sim(x_1 + x_1) \\ &= x_1 \downarrow x_1 \end{aligned} \tag{5}$$

To obtain + from the NOR function, we apply the involution law and use the definition of the NOR function and equation (5):

$$\begin{aligned} x_1 + x_2 &= \sim[\sim(x_1 + x_2)] \\ &= \sim[x_1 \downarrow x_2] \\ &= [x_1 \downarrow x_2] \downarrow [x_1 \downarrow x_2] \end{aligned} \tag{6}$$

Hence, $S_4 = \{\downarrow\}$ is functionally complete. The gate representation of equation (5) is shown in fig. 5

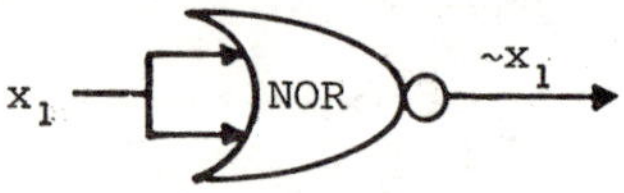

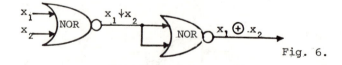

x1
NOR
~x1

Fig. 5.

The gate representation of equation (6) is shown in fig. 6

x1
x2
NOR
x1↓x2
NOR
x1 ⊕ x2

Fig. 6.

Every gating function can be implemented by a network of NOR gates only: Though, for simplicity, only two variables were used in showing the functional completeness of $S_1 - S_4$, the results can be shown to be true for any number of variables by a simple induction argument.

Find the prime implicants of the following Boolean expressions by using the consensus method.

(1) $xyz + \bar{x}y + \bar{z}y + \bar{x}\,\bar{y}$

(2) $(x+y)(y+z)(\bar{z}+\bar{x})(\bar{x}+\bar{y}+z)(x+y+z)$

Solution: The consensus method utilizes the following theorems:

Theorem 1: $XY + \bar{X}Z = XY + \bar{X}Z + YZ$

$(X+Y)(\bar{X}+Z) = (X+Y)(\bar{X}+Z)(Y+Z)$

Theorem 2: $X + X = X$, $X \cdot X = X$

Theorem 3: $X + XY = X$, $X(X+Y) = X$

The consensus method can be stated as the following:

Using theorem 2 and theorem 3, delete terms which include any other terms in a given Boolean expression. At the same time, apply theorem 3 systematically to all pairs of terms to obtain all possible included terms which are added to the expression. An included term that can be immediately eliminated by the use of theorem 2 or theorem 3 is not added. Continue this process until no more included terms can be formed, or until the only included terms that can be formed would be immediately eliminated by using theorem 2 or theorem 3. The final Boolean expression obtained is the sum of all the prime implicants of the given Boolean expression. Applying the consensus method, one has:

(1) $xyz + \bar{x}y + \bar{z}y + \bar{x}\,\bar{y}$

$= xyz + \bar{x}y + yz \cdot y + \bar{z}y + \bar{x}\,\bar{y}$

Consensus of $x \cdot (yz)$ and $\bar{x} \cdot y$ (theorem 1)

$= xyz + \bar{x}y + yz + \bar{z}y + \bar{x}\,\bar{y}$

$X \cdot X = X$

$= \bar{x}y + yz + \bar{z}y + \bar{x}\,\bar{y}$

$X + XY = X$
xyz includes yz

$= \bar{x}y + zy + \bar{z}y + y \cdot y + \bar{x}\,\bar{y}$

Consensus of zy and $\bar{z}y$ (theorem 1)

$= \bar{x}y + zy + \bar{z}y + y + \bar{x}\,\bar{y}$

$= y + \bar{x}\,\bar{y}$

$\bar{x}y$, zy and $\bar{z}y$ include y, (theorem 3)

$= y + \bar{x}\,\bar{y} + 1 \cdot \bar{x}$

Consensus of $y \cdot 1$ and $\bar{y} \cdot \bar{x}$, (theorem 1)

$= y + \bar{x}\,\bar{y} + \bar{x}$

$1 \cdot X = X$

$= y + \bar{x}$ (Theorem 3) $\bar{x}$ is included in $\bar{x}\,\bar{y}$

Thus, the prime implicants for

$xyz + \bar{x}y + y\bar{z} + \bar{x}\,\bar{y}$ are y and $\bar{x}$. That is,

$xyz + \bar{x}y + y\bar{z} + \bar{x}\,\bar{y} = \bar{x} + y$.

(2) $(x+y)(y+z)(\bar{z}+\bar{x})(\bar{x}+\bar{y}+z)(x+y+z)$

   $= (x+y)(y+z)(\bar{z}+\bar{x})(\bar{x}+\bar{y}+z)$ $(x+y+z)$ includes $(x+y)$ (theorem 3)

   $= (x+y)(y+z)(\bar{x}+\bar{z})(y+\bar{z})(\bar{x}+\bar{y}+z)$ Consensus of $(x+y)$ and $(\bar{x}+\bar{z})$

   $= (x+y)(z+y)(\bar{x}+\bar{z})(\bar{z}+y)(y+y)(\bar{x}+\bar{y}+z)$

                                     Consensus of $(z+y)$ and $(\bar{z}+y)$

   $= (x+y)(z+y)(\bar{x}+\bar{z})(\bar{z}+y)y(\bar{x}+\bar{y}+z)$

                                     $X + X = X$

   $= (\bar{x}+\bar{z})y(\bar{x}+\bar{y}+z)$ Theorem 3, y is included in $(x+y)$, $(z+y)$ and $(\bar{z}+y)$

   $= (\bar{x}+\bar{z})y(\bar{x}+\bar{y}+z)(\bar{x}+(\bar{x}+\bar{y}))$ Consensus of $(\bar{z}+\bar{x})$ and $(z+\bar{x}+\bar{y})$

   $= (\bar{x}+\bar{z})y(\bar{x}+\bar{y}+z)(\bar{x}+\bar{y})$ $X + X = X$

   $= (\bar{x}+\bar{z})y(\bar{x}+\bar{y})$ $(\bar{x}+\bar{y}+z)$ includes $(\bar{x}+\bar{y})$

   $= (\bar{x}+\bar{z})(y+0)(\bar{x}+\bar{y})$ $X = X + 0$

   $= (\bar{x}+\bar{z})y(\bar{x}+\bar{y})(0+\bar{x})$ Consensus of $(y+0)$ and $(\bar{y}+\bar{x})$

   $= (\bar{x}+\bar{z})y(\bar{x}+\bar{y})\bar{x}$ $X + 0 = X$

   $= \bar{x}y$ Theorem 3, $\bar{x}$ is included in both $(\bar{x}+\bar{z})$ and $(\bar{x}+\bar{y})$

$(x+y)(y+z)(\bar{z}+\bar{x})(\bar{x}+\bar{y}+z)(x+y+z) = (\bar{x}) \cdot (y)$

Therefore, the prime implicant is $\bar{x}$ and y. Note that the given Boolean expression is in the product-of-sums form. Therefore, $\bar{x}y$ is not a single term but the multiplication of two prime implicants, i.e., $(\bar{x}) \cdot (y)$.

In addition, note that since the Boolean expression in (2) is the dual of the Boolean expression in (1), the final results in (1) and (2) are also dual of each other.

(a) Find all the prime implicants for each of the following Boolean expressions:

(1) $ab\bar{c}d + a\bar{b}c\bar{d} + ab\bar{d} + c\bar{d} + a\bar{c}$

(2) $abcd + \bar{a}\,\bar{b}\,c\,d + a\bar{b} + ac + ad$

$+ \bar{a}\,\bar{b}\,\bar{c}\,\bar{d} + \bar{a}\,\bar{d} + abc + dc$

(b) Find the essential prime implicants for the Boolean expressions in (1) and (2).

(c) Minimize the Boolean expressions in (1) and (2).

Solution: The consensus method is used here. The method can be stated as follows. Use the theorems:

$$"X + X = X, \quad X \cdot X = X" \quad \text{and}$$

$$"X + XY = X"$$

to eliminate terms that include other terms in a Boolean expression. At the same time, the theorem

$$"XY + \bar{X}Z = XY + \bar{X}Z + YZ"$$

is used to obtain all possible included terms (which are added to the expression) for all pairs of terms.

Repeat this procedure until no more included terms can be formed, or until the only included terms that can be formed would be immediately eliminated by the use of the first theorem given in the procedure.

(1) $ab\bar{c}d + a\bar{b}c\bar{d} + ab\bar{d} + c\bar{d} + a\bar{c}$

$= ab\bar{c}d + ab\bar{d} + c\bar{d} + a\bar{c}$ $X + XY = X$

$= ab\bar{d} + c\bar{d} + a\bar{c}$ $X + XY = X$

$= ab\bar{d} + c\bar{d} + a\bar{c} + a\bar{d}$ $XY + \bar{X}Z = XY + \bar{X}Z + YZ$

$= c\bar{d} + a\bar{c} + a\bar{d}$ $X + XY = X$

Thus, the prime implicants are $c\bar{d}$, $a\bar{c}$ and $a\bar{d}$.

(2) $abcd + \bar{a}\,\bar{b}\,c\,d + a\bar{b} + ac + ad + \bar{a}\,\bar{b}\,\bar{c}\,\bar{d} + \bar{a}\,\bar{d} + abc + dc$

$= \bar{a}\,\bar{b}\,c\,d + a\bar{b} + ac + ad + \bar{a}\,\bar{b}\,\bar{c}\,\bar{d} + \bar{a}\,\bar{d} + abc + dc$

                               $X + XY = X$

$$= a\overline{b} + ac + ad + \overline{a}\,\overline{b}\,\overline{c}\,\overline{d} + \overline{a}\,\overline{d}$$

$$+ abc + dc \qquad\qquad X + XY = X$$

$$= a\overline{b} + ac + ad + \overline{a}\,\overline{d} + abc + dc \qquad X + XY = X$$

$$= a\overline{b} + ac + ad + \overline{a}\,\overline{d} + dc \qquad X + XY = X$$

$$= a\overline{b} + ac + ad + \overline{a}\,\overline{d} + dc + \overline{a}c \qquad XY + \overline{X}Z$$

$$= XY + \overline{X}Z + YZ$$

$$= a\overline{b} + ac + ad + \overline{a}\,\overline{d} + dc + \overline{a}c + cc \qquad XY + \overline{X}Z = XY + \overline{X}Z + YZ$$

$$= a\overline{b} + ac + ad + \overline{a}\,\overline{d} + dc + \overline{a}c + c \qquad X \cdot X = X$$

$$= a\overline{b} + ad + \overline{a}\,\overline{d} + dc + \overline{a}c + c \qquad X + XY = X$$

$$= a\overline{b} + ad + \overline{a}\,\overline{d} + dc + c \qquad X + XY = X$$

$$= a\overline{b} + ad + \overline{a}\,\overline{d} + c \qquad X + XY = X$$

$$= a\overline{b} + ad + \overline{a}\,\overline{d} + c + \overline{b}\,\overline{d} \qquad XY + \overline{X}Z = XY + \overline{X}Z + YZ$$

Therefore, the prime implicants are $a\overline{b}$, ad, $\overline{a}\,\overline{d}$, c, and $\overline{b}\,\overline{d}$.

(b) The essential prime implicants for (1) are $c\overline{d}$ and $a\overline{c}$. The essential prime implicants for (2) are $a\overline{b}$, ad, $\overline{a}\,\overline{d}$, and c. Observe that the prime implicants of a Boolean expression are not necessarily the essential prime implicants.

(c) The minimized Boolean expression contains essential prime implicants only. Hence, the minimized result for (1) is $c\overline{d} + a\overline{c}$ whereas the minimized expression for the Boolean expression in (2) is $a\overline{b} + ad + \overline{a}\,\overline{d} + c$.

MINIMIZATION

● **PROBLEM** 10-29

A student is interested in joining Club A of a university. After consulting with a member of Club A, the student finds that a student can join the club if anyone of the following conditions is satisfied:

(a) A student has completed 40 credits or more and has a GPA above 3.5.

(b) The student completed at least 40 credits with a GPA below 3.5, but he or she majors in English.

(c) The student majors in English and has completed at least 40 credits. In addition, he or she must be a second-degree student.

(d) A student majors in English with a GPA above 3.5 but is not a second-degree student.

(e) A second-degree student majors in English, and has completed less than 40 credits.

(f) A student majors in English, has completed less than 40 credits with a GPA above 3.5.

(g) A student majors in English.

Find a simpler statement which is equivalent to the statements above for the requirements to join Club A.

Solution: First introduce the following logic variables to represent the statements (a) through (g).

A = A student completed at least 40 credits.

B = A student whose GPA is above 3.5.

C = A student majors in English.

D = A second-degree student.

E = A student who can join the club.

Then, $A = T$ or $A = 1$ means that the student has completed 40 credits or more, whereas $A = F$ (or $A = 0$), or $\overline{A} = 1$ means that the student has completed less than 40 credits. Therefore, the given conditions for a student to be qualified to join the club can be expressed as the following logical algebraic equation:

$$E = AB + A\overline{B}C + ADC + B\overline{D}C + \overline{A}CD + \overline{A}BC + C$$

Note that each term in the above equation represents one of the seven conditions, and E is true if any of the terms on the right side of the equation are true.

Now, the equation is simplified as follows:

$$E = AB + A\overline{B}C + ADC + B\overline{D}C + \overline{A}CD + \overline{A}BC + C$$

$$= AB + (A\overline{B} + AD + B\overline{D} + \overline{A}D + \overline{A}B + 1)C$$

$$= AB + (1) \cdot C$$

$$= AB + C$$

Hence, the final result is $E = AB + C$ which "says" that a student can join Club A ($E = $ true) if he or she majors in English ($C = $ true) or he or she has completed 40 credits and has a GPA above 3.5 ($A = B = $ true, i.e. $AB = $ true).

(1)   Plot the following Boolean expression on Karnaugh maps (K-maps):

(a) $x\bar{y} + \overline{\bar{x}y} + xy$

(b) $xyz + \bar{x}y\bar{z} + \bar{x}\,\bar{y} + xyzq$

(c) $xy + \bar{z} + \bar{x}\,\bar{z}$

(2)   Simplify the following Boolean expression using a Karnaugh map:

$E = x\bar{y}z + \overline{xz} + x\,\bar{y}\,\bar{z} + (x+y)\bar{z}$

Solution: (a) $x\bar{y} + \overline{\bar{x}y} + xy$

$$= x\bar{y} + x + \bar{y} + xy$$

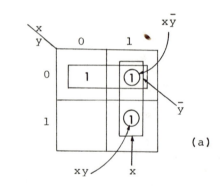

(a)

(b)

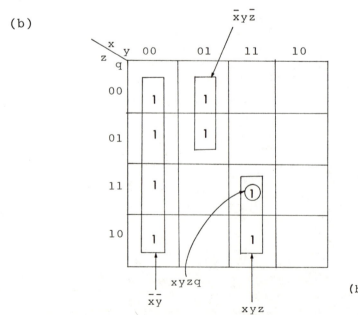

(b)

(c)

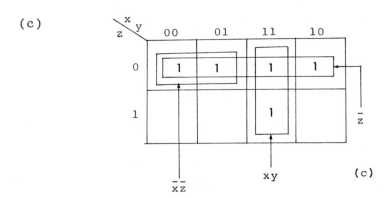

(c)

(2) $E = x\bar{y}z + \overline{xz} + x\,\bar{y}\,\bar{z} + (x+y)\bar{z}$

$\quad = x\bar{y}z + \bar{x} + \bar{z} + x\,\bar{y}\,\bar{z} + x\bar{z} + y\bar{z}$

The Karnaugh map for E is shown in Fig. 1.

Read the map in Fig. 1 by looping the maximum number of 1's in each group as shown.

The simplified E is

$\quad\quad E = \bar{x} + \bar{y} + \bar{z}.$

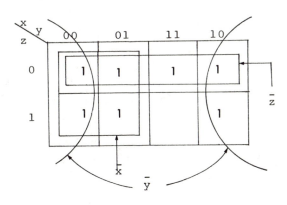

Fig. 1

(a)   Which theorem allows us to use the Karnaugh maps to simplify Boolean expressions?

(b)   A possible three-variable Karnaugh map is shown in Fig. 1. Explain why it is never used.

(c)   Simplify the Boolean expression given in the Karnaugh map in Fig. 2.

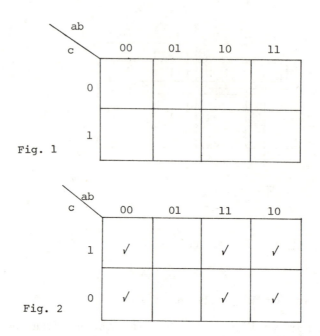

Fig. 1

Fig. 2

Solution: (a) The theorem we use to simplify Boolean expressions with K-maps is the following:

$$XY + X\overline{Y} = X,$$

$$(X+Y)(X+\overline{Y}) = X$$

(b) The given K-map is never used because it's inconvenient to apply the theorem $XY + X\overline{Y} = X$ for certain cases (whereas it works well for other cases). For example, the expression $\overline{a}\,\overline{b}\,\overline{c} + \overline{a}b\overline{c} + ab\overline{c} + abc$ can be simplified by the map as shown in Fig. 3.

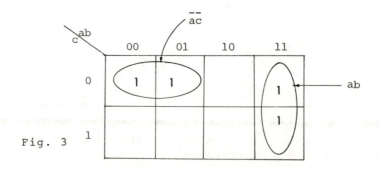

Fig. 3

However, for the expression $\overline{a}b + a\overline{b}c + a\overline{b} + a\overline{b}\,\overline{c}$, the K-map will not work well as indicated in Fig. 4. This is because some of the adjacent boxes in the map represent combinations that differ by more than two variables.

758

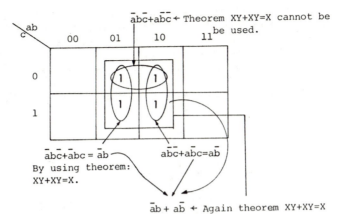

$\overline{abc}+\overline{abc}$ ← Theorem XY+XY=X cannot be be used.

$\overline{abc}+\overline{abc} = \overline{ab}$

By using theorem:
XY+XY=X.

$\overline{abc}+a\overline{bc}=\overline{ab}$

$\overline{ab} + a\overline{b}$ ← Again theorem XY+XY=X cannot be used since this expression is in the form of $X\overline{Y}+\overline{X}Y$.

Fig. 4

On the other hand, if we use the K-map as shown in Fig. 5, the expression $\overline{ab} + \overline{abc} + a\overline{b} + a\overline{b}\,\overline{c}$ can be easily simplified (see Fig. 5). Furthermore, there is no problem for any possible Boolean expression of three variables to be simplified by the K-map in Fig. 5. An example is shown in Fig. 6.

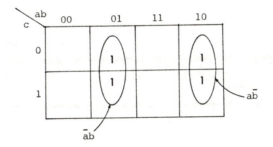

$\overline{ab}$

$a\overline{b}$

Fig. 5 $\overline{ab} + \overline{abc} + a\overline{b} + a\overline{bc} = \overline{ab} + a\overline{b}$

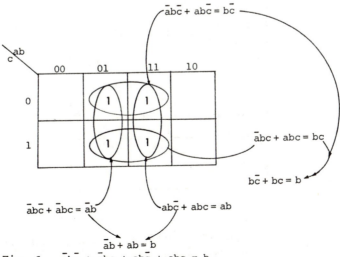

$\overline{abc} + abc = b\overline{c}$

$\overline{abc} + abc = bc$

$b\overline{c} + bc = b$

$\overline{abc} + \overline{abc} = \overline{ab}$

$a\overline{bc} + abc = ab$

$\overline{ab} + ab = b$

Fig. 6 $\overline{abc} + \overline{abc} + a\overline{bc} + abc = b$

759

(c) The simplified expression is $a + \bar{b}$. Note that though the K-map in this problem is different from the K-maps we used before, it still works well for simplifying a Boolean expression. This is because the map guarantees that the terms represented by any adjacent boxes differ by only one variable. This guarantees that the theorem $XY + \bar{X}Y = X$ can always be applied to any valid grouping of 2^n adjacent boxes, where n is a positive integer.

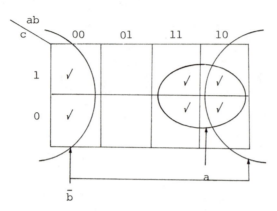

Fig. 7

● **PROBLEM** 10-32

Simplify the following Boolean expression given by the Karnaugh map in Fig. 1.

cd＼ab	00	01	11	10
00	0	4 1	12 1	8
01	1	5	13 1	9 1
11	3 1	7	15 1	11 1
10	2 1	6	14	10

Fig. 1

Solution: The process of simplifying the given Boolean expression is shown in Fig. 2.

760

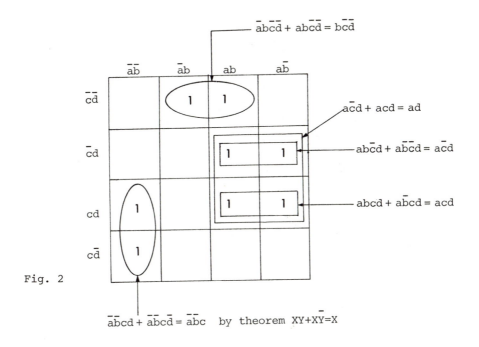

$\overline{abcd} + abc\overline{d} = b\overline{c}\overline{d}$

$\overline{ab}$ $\overline{ab}$ ab $a\overline{b}$

$\overline{c}\overline{d}$

$a\overline{c}d + acd = ad$

$\overline{c}d$

$ab\overline{c}d + a\overline{b}\overline{c}d = a\overline{c}d$

cd

$abcd + a\overline{b}cd = acd$

$c\overline{d}$

Fig. 2

$\overline{a}bc\overline{d} + \overline{a}bcd = \overline{a}bc$ by theorem XY+X$\overline{Y}$=X

The simplified expression is

$$E = ad + \overline{a}\,\overline{b}\,c + b\,\overline{c}\,\overline{d}.$$

● **PROBLEM** 10-33

Simplify the following Boolean expressions whose
Karnaugh maps are given in Figs. 1, 2, 3, and 4.

c \ ab	00	01	11	10
0	0 1	2	6 1	4 1
1	1 1	3 1	7	5 1

Fig. 1

c \ ab	00	01	11	10
0	0		0	0
1	0	0		0

Fig. 2

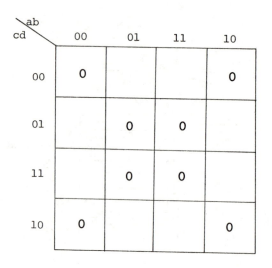

Fig. 3

Fig. 4

Solution: (1) Since in the K-map the boxes are filled with 1's, the original expression is in the sum-of-products form. The expanded sum-of-products of the given expression is:

$$\overline{a}\,\overline{b}\,\overline{c} + \overline{a}\,\overline{b}\,c + \overline{a}bc + a\overline{b}\,\overline{c} + a\overline{b}\,\overline{c} + abc$$

The given Boolean expression is simplified as shown in Fig. 5. The simplified expression is $a\overline{c} + \overline{a}c + \overline{b}$.

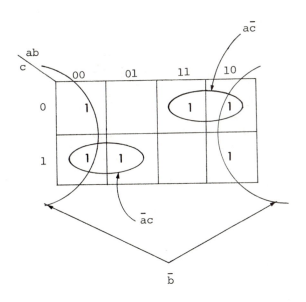

Fig. 5

(2) Since the boxes are filled with 0's the original expression is in the product-of-sums form. The expanded
product-of-sums of the given expression is:

(a+b+c)(a+b+$\overline{c}$)(a+$\overline{b}$+$\overline{c}$)($\overline{a}$+$\overline{b}$+c)($\overline{a}$+b+c)($\overline{a}$+b+$\overline{c}$).

The expression is simplified as shown in Fig. 6.

The simplified expression is (a+$\overline{c}$)($\overline{a}$+c)b.

We can obtain the same result by taking the complement
of the simplified result from (1), that is,

$$\overline{a\overline{c} + \overline{a}c + \overline{b}}$$

$$= (\overline{a} + c)(a + \overline{c})b$$

(using DeMorgan's theorem). This is a valid procedure because
the K-maps in Fig. 1 and Fig. 2 are the same except the same
boxes that are filled with 1's in the K-map in Fig. 1 are
filled with 0's in Fig. 2. This implies that the Boolean expression represented by the K-map in Fig. 1 is the complement
of the expression represented by the K-map in Fig. 2 and
vice versa.

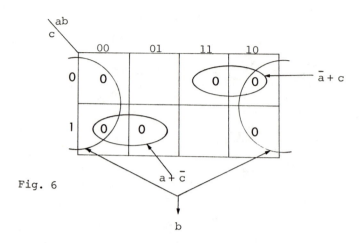

Fig. 6

(3)

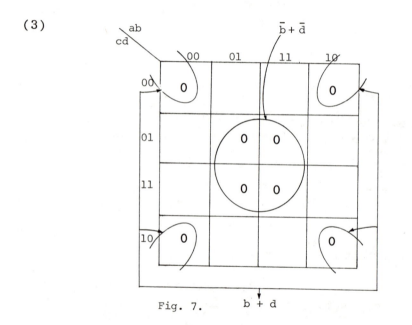

Fig. 7.   b + d

The simplified expression is

$(\bar{b}+\bar{d})(b+d)$.

(4) Observe that the boxes of the K-map in Fig. 4 are
filled with 1's whereas the boxes of the K-map in Fig. 3 are
filled with 0's. Furthermore, note that the boxes filled
with 1's in Fig. 4 are those empty boxes in Fig. 3. Therefore, the original expression of (3) must be equivalent to
the original expression in (4).

Hence, the simplified expression in (3) must also equal
the simplified expression in (4). That is

$$(\overline{b}+\overline{d})(b+d) = \overline{b}b + \overline{b}d + \overline{d}b + \overline{d}d$$

$$= 0 + \overline{b}d + \overline{d}b + 0$$

$$= \overline{b}d + b\overline{d}$$

Indeed, we do get the same result by simplifying the given expression of (4) as shown in Fig. 8.

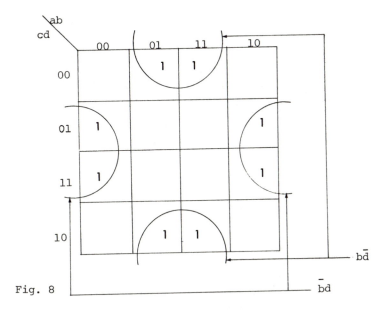

Fig. 8

● PROBLEM 10-34

Minimize the Boolean expressions given in Figs. 1, 2, 3, and 4.

cd \ ab	00	01	11	10
00	0 (0)	4 (0)	12 (0)	8 (0)
01	1 (0)	5	13	9 (0)
11	3 (0)	7	15	11 (0)
10	2 (0)	6 (0)	14 (0)	10 (0)

Fig. 1

	$\overline{a}\overline{b}$	$\overline{a}b$	ab	$a\overline{b}$
$\overline{c}\overline{d}$	✓	✓		✓
$\overline{c}d$		✓	✓	
cd		✓	✓	
$c\overline{d}$	✓		✓	✓

Fig. 2

	$a+b$	$a+\overline{b}$	$\overline{a}+\overline{b}$	$\overline{a}+b$
$c+d$	✓			✓
$c+\overline{d}$	✓	✓	✓	✓
$\overline{c}+\overline{d}$	✓	✓	✓	✓
$\overline{c}+d$	✓			✓

Fig. 3

bc / de	00	01	11	10
00	0 1	4	12	8 1
01	1	5	13	9
11	3	7	15	11
10	2 1	6	14	10 1

Fig. 4   a = 0

bc / de	00	01	11	10
00	16 1	20	28	24 1
01	17 1	21 1	29 1	25 1
11	19	23	31	27
10	18 1	22	30	26 1

a = 1

: (a) The detailed simplification process is given in Fig. 5.

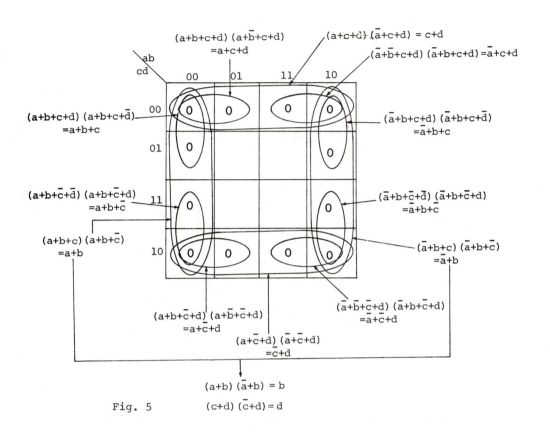

Fig. 5

The grouping is shown in Fig. 6.

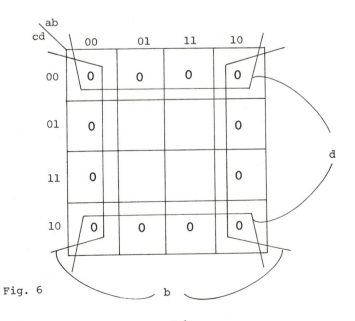

Fig. 6

767

Therefore, the simplified expression is $E = b \cdot d$.

(b) The simplified expression is

$$E = bd + \overline{b}\,\overline{d} + ac\overline{d} + \overline{a}\,\overline{c}\,\overline{d}.$$

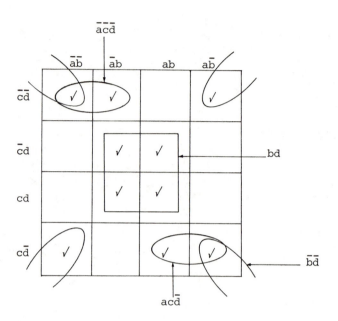

(c) The simplified expression is

$$E = b + \overline{d}.$$

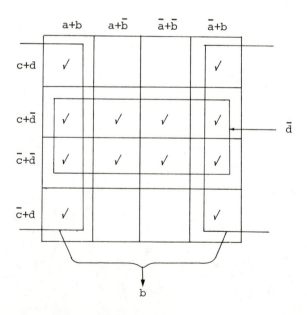

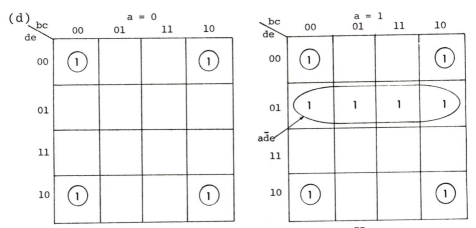

(d)

a = 0

bc de	00	01	11	10
00	①			①
01				
11				
10	①			①

a = 1

bc de	00	01	11	10
00	①			①
01	1	1	1	1
11				
10	①			①

$a\bar{d}e$

Fig. 8 All the circled terms added together give $\bar{c}\bar{e}$. Thus, the
simplified result is
$$a\bar{d}e + \bar{c}\bar{e}.$$

● **PROBLEM** 10-35

Simplify the following Boolean expressions given by the
Karnaugh maps of Figs. 1, 2 and 3.

ab cd	00	01	11	10
00	1	1	1	1
01	1	1	1	1
11	1	1	1	1
10	1	1	1	1

Fig. 1

ab cd	00	01	11	10
00	0		0	0
01			0	0
11			0	0
10	0		0	0

Fig. 2

769

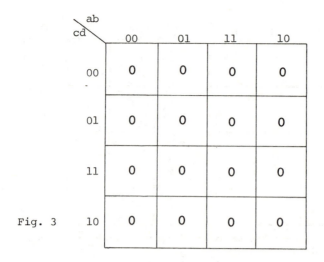

Fig. 3

Solution: (a) The given expression is simplified as shown
in Fig. 4. One can obtain the same result by inspection
since the K-map of the Boolean expression indicates that the
expression includes ALL 16 possible combinations of all four
variables a, b, c and d. A Boolean expression of n variables
which includes all 2^n possible combinations always has the
truth value of 1.

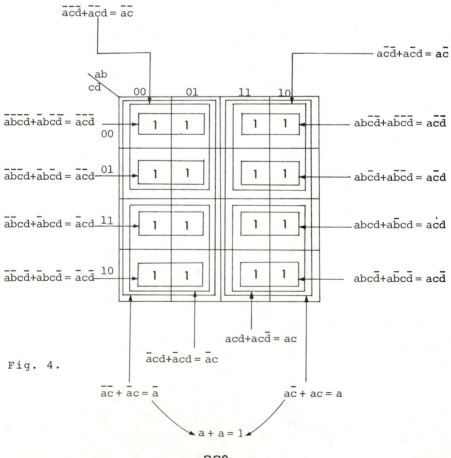

Fig. 4.

770

Thus, the simplified expression is E = 1.

(b)   Thus, the simplified Boolean expression is $\bar{a}(b+d)$.

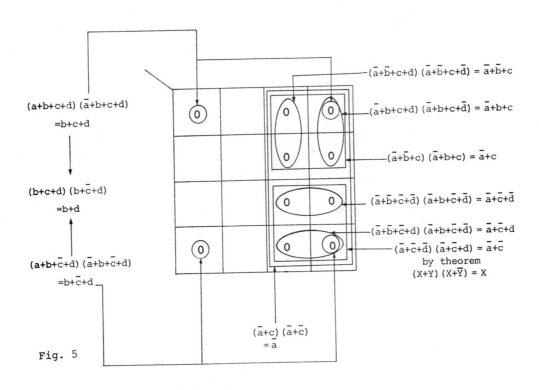

$(a+b+c+d)\ (\bar{a}+b+c+d)$

$=b+c+d$

$(b+c+d)\ (b+\bar{c}+d)$

$=b+d$

$(a+b+\bar{c}+d)\ (\bar{a}+b+\bar{c}+d)$

$=b+\bar{c}+d$

$(\bar{a}+\bar{b}+c+d)\ (\bar{a}+b+c+\bar{d}) = \bar{a}+\bar{b}+c$

$(\bar{a}+b+c+d)\ (\bar{a}+b+c+\bar{d}) = \bar{a}+b+c$

$(\bar{a}+\bar{b}+c)\ (\bar{a}+b+c) = \bar{a}+c$

$(\bar{a}+\bar{b}+\bar{c}+\bar{d})\ (\bar{a}+b+\bar{c}+\bar{d}) = \bar{a}+\bar{c}+\bar{d}$

$(\bar{a}+\bar{b}+\bar{c}+d)\ (\bar{a}+b+\bar{c}+d) = \bar{a}+\bar{c}+d$

$(\bar{a}+\bar{c}+\bar{d})\ (\bar{a}+\bar{c}+d) = \bar{a}+\bar{c}$

by theorem

$(X+Y)\ (X+\bar{Y}) = X$

$(\bar{a}+c)\ (\bar{a}+\bar{c})$

$= \bar{a}$

Fig. 5

(c)   Since the K-map in Fig. 3 indicates that the given expression is a product of all the possible combinations (sums) of four variables, as explained in Part (a) of this problem, such a Boolean expression always has the truth value of 1. Therefore the given expression E is simplified to E = 1.

● **PROBLEM** 10-36

Use a K-map to minimize the following function

$f(x,y,z,p) = \Pi(0,1,2,4,8,9,10,12)$

Solution: The following algorithm applied to a K-map will lead to a minimal expression for a logical function:

   Step 1:   Check any box that cannot be combined with any other; it will be accepted as an essential prime implicant.

   Step 2:   Identify the boxes that can be combined with a single other box in only one way, and group such two-box

771

combinations. A box that can be combined into a grouping
of two but can be so combined in more than one way is
not considered here.

Step 3: Identify the boxes that can be combined with
three other boxes in only one way. Encircle such four-
box combinations if all four boxes are not already
covered in groupings of two. Temporarily bypass boxes
that can be encompassed in a group of four in more than
one way.

Step 4: Repeat the preceding for groups of eight, etc.

Step 5: After the above procedure, if there still remain some uncovered boxes, they should be combined with
each other and/or with other already covered boxes in
as few groupings as possible, and each grouping should
be as large as possible.

Using the algorithm just described, one can obtain the
K-map as shown in Fig. 1.

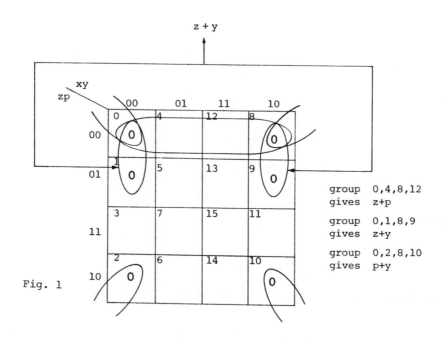

Fig. 1

group 0,4,8,12
gives z+p

group 0,1,8,9
gives z+y

group 0,2,8,10
gives p+y

The final result is f(x,y,z,p) = (z+p)(z+y)(p+y)

● PROBLEM 10-37

Minimize the given minterm function f(A,B,C,D) via the Karnaugh map.

   f(A,B,C,D) = Σm(0,1,3,8,9,11,13,14)

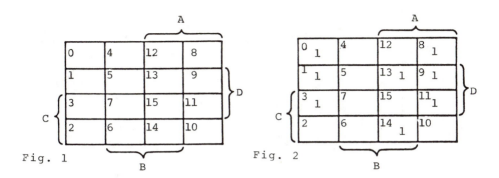

Fig. 1

Fig. 2

Solution: We are given the function $f(A,B,C,D) = \Sigma m(0,1,3,8,9,11,13,14)$ which consists of minterms to be minimized. The first step in obtaining the solution is to plot the function on the K-map.

In figure #1 a K-map for four variables is given. These maps are standard for each problem, however they change size according to the number of variables in the function.

To plot the given function into the K-map, 1's are put into the boxes which number matches with the given minterm variable. The final K-map with the variables plotted looks as shown in figure 2. Secondly, the number of adjacencies for each minterm is counted.

The number of adjacencies for each minterm is shown in the lower right-hand corner of the minterm block of figure 3.

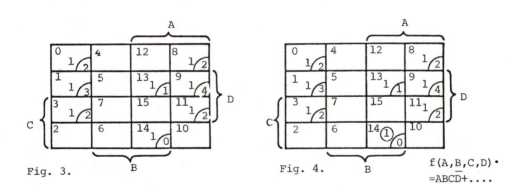

Fig. 3.

Fig. 4.

$f(A,B,C,D) \cdot$
$=ABC\bar{D}+\ldots$

Now the map is ready for the simplification process. This is done by choosing first the minterm m_{14} which has no adjacencies, as in figure 4.

This minterm must be taken as a group itself; hence the first term in the minimized function is $ABC\bar{D}$ since minterm box #14 is not covered by variable D.

Next, minterms with one adjacency are examined; m_{13} is the only one. Consequently, m_{13} has one way of being grouped and that is with m_9. This is illustrated in figure 5.

773

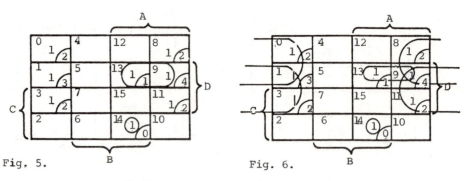

Fig. 5.

Fig. 6.

$$f(A,B,C,D) = ABC\overline{D} + A\overline{C}D$$

Note that minterms m_{13} and m_9 have the following variables:

$m_{13} = AB\overline{C}D$, $m_9 = A\overline{B}\overline{C}D$, therefore, B (in m_{13}), and $\overline{B}$(in m_9) may be cancelled according to Boolean algebra theorems. Continuing the solution; four minterms with two adjacencies are taken into account. These are m_0, m_3, m_8 and m_{11}. One of these is picked to group at random. If m_3 is picked (it is adjacent to m_1, and m_{11}), since m_9) is also available (any minterm may be used as many times as it is needed), a large group of four minterms may be formed. Note that

$$m_1 = \overline{A}\,\overline{B}\,\overline{C}\,D$$
$$m_3 = \overline{A}\,\overline{B}\,C\,D$$
$$m_{11} = A\,\overline{B}\,C\,D$$
$$m_9 = A\,\overline{B}\,\overline{C}\,D$$

in these four minterms A, and $\overline{A}$ terms also C, and $\overline{C}$ terms cancel each other, using Boolean theorems $a \cdot \overline{a} = 0$, therefore the minimized form of minterm m_1, m_3, m_{11}, and m_9 is $\overline{B}D$. The function now becomes $f(A,B,C,D) = ABC\overline{D} + A\overline{C}D + \overline{B}D$.

Finally there are still two minterms which have not been accounted for. These two, m_0 and m_8, can be grouped with m_1 and m_9 respectively, to form a last group of four minterms as shown in figure 6. This last group adds the final product term $\overline{B}\,\overline{C}$ to the function $f(A,B,C,D)$, and hence: $f(A,B,C,D) = ACB\overline{D} + A\overline{C}D + \overline{B}D + \overline{B}\,\overline{C}$.

● **PROBLEM** 10-38

Minimize the following minterm function containing 'don't cares' using the Karnaugh-Map.

$$f(A,B,C,D) = \Sigma m(5,6,7,8,9) + d(10,11,12,13,14,15)$$

Solution: $f(A,B,C,D) = \Sigma m(5,6,7,8,9) + d(10,11,12,13,14,15)$. In the design of digital circuits one often encounters cases in which the switching function is not completely specified. In other words, a function may be required to contain certain minterms, omit certain minterms, with the remaining minterms being optional; that is certain minterms may be included in the logic design if they help simplify the logic circuit. A minterm which is optional is called a don't care minterm.

The plotting of the given function $f(A,B,C,D)$ is shown in fig. 1.

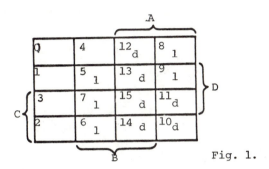

Fig. 1.

In the use of don't cares there is one additional rule which may be used in mapping. Recall that the don't cares by definition can be either 0 or 1. Hence in minimizing terms in Sum of Product form, the don't cares may be chosen to be 1 if in doing so the set of blocks on the map which can be grouped together is larger than would otherwise be possible without including the don't cares. In other words, with regard to don't cares one can take it or leave it, depending on whether they do or do not aid in the simplification of the function.

In the map of fig. 1 the following minterms and don't cares may be grouped.

$$m_8 = A \bar{B} \bar{C} \bar{D}$$
$$m_9 = A \bar{B} \bar{C} D$$
$$d_{11} = A \bar{B} C D$$
$$d_{10} = A \bar{B} C \bar{D}$$
$$d_{12} = A B \bar{C} \bar{D}$$
$$d_{13} = A B \bar{C} D$$
$$d_{15} = A B C D$$
$$d_{14} = A B C \bar{D}$$

and by the use of Boolean theorem $A \cdot \bar{A} = 0$, terms $B, C,$ and D are eliminated. The result from this group would be A.

The next group is minterms 5 and 7, and don't cares 13 and 15.

775

$$m_5 = \bar{A} \ B \ \bar{C} \ D$$
$$m_7 = \bar{A} \ B \ C \ D$$
$$d_{13} = A \ B \ \bar{C} \ D$$
$$d_{15} = A \ B \ C \ D$$

With the same argument, the eliminated terms are A and C. Leaving the resulting terms BD.

The third and the last group consists of minterms 7 and 6, and don't cares 15 and 14.

$$m_6 = \bar{A} \ B \ C \ \bar{D}$$
$$m_7 = \bar{A} \ B \ C \ D$$
$$d_{14} = A \ B \ C \ \bar{D}$$
$$d_{15} = A \ B \ C \ D$$

Giving the result BC. The resulting total function can now be written as the sum of reduced terms that are found, namely;

$$f(A,B,C,D) = A + BD + BC \ .$$

Note that this function is much simpler to deal with than it would have been without the inclusion of don't cares.

● **PROBLEM** 10-39

Simplify F together with its don't care condition d in
(a)   sum-of-products form and
(b)   product-of-sums form.

$$F(A,B,C,D) = \Sigma \ (0,1,2,8,9,12,13)$$
$$d(A,B,C,D) = \Sigma \ (10,11,14,15)$$

Solution: F is simplified with a four-variable K map. Each don't care minterm can be treated as a 0 or a 1, whichever can help minimize F the most.

(a)   The K-map is drawn in fig. 1 with X's representing don't cares. F minterms of 8,9,12 and 13 and don't cares treated as 1's at 10,11,14 and 15 combine to form A. F minterms at 0,1,8, and 9 combine to form $\bar{B}\bar{C}$. F minterms at 0,2, and 8 and a don't care treated as a 1 at 10 combine to form $\bar{B}\bar{D}$. Thus using sum-of-products:

$$F = A + \bar{B}\bar{C} + \bar{B}\bar{D}$$

776

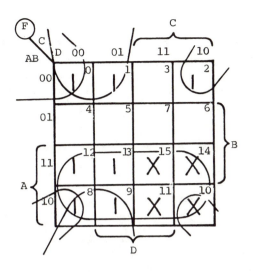

Fig. 1

(b) Use the K-map of fig. 2, which is the same as fig. 1, but now minimize $\bar{F}$ with the aid of the don't cares and then use DeMorgan's Law to change F to the product-of-sums form. $\bar{F}$ minterms which are 0's for the product-of-sums case, at 4,5,6, and 7 combine to form $\bar{A}B$. $\bar{F}$ minterms at 3 and 7 and don't cares treated as 0's at 11 and 15 combine to form CD. Thus

$$\bar{F} = \bar{A}B + CD$$

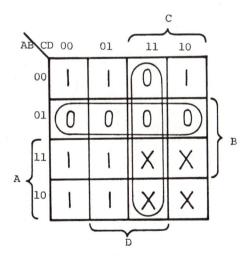

Fig. 2

Applying DeMorgan's Law gives the product-of-sums form

$$F = (A + \bar{B}) \cdot (\bar{C} + \bar{D})$$

Plot the following function on a K-map and simplify in SOP and POS forms.

$$F(A,B,C,D) = \Sigma(0,2,5,8,10,13,14,15) + X(1,11,12).$$

By using the simplified expression, determine the output when a redundant input occurs.

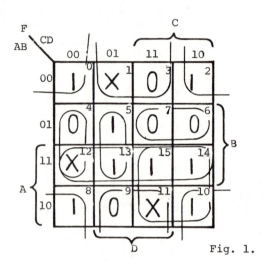

Fig. 1.

Solution: The K-map is shown in figure 1. The "X" mark at locations one, eleven, and twelve denotes the redundancies.

First simplify in the SOP form. Minterms at 13,14, and 15 and the redundancy at 12 combine to form AB. Minterms at 0,2,8, and 10 combine to form $B\bar{D}$. Minterms at 5 and 13 combine to form $B\bar{C}D$. The SOP form is:

$$F = AB + \bar{B}\bar{D} + B\bar{C}D$$

To simplify F into the POS form, simplify $\bar{F}$ in SOP form and then use DeMorgan's Law. From the K-map it is seen that the 0's at 6 and 7 combine to form $\bar{A}BC$. The 0 at 4 and the redundancy at 12 combine to form $B\bar{C}\bar{D}$. The 0's at 3 and 9 and the redundancies at 1 and 11 combine to form $\bar{B}D$. The SOP of $\bar{F}$ is

$$\bar{F} = \bar{A}BC + B\bar{C}\bar{D} + \bar{B}D$$

	A	B	C	D	F SOP	POS
X_1	0	0	0	1	1	0
X_{11}	1	0	1	1	1	0
X_{12}	1	1	0	0	1	0

Fig. 2.

778

Apply DeMorgan's Law to $\bar{F}$ gives the POS form of F.

$$F = (A+\bar{B}+\bar{C})(\bar{B}+C+D)(B+\bar{D})$$

The table of fig. 2 shows the output when redundancies occur.

● **PROBLEM** 10-41

(a) Use a K-map to minimize the following function.

$f(a,b,c,d,e) = \Sigma(0,2,4,5,7,8,10,13,15,16,18.20,$

$21,23,24,26,29,31)$

(b) Assume that all 1's in the K-map in part (a) are 0's now, thus the new map represents the complement of the function in part (a). Simplify this complementary function and express the final result as a product of sums.

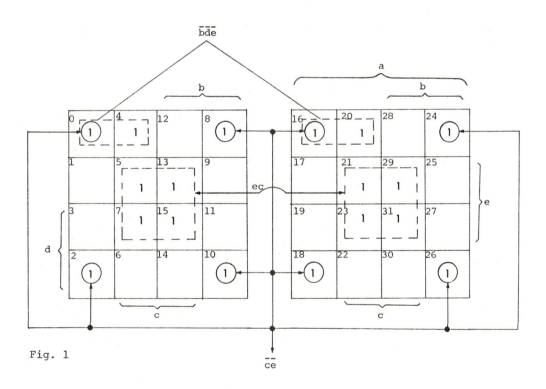

Fig. 1

Solution: The following algorithm applied to a K-map will lead to a minimal expression for a logic function.

Step 1: Check any box that cannot be combined with any other; it will be accepted as an essential prime impli‐cant.

Step 2: Identify the boxes that can be combined with a single other box in only one way, and group such two-

779

box combinations. A box that can be combined into a grouping of two but can be so combined in more than one way is not considered here.

Step 3: Identify the boxes that can be combined with three other boxes in only one way. Encircle such four-box combinations if all four boxes are not already covered in groupings of two. Temporarily bypass boxes that can be encompassed in a group of four in more than one way.

Step 4: Repeat the preceding for groups of eight, etc.

Step 5: After the above procedure, if there still remain some uncovered boxes, they should be combined with each other and/or with other already covered boxes in as few groupings as possible and each grouping should be as large as possible.

(a) The K-map for the given function is shown in Fig. 1.

By following the algorithm given (see Fig. 1) one obtains

$$f(a,b,c,d,e) = ce + \overline{c}\,\overline{e} + \overline{b}\,\overline{d}\,\overline{e}.$$

(b) Replacing 1's in the K-map in Fig. 1 by 0's, one obtains the function

$$f*(a,b,c,d,e) = \Pi(0,2,4,5,7,8,10,13,15,16,18,20,21,23,24,29,31)$$

As in part (a), this function is simplified by using a K-map (use the method given at the beginning of this problem). The result is

$$f*(a,b,c,d,e) = (\overline{c}+\overline{e})(c+e)(b+d+e)$$

Note that in simplifying the K-map, the grouping of 0's is exactly the same as the grouping of 1's in part (a). However, in this case, a box within the range of a is associated with $\overline{a}$, rather than with a (and vice versa). The same applies for the other variables.

Observe that $f(a,b,c,d,e) = ce + \overline{c}\,\overline{e} + \overline{b}\,\overline{d}\,\overline{e}$ and $f*(a,b,c,d,e) = (\overline{c}+\overline{e})(c+e)(b+d+e)$ are complements of each other, i.e. $f = \overline{f*}$ or $\overline{f} = f*$.

This is the expected result since after replacing 1's in the K-map in Fig. 1 by 0's, the new K-map represents the complement of the original function.

Therefore, one can find the result of (b) by simply finding the complement of the final result from part (a).

That is $f*(a,b,c,d,e) = \overline{f(a,b,c,d,e)}$

$$= \overline{ce + \overline{c}\,\overline{e} + \overline{b}\,\overline{d}\,\overline{e}}$$

(by DeMorgan's theorem) $= (\overline{c}+\overline{e})(c+e)(b+d+e)$.

● **PROBLEM** 10-42

Read the following K-map.

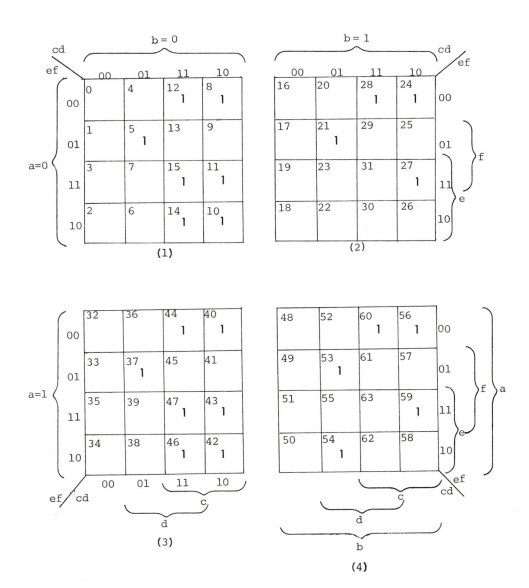

Solution:

The K=map is shown in Fig. 1. The given function is

$$f(a,b,c,d,e,f) = \Sigma(5,8,10,11,12,14,15,21,24,27,$$
$$28,37,40,42,43,44,46,47,53,$$
$$54,56,59,60)$$

The simplified f is

$$f(a,b,c,d,e,f) = ab\bar{c}de\bar{f} + bc\bar{d}ef + \bar{c}d\bar{e}f + \bar{b}ce + c\,\bar{e}\,\bar{f}.$$

781

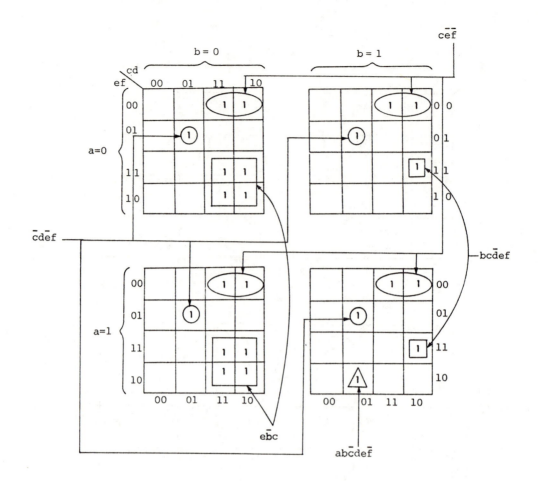

b = 0   b = 1

cef̄

- PROBLEM 10-43

Use Quine-McCluskey Method to minimize the function f(A,B,C,D).

$$f(A,B,C,D) = \Sigma m(2,4,6,8,9,10,12,13,15)$$

Solution: To begin the Quine-McCluskey minimization technique the minterms are grouped according to the number of 1's in the binary representation of the minterm number. This grouping of items is illustrated in fig. 1.

Minterms	A	B	C	D
2	0	0	1	0
4	0	1	0	0
8	1	0	0	0
6	0	1	1	0
9	1	0	0	1
10	1	0	1	0
12	1	1	0	0
13	1	1	0	1
15	1	1	1	1

Fig. 1

Minterms 2,4, and 8 form the first group since they both contain a
single 1 in their binary representation form. Minterms 6,9,10 and
12 form the second group hence they contain two 1's in their binary
representations. The third group which consists of three 1's is the
minterm 13. And the fourth group with four 1's, minterm 15. Once
this method is formed, using the table obtained, adjacent minterms
are found and combined into minterm lists in a minimizing table.

   In the following tables this procedure is illustrated;

Minterm	A	B	C	D	
2	0	0	1	0	Group 1
4	0	1	0	0	Group 1
8	1	0	0	0	Group 1
6	0	1	1	0	Group 2
9	1	0	0	1	Group 2
10	1	0	1	0	Group 2
12	1	1	0	0	Group 2
13	1	1	0	1	Group 3
15	1	1	1	1	Group 4

Fig. 2

In fig. 2 all 4 groups are given. Note here that two terms can be
combined if and only if they differ in a single literal. Hence, in
Fig. 2 group 1 terms can only be combined with group 2 terms, to form
the list given in fig. 3.

Minterms	A	B	C	D	
2,6	0	-	1	0	PI_2
2,10	-	0	1	0	PI_3
4,6	0	1	-	0	PI_4
4,12	-	1	0	0	PI_5
8,9	1	0	0	-	✓
8,10	1	0	-	0	PI_6
8,12	1	-	0	0	✓
9,13	1	-	0	1	✓
12,13	1	1	0	-	✓
13,15	1	1	-	1	PI_7

Fig. 3

Note here that in forming fig. 3 several minterms are combined together.
Such as m_2 of group 1, and m_6 of group 2. In these two minterms
only the term B differs, and is therefore dropped.

   Minterm 2 of group 1, and 10 of group 2 differ in A, Minterms 4 of
group 1, and 6 of group 2 differ in C; Minterms 4 of group 1, and 12 of
group 2 differ in A; Minterms 8 of group 1, and 9 of group 2 differ in
D; Minterms 8 of group 1, and 10 of group 2 differ in C; Minterms 8 of
group 1, and 12 of group 2 differ in B.

When all the combinations between the groups 1 and 2 have been made
and they have been entered in the list, a line is drawn under these
combinations, and combination of terms in group 2 with those in group
3 starts, following the same logic.

   Next, using the list of fig. 3, the list in fig. 4 is obtained.

Minterms	A	B	C	D	
8,9,12,13	1	-	0	-	PI_1

Fig. 4

As before, two terms in list 2 can be obtained only if they differ in a single literal, only terms which have the same literal missing can possibly be combined. Note that in fig. 3 minterm combinations 8, 12 and 9,13 also 8,9 and 12,13 can be combined to yield terms 8,9,12,13 in fig. 4. These terms are checked off in the table of fig. 3 and all the other terms are labeled as Prime Implicants.

To determine the smallest number of Prime Implicants required to realize the function, a Prime Implicant Chart is formed as in fig. 5.

	2	4	6	8 ✓	9 ✓	10	12 ✓	13 ✓	15 ✓
PI_1				X	Ⓧ		X	X	
PI_2	X		X						
PI_3	X					X			
PI_4		X	X						
PI_5		X					X		
PI_6					X	X			
PI_7								X	Ⓧ

Fig. 5.

The double line through the chart between PI_1 and PI_2 is used to separate prime implicants which contain different number of literals.

An examination of the minterm columns indicates that minterms 9 and 15 are covered by only one prime implicant. Therefore prime implicants PI_1 and PI_7 must be chosen, and hence they are essential prime implicants. Note that in choosing these two prime implicants, minterms 8,9,12,13, and 15 are also covered. These minterms are shown checked in the table.

To cover the remaining minterms 2,4,6, and 10 a reduced prime implicant chart is formed as in fig. 6.

Prime implicants PI_5 and PI_6 may be omitted because they are covered by PI_4 and PI_3 respectively. Hence ignoring PI_5 and PI_6 for the moment minterms 2,4,6, and 10 can be most efficiently covered by choosing PI_3 and PI_4.

Therefore a minimal realization of the original function would be;

$$f(A,B,C,D) = PI_1 + PI_3 + PI_4 + PI_7$$

$$f(A,B,C,D) = 1 - 0 - + - 0\ 1\ 0 + 0\ 1 - 0 + 1\ 1 - 1$$

and using the variable coding;

$$f(A,B,C,D) = A\bar{C} + \bar{B}C\bar{D} + \bar{A}B\bar{D} + ABD$$

784

	2	4	6	10
PI_2	X		X	
PI_3	X			X
PI_4		X	X	
PI_5		X		
PI_6				X

Fig. 6.

● PROBLEM 10-44

Given $f(a,b,c,d) = \overline{a}\,\overline{b}\,\overline{c}\,\overline{d} + \overline{a}\,\overline{b}\,c\,d + ab + a\,\overline{b}\,\overline{c}$, and optional combinations $\overline{a}bcd$, $a\overline{b}c\overline{d}$ and $a\overline{b}cd$.

Simplify f by the tabular method.

Solution: Step 1. Expand f into the form of standard sum-of-products.

$$f(a,b,c,d) = \overline{a}\,\overline{b}\,\overline{c}\,\overline{d} + \overline{a}\,\overline{b}\,c\,d + a\,\overline{b}\,\overline{c}(d+\overline{d}) + ab(c+\overline{c})(d+\overline{d})$$

$$= \overline{a}\,\overline{b}\,\overline{c}\,\overline{d} + \overline{a}\,\overline{b}\,c\,d + a\,\overline{b}\,\overline{c}\,d + a\,\overline{b}\,\overline{c}\,\overline{d}$$

$$+ abcd + abc\overline{d} + ab\overline{c}d + a b \overline{c}\,\overline{d}$$

$$= \Sigma(0,3,8,9,12,13,14,15)$$

Thus, f with the optional combinations can be expressed as

$$f(a,b,c,d) = \Sigma(0,3,8,9,12,13,14,15) + \Sigma_\phi(7,10,11)$$

Step 2. Express the result in step 1 into a table as shown below.

	a	b	c	d	
	0	0	0	0	0
	0	0	1	1	3
	1	0	0	0	8
original terms	1	0	0	1	9
	1	1	0	0	12
	1	1	0	1	13
	1	1	1	0	14
	1	1	1	1	15
optional combinations	0	1	1	1	7
	1	0	1	0	10
	1	0	1	1	11

785

Next, rearrange the above table so that the rows are listed according to the number of 1's in each row. Furthermore, rows with the same number of 1's are grouped together. (see table below).

	a	b	c	d
no 1's	0	0	0	0
one 1 per row	1	0	0	0
two 1's per row	0	0	1	1
	1	0	0	1
	1	1	0	0
	1	0	1	0
three 1's per row	0	1	1	1
	1	1	0	1
	1	1	1	0
	1	0	1	1
four 1's per row	1	1	1	1

Table 1

Step 3. Two rows from two adjacent groups will be combined if they differ in only one column; in one row that column must contain a zero, and in the other row that column must contain a 1. The rest of the corresponding columns in both rows must be identical. If a column in one row contains a "–", for instance, the other row's corresponding column must also contain a "–".

The theorem used for the combination is $AB + A\overline{B} = A$. For example, row 2 can be combined with row 4 to yield

$$a\,\overline{b}\,\overline{c}\,\overline{d} + a\,\overline{b}\,\overline{c}\,d = a\,\overline{b}\,\overline{c}, \text{ or } 1000 + 1001 = 100-.$$

The position of the eliminated variable is replaced by a bar, and the result is listed in a new table. Note that a row may be combined with more than one row. Furthermore, because of the nature of table 1, one needs to consider only the rows in the adjacent group(s). Each row in a group must be compared with all the rows in its adjacent group(s).

A row that cannot be combined with any other rows is marked with "*"; it's called a prime implicant -- an irreducible term.

The result of this step is shown in Table 2.

Table 1			Table 2		

Table 1

Row No.	abcd
1	0000
2	1000
3	0011
4	1001
5	1100
6	1010
7	0111
8	1101
9	1110
10	1011
11	1111

Table 2

Rows Combined	abcd
1 + 2	-000
2 + 4	100-
2 + 5	1-00
2 + 6	10-0
3 + 7	0-11
3 + 10	-011
4 + 8	1-01
4 + 10	10-1
5 + 8	110-
5 + 9	11-0
6 + 9	1-10
6 + 10	101-
7 + 11	-111
8 + 11	11-1
9 + 11	111-
10 + 11	1-11

Step 4. Rearrange Table 2 so that the rows in table 2 are listed according to the number of 1's per row. The new table is shown below.

Table 3

a	b	c	d
-	0	0	0
1	0	0	-
1	-	0	0
1	0	-	0
0	-	1	1
1	-	0	1
-	0	1	1
1	0	-	1
1	1	0	-
1	1	-	0
1	-	1	0
1	0	1	-
-	1	1	1
1	1	-	1
1	1	1	-
1	-	1	1

Step 5. Repeat step 3 and step 4 until one obtains a table in which all rows are prime implicants.

Table 3			Table 4			Table 5		Table 6		
	abcd			abcd			abcd		abcd	
1	-000*		2 + 9	1-0-		1	1-0-	1 + 7	1---	
2	100-		2 + 12	10--		2	10--	2 + 6	1---	
3	1-00		3 + 6	1-0-		3	1--0	3 + 5	1---	
4	10-0		3 + 11	1--0		4	--11*			
5	0-11		4 + 8	10--		5	1--1			
6	1-01		4 + 10	1--0		6	11--			
7	-011		5 + 16	--11		7	1-1-			
8	10-1		6 + 16	1--1						
9	110-		7 + 13	--11						
10	11-0		8 + 14	1--1						
11	1-10		9 + 15	11--						
12	101-		10 + 14	11--						
13	-111		11 + 16	1-1-					Table 7	
14	11-1		12 + 15	1-1-					abcd	
15	111-								1--- *	
16	1-11									

Therefore, the prime implicants are -000, --11 and 1---.

Observe that by the theorem X + X = X, when a row appears in a table more than once, all but one of the repeated rows are dropped.

Step 6. Construct a prime implicant table (table 8) as shown.

TABLE 8

Row No.	0000	0011	1000	1001	1100	1101	1110	1111		
1	✓		✓						-000	A
2		✓						✓	--11	B
3			✓	✓	✓	✓	✓	✓	1---	C

original terms in f　　　　　　prime implicants

For each prime implicant (each row in Table 8), check marks
are placed in the columns of the terms accounted by that
prime implicant. For example, 0000 is accounted for by
prime implicant -000, so a check mark is placed in column 1,
row 1. Note that a "-" in a prime implicant can be either a
zero or a one.

Step 7. For the sake of simplicity, each prime
implicant is named by a single letter as shown in Table 8.

A Boolean expression is written for each original term
in f to indicate which prime implicants can account for them.

0000 is accounted by A

0011 is accounted by B

1000 is accounted by A or C, i.e. (A+C)

1001 is accounted by C

1100 is accounted by C

1101 is accounted by C

1110 is accounted by C

1111 is accounted by B or C, i.e. (B+C)

Thus, the given function is accounted for by A and B and
(A or C) and C and C and C and C and (B or C), or AB(A+C)CCC
(B+C) = AB(A+C)C(B+C).

Step 8. Express the Boolean expression AB(A+C)C
(B+C) as a product of sums (make simplification wherever possible).

$$AB(A+C)C(B+C)$$

$$= (ABA+ABC)C(B+C)$$

$$= (AB+ABC)C(B+C)$$

$$= (AB)C(B+C)$$

$$= ABC(B+C)$$

$$= ABC + ABC = ABC$$

Hence, the given function can be accounted for by the
prime implicants A and B anc C (i.e. ABC). Therefore,

$$f(a,b,c,d) = \bar{b}\,\bar{c}\,\bar{d} + cd + a$$

since $A = -000 = \bar{b}\,\bar{c}\,\bar{d}$

$B = --11 = cd$

and $C = 1--- = a.$

Note that very often, unlike in this problem, the final result of step 8 is a sum of several terms, each of which is a
product of some or all of the prime implicants of a given
function. If this were the case, then the given function could
could be accounted for by any one of the terms in the sum.

In other words, by the tabular method one not only obtains a minimum sum of products for the given function, but one also obtains all the possible minimums (if there are more than one) of a given f. In addition, the method gives all irredundant solutions (if there are more than one), i.e. all solutions from which no prime implicant may be removed and still have all output combinations accounted for.

Finally, note that when there is more than one solution, often the final result is selected such that it consists of the least number of prime implicants.

SWITCHING CIRCUITS

• **PROBLEM** 10-45

Implement the following Boolean expressions using logic gates:

a. $Y_1 = ab + \overline{c}da + \overline{\overline{c}\overline{a}}$

b. $Y_2 = (\overline{a+b})(\overline{a}+c\overline{b})$

c. $Y_3 = (a+b\overline{c})(b+d) + ab\overline{c}$

Solution: (a) Y_1 can be simplified as follows:

$$Y_1 = ab + \overline{c}da + \overline{\overline{c}\overline{a}}$$

$$= ab + \overline{c}da + \overline{c} + a$$

$$= a(b+\overline{c}d+1) + \overline{c}$$

$$= a \cdot 1 + \overline{c} = a + \overline{c}$$

Y_1 is implemented as shown in Fig. 1.

The implementation of Y_1 which is not simplified (or changed in any way) is shown in Fig. 2.

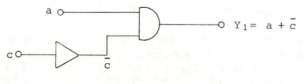

Fig. 1.

790

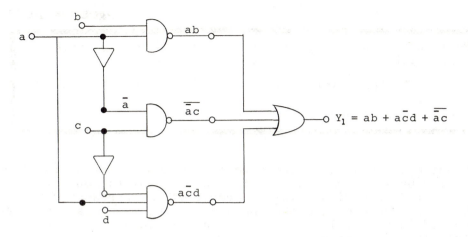

Fig. 2.

(b)

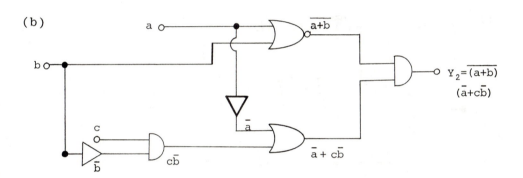

Fig. 3

(c)

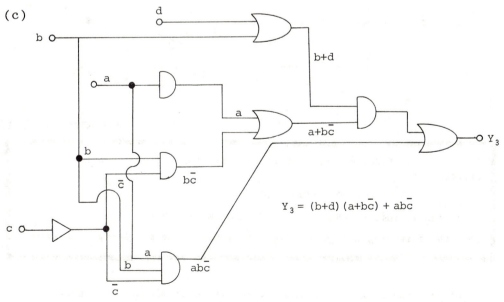

$$Y_3 = (b+d)(a+b\bar{c}) + ab\bar{c}$$

Fig. 4

Construct a switching circuit for the following Boolean expression:

(a) $(\overline{A}BC+C\overline{B})D + A$

(b) $(A+B+C)(\overline{B}+C)$

Solution: In a switching circuit the AND operation is represented by switches representing operands of the operation connected in series.   The OR operation is represented by switches representing the operands of the OR operations connected in parallel.   Therefore, the switching circuits for the given Boolean expressions are:

(a)

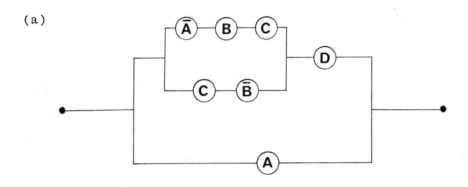

(b)

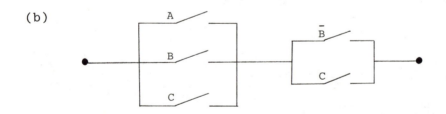

Use the laws of Boolean algebra to simplify the following Boolean expression:

$$[(A+B) + (A+C) + (A+D)] \cdot [A \cdot \sim B] \cdot (1)$$

+ is the inclusive OR, · is AND and N is NOT.

Show the switching and gate representations of the resulting expression.

Solution: One suggested plan of attack is to eliminate redundancy in the expression.   Since A appears four times, we try to simplify the terms using A. The law that allows each step is given on the right:

$[(A+B) + (A+C) + (A+D)] \cdot [A \cdot \sim B]$

$= [(B+A) + (A+C) + (A+D)] \cdot [A \cdot \sim B]$ Comm. of +

$= [B + (A+A) + C + (A+D)] \cdot [A \cdot \sim B]$ Assoc. of +

$= [B + A + C + (A+D)] \cdot [A \cdot \sim B]$ Idempotent law

$= [B + C + A + (A+D)] \cdot [A \cdot \sim B]$ Comm. of +

$= [B + C + (A+A) + D] \cdot [A \cdot \sim B]$ Assoc. of +

$= [B + C + A + D] \cdot [A \cdot \sim B]$ Idempotent law (1)

Working on B and $\sim$ B,

$(1) = [B + C + A + D] \cdot [\sim B \cdot A]$ Commutativity of $\cdot$

$= [(B + C + A + D) \cdot \sim B] \cdot A$ Associativity of $\cdot$

$= [(B \cdot \sim B) + (C \cdot \sim B) + (A \cdot \sim B) + (D \cdot \sim B)] \cdot A$

Distributive law of $\cdot$ over +

$= [0 + (C \cdot \sim B) + (A \cdot \sim B) + (D \cdot \sim B)] \cdot A$ Law of complement

$= [(C \cdot \sim B) + (A \cdot \sim B) + (D \cdot \sim B)] \cdot A$ Identity of $\cdot$

$= [(C + A + D) \cdot \sim B] \cdot A$ Dist. law of $\cdot$ over + (2)

The switching representation of (2) would be as shown in Fig. 1.

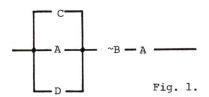

Fig. 1.

Note that the circuit conducts if and only if both A and $\tilde{B}$ are closed, regardless of the settings of C and D.
The gate representation of (2) would be as shown in Fig. 2.

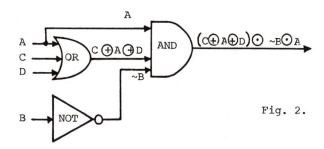

Fig. 2.

793

The reduction of statement (1) to statement (2) was tedious but straightforward. It allows an eight-switch circuit to be replaced by a five-switch circuit, or a seven-gate circuit to be replaced by a three-gate circuit. However, the reduction method depended on observation and a "feel" for algebraic manipulation. Fortunately, minimization algorithms exist which do not depend on such subjective methods and can, in fact, be implemented on a computer.

● **PROBLEM** 10-48

Determine the Boolean expression of the following circuit, simplify the expression obtained, then represent the simplified expression by its gate representation.

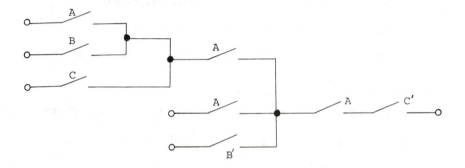

Solution: The Boolean expression for the given circuit is

$(((A+B)+C)A' + A + B')AC'$

$$= ((A+B+C)A' + A + B')AC'$$

$$= (AA' + A'B + A'C + A + B')AC'$$

$$= (0 + A'B + A'C + A + B')AC'$$

$$= AA'BC' + AA'CC' + AAC' + AB'C'$$

$$= 0 + 0 + AC' + AB'C'$$

$$= AC'(1+B')$$

$$= AC'$$

Therefore, the gate representation is:

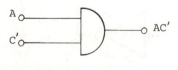

794

(a) Evaluate the following statement by using a truth table:

$$[(A + B) + (A + C) + (A + D)] \cdot [A \cdot \sim B] \qquad (1)$$

The + operator denotes the inclusive OR function and the ·
operator denotes the AND function. NOT is represented by $\sim$.

(b) Change statement (1) into its equivalent switching circuit.

(c) Show the gate representation of statement (1).

Solution: (a) Set up the truth table by specifying all possible
combinations of A,B,C, and D and evaluating the terms of the statement for each combination. Since there are 4 variables, the truth
table has 2^4 = 16 rows of entries as shown in figure 1. First,
evaluate the innermost terms:

A	B	~B	C	D	A + B	A · ~ B	A + C	A + D
0	0	1	0	0	0	0	0	0
0	0	1	0	1	0	0	0	1
0	0	1	1	0	0	0	1	0
0	0	1	1	1	0	0	1	1
0	1	0	0	0	1	0	0	0
0	1	0	0	1	1	0	0	1
0	1	0	1	0	1	0	1	0
0	1	0	1	1	1	0	1	1
1	0	1	0	0	1	1	1	1
1	0	1	0	1	1	1	1	1
1	0	1	1	0	1	1	1	1
1	0	1	1	1	1	1	1	1
1	1	0	0	0	1	0	1	1
1	1	0	0	1	1	0	1	1
1	1	0	1	0	1	0	1	1
1	1	0	1	1	1	0	1	1

Fig. 1

Now, transfer the results in columns 6,8, and 9 to a second truth table,
shown in figure 2 which evaluates the first bracketed term.

A + B	A + C	A + D	(A + B)+(A + C) + (A + D)
0	0	0	0
0	0	1	1
0	1	0	1
0	1	1	1
1	0	0	1
1	0	1	1
1	1	0	1
1	1	1	1
1	1	1	1
1	1	1	1
1	1	1	1
1	1	1	1
1	1	1	1
1	1	1	1
1	1	1	1
1	1	1	1

Fig. 2

795

Finally, transfer the result in the last column of this truth table and the result in column 7 of the first truth table to a third truth table, shown in figure 3, which evaluates the entire statement (1).

A · ~ B	(A + B) + (A + C) + (A + D)	Statement (1)
0	0	0
0	1	0
0	1	0
0	1	0
0	1	0
0	1	0
0	1	0
0	1	0
1	1	1
1	1	1
1	1	1
1	1	1
0	1	0
0	1	0
0	1	0
0	1	0

Fig. 3

The entry in each row of column 3 is the result of performing the operation on the entries in columns 1 and 2 of that row.

(b) The switching circuit equivalent to statement (1) is obtained by connecting, in parallel, variables which are operands of the + operation and connecting, in series, variables which are operands of the · operation. First, find equivalent switching circuits for the innermost terms. These are shown in figure 4.

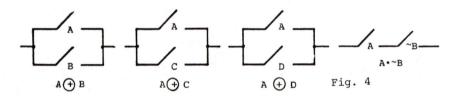

A ⊕ B   A ⊕ C   A ⊕ D   Fig. 4

The switching circuit equivalent to the first bracketed term is a parallel connection of the three parallel circuits A + B, A + C, A + D, as shown in figure 5.

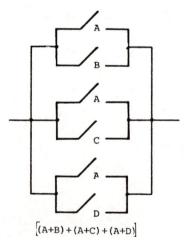

[(A+B) + (A+C) + (A+D)]   Fig. 5

796

The switching circuit equivalent to statement (1) is a series connection of the parallel circuit

$$[(A + B) + (A + C) + (A + D)]$$

and the series circuit $[A \cdot \sim B]$. The complete switching circuit is shown in figure 6.

$$[(A + B) + (A + C) + (A + D)] \cdot [A \cdot \sim B]$$

Note that the circuit conducts if and only if A is closed and B is open, regardless of the settings of C and D. This agrees with the truth table, since statement (1) has value if and only if A = 1 and B = 0.

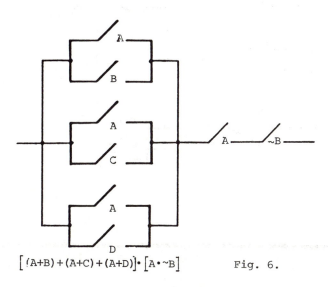

$$\left[(A+B) + (A+C) + (A+D) \right] \cdot \left[A \cdot \sim B \right] \qquad \text{Fig. 6.}$$

(c) Variables which are operands of the + operation are inputs to an OR gate. Variables which are operands of the · operation are inputs to an AND gate. A variable that is the operand of the $\sim$ operation is the input to a NOT gate. The gate representation of the terms in parentheses are shown in figure 7.

Fig. 7.

The composition of 2 operations is done by plugging the output of 1 gate into the input of another, so that the gate representation of $[A \cdot \sim B]$ is shown in figure 8.

Fig. 8.

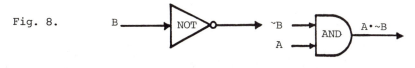

Finally, by the same reasoning, the complete gate representation of

$$[(A + B) + (A + C) + (A + D)] \cdot [A \cdot \sim B]$$

is shown in figure 9.

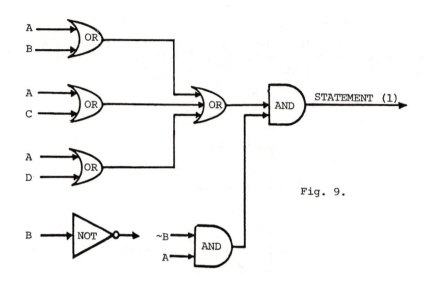

Fig. 9.

(1) Write the Boolean expression for the following circuit.

(2) Find the output Y of the following circuit.

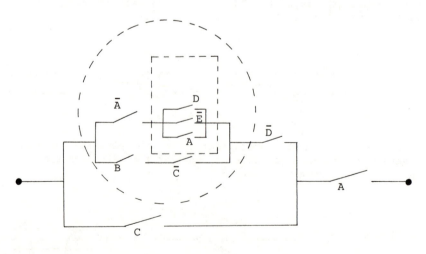

Fig. 1

798

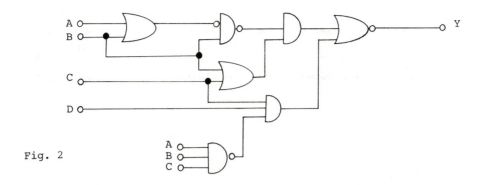

Fig. 2

Solution: (1) In a switching circuit, variables that are operands of the OR operation are connected in parallel, and variables which are operands of the AND operation are connected in series. For instance, the three switches representing variables D, $\overline{E}$ and A (as indicated by the boxed portion in Fig. 1) are connected in parallel. They can therefore be represented by the Boolean expression $D + \overline{E} + A$. The boxed portion of the circuit in Fig. 1 is connected with $\overline{A}$ in series then connected with $B \cdot \overline{C}$ in parallel. Hence the Boolean expression for the circled portion in Fig. 1 is

$$\overline{A} \cdot (D + \overline{E} + A) + B\overline{C}$$

By the same token, after examining the entire circuit one obtains the expression for the given switching circuit as

$$A \cdot \left[\left[(\overline{A} \cdot (D+\overline{E}+A)+B\overline{C}) \cdot \overline{D} \right] + C \right]$$

(2) By finding the inputs and the output of each logic gate in the given circuit as shown in Fig. 3, one has

$$Y = \overline{\{ [(\overline{\overline{A+B}})B](B+C) + DC\ \overline{ABC} \}}$$

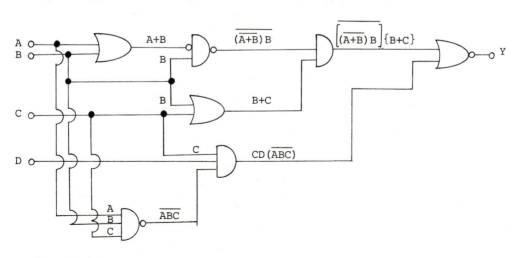

Fig. 3

799

Obtain expressions for the exclusive-OR and the equivalence functions
of two variables A and B.

Solution: The Exclusive-OR function is defined as a function which gives
an output 1 whenever either A or B is 1, but gives an output 0
whenever A and B are both 1 or both 0.
   The equivalence function of two variables A and B is defined as a
function which gives an output 1 whenever both A and B are equal
(both 1 or both 0), but gives an output 0 whenever A and B are
unequal.
   These two functions can be represented by a Truth Table as follows:
If A and B are binary variables, they can be represented in 4
different ways, giving us the 4 rows in the Truth Table. The Truth
Table is shown in figure 1.

Col. 1	Col. 2	Col. 3
VARIABLES A B	EXCLUSIVE-OR A × B	EQUIVALENCE A ≡ B
0 0	0	1
0 1	1	0
1 0	1	0
1 1	0	1

We note from the Truth Table of fig. 1 that the exclusive-OR:and the
equivalence functions are complements of each other.
   Also, from the Truth Table, we can write the equations:

a)   Expressions for exclusive-OR

   $A \times B = A' \cdot B + A \cdot B'$.   (This is a sum of products form of the

expression obtained by considering entries of column 2 which are equal
to 1). Also,

   $A \times B = (A+B) \cdot (A'+B')$.   (This is a product of sums form, obtained
by considering entries of column two which are equal to 0.)

b)   Expressions for Equivalence:

   $A \equiv B = A' \cdot B' + A \cdot B$ (Sum of products form)

   $A \equiv B = (A+B') \cdot (A'+B)$.   (Product of sums form)

   We shall now realize the exclusive-OR and the equivalence functions,
using a)   NAND gates only, and, **b)**   NOR gates only.

(a)

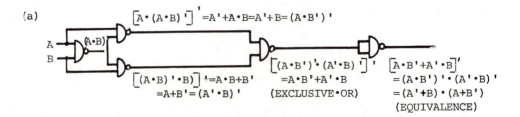

(b)

$[A+(A+B)']'=A' \cdot (A+B)=A' \cdot B$

$[(A+B)'+B)]'=(A+B) \cdot B'$
$= A \cdot B'$

$[A' \cdot B+A \cdot B']'$
$=(A' \cdot B)' \cdot (A \cdot B')'$
$=(A+B') \cdot (A'+B)$
(EQUIVALENCE)

$[A' \cdot B+A \cdot B']$
(EXCLUSIVE·OR)

● **PROBLEM** 10-52

The equivalence (also called biconditional) function of x_1, x_2, written $x_1 \equiv x_2$ (or $x_1 \Leftrightarrow x_2$), is defined as:

$$x_1 \equiv x_2 = (\sim x_1 + x_2) \cdot (x_1 + \sim x_2) \tag{1}$$

a) The $+$ operator denotes the inclusive OR function here; $\cdot$ represents AND and $\sim$ representes the negation operator. Set up the truth table for the equivalence function.

b) Use AND, OR and NOT gates to evaluate the equivalence function.

Solution: (a) The truth table is set up by first specifying the possible combinations of values of x_1 and x_2. There are $2^2 = 4$ of them and they represent the numbers 0 to 3 in the binary system. The next step is to evaluate the terms of statement (1):

x_1	x_2	$\sim x_1$	$\sim x_2$	$\sim x_1 + x_2$	$x_1 + \sim x_2$
0	0	1	1	1	1
0	1	1	0	1	0
1	0	0	1	0	1
1	1	0	0	1	1

Now transfer columns 1,2,5, and 6 to a truth table evaluating statement (1):

x_1	x_2	$\sim x_1 + x_2$	$x_1 + \sim x_2$	$x_1 \equiv x_2 = (\sim x_1 + x_2) \cdot (x_1 + \sim x_2)$
0	0	1	1	1
0	1	1	0	0
1	0	0	1	0
1	1	1	1	1

Note that $x_1 \equiv x_2 = 1$ if and only if $x_1 = x_2$, which is why it is called the equivalence function.

(b) Recall that variables which are operands of the + operation are
inputs to an OR gate and a variable that is the operand of the ~
operation is the input to a NOT gate. Thus, the gate representations
of the terms in parentheses are as shown in Fig. 1.

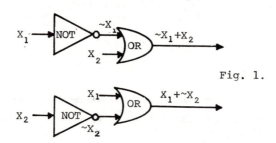

Fig. 1.

Since each term is an operand of the · operation, each of the above
gate circuits is an input to an AND gate. Thus, the gating circuit
of statement (1) is as shown in Fig. 2.

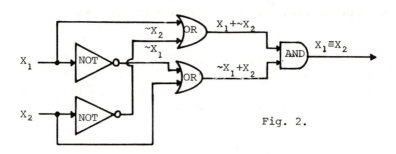

Fig. 2.

Note that the output of this network is 1 if and only if $x_1 = x_2$.
Thus, the network acts as a comparing device. Specifically, it is the
circuit for an equality comparator. An equality comparator compares
one bit, x_1, with another bit, x_2. If two strings of bits are to be
compared, they could be compared serially. That is, compare a bit in
the first position of the first string with a bit in the first position
of the second string, then compare a bit in the next position of the
first string with a bit in the next position if the second string and
continue until either a zero is output, indicating the strings are not
equal, or all positions have been compared, indicating the strings are
equal. Or they could be compared in parallel. That is, a parallel
comparing device that inputs two strings of length n would have n
equality comparators. An output of a string of ones indicates that the
two inputs are equal. The advantage of parallel comparing is its higher
speed compared with series comparison. It will always require k time
units to compare strings where k is the operation time of the equality
comparator. The series comparator requires n·k time units at worst
and k time units at best. The advantage of the series comparator is
its economy. But since gates can be manufactured inexpensively, no commercial computer would use a series comparator, since every instruction
must be identified by using the comparator and hence, there would be a
serious loss of speed.

802

Represent the equivalence function circuit realizations, using:
a)   NAND gates only
b)   NOR gates only

Solution: a) Using NAND gates:
   The equivalence function can be written as:

$A \equiv B = A \cdot B + A' \cdot B'$ [Using the Sum of Products form]

$= [(A \cdot B + A' \cdot B')']'$ [Complementing twice]

$= [(A \cdot B)' \cdot (A' \cdot B')']'$ [Applying DeMorgan's Rules]

$= (A \uparrow B) \uparrow (A' \uparrow B')$ [Using the Shaeffer stroke notation]

The circuit realization is shown in fig. 1 .

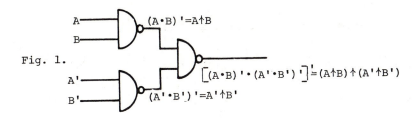

Fig. 1.

b) Using NOR gates:

   We can write

$A \equiv B = (A+B') \cdot (A'+B)$ [Using the Product of Sums form]

$= [((A+B') \cdot (A'+B))']'$ [Complementing twice]

$= [(A+B')'; + (A'+B)']'$ [Applying DeMorgan's rule to the inner primed bracket]

$= [(A \downarrow B') \downarrow (A' \downarrow B)]$.

The circuit realization is shown in figure 2.

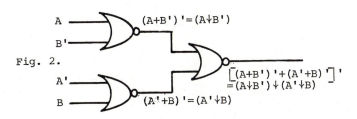

Fig. 2.

Explain the properties of NAND gates, and use NAND gates to generate AND, OR, and NOT operations. Draw the symbol diagrams, and give truth tables for each operation.

Fig. 1.

a
b
$f(a,b)$

Solution: The NAND gate is a combination of an AND gate followed by a NOT gate. In fig. 1 the structure of a NAND gate is shown. The key to understanding a NAND gate is to notice that the output is 0 if and only if its inputs are simultaneously 1.

The Truth table of the NAND gate is given in fig. 2.

a	b	a b	f(a,b)
0	0	0	1
0	1	0	1
1	0	0	1
1	1	1	0

Fig. 2

$f(a,b) = \overline{ab}$

NAND gates are usually represented as in fig. 3 when used in circuit diagrams;

Several other interesting properties of NAND gates are;

$$f(a,b) = \overline{ab} = \bar{a} + \bar{b} \qquad \text{(DeMorgan's Law)}$$

$$f(a,a) = \bar{a} + \bar{a} = \bar{a}$$

$$f(\bar{a},\bar{b}) = \overline{\bar{a}\bar{b}} = a + b \qquad \text{(DeMorgan's Law)}$$

(NAND GATE)

Fig. 3.

a
b
$f(a,b)$

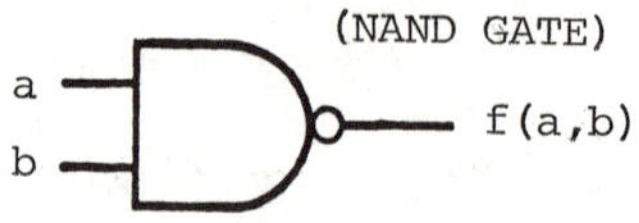

A NAND gate with both of its inputs the same acts like a NOT gate. This idea is illustrated in fig. 4. Truth table for the gate illustrated in fig. 4 is shown in fig. 5.

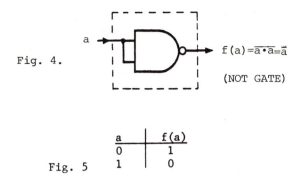

Fig. 4.

$f(a)=\overline{a\cdot a}=\overline{a}$

(NOT GATE)

a	f(a)
0	1
1	0

Fig. 5

Use of NAND gates to generate AND operations;

Two NAND gates connected as shown in fig. 6 generate an AND operation at the output. Since AND gates have the truth table shown in fig. 7.

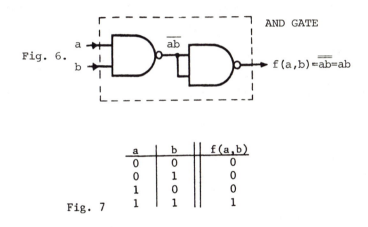

AND GATE

Fig. 6.   a   b $\overline{ab}$ $f(a,b)=\overline{\overline{ab}}=ab$

a	b	f(a,b)
0	0	0
0	1	0
1	0	0
1	1	1

Fig. 7

It can easily be verified that the output in fig. 4 has the same operation. When inputs a and b are both 0, output of the first NAND gate is 1. This output is the input of the second NAND gate, therefore both inputs to the second NAND gate are 1. The output of this NAND gate then is found to be 0, from the definition of NAND gates.

The OR gate can be obtained from three NAND gates, connected as shown in fig. 8.

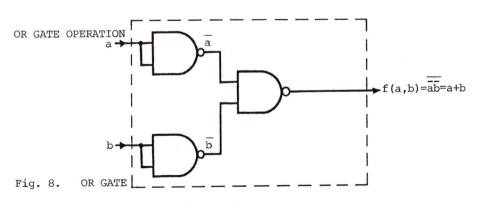

OR GATE OPERATION

a $\overline{a}$ b $\overline{b}$ $f(a,b)=\overline{\overline{a}\overline{b}}=a+b$

Fig. 8.   OR GATE

805

When inputs a and b are both 0, the input to the third NAND gate is 1 1 and its output is 0. In all the other input combinations the output will be 1. The truth table of the OR gate is given in fig. 9.

a	b	f(a,b)
0	0	0
0	1	1
1	0	1
1	1	1

Fig. 9

● **PROBLEM** 10-55

Explain the operation of NOR gates, and generate OR, NOT, and AND operations using NOR gates. Draw symbol diagrams, and explain the functions of each system with the help of truth tables.

<u>Solution</u>: The NOR gate is derived by combining the functions of an OR operator followed by a NOT operator. The key to remembering the function of a NOR gate is the first row of the truth table of fig. 1;

a	b	a + b	f(a,b)
0	0	0	1
0	1	1	0
1	0	1	0
1	1	1	0

$$f(a,b) = \overline{a+b}$$

Fig. 1

The output of a NOR gate is 1 if and only if both inputs are simultaneously 0. The block diagram of a NOR gate is shown in fig. 2.

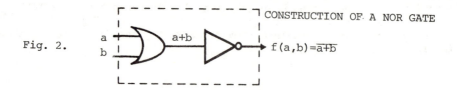

Fig. 2.

CONSTRUCTION OF A NOR GATE

$f(a,b) = \overline{a+b}$

NOR gates may also be used to generate AND, OR, and NOT operations. As seen in fig. 3, an OR gate can be constructed using two NOR gates.

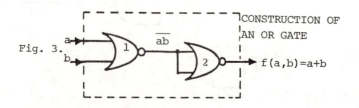

Fig. 3.

CONSTRUCTION OF AN OR GATE

$f(a,b) = a+b$

When the inputs a and b into gate 1 are both 0, the output of
gate 1 (which is the input of gate 2) is 1. Therefore, the output of
gate 2 is 0, following the truth table given for NOR gates. When
one or both inputs are 1 the output of the OR gate is 0. The truth
table of fig. 3(a) illustrates this principle throughly;

a	b	f(a,b)
0	0	0
0	1	1
1	0	1
1	1	1

$$f(a,b) = a + b$$

Fig. 3(a)

Construction of a NOT gate using NOR gates:

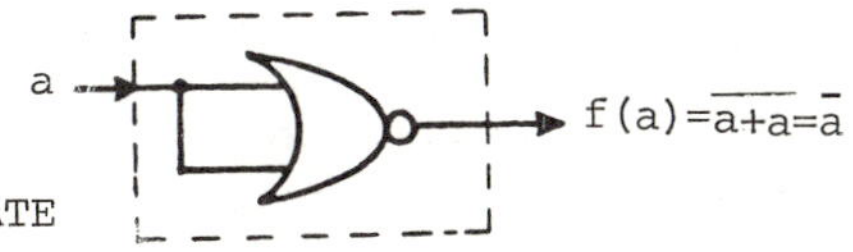

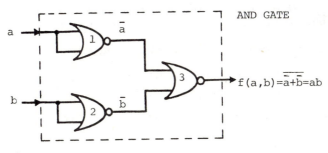

Fig. 4.   NOT GATE $f(a)=\overline{a+a}=\bar{a}$

Operation of the circuit in fig. 4 is illustrated by its truth table
given in fig. 5;

a	f(a)
0	1
1	0

Fig. 5

Fig. 6 shows the construction of an AND gate using NOR gates. 1st
and 2nd NOR gates act as NOT gates and supply $\bar{a}$, and $\bar{b}$ into the
third NOR gate. The output of this gate is then; $f(a,b) = \overline{\bar{a} + \bar{b}} = ab$,
by the use of DeMorgan's Law.

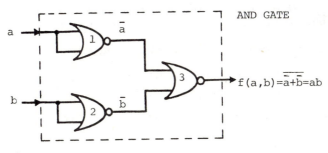

AND GATE

$f(a,b)=\overline{\bar{a}+\bar{b}}=ab$

Fig. 6.

Construct a gate representation for

$$f(x,y,z) = (xy+\overline{z})(x+y)z + xy\overline{z} + (\overline{x+z})$$

with inputs x, y and z.

Solution:

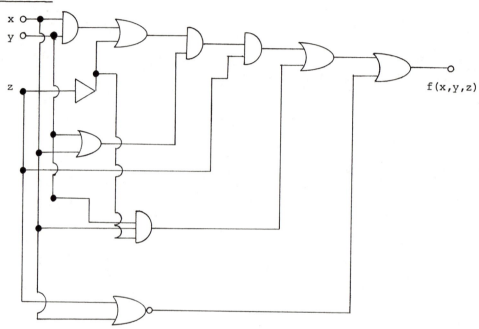

$f(x,y,z)$

Implement the following logic expression using NAND gates only.

$$Y = abc + \overline{a}(b+\overline{c}) + \overline{ab}\overline{c}d + (\overline{cb+d})$$

Solution: One of the methods for solving this kind of problem is the following:

    Step 1: Simplify the expression given.

    Step 2: Express the result from step 1 as a sum of products.

    Step 3: Double invert the result from step 2.

$$Y = abc + \overline{a}(b+\overline{c}) + \overline{ab}\overline{c}d + (\overline{cb+d})$$

$$Y = abc + \overline{ab} + \overline{a}\,\overline{c} + \overline{abc}d + \overline{cb} \cdot \overline{d}$$

$$Y = abc + \overline{ab} + \overline{a}\,\overline{c} + \overline{abc}d + \overline{c}\,\overline{d} + \overline{b}\,\overline{d}$$

Y is simplified by using a K-map as shown in Fig. 1. The result is

$$Y = \overline{a}\,\overline{c} + cb + \overline{d}.$$

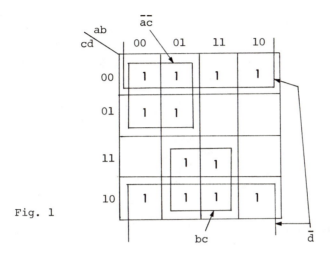

Fig. 1

$$\overline{\overline{Y}} = \overline{\overline{\overline{a}\,\overline{c} + cb + \overline{d}}}$$

$$= \overline{\overline{(\overline{a}\,\overline{c})\,(\overline{cb})\overline{\overline{d}}}}$$

$$= \overline{(\overline{a}\,\overline{c})\,(\overline{cb})d}$$

Observe that the above expression contains only NAND operations. It's implemented as shown in Fig. 2.

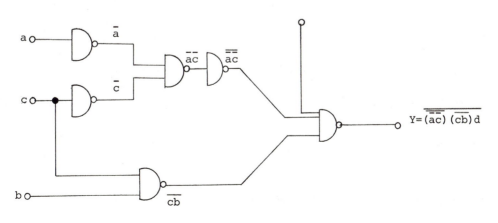

Fig. 2

Implement the following logic expression using only NOR gates:

$$Y = bc + \overline{a}(\overline{b}+c) + abc$$

Solution: One of the methods for solving this kind of problem is the following:

Step 1: Simplify the expression given.

Step 2: Express the result from step 1 as a product-of-sums.

Step 3: Double invert the result from step 2.

Using a K-map to simplify the given expression, one obtains

$$Y = \overline{a}\,\overline{b} + bc$$

Y can be expressed as a product-of-sums as follows:

$$Y = \overline{a}\,\overline{b} + bc$$

$$= (\overline{a}\,\overline{b}+b)(\overline{a}\,\overline{b}+c) \qquad X+YZ = (X+Y)(X+Z)$$

$$= (b+\overline{a})(b+\overline{b})(\overline{a}+c)(\overline{b}+c) \qquad X+YZ = (X+Y)(X+Z)$$

$$= (b+\overline{a})(1)(\overline{a}+c)(\overline{b}+c) \qquad X+\overline{X} = 1$$

$$= (b+\overline{a})(\overline{a}+c)(\overline{b}+c) \qquad X \cdot 1 = X$$

Double invert Y yielding:

$$Y = \overline{\overline{Y}} = \overline{\overline{(b+\overline{a})(\overline{a}+c)(\overline{b}+c)}}$$

$$= \overline{\overline{(b+\overline{a})} + \overline{(\overline{a}+c)} + \overline{(\overline{b}+c)}}$$

Note that expression Y contains NOR operations only.

Y is implemented as shown in the figure.

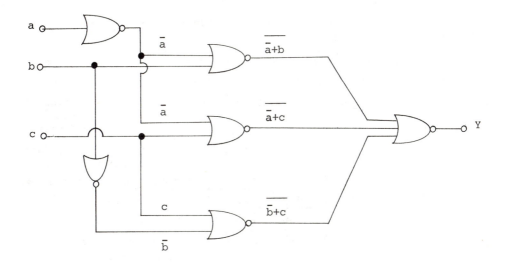

a) Carefully examine the given switching network, find its output, and explain each operation.
b) Draw an equivalent network which will give the same output function as in part (a), using the minimum amount of gates.

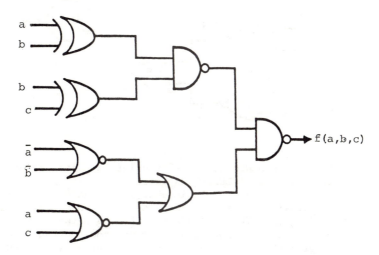

Solution: The circuit and the output of each gate is shown in fig. 1. Gates 1 and 2 are EXCLUSIVE OR gates, 3 and 4 are NOR gates, 5 is an OR gate, 6, and 7 NAND gates.

The output of gate 7 is determined by the use of outputs of gates 5, and 6, and according to the definition of NAND gates (i.e., $f(a,b) = \overline{ab}$). Therefore the output function $f(a,b,c)$ is:

$$f(a,b,c) = \overline{\overline{(a \oplus b)(b \oplus c)} \cdot \overline{\overline{a} + \overline{b} + \overline{a + c}}}$$

811

By the use of Boolean algebra theorems, this function can be reduced as follows:

$$f(a,b,c) = (a \oplus b)(b \oplus c) + \overline{\overline{\overline{a} + \overline{b}} + \overline{a + c}} \quad \text{(DeMorgan's Law)}$$

$$= (a \oplus b)(b \oplus c) + (\overline{a} + \overline{b}) \cdot (a + c) \quad \text{(DeMorgan's Law)}$$

$$= (a\overline{b} + \overline{a}b)(b\overline{c} + \overline{b}c) + (\overline{a} + \overline{b})(a + c) \quad \begin{array}{l}\text{(Definition} \\ \text{of Exclusive} \\ \text{OR function)}\end{array}$$

$$= a\overline{b}b\overline{c} + a\overline{b}\overline{b}c + \overline{a}bb\overline{c} + \overline{a}b\overline{b}c + (\overline{a} + \overline{b})(a + c) \quad \begin{array}{l}\text{(Dis-} \\ \text{tributive Property)}\end{array}$$

$$= 0 + a\overline{b}c + \overline{a}b\overline{c} + 0 + \overline{b}a + \overline{b}c + \overline{a}c \quad \text{(Negation Laws)}$$

$$= a\overline{b}c + \overline{a}b\overline{c} + \overline{a}c + a\overline{b} \quad \text{(Identity Laws)}$$

$$= \overline{b}c + \overline{a}b\overline{c} + \overline{a}c + a\overline{b} \quad \text{(DeMorgan's Law)}$$

$$= \overline{b}c + \overline{a}b + \overline{a}c + a\overline{b} \quad \text{(DeMorgan's Law)}$$

$$= \overline{b}c + \overline{a}b + a\overline{b} \quad \text{(DeMorgan's Law)}$$

$$= \overline{b}c + a \oplus b \quad \text{(Definition of exclusive OR)}$$

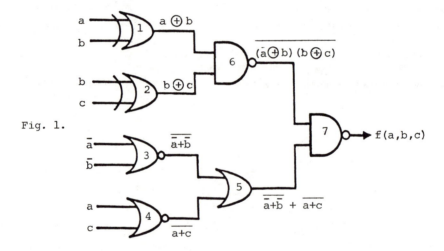

Fig. 1.

This final function can now be translated into a switching network by the use of AND, OR, and EXCLUSIVE OR gates.

The first term $\overline{b}c$ can be obtained by an AND gate, and the second term with an Exclusive OR gate. These functions then must be added by the use of an OR gate. The equivalent switching network and its truth table is given in fig. 2. Note that original and the equivalent circuits both give the same results, only the latter is in reduced form.

a	b	c	$\overline{b}c$	$a \oplus b$	$f(a,b,c)$
0	0	0	0	0	0
0	0	1	1	0	1
0	1	0	0	1	1
0	1	1	0	1	1
1	0	0	0	1	1
1	0	1	1	1	1
1	1	0	0	0	0
1	1	1	0	0	0

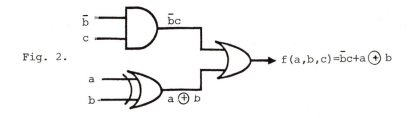

Fig. 2.

$f(a,b,c)=\bar{b}c+a\oplus b$

● **PROBLEM** 10-60

Perform the following conversion :
$1\ 1\ 0\ 1\ 1\ 1\ 0\ 1\ 1_2$ into base 10

Solution:

$1\ 1\ 0\ 1\ 1\ 1\ 0\ 1\ 1_2$ may be converted into base 10 by considering
this fact: Each digit in a base two number may be thought of as a switch,
a zero indicating "off", and a one indicating "on". Also note that each
digit corresponds, in base 10, to a power of two. To clarify, look at
the procedure:

2^8	2^7	2^6	2^5	2^4	2^3	2^2	2^1	2^0
1	1	0	1	1	1	0	1	1_2

$(1\times256)+(1\times128)+(0\times64)+(1\times32)+(1\times16)+(1\times8)+(0\times4)+(1\times2)+(1\times1)\ =443.$

If the switch is "on", then you add the corresponding power of two. If
not, you add a zero.

CHAPTER 11

LINEAR PROGRAMMING AND THE THEORY OF GAMES

SYSTEMS OF LINEAR INEQUALITIES

● **PROBLEM** 11-1

Determine the graph of

(a) x + 4y < 9

(b) 4x + 3y ≥ 20

Solution:

(a) The graph will be a half plane on one side of the line
x + 4y = 9. Let's examine the location of any convenient point
which is not on the line x + 4y = 9, say, the origin (0, 0).
Observe that: at origin x = y = 0, x + 4y = 0 + 4·0 = 0 < 9.
Hence, the graph of x + 4y < 9 is represented by the half
plane under the line as shown in Fig. 1.

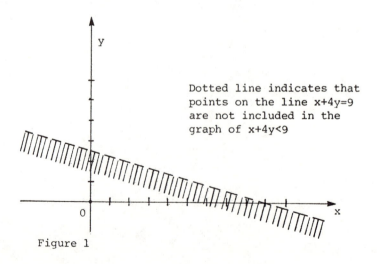

Dotted line indicates that
points on the line x+4y=9
are not included in the
graph of x+4y<9

Figure 1

(b) By the similar procedure one can determine the graph of
4x + 3y ≥ 20 as shown in Fig. 2 .

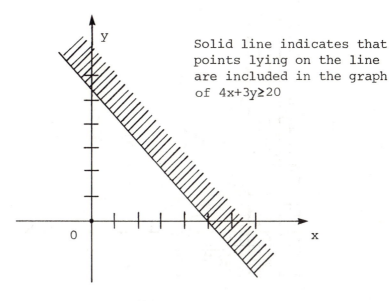

Solid line indicates that
points lying on the line
are included in the graph
of 4x+3y≥20

Figure 2

● **PROBLEM 11-2**

Graph the system

$$\begin{cases} x \geq 4 \\ 2x \leq 18 \end{cases}$$

Solution: $2x \leq 18$ is $x \leq \dfrac{18}{2}$, i.e. $x \leq 9$ hence, the solution
for x is $4 \leq x \leq 9$.

The graph is shown in the figure.

● **PROBLEM 11-3**

Describe geometrically the solutions to the following
system of inequalities:

$$\begin{cases} 4x + y \leq 8 \\ x + y < 4 \end{cases}$$

815

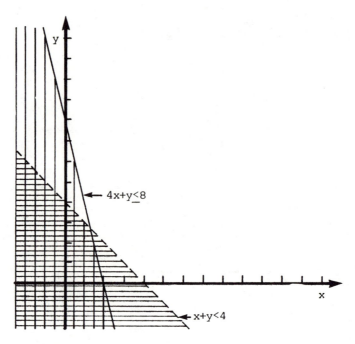

Figure 1

Solution: The graphs of $4x + y \leq 8$ and $x + y < 4$ are shown in Fig. 1, indicated by regions shaded $||||$ and $\equiv$ respectively. The points that satisfy both inequalities simultaneously are the points that make up the solutions to the given system. These are the points that lie in both regions (shaded by $\#\#$). The graph for the solutions are shown in Fig. 2.

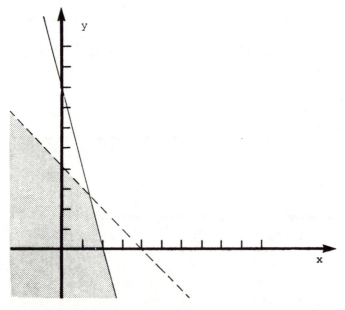

Figure 2

Graph the solutions for the following system

$$\begin{cases} x + 2y \geq 8 \\ x - 2y \geq 2 \\ \qquad x \leq 9 \end{cases}$$

<u>Solution</u>:

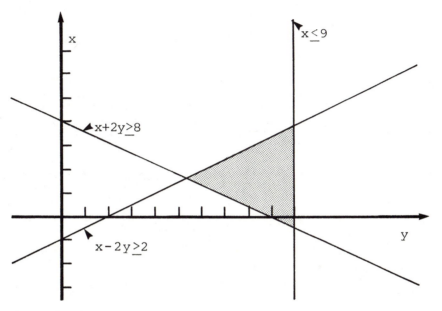

The solutions are indicated by the shaded region.

● **PROBLEM** 11-5

Graph the solution set of the following system of inequalities.

$$\begin{cases} x + y \leq 8 \\ x + y \geq 4 \\ \qquad x \geq 0 \\ 0 \leq y \leq 4 \end{cases}$$

<u>Solution</u>: First draw lines $x + y = 8$, $x + y = 4$, $x = 0$, $y = 0$, and $y = 4$ as shown in Fig. 1.

The common area of the graphs of $x + y \leq 10$, $x + y \geq 4$, $x \geq 0$, and $0 \leq y \leq 4$ is then the graph of the solution set of the given system.

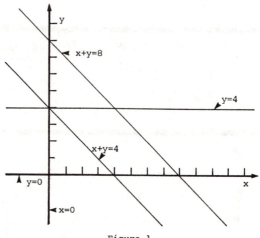

Figure 1

The graph of x + y ≤ 8 is the region below the line x + y = 8, the graph for x + y ≥ 4 is the area above the line x + y = 4, the region on the right of x axis is the graph for x ≥ 0, and the region between the lines y = 4 and y = 0 gives the graph of 0 ≤ y ≤ 4

The graph of the solution of the given system is the shaded region as shown in Fig. 2.

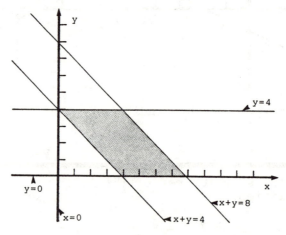

Figure 2

● **PROBLEM** 11-6

Graph the solutions of the following system

$$\begin{cases} y \geq 0 \\ x - y + 8 \geq 0 \\ x + y \geq 0 \\ 11x - y - 44 \leq 0 \end{cases}$$

The graph of the solution set is shown in Fig. 1.

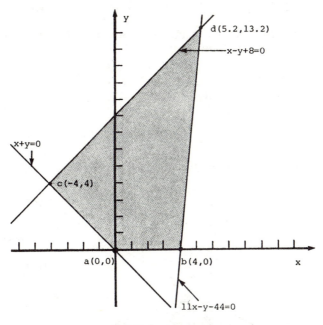

Figure 1

Each given inequality has a graph of plane area, the region that is common to all four plane areas described by the inequalities is the solution set. The solution set is a convex polygon with 4 vertices a, b, c, and d as shown in Fig. 1. These vertices are found by solving systems of equations

(a) $\begin{cases} y = 0 \\ x + y = 0 \end{cases}$ (b) $\begin{cases} y = 0 \\ 11x - y - 44 = 0 \end{cases}$

(c) $\begin{cases} x + y = 0 \\ x - y + 8 = 0 \end{cases}$ and (d) $\begin{cases} x - y + 8 = 0 \\ 11x - y - 44 = 0 \end{cases}$

The solutions are indicated in Fig. 1.

● **PROBLEM 11-7**

(a) Sketch the set of solutions to the following system:

$$2x + y \le 2$$
$$-x + 3y \le 4$$
$$x \ge 0$$
$$y \ge 0$$

(b) Find the vertices of the solution set.

Solution: One may first sketch the lines

819

$$2x + y = 2, \quad -x + 3y = 4, \quad x = 0 \text{ and } y = 0$$

(clearly $x = 0$ is y - axis and $y = 0$ is x - axis).

The set of solutions for this system is the area common to all areas that are defined by

$$2x + y \leq 2$$
$$-x + 3y \leq 4$$
$$x \geq 0$$
$$y \geq 0$$

This is indicated in the following figure.

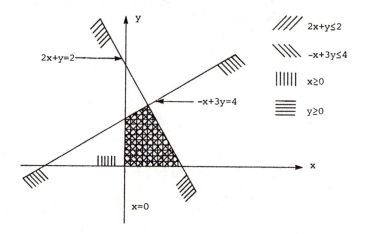

(b) By solving the following system of equations:

$$\begin{cases} -x + 3y = 4 \\ \quad x = 0 \end{cases}$$

one can determine that vertex A is the point $y = \frac{4}{3}$ and $x = 0$, $(0, \frac{4}{3})$.

Similarly by solving $\begin{cases} 2x + y = 2 \\ -x + 3y = 4 \end{cases}$ $\begin{cases} x = \frac{2}{7} \\ y = \frac{10}{7} \end{cases}$

one obtains the following coordinates for point B:

$(\frac{2}{7}, \frac{10}{7})$.

Finally, vertex C can be found by solving

$$\begin{cases} 2x + y = 2 \\ \quad y = 0 \end{cases} \qquad \begin{cases} x = 1 \\ y = 0 \end{cases}$$

Thus, C is the point $(1, 0)$.

820

Therefore, the vertices of the solution set are

A$(0, \frac{4}{3})$, B$(\frac{2}{7}, \frac{10}{7})$, C$(1, 0)$ and 0 $(0, 0)$.

GEOMETRIC SOLUTIONS AND DUAL OF LINEAR PROGRAMMING PROBLEMS

● **PROBLEM** 11-8

Give an example of a problem that is amenable to linear programming methods.

Solution: An appropriate description of linear programming would be the maximization of a linear function of n non-negative variables, subject to m linear constraints. The key word is linear. For maximizing a non-linear function subject to constraints, the method of Lagrange multipliers is available.

One of the first problems to be posed in linear programming is known as the diet problem. The essential idea is that a nutritionist wishes to find the cheapest diet that meets fixed nutritional requirements.

Suppose only two goods, A and B, are available. Let each ounce of food A contain 2 units of protein, 1 unit of iron and 1 unit of thiamine. Each ounce of food B contains 1 unit of protein, 1 unit of iron and three units of thiamine. Suppose that each ounce of A costs 30 cents while an ounce of B costs 40 cents. The nutritionist wants the meal to provide at least 12 units of protein, at least 9 units of iron and at least 15 units of thiamine. How many ounces of each of the foods should be used to minimize the cost of the meal?

We can cast the above problem into mathematical form. Notice that the two foods A and B have constant marginal cost.

Let x and y denote the number of ounces of foods A and B respectively. Then the number of units of protein supplied by the meal is
$$2x + y .$$
Thus, the first constraint is
$$2x + y \geq 12 \qquad (1)$$
The inequality in (1) indicates that consumption of protein can exceed 12 units without destroying the solution of the problem.

Similarly we obtain the iron and thiamine constraints:
$$x + y \geq 9$$
$$x + 3y \geq 15.$$
Finally, we require that the amounts of food A and food B bought be non-negative, i.e.,
$$x \geq 0 , y \geq 0 .$$
We wish to minimize the cost of the meal which is
$$z = 30x + 40y \qquad (2)$$
The equation (2) is the objective that we wish to minimize. It is called the objective function.

The linear programming problem is therefore: Find values of x and y that minimize
$$z = 30x + 40y$$

subject to the restrictions
$$2x + y \geq 12$$
$$x + y \geq 9$$
$$x + 3y \geq 15$$

x ≥ 0 , y ≥ 0 .

We could solve the above problem graphically but if the number of unknowns were higher, a numerical method of solution known as the simplex
algorithm would be used.

● **PROBLEM 11-9**

Show algebraically that $S = \{(x_1, x_2) \in R^2 | x_1 + x_2 \geq 2\}$
is convex.

Solution: If S is convex, then all points on the line
that connect any two points in S are also in S.

S is convex, a fact obvious from a graph of S. To
prove this algebraically, take any two points $P = (p_1, p_2)$
and $Q = (q_1, q_2)$ in S. Then $p_1 + p_2 \geq 2$ and $q_1 + q_2 \geq 2$.
Take any point:

$$tP + (1 - t)Q = [tp_1 + (1 - t)q_1, tp_2 + (1 - t)q_2],$$

$$0 \leq t \leq 1$$

on the line segment between P and Q. Then,

$$tp_1 + (1 - t)q_1 + tp_2 + (1 - t)q_2$$

$$= t(p_1 + p_2) + (1 - t)(q_1 + q_2)$$

$$\geq 2t + 2(1 - t) = 2$$

using the fact that t and 1 - t are nonnegative. Thus
$tP + (1 - t)Q$ is in S and, therefore, this is an algebraic
proof that S is convex.

● **PROBLEM 11-10**

Is the function $f(x) = 7x + 4$ convex or concave?

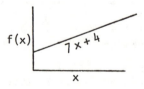

Solution:

$$\frac{df}{dx} = +7$$

$$\frac{d^2f}{dx^2} = 0$$

Since $\frac{d^2f}{dx^2} = 0$, the function is both convex and concave.

Solve the following linear programming problem:

$$\text{Maximize} \quad 6L_1 + 11L_2 \tag{1}$$

subject to:

$$2L_1 + L_2 \leq 104$$
$$L_1 + 2L_2 \leq 76 \tag{2}$$

and $L_1 \geq 0$, $L_2 \geq 0$.

Solution: Since there are only two variables $(L_1$ and $L_2)$, we can portray the above problem geometrically.

The two constraints $L_1 \geq 0$, $L_2 \geq 0$ generate the first quadrant of the Euclidean plane. The lines $L_1 = 0$, $L_2 = 0$ form the boundaries of this region.

Instead of considering the inequalities directly, we first find the boundaries of the constraint region. This is equivalent to graphing the two lines $2L_1 + L_2 = 104$ and $L_1 + 2L_2 = 76$.

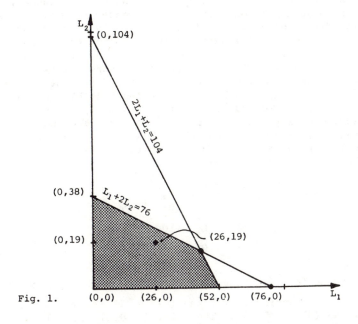

Fig. 1.

The feasible region is the area that simultaneously satisfies all four constraints. In Fig. 1, it is the darkened region. The feasible region represents possible solutions to the maximization problem. For example, (26,19) is a feasible solution, and when substituted into the objective function yields a value of

$$6(26) + 11(19) = 365.$$

Yet a better solution is the point (0,38) on the boundary of the feasible region which yields a value of

$$6(0) + 11(38) = 418.$$

What we require is a method that will reduce the infinite set of feasible points to a finite number of points. Any point not on the boundary of the feasible region is non-optimal, since we can increase (1) by increasing one variable while keeping the other constant and still satisfy the constraints.

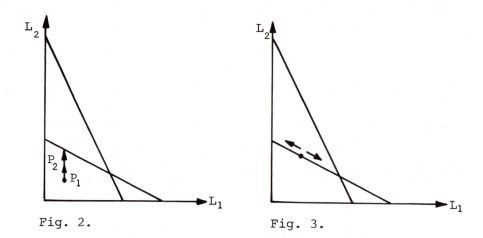

Fig. 2.   Fig. 3.

We can continue to do this until we meet a constraint line as in Fig. 2. Thus, the set of optimal points consists of points on the boundary of the feasible region. Now consider a point on the boundary. If movement along the boundary in one direction increases (1), then movement in the opposite direction decreases (1). Now we can keep on moving in the direction of increasing (1) until we meet another constraint. Thus the points at which (1) is locally maximized occur at the intersection of two or more constraints. Since the number of constraints is finite the set of such points is finite and every point in this set yields a greater value for (1) than any other point in the feasible region.

We now have a method for solving the given problem. First we find the intersection points of ·pairs of constraints. Then we check which points are feasible. Finally, from among these basic feasible points we find that point which maximizes the objective function.

The four constraints are:

$$L_1 = 0 \tag{3}$$
$$L_2 = 0 \tag{4}$$
$$2L_1 + L_2 = 104 \tag{5}$$
$$L_1 + 2L_2 = 76 \tag{6}$$

Taking these four equations two at a time we have $\binom{4}{2}$ or 6 points. For example, the solution to (3) and (4) is the point $(0,0)$. The solution to (4) and (6) is $(76,0)$. The point $(0,0)$ is feasible but the point $(76,0)$ is not. In this way we obtain the following table:

System of equations	(L_1, L_2)	Feasible Point
(3) and (4)	(0,0)	yes
(3) and (5)	(0,104)	no
(3) and (6)	(0,38)	yes
(4) and (5)	(52,0)	yes
(4) and (6)	(76,0)	no
(5) and (6)	(44,16)	yes

Next we compute the value of (1) for each of the above feasible points.

Feasible point	$f(L_1, L_2) = 6L_1 + 11L_2$
(0,0)	0
(0,38)	418
(52,0)	312
(44,16)	440

Thus, the objective function is maximized when $L_1 = 44$ and $L_2 = 16$. The value of the objective function is then 440.

● **PROBLEM 11-12**

A company makes desk organizers. The standard model requires 2 hours of the cutter's and one hour of the finisher's time. The deluxe model requires 1 hour of the cutter's time and 2 hours of the finisher's time. The cutter has 104 hours of time available for this work per month , while the finisher has 76 hours of time available for work. The standard model brings a profit of $6 per unit, while the deluxe one brings a profit of $11 per unit. The company, of course, wishes to make the most profit. Assuming they can sell whatever is made, how much of each model should be made in each month?

Solution: The company wishes to make the most profit within the given constraints. We graph the constraints and within the defined region we pick the point with the most profit. The profit is found by the formula:

Profit = $6 X + $11 Y

where X stands for the number of standard desk organizers and Y stands for the number of deluxe ones.

The constraints for this problem are:

(1) $X \geq 0$; we cannot have a negative number of standard units.

(2) $Y \geq 0$; we cannot have a negative number of deluxe units.

(3) The finisher has only 76 hours of time available. Since a standard model takes one hour of the finisher's time, and a deluxe model takes 2 hours of the finisher's time, we get the constraint

$$X + 2 Y \leq 76.$$

(4) The cutter has only 104 hours of time available. A standard unit takes two hours of the cutter's time, and a deluxe unit takes one hour of the cutter's time, thus, we get the constraint

$$2 X + Y \leq 104.$$

We can now graph these constraints to get the region in

825

which we can choose our point of maximum profit.

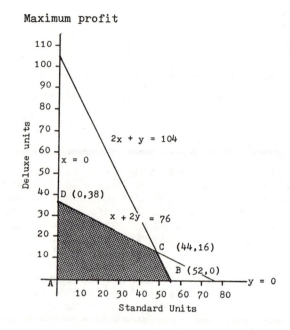

Maximum profit

The shaded area of the graph is the area which conforms to the constraints. Within this region we must pick the point with the maximum profit. By a theorem of linear programming we know that the point of maximum profit occurs at a corner of the region. Thus, we need only check the corners and take the point with the most profit.

A)   (0,0)   Profit = $6 (0) + $11 (0) = $0
B)   (52,0)   Profit = $6 (52) + $11 (0) = $312
C)   (44,16) Profit = $6 (44) + $11 (16) = $440
D)   (0,38)   Profit = $6 (0) + $11 (38) = $418

By observation, we note that the point with the largest profit is (44,16) (Point C). Thus, for the company to make the maximum profit of $440, they must produce 44 standard units and 16 deluxe ones.

● **PROBLEM 11-13**

A marketing manager wishes to maximize the number of people exposed to the company's advertising. He may choose television commercials, which reach 20 million people per commercial, or magazine advertising, which reaches 10 million people per advertisement. Magazine advertisements cost $40,000 each while a television advertisement costs $75,000. The manager has a budget of $2,000,000 and must buy at least 20 magazine advertisements. How many units of each type of advertising should be purchased?

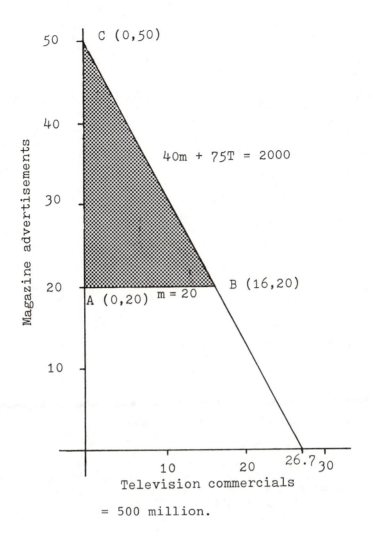

Solution: We find the constraints of the problem, and graph them to find the region defined by them. From this region we will pick the point which maximizes the number of people exposed to the advertisements.

The constraints are:

Let T stand for the number of television commercials and M stand for the number of magazine advertisements.

(1) $T \geq 0$ We cannot have a negative number of television commercials

(2) $M \geq 20$ We must have at least twenty magazine advertisements.

(3) This constraint comes from the costs. In thousands the cost of a television commercial is $75 and the cost of a magazine advertisement is $40. He is budgeted to $2,000,000 so we get the constraint

$$40M + 75T \leq 2,000.$$

We now graph these constraints to find the region which is defined by them.

The shaded area is the region which is defined by the constraints. To find the point which yields the highest number of people exposed to the advertisement can be found by a theorem of linear programming which states that the point must be one of the corners of the region.

A) (0,20) number of people = 20 million × T + 10 million × M = 20 million × 0 + 10 million × 20 = 200 million.

B) (16,20) number of people = 20 million × 16 + 10 million × 20 = 520 million.

C) (0,50) number of people = 20 million × 0 + 10 million × 50

Thus, the best thing for the manager to do is have 16 television commercials and 20 magazine advertisements.

● **PROBLEM 11-14**

A certain textile mill finishes cotton cloth obtained from weaving mills. The mill turns out two styles of cloth, a lightly printed style and a heavily printed one. The mill's output during a week is limited only by the capacity of its equipment for two of the finishing operations -- printing and bleaching -- and not by demand considerations. The maximum weekly output of the printing machinery is 800 thousand yards of cloth if the light pattern is printed exclusively, 400 thousand yards if the heavy pattern is printed exclusively, or any combination on the printing line L + 2H = 800 (where L represents light pattern and H heavy pattern). In a week, the maximum the bleaching equipment can handle is 500 thousand yards of the light-patterned cloth exclusively, 550 thousand yards of the heavy-patterned cloth exclusively, or any combination on the bleaching line, 1.1 L + H = 550. The mill gained $300 and $290 per thousand yards of the light -- and heavy-patterned cloths, respectively.
1) Draw the graph of the two lines described above.
2) Solve the linear programming problem of maximizing the gain from a week's production.

Solution: 1)

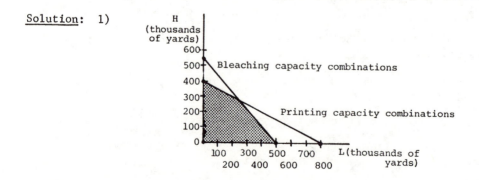

828

2) We will use the graphical approach to solve this linear programming problem. We will utilize the graph above. Looking at the graph, we notice that any combination of L and H that falls in the area between the two axes and the printing line can be printed during a week. Again, any combination of L and H that falls in the area between the two axes and the bleaching line can be bleached in one week.

The rules for solving a linear programming problem through a graphical approach are:

i) Graph all the constraining inequalities to obtain a picture of the feasible region.
ii) Solve the corresponding equations to find the vertices, or corners of the feasible area.
iii) Find the vertex point which yields the maximum (or minimum) value of the function under consideration.

Thus, the vertices are:

Output (Thousands of Yards)

Light Pattern	Heavy Pattern
0	400
250	275
500	0
0	0

The vertices $(0,400)$, $(500,0)$, $(0,0)$ are obtained easily from the graph. The fourth vertex $(250, 275)$ is obtained from the simultaneous equations:

$$L + 2H = 800$$
$$1.1L + H = 550$$

From the first equation, $L = 800 - 2H$. Substituting this into the second equation yields

$$1.1(800 - 2H) + H = 550$$

or

$$880 - 2.2H + H = 550$$
$$- 1.2H = -330$$
$$1.2H = 330$$
$$H = 330/1.2$$
$$H = 275 \quad .$$

Thus, $L + 2H = 800$ becomes $L + 2 \times 275 = 800$, $L = 250$.
The mill gained \$300 and \$290 per thousand yards of the light- and heavy-patterned cloths, respectively. Thus, for the four vertices, the gain from the week's production is:

at $(0,400)$: $\$300.0 + \$290 \cdot 400 = \$116,000$
at $(250,275)$: $\$300 \cdot 250 + \$290 \cdot 275 = \$155,000$
at $(500,0)$: $\$300 \cdot 500 + \$290 \cdot 0 = \$150,000$
at $(0,0)$: $\$300 \cdot 0 + \$290 \cdot 0 = \$ 0.$

Thus, if the textile mill wants to obtain the highest possible gain from the week's operations, the combination 250 light pattern, 275 heavy pattern should be selected.

A businessman needs 5 cabinets, 12 desks, and 18 shelves cleaned out. He has two part time employees Sue and Janet. Sue can clean one cabinet, three desks and three shelves in one day, while Janet can clean one cabinet, two desks and 6 shelves in one day. Sue is paid $25 a day, and Janet is paid $22 a day. In order to minimize the cost how many days should Sue and Janet be employed?

<u>Solution:</u>   The businessman wishes to minimize the cost of cleaning out the office. To do this we must graph the constraints, and take the point which gives the minimum cost. To find the cost we use the formula

$$\text{Cost} = \$25\ X + \$22\ Y,$$

where X is the number of days Sue is employed and Y is the number of days Janet is employed. We now will find the constraints.

(1) Since Sue can do a cabinet in one day, and Janet can do a cabinet in one day, and we must have at least 5 cabinets cleaned, we get the constraint

$$X + Y \geq 5.$$

(2) Since Sue can do 3 desks in one day, and Janet can do 2 desks in one day, and we require 12 desks to be cleaned, we have the constraint

$$3X + 2Y \geq 12.$$

(3) Similarly the constraint $3X + 6Y \geq 18$ comes from Sue being able to clean 3 shelves and Janet being able to clean 6 shelves in one day.

(4) $X \geq 0$ Sue cannot work a negative number of days

(5) $Y \geq 0$ Janet cannot work a negative number of days.

We now graph the constraints to find the region described by them.

The shaded area is the region described by the constraints. Note that this is an infinite region, because they can work more days than is needed. To find the minimum point, we refer to a theorem of linear programming which states that a minimum cost must occur in one of the corners. Thus, we need only check which of the four corners has the smallest cost, and we will have the answer.

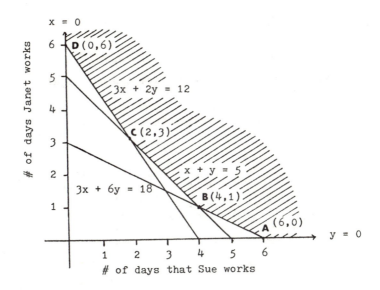

x = 0

3x + 2y = 12

C(2,3)

D(0,6)

x + y = 5

3x + 6y = 18

B(4,1)

A(6,0)

y = 0

6
5
4
3
2
1

1 2 3 4 5 6

# of days that Sue works

# of days Janet works

A)   (6,0) $25 (6)   + $22 (0) = $150

B)   (4,1) $25 (4)   + $22 (1) = $122

C)   (2,3) $25 (2)   + $22 (3) = $116

D)   (0,6) $25 (0)   + $22 (6) = $132

We can now see that the minimum cost is, with Janet working 3 days and Sue working 2 days, equal to $116. (Point C on the graph.)

● **PROBLEM 11-16**

Convert the following standard linear programming problem to canonical form:

Maximize

$$8x_1 + 15x_2 + 6x_3 + 20x_4$$

subject to:

$$x_1 + 3x_2 + x_3 + 2x_4 \leq 9$$
$$2x_1 + 2x_2 + 2x_3 + 3x_4 \leq 12$$
$$3x_1 + 3x_2 + 2x_3 + 5x_4 \leq 16$$

$x_1, x_2, x_3, x_4 \geq 0$.

Solution: The general standard linear programming problem takes the form

$$\text{Max } f(x_1, \ldots, x_n) = b_1 x_1 + b_2 x_2 + \ldots + b_n x_n \qquad (1)$$

subject to:

$$a_{11} x_1 + a_{12} x_2 + \ldots + a_{1n} x_n \leq c_1$$
$$a_{21} x_1 + a_{22} x_2 + \ldots + a_{2n} x_n \leq c_2 \qquad (2)$$
$$\vdots \qquad \vdots \qquad \qquad \vdots \qquad \vdots$$
$$a_{m1} x_1 + a_{m2} x_2 + \ldots + a_{mn} x_n \leq c_n$$

831

and $x_i \geq 0 \quad (i = 1, 2, \ldots, n)$. $\qquad\qquad$ (3)

The function $f(x_1, \ldots, x_n)$ is the objective function. It is a real-valued function whose argument comes from the set of all n-tuples that satisfy (2) and (3).

We can rewrite the above program in matrix form. Thus,

$\qquad$ Max $\qquad f(X) = (B \cdot X)$

$\qquad$ subject to: $\quad AX \leq C$, $X \geq 0$,

where $B = (b_1, \ldots, b_n)$, $X = (x_1, x_2, \ldots, x_n)$, $C = (c_1, \ldots, c_n)$ and

$$A = \begin{bmatrix} a_{11} & a_{12} & \cdots & \cdots & a_{1n} \\ a_{21} & & & & \\ \cdot & & & & \cdot \\ \cdot & & & & \cdot \\ \cdot & & & & \cdot \\ a_{m1} & a_{m2} & \cdots & \cdots & a_{mn} \end{bmatrix}$$

The solution of the problem is facilitated by converting the inequalities to equalities. We do this by introducing m new variables, one for each of the inequalities. Thus, let

$$x_{n+1} = c_1 - (a_{11}x_1 + a_{12}x_2 + \ldots + a_{1n}x_n)$$
$$\vdots$$
$$x_{n+m} = c_m - (a_{m1}x_1 + a_{m2}x_2 + \ldots + a_{mn}x_n)$$

Then the system of inequalities (2) becomes the system of equalities

$$a_{11}x_1 + a_{12}x_2 + \ldots + a_{1n}x_n + x_{n+1} = c_1$$
$$a_{21}x_1 + a_{22}x_2 + \ldots + a_{2n}x_n + x_{n+2} = c_2$$
$$\vdots$$
$$a_{m1}x_1 + a_{m2}x_2 + \ldots + a_{mn}x_n + x_{n+m} = c_m$$

By increasing the number of unknowns from n to $n+m$, we obtain a system of equalities: the analysis of such systems is much more developed than the study of systems of inequalities. When the inequalities in the constraint inequations are converted to equalities, the program is in canonical form.

The given problem in canonical form is:

$\qquad$ Maximize

$$8x_1 + 15x_2 + 6x_3 + 20x_4$$

$\qquad$ subject to

$$x_1 + 3x_2 + x_3 + 2x_4 + x_5 = 9$$
$$2x_1 + 2x_2 + 2x_3 + 3x_4 + x_6 = 12$$
$$3x_1 + 3x_2 + 2x_3 + 5x_4 + x_7 = 16$$
$$x_i \geq 0 \quad (i = 1, 2, \ldots, 7).$$

In order to produce 1000 tons of non-oxidizing steel for engine valves, at least the following units of manganese, chromium and molybdenum, will be needed weekly: 10 units of manganese, 12 units of chromium, and 14 units of molybdenum (1 unit is 10 pounds). These metals are obtainable from dealers in non-ferrous metals, who, to attract markets make them available in cases of three sizes, S, M and L. One S case costs \$9 and contains 2 units of manganese, 2 units of chromium and 1 unit of molybdenum. One M case costs \$12 and contains 2 units of manganese, 3 units of chromium, and 1 unit of molybdenum. One L case costs \$15 and contains 1 unit of manganese, 1 unit of chromium and 5 units of molybdenum.

How many cases of each kind should be purchased weekly so that the needed amounts of manganese, chromium and molybdenum are obtained at the smallest possible cost? What is the smallest possible cost?

Solution: The general linear programming problem has the following form:

$$\text{Minimize} \quad z = c_1 x_1 + c_2 x_2 + \ldots + c_n x_n \tag{1}$$

Subject to:

$$a_{11} x_1 + a_{12} x_2 + \ldots + a_{1n} x_n \geq b_1$$
$$\vdots \qquad \vdots \qquad \qquad \vdots \qquad \vdots \tag{2}$$
$$a_{m1} x_1 + a_{m2} x_2 + \ldots + a_{mn} x_n \geq b_m$$

$x_i \geq 0$ $(i = 1, 2, \ldots, n)$.

The linear function (1) is called the objective function while the inequalities (2) represent the constraints.

To cast the above problem into this form, we first find the objective function. Since we are buying cases we want to minimize

$$z = 9S + 12M + 15L . \tag{3}$$

To obtain the constraints, we see that at least 10 units of manganese are required. Each small case contains 2 units, each medium case contains 2 units and each large case contains 1 unit. Thus, the manganese constraint is

$$2S + 2M + L \geq 10 .$$

Similarly we obtain the constraints on chromium and molybdenum:

$$2S + 3M + L \geq 12$$
$$S + M + 5L \geq 14 .$$

Thus the problem is:

$$\text{Minimize} \quad z = 9S + 12M + 15L \tag{3}$$

Subject to:

$$2S + 2M + L \geq 10 \tag{4}$$
$$2S + 3M + L \geq 12 \tag{5}$$
$$S + M + 5L \geq 14 \tag{6}$$
$$S \geq 0 , M \geq 0 , L \geq 0 . \tag{7}$$

The region common to the constraints (4)-(7) is called the feasible region. From the theory of linear programming we know that a point at which the objective function is optimized must be on the boundary. Hence we can convert the inequalities in (4)-(7) to equations. A fundamental theorem of linear programming is that if a solution exists it must be at a point of intersection of three equations. We have 6 constraints and three unknowns; hence the number of possible solutions is $\binom{6}{3}$ or 20. For example, the point (0,0,0), corresponding to the solution set of S = M = L = 0, is a possible solution. But since it fails to satisfy the constraints (4)-(6) it is not a feasible solution. On the other hand, the solution to S = 0, 2S + 2M + L = 10 and 2S + 3M + L = 12 is (0,2,6) which is a feasible point. Substituting into the objective function (3):

$$z = 9(0) + 12(2) + 15(6) = \$114.$$

Proceeding in this way for every triplet of equations we find that the solution to (4)-(6) is (2,2,2). This point satisfies all the constraints, i.e., it is feasible. Substituting S = 2, M = 2 and L = 2 in (3):

$$z = 9(2) + 12(2) + 15(2) = \$72.$$

This is the minimum cost and is the solution to the problem. We should buy 2 small cases, 2 medium cases and 2 large cases to obtain this minimum cost.

● **PROBLEM 11-18**

A manufacturer of electronic instruments produces two types of timer: a standard and a precision model with net profits of $10 and $15, respectively. His work force cannot produce more than 50 instruments per day. Moreover, the four main components used in production are in short supply so that the following stock constraints hold:

Component	Stock	Number used per timer	
		Standard	Precision
a	220	4	2
b	160	2	4
c	370	2	10
d	300	5	6

Graphically determine the point of optimum profit. If profits on the standard timer were to change, by how much could they change without altering the original solution?

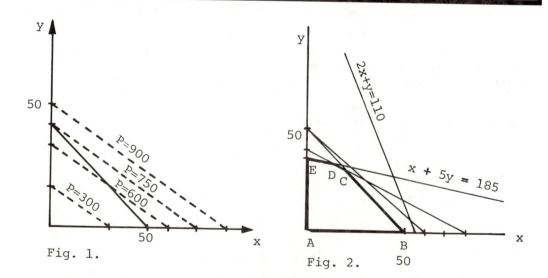

Fig. 1.

Fig. 2.

Solution: Let x and y denote the number of standard and precision timers produced respectively. Then the manufacturer's problem is to

Maximize 10x + 15y

subject to x + y ≤ 50

834

$$4x + 2y \leq 220$$
$$2x + 4y \leq 160$$
$$2x + 10y \leq 370$$
$$5x + 6y \leq 300$$

$x, y \geq 0$.

Begin the analysis by considering only the first constraint and the profit line as in Fig. 1. The feasible region is that portion of the xy plane that satisfies $x \geq 0$, $y \geq 0$ and $x+y \leq 50$. Now consider $P = 10x + 15y$. $P(x,y)$ is a function of two variables and we would require a three dimensional graph to plot it. But if we let P be a constant P_0, we obtain the linear equation

$$P_0 = 10x + 15y ,$$

(1)

or,

$$y = \frac{P_0}{15} - \frac{2}{3} x .$$

Lines of the form (1) are isoprofit lines, i.e., at all points on this line, the profit is the same. By substituting various values for P_0 we obtain a series of parallel lines cutting the x and y axis at $P_0/10$ and $P_0/15$ respectively, and having a slope of $-10/15 = -2/3$. Note that as we move further away from the origin, P increases. If $P = 300$, all combinations of x and y are within the feasible region. This suggests increasing P until it touches the feasible region at as few points as possible. When $P = 750$, the profit line is tangent to the feasible region at $y = 50$. Hence profit is maximized when 50 precision timers and 0 standard timers are produced. Now, add in the other constraints to obtain Fig. 2.

The feasible region now consists of the polygon ABCDE. We would obtain the optimum profit point by again drawing in the isoprofit lines. Maximum profit occurs at C.

Now, suppose the profit on the standard timer is changed. Examining Fig. 2 we can see that the profit line will always pass through C provided its slope lies between those of the lines DC and CB. Suppose it becomes the same as that of DC($-\frac{1}{2}$). The profit line is then $P = 7.5x + 15y$ and the optimum occurs anywhere along DC so that $x = 20$, $y = 30$ and $x = 10$, $y = 35$ both give the same maximum value for P of 600. Suppose the slope is varied again by reducing the profit on the standard model to 7, giving $P = 7x + 15y$ (slope $-7/15$); we find the optimum has now moved to the point $D(x = 10, y = 35)$ with a corresponding maximum profit of \$595.

● **PROBLEM** 11-19

The solution to the problem:

Minimize $45I + 50L + 60A$ (1)

subject to

$$I + 2L + 4A \geq 6$$
$$3I + 3L + 2A \geq 3$$
$$I + 3L + A \geq 2$$

$I \geq 0, L \geq 0, A \geq 0,$ is

$$I = 0, L = 1/5, A = 7/5 .$$

This yields a value of 94 when substituted into the objective function (1). The solution to the program

Maximize $6B + 3E + 2M$ (3)

subject to

$$B + 3E + M \leq 45$$
$$2B + 3E + 3M \leq 50$$
$$4B + 2B + B \leq 60 \qquad (4)$$

$B \geq 0,\ E \geq 0,\ M \geq 0,$ is

$$B = 13,\ E = 0 \text{ and } M = 8.$$

Here again the value of the function is 94.

What is the relationship between the two programs?

Solution: The minimization problem (1), (2) and the maximization problem (3),(4) are called mathematical duals of each other. To see how they are related, we rewrite the problems in matrix form. Thus,

$$\text{Minimize} \qquad \begin{bmatrix} 45 & 50 & 60 \end{bmatrix} \begin{bmatrix} I \\ L \\ A \end{bmatrix} \qquad (1')$$

subject to:

$$\begin{pmatrix} 1 & 2 & 4 \\ 3 & 3 & 2 \\ 1 & 3 & 1 \end{pmatrix} \begin{pmatrix} I \\ L \\ A \end{pmatrix} \geq \begin{pmatrix} 6 \\ 3 \\ 2 \end{pmatrix} \qquad (2')$$

$I \geq 0,\ L \geq 0,\ A \geq 0.$ The dual is:

$$\text{Maximize} \qquad \begin{bmatrix} 6 & 3 & 2 \end{bmatrix} \begin{pmatrix} B \\ E \\ M \end{pmatrix} \qquad (3')$$

subject to:

$$\begin{pmatrix} 1 & 3 & 1 \\ 2 & 3 & 3 \\ 4 & 2 & 1 \end{pmatrix} \begin{pmatrix} B \\ E \\ M \end{pmatrix} \leq \begin{pmatrix} 45 \\ 50 \\ 60 \end{pmatrix} \qquad (4')$$

The coefficients of the objective function, (1') in the minimization problem have become the constraint values for the maximization problem, in (4'). Conversely, the coefficients of the objective function of the maximization problem, (3') are the constraint values for the minimization problem. Notice, also, that the matrix of coefficients (4') is just the transpose of the matrix of coefficients in (2').

From the above example, we see that every linear program has a dual. In general, the dual of

$$\text{Maximize} \qquad b_1 x_1 + b_2 x_2 + \ldots + b_n x_n$$

subject to:

$$\begin{pmatrix} a_{11} x_1 + a_{12} x_2 + \ldots + a_{1n} x_n \\ a_{21} x_1 + a_{22} x_2 + \ldots + a_{2n} x_n \\ \cdot \\ \cdot \\ \cdot \\ a_{m1} x_1 + a_{m2} x_2 + \ldots + a_{mn} x_n \end{pmatrix} \leq \begin{pmatrix} c_1 \\ c_{.2} \\ \cdot \\ \cdot \\ \cdot \\ c_m \end{pmatrix}$$

$x_i \geq 0 \quad (i = 1, 2, \ldots, n)$, is

$$\text{Minimize} \quad c_1 y_1 + c_2 y_2 + \ldots + c_m y_m$$

subject to:

$$\begin{pmatrix} a_{11} y_1 + a_{21} y_2 + \ldots + a_{m1} y_1 \\ \cdot \\ \cdot \\ \cdot \\ a_{1n} y_1 + a_{2n} y_2 + \ldots + a_{mn} y_m \end{pmatrix} \geq \begin{pmatrix} b_1 \\ \cdot \\ \cdot \\ \cdot \\ b_n \end{pmatrix}$$

Since a linear program and its dual are so intimately related, we can expect their solutions to at least be acquaintances. According to the Fundamental Theorem of linear programming, if a linear program and its dual both have a feasible point, then they both have the same optimal (= best feasible) point. If one has no feasible point, the other has no optimal point, i.e., the dual cannot be solved.

● **PROBLEM 11-20**

According to the Fundamental Theorem of linear programming, if either a linear program or its dual has no feasible point, then the other one has no solution. Illustrate this assertion with an example.

<u>Solution</u>: Consider the linear program

$$\text{Maximize} \quad x + y \tag{1}$$

subject to

$$x - y \le 1 \tag{2}$$

$$x + y \ge 4 \tag{3}$$

$$x \ge 0 \, , \, y \ge 0 \, . \tag{4}$$

We first convert the program into standard form. The second inequality, (3), needs to be reversed. We then obtain

$$-x - y \le -4 \, . \tag{5}$$

The dual to the program (1), (2), (4) and (5) is

$$\text{Minimize} \quad u - 4v \tag{6}$$

subject to
$$u - v \ge 1$$
$$-u - v \ge 1 \tag{7}$$
$$u \ge 0 \, , \, v \ge 0 \, .$$

But for non-negative u and v, the constraint $-u - v \ge 1$ can never be satisfied. Thus the minimization problem has no feasible solution. Hence the maximization problem has no best feasible solution.

   We can see that the last statement is true without appealing to the Fundamental Theorem. Graphing the constraints to the maximization problem: (See fig.)
   The feasible region is unbounded from above. This means that we can keep on increasing the value of the objective function while satisfying the constraints. Thus no maximum exists, i.e., there is no best feasible solution.

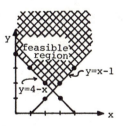

● **PROBLEM 11-21**

Consider the problem:

$$\text{maximize } x_1 + 3x_2$$

subject to:

$$6x_1 + 19x_2 \leq 100$$

$$3x_1 + 5x_2 \leq 40$$

$$x_1 - 3x_2 \leq 33$$

$$x_2 \leq 25$$

$$x_1 \leq 42$$

$$x_1, x_2 \geq 0 .$$

Find its dual problem.

Solution: The dual problem is constructed from the primal problem (the primal can be constructed from the dual similarly) as follows:

1. Each constraint in one problem corresponds to a variable in the other problem.

2. The elements of the right-hand side of the constraints in one problem are equal to the respective coefficients of the objective function in the other problem.

3. One problem seeks maximization and the other seeks minimization.

4. The maximization problem has ($\leq$) constraints and the minimization problem has ($\geq$) constraints.

5. The variables in both problems are nonnegative.

The dual for this problem is constructed as follows.

Designate y_1, y_2, y_3, y_4, and y_5 as the dual variables associated with the first, second, third, fourth and fifth primal constraints. The dual problem is:

minimize

$$100y_1 + 40y_2 + 33y_3 + 25y_4 + 42y_5$$

subject to:

$$6y_1 + 3y_2 + y_3 \qquad + y_5 \geq 1$$

$$19y_1 + 5y_2 - 3y_3 + y_4 \qquad \geq 3$$

$$y_1 , y_2 , y_3 , y_4 , y_5 \geq 0$$

In this case, the number of constraints in the dual problem are less than those in the primal problem. Thus, it is easier to solve the dual problem computationally, for an optimal solution. In LP problems computational difficulty depends on the number of constraints rather than the number of variables.

Find the dual to:

$$\max \quad 2x_1 + x_2 + x_3 - x_4$$

subject to:

$$x_1 - x_2 + 2x_3 + 2x_4 \leq 3$$
$$2x_1 + 2x_2 - x_3 \qquad = 4$$
$$x_1 - 2x_2 + 3x_3 + 4x_4 \geq 5$$
$$x_1, x_2, x_3 \qquad \geq 0,$$
$$x_4 \text{ unrestricted.}$$

<u>Solution</u>: $\min 3y_1 + 4y_2 + 5y_3$

subject to:

$$y_1 + 2y_2 + y_3 \geq 2$$
$$-y_1 + 2y_2 - 2y_3 \geq 1$$
$$2y_1 - y_2 + 3y_3 \geq 1$$
$$2y_1 \qquad + 4y_3 = -1$$
$$y_1 \geq 0$$
$$y_2 \text{ unrestricted}$$
$$y_3 \leq 0 .$$

Find the dual to:

$$\text{Maximize} \quad P = x_1 + 2x_2$$

where:

$$x_1 \geq 0, \ x_2 \geq 0$$

and

$$x_1 + 2x_2 \leq 10$$
$$-x_1 - x_2 \leq -30 .$$

Minimize $C = 10y_1 - 30y_2$

where:

$y_1 \geq 0, \; y_2 \geq 0$

and

$y_1 - y_2 \geq 1$

$2y_1 - y_2 \geq 2 \; .$

These constraints put no restrictions on y_2, and hence C can be made arbitrarily small.

● **PROBLEM** 11-24

Find the dual to the following primal problem:

Maximize $\quad z = x_1 + 1.5x_2$

subject to: $\quad 2x_1 + \; 3x_2 \leq 25$

$\quad\quad\quad\quad\quad x_1 + \quad x_2 \geq 1$

$\quad\quad\quad\quad\quad x_1 - \; 2x_2 = 1$

$\quad\quad\quad\quad\quad\quad x_1, x_2 \geq 0$

Solution: It may be rewritten as:

Maximize $z = x_1 + 1.5x_2$

subject to: $\quad 2x_1 + \; 3x_2 \leq 25$

$\quad\quad\quad\quad -x_1 - \quad x_2 \leq -1$

$\quad\quad\quad\quad\quad x_1 - \; 2x_2 \leq 1$

$\quad\quad\quad\quad -x_1 + \; 2x_2 \leq -1$

$\quad\quad\quad\quad\quad\quad x_1, x_2 \geq 0$

by multiplying the second constraint by -1 and representing the third constraint as two inequalities. The dual of the rewritten problem is then the following:

Minimize $z = 25y_1 - y_2 + \; y_3 - y_4$

Subject to:
$$2y_1 - y_2 + y_3 - y_4 \geq 1$$
$$3y_1 - y_2 - 2y_3 + 2y_4 \geq 1.5$$
$$y_1, y_2, y_3, y_4 \geq 0 .$$

Since $y_3 - y_4$ appears in every constraint and objective function, it can be replaced with an unrestricted variable (y_5), as follows:

Minimize $z = 25y_1 - y_2 + y_5$

subject to: $2y_1 - y_2 + y_5 \geq 1$
$$3y_1 - y_2 - 2y_5 \geq 1.5$$
$$y_1, y_2 \geq 0,$$

y_5 unrestricted in sign.

● **PROBLEM** 11-25

A: Maximize $f = 4x_1 + 3x_2$

subject to: $-2x_1 - x_2 \leq -4$
$$2x_1 - 2x_2 \leq 5$$
$$x_1, x_2 \geq 0.$$

This problem is unbounded.

B: Maximize $f = 3x_1 + 5x_2$

subject to: $x_1 - x_2 \leq -2$
$$-x_1 + x_2 \leq -2$$
$$x_1, x_2 \geq 0.$$

Now the second constraint can be rewritten as:

$$x_1 - x_2 \geq 2$$

(by multiplying through by -1). There are no points which can possibly satisfy these two conflicting constraints, so the primal is infeasible.

Give the dual problems and their optimal solutions.

Solutions:

A: The dual is:

Minimize $g = -4y_1 + 5y_2$

subject to: $-2y_1 + 2y_2 \geq 4$

$- y_1 - 2y_2 \geq 3$

$y_1, y_2 \quad \geq 0.$

Rewrite the second constraint as:

$$y_1 + y_2 \leq -3$$

which is clearly impossible to satisfy for nonnegative y_1 and y_2. Thus, the dual is infeasible.

B: The dual is:

Minimize $g = -2y_1 - 2y_2$

subject to: $y_1 - y_2 \geq 3.$

$-y_1 + y_2 \geq 5$

$y_1, y_2 \geq 0.$

If one rewrites the second constraint as

$$y_1 - y_2 \leq -5$$

it is apparent that the dual is also infeasible.

THE SIMPLEX METHOD

● **PROBLEM** 11-26

Consider the linear programming problem:

minimize $x_0 = 3x_1 - 3x_2 + 7x_3$

subject to

$$x_1 + x_2 + 3x_3 \leq 40$$

$$x_1 + 9x_2 - 7x_3 \geq 50$$

$$5x_1 + 3x_2 \qquad = 20$$

$$|5x_2 + 8x_3| \qquad \leq 100$$

$$x_1 \geq 0, x_2 \qquad \geq 0$$

x_3 is unconstrained in sign.

Find its canonical form.

<u>Solution:</u> The characteristics of canonical form are:

1. All decision variables are nonnegative.

2. All constraints are of the $\leq$ type.

3. The objective function is of the maximization type.

This problem can be put in the canonical form as follows:

$$|5x_2 + 8x_3| \leq 100$$

is equivalent to

$$5x_2 + 8x_3 \leq 100$$

and

$$5x_2 + 8x_3 \geq -100.$$

Also,

$$x_3 = x_3^+ - x_3^-$$

where $x_3^+ \geq 0$ and $x_3^- \geq 0$. Finally, if the objective function is transformed to maximization, the canonical form becomes:

$$\text{maximize} \quad g_0 = (-x_0) = -3x_1 + 3x_2 - 7(x_3^+ - x_3^-)$$

subject to

$$x_1 + x_2 + 3(x_3^+ - x_3^-) \leq 40$$
$$-x_1 - 9x_2 + 7(x_3^+ - x_3^-) \leq -50$$
$$5x_1 + 3x_2 \leq 20$$
$$-5x_1 - 3x_2 \leq -20$$
$$5x_2 + 8(x_3^+ - x_3^-) \leq 100$$
$$-5x_2 - 8(x_3^+ - x_3^-) \leq 100$$

$$x_1 \geq 0, \quad x_2 \geq 0, \quad x_3^+ \geq 0, \quad x_3^- \geq 0.$$

The only difference between the original and the canonical forms in the above problem occurs in the objective function where x_0 in the original problem becomes equal to $(-g_0)$ in the canonical form. The values of the variables are the same in both cases, since the constraints are mathematically equivalent.

Consider the following problem:

minimize $\qquad z = 2x_1 + 4x_2$

subject to

$$x_1 + 5x_2 \leq 80 \qquad\qquad (1)$$

$$4x_1 + 2x_2 \geq 20 \qquad\qquad (2)$$

$$x_1 + x_2 = 10 \qquad\qquad (3)$$

$$x_1, x_2 \geq 0$$

Find the canonical form.

Solution: A function will be minimized if one maximizes the negative of the function. Thus, the objective function above will be minimized if the canonical objective function below is maximized:

$$z_c = -z = -2x_1 - 4x_2.$$

An equality constraint can be replaced by two inequality constraints of opposite sense. Constraint (3) can be replaced by the two constraints:

$$x_1 + x_2 \leq 10 \qquad\qquad (4)$$

$$x_1 + x_2 \geq 10 \qquad\qquad (5)$$

Constraint inequalities can be reversed by multiplying both sides of the inequality by minus one. Constraints (5) and (2) can be converted to ($\leq$) types by multiplying both sides by (-1), thus,

$$- x_1 - x_2 \leq -10 \qquad\qquad (6)$$

$$-4x_1 - 2x_2 \leq -20 \qquad\qquad (7)$$

Note that the simplex method requires that all right-hand-side constants be nonnegative. Thus, canonical form is not suitable for use in the simplex algorithm.

The statement of the original problem in canonical form is given below:

maximize $\qquad\qquad z_c = - 2x_1 - 4x_2$

subject to

$$x_1 + 5x_2 \leq 80$$

$$-4x_1 - 2x_2 \leq -20$$

$$x_1 + x_2 \leq 10$$

$$-x_1 - x_2 \leq -10$$

$$x_1, x_2 \geq 0 .$$

● **PROBLEM** 11-28

Put the following linear programming problem into canonical form:

Minimize $4M + 4T + W$

subject to:

M ≥ 0

T ≥ 0

W ≥ 0

M + T + W ≥ 10

M + T + 2W ≥ 6 .

Solution: The corresponding system of equations in canonical form is:

Maximize $-4M - 4T - W$

subject to:

$$-M - T - W \leq -10$$

$$-M - T - 2W \leq -6$$

$$M, T, W \geq 0, \qquad S_1, S_2, S_3 \geq 0 .$$

● **PROBLEM** 11-29

Maximize $x_0 = 4x_1 + 3x_2$

subject to: $2x_1 + 3x_2 \leq 6$

$$-3x_1 + 2x_2 \leq 3$$

$$2x_2 \leq 5$$

$$2x_1 + x_2 \leq 4$$

$$x_1, x_2 \geq 0 \ .$$

Find the standard form of this linear programming problem.

Solution: The characteristics of the standard form are:

1. All constraints are equations except for the nonnegative constraints which remain inequalities (≥ 0).

2. The right-hand side element of each constraint equation is nonnegative.

3. All variables are nonnegative.

4. The objective function is of the maximization or the minimization type.

Thus the standard form of the above problem is:

Maximize $x_0 = 4x_1 + 3x_2$

subject to:

$$2x_1 + 3x_2 + S_1 \qquad\qquad = 6$$
$$-3x_1 + 2x_2 \qquad + S_2 \qquad = 3$$
$$2x_2 \qquad\quad + S_3 \quad = 5$$
$$2x_1 + x_2 \qquad\qquad + S_4 = 4$$

$$x_1, x_2, S_1, S_\sim, S_3, S_4 \geq 0 \ .$$

● **PROBLEM 11-30**

Consider the system: Minimize
$$x_1 + x_2 + \frac{1}{2}x_3 - \frac{13}{3}x_4$$

subject to:

$$2x_1 - \frac{1}{2}x_2 + x_3 + x_4 \qquad\quad \leq 2$$
$$x_1 + 2x_2 + 2x_3 - 3x_4 + x_5 \geq 3$$
$$x_1 \qquad\quad - x_3 + x_4 - x_5 \geq \frac{2}{3}$$
$$3x_1 - x_2 \qquad\quad + 2x_4 - \frac{3}{2}x_5 = 1$$

$$x_i \geq 0, i = 1, \ldots, 5$$

and put into standard form.

<u>Solution</u>:   Let x_6 be a slack variable, and x_7 and x_8 surplus variables; then the system becomes:

Minimize

$$x_1 + x_2 + \frac{1}{2}x_3 - \frac{13}{3}x_4$$

subject to:

$$2x_1 - \frac{1}{2}x_2 + x_3 + x_4 \qquad + x_6 \qquad = 2$$

$$x_1 + 2x_2 + 2x_3 - 3x_4 + x_5 \qquad -x_7 \qquad = 3$$

$$x_1 \qquad - x_3 + x_4 - x_5 \qquad -x_8 = \frac{2}{3}$$

$$3x_1 - x_2 \qquad + 2x_4 - \frac{3}{2}x_5 \qquad = 1$$

$$x_i \geq 0, i = 1, \ldots, 8.$$

● **PROBLEM** 11-31

Put into standard form:

Maximize $3x + y$

subject to: $x \geq 0$

$y \geq 0$

$2x - y \leq -10$

$x + 2y \leq 14$

$x \qquad \leq 12.$

<u>Solution</u>:   Maximize

$$3x + y + 0S_1 + 0S_2 + 0S_3$$

subject to:

$$-2x + y - S_1 \qquad = 10$$

$$x + 2y \qquad + S_2 \qquad = 14$$

$$x \qquad + S_3 = 12$$

$$x \geq 0, \quad 4 \geq 0, \quad S_1 \ S_2 \ S_3 \geq 0$$

where S_1, S_2, S_3 are the slack variables.

● PROBLEM 11-32

Maximize

$$z = 7x_1 + 10x_2$$

subject to:

$$5x_1 + 4x_2 \leq 24$$

$$2x_1 + 5x_2 \leq 13$$

$$x_1, x_2 \geq 0.$$

Find an initial feasible solution.

<u>Solution:</u>

$$1z - 7x_1 - 10x_2 - 0S_1 - 0S_2 = 0$$

$$0z + 5x_1 + 4x_2 + 1S_1 + 0S_2 = 24$$

$$0z + 2x_1 + 5x_2 + 0S_1 + 1S_2 = 13.$$

In tableau form this is expressed as:

Basis	z	x_1	x_2	S_1	S_2	b_i
—	1	−7	−10	0	0	0
S_1	0	5	4	1	0	24
S_2	0	2	5	0	1	13

The variables in the basis are indicated in the first column of the tableau. The selection of S_1 and S_2 for the basis is equivalent to setting x_1 and x_2 to zero; hence, the system of three equations is reduced to:

$$1z + 0S_1 + 0S_2 = 0$$

$$0z + 1S_1 + 0S_2 = 24$$

$$0z + 0S_1 + 1S_2 = 13$$

which provides an immediate solution of $z = 0$, $S_1 = 24$, and $S_2 = 13$.

● PROBLEM 11-33

Consider the following standard maximum problem:

Maximize $u = 4x + 2y + z$ (1)

subject to: $x + y \leq 1$

$$x + z \leq 1 \qquad\qquad (2)$$

and

$$x \geq 0, \; y \geq 0, \; z \geq 0. \tag{3}$$

Identify the basic feasible points (extreme points) of the constraint set. Determine which ones, if any are degenerate.

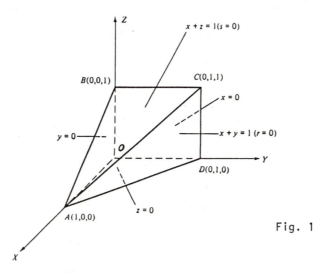

Fig. 1

<u>Solution</u>: Note that the function to be maximized is really irrelevant to the solution of the problem, i.e., the constraint set is determined solely from (2) and (3).

If slack variables r and s are introduced, then rewrite (2) and (3) as:

$$x + y - 1 = -r \tag{4}$$

$$x + z - 1 = -s \tag{5}$$

and

$$x \geq 0$$

$$y \geq 0$$

$$z \geq 0$$

$$r \geq 0$$

$$s \geq 0.$$

The sketch of the constraint set for this problem is given in Figure 1. Plot the bounding planes x + y = 1 (or r = 0) and x + z = 1 (or s = 0) as well as x = 0, y = 0, and z = 0. The constraint set consists of those points within and on the surface of the pyramid shown. This is a case of a bounded constraint set.

Note that x + y = 1 is parallel to the Z axis because
the variable z is missing from it, while x + z = 1 is
parallel to the Y axis because here the variable y is
missing. The plane labeled x = 0 is none other than
the yz plane. Similarly, y = 0 and z = 0 are coordinate
planes.

The extreme points (or basic feasible points) can
be determined by inspection of Figure 1 and by using
Equations (4) and (5). Thus, point B is located at the
intersection of x + z = 1 (s = 0), y = 0, and x = 0.
Thus, setting x = 0, y = 0, and s = 0 in Equations (4)
and (5) yields

$$-1 = -r \text{ and } z - 1 = 0.$$

This results in r = 1 and z = 1 for point B. Thus, B
is a nondegenerate basic feasible point because the
solved-for variables are both positive. Notice that
n = 3 variables are set equal to zero and the remaining
m = 2 variables are solved from equations (4) and (5).

Point C is located at the intersection of x + z =
1 (s = 0), x + y = 1 (r = 0), and x = 0. Thus, setting
x = 0, r = 0, and s = 0 in Equations (4) and (5) yields:

$$y - 1 = 0 \text{ and } z - 1 = 0.$$

This results in y = 1 and z = 1 for point C. Thus,
point C is also a nondegenerate basic feasible point.

As a final calculation, consider point A, which lies
at the intersection of x + y = 1 (r = 0), y = 0, and z = 0.
Thus, setting r = 0, y = 0, and z = 0 in Equations (4)
and (5) yields:

$$x - 1 = 0 \text{ and } x - 1 = -s.$$

This results in x = 1 and s = 0 for point A. Thus, note
that point A is a degenerate basic feasible point because
one of the solved-for variables is equal to zero.

The following table indicates the values of the
five variables (three original plus two slack variables)
for the five basic feasible points of this problem.
Only one--point A--is a degenerate basic feasible point:

Basic feasible point	x	y	z	r	s	u
O	0	0	0	1	1	0
A (degenerate)	1	0	0	0	0	4
B	0	0	1	1	0	1
C	0	1	1	0	0	3
D	0	1	0	0	1	2

There are geometrical interpretations to nondegenerate
and degenerate basic feasible points. For a standard
problem a bounding hyperplane (in the case just considered,
an ordinary plane) can be represented by an equation of
the form:

Variable $= 0$.

In a three-dimensional problem such as the problem above, (see Figure 1), a nondegenerate basic feasible point occurs at a point where exactly three of the bounding planes intersect, i.e., at a point where exactly three of the variables equal zero, the other two variables being positive at this point. This is what happens at points O, B, C, and D of the problem.

On the other hand, basic feasible point A occurs at the intersection of four bounding planes, i.e., four variables equal zero, the remaining variable being positive. Thus, for the case where $n = 3$, a degenerate basic feasible point is a point of intersection of more than three bounding planes.

In a standard linear programming problem, a degenerate basic feasible point may be interpreted as a point solution in the constraint set where more than n bounding hyperplanes intersect. If exactly n bounding hyperplanes intersect at a solution in the constraint set, then such a solution would be a nondegenerate basic feasible solution.

● **PROBLEM** 11-34

Minimize $x_1 + x_2$

Subject to: $x_1 + 2x_2 \leq 4$

$x_2 \leq 1$

$x_1, x_2 \geq 0$.

Find a basic feasible solution to the above problem, starting from a b.f.s. with x_1 and x_2 in the basis.

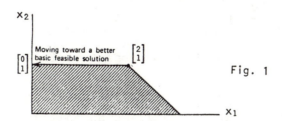

Fig. 1

Improving a basic feasible solution

Solution: Introduce the slack variables x_3 and x_4 to put the problem in a standard form. This leads to the following constraint matrix $\vec{A}$:

$$\vec{A} = [\vec{a}_1, \ \vec{a}_2, \ \vec{a}_3, \ \vec{a}_4] = \begin{bmatrix} 1 & 2 & 1 & 0 \\ 0 & 1 & 0 & 1 \end{bmatrix}.$$

851

Considering the basic feasible solution corresponding to $\vec{B} = [\vec{a}_1, \vec{a}_2]$ (in other words, x_1 and x_2 are the basic variables while x_3 and x_4 are the nonbasic variables), one gets:

$$\vec{x}_B = \begin{bmatrix} x_1 \\ x_2 \end{bmatrix} = \vec{B}^{-1}\vec{b} = \begin{bmatrix} 1 & 2 \\ 0 & 1 \end{bmatrix}^{-1} \begin{bmatrix} 4 \\ 1 \end{bmatrix} = \begin{bmatrix} 1 & -2 \\ 0 & 1 \end{bmatrix} \begin{bmatrix} 4 \\ 1 \end{bmatrix} = \begin{bmatrix} 2 \\ 1 \end{bmatrix}$$

$$\vec{x}_N = \begin{bmatrix} x_3 \\ x_4 \end{bmatrix} = \begin{bmatrix} 0 \\ 0 \end{bmatrix}.$$

This point is shown in Figure 1. In order to improve this basic feasible solution, calculate $z_j - c_j$ for the nonbasic variables:

$$z_3 - c_3 = \vec{c}_B \vec{B}^{-1} \vec{a}_3 - c_3$$

$$= (1,1) \begin{bmatrix} 1 & -2 \\ 0 & 1 \end{bmatrix} \begin{pmatrix} 1 \\ 0 \end{pmatrix} - 0$$

$$= (1,1) \begin{pmatrix} 1 \\ 0 \end{pmatrix} - 0$$

$$= 1$$

$$z_4 - c_4 = \vec{c}_B \vec{B}^{-1} \vec{a}_4 - c_4$$

$$= (1,1) \begin{bmatrix} 1 & -2 \\ 0 & 1 \end{bmatrix} \begin{pmatrix} 0 \\ 1 \end{pmatrix} - 0$$

$$= (1,1) \begin{pmatrix} -2 \\ 1 \end{pmatrix} - 0$$

$$= -1$$

Since $z_3 - c_3 > 0$, then the objective improves by increasing x_3.

The criterion $z_k - c_k > 0$ for a nonbasic variable x_k to enter the basis is justified as follows. Note that $z = \vec{c}_B \vec{b} - (z_k - c_k) x_k$, where

$$z_k = \vec{c}_B \vec{B}^{-1} \vec{a}_k = \vec{c}_B \vec{y}_k = \sum_{i=1}^{m} c_{B_i} y_{ik} \tag{1}$$

and c_{B_i} is the cost of the ith basic variable. Note that if x_k is raised from zero level, while the other nonbasic variables are kept at zero level, then the basic variables x_{B_1}, x_{B_2}, . . . , x_{Bm} must be modified. In other words, if x_k is increased by 1 unit, then x_{B_1}, x_{B_2}, . . . , and x_{B_m} will be decreased respectively by y_{1k}, y_{2k}, . . . , y_{mk} units (if $y_{ik} < 0$, then x_{B_i} will be increased). The saving (a negative saving means more cost) that results from the modification of the basic variables, as a result of increasing x_k by 1 unit, is therefore, $\Sigma_{i=1}^{m} c_{B_i} y_{ik}$, which is z_k (see Equation 1). However, the cost of increasing x_k itself by 1 unit is c_k. Hence $z_k - c_k$ is the saving minus the cost of increasing x_k by 1 unit. Naturally, if $z_k - c_k$ is positive, it will be advantageous to increase x_k. For each unit of x_k, the cost will be reduced by an amount $z_k - c_k$ and hence it will be advantageous to increase x_k as much as possible. On the other hand, if $z_k - \dot{c}_k < 0$, then by increasing x_k, the net saving is negative, and this action will result in a larger cost. So this action is prohibited. Finally, if $z_k - c_k = 0$, then increasing x_k will lead to a different solution, with the same cost. So whether x_k is kept at zero level, or increased, no change in cost takes place.

Now suppose that x_k is a basic variable. In particular, suppose that x_k is the tth basic variable, that is, $x_k = x_{B_t}$, $c_k = c_{B_t}$, and $\vec{a}_k = \vec{a}_{B_t}$. Recall that $z_k = \vec{c}_B \vec{B}^{-1} \vec{a}_k = \vec{c}_B \vec{B}^{-1} \vec{a}_{B_t}$. But $\vec{B}^{-1} \vec{a}_{B_t}$ is a vector of zeros except for one at the tth position. Therefore, $z_k = c_{B_t}$, and hence $z_k - c_k = c_{B_t} - c_{B_t} = 0$. The modified solution is given by:

$$\vec{X}_B = \vec{B}^{-1} \vec{b} - \vec{B}^{-1} \vec{a}_3 x_3$$

$$\begin{bmatrix} x_1 \\ x_2 \end{bmatrix} = \begin{bmatrix} 2 \\ 1 \end{bmatrix} - \begin{bmatrix} 1 \\ 0 \end{bmatrix} x_3$$

The maximum value of x_3 is 2 (any larger value of x_3 will force x_1 to be negative). Therefore the new basic feasible solution is:

$$(x_1, x_2, x_3, x_4) = (0,1,2,0).$$

Here x_3 enters the basis and x_1 leaves the basis. Note that the new point has an objective value equal to 1, which is an improvement over the previous objective value of 3. The improvement is precisely $(z_3 - c_3)x_3 = 2$.

● **PROBLEM** 11-35

Assume the following system of equations:

$$\begin{cases} 5x_1 + 4x_2 + 3x_3 = 8 \\ \\ 2x_1 + 7x_2 + 5x_3 = 5 \\ \\ 4x_1 + 4x_2 + 2x_3 = 4. \end{cases}$$

Solve for x_1, x_2, x_3 using elementary row operations.

Solution:

Step 1. Select the variable farthest to the right, x_3.

Step 2. Multiply the last equation by 1/2:

$$\begin{cases} 5x_1 + 4x_2 + 3x_3 = 8 \\ \\ 2x_1 + 7x_2 + 5x_3 = 5 \\ \\ 2x_1 + 2x_2 + x_3 = 2 \ . \end{cases}$$

Step 3. Eliminate x_3 from all equations above it. Multiply the third equation by five and subtract it from the second:

$$\begin{cases} 5x_1 + 4x_2 + 3x_3 = 8 \\ \\ -8x_1 - 3x_2 \qquad = -5 \\ \\ 2x_1 + 2x_2 + x_3 = 2 \ . \end{cases}$$

Multiply the third equation by three and subtract it from the first:

$$\begin{cases} -\ x_1\ -\ 2x_1 & = 2 \\[1em] -8x_1\ -\ 3x_2 & = -5 \\[1em] 2x_1\ +\ 2x_2\ +\ x_3 & = 2\ . \end{cases}$$

Step 4. Temporarily strike out the last equation and the last variable (which already has been eliminated from the first two equations):

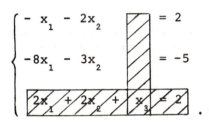

Now proceed by Steps 1-3 to deal with the smaller system. First multiply the second equation by -1/3:

$$\begin{cases} -x_1\ -\ 2x_2 & = 2 \\[1em] +8/3x_1\ +\ x_2 & = 5/3 \\[1em] \cancel{2x_1\ +\ 2x_2\ +\ x_3\ =\ 2}\ . \end{cases}$$

Multiply the second equation by two and add to the first:

$$\begin{cases} 13/3x_1 & = 16/3 \\[1em] 8/3x_1\ +\ x_2 & = 5/3 \\[1em] \cancel{2x_1\ +\ 2x_2\ +\ x_3\ =\ 2}\ . \end{cases}$$

Step 5. Strike out the second equation:

$$\begin{cases} 13/3x_1 & = 16/3 \\[1em] \\[1em] \cancel{8/3x_1\ +\ x_2\ =\ 5/3} \\[0.5em] \cancel{2x_1\ +\ 2x_2\ +\ x_3\ =\ 2}\ . \end{cases}$$

Multiply the first equation by 3/13:

$$\left\{ \begin{array}{l} x_1 = 16/13 \\ 8/3x_1 + x_2 = 5/3 \\ 2x_1 + 2x_2 + x_3 = 2 \end{array} \right.$$

Now all coefficients above the main diagonal have been eliminated and all coefficients on the main diagonal are equal to one.

Step 6. Eliminate all coefficients below the main diagonal. Multiply the first equation by 8/3 and subtract from the second equation:

$$\left\{ \begin{array}{l} x_1 = 16/13 \\ x_2 = -21/13 \\ 2x_1 + 2x_2 + x_3 = 2 . \end{array} \right.$$

Multiply the first equation by two and subtract it from the third equation:

$$\left\{ \begin{array}{l} x_1 = 16/13 \\ x_2 = -21/13 \\ 2x_2 + x_3 = -6/13 . \end{array} \right.$$

Multiply the second equation by two and subtract from the third:

$$\left\{ \begin{array}{l} x_1 = 16/13 \\ x_2 = -21/13 \\ x_3 = 36/13 \end{array} \right.$$

The solution is now complete.

Use row operations to solve:

$$x_1 + 4x_2 + x_3 = 2$$

$$2x_1 + 3x_2 = -1$$

$$8x_1 + 2x_3 = 0 .$$

Solution: Denote the first row by R_1, the second by R_2, and the third by R_3 .

Step	Operation	Result
1	$R_2 - 2R_1$	$x_1 + 4x_2 + x_3 = 2$ $-5x_2 - 2x_3 = -5$ $8x_1 + 2x_3 = 0$
2	$R_3 - 8R_1$	$x_1 + 4x_2 + x_3 = 2$ $-5x_2 - 2x_3 = -5$ $-32x_2 - 6x_3 = -16$
3	$R_1 + \frac{4}{5}R_2$	$x_1 - \frac{3}{5}x_3 = -2$ $-5x_2 - 2x_3 = -5$ $-32x_2 - 6x_3 = -16$
4	$R_2 - \frac{5}{32}R_3$	$x_1 - \frac{3}{5}x_3 = -2$ $-\frac{17}{16}x_3 = -\frac{5}{2}$ $-32x_2 - 6x_3 = -16$
5	$R_3 - \frac{96}{17}R_2$	$x_1 - \frac{3}{5}x_3 = -2$ $-\frac{17}{16}x_3 = -\frac{5}{2}$ $-32x_2 = -\frac{32}{17}$
6	$R_1 - \frac{48}{85}R_2$	$x_1 = -\frac{10}{17}$ $-\frac{17}{16}x_3 = -\frac{5}{2}$ $-32x_2 = 16$

TABLE 1

Solve, using row operations:

$$\begin{pmatrix} 1 & 4 & 3 \\ 2 & 5 & 4 \\ 1 & -3 & -2 \end{pmatrix} \begin{pmatrix} x_1 \\ x_2 \\ x_3 \end{pmatrix} = \begin{pmatrix} 1 \\ 4 \\ 5 \end{pmatrix} \quad \text{and} \quad = \begin{pmatrix} -1 \\ 0 \\ 2 \end{pmatrix}$$

<u>Solution</u>: It is possible to solve both sets of simultaneous equations at once. The following series of tableaus illustrate this:

$$\left(\begin{array}{ccc|cc} 1 & 4 & 3 & 1 & -1 \\ 2 & 5 & 4 & 4 & 0 \\ 1 & -3 & -2 & 5 & 2 \end{array} \right)$$

$$\left(\begin{array}{ccc|cc} 1 & 4 & 3 & 1 & -1 \\ 0 & -3 & -2 & 2 & 2 \\ 0 & -7 & -5 & 4 & 3 \end{array} \right)$$

$$\left(\begin{array}{ccc|cc} 1 & 0 & \frac{1}{3} & \frac{11}{3} & \frac{5}{3} \\ 0 & 1 & \frac{2}{3} & -\frac{2}{3} & -\frac{2}{3} \\ 0 & 0 & -\frac{1}{3} & -\frac{2}{3} & -\frac{5}{3} \end{array} \right)$$

$$\left(\begin{array}{ccc|cc} 1 & 0 & 0 & 3 & 0 \\ 0 & 1 & 0 & -2 & -4 \\ 0 & 0 & 1 & 2 & 5 \end{array} \right)$$

Thus, these answers are found:

$$x_1 = 3, \qquad x_2 = -2, \qquad x_3 = 2$$

to the first set of equations and the answers:

$$x_1 = 0, \qquad x_2 = -4, \qquad x_3 = 5$$

to the second set of equations.

Interchange the variables y and r in Tableau 1 by means of the pivot transformation. Interpret the resulting tableau. Determine what basic point corresponds to the resulting tableau. Do a check calculation.

x	y	z	1	
1	$\frac{2}{3}^*$	$\frac{1}{4}$	-900	$= -r$
0	$\frac{1}{3}$	$\frac{3}{4}$	-600	$= -s$
$\frac{1}{4}$	$\frac{2}{5}$	$\frac{1}{2}$	0	$= u$

(T.1)

Solution: Since it is desired to interchange the nonbasic variable y and the basic variable r, the pivot entry is thus

$p = \frac{2}{3}$, i.e., the entry label a_{12} with an asterisk next to it.

The resulting tableau-Tableau 2-is as follows

x	y	z	1	
$\frac{3}{2}$	$\frac{3}{2}$	$\frac{3}{8}$	-1350	$= -y$
$-\frac{1}{2}$	$-\frac{1}{2}$	$\frac{5}{8}$	-150	$= -s$
$-\frac{7}{20}$	$-\frac{3}{5}$	$\frac{7}{20}$	540	$= u$

(T.2)

Notice that the variables y and r have been interchanged, but that all other variables remained unchanged.

Pivot entry: $\frac{2}{3}$ to $\frac{3}{2}$.

Other entries in pivot row (row 1):

$$1 \text{ to } \frac{1}{\frac{2}{3}} = \frac{3}{2} , \quad \frac{1}{4} \text{ to } \frac{\frac{1}{4}}{\frac{2}{3}} = \frac{3}{8} ,$$

and

$$-900 \text{ to } \frac{-900}{\frac{2}{3}} = \frac{-2700}{2} = -1350.$$

Note that dividing by $\frac{2}{3}$ is equivalent to multiplying by $\frac{3}{2}$. Thus, for example,

$$\frac{1}{4} \cdot \frac{3}{2} = \frac{3}{8} .$$

Other entries in pivot column (column 2):

$$\frac{1}{3} \text{ to } \frac{\frac{1}{3}}{-\frac{2}{3}} = -\frac{1}{2}$$

and

$$\frac{2}{5} \text{ to } \frac{\frac{2}{5}}{-\frac{2}{3}} = -\frac{3}{5} .$$

If instead of dividing by $-\frac{2}{3}$ the pivot column (column 2) is multiplied by $-\frac{3}{2}$, the calculations would give the same results. For example, with respect to the second calculation it would be:

$$\frac{2}{5} \text{ to } \frac{2}{5} \cdot \frac{-3}{2} = \frac{-3}{5} .$$

Now work on the remaining entries, which are all type s entries:

Row 2:

$$0 \text{ to } 0 - \frac{1 \cdot \frac{1}{3}}{\frac{2}{3}} = -1 \cdot \frac{1}{3} \cdot \frac{3}{2} = -\frac{1}{2} .$$

$$\frac{3}{4} \text{ to } \frac{3}{4} - \frac{\frac{1}{4} \cdot \frac{1}{3}}{\frac{2}{3}} = \frac{3}{4} - \frac{1}{4} \cdot \frac{1}{3} \cdot \frac{3}{2} = \frac{6}{8} - \frac{1}{8} = \frac{5}{8} .$$

$$-600 \text{ to } -600 - \frac{(-900) \cdot \frac{1}{3}}{\frac{2}{3}} = -600 + 450 = -150.$$

Bottom Row:

$$\frac{1}{4} \text{ to } \frac{1}{4} - \frac{1 \cdot \frac{2}{5}}{\frac{2}{3}} = \frac{1}{4} - 1 \cdot \frac{2}{5} \cdot \frac{3}{2} = \frac{1}{4} - \frac{3}{5} = \frac{5-12}{20} = \frac{-7}{20} .$$

$$\frac{1}{2} \text{ to } \frac{1}{2} - \frac{\frac{1}{4} \cdot \frac{2}{5}}{\frac{2}{3}} = \frac{1}{2} - \frac{1}{4} \cdot \frac{2}{5} \cdot \frac{3}{2} = \frac{1}{2} - \frac{3}{20} = \frac{7}{20} .$$

$$0 \text{ to } 0 - \frac{(-900) \cdot \frac{2}{5}}{\frac{2}{3}} = +900 \cdot \frac{2}{5} \cdot \frac{3}{2} = 90 \cdot 2 \cdot 3 = 540.$$

Now translate the shorthand of Tableau 2 into equations:

$$\frac{3}{2}x + \frac{3}{2}r + \frac{3}{8}z - 1350 = -y \qquad\qquad (2')$$

$$-\frac{1}{2}x - \frac{1}{2}r + \frac{5}{8}z - 150 = -s \qquad\qquad (3')$$

$$-\frac{7}{20}x - \frac{3}{5}r + \frac{7}{20}z + 540 = u . \qquad\qquad (1')$$

Note that the Tableau 2 equations are indicated by primes.

It is now easy to determine coordinates of another basic point by working with the Tableau 2 equations. Set the non-basic variables x, r, and z equal to zero. Immediately obtain:

$$y = 1350, \qquad s = 150, \qquad \text{and } u = 540.$$

The point is a basic feasible point, since the two variables y and s are positive.

It is easy to make arithmetical errors in carrying out the pivot transformation calculations. Thus, a check procedure is useful. One such check is obtained by seeing if the coordinates of the basic point corresponding to Tableau 1 satisfy the equations of Tableau 2. From Tableau 1

if

$$x = y = z = 0,$$

then

$$r = 900, \ s = 600, \text{ and } u = 0.$$

Substituting these into the equations of Tableau 2, the following is obtained:

$$\frac{3}{2} \cdot 0 + \frac{3}{2} \cdot 900 + \frac{3}{8} \cdot 0 - 1350 = 1350 - 1350 = 0 = -y. \qquad (2')$$

Thus,

$$y = 0.$$

$$-\frac{1}{2} \cdot 0 - \frac{1}{2} \cdot 900 + \frac{5}{8} \cdot 0 - 150 = -450 - 150 = -600 = -s. \qquad (3')$$

Thus,

$$s = 600.$$

$$-\frac{7}{20} \cdot 0 - \frac{3}{5} \cdot 900 + \frac{7}{20} \cdot 0 + 540 = -3 \cdot 180 + 540 = 0 = u. \qquad (1')$$

Thus,

$$u = 0.$$

Hence, notice that all the equations of Tableau 2 check out, and this is a partial confirmation that the numbers in Tableau 2 are correct.

● **PROBLEM** 11-39

Consider the system in standard form:

$$x_1 \qquad + x_4 + x_5 - x_6 = 5$$
$$x_2 \qquad + 2x_4 - 3x_5 + x_6 = 3$$
$$x_3 - x_4 + 2x_5 - x_6 = -1$$

Set up the system of equations in tableau form, obtain a basic feasible solution having x_4 x_5, x_6 as basic variables.

Solution: Set up the coefficient array below:

x_1	x_2	x_3	x_4	x_5	x_6	
1	0	0	①	-1	1	5
0	1	0	2	-3	1	3
0	0	1	-1	2	-1	-1

The circle indicated is the first pivot element and corresponds to the replacement of x_1 by x_4 as a basic variable. After pivoting the following array is obtained:

862

x_1	x_2	x_3	x_4	x_5	x_6	
1	0	0	1	-1	-1	5
-2	1	0	0	(-5)	3	-7
1	0	1	0	3	-2	4

and again the next pivot element is circled, indicating the intention to replace x_2 by x_5. Then the following is obtained:

x_1	x_2	x_3	x_4	x_5	x_6	
3/5	1/5	0	1	0	-2/5	18/5
2/5	-1/5	0	0	1	-3/5	7/5
-1/5	3/5	1	0	0	(-1/5)	-1/5

Continuing, there results:

x_1	x_2	x_3	x_4	x_5	x_6	
1	-1	-1	1	0	0	4
1	-2	-3	0	1	0	2
1	-3	-5	0	0	1	1

From this last standard form the new basic solution is:

$$x_4 = 4 \qquad x_5 = 2 \qquad x_6 = 1 .$$

● **PROBLEM** 11-40

Find a basic feasible solution to:

$$2x_1 + x_2 + 2x_3 = 4$$

$$3x_1 + 3x_2 + x_3 = 3$$

$$x_1 \geq 0, \qquad x_2 \geq 0, \qquad x_3 \geq 0 .$$

<u>Solution</u>: Introduce artificial variables $x_4 \geq 0$, $x_5 \geq 0$ and an objective function $x_4 + x_5$. The initial tableau is:

x_1	x_2	x_3	x_4	x_5	b
2	1	2	1	0	4
3	3	1	0	1	3
0	0	0	1	1	0

Initial Tableau

863

A basic feasible solution to the expanded system is given by the artificial variables. To initiate the simplex procedure, update the last row so that it has zero components under the basic variables. This yields:

2	1	2	1	0	4
③	3	1	0	1	3
-5	-4	-3	0	0	-7

First Tableau

Pivoting in the column having the most negative bottom row component as indicated, the following is obtained:

0	-1	④/3	1	-2/3	2
1	1	1/3	0	1/3	1
0	1	-4/3	0	5/3	-2

Second Tableau

In the second tableau there is only one choice for pivot, and it leads to the final tableau shown:

0	-3/4	1	3/4	-1/2	3/2
1	5/4	0	-1/4	1/2	1/2
0	0	0	1	1	0

Final Tableau

Both of the artificial variables have been driven out of the basis, thus reducing the value of the objective function to zero and leading to the basic feasible solution to the original problem:

$$x_1 = 1/2 , \qquad x_2 = 0 , \qquad x_3 = 3/2 .$$

● PROBLEM 11-41

Maximize $3x_1 + x_2 + 3x_3$ subject to:

$$2x_1 + x_2 + x_3 \leq 2$$

$$x_1 + 2x_2 + 3x_3 \leq 5$$

$$2x_1 + 2x_2 + x_3 \leq 6$$

$$x_1 \geq 0, \quad x_2 \geq 0, \quad x_3 \geq 0.$$

Find the optimal solution to this linear program by using the simplex method.

Solution: To transform the problem into standard form so that the simplex procedure can be applied, change the maximization to minimization by multiplying the objective function by minus one, and introduce three nonnegative slack variables x_4, x_5, x_6.   Thus, the initial tableau is obtained.

a_1	a_2	a_3	a_4	a_5	a_6	b
②	①	1	1	0	0	2
1	2	③	0	1	0	5
2	2	1	0	0	1	6
-3	-1	-3	0	0	0	0

First tableau

The problem is already in standard form with the three slack variables serving as the basic variables.   There is at this point $r_j = c_j - z_j = c_j$, since the costs of the slacks are zero.   Application of the criterion for selecting a column in which to pivot shows that any of the first three columns would yield an improved solution.   In each of these columns the appropriate pivot element is determined by computing the ratios y_{i0}/y_{ij} and selecting the smallest positive one.   The three allowable pivots are all circled on the tableau.   Select a pivot that will minimize the amount of division required. Thus, for this problem select ①.

2	1	1	1	0	0	2
-3	0	①	-2	1	0	1
-2	0	-1	-2	0	1	2
-1	0	-2	1	0	0	1

Second tableau

Note that the objective function--the negative of the original one--has decreased from zero to minus two.   Again pivot on ①.

⑤	1	0	3	-1	0	1
-3	0	1	-2	1	0	1
-5	0	0	-4	1	1	3
-7	0	0	-3	2	0	4

Third tableau

The value of the objective function has now decreased to minus four and may pivot in either the first or fourth column. Select $\circled{5}$.

1	1/5	0	3/5	-1/5	0	1/5
0	3/5	1	-1/5	2/5	0	8/5
0	1	0	-1	0	1	4
0	7/5	0	6/5	3/5	0	27/5

Fourth tableau

Since the last row has no negative elements, the solution corresponding to the fourth tableau is optimal. Thus, $x_1 = \frac{1}{5}$, $x_2 = 0$, $x_3 = \frac{8}{5}$, $x_4 = 0$, $x_5 = 0$, $x_6 = 4$ is the optimal solution with a corresponding value of the negative objective of $-\frac{27}{5}$.

● **PROBLEM 11-42**

Find an optimal solution to this maximization problem using the simplex technique.

$$\text{maximize } x_0 = 3x_1 + 2x_2 + 5x_3$$

subject to:

$$x_1 + 2x_2 + x_3 \leq 430$$

$$3x_1 \qquad + 2x_3 \leq 460$$

$$x_1 + 4x_2 \qquad \leq 420$$

$$x_1, x_2, x_3 \geq 0$$

Solution: This is expressed in tableau form as follows:

Starting Tableau:

Basic	x_0	x_1	x_2	x_3	S_1	S_2	S_3	Solution
x_0	$\circled{1}$	-3	-2	-5	0	0	0	0
S_1	0	1	2	1	$\circled{1}$	0	0	430
S_2	0	3	0	2	0	$\circled{1}$	0	460
S_3	0	1	4	0	0	0	$\circled{1}$	420

First Iteration. x_3 is the entering variable. By taking ratios,

866

Current basic Solution	Ratios to coefficients of x_3
$S_1 = 430$	$430/1 = 430$
$S_2 = 460$	$460/2 = 230 \leftarrow S_2 = 0, x_3 = 230$
$S_3 = 420$	$420/0 = -$

S_2 becomes the leaving variable. The new tableau is thus given by

Basic	x_0	x_1	x_2	x_3	S_1	S_2	S_3	Solution
x_0	①	9/2	-2	0	0	5/2	0	1150
S_1	0	-1/2	2	0	①	-1/2	0	200
x_3	0	3/2	0	①	0	1/2	0	230
S_3	0	1	4	0	0	0	①	420

Second iteration. x_2 is the entering variable. By taking ratios,

Current basic Solution	Ratios to coefficients of x_2
$S_1 = 200$	$200/2 = 100 \leftarrow S_1 = 0, x_2 = 100$
$x_3 = 230$	$230/0 = -$
$S_3 = 420$	$420/4 = 105$

S_1 leaves the solution. The new tableau is

Basic	x_0	x_1	x_2	x_3	S_1	S_2	S_3	Solution
x_0	①	4	0	0	1	2	0	1350
x_2	0	-1/4	①	0	1/2	-1/4	0	100
x_3	0	3/2	0	①	0	1/2	0	230
S_3	0	2	0	0	-2	1	①	20

This is optimal since all the coefficients in the x_0-equation are nonnegative. The optimal solution is $x_1 = 0$, $x_2 = 100$, $x_3 = 230$, $S_1 = 0$, $S_2 = 0$, $S_3 = 20$, and $x_0 = 1350$.

● **PROBLEM 11-43**

Maximize $x_0 = 3x_1 + 9x_2$

subject to:

$$x_1 + 4x_2 \leq 8$$
$$x_1 + 2x_2 \leq 4$$
$$x_1, x_2 \geq 0$$

Use the simplex technique to solve.

Solution:

Starting Tableau:

Basic	x_0	x_1	x_2	S_1	S_2	Solution
x_0	①	-3	-9	0	0	0
S_1	0	1	4	①	0	8
S_2	0	1	2	0	①	4

First Iteration: Introduce x_2 and drop S_1.

Basic	x_0	x_1	x_2	S_1	S_2	Solution
x_0	①	$-3/4$	0	$9/4$	0	18
x_2	0	$1/4$	①	$1/4$	0	2
S_2	0	$1/2$	0	$-1/2$	①	0

Second Iteration: Introduce x_1 and drop S_2.

Basic	x_0	x_1	x_2	S_1	S_2	Solution
x_0	①	0	0	$3/2$	$3/2$	18
x_2	0	0	①	$1/2$	$-1/2$	2
x_1	0	①	0	-1	2	0

The optimal solution is $x_1 = 0$, $x_2 = 2$, and $x_0 = 18$.

● **PROBLEM 11-44**

Consider the following minimization problem:

$$\min z = 2x_1 + x_2 - x_3$$

subject to:

$$x_1 + x_2 + x_3 \leq 3 \tag{1}$$

$$x_2 + x_3 \geq 2$$

$$x_1 + x_3 = 1$$

$$x_1, x_2, x_3 \geq 0$$

Find an initial basic feasible solution.

Solution: Converting to the equivalent equality-con-strained problem, there follows:

$$\min z = 2x_1 + x_2 - x_3$$

subject to:

$$x_1 + x_2 + x_3 + x_4 = 3 \tag{2}$$

$$x_2 + x_3 - x_5 = 2$$

$$x_1 + x_3 = 1$$

868

$$x_1, \ldots, x_5 \geq 0$$

Add a different nonnegative variable, x_6 and x_7, not appearing elsewhere, in each of the last two equations of problem (2). Thus:

$$
\begin{array}{rcl}
x_1 + x_2 + x_3 + x_4 & = & 3 \qquad (3) \\
x_2 + x_3 \quad - x_5 + x_6 & = & 2 \\
x_1 \quad + x_3 \quad + x_7 & = & 1 \\
\end{array}
$$

$$x_1, \ldots, x_7 \geq 0$$

Now, an initial basic solution is easily obtained: $x_1 = x_2 = x_3 = x_5 = 0$, $x_4 = 3$, $x_6 = 2$, $x_7 = 1$. The difficulty, however, is that this solution is not feasible, for it implies:

$$
\begin{array}{rcl}
x_1 + x_2 + x_3 & = & 0 \\
x_2 + x_3 & = & 0 \\
x_1 \quad + x_3 & = & 0 \\
\end{array}
$$

and, therefore, the constraints given in (1) are not satisfied, and neither are those of (2). Since this is not feasible, artificial variables must be added to the program and then their values reduced to zero.

To achieve removal of artificials from the basic solution, one associates with each variable a cost so undesirable that the technique will drive these variables to zero, since it is an optimization technique. Thus in a max (min) problem one assigns a large negative (positive) cost to each artificial variable. These costs will be sufficiently large to dominate any expression in which they appear in the computation. It is conventional to employ a symbol as this cost, and M or -M, is frequently used for the min (max) problem, where M is taken to be positive.

Thus the problem becomes:

$$\min z = 2x_1 + x_2 + x_3 + Mx_6 + Mx_7$$

subject to:

$$
\begin{array}{rcl}
x_1 + x_2 + x_3 + x_4 & = & 3 \\
x_2 + x_3 \quad -x_5 + x_6 & = & 2 \\
x_1 \quad + x_3 \quad + x_7 & = & 1 \\
\end{array}
$$

$$x_1, \ldots, x_7 \geq 0$$

The initial tableau becomes:

Tableau 1

c_j	Basis	P_1	P_2	P_3	P_4	P_5	P_6	P_7	b
0	P_4	1	1	1	1	0	0	0	3
M	P_6	0	1	1	0	-1	1	0	2
M	P_7	1	0	1	0	0	0	1	1
	$c_j - z_j$	2- M	1 -M	-1-2M	0	M	0	0	

P_3 should come into the basis, since $-1 -2M$ is the most negative $c_j - z_j$ available, by virtue of M's dominance. P_7, then, will leave the basis, yielding the tableau:

Tableau 2

c_j	Basis	P_1	P_2	P_3	P_4	P_5	P_6	P_7	b
0	P_4	1	1	0	1	0	0	-1	2
M	P_6	-1	1	0	0	-1	1	-1	1
-1	P_3	1	0	1	0	0	0	1	1
	$c_j - z_j$	3 + M	1 -M	0	0	M	0	1+2M	

P_2 should enter and P_6 should be removed, yielding the tableau:

Tableau 3

c_j	Basis	P_1	P_2	P_3	P_4	P_5	P_6	P_7	b
0	P_4	1	0	0	1	1	-1	0	1
1	P_2	-1	1	0	0	-1	1	-1	1
-1	P_3	1	0	1	0	0	0	1	1

In tableau 3 both artificial variables are zero, and $x_4 = 1$, $x_2 = 1$, $x_1 = x_5 = 0$, which is a feasible solution for (2); this then is selected as the initial basic feasible solution.

● **PROBLEM 11-45**

Maximize $\qquad f = 2x_1 + 3x_2$

subject to: $\qquad x_1 + x_2 \geq 3$

$\qquad\qquad x - 2x \leq 4$

$\qquad\qquad x_1, x_2 \geq 0$

Solve by the Simplex Method, applying the Big-M technique.

Maximize			2	3	0	0	-M			
i	c_B	x_B	x_1	x_2	x_3	x_4	x_5	b_i	θ_i	
1	-M	x_5	1	①	-1	0	1	3	3	Out →
2	0	x_4	1	-2	0	1	0	4	-	
		z_j	-M	-M	M	0	-M	-3M		
		c_j-z_j	2+M	3+M	-M	0	0			

     ↑In   (a) Initial Tableau

Maximize			2	3	0	0			
i	c_B	x_B	x_1	x_2	x_3	x_4	b_i	θ_i	
1	3	x_2	1	1	-1	0	3	-	?
2	0	x_4	3	0	-2	1	10	-	?
		z_j	3	3	-3	0	9		
		c_j-z_j	-1	0	3	0			

Fig. 1   ↑In   (b) Second Tableau

Solution: The first and second simplex tableaus are shown in Fig. 1 a and b. Consider the second tableau. The basic solution shown cannot be optimal because x_3 should be brought into the basis. However, there is no variable to leave: both $a_{1,3}$ and $a_{2,3}$ are negative, so neither can be the pivot element. In practical terms, as x_3 increases, so do x_2 and x_4. Thus, nonnegativity does not identify a variable to leave the basis; in fact, neither variable (x_2 or x_4) can be pivoted out without violating nonnegativity. The reason is that there is no limit to the size of x_3 because there is no limit to either x_2 or x_4. Therefore, maximum profit can be made.

This can be demonstrated simply by writing the equations that correspond to rows 1 and 2 in Fig. 1b. Row 1 tells us:

$$x_1 + x_2 - x_3 = 3$$

or

$$x_2 = 3 + x_3 - x_1$$

and row 2 yields

$$3x_1 - 2x_3 + x_4 = 10$$

or

$$x_4 = 10 + 2x_3 - 3x_1$$

Now x_1 is non-basic, so it has a value of zero. Eliminating x_1 from the above leaves:

$$x_2 = 3 + x_3$$

$$x_4 = 10 + 2x_3$$

Thus, as x_3 increases, so do x_2 and x_4.

Of course, the "real world" does not permit an infinite profit. When such a result occurs in a linear programming model, it invariably indicates a defect in the model. Perhaps one or more constraints were left out, or the situation may require a nonlinear model.

Solve the problem below by the two-phase method.

$$\text{Maximize } z = x_1 + x_2$$

$$\text{subject to: } 3x_1 + 2x_2 \leq 20$$

$$2x_1 + 3x_2 \leq 20$$

$$x_1 + 2x_2 \geq 2$$

$$x_1, x_2 \geq 0.$$

Solution: Add the slack and artificial variables as follows:

Phase 1: Maximize $z_1 = \qquad\qquad\qquad - x_a$

Phase 2: Maximize $z_2 = x_1 + x_2$

$$\text{subject to: } 3x_1 + 2x_2 + x_3 \qquad\qquad = 20$$

$$2x_1 + 3x_2 \qquad + x_4 \qquad = 20$$

$$x_1 + 2x_2 \qquad\qquad -x_5 + x_a = 2$$

Putting the problem into tableau form yields Tableau 0.

Tableau 0

		x_1	x_2	x_3	x_4	x_5	x_a
z(phase 1)	0	0	0	0	0	0	1
z(phase 2)	0	-1	-1	0	0	0	0
x_3	20	3	2	1	0	0	0
x_4	20	2	3	0	1	0	0
x_a	2	1	2	0	0	-1	[1]

The basic representation must be made complete; hence, the partial iteration on the dashed pivotal element must be undertaken to yield Tableau 1.

Tableau 1

		x_1	x_2	x_3	x_4	x_5	x_a
z(phase 1)	-2	-1	-2	0	0	1	0
z(phase 2)	0	-1	-1	0	0	0	0
x_3	20	3	2	1	0	0	0
x_4	20	2	3	0	1	0	0
x_a	2	1	$\boxed{2}$	0	0	-1	1

Using the phase 1 objective function, x_2 is selected to enter the basis replacing x_a and yielding Tableau 2.

Tableau 2

		x_1	x_2	x_3	x_4	x_5	x_a
z(phase 1)	0	0	0	0	0	0	1
z(phase 2)	1	$-\frac{1}{2}$	0	0	0	$-\frac{1}{2}$	$\frac{1}{2}$
x_3	18	2	0	1	0	1	-1
x_4	17	$\frac{1}{2}$	0	0	1	$\boxed{\frac{3}{2}}$	$-\frac{3}{2}$
x_2	1	$\frac{1}{2}$	1	0	0	$-\frac{1}{2}$	$\frac{1}{2}$

Variable x_a can now be dropped. Since the phase 1 objective function is zero, the phase 1 objective function can also be dropped.

Proceed to phase 2. Introducing x_5 (choosing x_5 arbitrarily instead of x_1) into the basis to replace x_4, and then introducing x_1 into the basis to replace x_3, leads to the following tableaus.

Tableau 3

		x_1	x_2	x_3	x_4	x_5
z(phase 2)	$\frac{20}{3}$	$-\frac{1}{3}$	0	0	$\frac{1}{3}$	0
x_3	$\frac{20}{3}$	$\boxed{\frac{5}{3}}$	0	1	$-\frac{2}{3}$	0
x_5	$\frac{34}{3}$	$\frac{1}{3}$	0	0	$\frac{2}{3}$	1
x_2	$\frac{20}{3}$	$\frac{2}{3}$	1	0	$\frac{1}{3}$	0

Tableau 4

		x_1	x_2	x_3	x_4	x_5
z(phase 2)	8	0	0	$\frac{1}{5}$	$\frac{1}{5}$	0
x_1	4	1	0	$\frac{3}{5}$	$-\frac{2}{5}$	0
x_5	10	0	0	$-\frac{1}{5}$	$\frac{4}{5}$	1
x_2	4	0	1	$-\frac{2}{5}$	$\frac{3}{5}$	0

Tableau 4 is optimal.

Consider:

Minimize $- x_1 + 2x_2 - 3x_3$

Subject to:
$$x_1 + x_2 + x_3 = 6$$
$$-x_1 + x_2 + 2x_3 = 4$$
$$2x_2 + 3x_3 = 10$$
$$x_3 \leq 2$$
$$x_1, x_2, x_3 \geq 0.$$

Solve by the two-phase method.

Solution: A slack variable x_4 needs to be introduced. The constraint matrix $\vec{A}$ is given below:

$$\vec{A} = \begin{bmatrix} 1 & 1 & 1 & 0 \\ -1 & 1 & 2 & 0 \\ 0 & 2 & 3 & 0 \\ 0 & 0 & 1 & 1 \end{bmatrix}$$

Note that the matrix is of full rank. The sum of the first two rows of $\vec{A}$ is equal to the third row; that is,

any one of the first three constraints is redundant and can be eliminated. Assume this fact is not known, however, and introduce the artificial variables x_5, x_6, and x_7. The phase I objective is: Minimize $x_0 = x_5 + x_6 + x_7$. Phase I proceeds as follows.

Phase I

x_0	x_1	x_2	x_3	x_4	x_5	x_6	x_7	RHS
1	0	0	0	0	-1	-1	-1	0
1	1	1	1	0	1	0	0	6
0	-1	1	2	0	0	1	0	4
0	0	2	3	0	0	0	1	10
0	0	0	1	1	0	0	0	2

Add rows 1, 2, and 3 to row 0, to display $z_5 - c_5 = z_6 - c_6 = z_7 - c_7 - 0$.

	x_0	x_1	x_2	x_3	x_4	x_5	x_6	x_7	RHS
x_0	1	0	4	6	0	0	0	0	20
x_5	0	1	1	1	0	1	0	0	6
x_6	0	-1	1	2	0	0	1	0	4
x_7	0	0	2	3	0	0	0	1	10
x_4	0	0	0	①	1	0	0	0	2

	x_0	x_1	x_2	x_3	x_4	x_5	x_6	x_7	RHS
x_0	1	0	4	0	-6	0	0	0	8
x_5	0	1	1	0	-1	1	0	0	4
x_6	0	-1	①	0	-2	0	1	0	0
x_7	0	0	2	0	-3	0	0	1	4
x_4	0	0	0	1	1	0	0	0	2

	x_0	x_1	x_2	x_3	x_4	x_5	x_6	x_7	RHS
x_0	1	4	0	0	2	0	-4	0	8
x_5	0	②	0	0	1	1	-1	0	4
x_2	0	-1	1	0	-2	0	1	0	0
x_7	0	2	0	0	1	0	-2	1	4
x_3	0	0	0	1	1	0	0	0	2

	x_0	x_1	x_2	x_3	x_4	x_5	x_6	x_7	RHS
x_0	1	0	0	0	0	-2	-2	0	0
x_1	0	1	0	0	$\frac{1}{2}$	$\frac{1}{2}$	$-\frac{1}{2}$	0	2
x_2	0	0	1	0	$-\frac{3}{2}$	$\frac{1}{2}$	$\frac{1}{2}$	0	2
x_7	0	0	0	0	0	-1	-1	1	0
x_3	0	0	0	1	1	0	0	0	2

Since all the artificial variables are at level zero, proceed to phase II with a basic feasible solution of the original problem. Either proceed directly with the artificial x_7 into the basis at zero level, or attempt to eliminate x_7 from the basis. The only legitimate nonbasic variable is x_4, and it has zero coefficient in row 3 corresponding to x_7. This shows that the third row (constraint of the original problem) is redundant and can be eliminated. This will be done while moving to phase II.

PHASE II

Obviously $z_1 - c_1 = z_2 - c_2 = z_3 - c_3 = 0$. Thus x_5 and x_6 are nonbasic artificial variables and wll not be introduced in the phase II problem. In order to complete row 0 calculate $z_4 - c_4$:

$$z_4 - c_4 = \vec{c}_B B^{-1} \vec{a}_4 - c_4$$

$$= (-1, 2, -3) \begin{bmatrix} \frac{1}{2} \\ -\frac{3}{2} \\ 1 \end{bmatrix} - 0$$

$$= -\frac{13}{2}$$

Since the objective is minimization and $z_4 - c_4 \leq 0$ for the only nonbasic variable, the stop; the solution

875

obtained from phase I is optimal. The tableau below
displays the optimal solution.

	z	x_1	x_2	x_3	x_4	RHS
z	1	0	0	0	$-\frac{11}{2}$	-4
x_1	0	1	0	0	$\frac{1}{2}$	2
x_2	0	0	1	0	$-\frac{3}{2}$	2
x_3	0	0	0	1	1	2

● **PROBLEM** 11-48

Find a basic feasible solution to the problem:

$$\text{Maximize} \quad x_1 + 2x_2 + 3x_3 + 4x_4 \tag{1}$$

while satisfying the conditions

$$\begin{aligned} x_1 + 2x_2 + x_3 + x_4 &= 3 \\ x_1 - x_2 + 2x_3 + x_4 &= 4 \\ x_1 + x_2 - x_3 - x_4 &= -1 . \end{aligned} \tag{2}$$

$x_1, x_2, x_3, x_4 \geq 0$.

Solution: We can solve this problem using the simplex method. Note that
the number of equations, 3, is less than the number of unknowns, 4. The
system of linear equations can be written in vector form as:

$$x_1 \begin{pmatrix} 1 \\ 1 \\ 1 \end{pmatrix} + x_2 \begin{pmatrix} 2 \\ -1 \\ 1 \end{pmatrix} + x_3 \begin{pmatrix} 1 \\ 2 \\ -1 \end{pmatrix} + x_4 \begin{pmatrix} 1 \\ 1 \\ -1 \end{pmatrix} = \begin{pmatrix} 3 \\ 4 \\ -1 \end{pmatrix} \tag{3}$$

Any three of the vectors on the left can be used as a basis for R^3. Thus,
by setting either $x_1 = 0$, or $x_2 = 0$ or $x_4 = 0$ we can still find
x_i, x_j, x_k (i,j,k = 1 or 2 or 3 or 4) that satisfy (3). Such solutions
are called basic solutions, since they depend on the particular basis
chosen when one of the x_i's is set equal to zero. A basic feasible solution (non-negative x_i), when substituted into the objective function will
yield some value, not necessarily optimal. Now, the number of basic solutions is the number of ways of selecting a basis for R^3 from a set of
four vectors, i.e.,

$$\binom{4}{3} = \frac{4!}{3!\ 1!} = 4 .$$

According to the theory of linear programming, if an optimal solution to
(1) and (2) exists, then an optimal basic solution exists. Thus the optimal solution can be found by using one of the four bases of R^3 .
Let $v_1 = [1,1,1]$, $v_2 = [2,-1,1]$, $v_3 = [1,2,-1]$ and $v_4 = [1,1,-1]$.
Let $b = [3,4,-1]$. Then the system (3) becomes

$$x_1 v_1 + x_2 v_2 + x_3 v_3 + x_4 v_4 = b.$$

Letting $x_2 = 0$, we see that

$$x_1 = 1, \ x_2 = 0, \ x_3 = 1, \ x_4 = 1$$

is a basic feasible solution which depends on $\{v_1, v_3, v_4\}$. Corresponding to this solution we can construct a simplex tableau. The general form of the simplex tableau is as follows:

	v_1	v_2 . . . . . v_q . . . v_n	b
v_{k1}	s_{11}	s_{12} . . . . . s_{1q} . . . s_{1n}	x_{k1}
v_{k2}	s_{21}	s_{22} . . . . . s_{2q} . . . s_{2n}	x_{k2}
.	.		.
.	.		.
.	.		.
v_{kr}	s_{r1}	s_{r2} . . . . . s_{rq} . . . s_{rn}	x_{kr}
	d_1	d_2 . . . . . d_q . . . d_n	D

The column vectors $v_1, \ldots, v_n$ can all be expressed as linear combinations of the basic vectors $v_{k1}, v_{k2}, \ldots, v_{kr}$. The coordinates (s_{ij}), $i = 1, \ldots, r$, $j = 1, \ldots, n$ form the main body of the table. The numbers $d_1, d_2, \ldots, d_n$, D in the last row of the table are defined as follows:

$$d_j = (c_{k1}s_{1j} + c_{k2}s_{2j} + \ldots + c_{kr}s_{rj}) - c_j \, , \ j = 1, \ldots, n.$$

$$D = c_{k1}x_{k1} + c_{k2}x_{k2} + \ldots + c_{kr}x_{kr}$$

$$= c_1x_1 + c_2x_2 + \ldots + c_nx_n \, .$$

The c_i $(i = 1, 2, \ldots, n)$ represent the coefficients of the objective function, $z = c_1x_1 + c_2x_2 + \ldots + c_nx_n$. The c_{kr} are the coefficients of the x_{kr} which are the coefficients of the basis $\{v_{k1}, \ldots, v_{kr}\}$ chosen from $v_1, v_2, \ldots, v_n$. D is the quantity to be maximized. The d_j indicate when the solution is optimal or how to proceed to a more nearly optimal solution. Returning to the given problem,

$$b = v_1 + v_3 + v_4; \ v_1 = 1v_1 + 0v_3 + 0v_4;$$
$$v_2 = 3/2 \, v_1 - 3v_3 + 7/2 \, v_4; \ v_3 = 0v_1 + 1v_3 + 0v_4;$$
$$v_4 = 0v_1 + 0v_3 + 1v_4 \, .$$

$$d_1 = c_1 1 + c_3(0) + c_4(0) - c_1 = 0$$

$$d_2 = c_1(3/2) + c_3(-3) + c_4(7/2) - c_2 = 9/2$$

$$d_3 = c_1(0) + c_3(1) + c_4(0) - c_3 = 0$$

$$d_4 = c_1(0) + c_3(0) + c_4(1) - c_4 = 0$$

$$D = c_1x_1 + c_2x_2 + c_3x_3 + c_4x_4 = 8$$

The simplex tableau for this solution is:

	v_1	v_2	v_3	v_4	b
v_1	1	3/2	0	0	1
v_3	0	-3	1	0	1
v_4	0	7/2	0	1	1
	0	9/2	0	0	8

Next, we would choose another basis for the basic solution and construct another tableau. If the new value of D were greater than before we would proceed yet again using another new basis. This is the simplex method. We see that rules for choosing a new basis and for deciding when a solution is optimal are needed. It is important to note the following criteria:

(1) If the numbers $d_1, d_2, \ldots, d_n$ are all non-negative, then the given solution is optimal.

(2) If, for some index q, d_q is negative and $s_{1q}, s_{2q}, \ldots, s_{rq}$ are all non-positive, then the given problem does not have an optimal solution.

● **PROBLEM 11-49**

Find nonnegative numbers x_1, x_2, x_3, x_4 which maximize

$$4x_1 + 5x_2 + 3x_3 + 6x_4 \qquad (1)$$

and satisfy the inequalities

$$x_1 + 3x_2 + x_3 + 2x_4 \leq 2$$
$$3x_1 + 3x_2 + 2x_3 + 2x_4 \leq 4 \qquad (2)$$
$$3x_1 + 2x_2 + 4x_3 + 5x_4 \leq 6 .$$

Solution: We cannot use the simplex method directly to solve (1) subject to (2) because we need equalities in the constraint equations. Hence, introduce three additional variables, x_5, x_6, x_7 to convert each inequality to an equality. The system (2) then becomes

$$x_1 + 3x_2 + x_3 + 2x_4 + x_5 = 2$$
$$3x_1 + 3x_2 + 2x_3 + 2x_4 + x_6 = 4 \qquad (3)$$
$$3x_1 + 2x_2 + 4x_3 + 5x_4 + x_7 = 6 .$$

These new variables are called slack variables. We now have 3 equations in 7 unknowns. The system (3) can be written as

$$x_1 \begin{pmatrix} 1 \\ 3 \\ 3 \end{pmatrix} + x_2 \begin{pmatrix} 3 \\ 3 \\ 2 \end{pmatrix} + x_3 \begin{pmatrix} 1 \\ 2 \\ 4 \end{pmatrix} + x_4 \begin{pmatrix} 2 \\ 2 \\ 5 \end{pmatrix} + x_5 \begin{pmatrix} 1 \\ 0 \\ 0 \end{pmatrix} + x_6 \begin{pmatrix} 0 \\ 1 \\ 0 \end{pmatrix} + x_7 \begin{pmatrix} 0 \\ 0 \\ 1 \end{pmatrix}$$

$$= \begin{pmatrix} 2 \\ 4 \\ 6 \end{pmatrix} , \text{ or}$$

$$x_1 v_1 + x_2 v_2 + x_3 v_3 + x_4 v_4 + x_5 v_5 + x_6 v_6 + x_7 v_7 = [2,4,6] . \qquad (4)$$

By setting any four of the unknowns in (4) to zero, we obtain a system of three equations in three unknowns with a unique solution. Say we let $x_{\ell_1} = x_{\ell_2} = x_{\ell_3} = x_{\ell_4} = 0$. Then $v_{\ell_5}, v_{\ell_6}, v_{\ell_7}$ form a basis for R^3. The corresponding solution of x's is called a basic feasible solution. We can obtain

$$\binom{7}{3} = \frac{7!}{3! \; 4!} = 35$$

basis feasible solutions. According to the theory of linear programming if an optimal solution exists, one of these solutions is an optimal solution. The simplex method is a systematic way of changing the basis vectors and computing solutions. To get started on the simplex method we need a basic feasible solution. Let $x_1 = x_2 = x_3 = x_4 = 0$. Then, from (4),

$$x_5 \begin{pmatrix} 1 \\ 0 \\ 0 \end{pmatrix} + x_6 \begin{pmatrix} 0 \\ 1 \\ 0 \end{pmatrix} + x_7 \begin{pmatrix} 0 \\ 0 \\ 1 \end{pmatrix} = \begin{pmatrix} 2 \\ 4 \\ 6 \end{pmatrix}.$$

Thus $x_1 = x_2 = x_3 = x_4 = 0$, $x_5 = 2$, $x_6 = 4$, $x_7 = 6$ is a basic feasible solution. We form the simplex tableau using this solution.

	v_1	v_2	v_3	v_4	b	v_5	v_6	v_7
v_5	1	3	1	2	2	1	0	0
v_6	3	3	2	2	4	0	1	0
v_7	3	2	4	5	6	0	0	1
	-4	-5	-3	-6	0	0	0	0

The column under v_1 was derived as follows: $v_1 = [1,3,3] = 1v_5 + 3v_6 + 3v_7$. The last element was computed using the formula $d_1 = c_1(1) + c_2(3) + c_3(3) - c_1$, where c_5, c_6, c_7 are the coefficients of the slack variables in the objective function

$$c_1 x_1 + c_2 x_2 + \ldots + c_3 x_3 + c_4 x_4 + c_5 x_5 + c_6 x_6 + c_7 x_7 .$$

But $c_1 = 4$, $c_2 = 5$, $c_3 = 3$, $c_4 = 6$, $c_5 = c_6 = c_7 = 0$. Thus, $d_1 = -4$. The other columns of the simplex tableau were found in a similar manner. The value of the program, D, is zero. To find another basic feasible solution we use the following two rules to help us select another basis for R^3.

1) Choose the column that has the most negative value of d_j (here, v_4).

2) Choose the row which has the smallest ratio of the jth element of b to the corresponding element of the column chosen in 1). (Here v_5 is the row since 2/2 is smaller than 4/2, 6/5).

3) Remove the chosen basis vector and in its place put the chosen non-basic vector to obtain a new basis.
(Here, we remove v_5 and put v_4 in its place).
We must now find a new tableau associated with this new basis. To do this we use the following rule:

4) Divide the chosen row by the element in the chosen column. Then subtract multiples of the chosen row from each of the other rows so that the chosen column will have zeros everywhere except for a one in the chosen

row.

Carrying out these steps on the tableau we obtain the new tableau

	v_1	v_2	v_3	v_4	b	v_5	v_6	v_7
v_4	1/2	3/2	1/2	1	1	1/2	0	0
v_6	2	0	1	0	2	-1	1	0
v_7	1/2	-11/2	3/2	0	1	-5/2	0	1
	-1	4	0	0	6	3	0	0

Now, according to 1), the first column has the most negative value and is chosen. Of the ratios $1/\frac{1}{2}$ $2/2$ and $1/\frac{1}{2}$, v_6 is the minimum ratio. Thus we remove v_6 and put v_1 in its place. The new tableau is

	v_1	v_2	v_3	v_4	b	v_5	v_6	v_7
v_4	0	3/2	1/4	1	1/2	3/4	-1/4	0
v_1	1	0	1/2	0	1	-1/2	1/2	0
v_7	0	-11/2	5/4	0	1/2	-9/4	-1/4	1
	0	4	1/2	0	7	5/2	1/2	0

Since all $d_j \geq 0$, the solution $x_1 = 1$, $x_2 = x_3 = 0$, $x_4 = \frac{1}{2}$, $x_5 = x_6 = 0$, $x_7 = \frac{1}{2}$ is an optimal solution. The value of the problem is 7.

● **PROBLEM 11-50**

Find nonnegative numbers x_1, x_2, x_3, x_4 which maximize

$$3x_1 + x_2 + 9x_3 - 9x_4$$

and satisfy the conditions

$$x_1 + x_2 + x_3 - 5x_4 = 4$$
$$x_1 - x_2 + 3x_3 + x_4 = 0 .$$

Solution: Use the simplex algorithm to solve the problem. The calculations are conveniently set forth in the form of a table.

c_j	solution variables	solution values	3 x_1	1 x_2	9 x_3	-9 x_4
3	x_1	4	1	1	1	-5
1	x_2	0	1	-1	3	1
	z_j					
	$c_j - z_j$					

Under each column (x_1, x_2, x_3, x_4) are written the coefficients from the constraint equations of the variables found in the heading. For example, under x_3 is written $(1,3)$. Under the column headed solution

880

values, the constants of the constraints are written. The first row in the heading of the table contains the c_j's or the profit per unit (the coefficients of the variables in the profit equation).

Before filling in the rest of the table, we must identify an initial solution. A basic feasible solution is

$$x_1 = x_2 = 2, \ x_3 = x_4 = 0 \ .$$

The value associated with this solution is $3(2) + (1)2 + 9(0) - 9(0) = 8$. The terms x_1, x_2 are entered in the simplex table under the solution variables column, and their per-unit profits are entered in the first column under the c_j heading.

Finally, consider the computation of the z_j's and $c_j - z_j$'s. The z_j total of a column is the amount of profit which is given up by replacing some of the present solution mix with one unit of the item heading the column. It is found by multiplying the c_j of the row by the number in the row and j^{th} column (the substitution coefficient) and adding.

c_j	solution variables	solution values	3 x_1	1 x_2	9 x_3	-9 x_4
3	x_1	4	1	1	1	-5
1	x_2	0	1	-1	3	1
	z_j	8	4	2	6	-14
	$c_j - z_j$		-1	-1	3	5

The $c_j - z_j$ row represents the net profit that is added by one unit of the product. x_3 and x_4 are the only positive profits. That means we want to replace some of x_1 or x_2 with one or more units of x_3 . The next step is to determine which row (x_1 or x_2) is to be replaced by x_3 . Divide each amount in the "Solution values" column by the amount in the comparable row of the x_3 column:

$$\text{for } x_1 \text{ row: } \frac{-4}{5}$$

$$\text{for } x_2 \text{ row: } \frac{0}{1}$$

Since negative ratios don't count, we choose x_2 for elimination, i.e., pivot on 1. To obtain a new table, convert all other elements in the x_4 column to zero.

c_j	solution variables	solution values	3 x_1	1 x_2	9 x_3	-9 x_4
3	x_1	4	6	-4	16	0
9	x_3	0	1	-1	3	1
	z_j	12	27	-21	75	9
	$c_j - z_j$		-24	22	-66	-18

The z_j and $c_j - z_j$ are calculated as before. If all the $c_j - z_j$ were negative or zero, the solution would be optimal. But the column headed x_2 has a positive amount. On the other hand, both elements in this column are negative indicating that no pivoting is possible.

By the rules of the simplex algorithm the problem has no optimal solution.

Use the simplex algorithm to solve the following linear programming problem:

Maximize

$$z = 4x_1 + 8x_2 + 5x_3 \tag{1}$$

subject to

$$\begin{aligned} x_1 + 2x_2 + 3x_3 &\leq 18 \\ x_1 + 4x_2 + x_3 &\leq 6 \\ 2x_1 + 6x_2 + 4x_3 &\leq 15 \end{aligned} \tag{2}$$

$$x_1 \geq 0,\ x_2 \geq 0,\ x_3 \geq 0. \tag{3}$$

Solution: We first convert the inequalities to equalities by defining slack variables. Let

$$\begin{aligned} x_4 &= 18 - x_1 - 2x_2 - 3x_3 \\ x_5 &= 6 - x_1 - 4x_2 - x_3 \\ x_6 &= 15 - 2x_1 - 6x_2 - 4x_3 \ . \end{aligned}$$

Thus, the resource constraints (2) become:

$$\begin{aligned} x_1 + 2x_2 + 3x_3 + x_4 &= 18 \\ x_1 + 4x_2 + x_3 + x_5 &= 6 \\ 2x_1 + 6x_2 + 4x_3 + x_6 &= 15 \ . \end{aligned} \tag{4}$$

Now use the tableau method to solve (1), (4) and (3). Treating (4) as a matrix, form the coefficient tableau

$$\left[\begin{array}{ccc|ccc|c} 1 & 2 & 3 & 1 & 0 & 0 & 18 \\ 1 & 4 & 1 & 0 & 1 & 0 & 6 \\ 2 & 6 & 4 & 0 & 0 & 1 & 15 \\ \hline -4 & -8 & -5 & 0 & 0 & 0 & 0 \end{array}\right] \tag{5}$$

The bottom row in (5) is the coefficient vector of the objective function. Now convert the matrix

$$\begin{bmatrix} 1 & 2 & 3 \\ 1 & 4 & 1 \\ 2 & 6 & 4 \end{bmatrix} \quad \text{to a matrix of 0's and 1's .}$$

882

Simultaneously we will obtain the maximum possible value of the objective function subject to the constraints.

In the conversion process use the following rules:

1) Locate the most negative number in the extra row and select the column in which this number occurs. Here, -8 is the most negative number which occurs under column 2.

2) Now choose an element in this column as the pivot element. Recall that the pivot element is the element that is converted to 1 and is then used to eliminate other elements in the column (by elementary row operations). The choice of the pivot is a distinguishing feature of the simplex method. Form the ratios of the constraint constants to the positive elements of the chosen column and pivot on that column element which is the denominator of the smallest ratio. Here the ratios are 18/2, 6/4 and 15/6. Thus we choose the second element as the pivot.

3) Convert the remaining entries in the chosen column to zero.

4) Repeat 1)-3) for another column or until every column has a single 1 and the rest of its entries zero. The tableau (5) becomes

1/2	0	5/2	1	-1/2	0	15
1/4	1	1/4	0	1/4	0	3/2
1/2	0	5/2	0	-3/2	1	6
-2	0	-3	0	2	0	12

$$(6)$$

The most negative entry in the last row is -3. The ratios are
$$\frac{15}{(5/2)}, \quad \frac{3/2}{(1/4)} \quad \text{and} \quad \frac{6}{(5/2)}.$$

The lowest ratio is $6/(5/2)$. We choose this element as pivot and convert all other entries in this column to zero to obtain

0	0	0	1	1	-1	9
1/5	1	0	0	2/5	-1/10	9/10
1/5	0	1	0	-3/5	2/5	12/5
-7/5	0	0	0	1/5	6/5	96/5

The lowest ratio in the first column is $\frac{9}{10}/\frac{1}{5}$. Pivoting on this element

0	0	0	1	1	-1	9
1	5	0	0	2	-1/2	9/2
0	-1	1	0	-1	1/2	3/2
0	7	0	0	3	1/2	51/2

Hence an optimal solution is
$$x_1 = 9/2, \; x_2 = 0 \quad \text{and} \quad x_3 = 3/2.$$
The slack variables are
$$x_4 = 9, \; x_5 = 0 \quad \text{and} \quad x_6 = 0.$$
The optimal value of z is $51/2$.

883

Show through the simplex method and then graphically that the following linear program has no solution.

$$\text{Maximize} \quad 2x + y$$

subject to

$$-x + y \le 1$$
$$x - 2y \le 2$$

$x, y \ge 0$.

Solution: First convert the inequalities to equalities by the addition of slack variables. Thus

$$-x + y + s_1 = 1$$
$$x - 2y + s_2 = 2$$

where $s_1 = 1 - (-x+y)$ and $s_2 = 2 - (x-2y)$. This may be rewritten as

$$x \begin{bmatrix} -1 \\ 1 \end{bmatrix} + y \begin{bmatrix} 1 \\ -2 \end{bmatrix} + s_1 \begin{bmatrix} 1 \\ 0 \end{bmatrix} + s_2 \begin{bmatrix} 0 \\ 1 \end{bmatrix} = \begin{bmatrix} 1 \\ 2 \end{bmatrix} \qquad (1)$$

There are 2 equations in four unknowns. Letting two of the unknowns equal zero yields two equations in two unknowns which will have a unique solution only if the two vectors with non-zero coefficients form a basis for $\mathbb{R}^2$. The corresponding coefficients then are said to be a basic feasible solution. For example, suppose $x = 0$, $s_1 = 0$ in (1). Then

$$y \begin{bmatrix} 1 \\ -2 \end{bmatrix} + s_2 \begin{bmatrix} 0 \\ 1 \end{bmatrix} = \begin{bmatrix} 1 \\ 2 \end{bmatrix}$$

$y = 1$, $s_2 = 4$ is the unique solution. Thus $\begin{bmatrix} 1 \\ -2 \end{bmatrix}$ and $\begin{bmatrix} 0 \\ 1 \end{bmatrix}$ form a basis

for $\mathbb{R}^2$ and $y = 1$, $s_2 = 4$ is a basic feasible solution. Now form a simplex tableau using this solution. First find v_1, v_2, v_3 and v_4 in terms of the basis $\{v_2, v_4\}$. $[-1,1] = c_1[1,-2] + c_2[0,1]$ and $c_1 = -1$, $c_2 = -1$.

$$[1,-2] = 1[1,-2] + 0[0,1]$$
$$[1,0] = c_1[1,-2] + c_2[0,1]$$
$$\text{and} \quad c_1 = 1, \ c_2 = 2$$
$$[0,1] = 0[1,-2] + 1[0,1] \quad .$$

Let $b = [1,2] = c_1 v_2 + c_2 v_4 = c_1[1,-2] + c_2[0,1]$. Then $c_1 = 1$, $c_2 = 4$. The simplex tableau summarizes the above information:

	v_1	v_2	v_3	v_4	b
v_2	-1	1	1	0	1
v_4	-1	0	2	1	4
d	-2	-1	0	0	0

According to the rules, choose v_1 as the column for pivoting. But the minimum ratio rule cannot be applied to this column because both elements are nonpositive. Hence there is no solution to the maximum linear program

884

(and, therefore, by the duality theorem, no solution to the minimum dual). Since there are only two variables we can graphically demonstrate the above problem. (See fig.)

The feasible region extends upwards indefinitely. Thus, whatever combination of x and y chosen to substitute into the objective function, we can always find another pair that will yield a higher value while satisfying the constraints.

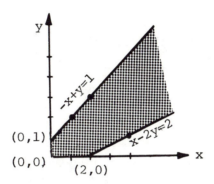

● **PROBLEM** 11-53

The following problem is an illustration of degeneracy.

$$\text{Maximize} \quad P = 4x_1 + 3x_2 \tag{1}$$

subject to

$$4x_1 + 2x_2 \le 10.0$$
$$2x_1 + 8/3x_2 \le 8.0 \tag{2}$$
$$x_1 \ge 0, \; x_2 \ge 1.8 \;.$$

What are the signs of degeneracy
a) in the simplex tableau
b) graphically?

Solution: To apply the simplex method convert the inequalities in (2) to equalities by the addition of artificial and slack variables. In maximization problems, artificial variables are introduced so as to facilitate the simplex method of solution. One of the requirements of the method is that every equation contain a variable whose coefficient is 1 in that equation and zero in every other equation. Thus, the system (2) becomes

$$4x_1 + 2x_2 + x_3 \qquad = 10$$
$$2x_1 + 8/3 \; x_2 + x_4 \qquad = 8$$
$$x_2 + \quad x_5 - x_6 = 1.8 \;.$$

Here x_3, x_4 and x_6 are slack variables, while x_5 is an artificial variable. To ensure that it will not appear in the final solution, assign it a profit factor coefficient of −M, where M is a very large number.

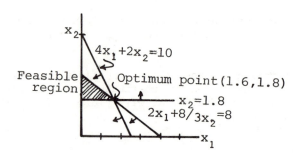

The initial simplex tableau with $x_1 = x_2 = x_6 = 0$ is

c_j	Solution variables	Solution values	4 x_1	3 x_2	0 x_3	0 x_4	-M x_5	0 x_6
0	x_3	10	4	2	1	0	0	0
0	x_4	8	2	8/3	0	1	0	0
-M	→ x_5	1.8	0	(1)	0	0	1	-1
	z_j	-1.8M	0	-M	0	0	-M	M
	c_j-z_j		4	M+3 ↑	0	0	0	-M

Replace x_5 by x_2 ; i.e., pivot on 1.

c_j	Solution variables	Solution values	4 x_1	3 x_2	x_3	x_4	x_5	x_6
0	x_3	6.4	4	0	1	0	-2	2
0	x_4	3.2	2	0	0	1	-8/3	8/3
3	x_2	1.8	0	1	0	0	1	-1
	z_j		0	3	0	0	3	-3
	c_j-z_j		4 ↑	0	0	0	-M-3	3

Here, both row x_3 and row x_4 are minimum ratios: $\frac{6.4}{4} = \frac{3.2}{2} = 1.6$. This is the signal that degeneracy exists. By the rules of the simplex algorithm, we can arbitrarily replace either row. If the chosen row eventually leads to no solution (the simplex tables begin to repeat themselves), then choose the other row at the point where the degeneracy was discovered.

   In the present problem, replacing row x_4 with x_1 leads immediately to the final simplex tableau and optimal solution $x_1 = 1.6$, $x_2 = 1.8$ and $P = 11.8$.

   Replacing row x_3 with x_1 yields the same result but after an additional iteration.

b) The degeneracy situation can also be identified by examining the graph of the problem. (See fig.)

   The optimum point occurs at the intersection of the three constraint equations. Since all the constraints are satisfied exactly, there is no slack in any constraint. In a nondegenerate case, at least one of these slack variables would be non-zero.

The optimum solution to the problem

$$\text{Maximize} \qquad P = 12x_1 + 9x_2 \qquad (1)$$

subject to

$$3x_1 + 2x_2 \leq 7$$
$$3x_1 + x_2 \leq 4 \qquad (2)$$
$$x_1 \geq 0, \; x_2 \geq 0$$

is $P = 9(7/2) = 31\frac{1}{2}$. The solution to the dual is $y_1 = 4\frac{1}{2}$, $y_2 = 0$.
Now assume the first constraint of (2) is changed from 7 to 8, i.e.,
$$3x_1 + 2x_2 \leq 8 .$$
Find the increase in P. What is the dual for this new problem?

Solution: The new problem is:

$$\text{Maximize} \qquad P = 12x_1 + 9x_2$$

subject to

$$3x_1 + 2x_2 \leq 8$$
$$3x_1 + x_2 \leq 4$$
$$x_1 \geq 0, \; x_2 \geq 0 .$$

The dual to this problem is

$$\text{Minimize} \qquad C = 8y_1 + 4y_2$$

subject to

$$3y_1 + 3y_2 \geq 12$$
$$2y_1 + y_2 \geq 9$$
$$y_1, y_2 \geq 0 .$$

We can graph the new program and its dual: (See fig.)

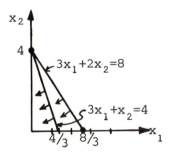
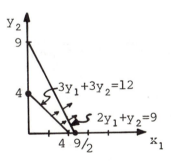

If a program has an optimal solution it has an optimal solution at the intersection of two constraints. From the graph of the primal, P is maximized when $x_1 = 0$ and $x_2 = 4$. The value of P is
$$P = 9(4) = 36.$$

Note that P has increased by $4\frac{1}{2}$ units as predicted by the dual value $y_1 = 4\frac{1}{2}$. But, examining the graph of the new dual there is no unique optimal solution, since any values for y_1 and y_2 on the line going from $y_1 = 4\frac{1}{2}$ to $y_2 = 9$ will satisfy the constraints.

887

Suppose that another unit is added to the first constraint in the primal which now becomes:

$$\text{Maximize} \quad P = 12x_1 + 9x_2$$

subject to

$$3x_1 + 2x_2 \leq 9$$
$$3x_1 + x_2 \leq 4$$
$$x_1 \geq 0, \ x_2 \geq 0 \ .$$

The optimum value of P is still 36. The value does not change, since the other unchanged restraint acts to prevent an improvement unless both restraints are changed. The solution is degenerate; of the two primal ordinary variables $(x_1$ and $x_2)$ and the two primal slack variables, only one of these four variables (x_2) is positive even though there are two constraints.

● **PROBLEM** 11-55

In a manufacturing process, the final product has a requirement that it must weigh exactly 150 pounds. The two raw materials used are A, with a cost of \$4 per unit and B, with a cost of \$8 per unit. At least 14 units of B and no more than 20 units of A must be used. Each unit of A weighs 5 pounds; each unit of B weighs 10 pounds.

How much of each type of raw material should be used for each unit of final product if we wish to minimize cost?

Solution: The objective function is

$$C = 4x_1 + 8x_2 \ . \tag{1}$$

The constraints are

$$5x_1 + 10x_2 = 150$$
$$x_1 \qquad \leq 20$$
$$x_2 \qquad \geq 14 \tag{2}$$
$$x_1 \qquad \geq 0$$

We will take the graphical approach to this linear programming problem. The constraints, (2) are graphed in the figure.

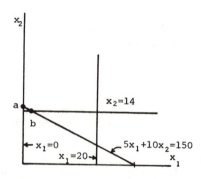

Since the pertinent region lies within $0 \leq x_1 \leq 20$, $x_2 \geq 14$, and on $5x_1 + 10x_2 = 150$, we can immediately find 2 solutions, points a and b where $a = (0,15)$ and $b = (2,14)$.

	Solution 1	Solution 2
Raw material A, (x_1)	0	2
Raw material B, (x_2)	15	14
Total cost, $4x_1 + 8x_2$	120	120

This is an example of a problem having multiple solutions. In such problems, two or more corner points have the same optimum value.

● **PROBLEM** 11-56

Assume that two products x_1 and x_2 are manufactured on two machines 1 and 2. Product x_1 requires three hours on machine 1 and one-half hour on machine 2. Product x_2 requires two hours on machine 1 and 1 hour on machine 2. There are six hours of available capacity on machine 1 and four hours on machine 2. Finally, each unit of x_1 produces a net increase in profit of $12.00 and each unit of x_2 an incremental profit of $4.00. 1)Maximize the profit. 2)Obtain a solution to this problem using the simplex method. 3)Apply sensitivity analysis to the final tableau.

Solution: In linear programming form the problem is:

$$\text{Maximize} \quad P = 12x_1 + 4x_2 \qquad (1)$$

subject to

$$3x_1 + 2x_2 \le 6$$
$$\tfrac{1}{2} x_1 + x_2 \le 4 \qquad (2)$$
$$x_1 \ge 0, \ x_2 \ge 0 .$$

Introduce the slack variables x_3, x_4:

$$3x_1 + 2x_2 + x_3 = 6$$
$$\tfrac{1}{2} x_1 + x_2 + x_4 = 4.$$

Let $v_1 = \begin{bmatrix} 3 \\ \tfrac{1}{2} \end{bmatrix}$, $v_2 = \begin{bmatrix} 2 \\ 1 \end{bmatrix}$, $v_3 = \begin{bmatrix} 1 \\ 0 \end{bmatrix}$, $v_4 = \begin{bmatrix} 0 \\ 1 \end{bmatrix}$. Let $x_1 = x_2 = 0$.

Then an initial feasible solution is $x_1 = x_2 = 0$, $x_3 = 6$, $x_4 = 4$. The initial simplex tableau is

	v_1	v_2	v_3	v_4	b	
v_3	③	2	1	0	6	
v_4	$\tfrac{1}{2}$	1	0	1	4	
d	-12	-4	0	0	0	D

Pivoting on 3, the next simplex tableau is

889

	v_1	v_2	v_3	v_4	b	
v_1	1	2/3	1/3	0	2	(3)
v_4	0	2/3	-1/6	1	3	
d	0	4	4	0	24	

Since all $d_j \geq 0$, the solution is $x_1 = 2$, $x_4 = 3$ and $P = 24$.

In order to carry out the sensitivity analysis we also need the dual of (1), (2). This is:

Minimize $U = 6y_1 + 4y_2$

subject to

$$3y_1 + \tfrac{1}{2} y_2 \geq 12$$

$$2y_1 + y_2 \geq 4$$

$$y_1 \geq 0, \ y_2 \geq 0 .$$

The optimum solution to the dual is $y_1 = 4$, $y_4 = 4$; $U = 24$. The dual variables are known as shadow prices. y_1 has a value of \$4 which means an hour of machine 1 has a value of \$4.00 . But y_1 is also the slack variable x_3 in the primal, i.e., unused hours on machine 1. Now if we could buy additional hours of machine 1 time for less than \$4.00, could we increase P indefinitely by adding more hours?

To answer this question, examine the v_3 column in (3). The meaning of the coefficients is that if we were to add one unit of x_3 it would replace one-third unit of x_1 and -1/6 unit of x_4 . But adding a unit of x_3 is equivalent to reducing hours of production by machine 1 on x_1. How far can production be reduced before a change in the solution mix occurs?

	Solution values	x_3	Solution values x_3
x_1	2	1/3	6
x_4	3	-1/6	-18

We see that six units of x_3 can be introduced before variable x_1 goes out of the solution. This means we can cut back a maximum of six hours of those available on machine 1 before production of x_1 stops.

On the other hand, adding negative x_3 means that we can make additional hours available on machine 1. Here we can add 18 additional hours of time on machine 1 before we run out of slack time on machine 2 (i.e., x_4 goes to zero and out of the solution).

Make the same kind of analysis relative to the amount of time available on machine 2. Here the slack variable is x_4 with a value of 3. Thus we can reduce x_4 only by 3 before there arises a shortage of machine 2 time. Also, since there already is slack, the additional time can be added indefinitely without altering the solution mix.

The above sensitivity analysis is summarized below:

	Machine-hrs. available	Shadow price	Shadow lower	Range for valid price upper
Machine 1	6	$4	0	24
Machine 2	4	0	1	no limit

Consider the following problem:

$$\text{Maximize} \quad P = 5x_1 + 8x_2 \tag{1}$$

subject to

$$2x_1 + x_2 \leq 14$$
$$x_1 + 3x_2 \leq 12 \tag{2}$$
$$x_2 \leq 3$$
$$x_1 \geq 0, \ x_2 \geq 0 .$$

Suppose that an additional constraint on x_1 and x_2 is imposed:

$$x_1 + x_2 \leq K \tag{3}$$

where K is some unspecified amount. How does the solution of (1), (2) and (3) change as K varies from zero to very large values?

Solution: We first graph the problem (1), (2), as in Fig. 1.

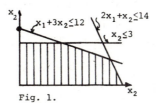

Fig. 1.

Assume the first two constraints represent time used on machine 1 and machine 2, respectively, to produce units of products x_1 and x_2 . The third constraint indicates that not more than three units of product x_2 can be sold. Now consider the additional constraint $x_1 + x_2 \leq K$. This states that the total amount of working capital used must be less than an unspecified amount, K. If $K = 0$ the only solution is that of no production and $x_1 = x_2 = 0$. As K increases to one, the first dollar of working capital is used to produce one unit of x_2 since x_2 is the more profitable product. Since each unit of K spent on x_2 produces $8 of profit as K increases, only x_2 is produced until the situation in Fig. 2 is reached.

Fig. 2.

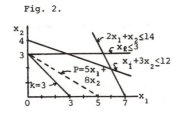

When $K > 3$, the market constraint $(x_2 \leq 3)$ becomes binding, and additional units of x_2 cannot be produced. Hence, production of x_1 now begins. Each dollar of working capital has an incremental value of $5, the profit associated with selling one unit of x_1. Production of x_1 continues until the situation in Fig. 3 is reached.

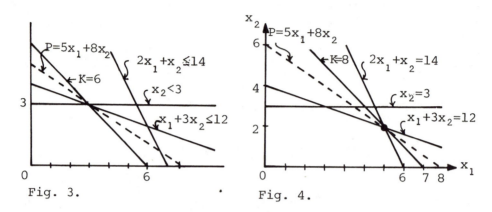

Fig. 3.   .   Fig. 4.

In Fig. 3, the constraint on time available on machine 2 also becomes binding, at the point at which six units of K are available. Now, three units of x_1 and 9 units of x_2 are produced for a total profit of $87. Note that P can still be increased by moving down the $x_1 + 3x_2 = 12$ constraint. But now each unit of working capital contributes only $3.50 profit. To see this, observe that one unit of x_2 can be substituted for three units of x_1 yielding an additional profit of $5(3) - $8(1) = $7. But this substitution requires three units for the new x_1 less one unit for reduced x_2. Hence the marginal value of K is $7/2 = $3.50. Finally, the situation portrayed in Fig. 4 is reached. Here the constraints on both machine times are binding. Increasing the working capital, K, beyond this point would have no effect on the solution. This sort of analysis is useful to a business man who is trying to decide how much working capital to invest in this production operation. He would not invest more than $8. However, he might invest even less if he had profitable alternative uses for his funds. For example, if he could use the working capital elsewhere to return him $4 per unit, then he would not invest more than six units. If he could get a return of $6 per unit, he would not use more than 3 units.

● **PROBLEM** 11-58

Assume there are three factories $(F_1, F_2$ and $F_3)$ supplying goods to three warehouses (W_1, W_2, W_3). The amounts available in each factory, the amounts needed in each factory and the costs of shipping from factory i to warehouse j are given in the table below:

Source / Destination	F_1	F_2	F_3	Units demanded
W_1	$.90	$1.00	$1.00	5
W_2	$1.00	$1.40	$ .80	20
W_3	$1.30	$1.00	$ .80	20
units available	20	15	10	45

Find the minimum cost of satisfying warehouse demands given that any factory may supply to any warehouse.

Solution: This is a transportation problem. It may be solved by iteration, i.e., we start with a solution and then use it to find more nearly optimal solutions.

An initial solution may be found by finding the box that has the lowest value in both its row and column. Place in that box the lower of demand or supply requirements. Next find the next lower value and repeat the placing of units shipped according to demand and supply requirements. We thus obtain

Source / Destination	F_1	F_2	F_3	units demanded
W_1	.90 / 5	$1.00 / 0	$1.00 / 0	5
W_2	1.00 / 10	1.40 / 0	.80 / 10	20
W_3	1.30 / 5	1.00 / 15	.80 / 0	20
units available	20	15	10	45

The total cost of this program is

$$(.90)(5) + (1.00)(10) + (1.30)(5) + (1.00)(15) + (.80)(10) = \$44.00.$$

Now check whether this solution is optimal. Pick a box with zero entry, say W_1F_2. This means that F_2 supplies no goods to W_1. The cost of F_2 directly supplying W_1 is $1.00 per unit. But F_2 is also indirectly supplying W_1 since by supplying W_3 it allows F_1 to supply W_1. What is this indirect cost?

The cost of shipping one unit from F_2 to W_1 by this indirect route is:

+ $1.00	charge for shipping from F_2 to W_3
- $1.30	every unit F_2 sends to W_3 saves the cost of supplying W_3 from F_1
+ $0.90	charge for shipping from F_1 to W_1
+ $0.60	

The indirect cost, i.e., the cost currently incurred is $0.60 whereas the direct cost is $1.00. Thus the current solution is cheaper.

The other zero boxes may be evaluated in a comparable manner. For example, the indirect shipment from F_2 to W_2 is the charge from F_2 to W_3 ($1.00) less the W_3F_1 charge ($1.30) plus the W_2F_1 charge ($1.00)= 70¢. Again, this is less than the cost of direct shipment ($1.40), so the current indirect route should be continued. In this way we obtain the following table:

Unused route	Cost of direct route	Cost of indirect route
W_1F_2	$1.00	$0.60
W_1F_3	$1.00	$0.70
W_2F_2	$1.40	$0.70
W_3F_3	$0.80	$1.10

By using W_3F_3 a saving of $.30 per unit can be obtained. But how many units can be shipped? The answer is the minimum number in any of the connections of the indirect route which must supply units for the transfer. This is five units, from box W_3F_1. Thus we ship 5 units by the direct route W_3F_3; since F_3 produces only 10 units, this imposes a reduction in the W_2F_3 box to 5. The new pattern is shown in the table below:

Source \ Destination	F_1	F_2	F_3	Units demanded
W_1	.90 / 5	1.00 / 0	1.00 / 0	5
W_2	1.00 / 15	1.40 / 0	.80 / 5	20
W_3	1.30 / 0	1.00 / 15	.80 / 5	20
units available	20	15	10	45

Once again we must compare the cost of using the direct route to the cost of using the indirect route.

Unused route	Cost of using direct route	Cost of using indirect route
W_3F_1	$1.30	$1.00
W_1F_2	$1.00	$0.90
W_2F_2	$1.40	$1.00
W_1F_3	$1.00	$0.70

In every case the cost of using the indirect route is less than the cost of the direct route, indicating that we are minimizing the shipment costs. The total cost of shipment from factories to warehouses is:

$$(5)(\$0.90) + (15)(\$1.00) + (15)(\$1.00) + 5(\$0.80) + 5(\$0.80) = \$42.50.$$

A mutual fund is deciding how to divide its investments among bonds, preferred stock, and speculative stock. It does not want to exceed a combined risk rate of 3 when the bonds have been assigned a risk rate of 1, the preferred stock, 3, and the speculative stock, 5. However, the fund does want a total annual yield of at least 10 percent. If the interest rate of the bonds is 8 percent, of the preferred stock is 12 percent, and of the speculative stock is 20 percent, how should the assets be distributed for the greatest annual yield?

Solution: The total amount to be invested does not matter. We let $x, y,$ and z be the fractions of the whole that will be invested in bonds, preferred stock, and speculative stock, respectively. Clearly these fractions must total the whole portfolio: $x + y + z = 1$. The other two constraints indicate the restrictions on the risk rate and total yield:

$$x + 3y + 5z \leq 3 \quad \text{(risk rate constraint)}$$

$$8x + 12y + 20z \geq 10 \quad \text{(annual yield constraint)}.$$

The annual yield constraint can be omitted, since whether it is there or not, we will make the left side of the inequality as large as possible. If it must be less than 10, there is no feasible solution to the problem as stated, and the fund must reconsider its goals. If it can equal or exceed 10 subject to the other constraints, that constraint is automatically satisfied. Thus we can summarize the problem:

$$\text{Maximize} \quad P = 8x + 12y + 20z$$

subject to $x + y + z = 1$ and $x + 3y + 5z \leq 3$. The non-negativity constraints, as usual, are assumed. The first constraint needs no slack variable, but it does need an artificial variable. Recall that a slack variable s_i is used to convert the ith inequality to an equality. An artificial variable is introduced to make the number of unknowns in each row the same. This makes it possible to start with the basic feasible solution $x = y = z = 0$ while not violating the constraints.

The second is given a slack variable as usual:

$$x + y + z + a = 1$$

$$x + 3y + 5z + s = 3.$$

Step 1: The first project is to rid the problem of a, since it has no real-world interpretation. We do this by finding the minimum of $A = a$ (which minimum had better be zero), or, equivalently, we find the maximum of $-A = -a$. We might write:

	x	y	z	s	a	Solution
a	①	1	1	0	1	1
s	1	3	5	1	0	3
-A	0	0	0	0	-1	0
-A	1	1	1	0	0	1

The row above the dashed line is a summary of the equation $-A = -a$. There is a non-zero number in the last row of the a-column. This is not allowed, because a is a basic variable, and each basic variable must have 1 in its column and the rest of the numbers 0. To obtain a 0 in the last row in the a-column, we add the a-row to the last row and write

the sum below; then we cross out the original row. Since all three positive numbers in the last row are equal, any one of the first three columns
may become the pivot column. We choose the first column to be the pivot
column. Now, divide the positive elements of the inner rectangle in the
pivot column into their corresponding elements in the "solutions" column.
Since $1/1 < 3/1$, then the first row is the pivot row. Thus, the 1 in
the upper left is the pivot. Now in the next table, replace the letter
to the left of the pivot row by the letter above the pivot column. Thus
x replaces a; a must leave. The pivot row is divided by the pivot.
Add the second row of the inner rectangle to -1x the first row of the
inner rectangle. Replace the second row with this result.

	x	y	z	s	a	Solution
x	1	1	1	0	1	1
s	0	2	(4)	1	-1	2
-A	0	0	0	0	-1	0

Thus, since our present objective function, A is 0, we can proceed
to Step 2 where the objective function in the original problem was
$$P = 8x + 12y + 20z .$$
Thus, our table is now:

	x	y	z	s	Solution
x	1	1	1	0	1
s	0	2	4	1	2
P	8	12	20	0	0

Since the largest entry in the last row is 20, the z-column is the pivot
column. Since $2/4 < 1/1$, the s-row is the pivot row and thus the 4 is
the pivot. Now we must replace the letter to the left of the pivot row
(s) by the letter above the pivot column (z). Divide the pivot row by the
pivot, 4.

	x	y	z	s	Solution
x	1	1	1	0	1
z	0	1/2	1	1/4	1/2
P	8	12	20	0	0

Now, it is left for us to add the appropriate multiples of the pivot
row to the other rows of numbers in such a way as to obtain zeros in the
rest of the pivot column. First of all, add the P row to $(-1) \cdot$x-row
to obtain:

	x	y	z	s	Solution
x	1	1	1	0	1
z	0	1/2	1	1/4	1/2
P	0	4	12	0	-8

896

Now add the P-row to (-12)·z-row. Add the x-row to (-1)·z-row.

	x	y	z	s	Solution
x	1	1/2	0	-1/4	1/2
z	0	1/2	1	1/4	1/2
P	0	-2	0	-3	-14

Since there are no positive numbers in the bottom row, we have obtained the optimum solution. It is P_{max} = 14 for $x = \frac{1}{2}$, $y = 0$, and $z = \frac{1}{2}$.

Thus the maximum possible yield under these conditions is 14 percent, well above the 10 percent desired minimum.

● **PROBLEM** 11-60

Suppose that we have 2 factories and 3 warehouses. Factory I makes 40 widgets. Factory II makes 50 widgets. Warehouse A stores 15 widgets. Warehouse B stores 45 widgets. Warehouse C stores 30 widgets. It costs $80 to ship one widget from Factory I to warehouse A, $75 to ship one widget from Factory I to warehouse B, $60 to ship one widget from Factory I to warehouse C, $65 per widget to ship from Factory II to warehouse A, $70 per widget to ship from Factory II to warehouse B, and $75 per widget to ship from Factory II to warehouse C.
1) Set up the linear programming problem to find the shipping pattern which minimizes the total cost.
2) Find a feasible (but not necessarily optimal) solution to the problem of finding a shipping pattern using the Northwest Corner Algorithm.
3) Use the Minimum Cell Method to find a feasible solution to the shipping problem.

Solution: 1) Let x_{11} = number of widgets shipped from Factory I to warehouse A, x_{12} = number of widgets shipped from Factory I to warehouse B, x_{13} = number of widgets shipped from Factory I to warehouse C, x_{21} = number of widgets shipped from Factory II to warehouse A, x_{22} = number of widgets shipped from Factory II to warehouse B, x_{23} = number of widgets shipped from Factory II to warehouse C. Thus, the linear programming problem can be formulated as follows:

Minimize $C = 80x_{11} + 75x_{12} + 60x_{13} + 65x_{21} + 70x_{22} + 75x_{23}$

subject to

$$x_{11} + x_{12} + x_{13} \leq 40 ,$$
$$x_{21} + x_{22} + x_{23} \leq 50 , \ x_{11} + x_{21} = 15,$$
$$x_{12} + x_{22} = 45 \ and \ x_{13} + x_{23} = 30 .$$

This linear programming problem may be solved using the simplex method.

2) The facts of the problem can be diagrammed in the following table, where the amounts the factories produce are written on the right, the amounts the warehouses can store are written on the bottom, and the numbers in the boxes are the costs of shipping from the factory on the left to the warehouse above.

Warehouse

	A	B	C	
I	80	75	60	40
II	65	70	75	50
	15	45	30	

Factory

The Northwest corner algorithm first allocates as many widgets as possible to the upper left box (the northwest box). Next, proceed to the nearest box into which something can still be placed, and allocate as much as possible to that one. Then the process continues, each time moving either one box to the right, or one down, or one diagonally down, depending on how the shipments can be made. Since the 15 at the bottom of the first column is less than the 40 at the right of the first row, Factory I can ship only 15 widgets to warehouse A, so we write a 15 in the upper left box. Then nothing else can go to warehouse A; that is, nothing else will be written in the boxes of the first column.

Warehouse

	A	B	C	
I	(15) 80	75	60	40
II	65	70	75	50
	15	45	30	

Factory

Thus, we move right from 15, making $x_{12} = 40 - 15 = 25$. Now the capacity of factory I has been exhausted, so there can be no more numbers written in the first row. Moving down, we next set $x_{22} = 45 - 25 = 20$ to fill warehouse B. Now we must move right and set $x_{23} = 50 - 20 = 30$. The results are as follows:

Warehouse

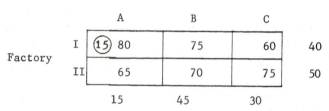

	A	B	C	
I	(15) 80	(25) 75	60	40
II	65	(20) 70	(30) 75	50
	15	45	30	

Factory

This table tells us we can ship 15 widgets from Factory I to warehouse A, 25 widgets from Factory I to warehouse B, 20 widgets from Factory II to warehouse B, and 30 widgets from Factory II to warehouse C. This is not the optimum solution (that is, it is not the cheapest), but it is feasible. The total cost is $C = 15 \cdot 80 + 25 \cdot 75 + 20 \cdot 70 + 30 \cdot 75 = 6725$.

3) The Northwest Corner Algorthm ignores the costs. The Minimum Cell Algorithm is another method of finding a feasible solution; unlike the Northwest Corner Algorithm, it does take the cost into account. The Minimum Cell Algorithm finds the cheapest possible rate, and it sends as much as possible at that rate. Again, the problem can be summarized in the following table:

Warehouse

		A	B	C	
Factory	I	80	75	60	40
	II	65	70	75	50
		15	45	30	

Since 60 is the cheapest possible rate, we decide to send as much as possible at that rate. Since warehouse C needs 30 widgets and Factory I has 40 widgets, we can send at most 30 widgets from Factory I to warehouse C; we do so and write it in the box. To show that we have accounted for all 30 items of warehouse C, we cross off the 30. To show that 30 of the 40 items in Factory I have been used, we cross off the 40 and write a 10 beside it to show there are 10 items left in Factory I.

Warehouse

		A	B	C		
Factory	I	80	75	(30) 60	~~40~~	10
	II	65	70	75	50	
		15	45	~~30~~		

The next cheapest shipping rate is 65. Since warehouse A needs only 15 widgets, we write a 15 in the box with the 65, showing that we will ship 15 widgets from Factory II to warehouse A. We cross out the 15 below the first column. And we cross out the 50 at the right of the second row and write a $50 - 15 = 35$ next to it.

Warehouse

		A	B	C		
Factory	I	80	75	(30) 60	~~40~~	10
	II	(15) 65	70	75	~~50~~	35
		~~15~~	45	~~30~~		

The next cheapest shipping rate is 70. We have only 35 of the widgets produced by Factory II remaining to send to warehouse B, so we write a 35 in the box with the 70. Then we cross out the 35 at the right of the second row, and replace the 45 below the second column with a $45 - 35 = 10$.

Warehouse

		A	B	C		
Factory	I	80	75	(30) 60	~~40~~	10
	II	(15) 65	(35) 70	75	~~50~~	~~35~~
		~~15~~	~~45~~ 10	~~30~~		

Since 75 is the lowest remaining rate, we ship the remaining 10 widgets from Factory I to warehouse B and indicate the feasible solution we obtain as follows:

	A	B	C		
Factory I	80	⑩ 75	㉚ 60	~~40~~ ~~10~~	
II	⑮ 65	㉟ 70	75	~~50~~ ~~35~~	

~~15~~   ~~45~~   ~~30~~
                 ~~10~~

This shipping pattern yields a cost of $15 \cdot 65 + 35 \cdot 70 + 10 \cdot 75 + 30 \cdot 60 =$ 5975.

● **PROBLEM** 11-61

A small-trailer manufacturer wishes to determine how many camper units and how many house trailers he should produce in order to make optimal use of his available resources. Suppose he has available 11 units of aluminum, 40 units of wood, and 52 person-weeks of work. (The preceding data are expressed in convenient units. We assume that all other needed resources are available and have no effect on his decision.) The table below gives the amount of each resource needed to manufactur each camper and each trailer.

	Aluminum	Wood	Person-weeks
Per camper	2	1	7
Per trailer	1	8	8

Suppose further that based on his previous year's sales record the manufacturer has decided to make no more than 5 campers. If the manufacturer realized a profit of $300 on a camper and $400 on a trailer, what should be his production in order to maximize his profit?

Solution: Letting x_1 represent the number of camper units, and x_2 the number of house trailers, we consider first the constraints. From the table we see that the manufacturer uses 2 units of aluminum per camper and 1 unit of aluminum per trailer. Thus he needs a total of $2x_1 + x_2$ units of aluminum. This fact, along with the fact that he has available only 11 units of aluminum, gives us the inequality $2x_1 + x_2 \leq 11$. Similarly, he needs a total of $x_1 + 8x_2$ units of wood. And since he has available only 40 units of wood, we get $x_1 + 8x_2 \leq 40$. The total number of person-weeks needed to build x_1 campers and x_2 trailers is $7x_1 + 8x_2$. And since only 52 weeks are available, $7x_1 + 8x_2 \leq 52$. Since he wants to produce no more than 5 campers, we have $x_1 \leq 5$. And finally, there exists a constraint that is unrelated to the numbers actually appearing in the statement of the problem. Certainly it is physically impossible for the manufacturer to produce a negative number of campers or trailers. Thus, we need that $x_1, x_2 \geq 0$. We want, of course, to maximize the total profit attained from x_1 campers and x_2 trailers, namely $300x_1 + 400x_2$. Thus, we have reduced the given problem to the following:

Maximize $300x_1 + 400x_2$

subject to the conditions that

$$2x_1 + x_2 \leq 11$$
$$x_1 + 8x_2 \leq 40 \tag{1}$$
$$7x_1 + 8x_2 \leq 52$$
$$x_1 \leq 5$$
$$x_1, x_2 \geq 0 \ .$$

The type of problem as above is called a maximum problem of linear programming. Now, we wish to determine the extreme points of the feasible solution set. One way to make the process of finding the extreme points efficient is to introduce slack variables. The purpose is to convert the inequalities of (1) to equalities. Specifically, we let $x_3 = 11 - (2x_1 + x_2)$, $x_4 = 40 - (x_1 + 8x_2)$, $x_5 = 52 - (7x_1 + 8x_2)$, and $x_6 = 5 - x_1$, and consider the system of equations:

$$2x_1 + x_2 + x_3 \qquad\qquad\qquad = 11$$
$$x_1 + 8x_2 \qquad + x_4 \qquad\qquad = 40 \tag{2}$$
$$7x_1 + 8x_2 \qquad\qquad +x_5 \qquad = 52$$
$$x_1 \qquad\qquad\qquad\qquad + x_6 = 5$$

and we still require that $x_1, x_2 \geq 0$. Moreover, the original inequality constraints will be satisfied if we require also that $x_3, x_4, x_5, x_6 \geq 0$. Observe, for example, that $x_3 = 11 - (2x_1 + x_2) \geq 0$ if and only if $2x_1 + x_3 \leq 11$. We first form the augmented matrix for (2). The function written below the matrix reminds us what we must maximize:

$$\begin{bmatrix} 2 & 1 & 1 & 0 & 0 & 0 & 11 \\ 1 & 8 & 0 & 1 & 0 & 0 & 40 \\ 7 & 8 & 0 & 0 & 1 & 0 & 52 \\ 1 & 0 & 0 & 0 & 0 & 1 & 5 \end{bmatrix}$$

$$300x_1 + 400x_2 + 0x_3 + 0x_4 + 0x_5 + 0x_6$$

Thus, our starting tableau is:

	x_1	x_2	x_3	x_4	x_5	x_6		
x_3	2	1	1	0	0	0	11	11/1 = 11
x_4	1	⑧	0	1	0	0	40	40/8 = 5
x_5	7	8	0	0	1	0	52	52/8 = 6.5
x_6	1	0	0	0	0	1	5	
	-300	-400	0	0	0	0	0	

To determine the pivot element: The elements of the last row of the tableau are called indicators. We begin by finding the negative indicator having the largest absolute value. In the tableau above the indicator is clearly -400, which appears in the second column. We therefore call the second column the pivot column. We now consider the ratio of each element in the last column to the corresponding element in the pivot column, if the pivot column is positive. The row associated with the smallest of these ratios is called the pivot row. In our example the pivot column contains three positive elements: a 1 in the first row, an 8 in the second row, and an 8 in the third row. Thus the ratios we must compare are $1/1 = 11$, $40/8 = 5$ and $52/8 = 6.5$. Since 5 is the smallest of the ratios, the second row is the pivot row. The pivot element is the element common to the pivot column and the pivot row, namely the 8 that is circled in the tableau above. Now, we must use elementary row operations to transform the tableau into one having a 1 in the place of the pivot element and 0's elsewhere in the pivot column. To accomplish this, we first multiply each element in the pivot row by the reciprocal of the pivot element to get:

	x_1	x_2	x_3	x_4	x_5	x_6	
x_3	2	1	1	0	0	0	11
x_4	1/8	1	0	1/8	0	0	5
x_5	7	8	0	0	1	0	52
x_6	1	0	0	0	0	1	5
	-300	-400	0	0	0	0	0

We then multiply the pivot row by -1 and add it to the first row, by -8 and add it to the third row, and by 400 and add it to the fifth row. The result is:

	x_1	x_2	x_3	x_4	x_5	x_6		
x_3	15/8	0	1	-1/8	0	0	6	$6 / \frac{15}{8} = 16/5$
x_2	1/8	1	0	1/8	0	0	5	$5 / \frac{1}{8} = 40$
x_5	6	0	0	-1	1	0	12	$12/6 = 2$
x_6	1	0	0	0	0	1	5	$5/1 = 5$
	-250	0	0	50	0	0	2000	

In the tableau above we replaced the x_4 in the notation column by x_2. This replacement indicates that the x_2 variable was brought into the solution and the x_4 variable was eliminated. Now, we must examine the last row of the tableau above. Since -250 is the only negative indicator, the first column is the pivot column. Comparing

$$6 / \left(\frac{15}{8}\right) = \frac{16}{5}, \ 5 / \left(\frac{1}{8}\right) = 40, \ \frac{12}{6} = 2, \ \text{and} \ 5/1 = 5,$$

we see that the third row is the pivot row. Thus the pivot element is 6, which is circled in the tableau above. We first multiply the pivot row by 1/6 so a 1 appears in the pivot position. We then multiply this new pivot row by -15/8, -1/8, -1, and 250, adding the results to the first, second, fourth, and fifth rows, respectively, to get the following tableau:

	x_1	x_2	x_3	x_4	x_5	x_6	
x_3	0	0	1	3/16	-5/16	0	9/4
x_2	0	1	0	7/48	-1/48	0	19/4
x_1	1	0	0	-1/6	1/6	0	2
x_6	0	0	0	1/6	-1/6	1	3
	0	0	0	25/3	125/3	0	2500

Note in the tableau above: Since our pivot element was in the first column and third row, we placed an x_1 in the third row of the notation column.

Since the tableau above include no negative indicators, we are done. Thus, $x_1 = 2$, $x_2 = 19/4$ is the point at which the function assumes its maximum value, namely, 2500.

LINEAR PROGRAMMING - ADVANCED METHODS

● **PROBLEM** 11-62

Consider the problem

maximize $\quad x_0 = 5x_1 + 12x_2 + 4x_3$

subject to $\quad x_1 + 2x_2 + x_3 \leq 5$

$\qquad\qquad 2x_1 - x_2 + 3x_3 = 2$

$\qquad\qquad x_1, x_2, x_3 \geq 0$

Solve the primal and dual problems by the Simplex method. Compare the results.

Solution: The dual is given by

minimize $\quad y_0 = 5y_1 + 2y_2$

subject to

$$y_1 + 2y_2 \geq 5$$

$$2y_1 - y_2 \geq 12$$

$$y_1 + 3y_2 \geq 4$$

$$y_1 \geq 0$$

y_2 unrestricted in sign

Notice that the last constraint, $y_1 \geq 0$, corresponds to S_1 in the primal. This same result could have been obtained from the original problem where y_1 corresponds to ($\leq$) constraint and y_2 corresponds to (=) constraint.

Inspection yields the solutions ($x_1 = 1$, $x_2 = 1$, $x_3 = $

1/3, and $y_1 = 7$, $y_2 = 2$) are feasible for the primal and dual problems. These give $x_0 = 18\frac{1}{3}$ and $y_0 = 39$ which shows that $x_0 < y_0$. From the property that $x_0 \leq y_0$, this result reveals the immediate information that the optimal value of the objective function lies between $18\frac{1}{3}$ and 39.

The standard form of this problem is

$$\text{maximize} \quad x_0 = 5x_1 + 12x_2 + 4x_3 + 0S_1$$

$$\text{subject to} \quad x_1 + 2x_2 + x_3 + S_1 = 5$$

$$2x_1 - x_2 + 3x_3 + 0S_1 = 2$$

$$x_1, \ x_2, \ x_3, \ S_1 \geq 0$$

Although the canonical and standard forms always yield equivalent duals, the standard form is more useful since it is the basis of all linear programming calculations, namely, the simplex method. The starting tableau for the primal problem is obtained from the standard form. Expressing the x_0-equation in terms of the nonbasic variables gives (R_1 is an artificial variable; artificial variables are nonnegative variables added to the left-hand side of each of the equations corresponding to constraints of the types ($>$) and ($=$) and their addition causes violation of the corresponding constraints. This difficulty is overcome by ensuring that the artificial variables will be zero ($=0$) in the final solution, (provided the solution of the problem exists. If the problem does not have a solution, at least one of the artificial variables will appear in the final solution at a positive level.)

Basic	x_0	x_1	x_2	x_3	S_1	R_1	Solution
x_0	①	$-5 - 2M$	$-12 + M$	$-4 - 3M$	0	0	$-2M$
S_1	0	1	2	1	①	0	5
R_1	0	2	-1	3	0	①	2

First Iteration: Introduce x_3 and drop R_1.

Basic	x_0	x_1	x_2	x_3	S_1	R_1	Solution
x_0	①	$-7/3$	$-40/3$	0	0	$4/3 + M$	$8/3$
S_1	0	$1/3$	$7/3$	0	①	$-1/3$	$13/3$
x_3	0	$2/3$	$-1/3$	①	0	$1/3$	$2/3$

Second Iteration: Introduce x_2 and drop S_1.

Basic	x_0	x_1	x_2	x_3	S_1	R_1	Solution
x_0	①	$-3/7$	0	0	$40/7$	$-4/7 + M$	$192/7$
x_2	0	$1/7$	①	0	$3/7$	$-1/7$	$13/7$
x_3	0	$5/7$	0	①	$1/7$	$2/7$	$9/7$

Third Iteration: Introduce x_1 and drop x_3.

Basic	x_0	x_1	x_2	x_3	S_1	R_1	Solution
x_0	①	0	0	$3/5$	$29/5$	$-2/5 + M$	$28\frac{1}{5}$
x_2	0	0	①	$-1/5$	$2/5$	$-1/5$	$8/5$
x_1	0	①	0	$7/5$	$1/5$	$2/5$	$9/5$

which is the optimal solution with $x_1 = 9/5$, $x_2 = 8/5$, $x_3 = 0$, and $x_0 = 28\frac{1}{5}$.

The dual problem will now be solved. Since y_2 is unrestricted in sign, it is replaced by $y_2^+ - y_2^-$ in the simplex tableau where $y_2^+ > 0$ and $y_2^- > 0$. Thus, adding the artificial variables $\bar{R}_1$, R_2, and $\bar{R}_3$ and expressing the $y_0{}^-$ equation in terms of the nonbasic variables, the starting tableau becomes

Basic	y_0	y_1	y_2^+	y_2^-	S_1	S_2	S_3	R_1	R_2	R_3	Solution
y_0	①	$-5+4M$	$-2+4M$	$2-4M$	$-M$	$-M$	$-M$	0	0	0	$21M$
R_1	0	1	2	-2	-1	0	0	①	0	0	5
R_2	0	2	-1	1	0	-1	0	0	①	0	12
R_3	0	1	3	-3	0	0	-1	0	0	①	4

Noting that this is a minimization problem, one obtains the optimal solution in five iterations. Its simplex tableau is given by

Basic	y_0	y_1	y_2^+	y_2^-	S_1	S_2	S_3	R_1	R_2	R_3	Solution
y_0	①	0	0	0	$-9/5$	$-8/5$	0	$9/5 - M$	$8/5 - M$	$-M$	$28\frac{1}{5}$
S_3	0	0	0	0	$-7/5$	$1/5$	①	$7/5$	$-1/5$	-1	$3/5$
y_2^-	0	0	-1	①	$2/5$	$-1/5$	0	$-2/5$	$1/5$	0	$2/5$
y_1	0	①	0	0	$-1/5$	$-2/5$	0	$1/5$	$2/5$	0	$29/5$

This yields $y_1 = 29/5$, $y_2^+ = 0$, $y_2^- = 2/5$ and $y_0 = 28\frac{1}{5}$.
Thus, $y_2 = y_2^+ - y_2^- = 0 - 2/5 = -2/5$. (Notice that y_2 is unrestricted in sign.)

A comparison of the primal and dual solutions shows that

$$\max x_0 = 28\frac{1}{5} = \min y_0$$

Notice that the optimal values of the objective function ($= 28\frac{1}{5}$) lies between the previously estimated values, $18\frac{1}{3}$ and 39.

Investigation of the optimal tableaus of the primal and the dual reveals the following results. Consider the variables of the starting solution in the primal. These are S_1 and R_1. The dual variables y_1 and y_2 correspond to the primal constraint equations containing S_1 and R_1, respectively. Now consider the coefficient of S_1 and R_1 in the x_0-equation of the optimal primal tableau. These are given by

Starting solution variables (primal)	S_1	R_1
x_0-equation coefficients	29/5	$(-2/5) + M$
Corresponding dual variables	y_1	y_2

Ignoring the constant M for the moment, one sees that the resulting coefficients 29/5 and -2/5 directly give the optimal solution of the dual problem. This means that optimal y_1 equals 29/5 and optimal y_2 equals -2/5, which is the same result obtained by solving the dual problem independently.

Similar investigation of the coefficients of the starting variables R_1, R_2, and R_3 in the y_0-equation of the optimal dual tableau gives

Starting solution variables (dual)	R_1	R_2	R_3
y_0-equation coefficients	$9/5 - M$	$8/5 - M$	$0 - M$
Corresponding dual variables	x_1	x_2	x_3

Again ignoring the constant M, one sees that these coefficients give directly the optimal primal solution $x_1 = 9/5$, $x_2 = 8/5$ and $x_3 = 0$. This is the same result obtained from the direct solution of the primal.

● **PROBLEM** 11-63

Find a solution to the following problem by solving its dual:

$$\text{Minimize} \quad 9x_1 + 12x_2 + 15x_3 \quad (1)$$

subject to

$$2x_1 + 2x_2 + x_3 \geq 10$$
$$2x_1 + 3x_2 + x_3 \geq 12 \quad (2)$$
$$x_1 + x_2 + 5x_3 \geq 14$$

$$x_1 \geq 0, \ x_2 \geq 0, \ x_3 \geq 0 . \quad (3)$$

Solution: The dual to (1), (2) and (3) is

$$\text{Maximize} \quad 10y_1 + 12y_2 + 14y_3 \quad (4)$$

subject to

$$2y_1 + 2y_2 + y_3 \leq 9$$
$$2y_1 + 3y_2 + y_3 \leq 12 \quad (5)$$
$$y_1 + y_2 + 5y_3 \leq 15$$

$$y_1 \geq 0, \ y_2 \geq 0, \ y_3 \geq 0 . \quad (6)$$

We solve this problem using the simplex method. Converting the inequalities to equalities by adding slack variables

$$2y_1 + 2y_2 + y_3 + y_4 = 9$$
$$2y_1 + 3y_2 + y_3 + y_5 = 12$$
$$y_1 + y_2 + 5y_3 + y_6 = 15 .$$

This may be rewritten as:

$$y_1 \begin{pmatrix} 2 \\ 2 \\ 1 \end{pmatrix} + y_2 \begin{pmatrix} 2 \\ 3 \\ 1 \end{pmatrix} + y_3 \begin{pmatrix} 1 \\ 1 \\ 5 \end{pmatrix} + y_4 \begin{pmatrix} 1 \\ 0 \\ 0 \end{pmatrix} + y_5 \begin{pmatrix} 0 \\ 1 \\ 0 \end{pmatrix}$$

$$+ y_6 \begin{pmatrix} 0 \\ 0 \\ 1 \end{pmatrix} = \begin{pmatrix} 9 \\ 12 \\ 15 \end{pmatrix} . \tag{7}$$

Since we have three equations in six unknowns, we can obtain a basic feasible solution by setting any three of the y_i's to zero. Let

$$v_1 = \begin{pmatrix} 2 \\ 2 \\ 1 \end{pmatrix}, \quad v_2 = \begin{pmatrix} 2 \\ 3 \\ 1 \end{pmatrix}, \quad v_3 = \begin{pmatrix} 1 \\ 1 \\ 5 \end{pmatrix}, \quad v_4 = \begin{pmatrix} 1 \\ 0 \\ 0 \end{pmatrix}, \quad v_5 = \begin{pmatrix} 0 \\ 1 \\ 0 \end{pmatrix}, \quad v_6 = \begin{pmatrix} 0 \\ 0 \\ 1 \end{pmatrix}$$

and

$b = \begin{pmatrix} 9 \\ 12 \\ 15 \end{pmatrix}$. Let $d_1 = 10$, $d_2 = 12$, $d_3 = 14$. Setting $y_1 = y_2 = y_3 = 0$,

a basic feasible solution is $y_4 = 9$, $y_5 = 12$, $y_6 = 15$. We start the simplex algorithm with v_4, v_5 and v_6 in the basis.

	v_1	v_2	v_3	b	v_4	v_5	v_6
v_4	2	2	1	9	1	0	0
v_5	2	3	1	12	0	1	0
v_6	1	1	5	15	0	0	1
d	-10	-12	-14	0	0	0	0

$$\underset{D}{\uparrow}$$

To increase the value of D, we must choose another vector for the basis.
We choose the column which has the most negative value in the row labelled 'd'. Here v_3 is the chosen column. Next we must decide which vector in the basis to discard. The rule here is to choose that row for which
the ratio of b to v_3 is the smallest. Here the ratios are 9/1,
12/1 and 15/5 . Hence we replace v_6 by v_3, by pivoting on 5. The
process of conversion is carried out by first converting the 5 to a 1
and then using elementary row operations to reduce every other element
under v_3 to zero. Thus we obtain the new tableau

	v_1	v_2	v_3	b	v_4	v_5	v_6
v_4	9/5	9/5	0	6	1	0	-1/5
v_5	9/5	14/5	0	9	0	1	-1/5
v_3	1/5	1/5	1	3	0	0	1/5
d	-36/5	-46/5	0	42	0	0	14/5

The value 42 was obtained by using $b = c_1 y_1 + c_2 y_2 + c_3 y_3 + c_4 y_4 +$

$c_5 y_5 + c_6 y_6$ where the y_i are the coefficients of the vectors in the
basis. Here the basis vectors are v_4, v_5, v_6. The coefficients are, from
(4), $c_1 = 10$, $c_2 = 12$, $c_3 = 14$, $c_4 = c_5 = c_6 = 0$ and the y_i are obtained
from the column labelled b. Hence

$$D = c_4(6) + c_5(9) + c_3(3) = 0(6) + 0(9) + 14(3) = 42.$$

Since the row d still contains negative entries we repeat the above procedure by pivoting on 14/5. We obtain

	v_1	v_2	v_3	b	v_4	v_5	v_6
v_4	9/14	0	0	3/14	1	-9/14	-1/14
v_2	9/14	1	0	45/14	0	5/14	-1/14
v_3	1/14	0	1	33/14	0	-1/14	3/14
d	-9/7	0	0	501/7	0	23/7	15/7

Next, we pivot on the first element in the first column, i.e., 9/14. The
final simplex tableau is

	v_1	v_2	v_3	b	v_4	v_5	v_6
v_1	1	0	0	1/3	14/9	-1	-1/9
v_2	0	1	0	3	-1	1	0
v_3	0	0	1	7/3	-1/9	0	2/9
d	0	0	0	72	2	2	2

Thus the solution to the maximization problem (the dual of the given problem) is

$$y_1 = 1/3, \ y_2 = 3, \ y_3 = 7/3$$

with value 72. We know from duality theory that if a program has an
optimal feasible point then so does its dual and both programs have the
same value. The dual of (4), (5), (6) is (1), (2), (3). Thus the solution to the minimization problem has value 72. But what are the values
of x_1, x_2, x_3 at this optimum point? From the Complementary Slackness
Theorem of duality theory, these values are the values of the slack variables in the final simplex tableau of the dual problem. Thus $x_1 = 2$,
$x_2 = 2$, $x_3 = 2$ is the required solution.

● **PROBLEM** 11-64

Minimize $z = 10x_1 + 5x_2 + 4x_3$

subject to

$3x_1 + 2x_2 - 3x_3 \geq 3$

$4x_1 \qquad + 2x_3 \geq 10$

$x_1, x_2, x_3 \geq 0$ 　　　　　　　　　　　(1)

Solve the primal problem by applying the Dual Simplex algorithm. Also, solve the dual problem by the Simplex method.

Solution: The primal problem: Adding slack variables x_4 and x_5, one gets

$$3x_1 + 2x_2 - 3x_3 - x_4 \quad\quad = 3$$
$$4x_1 \quad\quad + 2x_3 \quad\quad - x_5 = 10$$
$$10x_1 + 5x_2 + 4x_3 \quad\quad\quad\quad = z$$

To apply the Simplex process to this problem, add two artificial variables and proceed on. On the other hand, by multiplying the two constraints by (-1), the following is obtained.

$$-3x_1 - 2x_2 + 3x_3 + x_4 \quad\quad = -3$$
$$-4x_1 \quad\quad -2x_3 \quad\quad + x_5 = -10$$
$$10x_1 + 5x_2 + 4x_3 \quad\quad\quad\quad = z \quad\quad (2)$$

The system of constraints is in standard form with basic variables x_4 and x_5, the objective function is expressed in terms of the nonbasic variables x_1, x_2, and x_3, and the associated coefficients 10, 5, and 4 are nonnegative, but the associated basic solution $x_1 = x_2 = x_3 = 0$, $x_4 = -3$, $x_5 = -10$ is not feasible.

The basic step of the Dual Simplex algorithm which is intimately related to the dual problem of the linear programming problem, is the pivot operation. Dual Simplex algorithm differs from the standard Simplex process by the rules used to determine the pivot term at each step.

Consider the tableau presentation in Table 1 of the problem as stated in (2).

Table 1

	x_1	x_2	x_3	x_4	x_5	
x_4	-3	-2	3	1	0	-3
x_5	-4	0	-2	0	1	-10
	10	5	4	0	0	0

To apply the Dual Simplex algorithm, determine first the row to pivot in. According to the algorithm, the pivot term can be in any row with a negative constant term. In this tableau, $b_1 = -3$ and $b_2 = -10$; therefore the pivot term can come from either row. An arbitrary rule to use in such a case is to pivot in that row with the smallest b_i term, and so, here, to pivot in the second row, extracting x_5 from the basis.

Next determine what column to pivot in. The algorithm dictates that the pivot term be at a negative a_{ij} entry, and so, here, the pivot term will be either at $a_{21} = -4$ or $a_{23} = -2$. To determine at which entry one pivots, the ratios c_j/a_{rj} must be considered for those $a_{rj} < 0$ (where r is the pivoting row), and the pivot term be in that column, say column s, for which

$$\frac{c_s}{a_{rs}} = \text{Max} \left\{ \frac{c_j}{a_{rj}} \,\middle|\, a_{rj} < 0 \right\}$$

In this case compare $c_1/a_{21} = \frac{10}{-4} = -\frac{5}{2}$ with $c_3/a_{23} = \frac{4}{-2} = -2$. The maximum occurs in the third column, and therefore pivot at $a_{23} = -2$. (Note that here one is comparing two nonpositive ratios and seeking the maximum, and therefore is actually seeking that ratio of minimum absolute value. By nature of the algorithm, this will always be the case.)

Pivoting here, the tableaux of Table 2 is obtained.

Table 2

	x_1	x_2	x_3	x_4	x_5	
x_4	-3	-2	3	1	0	-3
x_5	-4	0	$\boxed{-2}$	0	1	-10
	10	5	4	0	0	0
x_4	-9	-2	0	1	$\frac{3}{2}$	-18
x_3	2	0	1	0	$-\frac{1}{2}$	5
	2	5	0	0	2	-20

Notice that the c_j^* entries, the 2, 5, 0, 0, and 2, remain nonnegative. The choice of pivoting column guarantees this. Now proceed on. In the second tableau, $b_1^* = -18$ is the only negative constant term, so pivot in the first row. Comparing those ratios corresponding to negative a_{rj}^* terms, one gets $c_1^*/a_{11}^* = -2/9 > c_2^*/a_{12}^* = -5/2$, and so pivot at the $a_{11}^* = -9$ term. The resulting tableau is in Table 3.

Note that after this step, the constant term column entries are nonnegative. In fact, with the original problem presented in this form, one has reached the resolution of the problem. The minimum value of the objective function is 24 and is attained at the point $(2,0,1,0,0)$.

The dual problem is to maximize v with

$$3y_1 + 4y_2 \leq 10$$
$$2y_1 \qquad\ \leq 5$$
$$-3y_1 + 2y_2 \leq 4$$
$$3y_1 + 10y_2 = v$$

910

Table 3

	x_1	x_2	x_3	x_4	x_5	
x_4	-9	-2	0	1	$\frac{3}{2}$	-18
x_3	2	0	1	0	$-\frac{1}{2}$	5
	2	5	0	0	2	-20
x_1	1	$\frac{2}{9}$	0	$-\frac{1}{9}$	$-\frac{1}{6}$	2
x_3	0	$-\frac{4}{9}$	1	$\frac{2}{9}$	$-\frac{1}{6}$	1
	0	$\frac{41}{9}$	0	$\frac{2}{9}$	$\frac{7}{3}$	-24

The solution is given by the tableaux in Table 4. The optimal result is 24 as for the primal problem.

Table 4

	y_1	y_2	y_3	y_4	y_5	
y_3	3	4	1	0	0	10
y_4	2	0	0	1	0	5
y_5	-3	2	0	0	1	4
	-3	-10	0	0	0	0
y_3	9	0	1	0	-2	2
y_4	2	0	0	1	0	5
y_2	$-\frac{3}{2}$	1	0	0	$\frac{1}{2}$	2
	-18	0	0	0	5	20
y_1	1	0	$\frac{1}{9}$	0	$-\frac{2}{9}$	$\frac{2}{9}$
y_4	0	0	$-\frac{2}{9}$	1	$\frac{4}{9}$	$\frac{41}{9}$
y_2	0	1	$\frac{1}{6}$	0	$\frac{1}{6}$	$\frac{7}{3}$
	0	0	2	0	1	24

● **PROBLEM** 11-65

Minimize $Y = 50Z_1 + 40Z_2$

subject to

$$3Z_1 + 2Z_2 \geq 35$$
$$5Z_1 + 6Z_2 \geq 60$$
$$2Z_1 + 3Z_2 \geq 30$$
$$Z_1, \quad Z_2 \geq 0$$

Solve by the Dual Simplex method.

Assume that a canonical form is generated through the introduction of slack variables $\bar{S}_1$, $\bar{S}_2$, and $\bar{S}_3$ such that

$$-3Z_1 - 2Z_2 + \bar{S}_1 = -35$$

$$-5Z_1 - 6Z_2 + \bar{S}_2 = -60$$

$$-2Z_1 - 3Z_2 + \bar{S}_3 = -30$$

so that an initial simplex tableau is as follows:

	50	40	0	0	0	
	Z_1	Z_2	$\bar{S}_1$	$\bar{S}_2$	$\bar{S}_3$	
$\bar{S}_1$	-3	-2	1	0	0	-35
$\bar{S}_2$	-5	-6	0	1	0	-60
$\bar{S}_3$	-2	-3	0	0	1	-30
$\bar{y}$	50	40				

Note that the optimality criteria ($v_1, v_2 \geq 0$) is satisfied but the solution is infeasible. Choose $\bar{S}_2$ to leave the basis. Note that Z_1 and Z_2 are both candidates for entering the basis. Since $-\dfrac{50}{5} < -\dfrac{40}{6}$,

choose Z_2 to enter the basis.

	50	40				
	Z_1	Z_2	$\bar{S}_1$	$\bar{S}_2$	$\bar{S}_3$	
$\bar{S}_1$	-4/3	0	1	-1/3	0	-15
Z_2	5/6	1	0	-1/6	0	10
$\bar{S}_3$	1/2	0	0	-1/2	1	0
$\bar{y}$	100/6	0	0	40/6	0	400

	50	40				
	Z_1	Z_2	$\bar{S}_1$	$\bar{S}_2$	$\bar{S}_3$	
Z_1	1	0	-3/4	1/4	0	45/4
Z_2	0	1	5/8	-9/24	0	5/8
$\bar{S}_3$	0	0	3/8	-5/8	1	-45/8
$\bar{y}$	0	0	25/2	5/2	0	1175/2

	50	40				
	Z_1	Z_2	$\bar{S}_1$	$\bar{S}_2$	$\bar{S}_3$	
Z_1	1	0	-3/5	0	2/5	9
Z_2	0	1	2/5	0	-3/5	4
$\bar{S}_2$	0	0	-3/5	1	-8/5	9
$\bar{y}$	0	0	14	0	4	610

The final solution is given by $Z_1 = 9$, $Z_2 = 4$, and $Y = 610$.

● **PROBLEM** 11-66

Apply the Dual Simplex algorithm to the following dual tableau to obtain a terminal tableau:

	s_1	s_2	1	
q_1	$-\frac{2}{3}$	$-\frac{1}{3}*$	12	$= -r_1$
q_2	$\frac{4}{3}$	$\frac{1}{3}$	-60	$= -r_2$
-1	$-\frac{20}{3}$	$-\frac{5}{3}$	200	$= u$
	$= p_1$	$= p_2$	$= -w$	

(T.k)

Solution: When there is an intermediate dual tableau where the maximum problem is lined up by the rows and the minimum problem by columns, as above, one can apply the Dual Simplex algorithm if all the c_j's are non-positive, but at least one of the $-b_i$'s is positive.

Focus on a row for which $-b_i$ is positive. Compute c_j/a_{ij} for all those a_{ij}'s in the row that are negative. These ratios will thus be positive or zero in value, since $c_j \leq 0$ for all j. Pick the a_{ij} for which c_j/a_{ij} is a minimum. This a_{ij} is then the pivot element (in case of ties, choose any of the eligible a_{ij}'s). In the typical pivot operation, s_j and r_i are interchanged in the maximum problem, while q_i and p_j are interchanged in the minimum problem (it is possible that all the a_{ij}'s in the row under focus are either positive or zero. This indicates that the minimum problem is un-bounded).

In the next tableau, after the pivot transformation has been effected, the c_j's will again be nonpositive, and usually the $-b_i$'s will be less positive in character. A terminal tableau that is a solution tableau is achieved when the $-b_i$'s have all become nonpositive.

Clearly, Dual Simplex algorithm applies because both c_1 and c_2 are negative, while $-b_1$ is positive. Tableau k corresponds to feasibility with respect to the minimum problem (at the b.f.p., $q_1 = q_2 = 0$, $p_1 = \frac{20}{3}$, $p_2 = \frac{5}{3}$, and $w = 200$). Compute the relevant c/a ratios for row 1:

$$\frac{c_1}{a_{11}} = \frac{(-\frac{20}{3})}{(-\frac{2}{3})} = +10$$

$$\frac{c_2}{a_{12}} = \frac{(-\frac{5}{3})}{(-\frac{1}{3})} = +5.$$

Thus, a_{12} is the pivot element in Tableau k. Carrying

913

out the pivot transformation leads to Tableau $k + 1$, in which s_2 and r_1 (maximum problem) and q_1 and p_2 (minimum problem), respectively, have been exchanged:

$$
\begin{array}{c c}
 & \begin{array}{ccc} s_1 & r_1 & 1 \end{array} \\
\begin{array}{c} p_2 \\ q_2 \\ -1 \end{array} &
\begin{array}{|cc|c|}
\hline
2 & -3 & -36 \\
\frac{2}{3} & 1 & -48 \\
\hline
-\frac{10}{3} & -5 & 140 \\
\hline
\end{array}
\begin{array}{l} = -s_2 \\ = -r_2 \\ = u \end{array}
\end{array}
\qquad (T.k+1)
$$

$$ = p_1 \quad = q_1 \quad = -w $$

This is a terminal tableau; the solutions of the respective problems are easily read off:

Maximum problem:

$$ s_1 = r_1 = 0, \quad s_2 = 36, \quad r_2 = 36, \text{ and Max } u = 140. $$

Minimum problem:

$$ p_2 = q_2 = 0, \quad p_1 = \frac{10}{3}, \quad q_1 = 5, \text{ and Min } w = 140. $$

● **PROBLEM** 11-67

Solve the following problem using the Dual Simplex method:

Maximize $z = -2x_1 - 2x_2$

subject to: $2x_1 + x_2 \geq 6$

$x_1 + 2x_2 \geq 6$

$x_1, x_2 \geq 0$

Solution: Denoting the slack variables as x_3 and x_4, one gets the starting tableau, Tableau 1.

Tableau 1

		x_1	x_2	x_3	x_4
z	0	2	2	0	0
x_3	-6	-2	-1	1	0
x_4	-6	-1	-2	0	1

The problem is dual feasible, i.e., all $a_{0j} \geq 0$, but not primal feasible, i.e., some $a_{i0} < 0$. Hence use the Dual Simplex method. Choose x_3 to leave the basis. The choice of variables to enter the basis is x_1 because

$$ \left|\frac{b_1}{a_{r1}}\right| = \left|\frac{2}{-2}\right| < \left|\frac{b_2}{a_{r2}}\right| = \left|\frac{2}{-1}\right| $$

914

Performing the iteration leads to Tableau 2.

Tableau 2

		x_1	x_2	x_3	x_4
z	-6	0	1	1	0
x_1	3	1	.5	-.5	0
x_4	-3	0	$\boxed{-1.5}$	-.5	1

In Tableau 2, x_4 is chosen to leave the basis and x_2 is the variable which must enter because

$$\frac{b_2}{|a_{r2}|} = \frac{1}{|-1.5|} < \frac{b_3}{|a_{r3}|} = \frac{1}{|-.5|}$$

The indicated iteration is performed, leading to Tableau 3, for which the solution is feasible and therefore optimal.

Tableau 3

		x_1	x_2	x_3	x_4
z	-8	0	0	$\frac{2}{3}$	$\frac{2}{3}$
x_1	2	1	0	$-\frac{2}{3}$	$-\frac{1}{3}$
x_2	2	0	1	$\frac{1}{3}$	$-\frac{2}{3}$

● **PROBLEM** 11-68

Maximize $3x_1 + x_2 + 3x_3$ subject to

$$2x_1 + x_2 + x_3 \leq 2$$

$$x_1 + 2x_2 + 3x_3 \leq 5$$

$$2x_1 + 2x_2 + x_3 \leq 6$$

$$x_1 \geq 0, \quad x_2 \geq 0, \quad x_3 \geq 0.$$

Solve by the Revised Simplex method.

Solution: Start with an initial basic feasible solution and corresponding $\vec{B}^{-1}$ as shown in the tableau below

Variable	$\vec{B}^{-1}$			$\vec{x}_B$
4	1	0	0	2
5	0	1	0	5
6	0	0	1	6

915

Compute

$$\vec{\lambda} = [0,0,0]\vec{B}^{-1} = [0,0,0]$$

and then

$$\vec{c}_D - \lambda\vec{D} = [-3,-1,-3].$$

Decide to bring $\vec{a}_2$ into the basis in order to simplify the calculation (although it violates the rule of selecting the most negative relative cost). Its current representation is found by multiplying $\vec{B}^{-1}$; thus one gets

Variable	$\vec{B}^{-1}$			$\vec{x}_B$	$\vec{y}_2$
4	1	0	0	2	①
5	0	1	0	5	2
6	0	0	1	6	2

After computing the ratios using the formula y_{io}/y_{ij} (where $y_{20}= 1$, $y_{21}=1$, $y_{22}=2$, $y_{23}=2$), select the pivot indicated. The updated tableau becomes

Variable	$\vec{B}^{-1}$			$\vec{x}_B$
2	1	0	0	2
5	-2	1	0	1
6	-2	0	1	2

then

$$\vec{\lambda} = [-1,0,0]\vec{B}^{-1} = [-1,0,0]$$

$$c_1 - z_1 = -1, \quad c_3 - z_3 = -2, \quad c_4 - z_4 = 1.$$

Select $\vec{a}_3$ to enter. The tableau obtained is

Variable	$\vec{B}^{-1}$			$\vec{x}_B$	$\vec{y}_3$
2	1	0	0	2	1
5	-2	1	0	1	①
6	-2	0	1	2	-1

Using the pivot indicated, obtain

Variable	$\vec{B}^{-1}$			$\vec{x}_B$
2	3	-1	0	1
3	-2	1	0	1
6	-4	1	1	3

916

Now

$$\vec{\lambda} = [-1,-3,0]\vec{B}^{-1} = [3,-2,0]$$

and

$$c_1 - z_1 = -7, \quad c_4 - z_4 = -3, \quad c_5 - z_5 = 2.$$

Select $\vec{a}_1$ to enter the basis. The tableau is

Variable	$\vec{B}^{-1}$			$\vec{x}_B$	$\vec{y}_1$
2	3	-1	0	1	⑤
3	-2	1	0	1	-3
6	-4	1	1	3	-5

Using the pivot indicated, obtain

Variable	$\vec{B}^{-1}$			$\vec{x}_B$
1	3/5	-1/5	0	1/5
3	-1/5	2/5	0	8/5
6	-1	0	1	4

Now

$$\vec{\lambda} = [-3,-3,0]\vec{B}^{-1} = [-6/5,-3/5,0]$$

and

$$c_2 - z_2 = 7/5, \quad c_4 - z_4 = 6/5, \quad c_5 - z_5 = 3/5.$$

Since the $c_i - z_i$'s are all nonnegative, the solution
$\vec{x} = (1/5,0,8/5,0,0,4)$ is optimal.

● **PROBLEM** 11-69

Minimize $-45x_1 - 80x_2$

subject to $x_1 + 4x_2 + x_3 \qquad = 80$

$2x_1 + 3x_2 \qquad + x_4 = 90$

$x_j \geq 0$

Solve by Revised Simplex method.

<u>Solution</u>: Here

$$\vec{c} = (-45, -80, 0, 0)$$

$$\vec{A} = (\vec{P}_1 \vec{P}_2 \vec{P}_3 \vec{P}_4) = \begin{pmatrix} 1 & 4 & 1 & 0 \\ 2 & 3 & 0 & 1 \end{pmatrix}$$

$$\vec{b} = \begin{pmatrix} 80 \\ 90 \end{pmatrix}$$

The first feasible basis $\vec{B}$ is given by the unit vectors $\vec{P}_3$ and $\vec{P}_4$; thus

$$\vec{B} = \begin{pmatrix} 1 & 0 \\ 0 & 1 \end{pmatrix} \quad \vec{B}^{-1} = \begin{pmatrix} 1 & 0 \\ 0 & 1 \end{pmatrix} \quad \vec{c}_0 = (0,0)$$

and

$$X_0 = \vec{B}^{-1}\vec{b} = \begin{pmatrix} 1 & 0 \\ 0 & 1 \end{pmatrix}\begin{pmatrix} 80 \\ 90 \end{pmatrix} = \begin{pmatrix} 80 \\ 90 \end{pmatrix} = (x_3, x_4)$$

The corresponding pricing vector is

$$\pi = c_0\vec{B}^{-1} = (0 \quad 0)\begin{pmatrix} 1 & 0 \\ 0 & 1 \end{pmatrix} = (0,0)$$

Price out the vectors $\vec{P}_1$ and $\vec{P}_2$ not in the basis by computing the $z_j - c_j$ using

$$\pi\vec{P}_1 - c_1 = (0,0)\begin{pmatrix} 1 \\ 2 \end{pmatrix} + 45 = 45$$

$$\pi\vec{P}_2 - c_2 = (0,0)\begin{pmatrix} 4 \\ 3 \end{pmatrix} + 80 = 80$$

Then select vector $\vec{P}_2$ to enter the basis because it corresponds to the maximum $z_j - c_j$. As the vector $\vec{X}_2$ is needed, which is the representation of $\vec{P}_2$ in terms of the current basis $\vec{B}$, compute

$$\vec{X}_2 = \vec{B}^{-1}\vec{P}_2 = \begin{pmatrix} 1 & 0 \\ 0 & 1 \end{pmatrix}\begin{pmatrix} 4 \\ 3 \end{pmatrix} = \begin{pmatrix} 4 \\ 3 \end{pmatrix}$$

Next, determine the vector to be eliminated by calculating the corresponding θ ratios of 80:4 and 90:3. As the former ratio is the minimum and corresponds to x_3, vector $\vec{P}_3$ is eliminated from the basis and is replaced by vector $\vec{P}_2$. To compute the inverse of the new basis $(\vec{P}_2\vec{P}_4)$ apply the elimination formula

$$\bar{b}_{ij} = b_{ij} - \frac{b_{ij}}{x_{\ell k}}x_{ik} \quad \text{for } i \neq 1, \quad \bar{b}_{\ell j} = \frac{b_{\ell j}}{x_{\ell k}}$$

to the matrix

$$\begin{array}{c|c}
\vec{X}_2 & \vec{B}^{-1} \\
\hline
\end{array}$$

$$\begin{pmatrix} \textcircled{4} & | & 1 & 0 \\ 3 & | & 0 & 1 \end{pmatrix}$$

to obtain the new inverse

$$(\vec{P}_2\vec{P}_4)^{-1} = \begin{pmatrix} 1/4 & 0 \\ -3/4 & 1 \end{pmatrix}$$

Repeat the process and denote the new basis by $\vec{B} = (\vec{P}_2\vec{P}_4)$ with

$$\vec{B} = \begin{pmatrix} 4 & 0 \\ 3 & 1 \end{pmatrix} \quad \vec{B}^{-1} = \begin{pmatrix} 1/4 & 0 \\ -3/4 & 1 \end{pmatrix} \quad \vec{c}_0 = (-80,0)$$

and

$$\vec{X}_0 = \vec{B}^{-1}\vec{b} = \begin{pmatrix} 1/4 & 0 \\ -3/4 & 1 \end{pmatrix}\begin{pmatrix} 80 \\ 90 \end{pmatrix} = \begin{pmatrix} 20 \\ 30 \end{pmatrix} = (x_2, x_4)$$

The corresponding pricing vector is

$$\pi = \vec{c}_0\vec{B}^{-1} = (-80,0)\begin{pmatrix} 1/4 & 0 \\ -3/4 & 1 \end{pmatrix} = (-20,0)$$

Price out vectors $\vec{P}_1$ and $\vec{P}_3$ which are not in the current basis and obtain

$$\pi\vec{P}_1 - c_1 = (-20,0)\begin{pmatrix} 1 \\ 2 \end{pmatrix} + 45 = 25$$

$$\pi\vec{P}_3 - c_3 = (-20,0)\begin{pmatrix} 1 \\ 0 \end{pmatrix} - 0 = -20$$

Thus, vector $\vec{P}_1$ is selected to enter the basis and compute

$$\vec{X}_1 = \vec{B}^{-1}\vec{P}_1 = \begin{pmatrix} 1/4 & 0 \\ -3/4 & 1 \end{pmatrix}\begin{pmatrix} 1 \\ 2 \end{pmatrix} = \begin{pmatrix} 1/4 \\ 5/4 \end{pmatrix}$$

with the corresponding θ ratios of $20:\frac{1}{4}$ and $30:\frac{5}{4}$; vector $\vec{P}_4$ is eliminated from the basis. The new basis is $\vec{B} = (\vec{P}_2\vec{P}_1)$, and the inverse is obtained by applying the elimination formulas to the matrix

$$\begin{array}{c|c} \vec{X}_1 & \vec{B}^{-1} \\ \hline \end{array}$$

$$\begin{pmatrix} \frac{1}{4} & \bigm| & \frac{1}{4} & 0 \\ \boxed{\frac{5}{4}} & \bigm| & -\frac{3}{4} & 1 \end{pmatrix}$$

i.e., the new $\vec{B}^{-1} = (\vec{P}_2\vec{P}_1)^{-1} = \begin{pmatrix} 2/5 & -1/5 \\ -3/5 & 4/5 \end{pmatrix}.$

919

Finally, using the new inverse one sees that the vectors $\vec{P}_3$ and $\vec{P}_4$ not in the basis price out negatively and the optimal solution is

$$\vec{X}_0 = \vec{B}^{-1}\vec{b} = \begin{pmatrix} 2/5 & -1/5 \\ -3/5 & 4/5 \end{pmatrix}\begin{pmatrix} 80 \\ 90 \end{pmatrix} = \begin{pmatrix} 14 \\ 24 \end{pmatrix} = (x_2, x_1)$$

with the minimum value of the objective function

$$z_0 = \vec{c}_0\vec{X}_0 = \vec{c}_0(\vec{B}^{-1}\vec{b}) = (\vec{c}_0\vec{B}^{-1})\vec{b} = \pi\vec{b}$$

$$= (-5, -20)\begin{pmatrix} 80 \\ 90 \end{pmatrix} = -2,200$$

INTEGER PROGRAMMING

● PROBLEM 11-70

The Brown Company has two warehouses and three retail outlets. Warehouse number one (which will be denoted by W_1) has a capacity of 12 units; warehouse number two (W_2) holds 8 units. These warehouses must ship the product to the three outlets, denoted by O_1, O_2, and O_3. O_1 requires 8 units. O_2 requires 7 units, and O_3 requires 5 units. Thus, there is a total storage capacity of 20 units, and also a demand for 20 units. The question is, which warehouse should ship how many units to which outlet? (The objective being, of course, to accomplish this at the least possible cost.)

Costs of shipping from either warehouse to any of the outlets are known and are summarized in the following table, which also sets forth the warehouse capacities and the needs of the retail outlets:

	O_1	O_2	O_3	Capacity
W_1	\$3.00	\$5.00	\$3.00	12
W_2	2.00	7.00	1.00	8
Needs (units)	8	7	5	

Solution: This seems to be a linear programming problem with three variables. However, the third variable can be computed from the previous two. Let X be the number of units sent from warehouse 1 to outlet 1. Since outlet 1 requires only 8 units, we have the constraint X < 8. Obviously X $\geq$ 0 means you either ship one or you don't. Let y be the number of units shipped from warehouse 1 to outlet 2. Similarly,

$$0 \overset{\le}{=} y \overset{\le}{=} 7$$

is another constraint. Because there are 12 units in warehouse 1, the number of units sent to outlet 3 is

$$12 - X - y.$$

Obviously, this must be larger or equal to zero. Thus,

$$12 - X - y \overset{\ge}{=} 0 \quad \text{or} \quad X + y \overset{\le}{=} 12$$

is a constraint.

The amount of units sent from warehouse 2 to outlet 1 is the original eight less the X that was sent from warehouse 1, or 8 - X. Similarly we find all of the others.

	O_1	O_2	O_3
W_1	x	y	$12 - x - y$
W_2	$8 - x$	$7 - y$	$x + y - 7$

Note that the quantities shipped from both warehouses to Outlet 3 have been determined by simply subtracting the quantities shipped to Outlets 1 and 2 from the total capacities of the warehouses.

So we have the following constraints:

(1) $X \overset{\ge}{=} 0$

(2) $y \overset{\ge}{=} 0$

(3) $X \overset{\le}{=} 8$

(4) $y \overset{\le}{=} 7$

(5) $X+y \overset{\le}{=} 12$

(6) $X+y \overset{\ge}{=} 7$ (from $O_3 - W_2 \overset{\ge}{=} 0$) .

We wish to minimize the cost function. It can be found by multiplying the number of units sent by their costs. We get

$$\text{Cost} = 3X + 5y + 3(12-X-y) + 2(8-X) + 7(7-y) +$$
$$1(X+y-7) = 94 - X - 4y.$$

To minimize this we plot the constraints to find the region that is defined by them.
The shaded region is the area that is defined by the constraints. By linear programming we know the minimum cost will appear at a corner of this region. We need only check these corners to pick the best point.

A   (8,0)   94 - 8 = \$86

B   (7,0)   94 - 7 = \$87

C   (0,7)   94 - 0 - (4 × 7) = \$66

921

| D | (5,7) | $94 - 5 - (4 \times 7) = \$61$ |
| E | (8,4) | $94 - 8 - (4 \times 4) = \$70$ |

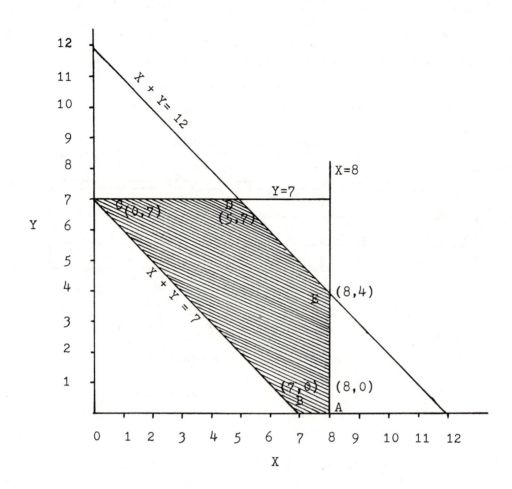

Thus the point with the least cost is point D. So for the lowest cost the shipping schedule should be

	O_1	O_2	O_3	
W_1	5	7	0	12
W_2	3	0	5	8
	8	7	5	

● **PROBLEM** 11-71

Suppose that in the course of solving a transportation problem, the following trial solution is computed:

Source \ Destination	F_1	F_2	F_3	Units demanded	
W_1	.90 / 0	1.00 / 5	1.00 / 0	5	
W_2	1.00 / 20	1.40 / 0	.80 / 0	20	(1)
W_3	1.30 / 0	1.00 / 10	.80 / 10	20	
	20	15	10	45	

How would you progress further to find the optimal solution?

Solution: When the number of boxes used in obtaining a trial solution is less than $F + W - 1$, (number of factories and warehouses minus 1) the problem of degeneracy appears. This problem arises as follows: To improve on a trial solution, consider alternative routes. An indirect route is the path a unit would have to follow from a factory to a given warehouse, using only established channels, (i.e., the shipment must avoid zero boxes; otherwise, we are shipping from a box which has no units, or introducing two new boxes into the solution instead of one). When a solution is degenerate, there are too many unused boxes.

To resolve the degeneracy case, record some very small amount, say d, in one of the zero boxes. This d is interpreted as a quantity of goods. The box with the d entry may either ship or receive goods but in the final solution the d is assigned a value of zero if it is still present in the calculations.

For the given trial solution, only 4 boxes are used while $W + F - 1 = 5$. Put d units in box W_1F_1. Total shipping cost is \$43. The zero boxes may now be evaluated in the standard manner.

Source \ Destination	F_1	F_2	F_3	Units demanded
W_1	.90 / d	1.00 / 5	1.00 / 0	5
W_2	1.00 / 20	1.40 / 0	.80 / 0	20
W_3	1.30 / 0	1.00 / 10	.80 / 10	20
Supply	20	15	10	45

Consider W_3F_1. The cost of the direct route is \$1.30. The indirect route is from F_1W_1, which permits a reduction of shipment from F_2 to W_1; but this in turn requires an increase in the shipment from F_2 to W_3. The costs are $W_1F_1 + W_3F_2 - W_1F_2$ or \$.90. Since the indirect route is cheaper than the direct route it should continue to be used. The other zero boxes may be evaluated similarly with the result below:

Source Destination	F_1	F_2	F_3	
W_1			1.00 .80	
W_2		1.40 1.10	.80 .90	
W_3	1.30 .90			

Shipping from F_3 to W_3 has an indirect cost in excess of the direct cost. The maximum amount which can be shifted is five units, since this is the minimum amount in a box which must be reduced (box W_1F_2). The new trial solution is optimal. In this example degeneracy disappeared in one step. It is possible for degeneracy to remain through several iterations; in fact, the optimal solution may be degenerate.

● **PROBLEM** 11-72

Solve the following problem in integer programming:
Find non-negative integers X_{ij} which will

$$\text{Minimize} \quad 200\ X_{11} + 300\ X_{12} + 250\ X_{21} + 100\ X_{22} + 250\ X_{31}$$
$$+ 250\ X_{32}$$

subject to

$$X_{11} \qquad + X_{21} \qquad + X_{31} \qquad \geq \quad 30$$
$$X_{12} \quad + X_{22} \qquad + X_{32} \geq \quad 20$$
$$-X_{11} -X_{12} \qquad\qquad\qquad \geq \quad -20$$
$$-X_{21} -X_{22} \qquad\qquad \geq \quad -20$$
$$-X_{31} -X_{32} \geq \quad -20$$

Solution: Since some of the constraints are negative we cannot apply the simplex algorithm to the given problem. However, the dual, which is a maximization problem, is receptive to the simplex technique. The dual is

$$\text{Maximize} \quad 30y_1 + 20y_2 - 20y_3 - 20y_4 - 20y_5$$

subject to:

$$y_1 \qquad - y_3 \qquad\qquad \leq \quad 200$$
$$y_2 - y_3 \qquad\qquad \leq \quad 300$$
$$y_1 \qquad\qquad - y_4 \qquad \leq \quad 250$$
$$y_2 \qquad - y_4 \qquad \leq \quad 100$$
$$y_1 \qquad\qquad\qquad -y_5 \leq \quad 250$$

$$y_2 \qquad\qquad -y_5 \quad\le\quad 250$$

$$y_1, y_2, \ldots, y_5 \qquad\quad \ge \quad 0$$

We can convert the inequalities to equalities by adding six slack variables

$$y_6, \ y_7, \ y_8, \ y_9, \ y_{10}, \ y_{11} \ .$$

An initial basic feasible solution is obtained by setting $y_1 = y_2 = y_3 = y_4 = y_5 = 0$. The simplex tableau is then

	v_1	v_2	v_3	v_4	v_5	v_6	v_7	v_8	v_9	v_{10}	v_{11}	b
v_6	1	0	-1	0	0	1	0	0	0	0	0	200
v_7	0	1	-1	0	0	0	1	0	0	0	0	300
v_8	1	0	0	-1	0	0	0	1	0	0	0	250
v_9	0	1	0	-1	0	0	0	0	1	0	0	100
v_{10}	1	0	0	0	-1	0	0	0	0	1	0	250
v_{11}	0	1	0	0	-1	0	0	0	0	0	1	250
d	-30	-20	20	20	20	0	0	0	0	0	0	0

Since -30 is the largest negative entry and $200/1$ is the minimum positive ratio we replace v_6 by v_1 in the basis on the left. Thus we pivot on the first element in the first column to obtain

	v_1	v_2	v_3	v_4	v_5	v_6	v_7	v_8	v_9	v_{10}	v_{11}	b
v_1	1	0	-1	0	0	1	0	0	0	0	0	200
v_7	0	1	-1	0	0	0	1	0	0	0	0	300
v_8	0	0	1	-1	0	-1	0	1	0	0	0	50
v_9	0	(1)	0	-1	0	0	0	0	1	0	0	100
v_{10}	0	0	1	0	-1	-1	0	0	0	1	0	50
v_{11}	0	1	0	0	-1	0	0	0	0	0	0	250
d	0	-20	-10	20	20	30	0	0	0	0	0	6,000

Now we replace v_9 by v_2 to obtain a new basic feasible solution. Pivoting on the encircled element:

	v_1	v_2	v_3	v_4	v_5	v_6	v_7	v_8	v_9	v_{10}	v_{11}	b
v_1	1	0	-1	0	0	1	0	0	0	0	0	200
v_7	0	0	-1	1	0	0	1	0	-1	0	0	200
v_8	0	0	1	-1	0	-1	0	1	0	0	0	50
v_2	0	1	0	-1	0	0	0	0	1	0	0	100
v_{10}	0	0	(1)	0	-1	-1	0	0	0	1	0	50
v_{11}	0	0	0	1	-1	0	0	0	-1	0	1	50
d	0	0	-10	0	20	30	0	0	20	0	0	8,000

The only negative entry is under v_3. Notice that the ratios of b to v_8 and of b to v_{10} are both the same, i.e., 50/1. By the rules of the simplex algorithm we can choose either v_8 or v_{10} for liquidation. We choose to replace v_{10} by v_3 and hence pivot on the encircled element. The next tableau is

	v_1	v_2	v_3	v_4	v_5	v_6	v_7	v_8	v_9	v_{10}	v_{11}	b
v_1	1	0	0	0	-1	0	0	0	0	1	0	250
v_7	0	0	0	1	-1	-1	1	0	1	1	0	250
v_8	0	0	0	-1	1	0	0	1	0	-1	0	0
v_2	0	1	0	-1	0	0	0	0	1	0	0	100
v_3	0	0	1	0	-1	-1	0	0	0	1	0	50
v_{11}	0	0	0	1	-1	0	0	0	-1	0	1	50
d	0	0	0	0	10	20	0	0	20	10	0	8,500

Since there are no more negative entries in the last row the algorithm has converged to a solution. The maximum feasible solution is 8,500. Therefore the minimum feasible solution is also 8,500. At this value, the values of the X_{ij} are read off from the slack variable values in the last row. Thus, $X_{11} = 20$, $X_{12} = 0$, $X_{21} = 0$, $X_{22} = 20$, $X_{31} = 10$, $X_{32} = 0$. Note that the main body of the simplex tableau consisted only of ones and zeros. This is a characteristic feature of integer programs. It makes pivoting easier and also ensures that the solution values will be integers.

Consider the following integer programming problem:

$$\text{Maximize} \quad P = 6x_1 + 3x_2 + x_3 + 2x_4 \quad (1)$$

subject to

$$
\begin{aligned}
x_1 + x_2 + x_3 + x_4 &\le 8 \\
2x_1 + x_2 + 3x_3 \phantom{{} + x_4} &\le 12 \\
5x_2 + x_3 + 3x_4 &\le 6 \\
x_1 &\le 1 \\
x_2 &\le 1 \\
x_3 &\le 4 \\
x_4 &\le 2
\end{aligned}
\quad (2)
$$

x_1, x_2, x_3, x_4 all non-negative integers. Use the branch and bound algorithm to solve this problem.

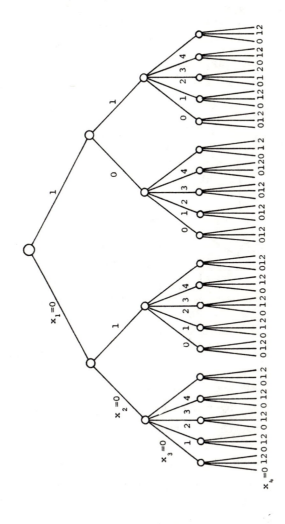

Solution: When a problem is required to have an integer solution, this means that there are a finite number of possible solution points. It is theoretically possible to enumerate and evaluate every possible solution to find the optimum.

We can list all possible solutions to the given problem by means of a tree diagram. The constraints $x_1 \leq 1$, $x_2 \leq 1$ imply that x_1 and x_2 can take on only the values zero or one. x_3 can take on five values while x_4 can take on 3 values. Thus, there are $2 \times 2 \times 5 \times 3 = 60$ possible solutions.

Some of the possible solutions will not satisfy the remaining constraints. For example, the solution $x_2 = 1$, $x_3 = 3$ and $x_4 = 2$ fails to satisfy the constraint $5x_2 + x_3 + 3x_4 \leq 6$. The branch and bound approach reduces the search by eliminating whole branches of the above tree. The principle used is: A branch can be eliminated if it can be shown to contain no feasible solution better than one already obtained.

Solving the problem as a linear programming problem (L.P.) yields an upper limit or bound on the possible integer solution. If the simplex method is used to solve the given problem we obtain the solution:

$$x_1 = 1; \ x_2 = 0; \ x_3 = 3.33; \ x_4 = 0.89$$

and $P = 11.11$. The integer solution must therefore be less than or equal to $P = 11$. An initial feasible integer solution is $x_1 = x_2 = x_3 = x_4 = 0$ and $P = 0$.

Now select an arbitrary variable and construct branches. Select x_4 for branching. Since the L.P. solution was $x_4 = .89$, we can let $x_4 = 0$ or $x_4 \geq 1$.

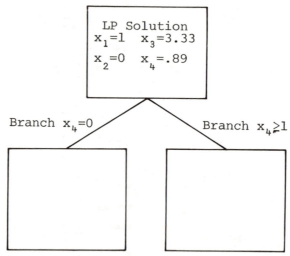

Consider the branch $x_4 = 0$. Replace the last constraint in the original problem ($x_4 \leq 2$) by the constraint $x_4 = 0$. Considered as a linear programming problem, the solution is $x_1 = 1$, $x_2 = .57$, $x_3 = 3.14$, $x_4 = 0$ and $P = 10.85$.

Now select x_2 as the branch variable. The two possible branches are $x_2 = 1$ and $x_2 = 0$. Let $x_2 = 1$ in the original problem (1),(2).

The solution to the L.P. problem is $x_1 = 1$, $x_2 = 1$, $x_3 = 1$, $x_4 = 0$ and
$P = 10$. This is better than the original feasible solution $P = 0$.
    Letting $x_2 = 0$ we obtain $P = 9.33$. Since the bound on profit in
this branch is less than 10, we eliminate this branch.
    Now move back to the branch $x_4 \geq 1$. The solution to the L.P. is
$$x_1 = 1, \ x_2 = 0, \ x_3 = 3, \ x_4 = 1 \ \text{ and } \ P = 11.$$
This is the optimal solution since all branches have been searched.

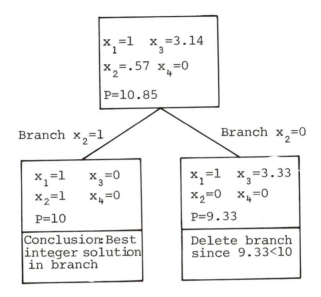

                                                        • PROBLEM 11-74

Use the branch and bound method to solve the integer programming problem
                Maximize $P = 2x_1 + 3x_2 + x_3 + 2x_4$
subject to
                $5x_1 + 2x_2 + x_3 + x_4 \leq 15$
                $2x_1 + 6x_2 + 10x_3 + 8x_4 \leq 60$
                $x_1 + x_2 + x_3 + x_4 \leq 8$
                $2x_1 + 2x_2 + 3x_3 + 3x_4 \leq 16$
$x_1 \leq 3$, $x_2 \leq 7$, $x_3 \leq 5$, $x_4 \leq 5$.

Solution: x_1 can take on any of the four values 0,1,2 or 3. Similarly, there are eight possibilities for x_2, six for x_3 and six for x_4.
By the Fundamental Principle of Counting there are $4 \times 8 \times 6 \times 6 = 1,152$
possible solutions. The branch and bound procedure eliminates non-optimal
solutions and thus reduces the amount of calculation. The steps in the
procedure are as follows:
1) Find an initial feasible integer solution.
2) Branch: Select a variable and divide the possible solutions into two
    groups. Select one branch for investigation.
3) Find an upper bound or maximum value for the problem defined by the

branch selected. This bound can be found by considering the problem as a linear programming problem.

4) **Compare:** Compare the bound obtained for the branch being considered with the best solution so far for the previous branches examined. If the bound is less, delete the whole new branch. If the bound is greater and an integer it becomes the new best solution so far. If the bound is greater but not an integer, continue in this same branch by branching further (Step 2).

5) **Completion:** When all branches have been examined, the best solution so far is the optimal solution.

The general approach is illustrated for the given problem in the figure.

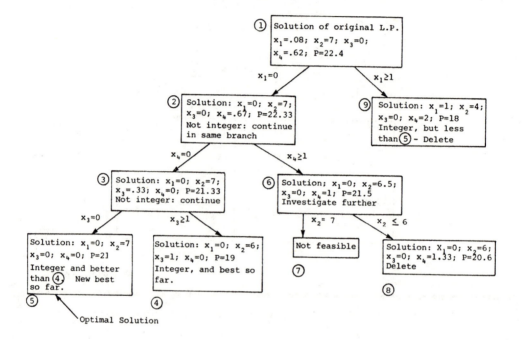

Thus the optimal solution to the integer program is $x_1 = 0$, $x_2 = 7$, $x_3 = 0$, $x_4 = 0$ and $P = 21$. When first selecting a variable for branching, a good rule to follow is to choose a variable whose linear programming solution is non-integer.

● **PROBLEM** 11-75

A travelling salesman must visit four towns. The distance between towns is given in the table below:

		To			
		A	B	C	D
From	A		1	4	5
	B	3		1	2
	C	2	4		3
	D	5	2	6	

930

The distance from town x to town y is not the same as from y to x
because of necessary detours (one-way streets, construction, etc.). What
is the minimum distance the salesman must travel if he is to touch every
town and finish back at the town he started from? Assume he can touch
each intermediate town only once.

Solution: This is a problem in integer programming. We start by reducing the entries in each row by the smallest entry in it. At least these
distances will have to be covered, whatever the circuit, since each town
will have to be left for some other town. Similarly, reduce the entries
in each column. The sum of all the reductions is a lower bound on the
length of any circuit.

Reduction by Rows

		A	B	C	d
(-1)	A		0	3	4
(-1)	B	2		0	1
(-2)	C	0	2		1
(-2)	D	3	0	4	

Reduction by Columns

	A	B	C	D
A		0	3	3
B	2		0	0
C	0	2		0
D	3	0	4	

(-1)

The total reduction is 7, and this is a lower bound for the length of a
circuit. Now we write against all zero entries the sum of the smallest
remaining entry in its row and the smallest remaining entry in its column.

	A	B	C	D
A		3		
B			3	0
C	2			0
D		3		

Take one of the largest of these values, 3, in AB. If AB is not a
link in the circuit (A̅B̅), then A will be left for some other node than
B, and B will be reached from some other node than A. Therefore, another 3 can be added to the lower bound in this case. But what happens
if AB is used? We exclude the A row and B column and also BA because this would bring us back to A, without having been to all towns.

	A	C	D
B		0	0
C	0		0
D	3	4	

Reduce the last row by 3 and add this to the lower bound, if we use
AB as a link.

	A	C	D
B		0	0
C	0		0
D	0	1	

which leads to

$$
\begin{array}{ccc}
 & A \quad C \quad D \\
B & \quad 1 \quad\ \ 0 \\
C & \quad 0 \quad\ \ 0 \\
D & \quad 1
\end{array}
$$

for the sum of the lowest remaining entries in row and column. If we do not take BC as a link, then the lower bound is increased by another 1. It is now 11. If we take BC, we omit row B, column C and CA (because this would close the short circuit ABCA).

$$
\begin{array}{cc}
 & A \quad D \\
C & \quad 0 \\
D & 0
\end{array}
$$

The circuit is closed by CD, DA, total length 10. If we do not take BC, then we know already that the length of any circuit is at least 11, longer than 10. Going one step further back, we see that a circuit not including AB has a length of at least 10.

Thus, the salesman must cover at least 10 miles if he is to complete a circuit touching every town. One such circuit is ABCDA .

● **PROBLEM** 11-76

An assembly line consisting of a collection of work stations has to perform a series of jobs in order to assemble a product. At each work station one or more of the jobs may be performed. Normally there are some restrictions on the order in which jobs may be done. These are called precedence relations, and there is a limit on the time a product can stay at any particular work station. Consider a product with 5 jobs. The decision involves allocating each job to a work station so that the number of work stations is minimized. Table 1 gives the jobs, any precedence relations that exist, and the time needed to complete each job.

Job i is either done at station j, or it is not done at station j. This is an either/or type situation which fits in well with zero-one variables.

$$
\text{Let } x_{ij} = \begin{cases} 1 & \text{if } i \text{ is done at station } j \\ 0 & \text{if } i \text{ is not done at station } j. \end{cases}
$$

Forumulate as an integer program.

Solution: Assume that there are 4 stations and the maximum time at each work station is 12 minutes. There is the following time constraint on each solution:

932

Table 1

Data for assembly line balancing.

Job i	Time (p_i) in minutes	Predecessors
1	6	—
2	5	—
3	7	—
4	6	3
5	5	2,4

$$\sum_{i=1}^{5} p_i x_{ij} \leq 12, \qquad j = 1, \ldots, 4 \tag{1}$$

(i.e., the time taken for jobs assigned to station j must be less than 12 minutes.) Equations (1) expand to

$$6x_{11} + 5x_{21} + 7x_{31} + 6x_{41} + 5x_{51} \leq 12$$

$$6x_{12} + 5x_{22} + 7x_{32} + 6x_{42} + 5x_{52} \leq 12$$

$$6x_{13} + 5x_{23} + 7x_{33} + 6x_{43} + 5x_{53} \leq 12 \tag{2}$$

$$6x_{14} + 5x_{24} + 7x_{34} + 6x_{44} + 5x_{54} \leq 12.$$

Next, handle precedence relations between jobs. By saying that job 3 must be done before job 4, that means that job 3 must be performed either at the same station as job 4 or at a prior station. Job i has been done at or before station k, if

$$\sum_{j=1}^{k} x_{ij} = 1,$$

and has not been done if

$$\sum_{j=1}^{k} x_{ij} = 0.$$

At station k, if

$$\sum_{j=1}^{k} x_{4j} \leq \sum_{j=1}^{k} x_{3j},$$

then job 4 cannot be done unless job 3 has been done because

$$\sum_{j=1}^{k} x_{4j} = 1$$

only if

$$\sum_{j=1}^{k} x_{3j} = 1.$$

For the precedence relations to be satisfied this must hold at all stations. Therefore

$$\sum_{j=1}^{k} x_{4j} \leq \sum_{j=1}^{k} x_{3j}, \qquad k = 1, \ldots, 4. \tag{3}$$

If neither job is done by station k, expression (3) holds trivially (i.e., $0 \leq 0$), and it also holds if both jobs have been done (i.e., $1 \leq 1$).

The precedence relations for job 5 are

$$\sum_{j=1}^{k} x_{5j} \leq \sum_{j=1}^{k} x_{2j}$$
$$\qquad\qquad\qquad k = 1, \ldots, 4. \tag{4}$$
$$\sum_{j=1}^{k} x_{5j} \leq \sum_{j=1}^{k} x_{4j}$$

It is also necessary to ensure that each job is done once and only once:

$$\sum_{j=1}^{k} x_{ij} = 1, \qquad i = 1, \ldots, 5. \tag{5}$$

The objective is to find the minimum number of stations to set up. This is achieved by allocating a lower "cost" to job i if it is done at station 1 than if it is done at station 2, etc. By minimizing these costs, the jobs are forced to the earliest possible work stations. The costs are arbitrary if a cost of j is given to x_{ij} (= job i done at station j). Thus

$$\text{minimize } z = \sum_{i=1}^{5} x_{i1} + 2 \sum_{i=1}^{5} x_{i2} + 3 \sum_{i=1}^{5} x_{i3} + 4 \sum_{i=1}^{5} x_{i4} \tag{6}$$

The collection of equations (2) to (6), together with the nonnegativity and integrality conditions on the variables

$$x_{ij} \geq 0 \text{ and integer}$$

make up the integer program for this problem. There is no need to put an upper limit of 1 on each x_{ij}; equation (4) does that implicitly.

Twenty-seven cells are arranged in a (3 x 3 x 3)-dimensional array as shown in Figure 1.

Three cells are regarded as lying in the same line if they are on the same horizontal or vertical line or the same diagonal. Diagonals exist on each horizontal and vertical section and connecting opposite vertices of the cube. (There are 49 lines altogether.)

Given 13 white balls (noughts) and 14 black balls (crosses), construct an integer programming model that would arrange them, one to a cell, so as to minimize the number of lines with balls all of one color.

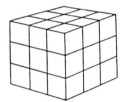

Fig. 1

Solution: This "pure" problem typifies the combinatorial character of quite a lot of integer programming problems. Clearly there are an enormous number of ways of arranging the balls in the three-dimensional array. Such problems often prove difficult to solve as integer programming models. There is an advantage to using a heuristic solution first. This solution can then be used to obtain a cut-off value for the branch and bound tree search. A formulation is described as follows:

Variables:

The cells are numbered 1 to 27. It is convenient to number sequentially row by row and section by section. Associated with each cell a 0-1 variable δ_j is introduced with the following interpretation :

$$\delta_j = \begin{cases} 1 & \text{if cell } j \text{ contains a black ball} \\ 0 & \text{if cell } j \text{ contains a white ball} \end{cases}$$

There are 27 such 0-1 variables.

There are 49 possible lines in the cube. With each of these lines a 0-1 variable γ_i is associated with the following interpretations:

$$
\gamma_i = \begin{cases} 1 & \text{if all the balls in the line i are of the same} \\ & \text{color} \\ 0 & \text{if there are a mixture of colors of ball in line i} \end{cases}
$$

There are 49 such 0-1 variables.

Constraints:

To ensure that the values of the variables γ_i truly represent the conditions above, we have to model the condition:

$$\gamma_i = 0 \rightarrow \delta_{i1} + \delta_{i2} + \delta_{i3} \geq 1 \qquad \text{and}$$

$$\delta_{i1} + \delta_{i2} + \delta_{i3} \leq 2$$

where i1, i2, and i3 are the numbers of the cells in line i. This condition can be modelled by the constraints

$$\delta_{i1} + \delta_{i2} + \delta_{i3} - \gamma_i \leq 2$$

$$\delta_{i1} + \delta_{i2} + \delta_{i3} + \gamma_i \geq 1$$

$$i = 1,2,\ldots,49$$

In fact these constraints do not ensure that if $\gamma_i = 1$ all balls will be of the same color in the line. When the objective is formulated it will be clear that this condition will be guaranteed by optimality.

In order to limit the black balls to 14 we impose the constraint:

$$\sum_j \delta_j = 14.$$

There are a total of 99 constraints.

Objective:

In order to minimize the number of lines with balls of a similar color we minimize

$$\sum_j \gamma_j.$$

In total this model has 99 constraints and 76 0-1 variables. The model is completed.

The computer solution to the model is:

The minimum number of lines of the same color is 4. There
are many alternative solutions one of which is given in
Figure 2, where the top, middle, and bottom sections of
the cube are given. Cells with black balls are shaded.

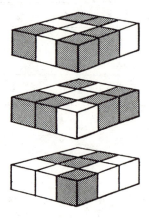

Fig. 2

● **PROBLEM** 11-7

Maximize $f(x_1, x_2) = 3x_1 + 13x_2$

subject to:

$2x_1 + 9x_2 \leq 40$

$11x_1 - 8x_2 \leq 82$

$x_1, x_2 \geq 0$ and integer.

Solve by the simplex method first, ignoring the integer requirements. Try to approximate an integer solution by using the result of the simplex algorithm. Also
solve by graphical means, and compare with the continuous
solution obtained from the simplex tableaux.

Solution: If the integer restrictions are ignored the
simplex algorithm can be applied (after adding slack
variables x_3 and x_4). From the tableaux of Table 1, the
maximum value of f for the simple linear programming
problem is $58 \frac{4}{5}$ and is attained at the point $(9 \frac{1}{5}, 2 \frac{2}{5})$.
Now, for the original problem with the integer restricted
variables, if would seem reasonable that the above solu-
tion point $(9 \frac{1}{2}, 2 \frac{2}{5})$ be rounded off to $(9, 2)$ or maybe
$(10, 2)$, or $(10, 3)$ or $(9, 3)$. However none of these
four points are feasible; the first three do not satisfy
the second inequality, and the last two do not satisfy
the first. Thus the simplex algorithm has provided us

937

with no useful information, and it is not all clear how one could proceed in general.

Table 1

	x_1	x_2	x_3	x_4	
x_3	2	⑨	1	0	40
x_4	11	-8	0	1	82
	-3	-13	0	0	0
x_2	$\frac{2}{9}$	1	$\frac{1}{9}$	0	$4\frac{4}{9}$
x_4	$\left(\frac{11\,5}{9}\right)$	0	$\frac{8}{9}$	1	$117\frac{5}{9}$
	$-\frac{1}{9}$	0	$\frac{13}{9}$	0	$57\frac{7}{9}$
x_2	0	1	$\frac{11}{113}$	$-\frac{2}{113}$	$2\frac{2}{5}$
x_1	1	0	$\frac{8}{113}$	$\frac{9}{113}$	$9\frac{1}{5}$
	0	0	$\frac{167}{113}$	$\frac{1}{113}$	$58\frac{4}{5}$

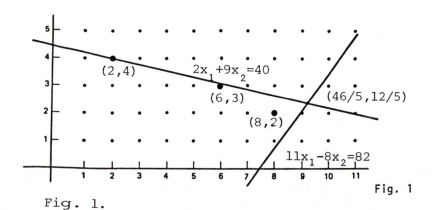

Fig. 1.

The graph for the problem is sketched in Figure 1. There are 36 lattice points (points with both coordinates integer) in the region bounded by the constraints. Since the coefficients of the objective finction are positive, htere can be no lattice points in the feasible region that are to the right or above the point at which the maximal value of the objectine finction is attained, and so the optimal value of f must occur at either (2, 4), (6, 3), or (8, 2). Now f(2, 4) = 58, f(6, 3) = 57, and f(8, 2) = 50, so the maximal value of f is 58 and is attained at the point (2, 4). Contrast here the proximites of the feasible lattice point (8, 2) and the actual solution point (2, 4) to the solution point of the noninteger restricted problem, $(9\frac{1}{5}, 2\frac{2}{5})$, and the difference in values of f at these two points, 50 and 58.

Consider the traveling salesman problem with the following cost data:

		To city				
		1	2	3	4	5
	1	M	20	4	10	25
	2	20	M	5	30	10
From city	3	4	5	M	6	6
	4	10	25	6	M	20
	5	35	10	6	20	M

Solve.

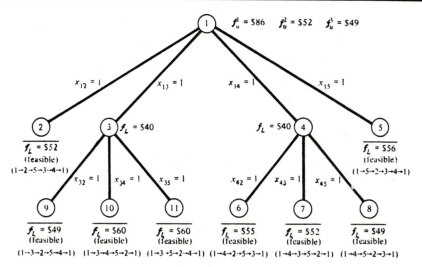

Fig. 1

<u>Solution</u>: The following algorithm is used to solve traveling salesman problems:

Step 1: Solve the original problem as an assignment problem. If the solution is a complete tour, it is optimal. Otherwise, go to step 2.

Step 2: Generate an upper bound on the minimum value of the objective function by finding any feasible tour. Denote this bound by f_U.

Generate the initial branches by setting $x_1 = 1$ for each city $j = 2, 3, \ldots, n$ (unless $c_{1j} = M$, indicating the route is infeasible). Compute a lower bound (f_L) on the minimum value of the objective function at each node as follows. From the original data, delete the first row and the jth column. Next, set $c_{j1} = +M$. Solve the resulting assignment problem and add its cost ($\hat{f}$) to c_{1j}, to yield f_L, that is, $f_L = c_{1j} + \hat{f}$. If $f_L > f_U$, fathom the node and go to step 3.

Step 3: If there are no active nodes, the current incumbent solution is optimal. Otherwise, choose the node with smallest value of f_L, and generate new branches by setting $x_{jk} = 1$ for each city k not previously visited on the partial tour. Go to step 4.

Step 4: Generate bounds f_L at each node by deleting row j and column k from the problem solved at the node immediately above the current node, setting $c_{kj} = M$ and then adding the solution cost $\hat{f}$ to c_{jk} and all earlier costs incurred in the partial tour.

The branch-and-bound tree is shown in Fig. 1.

Step 1: The solution to the assignment problem is $x_{13} = x_{34} = x_{41} = 1$ and $x_{25} = x_{52} = 1$, which contains subtours.

Step 2: Notice that the tour $x_{12} = x_{23} = x_{34} = x_{45} = x_{51} = 1$ is feasible with a cost of $20 + 5 + 6 + 20 + 35 = 86$, so we set $f_U^1 = \$86$.

Generate the initial branches $x_{12} = 1$ (node 2), $x_{13} = 1$ (node 3), $x_{14} = $ (node 4), and $x_{15} = 1$ (node 5).

At node 2 solve the assignment problem resulting by deleting the first row and second column from the original data data and then settting $c_{21} = M$, which yields the problem:

M	5	30	10
4	M	6	6
10	6	M	20
35	6	20	M

The solution is $x_{25} = x_{53} = x_{34} = x_{41} = 1$ with $\hat{f} = \$32$. Since it was assumed $x_{12} = 1$ with a cost of $20, there is a feasible tour with a cost $f_L = \$52$. Set $f_U^2 = \$52$, store this solution as the new incumbent and fathom node 2.

At node 3 delete the first row and third column from the original problem and set $c_{31} = M$:

20	M	30	10
M	5	6	6
10	25	M	20
35	10	20	M

This yields the problem with the solution $x_{34} = x_{41} = 1$ and $x_{25} = x_{52} = 1$ with $\hat{f} = \$36$. Thus, $f_L = c_{13} + \hat{f} = \$4 + \$36 = \$40$, so $f_L < f_U^2$ and node 3 cannot be fathomed.

At node 4 delete the first row and fourth column and set $c_{41} = M$ to obtain:

$$
\begin{array}{cccc}
20 & M & 5 & 10 \\
4 & 5 & M & 6 \\
M & 25 & 6 & 20 \\
35 & 10 & 6 & M
\end{array}
$$

The solution is $x_{43} = x_{31} = 1$ and $x_{25} = x_{52} = 1$ with $\hat{f} = \$30$. Thus, $f_L = \$10 + \hat{f} = \$40 < f_U^2$, so node 4 cannot be fathomed.

At node 5 delete the first row and fifth column and set $c_{51} = M$ to obtain:

$$
\begin{array}{cccc}
20 & M & 5 & 30 \\
4 & 5 & M & 6 \\
10 & 25 & 6 & M \\
M & 10 & 6 & 20
\end{array}
$$

The solution is $x_{52} = x_{23} = x_{34} = x_{41} = 1$ which, when coupled with $x_{15} = 1$, completes a tour with a cost of $f_L = \$56$. Thus node 5 can be fathomed.

Step 3: There are two active nodes (3 and 4) with $f_L = \$40$. Choosing node 4 arbitrarily, generate the branches $x_{42} = 1$ (node 6), $x_{43} = 1$ (node 7), and $x_{45} = 1$ (node 8).

Step 4: Generate the bounds at nodes 6, 7, and 8 by modifying the data at node 4, because these nodes are descendants of (derived from) node 4.

At node 6 delete the row for city 4 and the column for city 2, and set $c_{21} = M$, to obtain:

$$
\begin{array}{ccc}
M & 5 & 10 \\
4 & M & 6 \\
35 & 6 & M
\end{array}
$$

with solution $x_{25} = x_{53} = x_{31} = 1$. Coupled with $x_{14} = x_{42} = 1$, this completes a tour with $f_L = \$55$, so fathom node 6.

At node 7 delete the row for city 4 and the column for city 3, and set $c_{31} = M$, to obtain

$$
\begin{array}{ccc}
20 & M & 10 \\
M & 5 & 6 \\
35 & 10 & M
\end{array}
$$

with solution $x_{35} = x_{52} = x_{21} = 1$. Coupled with $x_{14} = x_{43} = 1$, this complete a tour with $f_L = \$52$, so fathom node 7.

At node 8 delete the row for city 4 and the column for city 5, and set $c_{51} = M$, to obtain:

$$
\begin{array}{ccc}
20 & M & 5 \\
4 & 5 & M \\
M & 10 & 6
\end{array}
$$

with solution $x_{52} = x_{23} = x_{31} = 1$. Since $x_{14} = x_{45} = 1$ above, there is a feasible tour with $f_L = \$49$. Since $f_L < f_U^2$, set $f_U^3 = \$49$ and store the new incumbent.

Step 3: The only remaining active node is node 3. Generate the branches $x_{32} = 1$ (node 9), $x_{34} = 1$ (node 10), and $x_{35} = 1$ (node 11).

Step 4: The calculations for the bounds result in feasible tours at all three nodes, with the values shown. Also find an alternate solution at node 9.

Step 3: There are no active nodes. There are two alternate optimal tours: $1 \to 4 \to 5 \to 2 \to 3 \to 1$, with a cost of $49, and $1 \to 3 \to 2 \to 5 \to 4 \to 1$, also with a cost of $49.

THE THEORY OF GAMES

● **PROBLEM** 11-80

Two players, A and B, each call out one of the numbers 1 and 2 simultaneously. If they both call 1, no payment is made. If they both call 2, B pays A $3.00. If A calls 1 and B calls 2, B pays A $1.00. If A calls 2 and B calls 1, A pays B $1.00.
What is the payoff matrix for this game? Is the game fair to both players?

Solution: This is a two person zero-sum game. It is zero-sum since whatever one player wins, the other must lose. Thus cooperation is impossible.

We construct the payoff matrix by listing A's strategies as rows and B's strategies as columns. The convention is to tabulate the payoffs to the row player.

A \ B	call 1	call 2
call 1	0	1
call 2	-1	3

The game is said to be fair if the value of the game is the same to both players To find the value of the game to A we assume he behaves rationally and see what conclusion he is forced to. A reasons that if he calls 1 his worst payoff is 0. If he calls 2, his worst payoff is $-1. Thus, he will always call 1, so as not to lose anything.
B thinks that: If I call 1 I can at worst draw with A, if I call 2 the worst payoff is $3.00. Thus I should call 1. Since the value of the game is the same to both A and B it is a fair game.

● **PROBLEM** 11-81

1) Two stores, R and C, are planning to locate in one of two towns. Town 1 has 60 percent of the population while town 2 has 40 percent. If both stores locate in the same town they will split the total business of

both towns equally, but if they locate in different towns each will get the business of that town. Where should each store locate?

2) Let us consider an extension of the above example. Stores R and C are trying to locate in one of three towns. The matrix game is:

		Store C locates in		
		1	2	3
Store R	1	50	50	80
locates in	2	50	50	80
	3	20	20	80

The entries in the matrix above represent the percentages of business that store R gets in each case. Where should each store locate?

Solution: 1) By the information given, we know that our payoff matrix is:

		Store C locates in	
		1	2
Store R	1	50	60
locates in	2	40	50

The entries of the matrix represent the percentages of business that store R gets in each case.

Definition: A game defined by a matrix is said to be strictly determined if and only if there is an entry of the matrix that is the smallest element in its row and is also the largest element in its column. This entry is then called a saddle point and is the value of the game.

Therefore, let us find the maxima of the columns and the minima of the rows.

		Store C		
		1	2	Row Minima
Store R	1	50	60	50
	2	40	50	40
Column Maxima		50	60	

Entry $a_{11} = 50$ is a saddle point. Hence it is the best strategy for both stores to locate in Town 1.

2) Let us examine the matrix game for column maxima and row minima.

		Store C locates in				
		1	2	3	Row Minima:	
Store R	1	50	50	80	50	
locates	2	50	50	80	50	(1)
in	3	20	20	50	20	
Column Maxima:		50	50	80		

Note that each of the four 50 entries in the 2 X 2 matrix in the upper left-hand corner of (1) above is a saddle value of the matrix, since each is simultaneously the minimum of its row and maximum of its column. Note the 50 entry in the lower right-hand corner is not a saddle value. The game is strictly determined with optimal strategies:

For store R: "Locate in either town 1 or town 2"
For store C: "Locate in either town 1 or town 2".

A new soda company, Super-Cola, recently entered the market. This company has three choices of advertising campaigns. Their major competitor, Cola-Cola, also has three counter campaigns of advertising to choose from in order to minimize the number of people switching from their soda to the new one. It has been found that their choices of campaign results in the following pay-off matrix:

Number, in 10,000's, of people switching from Cola-Cola to Super-Cola.			
	Cola-Cola		
Super-Cola	Counter-Compaign 1	Counter-Compaign 2	Counter-Compaign 3
Campaign 1	2	3	7
Campaign 2	1	4	6
Campaign 3	9	5	8

Find the best strategies for Super-Cola and Cola-Cola.

Solution: Each company wishes the strategy that is best for them. Super-Cola wishes to get the maximum amount of people from Cola-Cola, and Cola-Cola wishes to minimize their losses to Super-Cola. To do this, we use the minimax procedure. Super-Cola realizes that Cola-Cola will always look for the minimum losses, thus Super-Cola considers the minimum of what will happen for each choice of campaign. Super-Cola notices that the minimum gain for campaign 1 is 20,000 people, the minimum gain for campaign 2 is 10,000 people, but the minimum gain for campaign 3 is 50,000 people. Thus, Super-Cola will choose campaign 3 - the maximum of the minimums. Similarly Cola-Cola realizes that Super-Cola will want to choose the maximum for each of Cola-Cola's counter-campaigns. Thus, Cola-Cola only looks at the maximums. For counter-campaign 1 the maximum loss is 90,000 people, for counter-campaign 2 the maximum loss is 50,000 people, and for counter-campaign 3 the maximum loss is 80,000 people. Thus, Cola-Cola will choose counter-campaign 2 which is the minimum of the maximums. In this way they will minimize the losses. The point on the payoff matrix which they both choose is called the "saddle point". At this point neither company will change it's strategy for they

are doing the best that they can. This type of a
"game" is called a two-player zero-sum game, because whatever
one player wins, the other player loses. Thus, the
algebraic sum of the two is zero. Another way of looking
at this problem is "pure-strategy". Super-Cola will
look at the matrix and note that campaign 3 contains
the largest numbers in each column. Thus, campaign 3
is the best choice regardless of which counter-campaign
Cola-Cola chooses. Cola-Cola will notice this also.
They will choose counter-campaign 2, for that one
contains the minimum of all their choices, given Super-Cola
will choose campaign 3.

● **PROBLEM** 11-83

Players A and B simultaneously call out either of the numbers 1 and
2. If their sum is even, B pays A that number of dollars, if odd, A
pays B. What kind of strategy should both players adopt?

Solution: Games may be classified according to the following criteria:
1) Number of players
2) Number of moves
3) Whether or not they are zero-sum
4) Whether or not they are of full information

Here the number of players is two. There are two moves to the game,
A's move and B's move. The game is zero-sum since A's gain is B's loss
and vice-versa. Finally, a game is said to be of full-information if, at
each stage of the game, all previous moves are known to both players. The
given game is not of full-information since both players play simultaneously.
The analysis of a game is simplified by constructing its payoff matrix.
For the given game

	B:	1	2	Row Min.
A:				
	1	2	-3	-3
	2	-3	4	-3
col. max		2	4	

The optimal strategy for A is the maximum value of the game, i.e., the
maximum of the row minima. By calling out either 1 or 2 his maximum
value is -3. For B's optimum strategy we have the minimax value of the
game +2 (from calling 1). Thus A expects to lose 3 while B expects
to lose 2 (recall that positive entries in the payoff matrix are payments
by B to A). But since this is a zero-sum game, the above conclusion
cannot be true. The paradox arises because the game has no saddle point.
When there is one element that will clearly be chosen by both players, it
is called a saddle point. Thus, there is no predictable solution to a
single game. Over many games, both players play their strategies in a
random manner, the frequencies being chosen to give them their best payoffs over many games. We say that the players have moved from pure
strategies to mixed strategies. Note that if either player chose a pure
strategy he would be bound to lose over many games. Thus if B persists
in calling 1, A will continue to call 1.

Solve the following game

B:	B_1	B_2	B_3
A			
A_1	1	2	3
A_2	0	3	-1
A_3	-1	-2	4

Solution: From A's point of view, none of the strategies dominate each other. Similarly from B's point each strategy has some reward that cannot be exceeded by the other strategies. Thus, this is an irreducible 3 x 3 payoff matrix.

A knows that if he adopts strategy A_1 the worst that can happen is that he will receive 1 (if B plays B_1). Similarly the minimum returns from A_2 and A_3 are -1 and -2 respectively. Since A wants to maximize his payoff, he chooses the strategy that will yield him the maximum payoff amongst the minimum returns. Thus, he will play strategy A, where maximin = 1.

B, on the other hand wants to minimize the amount that he must pay A. If he plays B_1 the most he must pay A is 1 (if A plays A_1). Similarly, the maximum payments from playing B_2 and B_3 are 3 and 4 respectively (negative entries represent payments by A to B). B chooses the strategy that will yield him the minimum penalty amongst the maximum payments. Thus, he will play B_1 since minimax = 1.

The maximin and the minimax represent the values of the game to A and B, respectively. When they are equal, as in this case, the common value is the value of the game and is called a saddlepoint.

Every two-person zero-sum game with full information (i.e., each player knows the strategies of his opponent) has a saddlepoint. But if maximin $\neq$ minimax, the value of a single game is uncertain. By introducing the expected value of a game (i.e., the average value of the game when many games are played) even games without full information can be shown to have a unique value.

Consider the following payoff matrix:

	C_1	C_2	C_3	C_4
R_1	2	3	-3	2
R_2	1	3	5	2
R_3	9	5	8	10

Find the value of this game.

Solution: The game is in matrix form. Since there are two players, (R the row player and C the column player) and four choices for C with three choices for R, the matrix is of order 3 x 4. Each entry is considered as a payment by C to R if C pursues action C_j while R pursues action R_i. The value of the game is the choice that R and C make, provided this choice is common.

Assuming R and C behave rationally, we would observe the following behaviour. R, the row player reasons as follows: If I pick row 1, the worst that can happen is that I collect -$3 (this means R pays C, the column player, $3), if I choose row 2 the minimum I receive is $1.00 and if I select row 3 my minimum payoff will be $5.00. Thus, considering my three possible choices, selecting row 3 guarantees that I will receive $5.00 from C. This is the maximum of the minimum payoffs.

C, on the other hand, will reason as follows: If I pick column 1, it is possible that I may have to pay R $9.00. If I choose column 2, I may have to pay R $5.00, while choosing column 3 means the maximum I have to pay R is $8.00. Finally, if I decide to play column 4, my maximum payoff to R will be $10.00. Thus, to minimize the maximum payoff to R, I must choose column 2.

Hence R chooses row 3, C chooses column 2 and R receives a $5.00 payoff from C. The value of the game is 5.

● **PROBLEM** 11-86

Find the general solution to a 2 x 2 game using geometric methods.

Solution: The general 2 x 2 payoff matrix is:

$$
\begin{array}{ccc}
\text{B:} & B_1 & B_2 \\
\text{A:} & & \\
A_1: & a_{11} & a_{12} \\
A_2: & a_{21} & a_{22}
\end{array}
$$

The payoff matrix may or may not have a saddlepoint. Assume it does not have a saddlepoint. If B just plays B_1 and A plays A_1 and A_2 in proportions p and 1-p (not necessarily optimal), then A will receive, on average,

$$V = a_{11}p + a_{21}(1-p) \tag{1}$$

We can graph (1) treating the a's as fixed, but the p's as unknowns. When (1-p) = 0, $V = a_{11}$. When (1-p) = 1, $V = a_{21}$. All other points in Fig. 1 are mixed strategies. Let

$$
S_A^* = \begin{Bmatrix} A_1 & A_2 \\ p & (1-p) \end{Bmatrix} \qquad
S_B^* = \begin{Bmatrix} B_1 & B_2 \\ q & (1-q) \end{Bmatrix}
$$

be the optimal strategies for A and B. Thus, according to S_A^*, A plays A_1 with proportion p and A_2 the remaining proportion of the time. An important result in game theory is that if a player adopts an optimal mixed strategy, the value of the game to him does not depend on what strategy the other player chooses. Using this result we can solve for p and V from S_A^*:

947

$$V = a_{11}p + a_{21}(1-p) \qquad \text{(B plays } B_1 \text{ always)}$$

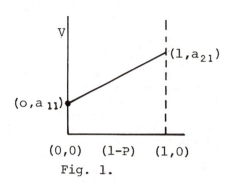

Fig. 1.

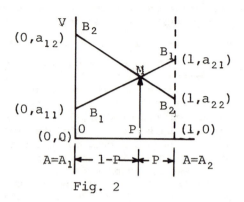

Fig. 2

$$V = a_{12}p + a_{22}(1-p) \qquad \text{(B plays } B_2 \text{ always)}$$

We plot both of these lines (Fig. 2). The solution set is the point of intersection, $p = PI$, $1-p = OP$, $V = PM$.

Suppose A were to choose strategy A_2 with probability less than $1-p$. Then B, by choosing strategy B_1B_1 will lose less than MP. Similarly if A chooses $1-p$ greater than OP, B can lessen $V(A)$ by playing on B_2B_2. Hence A will choose $(1-p) = OP$ since this will guarantee him most. A similar line of argument will give us B's optimum strategy from Fig. 3 (note the relabelling of the axes).

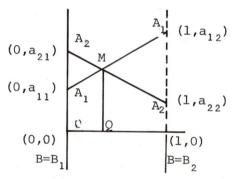

Fig. 3.

● **PROBLEM 11-87**

Simplify the following payoff matrix

A's strategies	B's strategies B_1	B_2	B_3	B_4
A_1	0	-1	2	-4
A_2	1	3	3	6
A_3	2	-4	5	1

948

Solution: It is usual to try to reduce the size of a game in order to solve it more easily. There are two basic ways in which this may be accomplished. We eliminate:
1)   Duplicate strategies
2)   Dominated strategies

We can use 2) in the given problem to reduce the size of the payoff matrix. The strategy A_i is said to be dominated by A_j if the payoffs to A in A_i are less than or equal to the payoffs to A in A_j. It is strictly dominated if the inequality is a strict inequality. We see that if A_j dominates A_i, then A will never play A_j, assuming rational behaviour.

   Examining the given payoff matrix, A_1 is strictly dominated by A_2 since every element in A_1 is less than the corresponding element in A_2. Hence A_1 may be eliminated from contention.

   Similarly, from B the column player's point of view, a strategy B_i is dominated by B_j if the payoffs to B in B_i are less than or equal to the payoffs to B in B_j. In the given payoff matrix B_2 dominates B_3 since the payoffs to B from playing B_2 are always greater or equal to the payoffs to B from playing B_3 (recall that a negative entry represents a payment by A to B and that positive elements are payments by B to A ; hence from B's point of view, smaller numbers are preferable to larger numbers). Thus we obtain the reduced payoff matrix.'

		B:	B_1	B_2	B_4
A:					
	A_2		1	3	6
	A_3		2	-4	1

But now, since A_1 was eliminated, B_2 strictly dominates B_4 and further reduction is possible:

	B:	B_1	B_2
A:			
A_2		1	3
A_3		2	-4

Thus the matrix has been reduced from one of order 3×4 to a payoff matrix of order 2×2 .

● **PROBLEM** 11-88

Consider the general 2×2 matrix game:

	B:	B_1	B_3
A:			
A_1		a_{11}	a_{12}
A_2		a_{21}	a_{22}

Solution: The game has a saddle point if the maximum of the minimum payoffs for A equals the minimum of the maximum payoffs for B. Let a_{11} be the saddle point for the game above. Thus a_{11} is less than or equal to $\max(a_{12}, a_{22})$. Further, a_{11} is the row minimum of A_1, giving

$$a_{11} \leq a_{12} . \tag{1}$$

Since a_{11} must be the column maximum of B_1,

$$a_{11} \geq a_{21} . \tag{2}$$

Also, $a_{11} \leq \max(a_{12}, a_{22})$, and $a_{11} \geq \min(a_{21}, a_{22})$. Now, either $a_{12} \geq a_{22}$ or $a_{12} \leq a_{22}$. If $a_{12} \geq a_{22}$, since $a_{11} \geq a_{21}$, A_1 dominates A_2. If $a_{12} \leq a_{22}$, since $a_{11} \leq a_{22}$ but $a_{11} \geq a_{21}$ we see that $a_{21} \leq a_{22}$. But from (1), $a_{11} \leq a_{12}$, hence B_1 dominates B_2. Thus the original payoff matrix reduces to the single saddle point a_{11}. For higher order games it is generally not true that a game with a saddle point must have a dominance.

● **PROBLEM** 11-89

Give an example of a two-person non-zero-sum game.

Solution: The fundamental theorem of two-person games is that every zero-sum game with mixed strategies has a unique and identical value to both opponents. However, when the game is non-zero sum, some solutions will yield more joint satisfaction to the participants than others. Thus co-operation becomes possible and the solutions are non-unique.

A typical two person non-zero sum game is the "prisoner's dilemma" a parable designed to illustrate that rational behaviour is not always the most satisfying.

Assume two people, A and B, are being accused of jointly engineering a crime. They are kept in separate cells and separately called for an interview with the Grand Inquisitor who presents them with the following alternatives: If you both admit responsibility for the crime you will both be sentenced to 5 years incarceration. If one of you denies complicity while the other admits guilt, then the one who claims to be innocent will receive 10 years while the confesee will be set free. Finally, if both of you deny having committed the crime, you will both receive 2 years each. Are you innocent or guilty?

This is a non-zero sum game without full information and has the payoff matrix given below:

	Admit (B_1)	Deny (B_2)
Admit (A_1)	-5 \\ -5	-10 \\ 0
Deny (A_2)	0 \\ -10	-2 \\ -2

B: (column header) A: (row header)

From A's point of view, the strategy A_1 dominates A_2. Hence A will choose A_1, i.e., he will admit that he and B committed the crime. From B's point of view, the strategy B_1 dominates B_2 and hence B will also admit his guilt. Thus, by acting rationally, A and B will receive 5 years each. But from the payoff table, this is non-optimal since they both could have obtained 2 years each.

● **PROBLEM** 11-90

Show how a game with the payoff matrix below can be converted to a linear programming problem.

	B's strategies	
A's strategies	2	4
	6	1

Solution: Since A's maximin strategy is to play row 1 while B's minimax strategy is to play column 2, we see that maximin ≠ minimax, i.e., no saddle point exists. However, if we let A choose row 1 with probability p and row 2 with probability 1-p then we can compute the expected value of the game to A. Now A has an expected value E_{c_1} against the column player playing column 1 of

$$E_{c_1} = 2p + 6(1-p) = -4p + 6.$$

Similarly, $E_{c_2} = 4p + (1)(1-p) = 3p + 1$. We can graph E_{c_i} versus p: (See fig.)

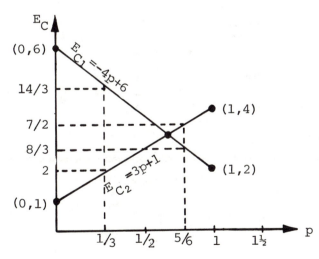

For any choice of p, $0 \le p \le 1$, $E_{c_1}, E_{c_2} > 0$. Let M denote the minimum expected value to the row player corresponding to his choosing row 1 with probability p and row 2 with probability 1-p. For example, if p = 1/3, $M = E_{c_2}$ since $E_{c_2} = 2$ is smaller than $E_{c_1} = 14/3$. Conversely, if

$p = 5/6$, $M = E_{c_1}$ since $E_{c_1} = 10/3$ is smaller than $E_{c_2} = 7/2$. Define s and t as follows:

$$s = p/M \qquad t = \frac{1-p}{M} . \tag{1}$$

Then, $s+t = \frac{p}{M} + \frac{1-p}{M} = 1/M$. Thus, the maximization of M is equivalent to the minimization of $s+t$ where s and t are both non-negative. Further restrictions on s and t are obtained by noting that for any p, E_{c_1} and $E_{c_2} \geq M$, i.e.,

$$2p + 6(1-p) \geq M \tag{2}$$
$$4p + (1)(1-p) \geq M .$$

But, from (1), $p = sM$ and $1-p = tM$. Hence, the inequalities (2) become

$$2sM + 6tM \geq M$$
$$4sM + tM \geq M .$$

Dividing through by M,

$$2s + 6t \geq 1$$
$$4s + t \geq 1 .$$

Thus, A's problem is:

Minimize $s + t$

subject to

$$2s + 6t \geq 1$$
$$4s + t \geq 1 \tag{3}$$
$$s \geq 0, t \geq 0 .$$

By analogous logic, B's problem is:

Maximize $x + y$

subject to

$$2x + 4y \leq 1$$
$$6x + y \leq 1 \tag{4}$$
$$x \geq 0, y \geq 0.$$

Note that (4) is the dual of (3). From the theory of duality we know that if they exist, the optimal values of (3) and (4) are the same. From game theory we know that the value of a game computed from either player's point of view must coincide.

Finally, George Dantizig, (the inventor of the simplex method), has shown that any game can be converted to a linear program and, conversely, any linear program can be converted to a game.

● **PROBLEM** 11-91

Consider a game with the following payoff matrix:

B:		B_1	B_2	min
A:	A_1:	2	-3	-3
	A_2:	-3	4	-3
max		2	4	

Find the value of the game when both players use mixed strategies.

Solution: The game has no saddle point since maximin = 2 $\neq$ minimax = -3.

Thus for a single game, there is no predictable value. Now a fundamental result in the theory of games is that by using mixed strategies every finite two-person zero-sum game has a solution, this solution being at the same time the best for both players. To find the common value of the game we reason as follows: Let

$$S_A^* = \begin{Bmatrix} A_1 & A_2 \\ p & 1-p \end{Bmatrix} \qquad S_B^* = \begin{Bmatrix} B_1 & B_2 \\ q & 1-q \end{Bmatrix} \qquad (1)$$

S_A^* in (1) is to be interpreted as saying that the strategy of A is to play A_1 with probability p and A_2 the remaining proportion of the time. S_B^* is the strategy of B. If S_A^* is optimal than V(A) does not depend on the frequencies with which B_1, B_2 are used, where V(A) denotes the value of the game to A. Hence,

$$V = p(2) + (1-p)(-3) \qquad \text{(if B uses } B_1 \text{ only)}$$
$$V = p(-3) + (1-p)(4) \qquad \text{(if B uses } B_2 \text{ only).}$$

Equating these we get

$$p = 7/12, \quad (1-p) = 5/12 \quad \text{and} \quad V = -1/12 .$$

Similarly, if B keeps to S_B^* , then

$$V = 2(2) + (1-q)(-3) \qquad \text{(A uses } A_1)$$
$$V = q(-3) + (1-q)(4) \qquad \text{(A uses } A_2)$$

giving $q = 7/12$, $(1-q) = 5/12$ and $V = -1/12$. Hence

$$S_A^* \begin{Bmatrix} A_1 & A_2 \\ 7/12 & 5/12 \end{Bmatrix} \qquad S_B^* = \begin{Bmatrix} B_1 & B_2 \\ 7/12 & 5/12 \end{Bmatrix}$$

and the value of the game is $-1/12$ to A, i.e., B will win $1/12$.

● PROBLEM 11-92

How would you solve a game with the following payoff matrix:

B:		B_1	B_2	B_3
A:				
	A_1:	-1	0	1
	A_2:	3	2	-1
	A_3:	-3	1	0

Solution: Since the maximin of A = 1 ≠ minimax of B = 1, the game has no saddle point. Furthermore, it has no dominance. Hence, it is an irreducible 3 × 3 game and cannot be solved by graphical methods. To find a method of solution, first generalize to the general solution of an m × n zero-sum game. Consider the general m × n zero-sum game: A has moves $A_1, \ldots, A_m$; B has moves $B_1, \ldots, B_n$. The payoff matrix to A is then

$$
\begin{array}{cccccccc}
\text{B:} & & B_1 & B_2 & \cdots & B_j & \cdots & B_n \\
\text{A:} & A_1 & a_{11} & a_{12} & \cdots & a_{1j} & \cdots & a_{1n} \\
& A_2 & a_{21} & a_{22} & \cdots & a_{2j} & \cdots & a_{2n} \\
& \vdots & \vdots & & & & & \vdots \\
& A_i & a_{i1} & a_{i2} & \cdots & a_{ij} & \cdots & a_{in} \\
& \vdots & \vdots & & & & & \vdots \\
& A_m & a_{m1} & a_{m2} & \cdots & a_{mj} & \cdots & a_{mn}
\end{array}
$$

Let S_A^* be A's optimum mixed strategy where the moves $A_1, A_2, \ldots, A_m$ are played with probability $p_1, p_2, \ldots, p_m$. Thus

$$
S_A^* = \left\{ \begin{array}{cccc} A_1 & A_2 & \cdots & A_m \\ p_1 & p_2 & \cdots & p_m \end{array} \right\}
$$

where $p_1 + p_2 + \cdots + p_m = 1$; $p_i \geq 0$ $(i = 1, 2, \ldots, m)$. Similarly,

$$
S_B^* = \left\{ \begin{array}{cccc} B_1 & B_2 & \cdots & B_n \\ q_1 & q_2 & \cdots & q_n \end{array} \right\}
$$

where

$$
q_1 + q_2 + \cdots + q_n = 1,
$$
$$
q_j \geq 0 \quad (j = 1, 2, \ldots, n).
$$

Assume that the value of the game to A is positive (this can be ensured by adding some constant to every element in the payoff matrix so that every element is positive. Let $V(A)$ be the value of the game to A. Assume B plays only the pure strategy B_j. Then the average payoff to A is:

$$
p_1 a_{1j} + p_2 a_{2j} + \cdots + p_m a_{mj} .
$$

This is not less than V since B_j need not necessarily be S_B^*, the optimal strategy for B. Thus,

$$
p_1 a_{1j} + p_2 a_{2j} + \cdots + p_m a_{mj} \geq V. \tag{1}
$$

The relationship (1) must be true for all j $(j = 1, \ldots, n)$. Put $x_1 = p_1/V$, $x_2 = p_2/V, \ldots, x_m = p_m/V$. Then (1) becomes

$$
a_{1j} x_1 + a_{2j} x_2 + \cdots + a_{mj} x_j \geq 1 \quad (j = 1, 2, \ldots, n).
$$

The condition $\sum\limits_{i=1}^{m} p_i = 1$ becomes

$$
x_1 + x_2 + \cdots + x_m = 1/V, \quad x_i \geq 0, \quad (i = 1, 2, \ldots, m).
$$

A wishes to maximize $V(A)$, i.e., minimize $1/V$. Thus the game problem becomes the standard linear programming problem.

$$
\text{Minimize} \quad (1/V) = x_1 + x_2 + \cdots + x_m \tag{2}
$$

subject to

$$a_{11}x_1 + a_{21}x_2 + \ldots + a_{m1}x_m \geq 1$$

$$a_{12}x_1 + a_{22}x_2 + \ldots + a_{m2}x_m \geq 1 \qquad\qquad (3)$$

$$\cdot \qquad\qquad\qquad \cdot$$
$$\cdot \qquad\qquad\qquad \cdot$$
$$\cdot \qquad\qquad\qquad \cdot$$

$$a_{1n}x_1 + a_{2n}x_n + \ldots + a_{mn}x_m \geq 1$$

$x_1 \geq 0$, $x_2 \geq 0, \ldots, x_m \geq 0$. Looking at the problem from B's point of view, the value of the game is $-V$ and, as before,

$$q_1 b_{1j} + q_2 b_{2j} + \ldots + q_n b_{nj} \geq - V$$

where the b's are the elements of B's payoff matrix ($b_{ij} = -a_{ji}$), and

$$q_1 + q_2 + \ldots + q_n = 1.$$

Putting $y_1 = q_1/V$, $y_2 = q_2/V, \ldots, y_m = q_m/V$ we obtain

$$y_1 b_{1j} + y_2 b_{2j} + \ldots + y_n b_{nj} \geq -1, \; (j = 1, \ldots, m), \qquad (4)$$

subject to $y_1 + y_2 + \ldots + y_n = 1/V$, $y_j \geq 0$ $(j = 1, \ldots, m)$. Now express the b_{ij} in terms of the payoff matrix elements a_{ji} . Then (4) becomes $-y_1 a_{j1} - y_2 a_{j2} - \ldots - y_n a_{jn} \geq -1$ or

$$a_{j1}y_1 + a_{j2}y_2 + \ldots + a_{jn}y_n \leq 1 \quad (j = 1, \ldots, m).$$

B wishes to minimize V, i.e., maximize $1/V$. Hence his problem is:

$$\text{Maximize} \quad (1/V) = y_1 + y_2 + \ldots + y_n \qquad\qquad (5)$$

subject to

$$a_{11}y_1 + a_{12}y_2 + \ldots + a_{1n}y_n \leq 1$$

$$a_{21}y_1 + a_{22}y_2 + \ldots + a_{2n}y_n \leq 1 \qquad\qquad (6)$$

$$\cdot$$
$$\cdot$$
$$\cdot$$

$$a_{m1}y_1 + a_{m2}y_2 + \ldots + a_{mn}y_n \leq 1$$

$y_1 \geq 0$, $y_2 \geq 0, \ldots, y_n \geq 0$. Observe that the two programs (2), (3) and (5), (6) are duals of each other. Thus the optimal solutions to these two linear programming problems must be such that they both give the same value for $1/V$. Turning to the given problem, we first ensure that V is positive by adding 4 to all terms in the original payoff matrix. We thus obtain the positive matrix:

$$
\begin{array}{c c c c c}
 & \text{B:} & B_1 & B_2 & B_3 \\
A = & A_1: & 3 & 4 & 5 \\
 & A_2: & 7 & 6 & 3 \\
 & A_3: & 1 & 5 & 4 \\
\end{array}
$$

Then the solution to the game is the solution to the linear programming problem

$$\text{Minimize} \quad 1/V = x_1 + x_2 + x_3$$

subject to

$$3x_1 + 7x_2 + x_3 \geq 1$$

$$4x_1 + 6x_2 + 5x_3 \geq 1$$

$$5x_1 + 3x_2 + 4x_3 \geq 1$$

$$x_1, x_2, x_3 \geq 0 .$$

● **PROBLEM** 11-93

Consider the game with the payoff matrix shown in Table 1. The payoff matrix has no saddle point; therefore in this case A and B do not have single best plans as their best strategies. Consequently, each player has to devise some mixed strategy in order to maximize his gain or minimize his loss.

What is the best mixed strategy for the players? Can A insure some minimum gain and what is this gain? Similarly, can B insure that he will not lose more than some maximum amount?

<u>Solution</u>: Let A play P with the frequency x, and Q with a frequency (1 - x). Then, if B plays S all the time, A's gain will be

$$g(A, S) = x(-3) + (1 - x)6 = 6 - 9x$$

If B plays T all the time, A's gain will be

$$g(A, T) = x(7) + (1 - x)1 = 1 + 6x$$

Table 1

		B	
	Plan	S	T
A	P	−3	7
	Q	6	1

It can be shown mathematically that if A chooses x, so that g (A, S) = g(A, T), then this will lead to the best strategy for him. Thus

$$6 - 9x = 1 + 6x$$

$$5 = 15x$$

i.e.

$$x = \frac{1}{3}$$

One gets the result

$$g(A) = \frac{1}{3}(-3) + \frac{2}{3}(6) = \$3.00$$

Thus, regardless of the frequency with which B plays either S or T, A's gain will be $3.00.

Therefore, by choice of frequencies $\frac{1}{3}$ and $\frac{2}{3}$, A can assure himself a gain of $3.00.

The same method can be applied by player B. Let frequency of choice of S be denoted by y and that of T be denoted by (1 - y). For best strategy one has

$$g(B, P) = y(-3) + (1 - y)7 = y(6) + (1 - y)1 = g(B, Q)$$

$$7 - 10y = 1 + 5y$$

$$6 = 15y$$

$$y = \frac{2}{5}$$

$$1 - y = \frac{3}{5}$$

$$g(B) = \frac{2}{5}(-3) + \frac{3}{5}(7) = \$3.00$$

Note that g(A) = g(B), as expected for a zero-sum game. A complete solution of the given game is:

1. A should play P and Q with frequencies $\frac{1}{3}$ and $\frac{2}{3}$ respectively.

2. B should play S and T with frequencies $\frac{2}{5}$ and $\frac{3}{5}$ respectively.

3. The value of the game is $3.00.

● **PROBLEM** 11-94

In a game of matching coins with two players, suppose A wins one unit of value when there are two heads, wins nothing when there are two tails, and loses ½ unit of value when there are one head and one tail. Determine the payoff matrix, the best strategies for each player, and the value of the game to A.

Solution: The payoff matrix (for A) is seen to be:

957

$$
\begin{array}{c}
\text{B}\\[2pt]
\begin{array}{cc}
\text{H} & \text{T}
\end{array}\\[2pt]
\text{A}\quad
\begin{array}{c|cc}
\text{H} & +1 & -\tfrac{1}{2}\\[6pt]
\text{T} & -\tfrac{1}{2} & 0
\end{array}
\end{array}
$$

Since there is no saddle point, it is known that the optimal strategies will be mixed strategies. The solution is obtained most easily by use of the formulas

$$
\frac{x_1}{x_2} = \frac{a_{22} - a_{21}}{a_{11} - a_{12}}, \qquad
\frac{y_1}{y_2} = \frac{a_{22} - a_{12}}{a_{11} - a_{21}},
$$

and

$$
v = \frac{a_{11}a_{22} - a_{12}a_{21}}{a_{11} + a_{22} - (a_{12} + a_{21})};
$$

where a_{ij}s refer to values in the payoff matrix for A as such:

$$
\begin{bmatrix}
a_{11} & a_{12}\\[6pt]
a_{21} & a_{22}
\end{bmatrix}
$$

and x_1: probability of A choosing row 1, x_2: probability of A choosing row 2, y_1: probability of B choosing column 1, y_2: probability of B choosing column 2, v: units of value that A will gain each time the game is played.

Thus

$$
\frac{x_1}{x_2} = \frac{1}{3}
$$

So

$$
x_1 = \frac{1}{4}, \qquad x_2 = \frac{3}{4}
$$

$$
\frac{y_1}{y_2} = \frac{1}{3}, \qquad y_1 = \frac{1}{4}, \qquad y_2 = \frac{3}{4}
$$

$$
v = \frac{0 - \frac{1}{4}}{1 + 1} = -\frac{1}{8}
$$

Thus each player should show heads 1/4 of the time and tails 3/4 of the time. The game is unfair to A, as he will lose on average 1/8 unit each time the game is played.

Consider the payoff table in which player I has only two pure strategies:

			II		
	Probability		y_1	y_2	y_3
Probability		Pure strategy	1	2	3
I	x_1	1	0	-2	2
	$1-x_1$	2	5	4	-3

Maximize the minimum expected payoffs of both players.

<u>Solution</u>: Since his mixed strategies are (x_1, x_2) and $x_2 = 1 - x_1$, it is only necessary for him to solve for the optimal value of x_1. However, it is straightforward to plot the expected payoff as a function of x_1 for each of his opponent's pure strategies. This graph can then be used to identify the point that maximizes the minimum expected payoff. The opponent's minimax mixed strategy can also be identified from the graph.

For each of the pure strategies available to player II the expected payoff for player I would be

(y_1, y_2, y_3)	Expected payoff
$(1, 0, 0)$	$0x_1 + 5(1 - x_1) = 5 - 5x_1$
$(0, 1, 0)$	$-2x_1 + 4(1 - x_1) = 4 - 6x_1$
$(0, 0, 1)$	$2x_1 - 3(1 - x_1) = -3 + 5x_1$

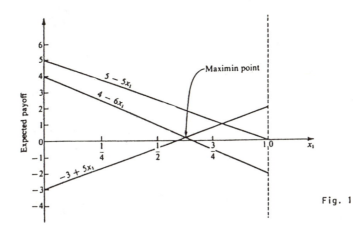

Fig. 1

Now plot these expected payoff lines on a graph, as shown in Fig. 1. For any given value of x_1 and of (y_1, y_2, y_3), the expected payoff will be the appropriate weighted average of the corresponding points on these three lines. In particular

$$\text{Expected payoff} = y_1(5 - 5x_1) + y_2(4 - 6x_1)$$
$$+ y_3(-3 + 5x_1).$$

Thus, given x_1, the minimum expected payoff is given by the corresponding point on the "bottom" line. According to the minimax (or maximin) criterion, player I should select the value of x_1 given the largest minimum expected payoff, so that

$$\underline{v} = v = \max_{0 \leq x_1 \leq 1} \{\min(-3 + 5x_1, 4 - 6x_1)\}.$$

Therefore the optimal value of x_1 is the one at the intersection of the two lines $(-3 + 5x_1, 4 - 6x_1)$. Solving algebraically,

$$-3 + 5x_1 = 4 - 6x_1,$$

so that $x_1 = 7/11$; thus $(x_1, x_2) = (7/11, 4/11)$ is the optimal mixed strategy for player I, and

$$\underline{v} = v = -3 + 5\left(\frac{7}{11}\right) = \frac{2}{11}$$

is the value of the game.

To find the corresponding optimal mixed strategy for player II, one would now reason as follows. According to the definition of upper value and the minimax theorem, the expected payoff resulting from this strategy $(y_1, y_2, y_3) = (y_1^*, y_2^*, y_3^*)$ will satisfy the condition,

$$y_1^*(5 - 5x_1) + y_2^*(4 - 6x_1) + y_3^*(-3 + 5x_1) \leq \overline{v} = v = \frac{2}{11}$$

for all values of $x_1 (0 \leq x_1 \leq 1)$; furthermore, when player I is playing optimally (that is, $x_1 = 7/11$), this inequality will be an equality, so that

$$\frac{20}{11}y_1^* + \frac{2}{11}y_2^* + \frac{2}{11}y_3^* = v = \frac{2}{11}.$$

Since (y_1, y_2, y_3) is a probability distribution, it is also known that

$$y_1^* + y_2^* + y_3^* = 1.$$

Therefore $y_1^* = 0$ because $y_1^* > 0$ would violate the next-to-last equation, i.e., the expected payoff on the graph at $x_1 = 7/11$ would be above the maximin point. (In general, any line that does not pass through the maximin must be given a zero weight to avoid increasing the expected payoff above this point.) Hence,

$$y_2^*(4 - 6x_1) + y_3^*(-3 + 5x_1) \begin{cases} \leq \dfrac{2}{11}, & \text{for } 0 \leq x_1 \leq 1 \\[2mm] = \dfrac{2}{11}, & \text{for } x_1 = \dfrac{7}{11}. \end{cases}$$

But y_2^* and y_3^* are numbers, so the left-hand side is the equation of a straight line, which is a fixed weighted average of the two "bottom" lines on the graph. Since the ordinate of this line must equal 2/11 at $x_1 = 7/11$, and it must never exceed 2/11, the line necessarily is horizontal. (This conclusion is always true, unless the optimal value of x_1 is either zero or 1, in which case player II also should use a single pure strategy.) Therefore

$$y_2^*(4 - 6x_1) + y_3^*(-3 + 5x_1) = \frac{2}{11}, \text{ for } 0 \leq x \leq 1.$$

Hence, to solve for y_2^* and y_3^*, select two values of x_1 (say, zero and 1), and solve the resulting two simultaneous equations. Thus

$$4y_2^* - 3y_3^* = \frac{2}{11},$$

$$-2y_2^* + 2y_3^* = \frac{2}{11},$$

so that $y_3^* = 6/11$ and $y_2^* = 5/11$. Therefore the optimal mixed strategy for player II is $(y_1, y_2, y_3) = (0, 5/11, 6/11)$.

● **PROBLEM** 11-96

Consider the following game. Two players A and B must each select a number out of 1, 2, or 3. If both have chosen the same number, A will pay B the amount of the chosen number. Otherwise A receives the amount of his own number from B. The payoff table for this game is shown in Table 1. What are the best strategies for A and B? Use LInear Programming to solve.

Table 1

A	1	2	3
		B	
1	-1^B	1	1
2	2	-2^B	2^A
3	3^A	3^A	-3^B

The linear program for A's best random strategy, x_1, x_2, x_3 (fractions of time during which row 1, 2, and 3 will be chosen so as to maximize A's gains) is

$$\max v$$

subject to

$$-x_1 + 2x_2 + 3x_3 \geq v$$

$$x_1 - 2x_2 + 3x_3 \geq v$$

$$x_1 + 2x_2 - 3x_3 \geq v$$

$$x_1 + x_2 + x_3 = 1$$

and

$$x_1, \; x_2, \; x_3 \geq 0$$

The optimal v could be negative because the payoff table does contain negative payoffs. However, if the number 3 is added to all a_{ij}'s, then all will be nonnegative. Then define $v' = v + 3$ or $v = v' - 3$. The problem can be rewritten with all non-negative variables as

$$\max v' - 3$$

subject to

$$2x_1 + 5x_2 + 6x_3 \geq v'$$

$$4x_1 + x_2 + 6x_3 \geq v'$$

$$4x_1 + 5x_2 \qquad \geq v'$$

$$x_1 + x_2 + x_3 = 1$$

$$x_1, \; x_2, \; x_3, \; v' \geq 0$$

It is best with hand calculation to eliminate the equality constraint by substituting

$$x_3 = 1 - x_1 + x_2$$

and replacing the nonnegativity condition $x_3 \geq 0$ by

$$x_1 + x_2 \leq 1$$

The resulting problem is

$$\max v' - 3$$

subject to

$$4x_1 + x_2 + v' \leq 6$$

$$2x_1 + 5x_2 + v' \leq 6$$

$$4x_1 + 5x_2 - v' \geq 0$$

$$x_1 + x_2 \leq 1$$

and

$$x_1, x_2, v' \geq 0$$

Slack variables are then used to convert the four inequality constraints to equalities. The problem becomes

$$\max v' - 3$$

subject to

$$4x_1 + x_2 + v' + s_1 = 6$$

$$2x_1 + 5x_2 + v' + s_2 = 6$$

$$4x_1 + 5x_2 - v' - s_3 = 0$$

$$x_1 + x_2 + s_4 = 1$$

The initial BFS will have basic variables s_1, s_2, s_3, s_4 and zero variables x_1, x_2, v'. This corner point is degenerate because the third constraint passes through the origin. However, the degeneracy gives no trouble. The first tableau of the simplex method is

	Const.	x_1	x_2	v'	Ratio	
E	-3	0	0	1		
s_1	6	-4	-1	-1	-6	
s_2	6	-2	-5	-1	-6	
s_3	0	4	5	-1	-0	Pivot
s_4	1	-1	-1	0	∞	

(Pivot, over v' column)

The next tableau shows no gain in objective because of the degeneracy:

	Const.	x_1	x_2	s_3	Ratio	
E	-3	$+4$	5	-1		
s_1	6	-8	-6	$+1$	-1	
s_2	6	-6	-10	$+1$	$-\dfrac{6}{10}$	Pivot
v'	0	$+4$	$+5$	-1	0	
s_4	1	-1	-1	0	-1	

(Pivot, over x_2 column)

The next tableau is

	Const.	Pivot x_1	s_2	s_3	Ratio	
E	0	1	$-\dfrac{1}{2}$	$-\dfrac{1}{2}$		
s_1	$\dfrac{24}{10}$	$-\dfrac{44}{10}$	$+\dfrac{6}{10}$	$+\dfrac{4}{10}$	$-\dfrac{24}{44}$	Pivot
x_2	$\dfrac{6}{10}$	$-\dfrac{6}{10}$	$-\dfrac{1}{10}$	$+\dfrac{1}{10}$	-1	
v'	3	1	$-\dfrac{1}{2}$	$-\dfrac{1}{2}$	$+3$	
s_4	$+\dfrac{4}{10}$	$-\dfrac{4}{10}$	$+\dfrac{1}{10}$	$-\dfrac{1}{10}$	-1	

The next tableau gives the optimal solution:

	Const.	s_1	s_2	s_3	Ratio
E	$\dfrac{6}{11}$	$-\dfrac{10}{44}$	$-\dfrac{4}{11}$	$-\dfrac{9}{22}$	
x_1	$\dfrac{6}{11}$	$-\dfrac{10}{44}$	$\dfrac{6}{44}$	$\dfrac{1}{11}$	
x_2	$\dfrac{3}{11}$	$+\dfrac{6}{44}$	$-\dfrac{2}{11}$	$\dfrac{1}{22}$	
v'	$\dfrac{39}{11}$	$-\dfrac{10}{44}$	$-\dfrac{4}{11}$	$-\dfrac{9}{22}$	
s_4	$\dfrac{2}{11}$	$+\dfrac{4}{44}$	$+\dfrac{1}{22}$	$-\dfrac{7}{110}$	

The result is

$$x_1 = \frac{6}{11}$$

$$x_2 = \frac{3}{11}$$

$$x_3 = \frac{2}{11}$$

and

$$v = \frac{6}{11}$$

The first three constraints hold as equalities in the optimal solution. Therefore it is known that B uses all three of his courses of action. The best random strategy for B can be found from these results by solving

three simultaneous equations in three unknowns. The ith equation represents the expected loss to B if A uses A_i. All these must equal the expected gain to A that is known. The equations are

$$-y_1 + y_2 + y_3 = \frac{6}{11}$$

$$2y_1 - 2y_2 + 2y_3 = \frac{6}{11}$$

$$3y_1 + 3y_2 - 3y_3 = \frac{6}{11}$$

The solution is

$$y_1 = \frac{5}{22}$$

$$y_2 = \frac{4}{11}$$

$$y_3 = \frac{9}{22}$$

This completes the solution to the game problem.

● **PROBLEM** 11-97

Suppose that two costume companies each make clown, skeleton, and space costumes. They all sell for the same amount
and use the same machinery and workmanship. Furthermore,
the market for costumes is fixed; a certain given number
of total costumes will be sold this Halloween. But each
company has its own individual styles which affect how
the costumes sell. On the basis of past experience, the
following matrix has been inferred. It indicates, for
example, that if both make clown outfits, then for every
20 that are sold, company I will lose 2 sales to company
II. Similarly, if company I makes clown outfits and
company II makes space suits, then for each 20 sold,
company I will sell 4 more than company II.

Company II

	Clown	Skeleton	Space
Clown	-2	0	4
Company I Skeleton	0	2	1
Space	-1	-4	0

How should each company plan its manufacturing?

Solution: Company I can quickly decide not to make space
outfits, because it will do better with skeleton costumes,
no matter what company II does. Similarly, company II
should not make space outfits, because it can always
be better off by making clown outfits, no matter what
the choice of company I. Thus the matrix has been
reduced as follows:

$$\begin{bmatrix} -2 & 0 & 4 \\ 0 & 2 & 1 \\ -1 & -1 & 0 \end{bmatrix}$$

965

With the remaining 2 by 2 matrix, it is clear that company I will profit most from playing the second row, and that company II should always play the first column. Thus this game also has a saddle point. It is $a_{21} = 0$.

The value of the game is 0; company I should always make skeleton outfits and company II should always make clown costumes.

P_1 and P_2 each extend either one, two, or three fingers, and the difference in the amounts put forth is computed. If this difference is 0, the payoff is 0; if the difference is 1, the player putting forth the smaller amount wins 1; and if the difference is 2, the player putting forth the larger amount wins 2.

Each player has three pure strategies. Let s_i denote P_1's pure strategy of extending i fingers, $1 \leq i \leq 3$, and similarly define t_j, $1 \leq j \leq 3$, for P_2. The payoff tableau is then

	t_1	t_2	t_3
s_1	0	1	-2
s_2	-1	0	1
s_3	2	-1	0

Formulate an equivalent linear programming model for determining an optimal strategy and security level for P_2. Solve by using the simplex method.

Solution: The simplex method is not directly applicable if the value of the game is not positive. For such a game a constant must first be chosen such that when this constant is added to each entry of the original payoff matrix, the game corresponding to this new matrix has a positive value. Then the simplex method can be applied to this new game, with the value of the original game equal to the value of the new game less the constant. Note that it may not be necessary to make all the entries in the modified payoff matrix positive; for example, if the matrix has at least one row with all positive entries, the value of the corresponding game is positive (the possibility of P_1 using the pure strategy of playing that particular row shows that his security level is positive).

By symmetry it is reasonable to expect the value of this game to be 0. To verify this and compute optimal strategies, first add 2 to each entry of the above matrix, giving the following matrix, which corresponds to a game with value at least 1 as all the entries in the last two rows are greater than or equal to 1.

$$\begin{bmatrix} 2 & 3 & 0 \\ 1 & 2 & 3 \\ 4 & 1 & 2 \end{bmatrix}$$

The associated linear programming problem corresponding to P_2's determination of an optimal strategy and security level is to

Maximize $y'_1 + y'_2 + y'_3$

subject to

$$2y'_1 + 3y'_2 \qquad\quad \leq 1$$

$$y'_1 + 2y'_2 + 3y'_3 \leq 1$$

$$4y'_1 + \ y'_2 + 2y'_3 \leq 1$$

$$y'_1,\ y'_2,\ y'_3 \geq 0$$

Adding three slack variables and solving leads to the tableaux of Table 1:

Table 1

	y'_1	y'_2	y'_3	y'_4	y'_5	y'_6	
y'_4	2	3	0	1	0	0	1
y'_5	1	2	③	0	1	0	1
y'_6	4	1	2	0	0	1	1
	-1	-1	-1	0	0	0	0
y'_4	2	③	0	1	0	0	1
y'_3	$\frac{1}{3}$	$\frac{2}{3}$	1	0	$\frac{1}{3}$	0	$\frac{1}{3}$
y'_6	$\frac{10}{3}$	$-\frac{1}{3}$	0	0	$-\frac{2}{3}$	1	$\frac{1}{3}$
	$-\frac{2}{3}$	$-\frac{1}{3}$	0	0	$\frac{1}{3}$	0	$\frac{1}{3}$
y'_2	$\frac{2}{3}$	1	0	$\frac{1}{3}$	0	0	$\frac{1}{3}$
y'_3	$-\frac{1}{9}$	0	1	$-\frac{2}{9}$	$\frac{1}{3}$	0	$\frac{1}{9}$
y'_6	㉜⁄₉	0	0	$\frac{1}{9}$	$-\frac{2}{3}$	1	$\frac{4}{9}$
	$-\frac{4}{9}$	0	0	$\frac{1}{9}$	$\frac{1}{3}$	0	$\frac{4}{9}$
y'_2	0	1	0	$\frac{5}{16}$	$\frac{1}{8}$	$-\frac{3}{16}$	$\frac{1}{4}$
y'_3	0	0	1	$-\frac{7}{32}$	$\frac{5}{16}$	$\frac{1}{32}$	$\frac{1}{8}$
y'_1	1	0	0	$\frac{1}{32}$	$-\frac{3}{16}$	$\frac{9}{32}$	$\frac{1}{8}$
	0	0	0	$\frac{1}{8}$	$\frac{1}{4}$	$\frac{1}{8}$	$\frac{1}{2}$

967

The value of the modified game is 2, and so the value of the original game is 0, as suggested. Since the optimal value of the above problem is attaned by $(y'_1, y'_2, y'_3) = \left(\frac{1}{8}, \frac{1}{4}, \frac{1}{8}\right)$, an optimal strategy for P_2 is $2\left(\frac{1}{8}, \frac{1}{4}, \frac{1}{8}\right)$ $= \left(\frac{1}{4}, \frac{1}{2}, \frac{1}{4}\right)$. Similarly, the solution to the dual problem, found in the bottom row in the slack variable columns, is $(x_1, x_2, x_3) = \left(\frac{1}{8}, \frac{1}{4}, \frac{1}{8}\right)$, and so an optimal strategy for P_1 is also $2\left(\frac{1}{8}, \frac{1}{4}, \frac{1}{8}\right) = \left(\frac{1}{4}, \frac{1}{2}, \frac{1}{4}\right)$.

● **PROBLEM** 11-99

Solve the game below for optimal strategies of the two players, by applying the simplex method.

$$
\begin{array}{ccc}
1 & 0 & 2 \\
0 & 2 & 0 \\
2 & 0 & -1
\end{array}
$$

Solution: The result is as follows: The optimal strategies are

$$\vec{u}* = \vec{y}* = (6,5,2)/13.$$

The optimal value of the program is

$$a* = 10/13,$$

and the pivot locations $k\ell = 13, 22$ or 31.

INDEX

Numbers on this page refer to <u>PROBLEM NUMBERS</u>, not page numbers

THE PROBLEM SOLVERS